Lecture Notes in Civil Engineering 634

Lecture Notes in Civil Engineering (LNCE) publishes the latest developments in Civil Engineering—quickly, informally and in top quality. Though original research reported in proceedings and post-proceedings represents the core of LNCE, edited volumes of exceptionally high quality and interest may also be considered for publication. Volumes published in LNCE embrace all aspects and subfields of, as well as new challenges in, Civil Engineering. Topics in the series include:

- Construction and Structural Mechanics
- Building Materials
- Concrete, Steel and Timber Structures
- Geotechnical Engineering
- Earthquake Engineering
- Coastal Engineering
- Ocean and Offshore Engineering; Ships and Floating Structures
- Hydraulics, Hydrology and Water Resources Engineering
- Environmental Engineering and Sustainability
- Structural Health and Monitoring
- Surveying and Geographical Information Systems
- Indoor Environments
- Transportation and Traffic
- Risk Analysis
- Safety and Security

To submit a proposal or request further information, please contact the appropriate Springer Editor:

- Pierpaolo Riva at pierpaolo.riva@springer.com (Europe and Americas);
- Swati Meherishi at swati.meherishi@springer.com (Asia—except China, Australia, and New Zealand);
- Wayne Hu at wayne.hu@springer.com (China).

All books in the series now indexed by Scopus and EI Compendex database!

Chih-Huang Weng
Editor

Proceedings of the 10th International Conference on Water Resource and Environment

Springer

Editor
Chih-Huang Weng
Department of Civil Engineering
I-Shou University
Kaohsiung City, Taiwan

ISSN 2366-2557 ISSN 2366-2565 (electronic)
Lecture Notes in Civil Engineering
ISBN 978-981-96-5094-1 ISBN 978-981-96-5095-8 (eBook)
https://doi.org/10.1007/978-981-96-5095-8

This Springer imprint is published by the registered company Springer Nature Singapore Pte Ltd.
The registered company address is: 152 Beach Road, #21-01/04 Gateway East, Singapore 189721, Singapore

Preface

As we stand at the threshold of the 10th International Conference on Water Resource and Environment (WRE 2024), it is both a moment to reflect on the advancements made in the field over the past decade and an opportunity to look forward to the challenges that lie ahead. Hosted in the vibrant city of Hong Kong, China, from December 15th to 18th, 2024, WRE 2024 promises to be a landmark event bringing together leading experts, researchers, policymakers, and practitioners from around the globe.

The conference proceedings presented here encapsulate the spirit of innovation, collaboration, and dedication to finding sustainable solutions for water resource management, environmental pollution, and sustainability. This compilation includes a diverse array of high-quality research papers, case studies, and insightful analyses contributed by scholars and professionals who are at the forefront of their fields. These contributions not only highlight the latest scientific findings but also offer practical guidance on addressing some of the most pressing issues facing our planet today.

WRE 2024 received a total of 193 submissions, resulting in 4 plenary speeches, 4 keynote speeches, and 10 invited presentations, alongside 48 oral presentations and 12 poster presentations. More than 100 Participants came from a diverse array of locations, including China, Egypt, Hong Kong, Hungary, India, Japan, Malaysia, Philippines, Qatar, Taiwan, and Turkey. From the 127 full paper submissions, 46 chapters were selected for publication in this volume following a rigorous and fair review process, yielding an acceptance rate of approximately 37%.

We are immensely grateful to all the authors who have shared their work, to the reviewers for their meticulous efforts, and to the attendees whose active participation makes this conference a success. Through such collective endeavors, we can aspire to achieve meaningful progress in safeguarding our water resources and the environment.

Kaohsiung City, Taiwan — Chih-Huang Weng

Organization

The 10th International Conference on Water Resource and Environment (WRE 2024)

Conference Honorary Chair

Advisor/Research Chair Prof. Ming-Hung Wong	The Education University of Hong Kong, Hong Kong

Conference General Chair

Distinguished Prof. Chih-Huang Weng	I-Shou University, Taiwan

Local Organizing Committee Chairs

Dr. Yiu Fai Tsang	The Education University of Hong Kong, Hong Kong
Dr. Yu Bon Man	The Education University of Hong Kong, Hong Kong

International Technical Program Committee Chairs

Prof. Cehui Mo	Jinan University, China
Vice-President/Prof. Shengdao Shan	Zhejiang University of Science and Technology, China
Prof. Tatiana Minkina	Southern Federal University, Russia

International Technical Program Committee

Lifetime Distinguished Prof. Yao-Tung Lin	National Chung Hsing University, Taiwan
Prof. Zhimin Qiang	Shanghai Jiao Tong University, China
Prof. Peiyue Li	Chang'an University, China
Prof. Foo Keng Yuen	Universiti Sains Malaysia, Malaysia
Dr. Anoop Kumar Shukla	Manipal Academy of Higher Education, India
Prof. Azza Moustafa Abdelaty	National Research Center, Egypt

Dr. Chunjiang An	Concordia University, Canada
Dr. Abdelkader Hamlat	University of Laghouat, Algeria
Prof. Fei Qi	Beijing Forestry University, China
Prof. George N. Zaimes	Democritus University of Thrace, Greece
Dr. Habib Bouzid	University of Mostaganem, Algeria
Dr. Hao Han	Institute of Earth Environment, Chinese Academy of Sciences, China
Prof. Hui Zhu	Chinese Academy of Science, China
Assoc. Prof. Ignacy Kitowski	University College of Applied Sciences in Chelm, Poland
Dr. Seyedmehdi Mohammadizadeh	Universidade Estadual de Campinas (UNICAMP), Brazil
Dr. Oleg D. Linnikov	Russian Academy of Sciences, Russia
Dr. Saad Mulahasan	University of Mustansiriyah, Iraq
Assoc. Prof. Porowska Dorota	University of Warsaw, Poland
Prof. Teik-Thye Lim	Nanyang Technological University (NTU), Singapore
Prof. R. S. Ajin	University of Florence, Italy
Dr. Narong Touch	Tokyo University of Agriculture, Japan
Dr. Pedro Antonio Guido Aldana	Mexican Institute of Water Technology, Mexico
Dr. Ackmez Mudhoo	University of Mauritius, Mauritius
Prof. Yuk Feng Huang	Universiti Tunku Abdul Rahman, Malaysia
Prof. Ricardo Gomes	Polytechnic of Leiria, Portugal
Prof. AbdAllah Tharwat Abdelkhalik	Jazan University, Saudi Arabia
Assoc. Prof. Xinxin Guo	Universiti Tunku Abdul Ralman, Malaysia
Assoc. Prof. Chai Hoon Koo	Universiti Tunku Abdul Rahman, Malaysia
ChM. Dr. Wong Lai Peng	Universiti Tunku Abdul Rahman, Malaysia
Prof. Hrouzková Svetlana	Slovak University of Technology in Bratislava, Slovak Republic
Assoc. Prof. Sébastien Déon	Université de Bourgogne Franche-Comté, France
Prof. Sarra Kitanou	Ibn Tofail University, Morocco
Dr. Reza Soleimanpour	Australian University, Kuwait

About This Book

Dive into the future of water resource management and environmental sustainability through the Conference Proceedings of the 10th International Conference on Water Resource and Environment (WRE 2024). This comprehensive collection showcases pioneering research, innovative solutions, and in-depth analyses from global experts across various disciplines, covering critical topics such as water resources and management, hydrology and hydrological engineering, water pollution and treatment technology, and environmental pollution and sustainability. Ideal for researchers, policymakers, engineers, and consulting firm leaders, these proceedings offer practical insights and actionable guidance to address current challenges and inspire new opportunities. Secure your copy today and join us in shaping a cleaner, safer, and more sustainable world for future generations. Visit our official website to learn more and order your copy now.

Contents

About the Editor

Emeritus Professor Chih-Huang Weng is affiliated with the Department of Civil Engineering at I-Shou University in Taiwan, where he is also serves as the President of North Kaohsiung Community University in Taiwan. He received his M.S. and Ph.D. degrees in 1990 and 1994, respectively, from the Department of Civil Engineering of The University of Delaware, USA. Currently, Prof. Weng is the Associate Editor of Environmental Geochemistry and Health (Springer) and a member of the Editorial Board Panel of Coloration Technology (Wiley). He has acted as a Guest Editor of SCI journals, such as Agricultural Water Management (Elsevier), Environmental Science and Pollution Research (Springer), and Lecture Notes in Civil Engineering. Additionally, he has organized and chaired numerous international conferences. Prof. Weng is listed in the World's Top 2% Scientists by Stanford University (2021–2024). His main research interests focus on using advanced oxidation processes and adsorption to treat wastewater and bacteria inactivation, groundwater remediation, and application of electrokinetic technologies to soil remediation/sludge treatment/activated carbon regeneration.

Water Resources and Management

Freshwater Salinization Syndrome: Global Public Health Threats Under the Anthropocene

Alice M. L. LI(✉)

Hong Kong Metropolitan University, Hong Kong, SAR, China
amlli@hkmu.edu.hk

Abstract. Freshwater salinization is an escalating water quality issue, with excess salt released through the anthropogenic-induced salt cycle, leading to freshwater salinization syndrome (FSS). This syndrome has become a worldwide concern, impacting safe drinking water, ecosystem health, biodiversity, corrosion of infrastructure, and food production. Epidemiological research has shown a significant positive link between drinking water salinity and elevated blood pressure or hypertension risk. Statistically, high systolic blood pressure is the leading risk factor for premature death globally. Hypertension is also the primary risk factor for coronary heart disease and stroke, as revealed in the Global Burden of Disease. Additionally, water and soil salinity can adversely affect nutritional health and food security, posing increasing challenges for sustainable health development for the present and future generations. Therefore, identifying gaps in knowledge from multiple perspectives of those arising global and ecological determinants of FSS will have significant implications for ecological interventions. Advocacy of "Planetary Health Diets" by using ecological public health approaches is highly recommended, especially in this era of eco-environmental and climate crises under this new epoch of the Anthropocene. This paper aims to provide scientific analyses from the perspective of FSS on how drinking water salinity increasingly poses risks of water-related, saline-contaminated health threats and salinity-related diseases, together with its domino and cascading effects on food security and nutritional health. Empirical analysis is conducted using epidemiological studies, plus citations of health facts and statistics regarding attributable risk factors.

Keywords: Freshwater Salinization Syndrome · Hypertension · Planetary Health Diets and Nutritional Health · Ecological Public Health

1 Introduction

During the past decades, freshwater salinization has increased in many parts of the world, posing significant risks to drinking water quality by leaching contaminants from soils, sediments, and water infrastructure [1]. This contamination not only reduces biodiversity but also threatens human health on global scale [2]. Evidence of research indicated that the salinity of freshwater ecosystems around the world is becoming increasingly saltier due to various anthropogenic-driven activities and factors contributing to freshwater salinization, including irrigation, oil extraction, potash mining, and road de-icing,

C. Weng (Ed.): WRE 2024, LNCE 634, pp. 3–16, 2025.
https://doi.org/10.1007/978-981-96-5095-8_1

which introduce salts entering into waterways. [3]. This process causes a phenomenon called "Freshwater Salinization Syndrome" (FSS), which also characterized by increased concentrations of contaminants such as heavy metals and nutrients in water [4]. This syndrome disrupts essential system-level functions and services, including crop production, drinking water, biodiversity infrastructures, ecosystems, and aquatic life and human health [5].

Freshwater is a vital resource for our planet, but its rising salinity is an escalating global environmental health concern [6]. And the risks of FSS are becoming more apparent in the current era of eco-environmental and climate mediated health threats as they are closely associated with the mediating effects of the Anthropocene. It indicates a range of components of the Earth's systems are now more controlled by anthropogenic forcing than by natural drivers; therefore, these are being characterized as a new epoch of the Anthropocene [7]. The term "Anthropocene" has, in fact, been coined three times historically, however only in the third, which is when this term gained popularity through the work of Dutch chemist and Nobel recipient Paul Crutzen [8]. This new epoch of the Anthropocene signifies a time when anthropogenic actions have become agents of geological change, shaping significant global chemical, physical and biological alterations to the atmosphere, landscape and oceans, as well as climate change [8]. This rising web of complex dynamic interaction includes the resultant effects of the anthropogenic salt cycle that are altering the natural balance of salt ions across Earth's surface systems, leading to such unprecedented effects across biophysical systems collectively recognized as FSS [9]. Examples of anthropogenic-induced salinization include the usage of road deicing salts, agricultural practices, mining operations, shale-gas extraction, brines used for secondary oil recovery, the application of fertilizers, and industrial wastes, plus climate change, all of which further poses major threats to the functioning of freshwater ecosystems and biodiversity [2, 10, 11]. Especially climate change, which is associated with rising temperatures, changing precipitation patterns, and rising sea levels that exacerbate the problems of water resources.

Thus far, it is important to understand the mechanisms interlinking FSS with chemical pollution, such as salts, which remains a significant public health issue, especially in industrialized nations. Ultimately, the magnitude of those underlying health threats posed by FSS to human health, together with water sustainability for food security and nutritional health, is now a major global concern. In addition, the complexity of FSS conditions is defined as risk factors, amongst other effects, including the risks of hypertension, which is currently set on a new perspective of health trajectory. And the World Health Organization (WHO) has already highlighted hypertension as the leading global risk factor for impacting morbidity and mortality [12].

Exposure to chemical contaminants, including salts in drinking water, can contribute to high blood pressure and a range of other chronic diseases, and these contaminants may also increase the risk of cancer and cardiovascular disease, adverse reproductive effects. They can affect children's health, particularly their neurodevelopment, among other health issues [13]. According to the hypertension guidelines issued by the European Society of Hypertension and the European Society of Cardiology, there are significant interrelationships between hypertension and the risks associated with cardiovascular, cerebral and renal health [12].

Beyond the risks of hypertension, the totality of issues on food security and water sustainability are closely interconnected. It is because different stages of nutritional transition historically reflect the shifts in dietary consumption or practices that coincide with economic, demographic, and epidemiological changes, as well as the impacts on environmental sustainability, including water sustainability [8]. Therefore, advocacy of "Planetary Health Diets" (PHD) is especially important in the context of the current eco-environmental and climate crises under the Anthropocene. Ecological public health approaches are therefore recommended for necessitating a process of change with synergistic interventions from different perspectives in need.

2 Freshwater Salinization Syndrome and Its Underlying Threats

Freshwater is a scarce commodity and a precious resource that is essential for sustaining life on Earth. Globally, less than 3% of usable freshwater, and the standardization of freshwater cycle serves multiple functions for food, energy, and other ecological services through ecosystems. Freshwater ecosystems are vital for human societies as they provide essential services such as drinking and irrigation water, food, climate regulation, and recreation opportunities [14].

Although dissolved salts and minerals are necessary for maintaining the health and vitality of organisms and all living beings that depend on this ecosystem service [15], salinization jeopardizes water resources for human needs and ecosystem services. However, the normal functioning services of ecosystems are being affected by several large-scale ecological disasters attributed to the salinization of freshwater ecosystems [10]. Freshwater salinization is an increasing water quality challenge that affects both surface and groundwater resources, which may have natural causes. Still, primarily this arising salinity occurs mostly due to human activities [16].

Primarily, the natural salt cycle is driven by relatively slow natural geological and hydrological processes such as rock weathering or the depositing of salt by wind and rain over thousands of years [17]. It is a matter of scientific fact that the salinity of freshwater is being affected by diverse anthropogenic and geological sources, as well as climate change, which can be grouped into three primary categories that drive the increasing severity of this FSS, including [18–20]:

1. Anthropogenic salt inputs, such as road salts, sewage, brines, sodic/saline irrigation discharge or runoff.
2. Human-accelerated weathering of natural geological materials by strong acids, including acid rain, fertilizers, acid mine drainage; resource extraction, water softeners, human use of easily weathered materials such as concrete and lime.
3. Saltwater intrusion and the evaporative concentration of ions due to hydrological alterations, sea level rise, and climate change.

Amongst others, the extensive use of road deicing salts to reduce vehicular accident rates has significantly increased freshwater salinity, posing threats to the biodiversity and functioning of freshwater ecosystems as well as to drinking water supplies and human health [21]. Deicing salts also mobilize harmful substances such as radon, mercury, and lead, further endangering freshwater biota and drinking water supplies [21], particularly with the added salinization from rising sea levels, which imposed additional constraints on groundwater resources [20]. Furthermore, the ongoing processes of salinization and alkalinization also added increasing uncertainty to the dynamics of terrestrial sequestration of atmospheric carbon dioxide, especially both of these changes can impair drinking water quality, harm freshwater biodiversity, accelerate infrastructure corrosion, and mobilize inorganic and organic pollutants [22].

Considering all these anthropogenic activities imposed by the growing salt production and usage disrupt, the natural balance of salt ions across the surface of Earth's systems, leading to those dynamical effects across biophysical systems collectively referred to as FSS [9]. Recent projections indicate that environmental salinization processes will continue to worsen due to global climate change [23]. Moreover, the combined sets of biological, physical, and chemical impacts commonly occur together in response to salinization, which further poses threats to aquatic life, sources of drinking water, infrastructure, agriculture, and critical ecosystem services [24]. There is no doubt that trajectories of human population growth, water use, and consequential environmental changes are expected to continue rising sharply [25]. Therefore, the proportion of salinized freshwater ecosystems is anticipated to increase due to intensified anthropogenic pressure and climate change [14].

3 Salinity-Related Health Risks of Hypertension Leading to Other Diseases

Salinity-related diseases are a growing concern due to the negative effect of high salt intake on the human body. Drinking water, often overlooked and underestimated source of salts, contributes significantly to the total salt burden in the body [26]. There are many epidemiological studies that found a significant positive correlation between different levels of salinity in drinking water and elevated blood pressure, or the risk of hypertension [26]. While sodium chloride is essential for our health, excessive consumption can lead to hypertension [27], which is a leading cause of premature death worldwide, as noted by the World Health Organization [28]. Data from 2019's global attributable deaths from Level 2 risk factors have been reported that high systolic blood pressure is now the top one and two most vulnerable factors following dietary risk factors, as shown in Fig. 1 [29].

Figure 1 reinforces the WHO's identification of hypertension as the most critical genesis of premature death globally. According to the National Organization of Public Health in 2021, hypertension accounts for about 50% of all deaths from stroke and heart disease worldwide [26]. Consequently, the risks of FSS associated with high salt intake, alongside the increased risks of hypertension and the possibilities of inducing other salinity-related diseases, are of growing concern.

Numerous studies have explored the epidemiological association between the health threats that arise or are aggravating from this FSS. Studies have supported that early onset (before 55yrs old) in parents was strongly associated with hypertension in their offspring, suggesting a familial trait that increases the risk for cardiovascular outcomes in adults with established hypertension [30]. For this reason, parental hypertension is used to assess and classify hypertension risk in young adults and has also been predicted

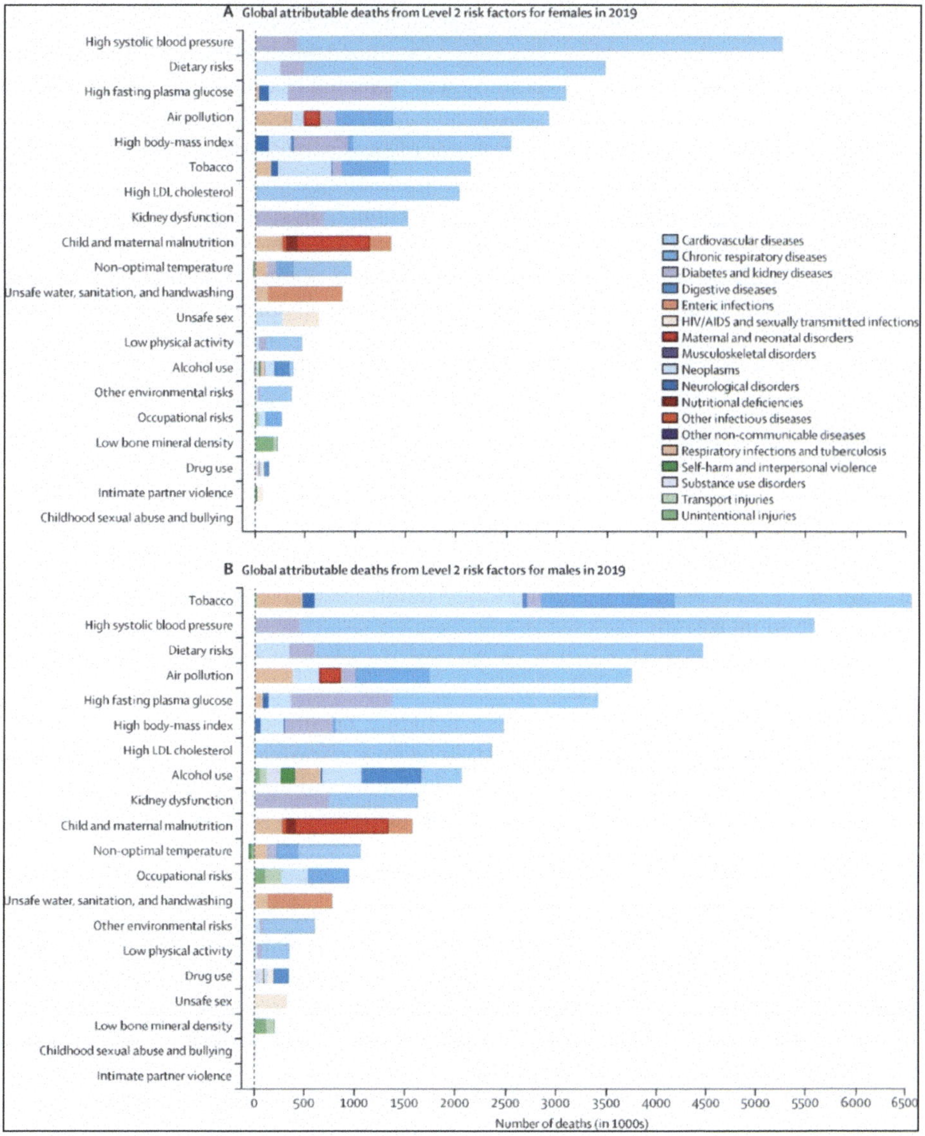

Fig. 1. Global attributable deaths from Level 2 risk factors [29, p. 1233].

for cardiovascular mortality and morbidity [31]. Amongst other risk factors for cardiovascular disease, high blood pressure has indeed the strongest evidence for causation, and a high prevalence of exposure [32, 33]. This echoes the studies from the Korea National Health and Nutrition Survey, which suggested that children with hypertensive parents are more likely to develop hypertension, indicating the familial aggregation of hypertension may be a useful screening tool for children in Korea [34].

In another study comparing drinking water salinity levels with the incidence of adult pre-hypertension and hypertension, researchers examined the relationship between water salinity, blood pressure, and age. This study was designed to categorize drinking water salinity into four groups based on its source, and the age group was also divided into two categories [28]. Despite the defining parameter of freshwater contains either $< 0.05\%$, or $< 0.1\%$ by some definition. And the caption of salinity levels in this study defined that freshwater ($<$ 1000 mg/L, i.e., less than 1,000 ppm (parts per million), or $< 0.10\%$), slightly saline (1000–2000 mg/L), moderately saline ($\geq$ 2000 mg/L). The findings indicated that the prevalence of elevated blood pressure, including both pre-hypertension and hypertension, significantly increased with higher levels of drinking water salinity concentration. This study illustrates a steep increase in the incidence of hypertension with rising salinity levels in drinking water, particularly among individuals aged 35 years and above. Whereas the prevalence of pre-hypertension is lower, suggesting there may be some other contributing factors involved [28].

Another cross-sectional research study identified a link between drinking water salinity associated with both hypertension and hyperdilute urine (i.e., urine specific gravity (USG) $<$ 1.003, whereas the normal range is defined clinically as 1.005–1.030), indicating altered kidney function [6]. This was significantly signified by kidney tubular epithelial damage or associated with either kidney failure or diabetes insipidus, polydipsia (excessive fluid intake) [6].

Etiologically, hypertension increases the risk of various chronic diseases, including cardiovascular diseases, myocardial infarction, stomach cancer, chronic respiratory disease, obesity, diabetes, and renal disease (including nephrosclerosis and retinopathy) [35]. Specifically, hypertension is a major risk factor for various types of cardiovascular diseases that may develop during a lifetime, including coronary disease, lift ventricular hypertrophy, valvular heart diseases, cardiac arrhythmias such as atrial fibrillation, cerebral stroke and renal failure [12].

Moreover, high salt intake through salinity of drinking water or dietary salt can lead to even more chronic diseases such as stroke, heart attack, pre-eclampsia, eclampsia, impaired liver function, kidney disease, low platelet count, skin diseases, diarrheal diseases, intrauterine growth retardation, preterm birth, and maternal health, such as miscarriage and/or premature birth and/or perinatal deaths among pregnant women [35, 36]. At the same time, other studies have reported that excess salinity has also been linked with increased infant mortality [37].

Most of the sodium in our diet comes from salt added to the food we eat, and a high-salt diet can elevate tissue sodium concentrations and affect immune responses in microenvironments, which in turn affects the development of immune-regulated diseases that include hypertension, multiple sclerosis, osteoporosis, cancer, and infections [38]. However, the source of salts in drinking water is most often unnoticed or underestimated. Thus far, elevated blood pressure associated with drinking water salinity due to FSS has become an emerging public health issue and global concern. Therefore, it is pivotal to be aware of the mechanism of salinity in drinking water that affects human health through the channel of FSS as an add-on high-salt intake. Additionally, the sequelae of FSS are increasingly being joined by the ailments of those salinity-related diseases, making the burden infinitely greater. Li et al. [38] have illustrated many examples of diseases that are salinity-related, including:

1. Cancer: Elevated sodium ion (Na^+) levels in solid tumors can alter cell metabolism and immune function. High salt intake is a potent inducer of inflammatory states, and has been linked to adverse effects on cancer. It also promotes Helicobacter pylori infection, gastric mucosa damage, hypergastrinemia, and cell proliferation, which is a causative source of gastric carcinogenesis.
2. Systemic lupus erythematosus (SLE): SLE is an autoimmune connective tissue disorder affecting multiple systems, and lupus nephritis is one of the most severe manifestations. Statistical data indicates that a high-salt intake accelerates lupus progression and apparently increases mortality.
3. Rheumatoid arthritis (RA): RA is a chronic inflammatory joint disease that can lead to cartilage and bone damage and disability. Studies suggest that reducing dietary salt intake can diminish the pro-inflammatory response.
4. Osteoporosis: Studies in humans have indicated that excessive salt intake may lead to calcium loss, which might increase the risk of osteoporosis.
5. Kidney diseases: Chronic kidney disease is one of the major global public health concerns, and high salt or sodium intake is potentially linked to an increased risk of kidney diseases [39]. Research studies indicated that drinking water salinity may cause kidney failure directly through excessive strain by increasing the workload for the sodium and other cellular pumps in the nephron or indirectly via its hypertensive effects [6]. The interaction between groundwater salinity and hypertension is significantly associated with chronic kidney diseases [39].
6. Consuming water with excessive salt concentrations poses health threats that can cause an imbalance of the sodium and potassium levels in our body [40].

There are still many more adverse effects of FSS related to drinking water salinity. This is especially so, as the climate crisis worsens along with global warming and the intrusion of water salinity, and saltwater intrusion into drinking water supplies also contributes to water scarcity. This phenomenon of seawater intrusion into coastal aquifers is a growing concern worldwide, particularly as surface water resources, like rivers and canals, are increasingly affected by saline water intrusion [36, 41]. This problem is especially pressing under climate change, sea level rise, and increasing salinity in drinking water sources [42].

Epidemiological association with ambient heat can be an important confounder that heavily influences drinking water salinity and sodium intake through a few mechanisms related to thermoregulation that affect blood pressure and cardiovascular risk [43]. Moreover, salts can facilitate the mobilization of toxic heavy metals into drinking water, further increasing the risks to human health [2].

Furthermore, higher salt concentrations in freshwater can also enhance the survival rates of *E. coli*. Freshwater ecosystem management especially relies upon the detection of *E. coli*, a key indicator as a sentinel for freshwater ecosystem management to assess infectious disease risk [44]. Annually, an estimated 90 million illnesses are attributed to human contact with waterborne pathogens in rivers, streams, and oceans, costing estimated from US$2.2 to $3.7 billion [44]. Urban freshwater salinization can also increase the prevalence of salt-tolerant disease vectors like mosquitoes, which can spread those mosquito-borne diseases, such as Zika virus, West Nile virus, Chikungunya virus, dengue, and malaria [45].

Evidently, FSS is not just a water crisis, but it is rather a significant global public health issue, as it poses an existential ecological determinant of sustainable health development globally and intergenerationally. Ultimately, the overarching goal of global public health is to facilitate sustainable health development across present and future generations through Sustainable Development Goals (SDGs). Henceforth, understanding the impacts of these transitional health risks is utterly crucial in this era of eco-environmental and climate mediated global public health crises in the Anthropocene.

4 Food Security for Nutritional Health

Water is a vital resource for all living beings, yet its availability is increasingly at risk due to the growing human population and the rising demands for more high-quality water for domestic and economic purposes [46]. The impacts of water and soil salinity are now becoming significant global concerns, impacting health and food security [41]. In fact, salinity often occurs alongside other environmental constraints and challenges, such as drought, waterlogging, pollution, acidity, and nutrient deficiency, which collectively and progressively exacerbate further threats to food security and general ecosystem resilience [23]. Factors like salinity, pH, and temperature are all critical factors for the survival of aquatic plants and animals, making them important parameters for water monitoring [47, 48]. Salinity, which measures the saltiness of water, refers to the sum of all ions dissolved in water, including calcium, magnesium, sodium, potassium, carbonate, and bicarbonate [49]. Because salinity is composed of these multiple major ions, the proportion of salinized freshwater ecosystems is expected to rise in the future due to a combination of factors such as anthropogenic intensification and climate change [14], plus growing water demands worldwide that leading to even higher salt concentrations [11, 50].

All in all, the current and potential water crisis is having a significant impact on food security due to both freshwater depletions and water salinity. Especially with the global population projected to reach nearly 10 billion people by 2050, ensuring a sustainable and healthy diet for everyone is indeed a major global challenge. In this era of widespread anthropogenic change, we need to better understand how species interact with multiple, interacting stressors [48]. For instance, agricultural activities pose a threat to many lake ecosystems worldwide [51].

According to Liu's study, agricultural irrigation discharges or runoff into Chagan Lake in Northeast China could potentially render the entire lake saline within three years [52]. The recommendations from this study are meant to assist authorities in recognizing the importance of balancing the risks and benefits of developing crop irrigation fields in areas surrounding large freshwater lakes [52]. Especially salinization can affect freshwater organisms across multiple trophic levels, thereby influencing lake ecosystems and trophic food webs [10]. Put simply, salinization can cause substantial zooplankton mortality, leading to a cascading effect that increases phytoplankton biomass and alters nutrient cycling and water clarity, ultimately triggering declines in fish production [11]. Moreover, the present structure of the global food system is unsustainable for the planet's natural systems [53], and the water crisis is deeply affecting food security in several ways, as follows:

1. Crop yields: High levels of salt accumulation in farmlands can cause dehydration of plants and loss of productivity [40]. And widespread land clearing and altered land use may induce the form of 'dryland salinity' or 'irrigation-induced salinity' also known as secondary salinity [17]. Furthermore, agro-ecosystems are very sensitive to salinization, which deteriorates food quality sharply [23].
2. Aquaculture: Salinity directly affects fish metabolism and osmoregulation, which can impact their growth and feed consumption. Some fish species, especially juveniles and fish eggs, are more susceptible to salinity, indirectly affecting food webs [54, 55]. The scientific understanding of how increasing salinization damages freshwater ecosystems, such as the fisheries collapse in the Aral Sea in Central Asia, complicates the development and implementation of necessary standards to protect freshwater life [10].
3. Livestock: Studies show that using saline water for livestock as an alternative source of drinking freshwater to avoid water scarcity can largely affect the productive aspects and health of that livestock [56].

Overall, salinity has direct and indirect effects on the environment and plant growth, making agricultural production nearly unworkable in severely saline areas [35]. In fact, the relationships between food security for nutritional health and water sustainability are two sides of the same coin.

5 Planetary Health Diets for Ecological Public Health

Public health has a long history of countering environmental threats to population health and well-being. Food systems play a crucial role in shaping our nutrition and health outcomes, as they are closely linked between human health and environmental well-being [53]. Therefore, it is crucial to identify knowledge gaps from an ecological perspective regarding the ecological determinants of food and environment, as which is essential for nutritional health and sustainable health development outcomes, as well as environmental sustainability. Research indicates that the current global dietary trends pose challenges from both a public health and a sustainability perspective [57].

In 2019, the EAT-Lancet Commission established the world's first scientific targets for healthy and sustainable food systems, introducing the recommended Planetary Health Diets (PHD). This initiative reinforces new nutritional knowledge that underscores the importance of maintaining human health and environmental sustainability, highlighting its relation to food production systems and their impacts on its prevailing climate conditions. PHD is an essential part of the nutritional transition for the reasons of climate change, as well as the rising Anthropocene impacts on water sustainability. The following graph illustrates the ways of creating a sustainable food future by reference to: (a) land use, (b) freshwater consumption, and (c) greenhouse gas emissions, from different food sources, which are specifically shown in Fig. 2 [57, 58].

Freshwater scarcity is one of the most urgent environmental challenges of this century already [46]. And this unprecedented phenomenon of FSS further adds to the difficulties of water sustainability as drinking water quality is paramount for sustainable

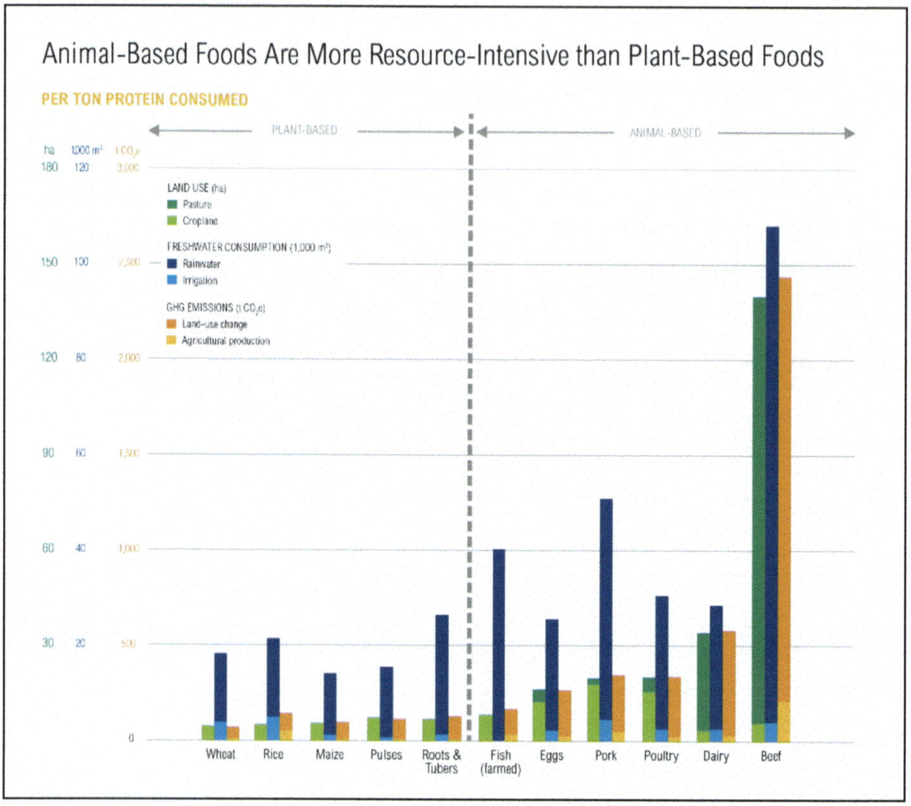

Fig. 2. Animal-based foods are more resource-intensive than plant-based foods [58].

health development from multiple perspectives of ecological public health *per se*. This is of significance to understand this FSS and its cumulative impacts affecting groups of biological, physical and chemical interactive dynamics of the natural salt cycle of Earth systems on a planetary scale by the existing and ongoing intense anthropogenic activities.

Henceforth, we must address the predicaments of environmental degradation and its significant ecological footprints on human health. Thus, advocacy for promoting planetary health literacy through the adoption of PHD is absolutely indispensable [57]. It connects environmental science and ecological public health by demonstrating how FSS affects not only drinking water quality and safety but also imposes other domino and cascading effects on the risks of hypertension and other salinity-related illnesses, ecosystems, food productivity and food security, as well as sustainability transition for global public health. In fact, PHD is the evolving knowledge of water conservation approach, as well as a sustainable way of nutritional health alongside the concern of environmental sustainability. This recommended nutritional transition of planetary health dietary trends by using an ecological public health approach is one of sustainable practices to safeguard and promote sustainable health development alongside food production systems and environmental sustainability, including the concern of water sustainability.

6 Conclusion

Human activities are causing an increase in salinity levels in Earth's air, soil, and freshwater, which could pose a significant threat or even catastrophic effects to the environment and human health if current trends continue. This anthropogenic salinization can disrupt ecosystem functions and services, degrade biological systems, and negatively impact human health and wellbeing [2]. The magnitude of these underlying threats of FSS is now a growing global water quality issue that has various causes and drastic consequences. It can have adverse effects on the body over time, causing to health problems such as high blood pressure, heart disease, stroke, kidney failure, skin diseases, diarrheal diseases, and many other salinity-related illnesses.

Therefore, global public health should be focused on addressing these increasingly crucial risk factors arising from the changing global and ecological determinants of health under the Anthropocene. This article takes innovative functional resilience and systems thinking by reconceptualizing these webs of dynamical complexities amongst anthropogenic activities, water functions, and its subsequent ecological processes on global public health perspectives by addressing those actual and potential emerging health threats, as well as the concern of food security for nutritional health and environmental sustainability across generations.

Salinity is a costly problem because drinking water quality guidelines and standards aim to ensure clean and safe water for human consumption [15]. When drinking water sources become more saline, costly treatments are often necessary to maintain salinity levels suitable for human usage [59]. Under this context, our food systems play a vital role in both human health and environmental sustainability. Our forward thinking is required from multiple perspectives alongside the advocacy of PHD as an integral part of these ecological public health approaches to tackle this issue. Simply because this recommended PHD serves as a sustainable diet for better conservation of water sustainability and land use, as well as reducing greenhouse gas emissions. As we are now facing the Anthropocene effects, where human activities are the dominant influence on the eco-environment and climate change, the contributions of ecological public health approaches to these unprecedented arising impacts on the threats of global public health are more important than ever.

References

1. Kaushal, S.S.: Increased salinization decreases safe drinking water. Environ. Sci. Technol. **50**, 2765–2766 (2016)
2. Schuler, M.S., Cañedo-Argüelles, M., Hintz, W.D., Dyack, B., Birk, S., Relyea, R.A.: Regulations are needed to protect freshwater ecosystems from salinization. Philosophical Transactions of The Royal Society B-biological Sciences **374**, 20180019 (2019)
3. Derry, A., Cañedo-Argüelles, M., Melles, S.J.: Freshwater ecosystems are becoming increasingly salty. Here's why this is a concern. The Conversation, March 22, (2023)

4. Environmental Protection Agency of United States (EPA).: Freshwater Salinization Syndrome: A Global Water Quality Threat. Accessed at EPA official website in 2024 (2022)
5. Kaushal, S.S., Mayer, P.M., Likens, G.E., et al.: Five state factors control progressive stages of freshwater salinization syndrome. Limnol. Oceanogr. **8**, 190–211 (2023)
6. Rosinger, A.Y., Bethancourt, H., Swanson, Z.S., Nzunza, R., Saunders, J., Dhanasekar, S., et al.: Drinking water salinity is associated with hypertension and hyperdilute urine among Daasanach pastoralists in Northern Kenya. Sci. Total. Environ. **770**, 144667 (2021)
7. Meybeck, M.: Global analysis of river systems: from Earth system controls to Anthropocene syndromes. Phil. Trans. R. Soc. Lond. B. **358**, 1935–1955 (2003)
8. Li, A.M.L.: Ecological Public Health for Nursing and Health Professionals in the Anthropocene. Cambridge Scholars Publishing, UK (2022)
9. Kaushal, S.S., Likens, G.E., Mayer, P.M., Shatkay, R.R., Shelton, S.A., et al.: The anthropogenic salt cycle. Nature Reviews Earth & Environment **4**, 770–784 (2023)
10. Cañedo-Argüelles, M., Hawkins, C.P., Kefford, B.J., Schäfer, R.B., Dyack, B.J., Brucet, S., et al.: Saving freshwater from salts: Ion-specific standards are needed to protect biodiversity. Science **351**(6276), 914–916 (2016)
11. Hintz, W.D., Arnott, S.E., Symons, C.C., Greoco, D.A., McClymont, A., et al.: Current water quality guidelines across North America and Europe do not protect lakes from salinization. PNAS **119**(9), 1–10 (2022)
12. Kjeldsen, S.E.: Hypertension and cardiovascular risk: General aspects. Pharmacol. Res. **129**, 95–99 (2018)
13. Levallois, P., Belmonte, C.V.: Drinking water quality and human health. MDPI. (2019)
14. Cañed-Argüelles, M., Kefford, B., Schäfer, R.: Salt in freshwaters: causes, effects and prospects – introduction to the theme issue. Philosophical Transactions of The Royal Society B-biological Sciences **374**, 20180002 (2018)
15. Carr, G.M., Neary, J.P.: Water Quality for Ecosystem and Human Health, 2nd ed. Canada: UNEP (United Nations Environment Programme Global Environment Monitoring System) (2008)
16. Thorslund, J., van Vliet, M.T.H.: A global dataset of surface water and groundwater salinity measurements from 1980–2019. Scientific Data **7**, 231 (2020)
17. Australian Government Initiative (AGI).: Salinity and water quality. Water Quality Australia (2024)
18. Kaushal, S.S., Likens, G.E., Pace, M.L., Utz, R.M., Haq, S., Gorman, J., Grese, M.: Freshwater salinization syndrome on a continental scale. PNAS, E574-E583. (2018)
19. Kaushal, S.S., et al.: Freshwater salinization syndrome: from emerging global problem to managing risks. Biogeochemistry **154**, 255–292 (2021)
20. Rakib, M.A., Sasaki, J., Matsuda, H., Fukunaga, M.: Severe salinity contamination in drinking water and associated human health hazards increase migration risk in the southwestern coastal part of Bangladesh. J. Environ. Manage. **240**, 238–248 (2019)
21. Hintz, W.D., Fay, L., Relyea, R.A.: Road salts, human safety, and the rising salinity of our fresh waters. Front. Ecol. Environ. **20**(1), 22–30 (2022)
22. E, B., Zhang, S., Driscoll, C.T., Wen, T.: Human and natural impacts on the U.S. freshwater salinization and alkalinization: A machine learning approach. Sci. Total. Environ. **889**, 164138 (2023)
23. Ondrasek, G., Rengel, Zed.: Environmental salinization processes: Detection, implications & solutions. Science of the Total Environment **754** (2021)
24. Maas, C.M., Kaushal, S.S., Rippy, M.A., Mayer, P.M., Grant, S. B., et al.: Freshwater salinization syndrome limits management efforts to improve water quality. Front. Environ. Sci. 1–20 (2023)
25. Dudgeon, D.: Multiple threats imperil freshwater biodiversity in the Anthropocene. Curr. Biol. **29**, R942–R995 (2019)

26. Xeni, C., Oliva, R., Jahan, F., Romaina, I., Naser, A.M., Rahman, M., et al.: Epidemiological evidence on drinking water salinity and blood pressure: a scoping review. Environ. Res. Health **1** (2023)
27. Wein, H.: How the body regulates salt levels. Natl. Inst (2017)
28. Ahmed, A., Al Nahian, M., Hutton, W., Lázár, A. N.: Chapter 27: Hypertension and Malnutrition as Health Outcomes Related to Ecosystem Services. In: Nicholls, R. J., Hutton, W., Adger, N., Hanson, S. E., Rahman, M. M., Salehin, M. (eds.) Ecosystem Services For Well-Being In Deltas: integrated Assessment for Policy Analysis, pp: 505–521. Switzerland: Palgrave MacMillan (2018)
29. Murray, C.J.L., et al.: Global burden of 87 risk factors in 204 countries and territories, 1990–2019: A systematic analysis for the Global Burden of Disease Study 2019. Lancet **2020**(396), 1223–49 (2020)
30. Niiranen, T.J., et al.: Heritability and risks associated with early onset hypertension: multigenerational, prospective analysis in the Framingham Heart Study. BMJ **357**, j1949 (2017)
31. Wang, N.Y., Young, J.H., Meoni, L.A., et al.: Blood pressure change and risk of hypertension associated with parental hypertension. Arch. Intern. Med. **168**(6), 643–648 (2008)
32. Fuchs, F.D., Whelton, P.K.: High blood pressure and cardiovascular disease. Hypertension **75**(2), 285–292 (2020)
33. Zhou, B., Perel, P., Mensah, G.A., Ezzati, M.: Global epidemiology, health burden and effective interventions for elevated blood pressure and hypertension. Nat. Rev. Cardiol. **18**, 785–802 (2021)
34. Jang, S., Kim, S.T., Kim, Y.K., Song, Y.H.: Association of blood pressure and hypertension between parents and offspring: The Korea National Health and Nutrition Examination Survey. Hypertens. Res. **46**, 368–376 (2023)
35. Rahman, M.S., Hossain, K.M., van Loenhout, J., wallemacq, P., Guha-Sapir, D.: Chapter 2: Effects of Salinity on Health Due to Environmental Exposure. In: Nasreen, M., Hossain, K.M., Khan, M.M. (eds.) Coastal Disaster Risk Management in Bangladesh: Vulnerability and Resilience, pp.15–43. London and New York: Routledge (2023)
36. Shammi, M., Rahman, M.M., Bondad, S.E., Bodrud-Doza, M.: Impacts of salinity intrusion in community health: a review of experiences on drinking water sodium from coastal areas of Bangladesh. Healthcare **7**(50), 1–19 (2019)
37. Araya, D., Podgorski, J., Berg, M.: Groundwater salinity in the Horn of Africa: Spatial prediction modeling and estimated people at risk. Environmental International, 176 (2023)
38. Li, X., Alu, A., Wei, Y., Wei, X., Luo, M.: The modulatory effect of high salt on immune cells and related diseases. Cell Prolif. **55**, e13250 (2022)
39. Mondal, R., Banik, P.C., Faruque, M., Mashreky, S.R., Ali, L.: Association of exposure to salinity in groundwater with chronic kidney disease among diabetic population in Bangladesh. PLoS ONE **18**(4), e0284126 (2023)
40. Ambulkar, A.: Salinity. The Encyclopaedia Britannica. Accessed at: https://www.britannica.com/science/salinity in July 2024 (2023)
41. Jabed, M.A., Paul, A., Nath, T.K.: People's perception of the water salinity impacts on human health: a case study in south-eastern coastal region of Bangladesh. Exposure and Health **12**, 41–50 (2020)
42. Talukder, M.R.R., Rutherford, S., Huang, C., Phung, D., Islam, M.Z., Chu, C.: Drinking water salinity and risk of hypertension: A systematic review and meta-analysis. Arch. Environ. & Occup. Health. **72**(3) (2017)
43. Naser, A.M., Clemmer, J.S., Mazumcer, H., Pickering, E.V., Caban-Martinez, A.J., et al.: Epidemiological association between water salinity & blood pressure in coastal populations: ambient temperature's role as a confounder. Wellcome Open Research **9**, 419 (2024)

44. DeVilbiss, S.E., Steele, M.K., Krometis, L.H., Badgley, B.D.: Freshwater salinization increases survival of *Escherichia coli* and risk of bacterial impairment. Water Res. **191**, 116812 (2021)
45. Centers for Disease Control and Prevention of United States (CDC).: Mosquito-borne Disease. Accessed at CDC official website in 2024 (2024)
46. UNEP (United Nations Environment Programme).: Water Quality for Ecosystem and Human Health, 2nd edn. United Nations (2008)
47. Marium, A., Chatha, A.M.M., Naz, S., et al.: Effect of temperature, pH, salinity and dissolved oxygen on fishes. J. Zool. Syst. (2023)
48. Walker, R.H., Smith, G.D., Hudson, S.B., French, S.S., Walters, A.W.: Warmer temperatures interact with salinity to weaken physiological facilitation to stress in freshwater fishes. Conserv. Physiol. (2020)
49. DataStream Initiative a Monitor's Guide to Water Quality.: The Gordon Foundation. Accessed at: Water: Our Source of Life - Water Quality Guide - DataStream, in 2024 (2021)
50. Cañedo-Argüelles, M., Sala, M., Peixoto, G., et al.: Can salinity trigger cascade effects on streams? A mesocosm approach. Sci. Total Environ. **540**, 3–10 (2016)
51. Guo, M., Zhou, X., Li, J., Wu, W., Chen, Y.: Assessment of the salinization processes in the largest inland freshwater lake of China. Stoch Environ Res Risk Assess **29**, 1823–1833 (2015)
52. Liu, X., Zhang, G., Zhang, J., et al.: Effects of irrigation discharge on salinity of a large freshwater lake: a case study in Chagan Lake, Northeast China. *Water* **12**, 2112 (2020)
53. Fanzo, J., Hood, A., Davis, C.: Eating our way through the Anthropocene. Physiol. Behav. **222**, 112929 (2020)
54. Prananingtyas, D., Prayogo, Rahardja, B.S.: Effect of different salinity level within water against growth rate, survival rate (FCR) of catfish (Clarias sp.). The 1st International Conference on Fisheries and Marine Science, IOP Conf. Ser.: Earth Environ. Sci. **236**, 012035 (2019)
55. Nielsen, D.L., Brock, M.A., Rees, G.N., Baldwin, D.S.: Effects of increasing salinity on freshwater ecosystems in Australia. Aust. J. Bot. **51**, 655–665 (2003)
56. Abdelsattar, M.M., Hussein, A.M.A., Abd El-Ati, M.N., Saleem, A.M.: Impacts of saline water stress on livestock production: A review. International Journal of Agricultural Science **2**(1), 1–12 (2020)
57. Li, A.M.L.: Chapter 8: Health Literacy of Planetary Diets: Key for nutritional Health, Obesity Epidemic and Environmental Sustainability. In: Disease and Health Research: New Insights, Vol. 2, pp. 98–124. UK: BP International (2024)
58. Ranganathan, J., Vennard, D., Waite, R., Dumas, P., Lipinski, B., et al.: Shifting Diets for a Sustainable Food Future. World Resources Institute (2016)
59. NSW Environment and Heritage.: Salinity. Australia: A NSW Government website (2024)

Study on Construction Planning of Water Network in Wenzhou City

Xiuxiu Chen, Rongsheng Yin, Bo Li(✉), and Yu Jiang

Zhejiang Design Institute of Water Conservancy and Hydroelectric Power Co., Ltd, Hangzhou 310000, Zhejiang, China
1295220640@qq.com

Abstract. The construction of a water network is an important measure to build a new development pattern, make up for the shortcomings of infrastructure, and promote the high-quality development of water conservancy. The basic conditions and existing problems of water network construction in Wenzhou City were analyzed from five aspects: water resources allocation, flood control and drainage, happy rivers and lakes, water conservancy informatization, and water management system. The overall layout of the Wenzhou water network was put forward: 'Three Horizontal and Two Vertical, Three Veins and Eight Pivots.' The construction tasks of the water resources guarantee system, flood control, and drainage system, happy rivers and lakes system, intelligent water conservancy system, and operation management system were clarified. The completion of the water network project can comprehensively enhance the water security guarantee ability of Wenzhou City, promote the high-quality development of water conservancy in the city, and provide references for the construction of the water network project in other cities and counties.

Keywords: Water network · Water resources allocation · Flood Control and Disaster Reduction · Happy Rivers and Lakes · Intelligent Water Conservancy · Water Management · Wenzhou City

1 Introduction

In the new stage of development in China, the problems of unbalanced and inadequate water conservancy development are becoming more prominent. Water conservancy work needs to change from 'have or not' to 'good or not' to meet people's growing need for a better life [1].

The implementation of the national water network project is a major task clearly put forward by the Fifth Plenary Session of the 19th CPC Central Committee. Different from the traditional river-lake water system connection project, the water network is a comprehensive system based on natural rivers and lakes, taking diversion and drainage projects as channels, regulation and storage projects as nodes, and intelligent regulation and control as a means, which integrates the functions of optimal allocation of water resources, flood control and disaster reduction in river basins, and protection of water

C. Weng (Ed.): WRE 2024, LNCE 634, pp. 17–28, 2025.
https://doi.org/10.1007/978-981-96-5095-8_2

ecosystems [2, 3]. The construction of a national water network can provide a solid water security guarantee for the comprehensive construction of a modern socialist country [4, 5]. Its strategic significance lies in solving the problem of uneven spatial and temporal distribution of water resources, adapting to the new situation and requirements of China's economic and social development, and promoting the high-quality development of water conservancy [6, 7]. The national water network planning and construction focus is to grasp the layout of three elements: outline, item, bond [8].

As a pioneer province of water conservancy modernization, Zhejiang Province actively promotes the construction of a water network in the province [9]. Zhejiang province has become the first batch of provincial water network pilot areas of the Ministry of Water Resources. Wenzhou is on the southeast coast of Zhejiang Province, which is the middle of the golden coastline of southeast China, with a land area of 12110 square kilometers and a sea area of 8649 square kilometers. Wenzhou is the economic, cultural and transportation center of southern Zhejiang, as well as one of the most important components of the Zhejiang water network. The research on the construction planning of the water network in Wenzhou can not only effectively support the construction of the provincial water network but also provide a strong guarantee for the high-quality development of Wenzhou's economy and society, provide a basis for the construction of Wenzhou's water network, and provide reference for the construction of other municipal water networks.

2 Data and Methods

The relevant data in this paper are mainly derived from our previous research or provided by the Wenzhou Water Conservancy Bureau [10–12].

According to Zhejiang City and County Water Network Construction Planning Guidelines [13], the water network comprises three elements: outline, item, and bond. 'Outline' refers to the main river mainstream water system, major regional water diversion projects, and key drainage channels. 'Item' refers to regional river networks, river connection projects, secondary channels of water resources allocation, urban and rural water supply pipelines, etc. 'Bond' refers to water regulation and storage projects such as important reservoirs, hubs, and flood detention areas. The overall layout of the water network was put forward according to these three elements.

Five systems need to be considered for the construction tasks: water resources guarantee system, flood control and drainage system, happy rivers and lakes system, intelligent water conservancy system, and operation management system. The technology roadmap was shown in Fig. 1.

3 Construction Foundation

3.1 Crisscrossed Water System

Wenzhou has abundant precipitation, with an average annual precipitation of 1834 mm, about three times the national average. Three rivers (Oujiang River, Feiyun River, Aojiang River) flow through Wenzhou into the sea among the eight major rivers in

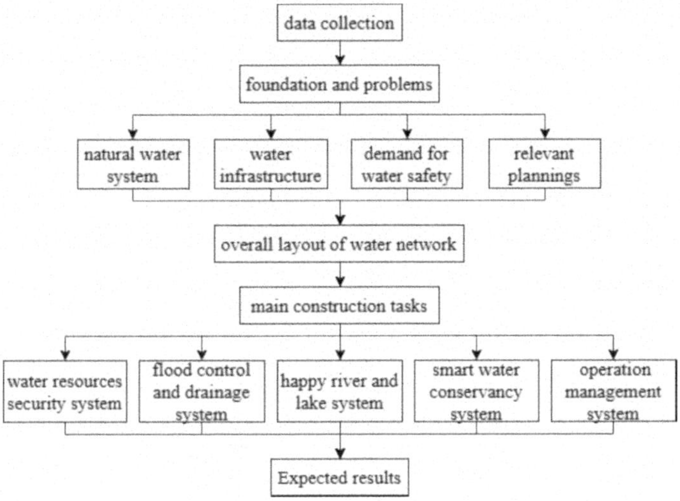

Fig. 1. Technology roadmap.

Zhejiang Province. There are 14465 rivers in Wenzhou, the total length of the river network is 18650 km, and the water area is 622.2 square kilometers, of which the river water area is 515.1 square kilometers.

3.2 More Perfect Water Conservancy Facilities

In terms of water resources protection, 183 reservoirs with water supply function and five key water diversion projects have been built in Wenzhou. The total water supply capacity of Wenzhou has reached 2.857 billion cubic meters. In terms of water disaster prevention, Wenzhou has built 251.2 km of embankments, 405.4 km of coastal seawalls, and 540 cubic meters per second of new enhanced drainage pumping stations in the main plains. The flood control capacity of 11 districts and counties except Yongjia is basically up to standard.

3.3 High-Quality Beautiful Rivers and Lakes

Taking the opportunity to build a beautiful water town in southern Zhejiang and a water ecological civilization city in Wenzhou, Wenzhou focuses on improving the water environment and restoring the water ecology. Combined with the actions of Joint Treatment of Water Banks, Big Demolition and Big Rectification, Big Construction and Big Beauty, Wenzhou has vigorously implemented seven types of projects: pollution source control, river and lake water system connection, beautiful river and lake construction, water source protection, comprehensive management of soil and water loss, hydropower ecological restoration and water culture promotion, to create a national water ecological civilization city.

3.4 Rapid-Developed Digital Water Conservancy

The Wenzhou water platform integrates six core business modules: water resources guarantee, water protection, water disaster prevention, water development planning, water affairs supervision, and water administration. The data resource and digital application systems have been constructed to drive the rapid development of digital twin water conservancy construction in Wenzhou. The construction of the digital twin basin of the Feiyun River has been included in the national pilot, laying a solid foundation for the construction of the Wenzhou digital water network.

3.5 Innovative Institutional Mechanisms

Wenzhou has vigorously explored the reform of the water conservancy project management system and created a series of 'Wenzhou experience' with distinct regional characteristics. Taking the lead in the implementing of water conservancy facilities insurance and agricultural drinking water disaster insurance in the country, the unpredictable disaster losses were dispersed into budgetary fixed insurance expenses, helping Wenzhou win the top ten experiences of national grass-roots water control in 2020 and 2022. The reform of multi-evaluation integration of water-related projects has been launched to effectively solve the simplification problem in demonstration and evaluation, winning the top ten experiences of national grass-roots water control in 2021.

4 Existing Problems

4.1 The Water Resources Security is Facing Challenges

In terms of water resource utilization, the spatial distribution of high-quality water resources and productivity in Wenzhou is inconsistent, the demand for high-quality water resources in most county-level cities is difficult to meet in the medium and long term. Some key areas lack emergency backup water sources, and there will be a situation of no water supply for a long time in the driest years in history. Yongjia has not yet achieved a multi-source water supply, and the ability to respond to sudden water supply safety incidents needs to be improved. In terms of water resources conservation, the utilization rate of reclaimed water is 19.03%, and the effective utilization coefficient of farmland irrigation water is 0.599, which needs to be improved. The average leakage rate of the pipe network is 9.53% in the urban area, but it has reached 20% in some areas or towns. It is urgent to remould the old and poor-quality pipes.

4.2 The Flood Control and Drainage System Still Has Shortcomings

In terms of flood control, the flood control projects of Oujiang River, Feiyun River, and Aojiang River are continuously constructed. However, the flood control system construction of Nanxi River, a tributary of Oujiang River, is relatively lagging. The flood control capacity of Yongjia County is insufficient to resist the flood with a return period of 20 years, and 27 km of embankments haven't been built in the mainstream of the Feiyun River. There is a lack of control projects in the upper reaches of the Aojiang

River, and the flood control pressure in the downstream towns is large. In terms of tide prevention, the upgrading and reinforcing of the seawall have been started but are not finished yet. The problems of seawall settlement and levee damage haven't been fully solved. In terms of waterlogging drainage, there are still problems such as insufficient river scale and numerous water-blocking places in Leliuhong Plain, eastern and southern areas of Wenrui Plain, Ruiping Plain, Jiangxiyang Plain and Jiangnanyang Plain, and some towns are still at risk of waterlogging.

4.3 The Modernization of the Water Governance System is Still Not Perfect

First, the ecological flow of important rivers cannot be guaranteed. The ecological flow of the Daitou section, which is in the mainstream of the Aojiang River, did not meet the requirement for 50 days in one year in history. The two major plains, Leliuhong Plain and Jiangnanyang Plain, lack a stable water supply, and the plain water body has poor fluidity and self-purification ability. Second, the control of river and lake shoreline space is insufficient, and problems such as disorderly occupation, dredging, stacking and construction, illegal fishing net laying, and illegal farming and planting on the shore still exist. Third, the happy rivers and lakes are still dominated by point-like advancement, which makes it difficult to meet the people's needs for a livable waterfront environment. Fourth, the comprehensive role of 'water conservancy plus' has not yet been fully realized. The integration of water conservancy with cultural tourism, sports, agriculture, and other industries is not enough, the degree of water space sharing and development is insufficient, and the economic potential of the waterfront needs to be stimulated.

4.4 The Intelligent Level of Water Conservancy Management Needs to Be Improved

First, the water conservancy monitoring and perception system must be improved, and the monitoring range can't cover the whole city. The monitoring frequency and accuracy can't meet the needs of digital twin water conservancy construction. The automatic monitoring rate of important hilly ponds and sluices is less than 40%. Problems such as low prediction accuracy, short forecast periods, and insufficient predictability still exist. Second, the digital twin system needs to be accelerated. The digital twin basin of the Feiyun River needs to be further optimized. The digital twin basins of Oujiang River, Aojiang River, and other important tributaries haven't been started, and most of the digital twins of large and medium-sized water conservancy projects haven't been started. Third, the application scenario of digital water conservancy is limited and needs to be further integrated with the actual water conservancy business. The intelligent application of water conservancy construction management, water space control, water resources management, and water-related public services needs to be strengthened.

4.5 The Legal System and Mechanism of Water Conservancy Management Need to Be Improved

First, the construction of the water conservancy legal system needs to be further accelerated. Only the 'Nanxi River Protection and Management Regulations' are issued at present. The legislative work of the Feiyun River, Aojiang River, and Shanxi Water Control Project progressed slowly, which can't meet the requirements of high-quality development and construction of common prosperity demonstration areas under the new situation [14]. Second, the advantages of active private capital in Wenzhou haven't been fully utilized, and the proportion of private capital participating in the construction and operation of water conservancy projects is still relatively low. The relevant supporting policies are insufficient, and the financing channels for water conservancy construction need to be further broadened. Third, the management of the water conservancy industry, project construction, and operation needs to be strengthened. The post-approval supervision of water-related projects is not strong enough. The professional personnel of the water conservancy industry are still insufficient. There are still great challenges in the completion and final acceptance of historical projects.

5 The Overall Layout of the Water Network

According to the distribution of river systems in Wenzhou City, the current situation of water conservancy projects and the demand for high-quality development of water conservancy in economic and social development, as well as related water conservancy plannings such as Overall Planning of Water Resources Conservation, Protection and Utilization in Wenzhou City (2023), The '14th Five-Year Plan' for Water Security in Wenzhou City (2021), this paper takes the provincial water resources allocation channel and the three major river projects in the city as the skeleton, takes the plain river management, regional river-lake water system connectivity and irrigation and drainage canal system as the vein, takes the main water sources and key hub projects as the nodes, and takes digitalization, networking and intelligent regulation as the means to coordinate the construction of water resources guarantee system, flood disaster prevention and control system, and water ecological protection system. As shown in Fig. 2, the overall layout of 'Three Horizontal and Two Vertical, Three Veins and Eight Pivots' of the Wenzhou water network was put forward to realize the integration and intercommunication of water networks at the provincial, municipal, and county level and form a modern, beautiful water network system in southern Zhejiang.

The 'Three Horizontal' refers to the three rivers flowing into the sea from west to east across Wenzhou: The Oujiang River, the Feiyun River, and the Aojiang River. These rivers have important functions such as flood discharge, water resources transportation, ecological conservation, transportation, and shipping.

The 'Two Vertical' refers to the coastal seawall and the river connecting channels. The coastal seawall channel realizes the connection of the coastal seawall in the whole city through the connection between the seawall project and the road networks. It creates a safe and convenient shoreline belt with Wenzhou characteristics. The river connecting channel realizes the water system connectivity of the three rivers and five plains, which can solve the drainage dilemma of coastal plains, and improve their drainage capacity.

The 'Three Veins' refers to the three main paths of water resources allocation in the city: the Shanxi-Zhaoshandu water diversion project, the Oujiang water diversion project, and the Wenzhou section of the water resources allocation channel in eastern Zhejiang.

'Eight Pivots' refers to eight large and medium-sized key water conservancy projects: Shanxi Water Control Project, Zeya Reservoir, Shunxi Reservoir, Qiaodun Reservoir, Nanxi River Water Supply Project, Xunzhai Reservoir, Feiyun River Estuary Gate, and Aojiang River Estuary Gate. These projects are important nodes for connecting major rivers and water conveyance channels.

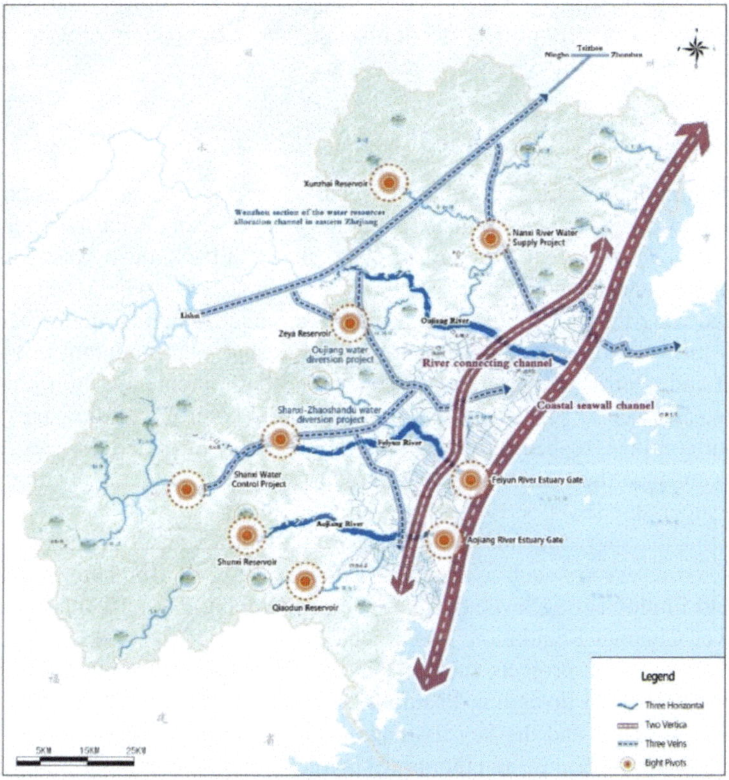

Fig. 2. Overall layout of the water network in Wenzhou.

6 Main Construction Tasks

6.1 Improve the Water Resources Security System

Guided by the demand and allocation of water resources within the city, we focus on solving the key problems, such as the shortage of total water resources and the imbalance of spatial allocation, highlight key areas, important users, and key links, implement the economic and intensive use of water resources, fully tap the potential of water resources in the three rivers and five plains, add high-quality water sources from outside the region, expand the scope of water supply for main water sources, improve the capacity of water resources allocation, and create a water resources guarantee system covering the whole region, to support the high-quality development of Wenzhou's economy and society.

Focusing on water conservation in agriculture, industry, urban life, and the use of unconventional water. Promoting water-saving actions. Building Yongjia Xunzhai, Pingyang Yuexi, Ruian Liuke, Yueqing Yinxi, and other medium-sized reservoirs to supplement high-quality water. Planning the estuary gate in Feiyun River, Aojiang River, and Nanxi River to solve the general water use problem in the coastal plains. Promoting the implementation of projects such as water resources allocation channels in eastern Zhejiang (Wenzhou section), Oujiang River water diversion project, and Nanxi River water supply improvement project to optimize the allocation pattern of water resources and increase water security.

6.2 Improve the Flood Control and Drainage System

Taking the three major basins (Oujiang River Basin, Feiyun River Basin and Aojiang River Basin) as the governance units, the relationship between the basin and the region, the upstream and downstream, the left bank and the right bank, the mainstream and the tributary, flood control, and drainage should be coordinated as a whole, and the way out of flood and waterlogging should be reasonably arranged. Engineering measures such as seawall protection, river management, plain drainage, and mountain flood disaster prevention and control in small watersheds should be implemented in a planned and step-by-step manner to solve the weak links of flood control and drainage, make up for the shortcomings, realize the 'full closure' of flood control and disaster reduction, effectively enhance the resilience of water security system, and enhance the ability to cope with super-standard floods and super-strong typhoons.

Promoting the upgrading and reconstruction of 200-km seawalls in the city. Planning medium-sized reservoirs such as Yueqing Longxi, Cangnan Dingbutou, and Taishun Xiyuan, and implementing levee projects such as Oujiang River Regulation Phase II, Shupu River Channel Regulation, and Jinchao Port Comprehensive Management, as well as flood diversion projects such as Guxi flood diversion, Dajing flood diversion, and Jinchao Port flood diversion. Promoting the management of drainage projects of the five coastal plains and the key drainage networks through measures such as river dredging, entrance widening, and pumping station drainage. Promoting small watershed management to prevent and resolve the risk of flash flood disasters.

6.3 Build a Happy River and Lake System

We should actively promote the construction of happy rivers and lakes in the whole region, fully excavate the connotation of water elements such as three rivers, wetlands, corridor bridges, et al., and construct scenes such as source area water resources protection, river waterfront space expansion, hydrophilic sharing, water ecological value conversion and Ouyue water culture, to make Wenzhou's river and lake water system become a livable space for urban and rural residents to live a happy life, to show the unique warmth of the happy rivers and lakes in southern Zhejiang, and help Wenzhou to realize the strategic positioning of the city of mountain, sea, poetry and painting.

Promoting the management of 186 km of small and medium-sized rivers. Promoting Pingyang County water system connectivity and water beauty rural construction. Accelerate the implementation of Happy River and Lake construction in counties like Ouhai,

Rui'an, Taishun, and Wencheng. Hydrophilic facilities such as recreational greenways, natural swimming pools, camping beaches, viewing platforms, water weirs, and dams should be built to meet the diverse needs of sports and leisure, tourism and health, education, and research. Several rural revitalization water conservancy demonstration projects, such as Baofuwan in Yantou Town, Taishi Village in Shatou Town, Qiongkou Village in Juyu Town, and Fuxi Village in Zaoxi Town, should be implemented.

6.4 Build a Smart Water Conservancy System

According to the principle of 'demand traction, the application first, emergency first, actual combat effectiveness', we will comprehensively promote the construction of Wenzhou's intelligent water conservancy system, striving to improve the intelligent level of Wenzhou's water network and promote the transformation of water safety management to active management and fine management. Focusing on the construction tasks of data, algorithms, and computing power, as well as the requirements of forecasting, early warning, rehearsal, and planning, we will promote the construction of a digital twin water conservancy in the whole city, consolidate the basic network of hydrological monitoring and perception, build digital twin basins, digital twin water networks, digital twin projects, and iterative intelligent business applications, to comprehensively improve the level of water conservancy modernization management and help the high-quality development of water conservancy in Wenzhou.

6.5 Deepen the Operation Management System

Focusing on the goal and task of high-quality development of water conservancy in the new stage, we will focus on solving the key and difficult problems and deep-seated contradictions in water conservancy development, establish and improve the legal system and mechanism of the water network by the idea of 'strengthening mechanism, emphasizing guarantee and improving capacity.' The mechanism includes the river and lake chief system with Wenzhou characteristics, the management methods of important water sources, comprehensive management of county water conservancy, and watershed protection legislation. The management ability of water conservancy construction and operation needs to be improved, including water conservancy construction management, water conservancy operation management, and water conservancy industry management. To promote the sustainable development of the water network, we should broaden the investment and financing channels of the water network, deepen the flood insurance mechanism of water conservancy disasters, promote water rights trading, build a high-level water conservancy operation management system, and promote the modernization of water conservancy governance capacity.

7 Expected Results

7.1 A Substantial Increase in Water Resources Security Capacity

After the completion of the Wenzhou water network, the supply capacity of high-quality water in Wenzhou can reach 1.262 billion cubic meters. The regulation capacity of the water resources allocation channel reach 1.313 billion cubic meters, ensuring the high-quality water demand of 11 million people and the agricultural water demand of 2.3 million acres of cultivated land, meeting the demand of social and economic development to the greatest extent, and making a typical demonstration for similar cities in the southern wet area.

7.2 Improvement in Flood Control and Disaster Reduction Capacity

After the completion of the Wenzhou water network, coastal areas can withstand tides with a return period of 50 to 200 years, county-level cities can withstand floods with a return period of 50 years, and the five major plains can withstand waterlogging with a return period of 20 years. The flood control capacity of the headwaters of the Oujiang River, Feiyun River, and Aojiang River can be enhanced, and the total flood control capacity of the reservoirs will reach 300 million cubic meters. The drainage capacity of the five major plains can be significantly improved, and the drainage capacity of pumping stations exceeds 700 cubic meters per second. The risk of urban waterlogging will be greatly reduced, and the high-quality development of the social economy can be guaranteed.

7.3 Build a Model of Happiness River and Lake

After the completion of the Wenzhou water network, the Happy River and Lake system, in line with Wenzhou's social and economic development, will be fully completed, which can effectively promote regional economic development and the construction of common prosperity. The functional orientation of the ecological pearl of Zhejiang and ecological industry agglomeration area will be clearer, and a model city for the construction of rivers and lakes in Jiangnan water town with shared benefits will be created.

7.4 Digital Twin Water Conservancy is at the Forefront

After the completion of the Wenzhou water network, the water network perception system covering flood and drought disasters, the whole life cycle of water conservancy projects, water quantity and quality, water use, river and lake water areas, wading activities, ecological flow, soil and water conservation will be fully completed. Aojiang Digital Twin Basin and Wenzhou water supply integrated dispatching platform operate efficiently, and 24 digital twin projects will be built. The data, model platform, and knowledge platform are rich, and the application scenarios of the water conservancy digital brain are extensive and diverse.

7.5 Create a Water Conservancy Innovation Vitality City

After the completion of the Wenzhou water network, it can adapt to the social and economic development of southern Zhejiang, fully stimulate the flexible private economic potential of Wenzhou, build a water economic model jointly built and shared by the whole society, create a more open, dynamic and innovative Wenzhou water network operation management system and mechanism, leading the modernization of water conservancy development in southern Zhejiang, and help the high-quality development of the region to build a common prosperity demonstration area.

8 Conclusion

Wenzhou water network is the connection between the provincial water network and the county water network. It plays a supplementary role in the provincial water network and a guiding role in the county water network. Based on the conditions of the natural water system, water resources endowment, and the current situation of water conservancy projects in Wenzhou, this paper analyzed the foundation and existing problems of water network construction in Wenzhou from five aspects: water resources allocation, flood control and disaster reduction, happy rivers and lakes, water conservancy informatization and water management system. The overall layout of the water network in Wenzhou was constructed, and the relevant construction tasks were put forward. The research on Wenzhou water network construction planning is conducive to promoting the construction of the Wenzhou water network and the high-quality development of the water conservancy. It can also provide ideas and references for the research of water network construction in other regions.

References

1. Wang, P., Li, J.Q., He, J., et al.: Strategic thinking of modern water network planning. Water Resour. Plan. Des. (9), 3–6+43 (2021)
2. Wang, P., Li, J.Q.: Strategic thinking of Xi'an modern water network construction planning. Water Resources Planning and Design **3**, 1–7 (2023)
3. Huang, J., Yan, T., Wu, M.J., et al.: Study on water network construction in Guiyang City. Express Water Resources & Hydropower Information **45**(7), 8–12 (2024)
4. Zuo, Q.T., Guo, J.H., Li, Q.W., et al.: Constructing the theoretical system of national water network by referring to the experience of South-to-North Water Diversion Project. China Water Resour. (11), 22–24+21 (2021)
5. Li, G.Y.: To provide a strong water security guarantee for comprehensively promoting the great cause of building a strong country and national rejuvenation with Chinese-style modernization. Water Resources Development and Management **10**(3), 1–2 (2024)
6. Party Group of the Communist Party of China of the Ministry of Water Resources.: Speed up the construction of the national water network, to provide a strong water security guarantee for national rejuvenation. China Water Resour. (13), 1–4 (2023)
7. Xia, J., Chen, J., Yu, D.X., et al.: Opportunities and challenges of China's modern water network construction under the changing environment. Acta Geogr. Sin. **78**(7), 1608–1617 (2023)
8. Liu, G.Q., Xu, C., Chen, X.Q., et al.: Thoughts on planning and constructing water network in Yangtze River Basin. Technol. Econ. Chang. **7**(1), 53–56+105 (2023)
9. Li, R.: Build a large water network in Zhejiang, strive for water conservancy modernization pioneer province. China Water Resour. (224), 66 (2023)
10. Zhejiang Provincial Water Conservancy Department, Zhejiang Provincial Development and Reform Commission: Zhejiang water network construction planning (2022)
11. Wenzhou Water Conservancy Bureau: Overall planning of the water resources conservation, protection, and utilization in Wenzhou City (2023)
12. Wenzhou Water Conservancy Bureau: The '14th Five-Year Plan' for water security in Wenzhou City (2021)
13. Zhejiang Provincial Water Conservancy Department: Zhejiang city and county water network construction planning guidelines (2022)

14. He, L.F.: Support the high-quality development of Zhejiang to build a demonstration area of common prosperity and provide a provincial example for the country to promote common prosperity. Macroecon. Manag. (7), 1–2+20 (2021)

Evaluation and Analysis of Water Quality of Five Rivers in Poyang Lake Based on CWQI and BCWQI

Chao Ji[1], Hongwei Yin[2], Jin Qian[2], and Yong Ji[2](✉)

[1] Jiangxi Provincial Port & Waterway Construction Investment Group Co., Ltd., Nanchang 330000, China
1934269652@qq.com

[2] College of Hydraulic Engineering, Nanchang Institute of Technology, Nanchang 330099, China
2907806226@qq.com, jiyong@nit.edu.cn

Abstract. The Poyang Lake basin serves as a natural regulator of the Yangtze River's water levels, which is crucial for water retention and purification. This study investigates the long-term temporal and spatial variations in water quality within the Wu Basin of Poyang Lake. Monthly monitoring data from five hydrological stations along the Rao, Xiu, Xinjiang, Gan, and Fu Rivers, connected to Poyang Lake, were utilized. Water quality was assessed using the Canadian Council of Ministers of the Environment Water Quality Index (CWQI) and the British Columbia Water Quality Index (BCWQI). Temporal trends were analyzed using the Mann-Kendall (M-K) trend test and time series analysis methods to explore spatial characteristics. The results include: (1) The CWQI and BCWQI spatial variability within the Wu Basin generally indicates good water quality, with an average evaluation score of 84. The western region exhibits better water quality than the eastern region. (2) Water Quality Index (WQI) values are predominantly influenced by dissolved oxygen (DO), chemical oxygen demand (CODMn), biochemical oxygen demand (BOD5), and ammonia nitrogen (NH3-N). (3) Significant temporal and spatial variations were observed in the Gan, Fu, and Xiu Rivers, while Xinjiang and Rao tributaries showed less pronounced differences. (4) Spatiotemporal disparities in water quality are primarily attributed to rainfall patterns, human activities, and land use. This research provides insights valuable for guiding water pollution control and comprehensive water quality management in the Wu Basin of Poyang Lake.

Keywords: CWQI · BCWQI · Poyang Lake · The Five Rivers · Water Quality Evaluation

C. Weng (Ed.): WRE 2024, LNCE 634, pp. 29–45, 2025.
https://doi.org/10.1007/978-981-96-5095-8_3

1 Introduction

Water quality monitoring is the basis for studying the change in water quality in a basin, and water quality parameters are the premise for evaluating the water quality in a basin. Timely and accurate monitoring of river water quality and obtaining water quality information of rivers and lakes are of great significance for basin water resources development, water environment protection, water ecological restoration, flood control and disaster reduction [1]. In recent years, China has intensified its efforts to control the water environment, and the surface water quality has generally improved throughout the country. However, the water quality in some basins is still unbalanced and uncoordinated [2, 3]. In order to take effective water environment management measures and prevent water pollution, it is imperative to carry out long-term monitoring and evaluation of water quality to understand the status of river water quality and its spatial-temporal changes [4, 5].

Water quality assessment is a basic method to study the evolution of watershed water, a basic work for rational utilization and protection of water resources, and a basis for water environment management and decision-making. Surface water quality assessment constitutes a complex process which needs to consider many samples and various water quality parameters [6]. At present, many water quality evaluation methods are widely used in China. The most commonly used water quality evaluation methods include the analytic hierarchy process [7], gray cluster evaluation [8], single factor evaluation, fuzzy mathematics evaluation, artificial neural network evaluation and comprehensive water quality index (WQI) method [9, 10].

Poyang Lake is situated on the south bank of the last reaches of the middle reaches of the Yangtze River. Influenced by the top support of the Yangtze River and the five rivers (Gan River, Fu River, Xinjiang River, Rao River and Xiu River) [10], the water period of Poyang Lake alternates periodically with the dry period. Gan River, Fuhe River, Xinjiang River, Rao River, Xiu River and the surrounding hills and mountains are important regions within the Poyang Lake basin. As the most extensive freshwater lake in China, it plays great special ecological functions such as flood water storage and biodiversity protection in China's Yangtze River Basin. Additionally, the Five Rivers area abounds with water resources, with numerous reservoirs, complex river networks and numerous waterways, and most of the basins are dominated by industry and agriculture, whose environmental quality influences are imported from the basin to the waters of Poyang Lake [11].

By studying the time-scale change characteristics of water quality in the five regions of Poyang Lake Basin, this study adopted the Canadian Council of Ministers of the Environment Water Quality Index (CWQI) and the British Columbia Water Quality Index Method (BCWQI) to comprehensively evaluate the water quality index of the five major rivers in the basin. Combined with trend test, spatial-temporal series analysis and linear regression analysis, the water quality data of 5 monitoring sections during 2008–2010 and 2021–2022 were evaluated, and trend analysis was conducted with the aim of comprehensively evaluate the overall status of regional water quality, the time change trend and spatial distribution characteristics of water quality. In order to provide targeted data support for the water environment quality management and water pollution prevention and control policies in this area, and provide an effective reference for the relevant research of river water quality evaluation.

2 Materials and Methods

2.1 Overview of the Study Area

Poyang Lake (east longitude 115°50'~116°44', north latitude 28°25'~29°45'), the largest freshwater lake in China, is located in the north of Jiangxi Province. It is one of the main tributaries of the middle stream basin of the Yangtze River. The Poyang Lake system is mainly comprised of the Gan River, Fu River, Xinjiang River, Rao River and Xiu River and their tributaries at all levels (Fig. 1). Its subtropical humid monsoon-influenced climate features hot and muggy summers, dry and cold in winter, and rarely frost in winter. According to the water system characteristics of the study area, the water quality of the five river systems of Poyang Lake was evaluated according to seven control hydrology stations: Ganjiang Waizhou, Fuhe Lijiadu, Xinjiang Meigang, Changjiang Dufeng Keng, Le'an River Hushan, Liaohe Wanjia Port and Xiuhe Qiujin.

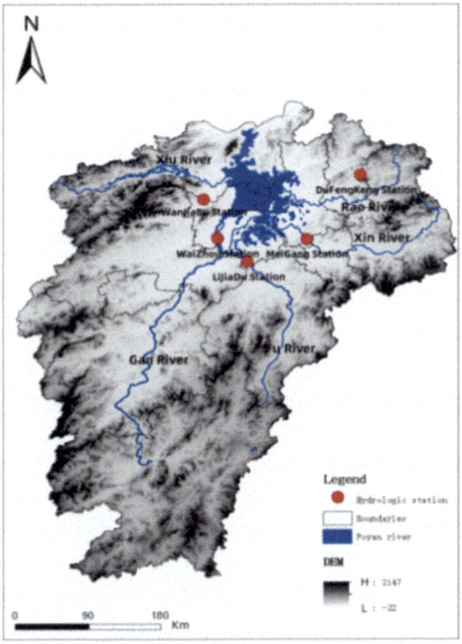

Fig. 1. Study area.

2.2 Data Sources

Monthly water quality data from 2008–2010 and 2021–2022 from 7 monitoring sections in Wu District of Poyang Lake Basin were used as water quality assessment indicators.

The other socio-economic data and pollution assessment data for analysis were collected from Jiangxi Provincial Environmental Bulletin and Jiangxi Provincial Environmental Statistical Annual Report. The monitoring stations cover seven river systems in the Poyang Lake basin, including the Ganjiang River Basin, Fuhe River Basin, Xinjiang River Basin, Changjiang River Basin, Le'an River Estuary and Liaohe River Basin. In this study, nine monitoring data of water temperature (T), pH value, dissolved oxygen (DO), ammonia nitrogen ($NH^{3-}N$), total phosphorus (TP), five-day biochemical oxygen demand (BOD_5), permanganate index (COD_{Mn}), sulfate (SO_4^{2-}) and nitrate (NO^{3-}) were selected as water quality evaluation objects [11]. The geographical location of each station is shown in the Fig. 1, and the water system information features are shown in the Table 1. In this study, Excel, SPSS 22.2, origin 2018 and RStudio integrated development environment with R language were used for data sorting, calculation and analysis, and ArcMap 10.2 software was used for terrain information visualization.

2.3 Evaluation Method

The water quality index method (WQI) calculates a large number of water quality monitoring data with a simple formula, and the quality of water is stratified into several levels, such as very good, good, good and poor. Water quality index evaluation method is simple to use, has intuitive evaluation results, and high accuracy, can reflect the degree of surface water pollution more completely, and has been widely used at home and abroad; For example, the National Sanitation Foundation Water Quality Index (NSFWQI), the Canadian Council of Ministers of the Environment Water Quality Index (CCMEWQI), and the Minimum Water Quality Index (MWQImin) model are commonly utilized in developing countries [10].

2.3.1 Canadian Council of Ministers of the Environment Water Quality Index Method

The Canadian Council of Ministers of the Environment Water Quality Index (CWQI) is grounded in the Water Quality Index published by the University of British Columbia in 1995 and is now used in many Canadian provinces. The indicator assesses if the water quality surveillance data surpasses the permissible water quality threshold in three dimensions: Scope, Frequency and Amplitude [12]. According to the value distribution grade of the Canadian Drinking Water Quality Composite Index (CWQI), the value range is 0 to 100, 0 means very poor water quality, 100 means excellent water quality. According to CWQI values, water quality is divided into five levels: very good (95-100), good (80-94), medium (60-79), Pass (45-59), and poor (0-44). The Canadian Council of Ministers of the Environment Water Quality Index is subjective in its classification [11].

2.3.2 Water Quality Index Method for British Columbia, Canada

The British Columbia Water Quality Index Method (BCWQI) was introduced by Environment Canada in 1995 as an additional water quality assessment index. The calculation procedures of this index method closely resemble those of the Canadian Council of Ministers of the Environment Water Quality Index (CWQI) method described earlier. The calculation steps are different from CWQI except for the final index calculation, and the formulas for F1 (range), F2 (frequency) and F3 (amplitude) are the same [12].

2.4 Statistical Methods

2.4.1 M-K Test

The Mann-Kendall (M-K) test was employed to investigate the WQI and assess the temporal trends of indicators. The statistical analysis was conducted using the MK test function in R Studio. At an alpha level of α, if $Z > Z(1\text{-}\alpha/2)$, the null hypothesis of no trend is rejected, indicating a significant change trend. Threshold values for Z at different confidence levels (90%, 95%, and 99%) are 1.65, 1.96, and 2.58, respectively. In this study, an alpha level of 0.05 ($\alpha = 0.05$) served as a metric for assessing the importance of variations. A Z-score of positive value underscores a significant upward tendency, in contrast, a Z-score of negative value underscores a significant downward tendency [13, 14].

2.4.2 Time Series Method

Temporal series analysis involves a sequence of random variables arranged over time, typically derived from observations of an underlying process at regular intervals. Long-term monitoring and evaluation of river water quality are essential for effective water environment management and pollution prevention. This approach helps in understanding the current status and spatiotemporal dynamics of water quality. Specifically, time series analysis is utilized to examine the temporal and spatial variations in WQI values within the Wu District. This method captures trends in random variables over time and extracts patterns from the data [16].

3 Results and Discussion

3.1 Evaluation of Water Quality Status Based on CWQI

3.1.1 Analysis of Variation Characteristics Between Years

An evaluation and comparison of the water quality in the Poyang Lake Basin's five rivers was conducted utilizing the Canadian Council of Ministers of the Environment's water quality index method for the years 2008–2010 and 2021–2022 [15]. Nine parameters, including pH, water temperature (T), DO, BOD_5, COD_{Mn}, TP, TN and NH_3^-, were utilized to calculate the CWQI value. The results show that the selected WuHe cross section data are 57.6–99.5. According to the grade of CWQI water quality evaluation, it is mainly divided into "very good", "good" and "medium", accounting for 16.5%, 63.8% and 14.6% of the total, respectively. From the perspective of spatial difference, the overall mean is 84.8, and the evaluation grade is "good". It is equivalent to the second level of water quality evaluation in our country.

Comparing the CWQI values of Wuhe rivers in Fig. 2, the water quality of Fu River has the lowest score compared with the other four tributaries, with an average value of 82.3, and the minimum CWQI value of 57.6, which is a "pass" level, indicating that the water quality has often been threatened or damaged. The water quality conditions often cannot maintain a natural or satisfactory level. Compared with the other four tributaries, the water quality of Gan River has the highest score. The average CWQI value is 86.7, with a value of minimum CWQI of 74.6, and the median level is 88.2, which is the best comprehensive water quality among the five major rivers in various indexes. There are spatial differences in the water quality of the five rivers in Poyang Lake, and the water quality scores of the western region (Gan River and Xiu River) are better than those of the eastern region (Rao River, Fu River and Xin River).

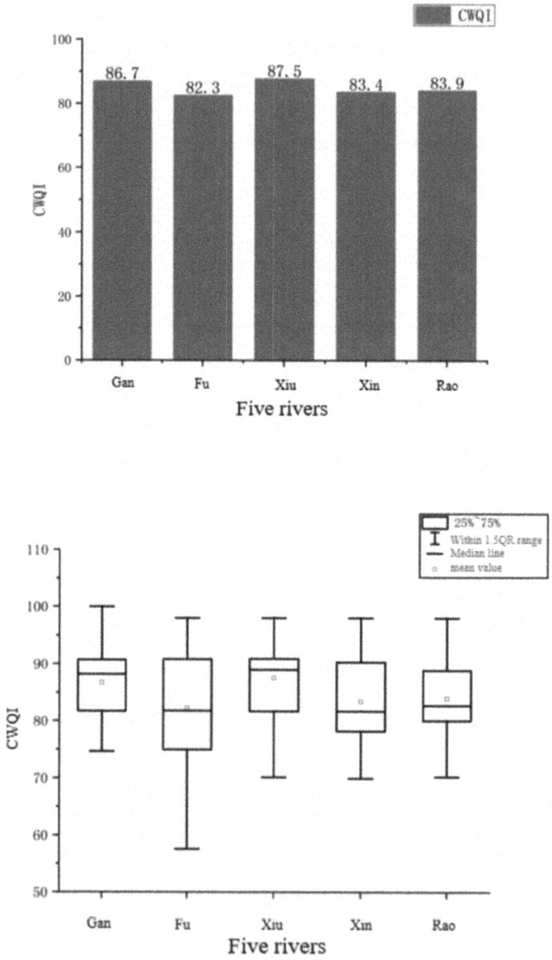

Fig. 2. CWQI values comprehensive statistical data of Wuhe District.

3.1.2 Analysis of General Characteristics During the Year

The monthly CWQI data of each river in the five time periods were statistically analyzed using the spatiotemporal series analysis method. The annual and monthly variation data of the CWQI value of each tributary river system in the Five Rivers area are shown in the following Fig. 3. Ganjiang River: From 2008 to 2010, the water quality of Ganjiang River did not change much and tended to be stable, and the overall average monthly water quality did not change greatly, and the average CWQI was 88. From the multi-year data analysis, the water quality grade in 2021–2022 is slightly lower than in 2008–2010, but the decline is relatively low.

The variation of CWQI of water quality in five rivers can be seen from the chart data. Fuhe River: In 2008, the water quality of Fu River was the best, with an average CWQI of 90, and in 2021, it was the worst, with an average CWQI of 72. After nearly 10 years of change, the water quality of Fu River shows an obvious decline trend from "good" to "average", but the water quality in 2022 is slightly improved compared with that in 2021. Xiu River: The water quality of the 5-year Xiu River is between 84–91, and the overall level shows a slow downward trend. The average annual level is above 80, which is at a "good" level. Xinjiang River: In the past five years, the water quality of Xiu River has been between 82–85, and the overall level shows a stable trend. The water quality has not changed much due to fewer human influence factors. Rao River: The change trend is consistent with that of Xinjiang River. In 5 years, the water quality of Rao River is between 81–85, and the overall level shows a stable trend without great change.

3.2 Evaluation of Water Quality Status Based on BCWQI

3.2.1 Analysis of variation characteristics between years

The BCWQI and British Columbia Water Quality Index results were compared and verified. The results show that the data of the five rivers section evaluated by the water quality index method of British Columbia, Canada, are between 56.6–99.9. The grades of water quality evaluation are mainly "very good", "good" and "medium", accounting for 15.5%, 68.6% and 13.8% of the total, respectively. From the perspective of spatial differences, the overall mean is 84.1, and the evaluation grade is "good". It is equivalent to the second level of water quality evaluation in our country.

Compared with the BCWQI values of the five major rivers from Fig. 4, the water quality of Fu River and Xin River has the lowest scores compared with the other three tributaries, with the average value of 82.1 and 82.0, and the minimum BCWQI value of 56.6, which is a "pass" level, indicating that the water quality has often been threatened or damaged. The water quality conditions are often unable to maintain a natural or satisfactory level. Compared with the other four tributaries, the water quality of Gan River has the highest score. The average CWQI value is 86, the minimum value is 71.4, and the median level is 88.8, the best comprehensive water quality among the five major rivers in all major indexes. According to the BCWQI value, the water quality of the five rivers of Poyang Lake has the same spatial difference, and the water quality score of the western region (Gan River and Xiu River) is better than that of the eastern region (Rao River, Fu River and Xinjiang River).

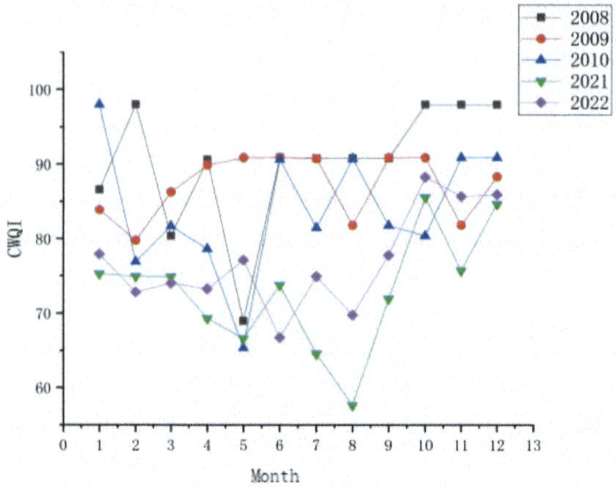

a-Fu River

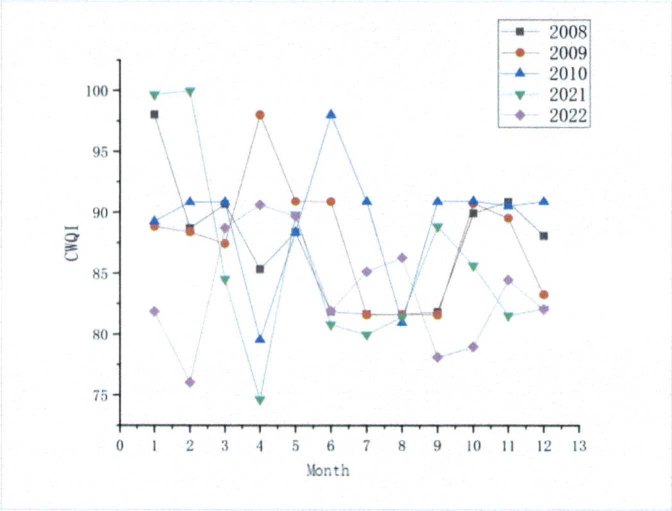

b-Gan River

Fig. 3. Variation curve of CWQI value of tributary.

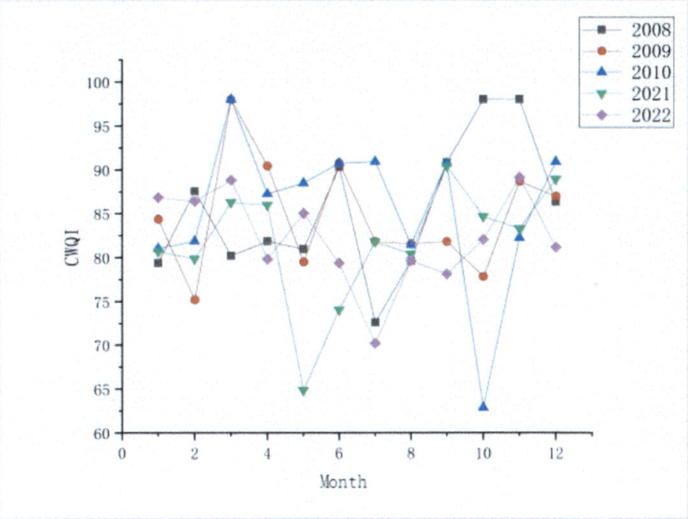

c-Rao River

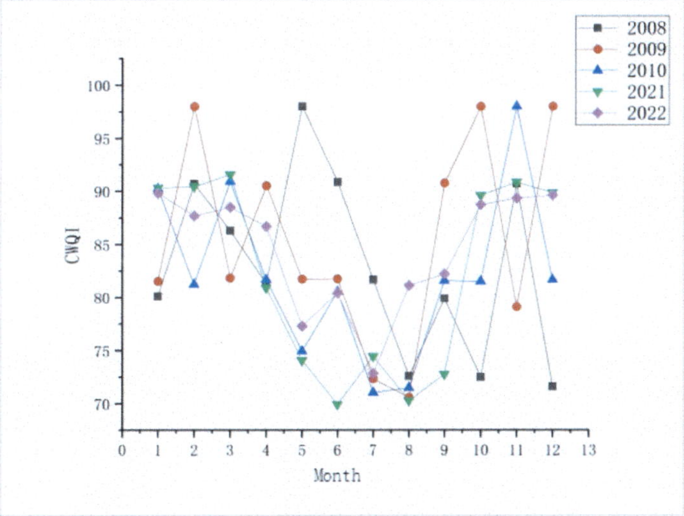

d-Xin River

Fig. 3. (*continued*)

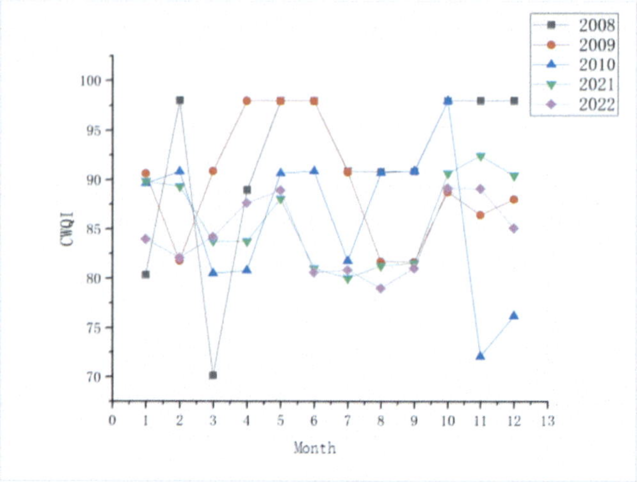

e-Xiu River

Fig. 3. (*continued*)

Compared with the comprehensive results of CWQI and BWQI, the overall spatial differences are close. The water quality scores of the western region of Poyang Lake region are better than the eastern level. The overall water quality evaluation results are between 80–94, which is equivalent to the second-class water quality level of China's surface water.

3.2.2 Analysis of General Characteristics during the Year

Using the spatio-temporal series analysis method, the BCWQI data of each river in 5 time periods were statistically analyzed. The CWQI value of each tributary river system in the Five rivers area was changed year by year and month as shown in the following Fig. 5 Gan River: 2010 Gan River water quality is the best, BCWQI average of 88, close to the median level of 89, the overall monthly average water quality did not change greatly, only in June to July reached "excellent" water quality. From the data analysis of many years, the water quality of Gan River has also evolved from "good" to "average", and reached the worst BCWQI score of 80 in 2022.

According to the data trends in Fig. 5, the changes in BCWQI of water quality for the five rivers indicate the trends in variation. Fu River: In 2008, the water quality of Fu River was the best, with the average BCWQI of 89, equal to the median level of 89, and the overall average monthly water quality parameters tended to be stable. 2022 is the worst year, with an average BCWQI level of only 72. In the past 10 years, the water quality of Fu River has changed from "good" to "average". Xiu River: The water quality of Xiu River in 2008 was the best, BCWQI average of 92, close to the "high quality" level. 2022 is the worst year, with an average BCWQI level of 81. It is consistent with the trend of water quality in Fu River, from "high quality" to "general". Xin River: The water quality parameters of Xinjiang River are between 79–84. The annual monthly average changes

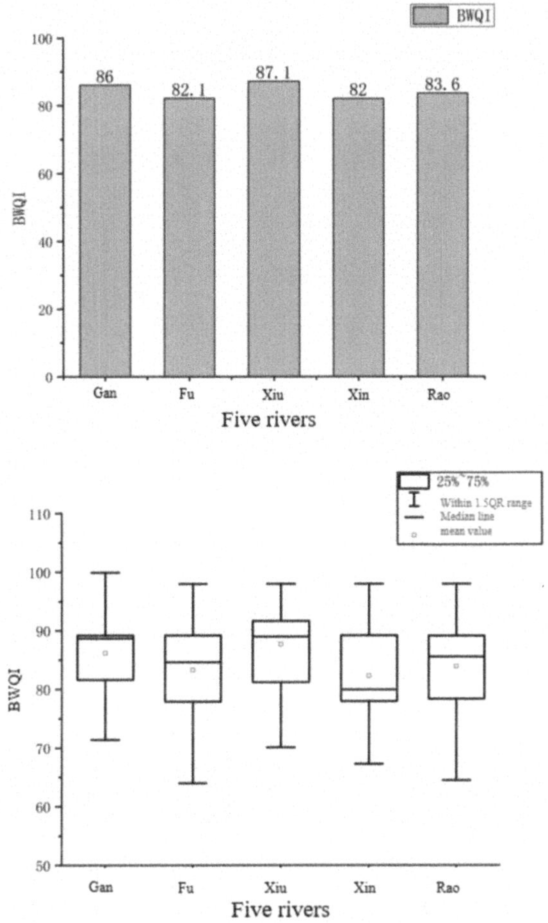

Fig. 4. BWQI value comprehensive statistical data of Wu District.

are similar, and there is no great change; Rao River: The water quality parameters of Rao River are between 78 and 85. Water quality did not change much from 2008 to 2010, but increased slightly in 2010. However, comparing the water quality parameters of 2021 and 2022, it can be observed that a substantial drop in water quality will occur in 2022.

3.3 Change Trend Analysis

To further investigate the trend of water quality change in five major rivers, linear regression analysis was used to analyze the annual mean value of CWQI and BWQI in five major rivers areas. The findings indicated that the water quality in the five major rivers area of Poyang Lake showed a funnel-like trend of first decreasing and then rising. However, the regression curve in Fig. 6 shows that the overall WQI value in five major rivers District has a significant downward trend, while the WQI value years in 2022 and 2021 show different upward trends, indicating that in the past ten years, China's

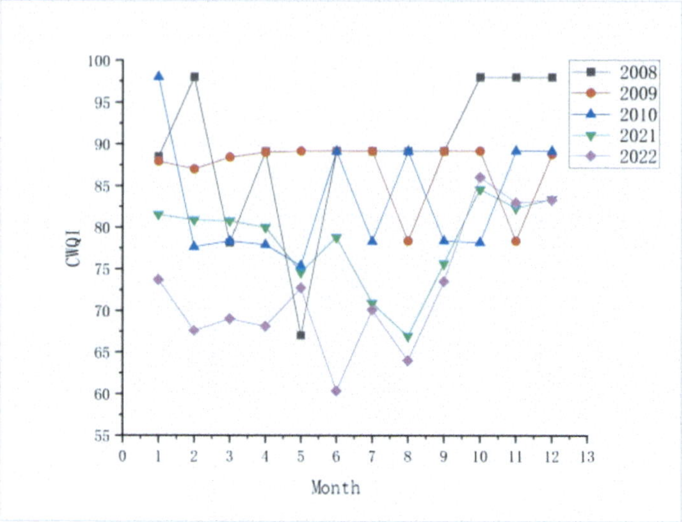

a-Fu River

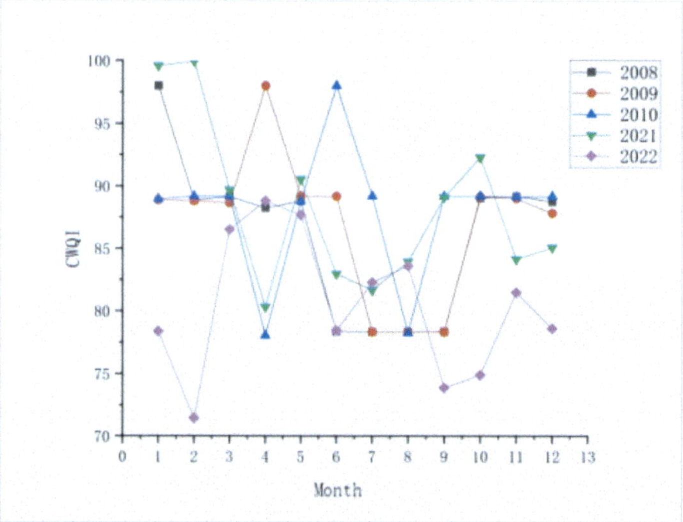

b-Gan River

Fig. 5. Changes of BCWQI value of tributary.

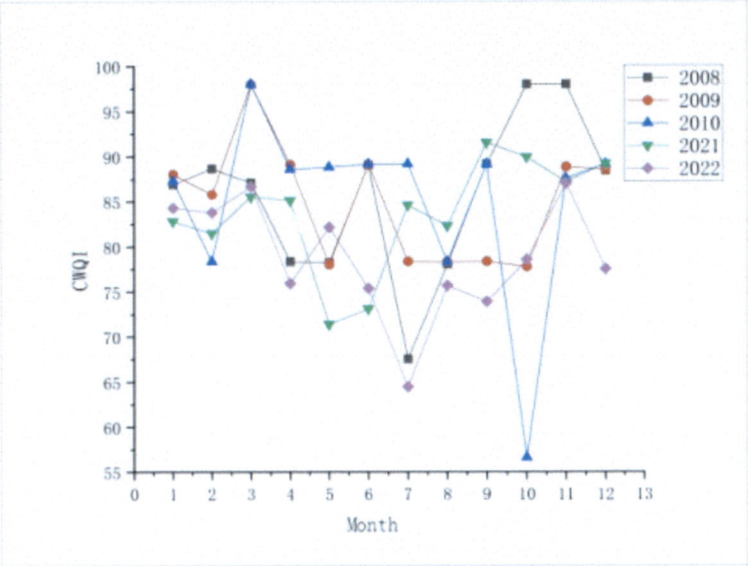

c-Rao River

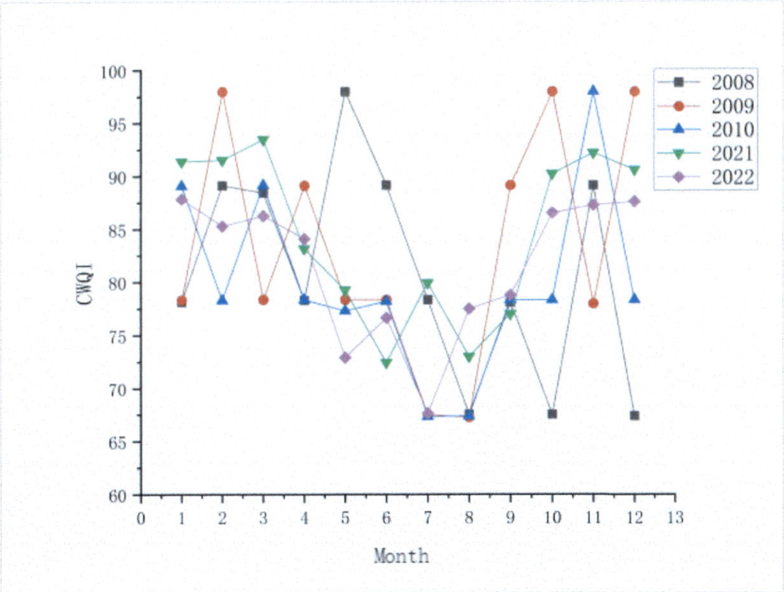

d-Xin River

Fig. 5. (*continued*)

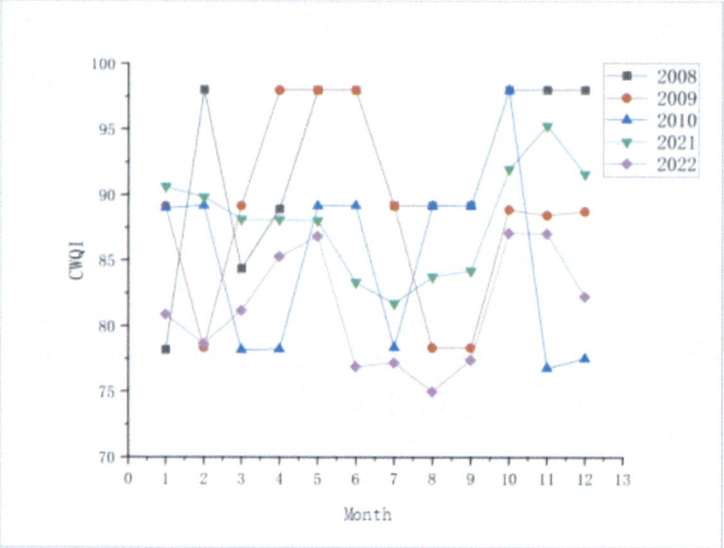

e-Xiu River

Fig. 5. (*continued*)

watershed governance in Poyang Lake region has achieved obvious effects, making the overall water quality grade in 2022 return to the overall average value of water quality. According to the statistical yearbook of Jiangxi Province in 2022, the proportion of excellent water quality in the monitoring section of the province was 93.6%, with Class V water quality accounting for 0.3%, and no poor Class V water quality recorded. In the national assessment of surface water sections within the province, the proportion of excellent water quality was 96.2%, marking a 0.7 percentage point increase from the previous year. The data from the yearbook aligns with the findings of this study.

3.4 Water Quality Index Analysis

The Spearman correlation matrix of 9 water quality indexes and WQI values is shown in the Table 1 below. As evident from Table 1, only 5 of the nine indicators are significantly correlated with the WQI value, and there is a positive correlation between DO and WQI. COD_{Mn}, BOD5, NH3-N, TP were negatively correlated with WQI. The correlation coefficients of DO, COD_{Mn}, BOD5 and NH3-N with WQI value $P<0.01$, that is, there is a high correlation. The results indicate that dissolved oxygen (DO), chemical oxygen demand (COD_{Mn}), biochemical oxygen demand (BOD_5), and ammonia nitrogen (NH3-N) are the primary water quality parameters influencing the water quality variation in the Wu area of Poyang Lake.

According to the Shapiro-Wilk normal test, at the significance level of 0.05, all the four indexes and WQI values in the five monitored cross sections did not follow a normal distribution, so the Spearman coefficient method was used for correlation analysis, and the Kruskal-Wallis test was used for the difference test.

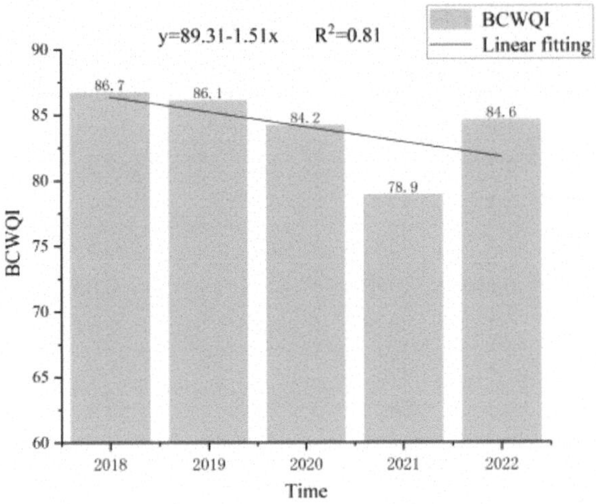

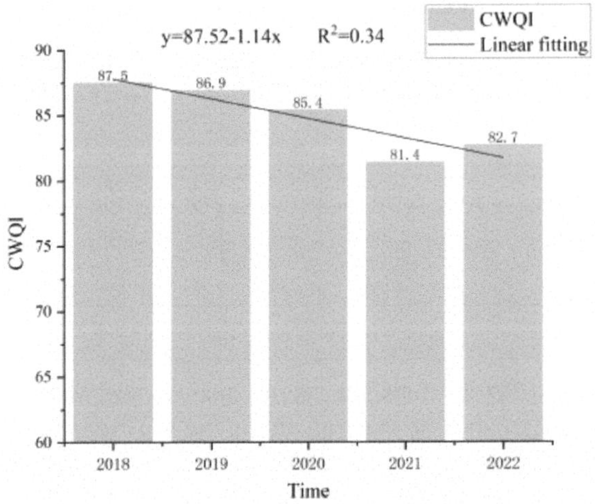

Fig. 6. Variation trend of WQI annual mean value in Wuhe District of Poyang Lake.

As evident from Table 2, the correlation data showed that the water quality of the five rivers in Poyang Lake was mainly polluted by organic matter and nutrients, and the pollution was caused by human activities, land use and industrial and agricultural pollution. The basin has the characteristics of a large construction land area, extensive agricultural planting, large population density, and dense industrial and economic development zones. The main pollution is industrial enterprises with point source pollution and rural living and agricultural planting pollution with non-point source pollution. Therefore, the implementation of basin regional management and water quality monitoring is an important measure at present, and on this basis, to improve the reasonable discharge of industrial and agricultural pollutants and the improvement of people's environmental

Table 1. Statistical data of overall water quality index and WQI value.

Index	Mean value	Median	Standard deviation	Minimum value	Maximum value
ρ(DO)/(mg/L)	5.3	5.2	1.8	2.3	11.6
ρ(CODMn)/(mg/L)	4.7	4.7	0.8	2.7	7.4
ρ(BOD5)/(mg/L)	2.4	2.1	1.1	0.5	7.6
ρ(NH3-N)/(mg/L)	0.56	0.47	0.41	0.02	2.2
ρ(TP)/(mg/L)	0.12	0.08	0.06	0.04	0.4
CWQI	84.8	82.7	7.5	57.6	99.5
BWQI	84.1	85.7	7.8	56.6	99.9

Table 2. Spearman correlation between water quality index and WQI value.

Index	DO	COD_{Mn}	BOD_5	NH^{3}-N	TP
CWQI	0.76	−0.53	−0.68	−0.62	−0.42
BWQI	0.74	−0.57	−0.69	−0.61	−0.45

awareness is to improve the long-term measures to enhance the water quality of the five rivers.

4 Conclusion

Based on CWQI and BCWQI values, time series analysis and M-K trend test, the temporal and spatial characteristics of water quality in the Wu area of Poyang Lake during 2008-2010 and 2021-2022 were evaluated and analyzed. The results showed that:

(1) Overall water quality results: The overall spatial difference between CWQI and BWQI is close, and the overall water quality evaluation results are both 84, which is a good level, equivalent to the second-class water quality level of surface water in China. The water quality grade is superior in the western region compared to the eastern region. The WQI value is primarily influenced by dissolved oxygen (DO), chemical oxygen demand (COD_{Mn}), biochemical oxygen demand (BOD_5), and ammonia nitrogen (NH3-N).
(2) Spatiotemporal difference results: The water quality of the three tributaries of Poyang Lake, Gan River, Fu River and Xiu River, showed significant spatiotemporal differences, changing from "good" to "general"; However, the spatiotemporal difference of Xinjiang and Rao tributaries is not significant, and the water quality has not changed greatly and is stable every year.
(3) Trend change findings: The water quality in the Wu area of the Poyang Lake exhibited an "inverted U-shaped" trend, initially declining before improving.

Pursuant to the conclusions of two kinds of water quality evaluation and the spatiotemporal change characteristics of the five rivers in Poyang area, that is suggested to implement the differential management among basins.

Enhancing regional management within the watershed, increasing the frequency of water quality monitoring, and rigorously ensuring the effectiveness of water quality control measures. We will strengthen efforts to improve river systems in eastern China, further promote the national environmental governance policy, strictly treat and prevent industrial and agricultural sewage, improve urban sewage pipes and non-point source pollution, and raise the technical level of river water environmental quality.

References

1. Liu, M., Fu, B.L., He, H.C., et al.: Water surface monitoring and water quality parameter inversion of the Li River based on multi-temporal active and passive remote sensing (2016–2020). Journal of Lake Sciences **33**(3), 687–705 (2019)
2. Ji, X. Y., Hou, H. H., Wang, S. S. et al.: Changes of surface water quality in China in recent years Ding. Environ. Sci. **202**, 1-16.]
3. Li, J., Tang, M., Liang, Y.X.: Evaluation of water quality improvement in Haihe River Basin, Henan Province from 2015 to 2018. Environmental Engineering **38**(5), 60–64 (2019)
4. Cristiano, E., ten Veldhuis, M., Van De Giesen, N.: Spatial and temporal variability of rainfall and their effects on hydrological response in urban areas–a review. Hydrology and Earth System Sciences **21**(7), 3859–3878 (2017)
5. Li, Y.D., Wu, J.: Current situation and change trend of water quality pollution in West Liaohe River Basin. Environmental Engineering **34**(S1), 807–809 (2016)
6. Liu, L. H., Wu, L. X., Wu, J. P. et al.: Review of foreign surface water quality index evaluation methods. Water Resour. Prot. **32**(01), 86–90+96 (2016)
7. Wang, F.Y., Zhang, S.Y., Hong, H.C., et al.: Water quality monitoring and evaluation of township centralized drinking water source. Water Supply Technology **5**(6), 1–5 (2011)
8. Zhao, X.C.: Application of grey correlation analysis in water quality assessment of Qingdao section of Dagu River. Water Ecology **2**(5), 115–118 (2009)
9. Wang, X.W., Liu, Z.Q., Guo, Q.Q., et al.: Spatial and temporal distribution of water quality and pollution sources in Huangzhou River Basin. Guizhou. Environmental Engineering **39**(9), 69–75 (2019)
10. Tian, H.X., Wang, J.J., Zhou, L., et al.: Water quality and pollution assessment in Lalu Wetland of Xizang Province. Environmental Engineering **39**(6), 198–206 (2019)
11. Li, J.: Distribution characteristics and analysis of typical plants in Nanjishan wetland. Nanchang Inst. Technol. (2017)
12. Su, C.J., Chen, X.H., Tan, Y.Q., et al.: Spatial and temporal variation of water quality in Tanjiang River and its driving factors. Water Resources Protection **37**(4), 109–116 (2021)
13. Zhang, S.Q., Zhang, H.B., Xin, C., et al.: Characterization of trend and morphological change of hydrological series. Water Resources Protection **35**(6), 58–67 (2019)
14. Wang, G., Wo, Y.B., Mao, J.Q., et al.: Spatial and temporal variation of water quality in urban gated rivers based on two-step clustering. Environmental Engineering **40**(1), 117–122 (2022)
15. Ye, F.Q., Ji, Y., Hui, Y., et al.: Analysis and evaluation of water quality in five rivers of Poyang Lake based on three water quality index methods. Renmin Zhujiang **27**(6), 18–26 (2024)
16. Zhao, S., Wu, Y.J., Cai, L., Zhang, Y.W., et al.: Spatial and temporal heterogeneities in water quality and their potential drivers in Lake Poyang (China) from 2009 to 2015. Limnologica - Ecology and Management of Inland Waters **24**(4), 109–116 (2018)

Estimation of Ecological Water Use in Typical Oasis Shelterbelts in Arid Region, Northwest China

Shuai Fu[1,2,3](✉), Yingxiang Feng[1], Bingxian Xu[1], Yuxin Peng[1], Jie Yu[1], Xiuxiang Li[3], and Lanhai Li[2]

[1] Land Resource Management, School of Finance and Public Administration, Jiangxi University of Finance and Economics, Nanchang 330013, China
fushuai1019@126.com

[2] Xinjiang Key Laboratory of Water Cycle and Water Utilization in Arid Zone, Xinjiang Institute of Ecology and Geography, Chinese Academy of Sciences, Urmqi 830011, China

[3] International Commodity Price Analysis and Forecast Research Innovation Team, Jiangxi University of Finance and Economics, Nanchang 330013, China

Abstract. Climate change and agricultural water resource development have constrained the water use and landscape pattern of shelterbelts. This study applied remote sensing technology to extract and evaluate the landscape pattern of shelterbelts, combined with the water consumption law of oasis shelterbelts, to calculate water consumption. The results show shelterbelts increased from 674.87 km^2 in 2017 to 784.68 km^2 in 2020. The landscape fragmentation index of shelterbelts also increased to 16.88, and the landscape disturbance degree in shelterbelts in 2020 was higher than that in 2017, with water consumption rising from 4.1 billion m^3 in 2017 to 5.2 billion m^3 in 2020. Understanding the development status of the landscape pattern of shelterbelts and scientifically calculating the water resources consumed by shelterbelts provide a scientific basis for water resource management in oasis shelterbelts in arid regions.

Keywords: Water Use · Landscape Pattern · Shelterbelts · Remote Sensing · Water Resource Management

C. Weng (Ed.): WRE 2024, LNCE 634, pp. 46–56, 2025.
https://doi.org/10.1007/978-981-96-5095-8_4

1 Introduction

Oases are the core areas for agricultural development in arid regions, where artificial shelterbelts act as barriers to the harmonious development of oasis agriculture and ecology, improving microclimates in farmland, preventing wind erosion, and enhancing oasis productivity. With the intensification of climate change and agricultural activities, the contradiction between the supply and demand of water resources for the oasis agriculture and shelterbelts in arid regions is becoming increasingly prominent. Extensive degradation of artificial shelterbelts in oases has led to landscape fragmentation, reducing the stability of protective forest networks, and restricting the stable development of oasis agricultural economy. Due to the unique geographical location and climatic conditions, the water use and water resource management of oasis crops and shelterbelts differ greatly from that of inland areas. Evaluating the landscape pattern of typical arid region oasis artificial shelterbelts and calculating ecological water consumption has become a pressing issue that needs to be addressed for the development of oasis agriculture and ecological construction in arid regions.

Oasis shelterbelts are usually composed of farmland shelterbelts located around farmland patches and windbreak and sand-fixation forests at the edges of oases [1, 2]. After long-term construction and management, the protective forest system in Xinjiang has gradually formed a spatial pattern characterized by belts, patches, and networks [3, 4]. Extensive research has been conducted on the development scale and structural configuration of shelterbelts for the protection benefits of oasis farmland systems, and the future development mode and water resource allocation in oases have also attracted scholars' attention [4–6]. Utilizing RS (remote sensing), GPS (global positioning), GIS (geographic information system), and FRAGSTATS tools can effectively assess the characteristics and problems of the landscape pattern at the patch level, landscape type, and landscape index in oasis shelterbelts, promoting regional landscape ecology assessment [7–9]. Various studies have evaluated the rationality of the landscape pattern of oasis shelterbelts by extracting landscape ecological indicators, such as fragmentation degree, isolation degree and isolation degree [4]. This rich diversity and spatial distribution of oasis shelterbelts play a crucial role in the structure and connectivity of oasis farmland landscapes [10, 11].

The water consumption of shelterbelts in arid regions is positively correlated with solar radiation, air temperature, wind speed, and vapor pressure deficit [12, 13], while air relative humidity is negatively correlated with leaf transpiration rate and stem flow rate [14]. Soil water stress and tree transpiration in arid region farmland shelterbelts exhibit a power exponent relationship [15]. The rise and fall of groundwater levels not only affect the stem flow rate of trees but also influence the leaf transpiration rate of trees, reducing the contribution of groundwater to tree transpiration [16–18]. The Penman-Monteith and the Jarvis models can better simulate the water consumption and canopy conductance changes in forests. The Jarvis model can adjust the combination of characteristic factors according to regional variations to simulate canopy conductance, providing convenience for simulating oasis shelterbelts in typical arid areas and offering a method for systematically exploring the relationships of various factors in the transpiration process [13]. By combining remote sensing image interpretation, field observations, landscape ecology principles, ArcGIS technology, and CA-Markov model, it is possible to effectively assess oasis land use changes, analyze and predict the development status and scale of artificial shelterbelts in oases [19–21]. Summary of previous research shows that through remote sensing technology, landscape ecology principles, GIS technology, and the transpiration water consumption model of artificial shelterbelts, it is possible to effectively analyze the suitable ratio and water resource regulation system of artificial shelterbelts in oases. Therefore, this study focused on the water consumption of typical oasis protection forests in arid areas. First, the landscape pattern of oasis protection forests in arid areas was analyzed by remote sensing and landscape ecology. Then the water consumption of shelter forests was scientifically estimated by combining the development area of shelter forests and the water consumption law in order to provide scientific basis and theoretical support for the landscape pattern development and water resources management of oasis protection forests in arid areas.

2 Materials and Methods

2.1 Study Area

The Manas River Basin (43°57′–45°16′N, 84°58′–86°32′E) is located on the southern edge of the Junggar Basin in Xinjiang, China. It is adjacent to parts of the Junggar Basin and the Tianshan Mountains and has a total area of approximately 9,474.42 km^2 (see Fig. 1).

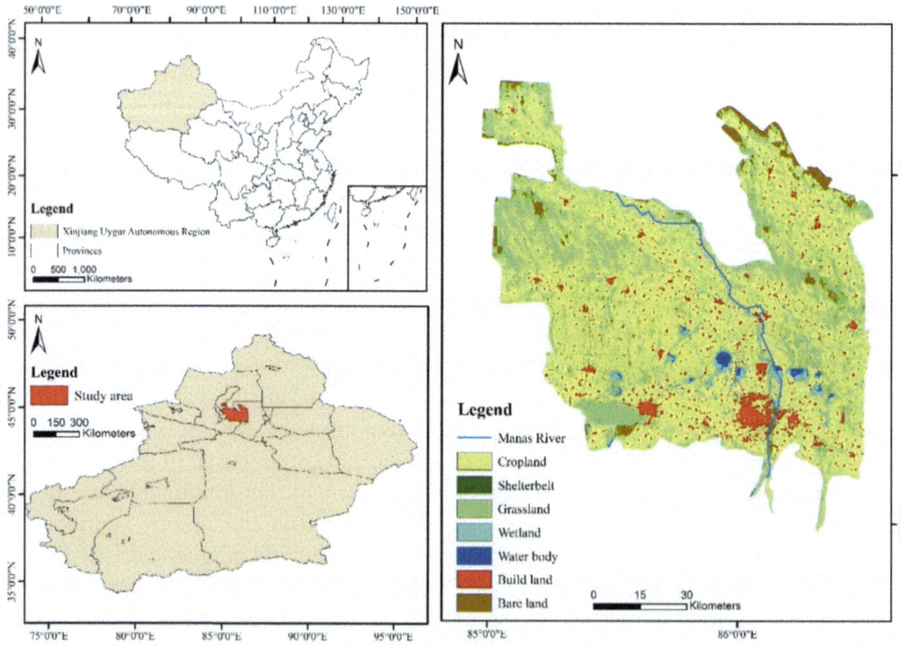

Fig. 1. Geographic location of the Manas River Basin.

Due to its location in the heart of the Eurasian continent, far from the sea, it exhibits typical temperate continental climate characteristics such as aridity, abundant sunlight, ample heat, scarce precipitation, and high evaporation (Table 1).

Table 1. Water Resources and shelterbelts in Manas oasis in 2020.

	Comment	Value
1	Annual average temperature	4.7 to 5.7 °C
2	Annual average	110 to 200 mm
3	Annual average evaporation	1,500 to 2,000 mm
4	Multi-year average runoff	$12.95 \times 108\ m^3$
5	Farmland area	$5{,}830.36\ km^2$
6	Shelterbelts area	$784.68\ km^2$
7	Agricultural irrigation water	$11.89 \times 108\ m^3$
8	Industrial, domestic, and urban greening water	$1.64 \times 108\ m^3$

2.2 Data Sources

Land-use data in the study area were derived from remote sensing image interpretation using 2020 Sentinel-2 satellite imagery with a resolution of 10 m and cloud cover of less than 10%. Data were obtained from the European Space Agency (https://dataspace.copernicus.eu/). The original data were radiometrically calibrated and atmospherically corrected using Sen2cor software, and specific bands B2, B3, B4, and B8 were selected with SNAP software for land cover identification. Based on the Land Use Classification Standard (GB/T 21010–2007) and visual interpretation on the Google Earth platform, the land cover categories within the study area were classified as cropland, protective forest, grassland, wetland, water bodies, built-up land, and barren land. As shelterbelts accounted for over 95% of the forested land in the study area, all forested areas were considered shelterbelts for simplicity. The interpretation accuracy exceeded 83%, meeting the research requirements.

2.3 Research Method

The water consumption of shelterbelts was calculated using a combination of the Penman-Monteith theory formula and Jarvis formula [22, 23]:

$$E_c^{pred} = \frac{\Delta(R_n - G) + \rho C_p VPD g_a}{\lambda\left[\Delta + \gamma\left(1 + g_a/g_c^{pred}\right)\right]} \tag{1}$$

where λ is the latent heat of vaporization of water (2465 kJ/kg), γ is the psychrometer constant (65.5 Pa/K), ρ is the density of the air (1.25 kg m^{-3}), and C_p is the specific heat of air (1010 Jkg^{-1} K^{-1}), VPD is vapor pressure deficit of the canopy air [1]. Δ (PaK^{-1}) is the slope of the relationship between saturation vapor pressure and temperature [24]. Rn is the net radiation above the forest canopy (Wm^{-2}), Rn is equal to net radiation, ra(sm^{-1}) is the aerodynamic resistance calculated from the logarithmic wind profile for neutral conditions as [25–27]; g_a represents aerodynamic conductance (ms^{-1}), and g_c^{pred} is the simulated canopy conductance value (ms^{-1}). Jarvis et al. assumed that different environmental variables have a relatively independent influence on canopy conductance, utilizing a factorial equation to describe the effects of microclimatic factors on canopy conductance [28]. The canopy conductance predictor model (g_c^{pred}, ms^{-1}) is a response equation (f, $0 \leq f \leq 1$). Comprising various environmental variables VPD, Rn, and REW affecting gc, determining the main factors affecting water consumption of shelterbelts through on-site observation data, composing the Jarvis [28, 29] model, and calibrating the Jarvis equation through measured data:

$$g_c = g_{cmax} \tag{2}$$

where gcmax is the maximum canopy conductance.

The expressions of $f(\theta), f(R_n) f(VPD)$, Rn, VPD, and θ vary, and their practicality also differs significantly between different regions and tree species. By combining experimental data with boundary line analysis, the types of and functions, as well as parameter values, were determined. The combination of the Jarvis equation and Penman-Monteith model not only allows for simulating changes in canopy conductance but also accurately predicts variations in water consumption of shelterbelts [30, 31]. By integrating plant growth models (canopy area, leaf area index, tree height at different growth stages) with meteorological data, the water consumption characteristics of shelterbelts under different moisture conditions and growth scenarios can be predicted.

3 Result and Discussion

3.1 Analysis of Oasis Shelterbelts Landscape Index in the Manas River Basin

Utilizing Fragstats 4.2 software and Excel, the landscape pattern index for shelterbelts in the study area for 2017 and 2020 was calculated (Table 2). From the table, it can be observed that the fragmentation of shelterbelt landscapes increased from 10.91 in 2017 to 16.88 in 2020, indicating a gradual intensification of fragmenting patches in the shelterbelts, posing higher ecological risks. The values for landscape separation and fractal dimensionality were higher in 2017 than in 2020, while disturbance exhibited higher values in 2020 than in 2017.

1. Landscape fragmentation: it represents the fragmentation degree of landscape segmentation, reflects the complexity of landscape spatial structure, and reflects the degree of human interference in landscapes to a certain extent.
2. Landscape separation: to measure the degree of separation of individual distribution of different patch numbers in a landscape type, reflects the degree of dispersion of patch distribution of landscape type.
3. landscape fractal dimension: an integer dimension that describes the geometric complexity of a patch or landscape Mosaic.

Table 2. Manas shelterbelts area and landscape index

	Landscape fragmentation	Landscape separation	landscape fractal dimension	Landscape disturbance index	Forest area /km^2
2017	10.91	0.86	4.73	0.68	674.87
2020	16.88	0.84	2.21	0.72	784.68

4. Landscape disturbance index: it is used to measure the extent to which a landscape is subjected to human or other forms of disturbance.

The current state of shelterbelt landscapes in the study area, as shown in Fig. 2, suggests that the fragmentation is primarily due to inadequate construction and unreasonable maintenance of the shelterbelts. This leads to low closure degrees and improper forest grid structure within the regional shelterbelt strips, directly resulting in severe fragmentation of patches. This phenomenon aligns with the analysis of habitat quality and degradation levels, indicating significant ecological risks in the area.

Furthermore, the high values of Landscape separation, fractal dimension, and disturbance in 2017 indicate severe disturbance to the landscape types, contributing to a higher corresponding index, which is a significant factor in the increased ecological risk in the region. The area of shelterbelts increased from 674.87 km^2 in 2017 to 784.68 km^2 in 2020, with an annual growth rate of 36 km^2, reflecting continuous planting and growth of shelterbelts, leading to an increase in their area. The continuous growth and planting of shelterbelts pose a challenge in terms of water demand and water resource allocation. Like previous studies: there are some problems in Mnasis Oasis shelterbelts, such as insufficient shelterbelts area, uneven distribution of shelterbelts network, a less closed mesh, poor molding condition, etc. [2, 4, 5, 10].

3.2 Assessment of Water Consumption by Oasis Shelterbelts in the Manas River Basin

By analyzing meteorological data from 2012 and 2016 (Data of Mosso Bay Meteorological Observation Station, Chinese Academy of Sciences), the water consumption patterns of different tree species in shelterbelts during various months were obtained. The transpiration water consumption of young trees ranged from 2.77 ± 1.31 cm to 0.75 ± 0.42 cm from May to September, while that of juvenile trees varied from 10.67 ± 1.36 cm to 2.87 ± 0.75 cm during the same period. As trees matured, their water consumption increased, reaching 21.33 ± 3 cm to 5.78 ± 0.61 cm from May to September. On a yearly basis, young trees consumed an average of 16.69 ± 5.1 cm, juvenile trees consumed 62.5 ± 8.2cm, and mature trees consumed up to 116.7 ± 16 cm of water. The simulated results align with measurements from nearby experimental sites [32]. The water consumption of small-leaf poplar trees in farmland shelterbelts increases according to a logistic curve as tree age progresses: $Tr = 847–844/[1 + (x/87.9)^{1.9}]$ ($R^2 = 0.98$, Tr: cm/year, x: tree ages) [14].

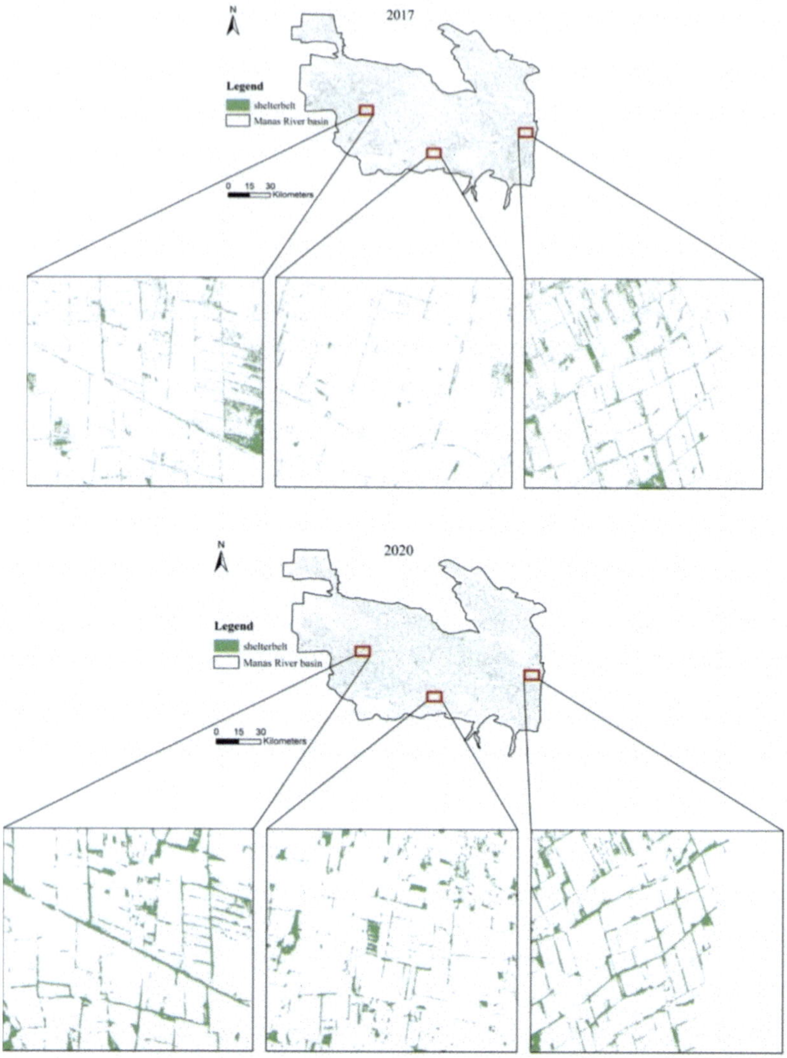

Fig. 2. Status of shelterbelts landscapes in the Manas River Basin in 2020.

According to Fig. 3 and Table 2, it is evident that not only has the area of shelterbelts increased, but also the growing years have also extended. This indicates a change in the water consumption of shelterbelts. By considering the water consumption pattern of shelterbelts and the changes in their area, we can estimate that the water consumption of shelterbelts in 2017 and 2020 reached between 410 million m^3 and 520 million m^3, increasing at a rate of 36 million m^3 per year.

The annual total water diversion in the Manas River Basin is 27.83 billion m^3 (including surface water, groundwater, and spring water), with 22.18 billion m^3 from surface water, and 3.25 billion m^3 from groundwater [33]. Agricultural water usage accounts for

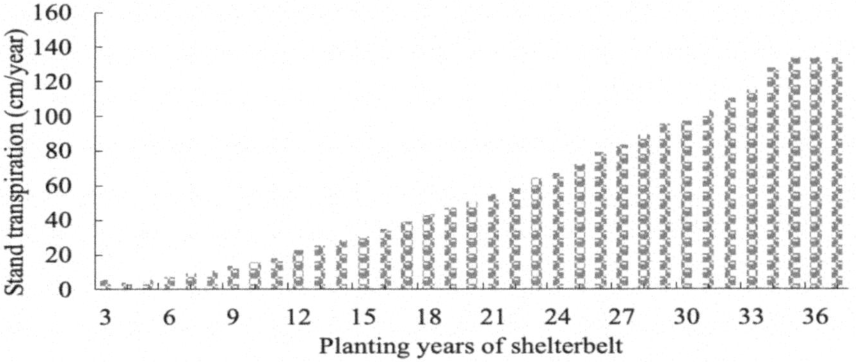

Fig. 3. Stand transpiration of shelterbelts in young, immature, and mature growth period.

26.22 billion m^3, representing 94.2%, industrial water usage is 0.64 billion m^3, accounting for 2.3%, and domestic water usage is 0.97 billion m^3 [34]. The water usage structure in the Manas basin is relatively simple, primarily composed of water consumption for crops and shelterbelts. With the rapid development of agriculture, total crop water consumption in oases has increased from 33.58 billion m^3 to 56.43 billion m^3, and the water requirement for farm shelterbelts has also increased from 410 million m^3 to 520 million m^3.

Under the rapid development of agriculture, the continuous increase in water demand for crops will inevitably squeeze the irrigation water allocated for shelterbelts on farmland. To achieve an optimal adjustment of water usage structure and maximize the comprehensive benefits of water resources, it is necessary for all levels of government to enhance their strategic understanding of water resources, recognize the role of farm shelterbelts, and not only maintain the agricultural and forestry system from the perspective of water-saving irrigation but also establish the concept of "ensuring ecological water use and reducing agricultural water use" to genuinely transform development concepts. This will make the water use in oasis agricultural and forestry systems more rational, thereby promoting the efficient and rational utilization of water resources.

4 Conclusion

Climate change, agricultural development and water resource allocation not only affect the landscape pattern of shelterbelts but also restrict the water consumption of shelterbelts. Through remote sensing, landscape ecology and hydrology analysis, the results show that: From 2017 to 2020, the area of shelter forest increased from 674.87 km^2 to 784.68 km^2, the landscape fragmentation degree increased from 10.91 to 16.88, the landscape disturbance degree increased from 0.68 to 0.72, and the landscape separation degree and fractal dimension decreased from 0.86 and 4.73 to 0.84 and 2.21, respectively; The water consumption of shelterbelts increased from 410 million m^3 in 2017 to 520 million m^3 in 2020, which brought great pressure on the allocation of agricultural water and ecological water in oasis. Finding out the development status of the shelterbelt landscape pattern and scientifically calculating the water consumption resources of shelterbelts are conducive to the rational layout of the oasis shelterbelt landscape and the distribution of water resources, and provide theoretical reference for the rational regulation of oasis water resources.

Acknowledgments. The research results were supported by the National Science Foundation (Grant No. 42361003); the Opening Foundation of Xinjiang Key Laboratory of Water Cycle and Utilization in Arid Zone, Xinjiang Institute of Ecology and Geography, Chinese Academy of Sciences (Grant No. XJYS0907–2023-22); 2022 Pure Zixi Carbon Neutral Practice Innovation Center project "Carbon neutral Fuzhou (Zixi) evaluation model construction and application(2022JDA07); Jiangxi University of Finance and Economics, International commodity Price analysis and forecast research innovation team; Science Foundation of Jiangxi Education Department (Grant No.GJJ190269).

References

1. Chang, X., Zhao, W., Zhang, Z., Su, Y.: Sap flow and tree conductance of shelter-belt in arid region of China. Agric. For. Meteorol. **138**, 132–141 (2006)
2. Sun, H., Liu, L.J., Li, X.Y., et al.: Effect of oasis protection forest network pattern on farmland evapotranspiration in arid region: A case study of oasis in Sangong River Basin. Xinjiang. Journal of Ecology **37**, 2436–2444 (2018)
3. Xie, F. B., Mi, T. T., Song, Y. Y.: Study on water consumption characteristics of non-irrigated afforestation trees. Water Saving Irrigation, 9–15 (2022)
4. Yang, S.Y., Yang, H.J., Liu, L.J., et al.: Landscape pattern evaluation of oasis farmland shelterbelt in Manas River basin. J. Ecol. **36**, 1690–1698 (2017)
5. Li, X.N., Xu, X.Y., Yang, X.M., et al.: Construction and application of farmland shelter forest health evaluation system in Guazhou County. Resources and Environment in Arid Land **36**, 187–194 (2022)
6. Li, Y., Zhang, R., Shi, Y., et al.: Water resource planning and ecological environment restoration in arid oasis: A case study of Minqin Oasis in Shiyang River Basin. Water Conservancy Planning and Design, 6–8(2019)
7. Li, F., Zhou, W. Z., Shao, Z. L., et al.: Changes of landscape pattern and ecosystem health assessment in the Western Qinling Mountains from 2000 to 2018. Journal of Ecology, 1–15 (2023)
8. Xiao, D. N., Gao, J., Shi, T. M.: Application of landscape ecology in urban planning and management. Advances in Earth Sciences, 813–820 (2001)
9. Yan, S.G., Li, H., Li, H.D., et al.: Redline ecosystem health assessment method for ecological protection based on land use and landscape pattern: a case study of Nanjing. J. Nat. Resour. **35**, 1109–1118 (2020)
10. Xue, L.Q., Ni, T., Liu, Y.H., et al.: Calculation of suitable scale of oasis in Manas River basin based on surface runoff prediction. Water Resources Conservation **37**, 7–13 (2021)
11. Zhu, T. Y., Zhou, G. S.: Study on the law of wind speed weakening in forest network area of farmland. Chinese Journal of Applied Ecology, 136–140 (1993)
12. Addington, R.N., Mitchell, R.J., Oren, R., et al.: Stomatal sensitivity to vapor pressure deficit and its relationship to hydraulic conductance in Pinus palustris. Tree Physiol. **24**, 561–569 (2004)
13. Fu, S., Sun, L., Luo, Y.: Combining sap flow measurements and modelling to assess water needs in an oasis farmland shelterbelt of Populus *simonii Carr* in Northwest China. Agric. Water Manag. **177**, 172–180 (2016)

14. Fu, S., Sun, L., Luo, Y.: Canopy conductance and stand transpiration of Populus *simonii Carr* in response to soil and atmospheric water deficits in farmland shelterbelt. Northwest China. Agroforestry Systems **91**, 1165–1180 (2016)
15. Gazal, R.M., Scott, R.L., Goodrich, D.C., et al.: Controls on transpiration in a semiarid riparian cottonwood forest. Agric. For. Meteorol. **137**, 56–67 (2006)
16. Brooks, J.R., Barnard, H.R., Coulombe, R., et al.: Ecohydrologic separation of water between trees and streams in a Mediterranean climate. Nat. Geosci. **3**, 100–104 (2010)
17. Lei, J., Songhao, S., and Xiaomin, M. Simulation on the impact of groundwater table on evapotranspiration of a riparian forest in arid regions. Journal of Tsinghua University, 601–612(2013)
18. Manzoni, S., Vico, G., Porporato, A., Katul, G.: Biological constraints on water transport in the soil-plant-atmosphere system. Adv. Water Resour. **51**, 292–304 (2013)
19. Wang, Q., Wang, H., Chang, R., et al.: Dynamic simulation patterns and spatiotemporal analysis of land-use/land-cover changes in the Wuhan metropolitan area. China. Ecological Modelling **464**, 109–131 (2022)
20. Liu, C., Zhang, F., Carl Johnson, V., et al.: Spatio-temporal variation of oasis landscape pattern in arid area: Human or natural driving. Ecol. Ind. **125**, 35–50 (2021)
21. Ge, G., Shi, Z., Zhu, Y., Yang, X., et al.: Land use/cover classification in an arid desert-oasis mosaic landscape of China using remote sensed imagery: Performance assessment of four machine learning algorithms. Global Ecology and Conservation **22**, 1–13 (2020)
22. Zhang, Z.: Effect of root-zone vertical soil moisture heterogeneity on water transport safety in soil-plant-atmosphere continuum in Robinia pseudoacacia. Agric. Water Manag. **246**, 35–46 (2021)
23. Zhou, H., Tong, S.Z., Ding, L., et al.: Improved application of the Penman-Monteith model using an enhanced Jarvis model that considers the effects of nitrogen fertilization on canopy resistance. Environ. Exp. Bot. **159**, 1–12 (2019)
24. Zeppel, M.J.B., Macinnis-Ng, C.M.O., Yunusa, I.A.M., et al.: Long-term trends of stand transpiration in a remnant forest during wet and dry years. J. Hydrol. **349**, 200–213 (2008)
25. She, D.L., Xia, Y.Q., Shao, M.A., et al.: Transpiration and canopy conductance of Caragana korshinskii trees in response to soil moisture in sand land of China. Agrofor. Syst. **87**, 667–678 (2013)
26. Sommer, R., Sá, T.D., Vielhauer, K., et al.: Transpiration and canopy conductance of secondary vegetation in the eastern Amazon. Agric. For. Meteorol. **112**, 103–121 (2002)
27. Whitley, R., Zeppel, M., Armstrong, N., et al.: A modified Jarvis-Stewart model for predicting stand-scale transpiration of an Australian native forest. Plant Soil **305**, 35–47 (2008)
28. Jarvis, P.G., Mcnaughton, K.G.: Stomatal control of transpiration: scaling up from leaf to region. Adv. Ecol. Res. **15**, 1–49 (1986)
29. Lsab, C., Jzab, C., Xiao, Z., et al.: Comparison of canopy transpiration between Pinus sylvestris var. mongolica and Pinus tabuliformis plantations in a semiarid sandy region of Northeast China. Agricultural and Forest Meteorology 314, 184–187 (2022)
30. Li, Q., Wang, L., Gul, H.N., Li, D.: Simulation and optimization of land use pattern to embed ecological suitability in an oasis region: A case study of Ganzhou district, Gansu province. China. Journal of Environmental Management **287**, 52–63 (2021)
31. Rkr, A., Ap, B., Hsj, C., et al.: Ecological aspects of the soil-water-plant-atmosphere system. Plant Perspectives to Global Climate Changes, 279–302 (2022)

32. Ce, S., Wenbo, F., Hongkai, Z., et al.: Preliminary study on water consumption process and irrigation system of poplar in Xinjiang under different irrigation gradients [J]. China Rural Water Resources and Hydropower **09**, 1–7 (2011)
33. FAN W.B. Relationship between spatial and temporal variation of crop water use structure and crop structure in Manas River Basin. Northwest A&F University, 56–78 (2014)
34. Wang H.J., Zhang, H.L., REN, S.T., et al.: Research on prediction model of soil water and salt status based on hyperspectral reflection characteristics. Transactions of the Chinese Society for Agricultural Machinery,45(07):133–138 (2014)

Climate Teleconnections Influencing Desertification in the Lancang-Mekong River Basin

Masoud Irannezhad[1,2](✉) and Zahrah Abdulghafour[1]

[1] Department of Civil Engineering, College of Engineering, Australian University, West Mishref, 13015 Safat, Kuwait
m.irannezhad@au.edu.kw

[2] Water, Energy and Environmental Engineering Research Unit, University of Oulu, 90014 Oulu, Finland

Abstract. The Lancang-Mekong River Basin (LMRB) is vital to ~ 70 million people in Mainland Southeast Asia, crucially reliant on primary freshwater-dependent sectors like agriculture and fishery. Hence, assessing water resources availability over the LMRB is important for local, regional, and global sustainability. As the ratio of precipitation to potential evapotranspiration, the Aridity Index (AI) reflects dryness and wetness patterns, which are naturally dependent on large-scale climate teleconnections, such as the Pacific Decadal Oscillation (PDO). The present study assessed the effects of these teleconnections on AI variability in the LMRB, using gridded (25° × 25°) daily precipitation and mean Surface Air Temperature (SAT) time series extracted from historical (1952–2005) NASA Earth Exchange Global Daily Downscaled Projections (NEX-GDDP). The Mann-Kendall nonparametric test and Spearman's rank correlation were applied to identify trends in AI and its relationships with teleconnections, respectively. Key findings include: (i) the LMRB was predominantly humid, except during January, February, March, and December; (ii) AI showed a drying trend across most time scales; (iii) the PDO strongly affected AI during the dry season (Nov-May), while (iv) the East Pacific/North Pacific (EP/NP) during the wet season (Jun-Oct); and (v) the PDO was also influential annually (Nov-Oct) and in most months. These results provide critical insights for enhancing water security and advancing the UN's 2030 Sustainable Development Agenda.

Keywords: Dryness · East Pacific/North Pacific (EP/NP) · NEX-GDDP · Pacific Decadal Oscillation (PDO) · Southeast Asia

1 Introduction

Mean global Surface Air Temperature (SAT) increased by about 1.1 °C in 2011–2020 compared to 1850–1900 [1]. In general, SAT was warmer over land (1.6 °C) than ocean (0.9 °C) [1]. Other studies also reported significant global mean Surface Sea Temperature (SST) increases during recent decades [2]. The primary cause behind warmer SAT and

C. Weng (Ed.): WRE 2024, LNCE 634, pp. 57–66, 2025.
https://doi.org/10.1007/978-981-96-5095-8_5

SST is extensive increases in atmospheric Greenhouse Gas (GHG) concentrations [1, 2]. Following the Clausius-Clapeyron (C-C) relationship, the considerable rises in SAT and SST intensify the water content of the atmosphere [3], change the global climate system, affect various components of the hydrological cycle, alter dryness and wetness patterns [4], and consequently impact water resources [5] around the world.

Not only vital for all forms of life, water also plays an imperative role in economic, social, and environmental sustainable development by supporting primary human activities such as agriculture, fishery, forestry, recreation, and energy production [6]. This is particularly true for the Lancang-Mekong River Basin (LMRB) in Mainland Southeast Asia (MSA), where approximately 40% of ~ 70 million inhabitants live in poverty and mostly rely on such various freshwater-dependent sectors [7]. Water shortages, hence, present serious sustainability risks to the six developing countries (Vietnam, Thailand, Myanmar, Laos, China, and Cambodia) within the LMRB due to their high-density human population, unsafe infrastructures, and poor land development and management policies [6].

Numerous previous studies have employed an Aridity Index (AI) to assess variations in regional dryness and wetness around the world [8]. The United Nations Environmental Program (UNEP) defined AI as the ratio of precipitation (P) to potential evapotranspiration (PET) across a region on a specific time scale, such as annual, seasonal, and monthly [9]. Along with analyzing alterations in regional AI, it is also crucial to explore their potential underlying physical processes. Both regional P and PET are naturally influenced by the well-known large-scale atmospheric-oceanic circulation patterns [10, 11]. Hence, such patterns can principally control AI variability over both space and time. These patterns are repeating and consistent modes of pressure and circulation anomalies across a wide geographical area, manifesting variations in the natural incidence of chaotic actions in climatic systems [12]. Their power and spatio-temporal effects are generally expressed by climate teleconnection indices, hereafter climate teleconnections [12]. For the LMRB, although there are already studies focusing on the relationships of different hydrological cycle components (P and PET) with such climate teleconnections, their role in AI variations across this region has received less attention.

This study aimed to examine significant changes in the historical aridity index across the LMRB in response to different phases of climate teleconnections. The particular intentions were to: (i) investigate historical variations and alterations in annual, seasonal, and monthly aridity indices over the entire LMRB; and (ii) identify the most dominant climate teleconnection for such dryness and wetness changes across the basin. The findings lay a foundation for acting towards achieving water security throughout the LMRB, playing a decisive role in local, regional, and global sustainable development.

2 Material and Methodology

With an area of ~ 795,000 km^2 in MSA (Fig. 1), the LMRB consists of the Lancang River Basin (LRB) and the Mekong River Basin (MRB) [7]. The Lancang-Mekong River (LMR) globally ranks as the 10th largest in mean annual discharge (475,000 m^3) and the 12th longest river (4,200 to 4,909 km, depending on measurement tools and methods) [7]. The LMRB is distributed across six countries, with Myanmar accounting for 3%,

Vietnam 8%, Cambodia 20%, China 21%, Thailand 23%, and Laos 25%. This basin generally experiences distinct dry and wet seasons that cover the Nov-May and Jun-Oct months, respectively. In this basin, precipitation typically increases from the northwest to the southeast [13]. During 1981–2010, annual mean precipitation ranged between 464 mm in the northern parts and 4,300 mm in the southeastern areas [10]. However, the annual mean SAT was about −4.8ºC in the north to 29.0ºC in the southwest [7].

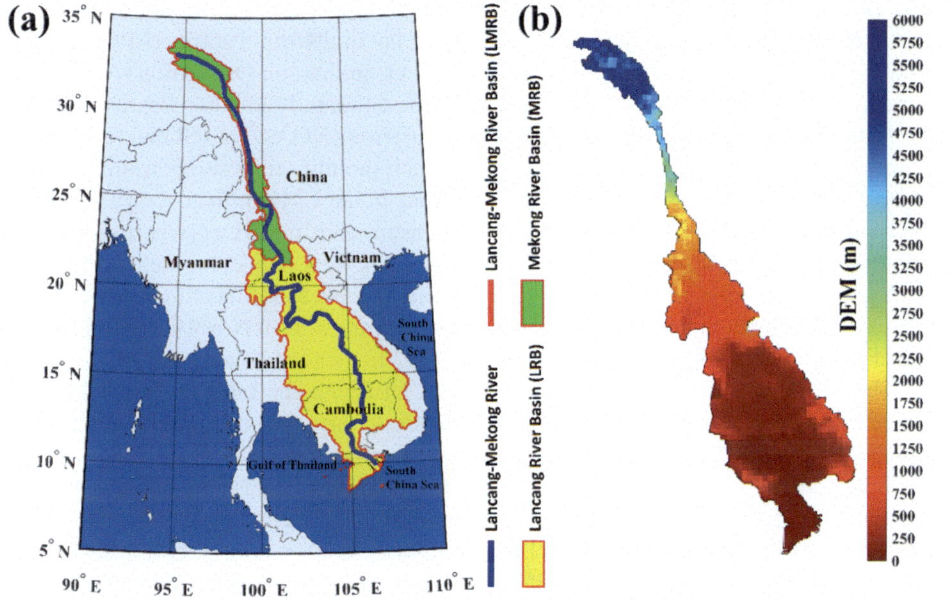

Fig. 1. (a) The geographical location of the Lancang-Mekong River Basin (LMRB), and (b) its Digital Elevation Model (DEM).

In the LMRB, the limited and unbalanced distribution of meteorological measurement stations hampers effective tracking of both spatial and temporal variations in different hydrological cycle components, particularly precipitation (P) [14]. Such in-situ P time series are often incomplete, uncertain, and not readily accessible, primarily due to transboundary conflicts among the six riparian nations over freshwater resource use [15]. As a result, previous studies have relied on gridded hydrometeorological datasets to detect climate change in Southeast Asia, including the LMRB [10, 16–19]. The present study employed high-resolution (0.25° × 0.25°) gridded daily P and mean SAT time series extracted from NASA Earth Exchange Global Daily Downscaled Projections (NEX-GDDP) product [20]. Such downscaled datasets from 21 models under the Climate Model Inter-comparison Project Phase 5 (CMIP5) were generated by applying the Bias-Corrected Spatial Disaggregation (BCSD) technique [21] to coarse-resolution Global Climate Models (GCMs) outputs. Using quantile mapping, the BCSD method corrects biases by comparing historical GCM simulations (1950–2005) with the gridded Global Meteorological Forcing Dataset (GMFD) from Princeton University's Terrestrial Hydrology Research Group [22]. For a more comprehensive assessment of uncertainties within the NEX-GDDP product, on the other hand, it is recommended to use the ensemble of all its 21 models [23]. Such ensemble mean of NEX-GDDP has commonly been used to assess regional hydrological alterations around the world [24, 25], including drought variability in the MRB [26]. Hence, such a historical period from 1950 to 2005 was selected as the temporal focus of this study.

The following large-scale climate teleconnections were selected: West Pacific (WP), Tropical/Northern Hemisphere (TNH), Southern Oscillation Index (SOI), Scandinavia (SCA), Polar/Eurasia (POL), Pacific Transition (PT), Pacific/North America (PNA), PDO, North Atlantic Oscillation (NAO), East Pacific/North Pacific (EP/NP), East Atlantic/West Russia (EA/WR), East Atlantic (EA), and Arctic Oscillation (AO). Their standardized monthly values are calculated by the Climate Prediction Center (CPC) at the National Oceanic and Atmospheric Administration (NOAA) based on climate data from 1981 to 2010. The present study utilized such monthly time series from Jan 1961 to Dec 2015. For seasonal climate teleconnections in the LMRB, the average values of these monthly datasets from Nov to May (dry season) and Jun to Oct (wet season) were calculated. Annual (Nov-Oct) climate teleconnections were then computed as the mean of these seasonal values. Thus, in this study, the annual and seasonal climate teleconnections for 1962 refer to the average of their monthly values from Nov 1961 to Oct 1962, Nov 1961 to May 1962 (dry season), and Jun 1962 to Oct 1962 (wet season).

This study applied the AI defined by the UNEP [9]: the ratio of P to PET. The Thornthwaite method [27] was used to calculate monthly, seasonal (wet and dry), and annual (Jun-May) PET. This method is an empirical formula that is relatively simple to implement, relying solely on regional SAT and latitude datasets [9]. Based on the AI values, UNEP categorizes regional climates into five different classes: humid ($AI > 0.65$), dry sub-humid ($0.50 < AI \leq 0.65$), semi-arid ($0.20 < AI \leq 0.50$), arid ($0.05 < AI \leq 0.20$), and hyper-arid ($AI < 0.05$) [9]. To determine the regional climate class for the LMRB, we computed the average value of the AI time series for all $0.25° \times 0.25°$ grid cells within the basin.

The Mann–Kendall (MK) non-parametric test [28] was utilized to identify statistically significant ($p < 0.05$) trends in different AI time series across the entire LMRB. The magnitude of trends was quantified by Sen's Slope estimator [29]. Additionally, Spearman's rank correlation coefficient (rho) was used to determine significant relationships between such AI time series and their corresponding monthly, seasonal, and annual climate teleconnections [30]. In cases where positive autocorrelation was present in the time series, the Trend-Free Pre-Whitening (TFPW) method [31] was employed to detect significant trends, while the Residual Bootstrap (RB) method [32], with 5,000 independent replications, to evaluate the standard deviation of the rho values.

3 Results

The long-term (1952–2005) average value (hereafter base value) for annual (Nov-Oct) aridity index was about 1.15 in the entire LMRB (Fig. 2). The highest and lowest annual (Nov-Oct) anomalies of aridity index were 0.06 in 1967 and –0.09 in 2005, respectively, during 1952–2005 (Fig. 2). Across the basin, a statistically significant decreasing trend, at the rate of 0.015 per decade, was detected in these historical annual anomalies of aridity index (Fig. 2). The PDO was the most influential climate teleconnection (with rho = −0.35, $p < 0.05$) for such variations and change in annual aridity index over the LMRB during 1952–2005 (Fig. 2).

For dry season (Nov-May), the AI was generally about 0.78 in the entire LMRB during 1952–2005 (Fig. 3). The historical dry season's AI anomalies ranged between –0.11 in 1990 to 0.09 in 1968 (Fig. 3). These anomalies substantially declined at the rate of 0.015 per decade over the basin during 1952–2005 (Fig. 3). Such variability and trend in historical AI during dry seasons were most strongly influenced by the PDO, with rho = −0.31 ($p < 0.05$) (Fig. 3).

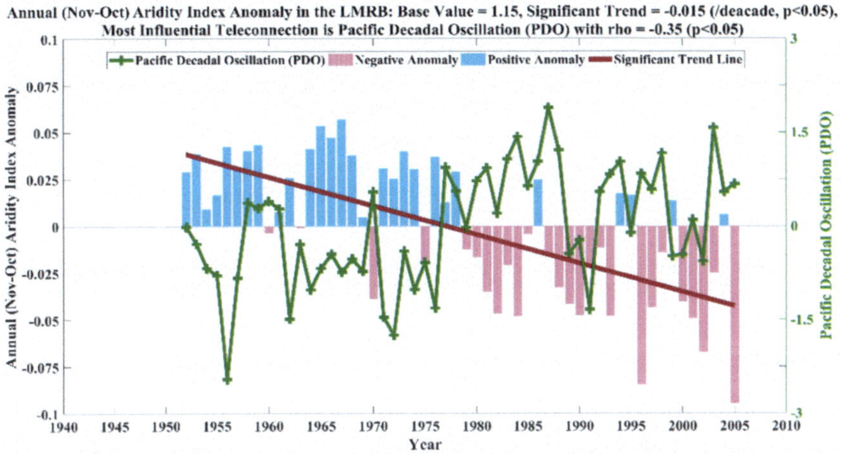

Fig. 2. Anomalies (positive and negative) with significant ($p < 0.05$) trend line and the most effective climate teleconnection for annual (Nov–Oct) Aridity Index (AI) across the entire LMRB during the historical period (1952–2005) of NEX-GDDP datasets.

On average, however, the AI was about 1.54 for the wet season (Jun–Oct) in the LMRB during 1952–2005 (Fig. 4). The wet season AI anomalies were from −0.11 in 1990 to 0.09 in 1968 (Fig. 4). These anomalies substantially declined at the rate of 0.015 per decade over the basin during 1952–2005 (Fig. 4). Such variability and trend in historical AI during dry seasons were most strongly influenced by the PDO, with rho = −0.31 ($p < 0.05$) (Fig. 4).

During 1952–2005, the base values for monthly AI ranged from 0.51 for Dec to 1.69 for Feb over the entire LMRB (Table 1). The lowest and highest interannual monthly AI anomalies were –0.25 in Oct 2005 and 0.96 in Feb 1960, respectively (Table 1). All statistically significant trends in interannual monthly AI values showed desertification across the entire LMRB during 1952–2005 (Table 1). Such substantial decreases were at a rate between –0.037 and –0.012 per decade for Oct and Mar, respectively, during 1952–2005 (Table 1). The climate teleconnection most significantly affecting interannual variations in monthly AI for most months was the PDO, with rho between –0.38 and –0.36 (Table 1). However, the PNA (NAO) most substantially controlled historical monthly AI values for Dec (Feb and Aug) in the LMRB (Table 1).

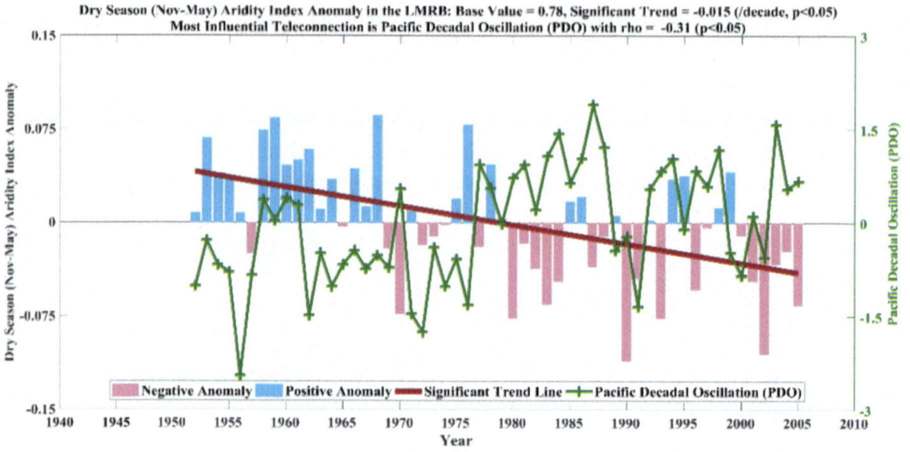

Fig. 3. Anomalies (positive and negative) with significant ($p < 0.05$) trend line and the most effective climate teleconnection for dry season (Nov–May) Aridity Index (AI) across the entire LMRB during the historical period (1952–2005) of NEX-GDDP datasets.

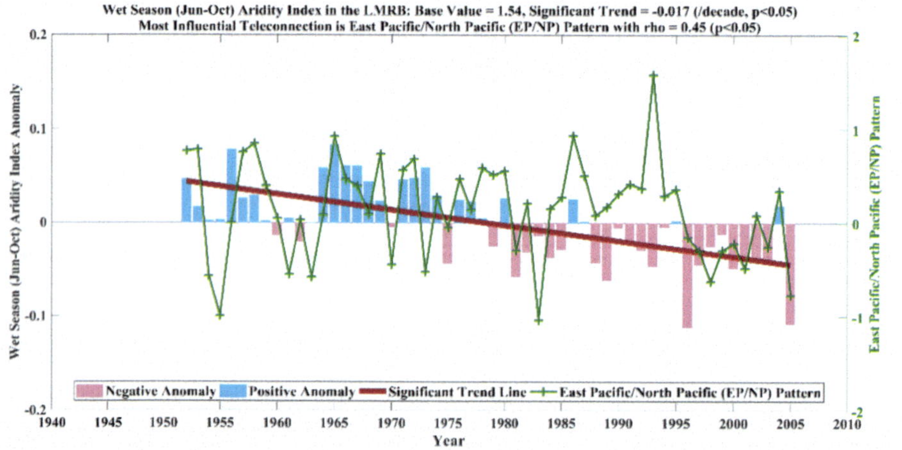

Fig. 4. Anomalies (positive and negative) with significant ($p < 0.05$) trend line and the most effective climate teleconnection for wet season (Jun-Oct) Aridity Index (AI) across the entire LMRB during the historical period (1952–2005) of NEX-GDDP datasets.

4 Discussion

On all different time scales, AI generally showed significant drying trends across the entire LMRB from 1952 to 2005. In contrast, many previous studies have reported increasingly wet climatic conditions in this basin over time [10, 13, 33]. However, most of these studies focused solely on analyzing historical P time series across the LMRB without investigating the effects of changes in SAT on dryness and wetness patterns. Even studies assessing drought variability based on the Standardized Precipitation Index (SPI)

Table 1. Interannual variations and trends in monthly Aridity Indices (AIs) and their most influential climate teleconnections across the entire LMRB during the historical (1952–2005) period of NEX-GDDP datasets.

Month	Base Value	Min Anomaly (Year)	Max Anomaly(Year)	Trend (/decade, $p < 0.05$)	Most Influential Teleconnection	Rho ($p < 0.05$)
Nov	0.82	−0.22 (1982)	0.28 (1965)	−0.018	PDO	−0.26
Dec	0.51	−0.17 (2005)	0.66 (1961)	−0.026	PNA	−0.27
Jan	0.52	−0.18 (2002)	0.79 (1960)	−0.021	–	–
Feb	0.56	−0.18 (1995)	0.96 (1969)	−0.021	NAO	−0.48
Mar	0.55	−0.17 (2002)	0.60 (1958)	−0.012	PDO	−0.38
Apr	0.68	−0.16 (1984)	0.16 (1956)	−0.013	PDO	−0.28
May	1.10	−0.21 (1993)	0.20 (1962)	−0.022	PDO	−0.31
Jun	1.58	−0.18 (2005)	0.16 (1956)	−0.023	PDO	−0.31
Jul	1.69	−0.12 (1994)	0.18 (1965)	–	–	–
Aug	1.68	−0.12 (1994)	0.20 (1964)	−0.013	NAO	−0.37
Sep	1.52	−0.18 (1996)	0.12 (1953)	−0.019	PDO	−0.28
Oct	1.22	−0.25 (2005)	0.27 (1976)	−0.037	PDO	−0.32

concluded with wetting trends across this basin from 1964 to 2005 [26]. Similar to our findings, however, studies employing the Standardized Precipitation-Evapotranspiration Index (SPEI), which considers the influence of SAT variability in drought assessment, reported more extended dry periods across the LMRB during 1985–1995 [34, 35]. This clearly highlights the importance of examining the impacts of SAT changes on the spatio-temporal analysis of regional dryness and wetness dynamics.

Variations and trends in annual, seasonal, and monthly AI time series for the LMRB were predominantly associated with the PDO and EP/NP patterns. Previous studies also reported strong relationships between precipitation [10, 13, 33] and SAT [36, 37] over the LMRB with these two climate teleconnections. Hence, both PDO and EP/NP play significant roles in controlling drought conditions throughout the LMRB [38].

As a long-term climate pattern in the North Pacific, the PDO shifts between warm (positive) and cool (negative) phases roughly every 20–30 years, affecting precipitation and SAT in Pacific bordering regions [12]. In its positive (negative) phase, warmer (cooler) waters near North America and cooler (warmer) waters in the central Pacific weaken (strengthen) the Asian monsoon, causing warmer (cooler) and drier (wetter) climatic conditions that increase the likelihood of drought (flood) in the LMRB [39, 40]. Similarly, the EP/NP pattern, marked by warmer SSTs in the eastern (northern) Pacific [12], significantly diminishes (intensifies) the Southeast Asian monsoon, leading to the warmer and drier (milder and wetter) climate that arises dryness (wetness) risks across the basin [41]. The EP/NP can amplify or moderate climate impacts, particularly when combined with broader phenomena like the PDO.

5 Conclusions

This study emphasized the critical role of large-scale climate teleconnections in variations in dryness and wetness across the LMRB, where approximately 70 million people rely on freshwater-dependent sectors, which crucial for overall sustainability. Although the LMRB is classified as humid for most of the year, dry sub-humid conditions dominated in January, February, March, and December. An analysis of the Aridity Index (AI) from 1962 to 2015 revealed significant drying trends over various time scales, with pronounced seasonality. The PDO had the strongest effect on the AI annually (Nov-Oct) and during the dry season (Nov-May). However, the EP/NP pattern was more influential during the wet season (Jun-Oct). Additionally, the PDO pattern notably affected the AI in most months. However, the AI for Dec (Feb and Aug) showed the strongest correlations with the PNA (NAO). These findings highlight the importance of understanding such effects of climate teleconnections for efficient regional water resource management, improving water security strategies, and achieving the United Nations' 2030 Sustainable Development Agenda.

Acknowledgement. This study was funded by the Kuwait Foundation for the Advancement of Sciences (KFAS), Grant No. CN23-44SE-2126.

References

1. IPCC; Masson-Delmotte, V., Zhai, P., Pirani, A., Connors, S.L., Péan, C., Berger, S., Caud, N., Chen, Y., Goldfarb, L., et al.: Climate Change 2021: The Physical Science Basis. Contribution of Working Group I to the Sixth Assessment Report of the Intergovernmental Panel on Climate Change (2021)
2. Iz, H.B.: Is the global sea surface temperature rise accelerating? Geod. Geodyn. **9**, 432–438 (2018). https://doi.org/10.1016/j.geog.2018.04.002
3. Ali, H., Mishra, V.: Contrasting response of rainfall extremes to increase in surface air and dewpoint temperatures at urban locations in India. Sci. Rep. **7**, 1228 (2017). https://doi.org/10.1038/s41598-017-01306-1
4. Pokhrel, Y., et al.: Global terrestrial water storage and drought severity under climate change. Nat. Clim. Chang. **11**, 226–233 (2021). https://doi.org/10.1038/s41558-020-00972-w

5. Gudmundsson, L., et al.: Globally observed trends in mean and extreme river flow attributed to climate change. Science **371**, 1159–1162 (2021). https://doi.org/10.1126/science.aba3996
6. Irannezhad, M., Ahmadi, B., Liu, J., Chen, D., Matthews, J.H.: Global water security: a shining star in the dark sky of achieving the sustainable development goals. Sustain. Horizons **1**, 100005 (2020). https://doi.org/10.1016/j.horiz.2021.100005
7. MRC Mekong River Commission: State of the Basin Report 2018, (2019)
8. Gao, Y., Li, X., Ruby Leung, L., Chen, D., Xu, J.: Aridity changes in the Tibetan Plateau in a warming climate. Environ. Res. Lett. **10**, 34013 (2015). https://doi.org/10.1088/1748-9326/10/3/034013
9. des Nations Unies pour l'environnement, P., Middleton, N.J., Middleton, N., Programme, U.N.E., Thomas, D.S.G. World Atlas of Desertification; A Hodder Arnold Publication; Arnold, (1997). ISBN 9780340691663
10. Irannezhad, M., Liu, J.: Evaluation of six gauge-based gridded climate products for analyzing long-term historical precipitation patterns across the Lancang-Mekong River Basin. Geogr. Sustain. (2022). https://doi.org/10.1016/j.geosus.2022.03.002
11. Sun, C., Zheng, Z., Chen, W., Wang, Y.: Spatial and temporal variations of potential evapotranspiration in the Loess Plateau of China during 1960–2017. Sustainability, 12 (2020). https://doi.org/10.3390/su12010354
12. Glantz, M.H., Katz, R., Nicholls, N. (eds.): Teleconnections Linking Worldwide Climate Anomalies : Scientific Basis and Societal Impact. Cambridge University Press, Cambridge, ISBN 9780521106849 0521106842 (2009)
13. Irannezhad, M., Liu, J., Chen, D.: Influential climate teleconnections for spatiotemporal precipitation variability in the Lancang-Mekong River Basin from 1952 to 2015. J. Geophys. Res. Atmos. (2020). https://doi.org/10.1029/2020jd033331
14. Wang, W., et al.: Modelling hydrologic processes in the Mekong River Basin using a distributed model driven by satellite precipitation and rain gauge observations. PLoS ONE (2016). https://doi.org/10.1371/journal.pone.0152229
15. Lutz, A., Terink, W., Droogers, P., Immerzeel, W., Piman, T.: Development of baseline climate data set and trend analysis in the Mekong Basin. 1–127 (2014). https://doi.org/10.1175/BAMS-88-9-1383
16. Sun, Q., Miao, C., Duan, Q., Ashouri, H., Sorooshian, S., Hsu, K.L.: A review of global precipitation data sets: data sources, estimation, and intercomparisons. Rev. Geophys. (2018). https://doi.org/10.1002/2017RG000574
17. Yatagai, A., Arakawa, O., Kamiguchi, K., Kawamoto, H.: A 44-year daily gridded precipitation dataset for Asia. Sola (2009). https://doi.org/10.2151/sola.2009
18. Yatagai, A., Kamiguchi, K., Arakawa, O., Hamada, A., Yasutomi, N., Kitoh, A.: APHRODITE: constructing a long-term daily gridded precipitation dataset for Asia based on a dense network of rain gauges. Bull. Am. Meteorol. Soc. **93**, 1401–1415 (2012). https://doi.org/10.1175/BAMS-D-11-00122.1
19. Chen, A., Chen, D., Azorin-Molina, C.: Assessing reliability of precipitation data over the Mekong River Basin: a comparison of ground-based, satellite, and reanalysis datasets. Int. J. Climatol. (2018). https://doi.org/10.1002/joc.5670
20. Thrasher, B., Maurer, E.P., McKellar, C., Duffy, P.B.: Technical note: bias correcting climate model simulated daily temperature extremes with quantile mapping. Hydrol. Earth Syst. Sci. **16**, 3309–3314 (2012). https://doi.org/10.5194/hess-16-3309-2012
21. Thrasher, B., Xiong, J., Wang, W., Melton, F., Michaelis, A., Nemani, R.: Downscaled climate projections suitable for resource management. Eos, Trans. Am. Geophys. Union **94**, 321–323 (2013). https://doi.org/10.1002/2013EO370002
22. Sheffield, J., Goteti, G., Wood, E.F.: Development of a 50-year high-resolution global dataset of meteorological forcings for land surface modeling. J. Clim. **19**, 3088–3111 (2006). https://doi.org/10.1175/JCLI3790.1

23. Knutti, R.: Should we believe model predictions of future climate change? Philos. Trans. R. Soc. A Math. Phys. Eng. Sci. **366**, 4647–4664 (2008). https://doi.org/10.1098/rsta.2008.0169
24. Musie, M., Sen, S., Srivastava, P.: Application of CORDEX-AFRICA and NEX-GDDP datasets for hydrologic projections under climate change in Lake Ziway Sub-Basin. Ethiopia. J. Hydrol. Reg. Stud. **31**, 100721 (2020). https://doi.org/10.1016/j.ejrh.2020.100721
25. Ahmadalipour, A., Moradkhani, H., Svoboda, M.: Centennial drought outlook over the CONUS using NASA-NEX downscaled climate ensemble. Int. J. Climatol. **37**, 2477–2491 (2017). https://doi.org/10.1002/joc.4859
26. Thilakarathne, M., Sridhar, V.: Characterization of future drought conditions in the lower Mekong River Basin. Weather Clim. Extrem. **17**, 47–58 (2017). https://doi.org/10.1016/j.wace.2017.07.004
27. Thornthwaite, C.W.: An approach toward a rational classification of climate. Geogr. Rev. **38**, 55–94 (1948). https://doi.org/10.2307/210739
28. Mann, H.B.: Non-Parametric Test Against Trend. Econometrica (1945)
29. Sen, P.K.: Estimates of the regression coefficient based on Kendall's Tau. J. Am. Stat. Assoc. (1968). https://doi.org/10.1080/01621459.1968.10480934
30. Helsel, D.R., Hirsch, R.M.: Statistical methods in water resources. Stat. Methods Water Resour. (1992). https://doi.org/10.2307/1269385
31. Yue, S., Pilon, P., Phinney, B., Cavadias, G.: The influence of autocorrelation on the ability to detect trend in hydrological series. Hydrol. Process. (2002). https://doi.org/10.1002/hyp.1095
32. Park, E., Lee, Y.J.: Estimates of standard deviation of spearman's rank correlation coefficients with dependent observations. Commun. Stat. Part B Simul. Comput. (2001). https://doi.org/10.1081/SAC-100001863
33. Irannezhad, M., Liu, J., Chen, D.: Extreme precipitation variability across the Lancang-Mekong River Basin during 1952–2015 in relation to teleconnections and summer monsoons. Int. J. Climatol. **42**, 2614–2638 (2022). https://doi.org/10.1002/joc.7370
34. Minh, H.V.T., et al.: Multi-scale characteristics of drought propagation from meteorological to hydrological phases: variability and impact in the upper Mekong Delta. Vietnam. Nat. Hazards (2024). https://doi.org/10.1007/s11069-024-06898-2
35. Quang, C.N.X., Hoa, H.V., Giang, N.N.H., Hoa, N.T.: Assessment of meteorological drought in the Vietnamese Mekong Delta in period 1985–2018. IOP Conf. Ser. Earth Environ. Sci. **652**, 12020 (2021). https://doi.org/10.1088/1755-1315/652/1/012020
36. Dong, X.: Influences of the Pacific Decadal Oscillation on the East Asian summer monsoon in Non-ENSO years. Atmos. Sci. Lett. **17**, 115–120 (2016). https://doi.org/10.1002/asl.634
37. Li, P., Zhou, B., Zhang, D., Huang, Y., Xie, W., Song, Z., Liu, Y.: On the association of pacific decadal oscillation with the interdecadal variability in Asian-Pacific oscillation. Atmos. Res. **310**, 107635 (2024). https://doi.org/10.1016/j.atmosres.2024.107635
38. Räsänen, T.A., Lindgren, V., Guillaume, J.H.A., Buckley, B.M., Kummu, M.: On the spatial and temporal variability of ENSO precipitation and drought teleconnection in Mainland Southeast Asia. Clim. Past (2016). https://doi.org/10.5194/cp-12-1889-2016
39. Tang, Q., Yun, X., Wang, J., Deng, H., Liu, B., Tran, T.C., Han, D., Fang, H., Shi, X., Pokhrel, Y., et al.: Water hazards: drought and flood BT. In: Chen, D., Liu, J., Tang, Q. (eds.) Water resources in the Lancang-Mekong River Basin: impact of climate change and human interventions, pp. 255–281. Springer Nature Singapore, Singapore (2024). ISBN 978-981-97-0759-1
40. Fan, L., Wang, Y., Cao, C., Chen, W.: Teleconnections of atmospheric circulations to meteorological drought in the Lancang-Mekong River Basin. Atmosphere (Basel). **15** (2024)
41. Li, R., Wang, S.Y., Gillies, R.R., Buckley, B.M., Truong, L.H., Cho, C.: Decadal oscillation of autumn precipitation in Central Vietnam Modulated by the East Pacific-North Pacific (EP-NP) teleconnection. Environ. Res. Lett. (2015). https://doi.org/10.1088/1748-9326/10/2/024008

Research on the Probabilistic Forecasting of Water Resources in the Huai River Basin Based on Multi-Model Ensemble

Kailei Liu[1,2], Jinglei Wang[3], Mei Yao[4], Gangyi Lin[5](✉), Guangye Yu[6], and Jiefeng Wu[7]

[1] Anhui Hydraulic Research Institute, Huai River Commission, Bengbu 233000, China
[2] Anhui Provincial Key Laboratory of Water Science and Intelligent Water Conservancy, Bengbu 233704, China
[3] China Water Resources Planning and Design Institute for the Huai River Basin, Hefei 230000, China
[4] Provincial Bureau of Hydrology, Hefei 230000, China
[5] Shanghai Electric Power Design Institute Co., Ltd, Shanghai 200000, China
exb25559@163.com
[6] Huaihe River Water Resources Protection Scientific Research Institute, Bengbu 233000, China
[7] Nanjing University of Information Science & Technology, Nanjing 210000, China
wujf27@nuist.edu.cn

Abstract. Water resources forecasting methods play crucial roles in long-term water resources management, while they have always been affected by unavoidable uncertainties from model structure, parameters, and input data sets. This study integrates 12 individual forecasting methods, such as kNN (k-nearest neighbor) and AR (Auto Regressive), based on the Bayesian Model Averaging approach to construct a probabilistic forecasting model BPF (Bayesian Probability Forecasting). After that, the numerical experimental simulation is issued in the Huai River Basin using data sets from 1997 to 2022. Through statistical and numerical analysis of the forecasting results, it is considered that: 1) there are significant differences among the forecasting results from the above 12 deterministic methods, and no single one can consistently perform more accurately than the others; 2) the mean forecasting results from the BPF model are always close to the most accurate one of the 12, and the BPF is considered to be stable and reliable; 3) the probabilistic forecasting results of the BPF perform with high coverage ratio (CR) and low bias width (BW) in both the calibration period and verification period, and thus considered to be able to describe the posterior probability distribution of water resources reliably.

Keywords: Annual Volume of Water Resources · Ensemble Members · Huai River Basin · Back Propagation Neural Network · K-Nearest Neighbor · Bayesian Model Averaging

C. Weng (Ed.): WRE 2024, LNCE 634, pp. 67–75, 2025.
https://doi.org/10.1007/978-981-96-5095-8_6

1 Introduction

The "Comprehensive Water Resources Planning of China, 2010–2030", approved by the State Council of China, clearly points out that water resources are the foundational natural resources on which the national economy and people's livelihood depend. The Huai River Basin is located in the north-south transition zone of China. Although the basin area only makes up 3.4% of the country's total area, its grain production constitutes 20.5% of the country's total grain output. In recent years, the Huai River Basin has experienced a surge in population growth and a swift economic boom. This rapid expansion has exacerbated the ongoing tension between the region's water supply and its escalating demand. Grasping the evolution trend of water resources in this area is of great significance for ensuring regional and national water supply security and food security.

Regression analysis, k-NN (the k-nearest neighbors), and BPNN (Back Propagation Neural Network) have been widely used for water resources forecasting applications. For example, Tian Lemeng [1] applies regression analysis methods to predict water scarcity in Beijing city. Kong Haojie [2] use regression analysis and BPNN methods to establish water resources forecasting models for the Jinan City. Jin Juliang [3] uses k-NN methods to predict daily runoff. Applying these methods has not only enriched the means of water resources forecasting, but also provided valuable information for comprehensive water resources planning in the region.

However, these traditional methods only provide deterministic forecasting results with limited forecasting information. Furthermore, the precision of deterministic forecast outcomes is inherently constrained due to the inherent uncertainties embedded within the input data and the predictive model's framework. As a result, the reliability and practical utility of such deterministic forecasts are somewhat diminished. This study advances the field by developing a comprehensive suite of twelve water resources forecasting methods. It builds upon the foundation of existing single-factor models, expanding to incorporate up to four-factor analyses. These methods are meticulously constructed using a blend of auto regression (AR) analysis, k-NN, and BPNN techniques. Furthermore, this study seamlessly integrates the forecasting outcomes from a diverse array of methodologies within the Bayesian Model Averaging (BMA) framework. It provides a nuanced depiction of the anticipated shifts in regional water resources, encapsulating these changes within the probabilistic contours of a distribution with a 90% confidence level.

Compared with traditional deterministic forecasting methods, the computational results of this study are richer in terms of presentation and numerical information, significantly enhancing the valuable content of information within the forecast outcomes. This, in turn, provides a more scientific and reliable basis for water resource management and planning.

2 Method Introduction

2.1 The Construction of Ensemble Members

The aggregate volume of Total volume of Water Resources (TWR) within a region is derived from the combined contributions of surface water and groundwater components. The annual TWR is ascertained by aggregating the yearly volumes of these two distinct water sources, with an adjustment to deduct any overlapping volumes that may have been counted twice. Additionally, due to the direct impact of precipitation on the volume of surface water resources, it is believed that there is a strong correlation between precipitation, surface water, groundwater and the TWR. Based on the identified strong correlations, it is possible to formulate a correlation equation that quantifies the relationship between each contributing factor.

According to the calculation formula of the TWR, it is possible to use historical water resources datasets to predict the future TWR of any specific year. This study attempts to filter out from one factor to four factors gradually, establishing the correlation between historical TWR and future ones. For the single-factor correlation, historical annual TWR data are used as the independent variable, and future annual TWR is used as the dependent variable to establish the single-factor correlation equation. A bivariate correlation relationship is established by using historical data on the volume of surface water and groundwater resources.

Three types of statistical mathematical models, i.e. AR, k-NN and BPNN, are used respectively to describe the aforementioned four correlation relationships. Subsequently, under the Bayesian theoretical framework, the 12 independently developed TWR forecasting models are integrated to contracture BMA ensemble forecast model with a membership of 12 (Fig. 1).

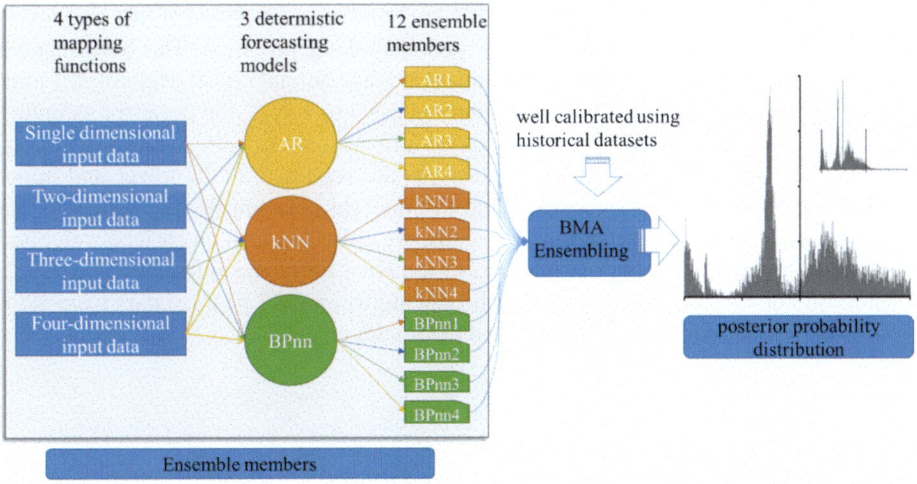

Fig. 1. Schematic diagram of the operational process of the probability forecast model based on the model cluster.

2.2 The Description for Posterior Distribution

Utilizing historical hydrological data, the parameters of the aforementioned models are meticulously calibrated to derive the parameters of BMA and the marginal probability distributions for both the observed and forecasted values of the TWR. The detailed calculation details of BMA can be found in Liu et al.' study [4], and are not elaborated here due to space limitations.

This study uses the Monte Carlo method to generate the posterior probability distribution of the forecasted TWR. The general steps are as follows: 1) Select randomly the optimal forecast method according to the weights of the 12 ensemble members; 2) Based on the sub-Gaussian model [4] constructed on the basis of the optimal method, the forecast value and corresponding probability value of the TWR in the normal space are generated; 3) The probability value is substituted into the marginal probability distribution function of the forecast value of the TWR in the original space to obtain a forecast value of the TWR; 4) Repeat the above steps multiple times to obtain a discrete distribution of the posterior forecast results of the TWR. The probability distribution of the TWR value obtained by the above steps is considered to reflect the posterior distribution characteristics of the TWR. Therefore, the mean of the posterior distribution can be published as the deterministic forecast result of the TWR, and the forecast values at the 5% and 95% percentiles can be published as the upper and lower limits of the 90% confidence forecast results.

3 Test Basin and Data Introduction

The Huai River Basin is located in the transitional zone of the north and south climates in China. It covers an area of about 27×10^4 km2 and spans 40 cities of five provinces, including Hubei, Henan, Anhui, Jiangsu, and Shandong province. The basin has an average annual precipitation of 875 mm, and an average annual runoff depth of 221 mm.

The forecast period for each forecast method is set to 1 year; the posterior sampling frequency is 50,000. After sorting the posterior forecast results from smallest to largest, the values at the 2000th and 38,000th positions are regarded as the lower and upper limits of the 90% confidence forecast, and the interval between the two values is the 90% confidence interval. To conduct this study, we collected data on surface water and groundwater resources, rainfall, TWR etc., of the Huai River Basin from 1997 to 2022. Considering the forecast period of 1 year for the algorithms, 13 years of data from 1998 to 2010 can be used for model structure and parameter calibration; 12 years of data from 2011 to 2022 can be used for verification.

4 Experimental Results Analysis

4.1 Deterministic Forecast Results

Using historical data from 1998 to 2010 for model structure and parameter calibration, and further using data from 2011 to 2022 to verify the structure and parameters of the established models. As shown in Fig. 2, the upper and lower ends of the box plot in the figure represent the distribution range of the forecast results of the 12 methods, the middle line of the box plot represents the median of the forecast results, the cross mark represents the mean of the forecast results.

From the forecasts shown in Fig. 2, the black dots representing the actual values of water resources are always within the maximum and minimum values of the box plot,

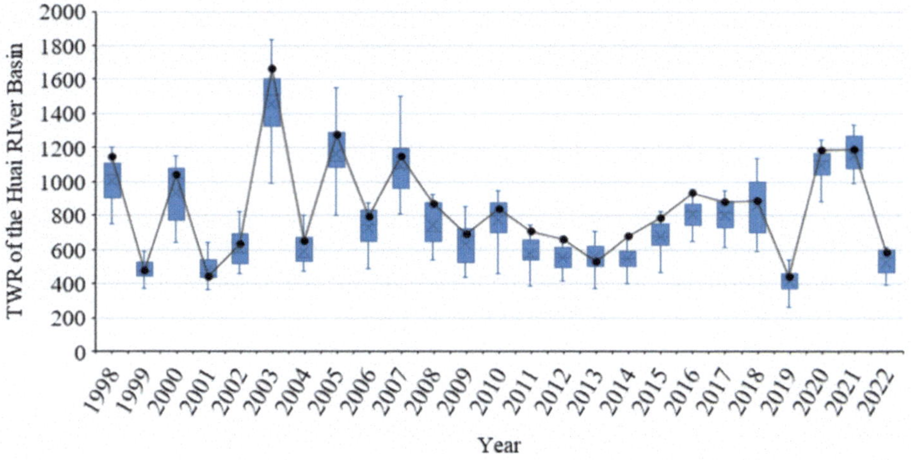

Fig. 2. Diagram for the Forecasts of TWR.

indicating that the deviation of the forecast results of each model from the observed value is not large. TWR can be roughly estimated based on the forecast results of the 12 methods. It can also be seen from Fig. 2 that there are significant differences in the forecast results of water resources between any two models in any year. For instance, in 2003, the BPNN2 and AR4 models yielded forecast results of 988 m^3 and 1833 m^3, respectively. Furthermore, we evaluated the relative superiority and stability of each model using the absolute error value of the forecast results. The larger absolute values indicate the worse performance of the analyzed method in the forecast of water resources for the year and vice versa. We counted the frequency of the relatively optimal performance of the 12 methods in 25 flood events and analyzed the relative superiority and stability of each model.

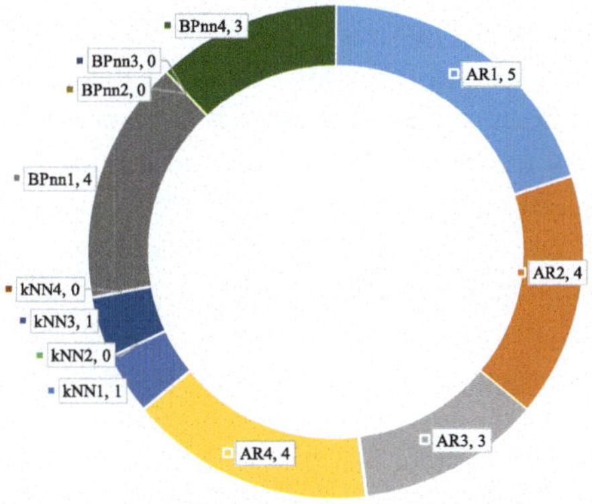

Fig. 3. Statistical chart of optimized ensemble members in 25 flood events.

Figure 3 illustrates the frequency with which each of the 12 methods demonstrated relatively superior performance across a series of 25 flood events. Examination of the Fig. 3 reveals that the AR1, AR2, AR4, and BPnn1 models have outperformed others in terms of accuracy, with the AR1 model showing the highest frequency of superior performance at five instances, closely followed by the AR2, AR4, and BPnn1 models, each with four instances of optimal performance. However, no single model has demonstrated optimal performance in 25 flood events or more than half of the years, leading to the conclusion that none of the individual models can consistently to maintain the highest level of accuracy. Therefore, when applying TWR forecasting, it is advisable to avoid relying solely on a single model for calculations.

4.2 Probability Forecast Results

As shown in the Fig. 4, we display the observed TWR, the forecast results of the 12 methods, and the BPF probability forecast results of each year on the same process line chart. The black dotted line represents the value of the TWR, the gray dots represent the forecast results of the 12 methods, and the blue band is the 90% confidence interval of the BPF. The closer the gray dotted line and the blue band align with the black dotted line, the greater the precision and reliability of the results. Additionally, a narrower blue band signifies lower uncertainty associated with the probabilistic forecast outcomes.

As illustrated by the forecast results in Fig. 4, the statistical outcomes for TWR from 1998 to 2022 are consistently contained within the 90% confidence interval. This inclusion indicates that the probabilistic forecasts, based on the ensemble members employed in this study, are robust enough to fully encompass the actual values of the TWR, thereby demonstrating a commendable level of accuracy and reliability.

Further precision analysis is presented in Table 1, contrasting the calibration and verification phases for the TWR forecasts. Among the ensemble of 12 forecast methods, the Nash-Sutcliffe Efficiency [5] (NSE) index ranges from 0.60 to 0.91 during the calibration period. However, this range extends from a lower limit of 0.41 to an upper limit of 0.94 during the verification period, which points to a discernible reduction in the collective performance of the models.

When evaluating the NSE and Root Mean Square Error [6] (RMSE) indicators of the BPF against those of the ensemble members, it is observed that the BPF's indicators fall within the spectrum of the ensemble's performance, closely aligning with the more favorable end. This positioning suggests that the BPF's aggregated forecast effectively mitigates the risk of suboptimal individual model outcomes, achieving a precision that is on par with and occasionally exceeds, the highest-performing method, coupled with a high reliability quotient. Especially noteworthy is the BPF's NSE indicator during the verification period, which reaches a commendable score of 0.83. This score not only meets but also exceeds the second-class standard as defined by the "Flood Forecasting Precision Evaluation Standard" in China. Consequently, the BPF forecasts are deemed sufficiently precise to be instrumental in guiding strategic decisions regarding medium and long-term water resource management.

Utilizing the statistical indicators of the CR and BW, we conducted an analysis to discern the performance variances in the BPF probabilistic forecast outcomes between the calibration and verification phases. A comparison reveals that the width of the BMA

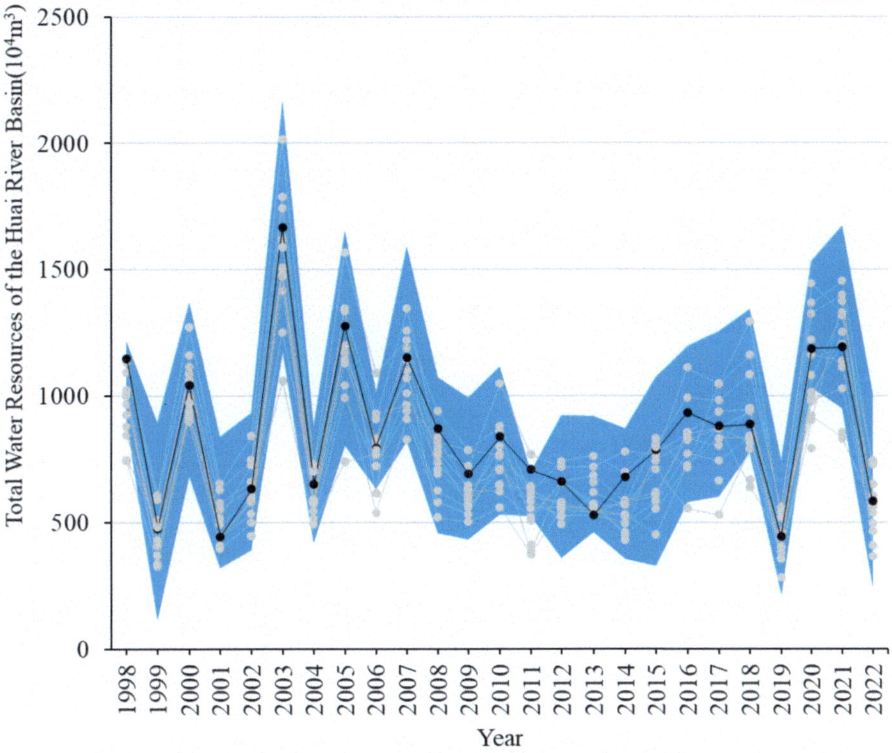

Fig. 4. Schematic diagram of the forecast results of the TWR in the Huai River Basin from 1998 to 2022.

confidence interval has expanded during the verification phase, increasing slightly from 0.76 to 0.77. Concurrently, there is a marginal decline in the CR indicator, dropping from 100% to 93%, which indicates a minor yet notable change. Despite these shifts, the aggregate findings affirm that the BPF ensemble forecasting methodology developed in this study can deliver annual-scale water resource forecasts with considerable reliability.

5 Probabilistic Forecasts of the 2022 TWR

In this study, we use the forecasted TWR for 2022 as a case study to illustrate the probability distribution of TWR in the Huai River Basin, represented through a histogram. This histogram delineates the forecast range of water resources into 100 distinct intervals, with the frequency value corresponding to the count of instances where the posterior probability forecast outcomes are situated within each interval. Given the sparse distribution of data points at the tails of the histogram, which tend to contain less reliable information, the presented histogram is truncated to encompass only the central interval, ranging from the 5th to the 95th percentile. This approach ensures that the visualization focuses on the most statistically significant portion of the data distribution.

Upon examining Fig. 5, it becomes evident that the posterior distribution of the projected water resources for 2022 exhibits a multimodal characteristic, suggesting a

Table 1. Forecast Results Statistics.

Indices		Ensemble Members	BPF		
			Ensemble Mean	Confidential Interval	
Calibration Period	NSE indices	0.60 ~ 0.91	0.97	CR	100%
	RMSE indices	101 ~ 293	60	BW	0.76
Validation Period	NSE indices	0.41 ~ 0.94	0.83	CR	93%
	RMSE indices	57 ~ 213	95	BW	0.77

complexity that defies straightforward representation by conventional distribution functions. Consequently, a precise probability distribution function for this dataset cannot be readily specified. To address this, we employ a discrete probability distribution to provide an approximation of the forecast's probabilistic nature. Based on the statistical analysis, the ensemble means of the forecasted water resources 2022 is determined to be 63.4 billion cubic meters. Additionally, the forecast values corresponding to the 5th and 95th percentiles are 24.2 billion m3 and 100.6 billion m3, respectively. These values offer a comprehensive view of the potential variability in the forecasted water resources.

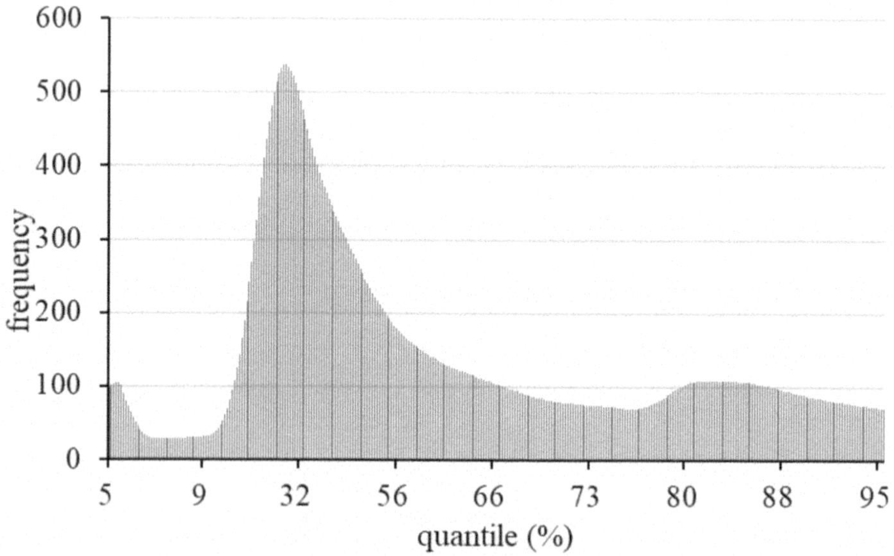

Fig. 5. Posterior distribution histogram of the 2022 forecasted water resources quantity.

6 Conclusion

This study takes the TWR as the research object and the Huai River Basin as the test basin. We introduce a probabilistic forecasting approach that leverages an ensemble of models. This method has been rigorously validated through numerical simulations spanning from 1998 to 2022, yielding high-reliability forecasts concerning the quantity of water resources and delineating the characteristics of their probability distribution. The BPF method, in particular, offers substantial technical support for water resources planning, certification, and other operational activities within the basin.

When juxtaposed with traditional deterministic forecasting techniques, the probabilistic forecasting method, underpinned by a model cluster, presents a distinct advantage. It furnishes a more comprehensive and dependable array of forecast information, effectively mitigating the uncertainty inherent in subsequent operational phases and markedly lowering the risks associated with decision-making processes. The BPF method merits further examination and broader adoption across water resources-related operations and other geographical areas.

It is important to note, however, that Total Water Resources are heavily influenced by a multitude of factors, including regional water conservancy projects, the exploitation of groundwater, and shifts in atmospheric patterns. Given that this study's scope is confined to historical data-based predictions of TWR, the constructed model ensemble may not account for all relevant influencing factors. Consequently, the model's architecture remains relatively simplistic, and the underlying mechanisms require further refinement. As such, the probabilistic forecast outcomes presented herein necessitate additional validation and are, at this stage, intended to serve as a reference for related research endeavors.

References

1. Tian, L.: Application of regression analysis in the prediction of water resource shortage. Statistics and Decision **2**(14), 84–86 (2012)
2. Kong, H.: A comparative analysis of the characteristics of sediment and water monitoring data of the yellow river based on multivariate nonlinear regression and BP neural network models. Journal of Zhejiang Business Technology Institute **23**(01), 18–22 (2024)
3. Jin, J., Wei, Y., Ding, J.: A nearest neighbor simulation model for predicting daily runoff processes. Journal of System Simulation **1**(11), 1494–1496 (2002)
4. Liu, K., Hu, Y., Wang, Y., et al.: Preliminary exploration of the application and parameter patterns of BMA ensemble forecasting in the Huai River Basin. Journal of Lake Sciences **29**(06), 1520–1527 (2017)
5. Shen, P., Ke, E., Liu, F., et al.: Practice of semi-distributed flood forecasting in the Fen River Basin. Zhejiang Water Conservancy Science and Technology **52**(02), 34–40 (2024)
6. Ye, J., Yu, H., Ge, J., et al.: Experimental analysis of deep learning to improve the accuracy of significant wave height forecasts. Journal of Ocean University of China (Natural Science Edition) **54**(03), 1–8 (2024)

Design of a Risk Management System for Water Diversion Project

Yu Feng[1,2,3], Weibing Du[4], Yongqiang Wang[1,2,3](✉), Zhe Chen[4], Taihe Zhang[4], and Jiang Wu[1,2,3]

[1] Changjiang Water Resources Commission, Changjiang River Scientific Research Institute, Wuhan 430010, Hubei, China
wangyongqiang1022@126.com

[2] Hubei Provincial Key Laboratory of Basin Water Resources and Ecological Environment, Changjiang River Scientific Research Institute, Wuhan 430010, Hubei, China

[3] Research Center On the Yangtze River Economic Belt Protection and Development Strategy, CWRC, Wuhan 430010, Hubei, China

[4] Henan Water Diversion Engineering Co., Ltd, Zhengzhou 450003, Henan, China

Abstract. The purpose of building the Yangtze River for the Huaihe River Diversion Project is to improve water supply in the Henan and Anhui Provinces of China. Risk management is an important task in this project, which aims to ensure safe water supply. This paper presents the design of a risk management system for water diversion projects. The system is developed based on a microservice architecture and is divided into three parts: data layer, model platform, and business application. The data layer stores all the business data of the system and provides data services. The model platform is the core component of the system. It integrates models related to the risk management of water diversion projects, including risk identification, risk evaluation, and contingency plan-matching models. Each model is published separately as a micro-service, and the model calculation service is managed by a service registry center. Model calculations are primarily executed in the form of asynchronous calls. In this case study, we used the Henan section of the Yangtze River to the Huaihe River Diversion Project as an example to demonstrate the risk-management system. The system implemented an identification and control process for the four risk categories, providing assistance in ensuring the construction and operational safety of the diversion project.

Keywords: Water Diversion Project · Risk Management · Decision Support System · Software Architecture

1 Introduction

Water supply is an essential service that any community needs to provide securely [1]; however, water resources have uneven spatial and temporal distribution [2]. Water diversion projects are an effective way to achieve spatial redistribution [3]. Various risks are involved during the construction and running period of water diversion projects, which

C. Weng (Ed.): WRE 2024, LNCE 634, pp. 76–89, 2025.
https://doi.org/10.1007/978-981-96-5095-8_7

impact project management [4], work trade in various project stages [5], and construction progress control [6]. Research has been conducted to address risk assessment and identification. For example, Monte Carlo probabilistic modeling has been used to evaluate the impact of pump stations [7]. The fuzzy analytic hierarchy process can play an effective role in assessing reservoir operational risks [8]. Fault tree analysis is a well-known technique for analyzing risks related to a specific system [9]. In addition to risks in engineering facilities, some natural factors, such as drought [10] and climate change [11, 12], can also affect the operation of water diversion facilities.

An information system is an effective means of assisting with the daily operational safety of water diversion projects [13]. Liu et al. (2021) built an integrated visualization system based on a building information model to support the safety management of water diversion projects [14]. The system realizes automatic safety management through processes such as data collection, analysis, early warning issuance, and emergency response. Lei et al. (2018) developed and implemented a decision support system that provides technical support for managing pollution accidents [15]. The decision support system integrates hydrodynamic and water quality simulations, source identification, emergency regulation, and evaluation of emergency measures to deal with sudden water pollution accidents. Gómez (2017) proposed a decision support system for operational risk management [16]. The system performs management through risk identification, evaluation, prioritization, and quantification.

The South-to-North water diversion project in China is a strategic inter-basin water transfer project implemented to alleviate the serious shortage of water resources in North China [17]. As part of the project, the Yangtze River to Huaihe River diversion project connects the two major water systems of the Yangtze and Huaihe Rivers, aiming to solve water shortages in counties and cities along the project. The Henan section of the project supplies water to several county-level administrative districts in Zhoukou and Shangqiu City in the Henan Province. Water diversion projects have various facilities, such as transmission channels, pipelines, storage reservoirs, and pumping stations. The failure of any of these facilities can adversely affect the water supply. Therefore, it is necessary to identify and analyze the risks in a project to reduce the losses caused by emergencies.

This study focused on integrating risk management models and conducting research on the development of risk assessment and management systems. Initially, we briefly described the models used for risk assessment and management. Subsequently, we introduced the design of the risk management system. We described the architecture and applied this design to deal with daily risk management in the Henan section of the water diversion project from the Yangtze River to the Huaihe River.

2 Materials and Methods

2.1 System Modules

The main goal of the risk management system is to deal with the following four types of risks: The risk modules are shown in Fig. 1. The risk points involved in the project were implemented in a data monitoring system using Geographic Information System (GIS) technology.

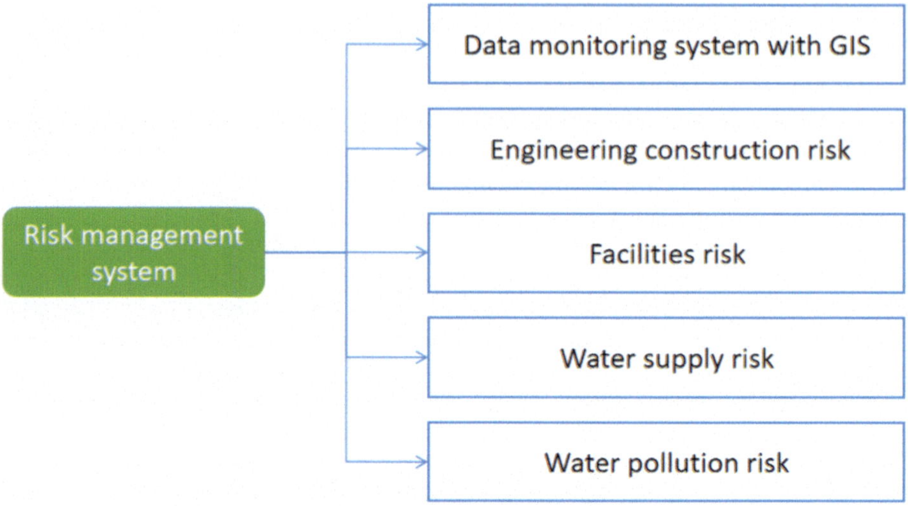

Fig. 1. System modules of the risk management system.

1) **Risks during the Engineering Construction Period**

Risks related to safety, progress, and investment during the construction period, including safety hazards, lagging progress, and insufficient funding.

2) **Risks of the Water Diversion Facilities**

Refers to the daily operational risks of pump stations, reservoirs, and embankments.

The risks associated with pump stations include operating conditions, device quality, and technical conditions. The operating conditions mainly consider the influences of sewage arrest, cleaning equipment, and inlet vortices. These include aging, sewage arrest and cleaning equipment damage, increased incoming dirt, and inlet vortex reverse rotation. Device quality mainly considers grid voltage fluctuations and the aging of the electrical equipment. The technical conditions mainly consider the pump characteristic errors and maintenance conditions, such as the reduces angle of the pump blades and the abnormal shape.

The first risk in embankments is a flood crash, which refers to the risk caused by a flood that overflows the embankment due to an insufficient height of the embankment or a high flood level in front of the embankment. The second is the risk of infiltration, which refers to the failure of the dike after the penetration or foundation of the embankment. The third is the risk of instability, which refers to the collapse and landslides of the river embankment, mainly owing to the loose foundation of the embankment.

The risk to reservoirs is mainly due to dam safety. These include the risk of embankment conditions, piercing gates and culverts, characteristics of embankment materials, and operation management.

3) **Risk to Water Supply**

Refers to the occurrence of insufficient water supply. This can be attributed to both natural and socioeconomic factors. Natural factors include floods, earthquakes, landslides, collapses, and subsidence. Socioeconomic factors include mass incidents, water disputes, water price increases, and changes in water diversion costs.

4) **Risk on Water Pollution**

Refers to water pollution. This can be caused by traffic accidents or sewage discharge from sewage treatment plants, landfills, and chemical plants. Once sewage enters a river, the contaminated water cannot be used as a water supply.

2.2 Risk Analysis

Figure 2 shows the flowchart of risk analysis, including three main steps:

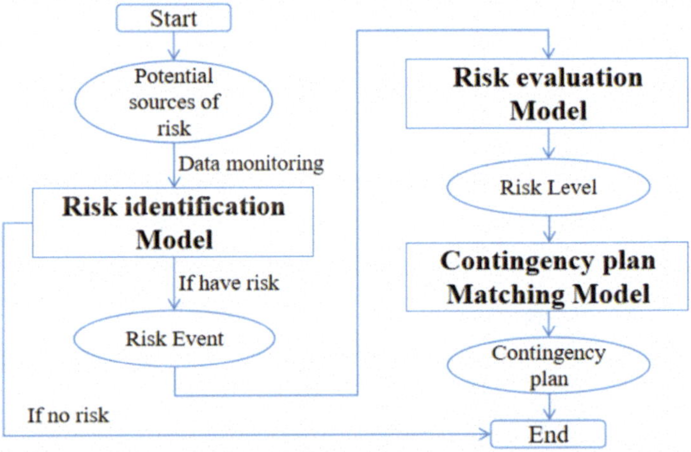

Fig. 2. Flowchart of risk analysis.

1) **Risk Identification**

In this project, data from the monitoring equipment were transmitted regularly to the risk management system. The system regularly activates the risk identification module to determine whether there is a risk. When a risk is identified using the risk identification model, a risk event is generated.

2) **Risk Level Evaluation**

Once a risk event is identified, a risk evaluation is initiated to analyze the risk level.

3) **Contingency Plans Generation**

Generates recommendations for corresponding contingency plans based on the risk level.

The core mathematical models in the system are the risk identification, risk evaluation, and contingency plan-matching models. It also integrates a water supply scheduling model to provide scheduling suggestions for the daily operation of the water diversion project and a water pollution simulation model to assist in the risk analysis.

2.3 System Architecture

The risk management system was designed based on the microservice architecture shown in Fig. 3. It contains five modules described in Sect. 2.1 and is divided into three layers. The data layer saves and manages all monitoring and business data used in the system. The service layer consists of a model platform and a web-based system. The web-based application system provides a user interface for convenient use of risk models to identify and analyze risks. The system also sets scheduled tasks that trigger the model calculations regularly.

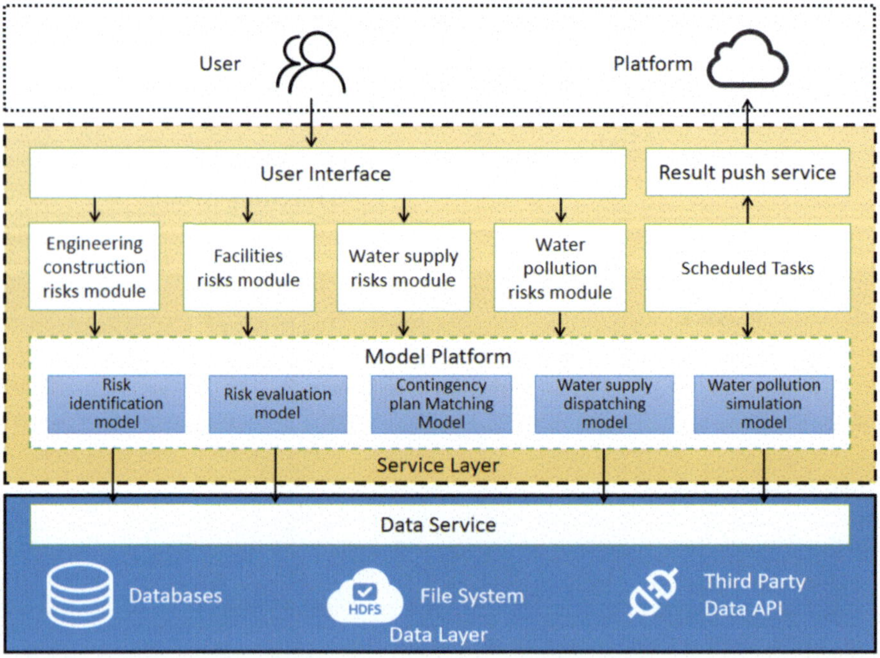

Fig. 3. Architecture of the platform.

The deployment structure of the system is shown in Fig. 4. When users access a risk-management system, requests are distributed through the gateway. Multiple web services are combined to form a service cluster. The Nacos service registration center, a platform for dynamic service discovery, configuration, and service management, is used to register microservices in the risk management system. These microservices include the data, models, and business services used in the system. When a service call occurs, the caller obtains the real IP and port from the registration center through the service name.

The configuration center hosts the configuration information within the platform to facilitate the management of microservice clusters. The authentication server implements login protection and permission control for system resources. Calling between services requires permission from an authorized server for authentication. Finally, the platform was monitored using a performance-monitoring module.

In this study, we implemented authorization based on OAuth 2.0. To implement a cluster of services, we adopted a stateless login. When the user logs successfully, the authentication server returns a bearer token to the user. The user then uses it to access the protected resource, and the authorization server does not store the user's login status.

The model platform provides interfaces for the model calculation. After the system receives a model calculation request, the model platform allocates the model and begins the calculation task with an asynchronous call. The model platform also uses scheduled tasks to perform rolling calculations.

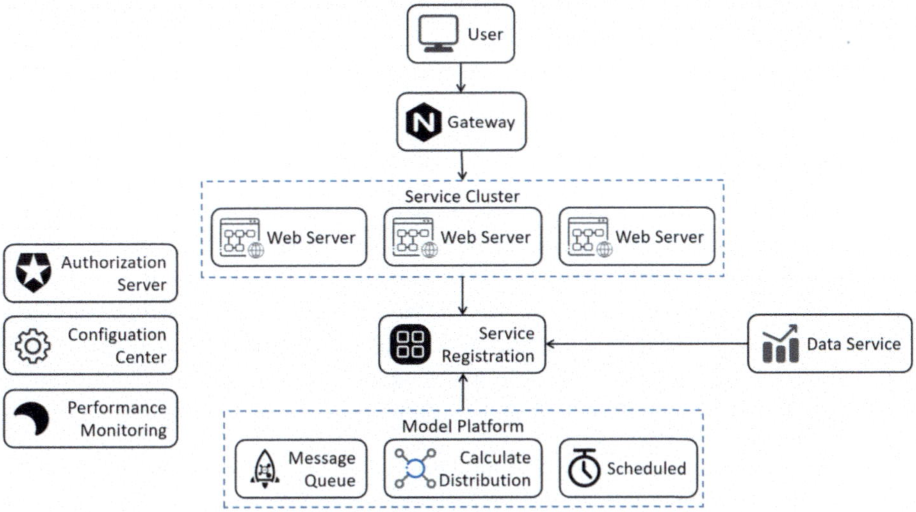

Fig. 4. Architecture of the platform.

2.4 Data Sources

The data layer unifies and stores data from various sources. The data sources for the risk assessment and management platform are divided into three main types:

(1) Relational Databases

The platform database and the databases from other third-party systems were included.

The database of the platform itself includes the following categories:

- Real-time monitoring database: This stores real-time monitoring data from objects such as rivers, gates, and pumping stations. The data items included water level, flow, rainfall, and opening.
- Basic information database: This stores the basic information of managed objects such as rivers, gates, and pumping stations.
- Business database: Stores various business data such as system configuration, user information, logs, and user operation.
- Spatial database: Stores GIS data and remote sensing image data, etc.
- Multimedia database: Stores audio and video data.

(2) Data from File System

Stored data is saved in a file format. This data type often contains a large amount of data and is compressed to reduce occupied storage space. We also used non-relational databases to store relational data.

(3) Data from Third-Party DataAPI

This platform requires data from other platforms, which are provided via a web-based application programming interface (API).

Based on the above data resources, we developed and registered data service in a service center.

2.5 Model Platform

The mathematical models were integrated into the model platform. The model platform implements a standardized definition and management of mathematical models, making their integration and management easier. The design of the model strictly follows the process of "input - > model - > output". The mathematical models were completely detached from the database, and the required variable data were provided in the input data. Before starting the model calculation, the input data were organized according to the requirements of the model using a JSON format. After the model was calculated, the result was output in the JSON format.

Here, we consider a risk evaluation model as an example. As shown in Fig. 5, the model does not read data directly through data services. The caller provides all input date required by the model, that is, the business back-end of the web system. When a user enters the model calculation module, the web system loads the required data for the model through the data service in advance and displays the input data on a graphical user interface. The system then waits until the user modifies and confirms the input data. In the risk analysis model, the input data primarily includes basic data, monitoring data, and conditions set by the user. After confirmation, the web system sends the data to the model platform, which completes the model calculations. After the model calculation is completed, the platform sends a signal to inform the web system that the calculation is complete, and the system reads the calculation result of the model from the model platform and displays it on a graphical user interface.

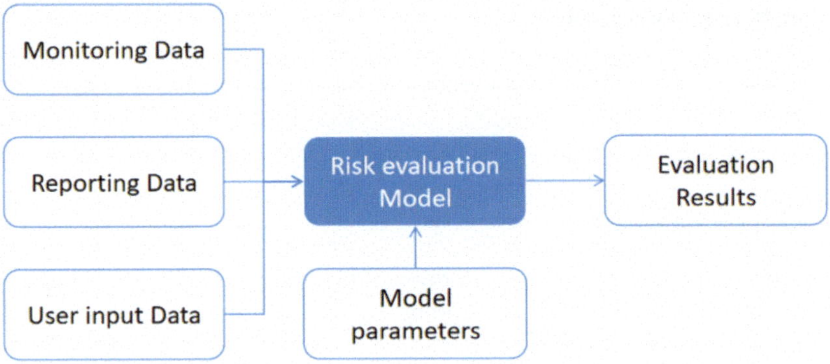

Fig. 5. Flowchart of the risk analysis model.

The model platform supports access to models written in different programming languages, such as Java, Python, and Fortran. Because of the varying requirements of computer hardware, models are often deployed separately on different server machines. The allocation of computing resources for the mathematical models on different server machines was scheduled using the model platform. Some models have input and output formats that do not conform to platform standards. These models require customized adapters that translate the input and output formats of the model platform into the format required by the model. The deployment structure of the model platform is illustrated in Fig. 6.

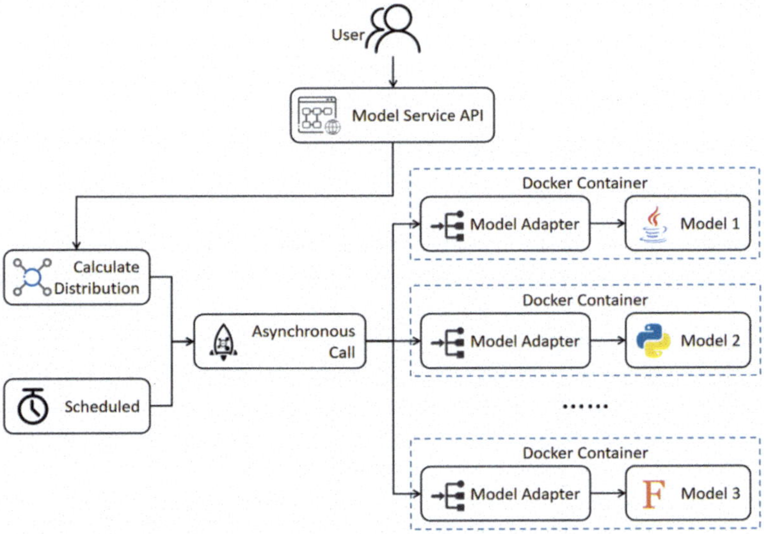

Fig. 6. Deployment structure of model platform.

3 Results and Discussion

In this section, we present the implementation of a risk management system for daily risk management of the diversion of the Yangtze River to the Huaihe River in the Henan section, China.

3.1 System Introduction

In this study, we focused on the Henan section of the Yangtze River to the Huaihe River diversion project, as well as the surrounding water-receiving areas along the project. The main task was to enhance urban and rural water supply capacities. Simultaneously, the improvement of the aquatic ecological environment should be considered. The construction project includes water conveyance channels, pumping stations, control gates, regulating reservoirs, and water conveyance pipelines, involving several counties in Zhoukou City and Shangqiu City in the Henan Province. A conceptual diagram of the project is presented in Fig. 7.

The main body of the water diversion project has been completed and will gradually be used in the future. In this study, we built a risk management system to help managers collect risk information. The system can identify, analyze, and manage risk events in a water diversion project.

As described in Sect. 2.1, the system has the corresponding features:

1) Provides a ledger of risk events, from which you can query the risk events that have occurred during the construction and operation periods.
2) Identify and analyze risk events that occur in engineering facilities. The main focus is on channels, pipelines, gates, pumps, and regulating reservoirs, as shown in Fig. 7.
3) Identify and analyze potential water shortage events in the water receiving area. As shown in Fig. 7, mainly focus on the county-level and above administrative regions.
4) Analyze and identify potential water pollution events.

Main modules of the system are as follows:

(1) **Data Monitoring Module**

The monitoring system displays basic and monitoring data for the following types of nodes: (1) reservoir, (2) pump, (3) channels, (4) gate, (5) water quality, (6) bridge, (7) pipe, and (8) risk information. Among them, risk information is not directly monitored but is calculated or manually reported by other risk-management modules. The positions of these nodes are marked on the GIS map. To facilitate the differentiation of the different types of nodes, we placed nodes of the same type in the same layer. After the user clicks on a node, the observation system displays more detailed information about the node, including basic information and real-time monitoring data.

The system regularly retrieved the latest monitoring data from the monitoring database and third-party data APIs. By comparing the latest monitoring data with a preset threshold in the system, data exceeding the threshold can be marked as a risk event. Simultaneously, the system regularly launches a risk identification model. If a risk was identified by the model, it was marked as a risk event. As shown in Fig. 8, the latest risk events are marked with red dots on the GIS map and users can click on

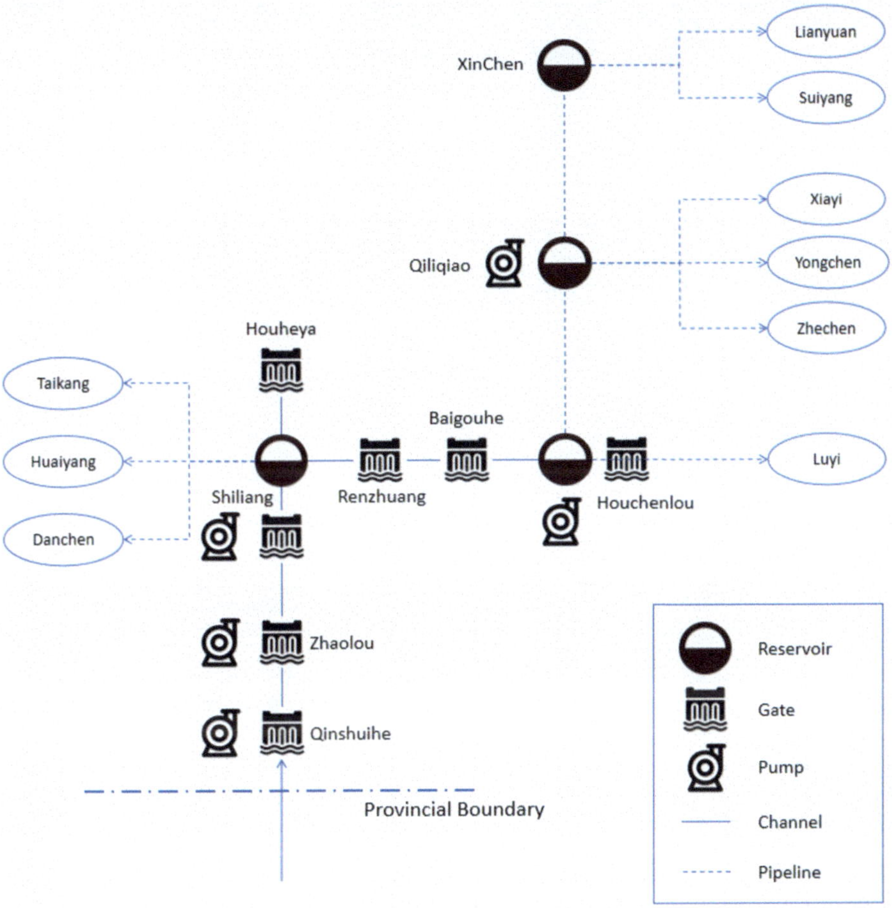

Fig. 7. Conceptual diagram of the project.

the relevant risk information. Only risk events within seven days are shown on the GIS map; earlier events need to be viewed in the list of historical events explained in the risk events display module.

(2) **Risk Events Display Module**

Historically, risk events have been collected and displayed in different subsystems. It contains all risk events after the system is put into use as well as historical risk events that are manually filled in. As explained in Sect. 2.1, we categorized the different risk events into four types: (1) risks during the engineering construction period, (2) risks of water diversion facilities, (3) risk of water supply, and (4) risk of water pollution. For example, Fig. 9 shows a table of historical events related to water pollution. The table contains the risk type, release time, and level. Other information, such as location, can be viewed by clicking on the details of the hyperlink. Manual reporting is also used to supplement and manage risk events that are not identified by the system.

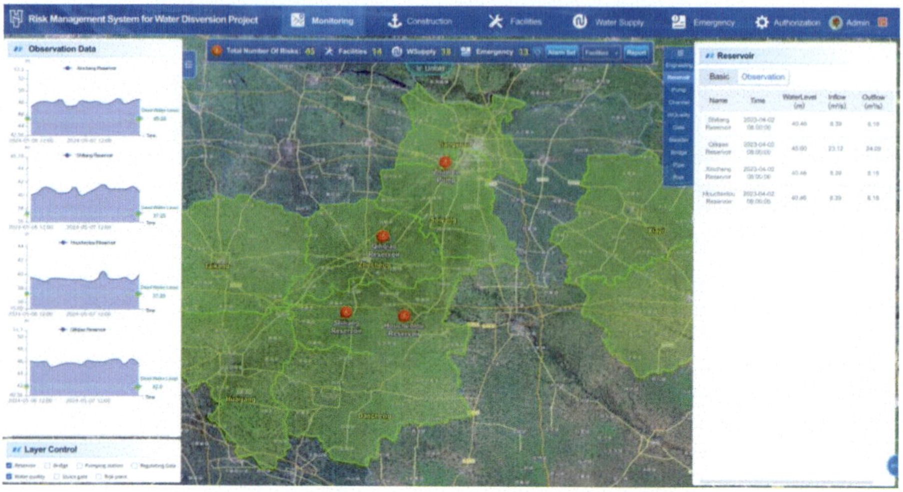

Fig. 8. User interface of the data monitoring system.

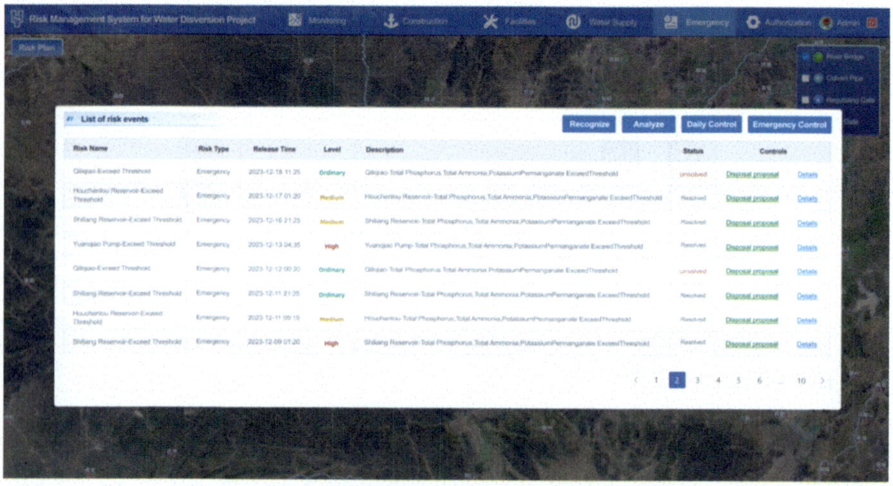

Fig. 9. Table of the historical risk events.

3.2 Case Study

Here, we provide an example to illustrate the risk management process. As shown in Fig. 10, the system automatically identifies risk events and displays them at the top of the page. We chose one of the water supply risk events named "Shiliang Pump Over Threshold." The panel on the left-hand side of the page displays information about the selected event. In this panel, we can enter the decrease in flow (default is 0) and duration (default is 0) for the Shiliang Pumping Station. The system calculates the number of days each reservoir can continue to supply water in this state and the number of days each receiving area experiences water shortages. Here, we enter the value of the decrease in flow as 18 and the value of the duration as 5, and then click the evaluation button. The results showed that the remaining water supply to the reservoir is 5.86 million cubic meters. Under this input state, the Shiliang and Qiliqiao reservoirs could only maintain water supply for four days. Taikang, Huayang, Dancheng, and Luyi will experience water shortages in two days; Zhecheng in three days; and Yongcheng in four days. Regions facing water shortages are marked in red on the GIS maps. This information provides managers with advice and suggestions.

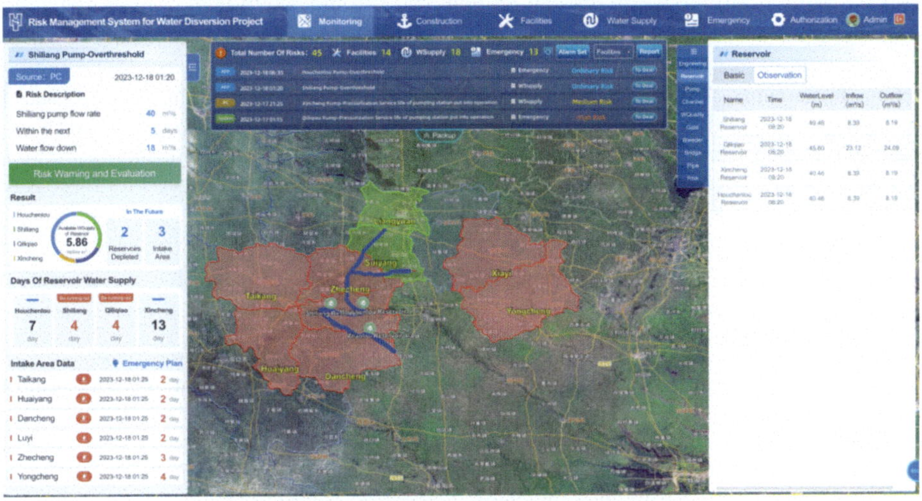

Fig. 10. Analysis of water supply risk event.

4 Conclusions

This study discusses the design of a risk management system. The data, model platform, and application layers are the main components of the platform. The data layer gathers all the data the platform requires during operation, which are stored separately in relational and non-relational databases classified by their characteristics. These stored data are provided to the application through data services that avoid direct access to the database by the application. The data service provides basic functions, such as data query, data management, statistical analysis, and data preprocessing, which can meet all the data requirements of the application. The model platform gathered the core mathematical models of the system, forming a standardized model calculation procedure. The model platform includes a series of basic functions related to the model, including model management, model operation, statistical analysis, asynchronous execution, and an API interface. When the model is connected to the model platform, customized adapters must be developed to satisfy the standardization requirements of the model platform. The application layer provides a graphical user interface for the management of four types of risk sources: (1) risks during the engineering construction period; (2) risks of water diversion facilities; (3) risks of water supply; and (4) risks of water pollution. The model platform provided computing resources for the model.

To demonstrate the potential of the proposed approach, we built a risk management system for the Henan section of the water transfer project from the Yangtze River to the Huaihe River. The platform collects monitoring data related to diversion projects. We can identify risk events in a water diversion project using the monitoring data and the risk identification model. The risk evaluation model helps us calculate the level of risk events and analyze their potential impacts. Finally, recommendations for the corresponding contingency plans can be matched through contingency plan-matching models, which assist in the security management of water diversion projects.

Acknowledgements. This research was funded by the Major Science and Technology Projects of the Ministry of Water Resources, grant number SKS-2022128; National Key Research and Development Program of China, grant number 2022YFC3202300; National Natural Science Foundation of China, grant number 42271044, 52009005; National Public Research Institutes for Basic R & D Operating Expenses Special Project, grant number CKSF2021486.

References

1. Fg, A., Ac, A., Eg, B., et al.: Dynamic, multi-objective optimal design and operation of water-energy systems for small, off-grid islands. Appl. Energy **250**, 605–616 (2019)
2. Zhang, C., Nong, X., Zhong, H., Shao, D., Chen, L., Liang, J.: A framework for exploring environmental risk of the longest inter-basin water diversion project under the influence of multiple factors: a case study in China. J. Environ. Manag. **322**, 116036 (2022), ISSN 0301-4797
3. Huang, W., Lin, H., Zhang, N.: RETRACTED ARTICLE: Cross basin water diversion project based on VR visual system and urban public art design. Arab. J. Geosci. **14**, 1733 (2021). https://doi.org/10.1007/s12517-021-08029-8
4. Zhao, Z.Y., Zuo, Z., Zillante, G.: Transformation of water resource management: a case study of the South-to-North Water Diversion project. J. Clean. Prod. **163**, 136–145 (2017), ISSN 0959-6526
5. Fung, I.W.H., Tam, V.W.Y., Lo, T.Y., Lu, L.L.H.: Developing a risk assessment model for construction safety. Int. J. Proj. Manag. **28**(6), 593–600 (2010), ISSN 0263-7863
6. Elbeltagi, E., Dawood, M.: Integrated visualized time control system for repetitive construction projects. Autom. Constr. **20**(7), 940–953 (2011), ISSN 0926-5805
7. Niu, Z.G., Xu, S.Y., Gong, Q.C.: Health risk assessment of odors emitted from urban wastewater pump stations in Tianjin. China. Environ. Sci. Pollut. Res. **21**, 10349–10360 (2014). https://doi.org/10.1007/s11356-014-2984-7
8. Jing, M., Jie, Y., Shou-yi, L., Lu, W.: Application of fuzzy analytic hierarchy process in the risk assessment of dangerous small-sized reservoirs. Int. J. Mach. Learn. Cybern. **9**(1), 113–123 (2015). https://doi.org/10.1007/s13042-015-0363-4
9. Yazdi, M., Zarei, E.: Uncertainty handling in the safety risk analysis: an integrated approach based on fuzzy fault tree analysis. J. Fail. Anal. Preven **18**, 392–404 (2018). https://doi.org/10.1007/s11668-018-0421-9
10. Haro, D., Solera, A., Paredes, J. et al.: Methodology for drought risk assessment in within-year regulated reservoir systems. Application to the Orbigo River System (Spain). Water Resour Manag. **28**, 3801–3814 (2014). https://doi.org/10.1007/s11269-014-0710-3

11. Shrestha, S., Shrestha, M., Babel, M.S.: Assessment of climate change impact on water diversion strategies of Melamchi Water Supply Project in Nepal. Theor. Appl. Climatol. **128**, 311–323 (2017). https://doi.org/10.1007/s00704-015-1713-6
12. Turner, S.W.D., Galelli, S.: Water supply sensitivity to climate change: An R package for implementing reservoir storage analysis in global and regional impact studies. Environ. Model. & Softw. **76**, 13–19 (2016), ISSN 1364-8152. https://doi.org/10.1016/j.envsoft.2015.11.007
13. Nematian, J., Movahhed, S.R.: An extended multi-objective mixed integer programming for water resources management through possibility theory. Eco. Inform. **54**, 100992 (2019)
14. Liu, T., Zhang, S., Wang, C.: A BIM-based safety management framework for operation and maintenance in water diversion projects. Water Resour. Manage **35**, 1619–1635 (2021). https://doi.org/10.1007/s11269-021-02813-7
15. Lei, X., Zheng, H., Shang, Y., Hao, W.: Assessing emergency regulation technology in the Middle Route of the South-to-North Water Diversion Project, China. Int. J. Water Resour. Dev. **34**(3), 405–417 (2018). https://doi.org/10.1080/07900627.2018.1441707
16. Osorio Gómez, J.C., Manotas Duque, D.F., Rivera, L., García-Alcaraz, J.L.: Decision support system for operational risk management in supply chain with 3PL Providers. In: Alor-Hernández, G., Valencia-García, R. (eds.) Current Trends on Knowledge-Based Systems. Intelligent Systems Reference Library, 120 (2017) Springer, Cham. https://doi.org/10.1007/978-3-319-51905-0_10
17. Tan, L., Luo, W., Yang, B., Huang, M., Shuai, S., Cheng, C., Zhou, X., Li, M., Hu, C.: Evaluation of landscape ecological risk in key ecological functional zone of South-to-North Water Diversion Project, China. Ecol. Indic. **147**, 109934 (2023), ISSN 1470-160X

A Bayesian Modified Bridging Model for Statistical Post-Processing of Ensemble Precipitation Forecasts in China

Yujie Li(✉), Weizhong Xu, Yunjie Ma, and Chaohao Zheng

Zhejiang Design Institute of Water Conservancy and Hydroelectric Power, Hangzhou 310002, Zhejiang, China
lyjzjhhu@126.com, {22112095,22012252}@zju.edu.cn

Abstract. Accurate and reliable precipitation forecasts are crucial for effective water resource management, disaster risk reduction, and agricultural planning. This paper proposes a Bayesian Modified Bridging (BMB) model, an improved statistical post-processing model based on the Bayesian Joint Probability (BJP) approach, to correct and bridge ensemble precipitation forecasts. The study evaluates the SEAS5 seasonal ensemble precipitation forecasts from ECMWF, covering precipitation across China. By comparing the correction effects of the BJP and BMB models, the aim is to provide more accurate and reliable precipitation forecasts. The BMB model incorporates both the mean and variance of the raw ensemble forecasts, enhancing the correction process's precision and reliability. Comprehensive evaluations using Continuous Ranked Probability Skill Score (CRPSS) and Probability Integral Transform Skill Score (PITSS) show that the BMB model significantly improves the accuracy and reliability of forecasts compared to the BJP model, with average CRPSS and PITSS improvements of 13.9% and 8.8%, respectively, for lead times of 0–5 months. By analyzing precipitation data across 3,781 grid points nationwide, the study finds that the BMB model consistently outperforms the BJP model in all forecast periods and months. These findings highlight the superior correction capabilities and robustness of the BMB model, providing a scientific basis for its widespread application in future research and operational forecasting. This study not only offers new methodologies for the meteorological and hydrological fields but also provides critical technical support for decision-makers in water resource management and disaster prevention.

Keywords: Bayesian Joint Probability Model · Bayesian Modified Bridging Model · Continuous Ranked Probability Skill Score · Ensemble Precipitation Forecast · Statistical Post-Processing

1 Introduction

The quest for accurate and reliable weather and climate forecasts has been a driving force in the advancement of meteorological science. Ensemble forecasting, which provides deterministic forecast results through the means or quantiles of ensemble members

C. Weng (Ed.): WRE 2024, LNCE 634, pp. 90–100, 2025.
https://doi.org/10.1007/978-981-96-5095-8_8

and quantifies forecast uncertainty using confidence intervals that represent the forecast states of all ensemble members, has gradually become a mainstream operational product for global meteorological and hydrological institutions [1–5]. These methods allow for a more comprehensive understanding of future weather conditions by not only predicting the most likely outcome but also providing a measure of the uncertainty associated with these predictions. This dual capability is particularly valuable in decision-making processes for water resource management, disaster risk reduction, and agricultural planning [6–9].

However, uncertainties in climate model inputs, structures, and parameters can lead to cumulative uncertainties in precipitation forecasts. These uncertainties arise from various sources, including errors in initial conditions, model parameterizations, and the representation of physical processes within the models. As a result, the accuracy and reliability of raw ensemble forecasts can be compromised, necessitating the application of post-processing techniques to enhance their utility. The primary method to address this issue is statistical post-processing correction. This technique aims to correct the systematic errors of the raw ensemble forecasts, generate ensemble members that maintain the spatiotemporal dependence structure among variables, and retain the original forecast skill. Statistical post-processing can significantly improve the quality of ensemble forecasts by adjusting for biases, reducing dispersion errors, and providing calibrated probabilistic information [1–3].

The Bayesian Joint Probability (BJP) post-processing approach is a well-established statistical model that has been successfully applied to correct ensemble forecasts of precipitation, runoff, evaporation, and other climate variables [10–13]. The BJP model uses Bayesian methods to combine information from different sources and derive posterior distributions that account for forecasts uncertainty. However, the traditional BJP approach only considers the mean of the raw ensemble forecasts, neglecting the impact of variance on the correction. This limitation can lead to suboptimal adjustments, particularly in cases where the ensemble spread provides valuable information about forecast uncertainty. This paper proposes a Bayesian Modified Bridging (BMB) model based on an improved BJP approach to address this shortcoming. The BMB model incorporates the mean and variance of the raw ensemble forecasts into the correction process, enhancing the model's ability to account for forecast uncertainty and improve accuracy. This advanced approach ensures that the corrected forecasts better reflect the underlying uncertainties in the climate models. Using the SEAS5 ensemble precipitation forecasts from ECMWF as an example, we perform statistical post-processing correction on China's 0.5-degree grid precipitation data, encompassing 3,781 grid points. The SEAS5 system, developed by the European Centre for Medium-Range Weather Forecasts, is one of the most advanced seasonal forecasting systems available today. It has demonstrated high skill in predicting seasonal climate variations and is widely used in operational forecasting. We compare and analyze the correction effects of the BJP and BMB models to comprehensively evaluate their performance. By rigorously assessing the improvements the BMB model brings, we aim to provide a scientific basis for its adoption in future research and operational forecasting. The results of this study have the potential to enhance the accuracy and reliability of ensemble precipitation forecasts, contributing to better-informed decision-making in various sectors affected by weather and climate.

2 Methodology

2.1 Observation and Raw Forecasts

This paper utilizes the advanced seasonal forecasting system SEAS5, developed by the European Centre for Medium-Range Weather Forecasts (ECMWF), to generate monthly ensemble precipitation forecasts. SEAS5 is considered one of the most advanced seasonal forecasting systems currently available, with its forecasting capabilities widely validated and applied in various fields. The system's ensemble consists of 25 members, each with a resolution of $0.4^\circ \times 0.4^\circ$. For this study, we evaluate reforecast data from 1981 to 2015 with a spatial grid resolution of 0.5°. Observed precipitation data is provided by the China Meteorological Administration, covering the entire country with a resolution of $0.5^\circ \times 0.5^\circ$, resulting in a total of 3,781 grid points [14]. Bilinear interpolation is employed to harmonize and unify the spatial resolution differences between the various data sources. Through a detailed analysis, we aim to assess and validate the forecasting capabilities of the SEAS5 system across different timescales.

2.2 Bayesian Modified Bridging Model

The Bayesian Joint Probability (BJP) post-processing method was initially developed to produce ensemble forecasts for runoff, precipitation, and other hydrometeorological variables [10–13]. In this model, the process begins by utilizing the raw prediction as the predictor and the observed values as the predictand. A log-sinh transformation is then applied to both variables to normalize their distributions and equalize variance. The BJP model's key feature is the use of Gibbs sampling, which operates within the framework of Markov Chain Monte Carlo to facilitate Bayesian parameter estimation. For more comprehensive details, refer to [11–13].

However, the BJP model, when handling raw ensemble forecasts, only considers the ensemble mean as the input for the correction model, neglecting the significance of each ensemble member within the raw ensemble forecast during the correction process. This paper proposes a Bayesian Modified Bridging (BMB) model, an improved post-processing model based on the BJP approach to address this limitation. Initially, the BJP model is independently applied to correct each member of the raw ensemble forecast (with the ensemble size being N), resulting in corrected ensemble forecasts. Subsequently, these corrected forecasts are bridged to obtain the probability density function of the final ensemble forecast. Using Gibbs sampling, a final probabilistic forecast with 1000 ensemble members is generated. The BMB model is defined by Eq. (1):

$$p_{BMB}(y|\theta) = \sum_{k=1}^{K} \alpha_k \phi(y|\theta_k) \tag{1}$$

where, α_k is the weight, $\alpha_k \geq 0$, $\sum_{k=1}^{K} \alpha_k = 1$ representing the contribution degree of a certain ensemble member to the forecast. $\phi(y|\theta_k)$ is the probability density function of the k-th ensemble member, $\theta_k = (\mu_k, \sigma_k^2)$, and the k-th member is expressed as:

$$\phi(y|\theta_k) = \frac{1}{\sqrt{2\pi}\sigma_k} \exp\left(-\frac{(y-\mu_k)^2}{2\sigma_k^2}\right) \tag{2}$$

The model relies on the following parameters $\theta_k = \{\alpha_k, \mu_k, \sigma_k^2, k = 1, 2, \ldots, K\}$ and employs the Expectation Maximization (EM) algorithm to estimate these parameters.

Step 1: Establish the complete data likelihood function, given by:

$$\log P(y, \gamma|\theta) = \sum_{k=1}^{K} \left\{ n_k \log \alpha_k + \sum_{n=1}^{N} \gamma_{nk} \left[\log\left(\frac{1}{\sqrt{2\pi}}\right) - \log \sigma_k - \frac{1}{2\sigma_k^2}(y_n - \mu_k)^2 \right] \right\} \tag{3}$$

where, the latent variable is denoted by γ_{nk}. If the n-th observation belongs to the k-th ensemble member, then $\gamma_{nk} = 1$, otherwise $\gamma_{nk} = 0$.

Step2: Estimate the expected value of the complete data likelihood, denoted as $Q(\theta, \theta^{(i)})$:

$$Q\left(\theta, \theta^{(i)}\right) = \sum_{k=1}^{K} \left\{ n_k \log \alpha_k + \sum_{n=1}^{N} \hat{\gamma}_{nk} \left[\log\left(\frac{1}{\sqrt{2\pi}}\right) - \log \sigma_k - \frac{1}{2\sigma_k^2}(y_n - \mu_k)^2 \right] \right\} \tag{4}$$

Step3: Maximize the expected complete data log-likelihood with respect to θ:

$$\hat{\mu}_k = \frac{\sum_{n=1}^{N} \hat{\gamma}_{nk} y_n}{\sum_{n=1}^{N} \hat{\gamma}_{nk}}, \ \hat{\sigma}_k^2 = \frac{\sum_{n=1}^{N} \hat{\gamma}_{nk}(y_n - \mu_k)^2}{\sum_{n=1}^{N} \hat{\gamma}_{nk}}, \ \hat{\alpha}_k = \frac{n_k}{N} = \frac{\sum_{n=1}^{N} \hat{\gamma}_{nk}}{N} \tag{5}$$

Step 4: Iterate Steps 2 and 3 until convergence. The resulting parameters are the estimates that maximize the likelihood of the corrected ensemble forecasts.

2.3 Evaluating Indicator

For monthly ensemble precipitation forecasting, skill evaluation usually includes accuracy, reliability, sharpness, and skill. Let y_{fct}^t and y_{obs}^t denote the values of the forecast series and the observed series at time t, F_{fct} and F_{obs} represent the cumulative distribution functions (CDFs) of the forecast and observed series. The ensemble forecast's predicted mean series is $\overline{y}_{fct}^t$, and $Q_{fct}^t(k\%)$ is the $k\%$ quantile of the forecast distribution, with T being the length of the sequence.

2.3.1 Accuracy Evaluation

Accuracy describes the discrepancy between the cumulative probability distribution function (CDF) of the forecast series and that of the observed series. It is a metric used to measure the similarity between different CDFs. The Continuous Ranked Probability Score (CRPS) is adopted as the metric for precision assessment. The smaller the CRPS value, the better the forecast result. The CRPS is defined as follows [1, 2]:

$$\begin{aligned} CRPS &= \frac{1}{T}\sum_{t=1}^{T}\int\left[F_{fct}\left(y_{fct}^{t}\right)-F_{obs}\left(y_{obs}^{t}\right)\right]^{2}dy^{t} \\ &= \frac{1}{T}\sum_{t=1}^{T}\int\left[F_{fct}\left(y_{fct}^{t}\right)-H\left(y_{fct}^{t}-y_{obs}^{t}\right)\right]^{2}dy^{t} \end{aligned} \tag{6}$$

where, $H\left(y_{fct}^{t}-y_{obs}^{t}\right)$ is the step function, defined as follows:

$$H\left(y_{fct}^{t}-y_{obs}^{t}\right)=\begin{cases}0 & y_{fct}^{t}<y_{obs}^{t}\\ 1 & y_{fct}^{t}\geq y_{obs}^{t}\end{cases} \tag{7}$$

2.3.2 Reliability Evaluation

Reliability assesses the dependability of the probability distribution of the forecast series, indicating whether the distribution is appropriately broad or narrow. The Probability Integral Transform Area is employed as a metric for evaluating reliability, and it is defined as follows [1, 2]:

$$PIT\ Area=\frac{2}{T}\sum_{t=1}^{T}\left|F_{fct}\left(y_{obs}^{t}\right)-\frac{t}{T+1}\right| \tag{8}$$

2.3.3 Skill Evaluation

Skill describes the improvement in forecast performance relative to a reference forecast. Since CRPS and PIT are absolute scores, skill evaluation often combines these metrics to convert absolute scores into relative scores, providing a more intuitive comparison of results. The skill evaluation is defined as follows [1, 2]:

$$Skill(\%)=\frac{Score_{ref}-Score_{fct}}{Score_{ref}}\times 100 \tag{9}$$

3 Results

Accuracy is the most critical metric for assessing the overall precision of ensemble forecasts. Figure 1 shows the spatiotemporal distribution characteristics of the continuous ranked probability skill score (CRPSS) of BMB-SEAS5 forecasts for LD = 0–5. From top to bottom, the subplots represent months 1–12, and from left to right, they represent six lead times. Each subplot displays the grid point and monthly CRPSS values, with the reference forecast being the BJP-SEAS5 forecast results. White grid points indicate that the model and the reference forecasts have similar accuracy. Blue grid points indicate that the model forecast has a smaller CRPS score than the reference forecast, demonstrating skill in accuracy. The darker the blue, the higher the forecast skill. Red grid points indicate that the model forecast has a larger CRPS score than the reference forecast, indicating no skill in accuracy. The darker the red, the lower the forecast skill. Overall, the BMB model's forecast skill significantly improved over the BJP model in almost all lead times and months. Although a small number of regions display a decline in forecast skill, these declines do not show consistent patterns. They are primarily concentrated in a few areas such as Xinjiang, Tibet, central North China, and some parts of Central South China.

Figure 2 further statistically analyzes the different CRPSS results: (a) shows the proportion of grid points with low (−25 to −15), relatively low (−15 to −5), equal (−5 to 5), relatively high (5 to 15), high (15 to 25), and very high (25 to 35) forecast skill at each lead time. The proportion of grid points with significantly improved accuracy skill is overwhelmingly high, with the proportion of grid points with forecast skill being relatively high and above increasing from 64.6% (LT = 0) to 79.5% (LT = 1), 80.5% (LT = 2–3), and 81.5% (LT = 4–5). (b) shows box plots of the average CRPSS per grid point for each lead time and month, indicating considerable differences in forecast skill across months with a relatively dispersed distribution. (c) shows the mean (blue) and median (red) CRPSS values per grid point for each lead time, averaged over all months. As the lead time extends from 0 months to 1 month, the forecast skill significantly improves and maintains this improvement as the lead time further extends.

Figure 3 shows the spatiotemporal distribution characteristics of the Probability Integral Transform Skill Score (PITSS) of BMB-SEAS5 forecasts for LT = 0–5. From top to bottom, the subplots represent months 1–12, and from left to right, they represent six lead times. Each subplot displays the grid point and monthly PITSS values, with the reference forecast being BJP-SEAS5. White grid points indicate that the model and reference forecasts have similar reliability. Blue grid points indicate that the model forecast has a smaller PIT Area score than the reference forecast, demonstrating skill in reliability. The darker the blue, the higher the forecast skill. Red grid points indicate that the model forecast has a larger PIT Area score than the reference forecast, indicating no skill in reliability. The darker the red, the lower the forecast skill. When the lead time is 0 months, BMB-SEAS5 shows a decline in reliability in parts of Northwest, Southwest, Central South, and East China, but the distribution is complex and does not show clear patterns. When the lead time extends to 1 month, reliability significantly improves, with red appearing only in very few parts of the Northwest. As the lead time extends to 5 months, the increase in PITSS is steadily maintained.

Figure 4 further statistically analyzes the different PITSS results: (a) shows the proportion of grid points with relatively low (−15 to − 5), equal (−5 to 5), relatively high (5 to 15), and high (15 to 25) forecast skill at each lead time. The proportion of grid points with significantly improved reliability skills shows an increasing trend, with the proportion of grid points with forecast skill being relatively high and above increasing from 38.7% (LT = 0) to 83.6% (LT = 1), 85.7% (LT = 2), and 87.2% (LT = 3–5). (b) shows box plots of the average PITSS per grid point for each lead time

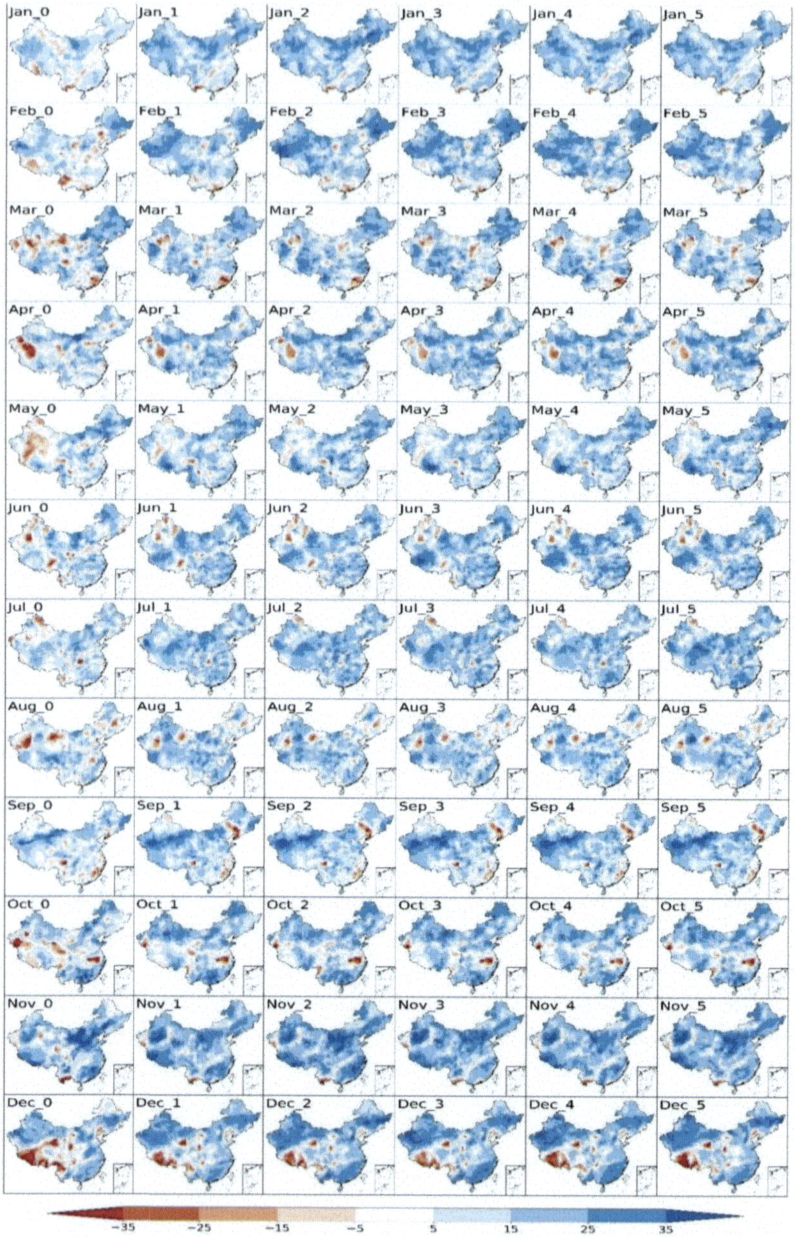

Fig. 1. Spatiotemporal distribution of the CRPSS of BMB-SEAS5 forecasts for LT = 0–5, with the reference forecast being BJP-SEAS5.

and month. Overall, when the lead time is 0 months, the forecast skill distribution is relatively dispersed across months, while for lead times of 1–5 months, the distribution is more concentrated. (c) shows the mean (blue) and median (red) PITSS values per grid

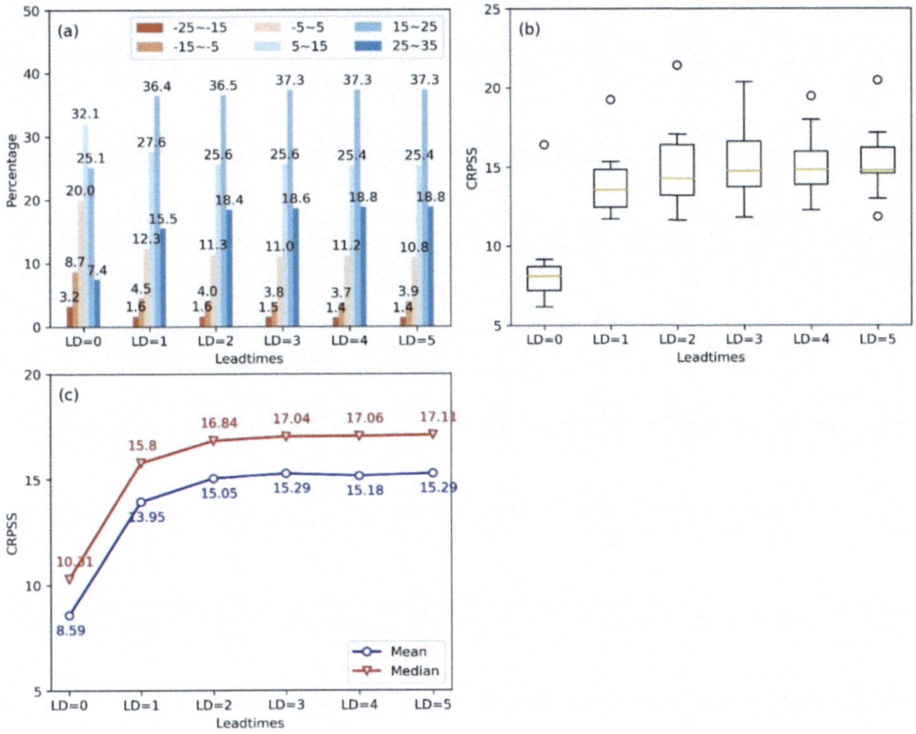

Fig. 2. Statistical analysis of the CRPSS of BMB-SEAS5 forecasts for LT = 0–5, with the reference forecast being BJP-SEAS5: (a) Percentage of grid points in each CRPSS interval; (b) Monthly distribution; (c) Trends of mean and median values.

point for each lead time, averaged over all months. Although the average PITSS is −1.09 for a lead time of 0 months, the median value is 2.28, indicating that the overall reliability of BMB-SEAS5 is higher and continues to improve as the lead time extends.

4 Conclusions

This paper proposes a Bayesian Modified Bridging (BMB) model, an improved statistical post-processing model based on the BJP approach, for correcting and bridging precipitation forecasts. The performance of the BMB model improvements is evaluated using accuracy skill scores (CRPSS) and reliability skill scores (PITSS). Initially, the SEAS5 precipitation forecast product, demonstrating high current forecast skill, is statistically post-processed using the BJP and BMB models, resulting in two sets of corrected ensemble forecasts: BJP-SEAS5 and BMB-SEAS5. A comprehensive evaluation is conducted using the CRPSS and PITSS metrics. The results indicate that for lead times of 0–5 months, the BMB-SEAS5 forecasts are consistently unbiased and reliable across all forecast times. The accuracy and reliability skill scores of BMB-SEAS5 are significantly higher than those of BJP-SEAS5. Specifically, the average CRPSS and PITSS values

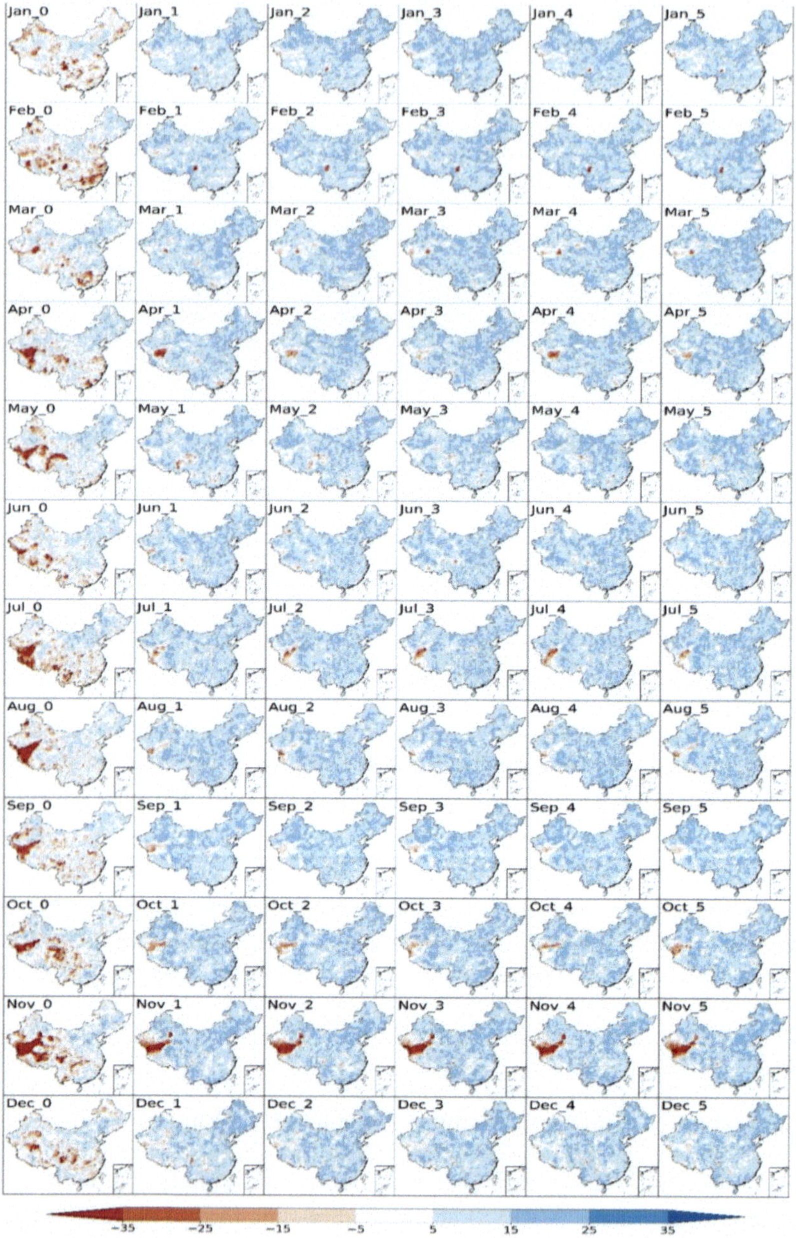

Fig. 3. Spatiotemporal distribution of the PITSS of BMB-SEAS5 forecasts for LT = 0–5, with the reference forecast being BJP-SEAS5.

for lead times of 0–5 months improved by 13.9% and 8.8%, respectively, compared to BJP-SEAS5. These findings demonstrate that the BMB model has superior correction capabilities and robustness compared to the BJP model.

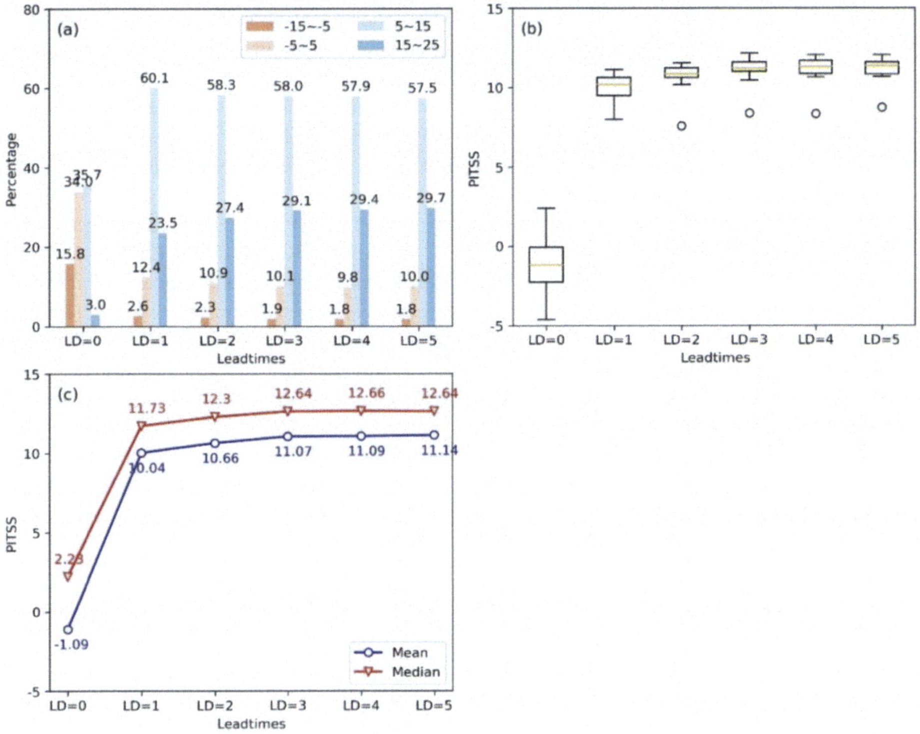

Fig. 4. Statistical analysis of the PITSS of BMB-SEAS5 forecasts for LT = 0–5, with the reference forecast being BJP-SEAS5: (a) Percentage of grid points in each PITSS interval; (b) Monthly distribution; (c) Trends of mean and median values.

References

1. Duan, Q., Pappenberger, F., Wood, A., Cloke, H.L., Schaake, J.C.: Handbook of hydrometeorological ensemble forecasting, 1st edn. Springer, Berlin/Heidelberg, Germany (2019)
2. Li, Y., Liang, Z., Hu, Y., Li, B., Xu, B., Wang, D.: A multi-model integration method for monthly streamflow prediction: modified stacking ensemble strategy. J. Hydroinf. **22**(2), 310–326 (2020)
3. Wang, Q.J., et al.: An evaluation of ECMWF SEAS5 seasonal climate forecasts for Australia using a new forecast calibration algorithm. Environ. Model. Softw. **122**, 104550 (2019)
4. Zhao, T., Wang, Q.J., Schepen, A.: A Bayesian modelling approach to forecasting short-term reference crop evapotranspiration from GCM outputs. Agric. For. Meteorol. **269**, 88–101 (2019)
5. Manzanas, R., et al.: Bias adjustment and ensemble recalibration methods for seasonal forecasting: A comprehensive intercomparison using the C3S dataset. Clim. Dyn. **53**, 1287–1305 (2019)
6. Peng, Z., et al.: Statistical calibration and bridging of ECMWF System4 outputs for forecasting seasonal precipitation over China. Journal of Geophysical Research: Atmospheres **119**(12), 7116–7135 (2014)

7. Bennett, J.C., Wang, Q.J., Li, M., Robertson, D.E., Schepen, A.: Reliable long-range ensemble streamflow forecasts: Combining calibrated climate forecasts with a conceptual runoff model and a staged error model. Water Resour. Res. **52**(10), 8238–8259 (2016)
8. Li, Y., et al.: A medium and long-term runoff forecast method based on massive meteorological data and machine learning algorithms. Water **13**(9), 1308 (2021)
9. Xu, Y., Gao, X., Shen, Y., Xu, C., Shi, Y., Giorgi, A.: A daily temperature dataset over China and its application in validating an RCM simulation. Adv. Atmos. Sci. **26**, 763–772 (2009)
10. Li, Y., et al.: Deterministic and probabilistic evaluation of raw and post-processing monthly precipitation forecasts: a case study of China. J. Hydroinf. **23**(4), 914–934 (2021)
11. Wang, Q.J., Robertson, D.E., Chiew, F.H.S.: A Bayesian joint probability modeling approach for seasonal forecasting of streamflows at multiple sites. Water Resour. Res. **45**(5), (2009)
12. Wang, Q.J., Robertson, D.E.: Multisite probabilistic forecasting of seasonal flows for streams with zero value occurrences. Water Resour. Res. **47**(2), (2011)
13. Wang, Q.J., Shrestha, D.L., Robertson, D.E., Pokhrel, P.: A log-sinh transformation for data normalization and variance stabilization. Water Resour. Res. **48**(5), (2012)
14. Wu, J., Gao, X., Giorgi, F., Chen, D.: Changes of effective temperature and cold/hot days in late decades over China based on a high-resolution gridded observation dataset. Int. J. Climatol. **37**, 788–800 (2017)

Characteristics of Vegetation Water Carbon Use Efficiency in the Middle Reaches of the Yellow River Basin by Remote Sensing Data

Yinge Liu[1,2](✉), Cheng Xu[2], Yujun Hou[2], Chuan Yao[2], Qianhui Liu[2], Maoqi Lin[2], and Dandan Li[2]

[1] Key Laboratory of Disaster Monitoring and Mechanism Simulating in Shaanxi Province, Baoji University of Arts and Sciences, Baoji 721013, China
Yingeliu@163.com

[2] College of Geography and Environment, Baoji University of Arts and Sciences, Baoji 721013, China

Abstract. Water use efficiency (WUE) of vegetation is an important indicator for assessing the adaptability of plant ecosystems. Taking the middle reaches of the Yellow River Basin as an example, trend analysis and univariate linear regression methods were employed to analyze the spatiotemporal characteristics of evapotranspiration (ET) and water use efficiency (WUE) based on remote sensing images. The results indicate that the average annual gross primary productivity (GPP) and ET show an increasing trend, whereas vegetation water use efficiency exhibits a decreasing trend. Among different vegetation types, the values of GPP, ET, and WUE are the highest for broad-leaved coniferous forests, while grassland vegetation has the lowest values for these parameters. The order of WUE size for different vegetation types is coniferous forest > broad-leaved forest > Grass bushes > Shrubs > Cultivate vegetation > swamp > meadow > Grassland. Only swamp vegetation shows an increasing trend of 0.01/10a in WUE, while other vegetation types show a decreasing trend in WUE. Among them, grass bushes and cultivated vegetation decreased the most, with trend rates of 0.13/10a and 0.1 /10a, respectively. These studies provide a basis for regional ecological environment restoration and water resource management.

Keywords: Vegetation Water Use Efficiency · Vegetation Type · Changing Characteristics

1 Introduction

The water use efficiency of vegetation is an important variable that describes the hydrological cycle in vegetation ecosystems. At the same time, the study of vegetation water use efficiency provides theoretical support for adapting to a special ecological environment. With the acceleration of global warming, some environmental problems have become increasingly prominent, and the water cycle process of vegetation has undergone significant changes. Therefore, WUE research on natural ecosystems such as grasslands,

C. Weng (Ed.): WRE 2024, LNCE 634, pp. 101–109, 2025.
https://doi.org/10.1007/978-981-96-5095-8_9

forests, and wetlands at the global and regional scales has become a hot topic [1, 2]. At the individual plant scale, WUE refers to the ratio of dry matter mass formed during long-term plant growth to water consumption. Dry matter mass can be characterized by indicators such as Gross Primary Production or Net Primary Production (NPP) of the region [3, 4]. However, due to differences in research scale, field, and data acquisition methods, regional WUE often has certain differences. In Europe, it was found that along the latitude gradient, WUE is higher in Northern Europe and lower in the Mediterranean region. While in the United States, most ecosystems have lower WUE during summer [5, 6]. Moreover, in arid areas, the seasonal and interannual variation trends of ecosystem WUE are consistent with the variation trends of GPP [7, 8]. Therefore, the spatiotemporal variation characteristics and mechanisms of WUE at the regional ecosystem scale are currently the focus of in-depth research. With the development of technology, satellite remote sensing data has the advantages of long time series and wide spatial range. The middle reaches of the Yellow River Basin have a significant impact on large-scale ecosystems due to their vast water resources and carbon cycling effects. Although numerous studies have been carried out on the carbon utilization rate of vegetation, they all have some limitations. Therefore, this article takes the middle Yellow River as an example to analyze the changes in vegetation water use efficiency by using MODIS series products and meteorological data and explore the spatial distribution differences of WUE of different vegetation types and their responses to precipitation and temperature, which helps to accurately understand the ecosystem adaptability and vulnerability to climate change. It also provides a scientific basis for water resources management, and offering scientific guidance for the formulation of ecological protection and restoration measures in response to climate change is of paramount significance for regional ecological environment planning.

2 Materials and Methods

2.1 Materials

The middle reaches of Yellow River Basin are located at 32°47′30"–40°58′40"N, 103°0′20"–113°42′0"E, with a total length of about 5464km and an area of 36.29 × 104 km^2. The terrain of the middle reaches is complex, encompassing loess hills, gullies, and plains, among others. Among them, the section of the Jin-Shaan Gorge is the longest continuous canyon of the mainstream of the Yellow River. The terrain is high in the west and low in the east, mainly composed of high mountains in the west, loess landforms in the middle, and alluvial plains in the east. It encompasses multiple climate types, including arid, semi-arid, and semi-humid regions. The annual average temperature in the middle reaches of the Yellow River Basin is -8.39 °C ~ 16.02 °C, the annual precipitation is between 200 mm ~ 700mm, and the annual sunshine hours generally reach 2000 ~ 3300 h. Various vegetation types include coniferous forests, broad-leaved forests, grasslands, shrubs, cultivated vegetation, swamps, meadows, and steppes (Fig. 1).

The GPP and ET remote sensing data required for this study are sourced from MOD17A2 and MOD16A2 from 2000 to 2022, with a spatial resolution of 1 km x 1 km and a temporal resolution of 8 days. The data was downloaded using the Google Earth Engine (GEE) platform (https://earthengine.google.com/), and outliers were removed

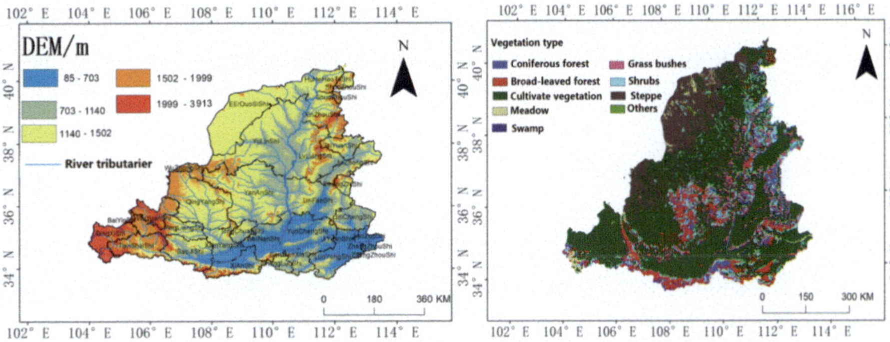

Fig. 1. DEM and vegetation type map of the Yellow River Basin.

from the original data. DEM data comes from SRTM (shuttle radar topography mission) data with a resolution of 90m, and precipitation and temperature data come from the Data Center of Resources and Environmental Sciences, Chinese Academy of Sciences (http://www.resdc.cn). The vegetation type data come from the 1:1 vegetation type atlas of China. Based on the IGBP (International Geosphere-Biosphere Programme) classification system, taking into account the actual situation, vegetation has been reclassified into nine distinct types: forest (including both coniferous and broadleaved varieties), grasslands, shrubs, cultivated vegetation, swamps, meadows, steppes, and others.

2.2 Methods

The WUE is calculated by the ratio of the gross primary productivity of vegetation photosynthesis to the water consumed by vegetation evapotranspiration. The formula is as follows:

$$WUE = \frac{\mathrm{GPP}}{\mathrm{ET}} \tag{1}$$

where *WUE* GPP and ET means water use efficiency, gross primary productivity and evapotranspiration, respectively.

We use univariate linear regression to calculate the trend rate, the Pearson method to calculate the correlation coefficient, and use two confidence levels of $p < 0.05$ and $p < 0.01$ to analyze the relationship between vegetation WUE and precipitation and temperature. The F-test method is used to analyze the significance of the complex correlation coefficient. The formula was as follows:

$$\gamma_{sl} = \frac{(n \times \sum_{i=1}^{n} i \times z_i - \sum_{i=1}^{n} i \times \sum_{i=1}^{n} z_{i})}{n \times \sum_{i=1}^{n} i^2 - \left(\sum_{i=1}^{n} i\right)} \tag{2}$$

$$r_{xy} = \frac{\sum (x - x)(y - y)}{\sqrt{\sum (x - x)^2 \sum (y - y)^2}} \tag{3}$$

where, γ_{sl} is the trend rate of each element, n is the number of samples, z is the value of the element in year, i^{th} and γ_{sl} indicate that element showed an increasing trend, vice versa. Where r_{xy} means the person-specific coefficient, x is the variable of WUE, y presents a meteorological element, and $\overline{x}$ and $\overline{y}$ present the average values of y in two different conditions or time periods, respectively. r_{xy}

3 Results and Discussion

3.1 Spatiotemporal Characteristics of Vegetation GPP, ET, and WUE

The calculation showed that the annual average vegetation GPP, ET, and WUE in the middle reaches of the Yellow River Basin were 602.83 gC $m^{-2}a^{-1}$, 319.11 mm, and 1.89 gC · $mm^{-1}m^{-2}$, respectively. As shown in Fig. 2, the changes in Gross Primary Productivity (GPP) and Evapotranspiration (ET) exhibit an increasing trend, with trend rates of 12.79 gC m^{-2} a^{-1}/a for GPP and8.07mm/a for ET, respectively. The WUE of vegetation showed a decreasing trend with a trend rate of −0.0079/a.

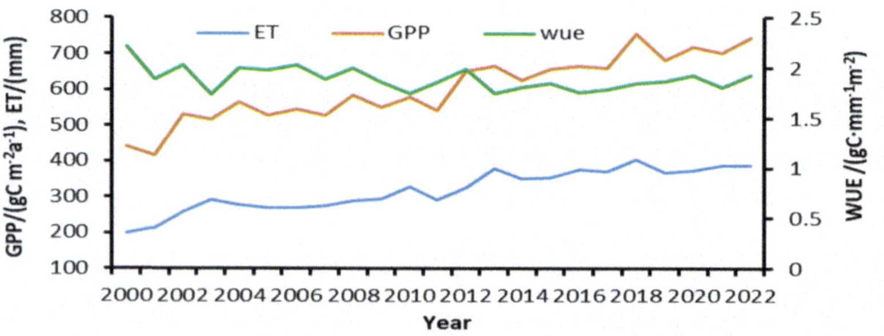

Fig. 2. Annual changes of GPP, ET, and WUE in the Yellow River Basin from 2000 to 2022.

The vegetation GPP, ET, and WUE exhibit spatial variability. From Figs. 3, 4, and 5, GPP and ET increased from northwest to southeast. Among them, GPP less than 300 gC $m^{-2}a^{-1}$ and ET less than 200 mm are mostly distributed in the northwest, these areas have harsh natural environments, dry climates, low vegetation coverage, and mainly grassland and desert vegetation types and meadows. The high GPP and ET areas (400–600 gC $m^{-2}a^{-1}$ and 200–400 mm), are mainly distributed in the southern region. The GPP of the entire watershed is generally on an increasing trend, with an increase area of 99.54% and 98.24% and a decrease area of only 0.46% and 1.76%. GPP and ET increased significantly after 2010. WUE in most areas is greater than 1.2, and the area with WUE higher than 1.6 accounts for 72.8% of the entire watershed. Overall, WUE showed an upward trend from 2010 to 2022 because returning farmland to forests and grasslands, as well as grassland restoration, are conducive to restoring the stability of the ecosystem and vegetation diversity.

3.2 WUE Variation of Different Vegetation Types

Among the vegetation types in the middle reaches of the Yellow River Basin, the largest proportion of cultivated vegetation is 55.53%, followed by grasslands at 17.38%, and

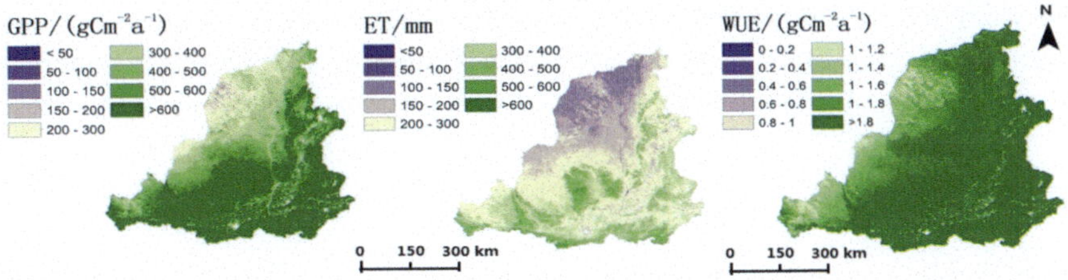

Fig. 3. The spatial distribution of annual average ET, GPP, and WUE in the middle reaches of the Yellow River.

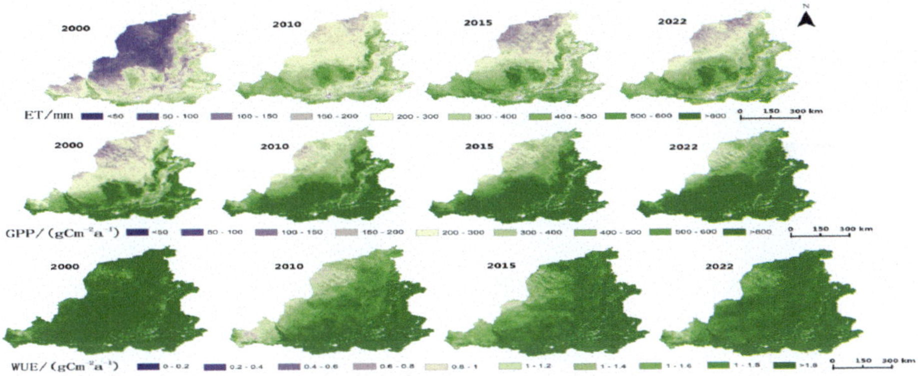

Fig. 4. Changes in annual average ET, GPP, and WUE from 2000 to 2022.

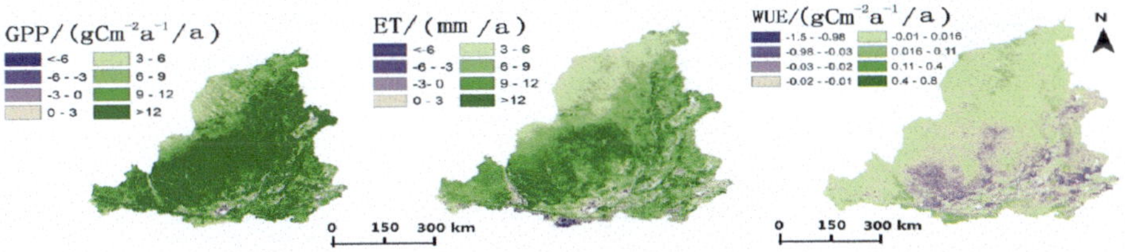

Fig. 5. The spatial distribution of annual average trend rates for ET, GPP and WUE.

swamps at the smallest at 0.02%. Coniferous forests, broad-leaved forests, shrubs, grasslands, and meadows account for 4.33%, 8.59%, 7.19%, 5.25%, and 1.54%, respectively. According to Figs. 6 and 7, As can be seen, that the average GPP and ET sizes of different vegetation types in the past 20 years are as follows: Broad-leaved forest (939.85 gC m^{-2}a-1) > Coniferous forest (864.89 gC $m^{-2}a-1$) > Shrubs (762.01 gC $m^{-2}a^{-1}$) > Grass bushes (698.82 gC $m^{-2}a^{-1}$) > Swamp (623.55 gC $m^{-2}a^{-1}$) > Cultivate vegetation (586.78 gC $m^{-2}a^{-1}$) > Meadow(542.01 gC $m^{-2}a^{-1}$) > Other (418.42 gC $m^{-2}a^{-1}$) > Steppe (337.04 gC $m^{-2}a^{-1}$) and Broad-leaved forest (460.11 gC $m^{-2}a^{-1}$) > Coniferous forest (420.11 gC $m^{-2}a^{-1}$), Shrubs (392.09 mm) > Grass bushes (351.76 mm) > Swamp (342.45 mm) > Meadow (325.19mm) > Cultivate vegetation (309.54 mm) > Other (231.73 mm) > Steppe (215.99 mm). All vegetation GPP and ET showed an increasing trend. Among them, the trend rates of GPP and ET in Grass bushes are the highest at 16.36 gC $m^{-2}a^{-1}$/a and 10.01 mm, followed by cultivated vegetation and Shrubs at 13.77 gC $m^{-2}a^{-1}$/a and 8.7 mm/a while the trend rates GPP and ET in meadow and swamp are the lowest at 10.06 gC $m^{-2}a^{-1}$/a and 5.49 mm, respectively.

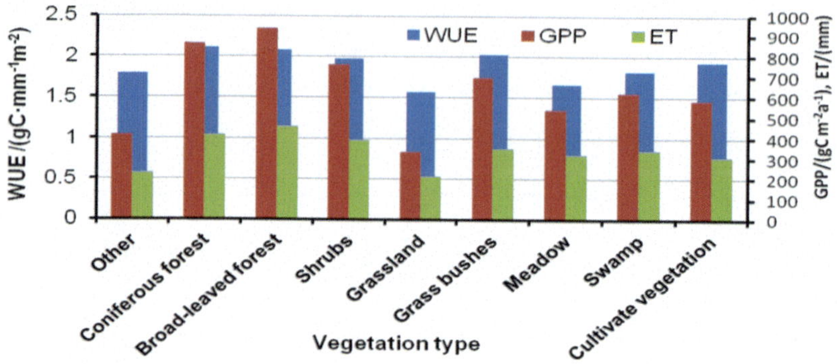

Fig. 6. Annual average WUE, ET, and GPP of different vegetation types.

The average WUE size order in different vegetation types over the past 20 years is as follows: coniferous forest (2.11 gC·$mm^{-1}m^{-2}$) > broad-leaved forest (2.08 gC·$mm^{-1}m^{-2}$) > Grass bushes (2.02 gC·$mm^{-1}m^{-2}$) > Shrubs (1.97 gC·$mm^{-1}m^{-2}$) > Cultivate vegetation (1.93 gC·$mm^{-1}m^{-2}$) > swamp (1.81 gC·$mm^{-1}m^{-2}$) > others (1.79 gC·$mm^{-1}m^{-2}$) > meadow (1.67 gC·$mm^{-1}m^{-2}$) and steppe (1.57 gC·$mm^{-1}m^{-2}$). All vegetation WUE shows a decreasing trend, with only swamps showing an increasing trend. The annual trend rates of WUE of coniferous forest, broad-leaved, forest, Grass bushes, Shrubs, cultivate vegetation, swamp, others, meadow, and steppe is −0.05 gC·$mm^{-1}m^{-2}$/10a, −0.07 gC·$mm^{-1}m^{-2}$/10a, −0.13 gC·$mm^{-1}m^{-2}$/10a, −0.09 gC·$mm^{-1}m^{-2}$/10a, −0.104 gC·$mm^{-1}m^{-2}$/10a, 0.01 gC·$mm^{-1}m^{-2}$10a, −0.065 gC·$mm^{-1}m^{-2}$/10a, −0.01 gC·$mm^{-1}m^{-2}$/10a, −0.019 gC·$mm^{-1}m^{-2}$/10a, respectively. The trend rate of grass clump is the highest, followed by cultivated vegetation, and meadow is the smallest.

3.3 Discussion

GPP and ET determine the ecosystem WUE. The spatial differentiation of vegetation GPP, ET, and WUE is related to the regional differences in vegetation distribution caused by heat and water in the middle reaches of the Yellow River Basin. The increase in temperature helps increase the photosynthetic rate of vegetation and CO_2. The production of fixed and other dry matter, thereby enhancing vegetation GPP. When the temperature surpasses the optimal range for vegetation, the plants consume increased amounts of water while simultaneously reducing their carbon fixation through photosynthesis. This, in turn, leads to a decrease in the overall water use efficiency (WUE) of the vegetation. High temperatures may lead to leaf dehydration and stomatal closure, limiting photosynthetic capacity and carbon sequestration [9]. The study area has the highest

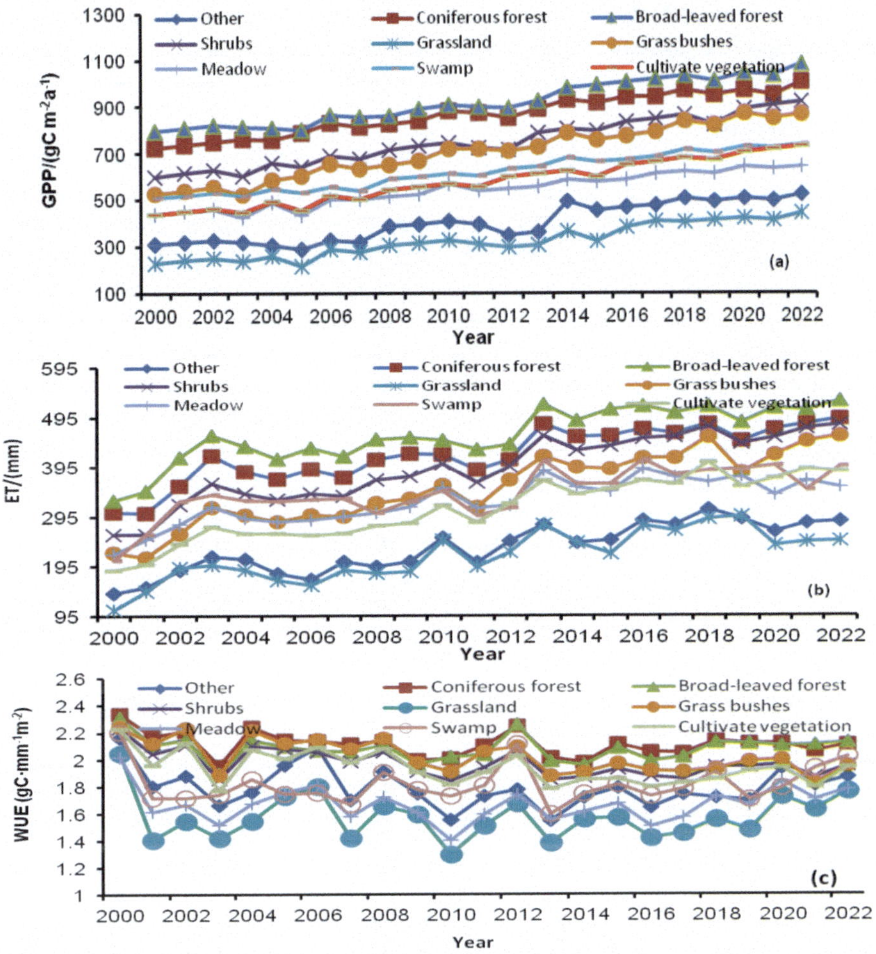

Fig. 7. Annual variation of average WUE for different vegetation types.

WUE in the coniferous forest, consistent with existing studies [10]. Higher precipitation can increase soil moisture, leading to poor soil drainage, limiting the photosynthetic capacity of vegetation, and correspondingly reducing WUE [11]. Meanwhile, the growth and development of meadows and steppe are limited in extreme environments of poor precipitation, resulting in lower WUE. The increase in precipitation is conducive to increasing vegetation GPP, thereby improving vegetation WUE, consistent with existing research results [12]. Changes in the underlying surface due to Land Use and Land Cover Change (LUCC) can significantly impact several key factors, including albedo, roughness, and the vegetation coverage ratio. These corresponding alterations can, in turn, lead to substantial changes in vegetation productivity and evapotranspiration processes. The difference in terrain characteristics and vegetation types may be the main reason for regional differences in GPP, ET, and WUE. This indicates that water heat matching improves vegetation carbon and water utilization efficiency.

4 Conclusion

From the above analysis, the following conclusion can be drawn: (1) Vegetation GPP and ET in the middle reaches of Yellow River Basin show an upward trend, while WUE shows a decreasing trend. However, the central ET and GPP increased significantly, while the WUE decreased significantly in central areas. The GPP and ET values of broad-leaved forests are the highest, and the WUE value of coniferous forests is the highest among different vegetation types, while the GPP, ET, and WUE values of grassland vegetation are the lowest. (2) All vegetation types showed an increasing trend in GPP and ET, while the highest GPP and ET change rates were observed in grass bushes, shrubs, and cultivated vegetation. Except for swamps, the WUE of all other vegetation shows a decreasing trend, and the WUE reduction rate of grass bushes and cultivated vegetation is relatively high. (3) The order of WUE change rates from highest to lowest is: grass bushes, cultivated vegetation, shrubs, others, coniferous forests, grasslands, broadleaved forests, and meadows. These studies provide a reference basis for regional vegetation restoration and water resource management. The efficiency of water use can be improved by selecting drought-tolerant vegetation types.

Acknowledgments. This study was supported by the National Natural Science Foundation Project of China (No. 41771048) and the Key R&D Program of Shaanxi Province (NO. 2022SF-364).

References

1. Ito, A., Inatomi, M.: Water-use efficiency of the terrestrial biosphere: A model analysis focusing on interactions between the global carbon and water cycles. J. Hydrometeorol. **13**(2), 681–694 (2012)
2. Ogutu, B.O., D'Adamo, F., Dash, J.: Impact of vegetation greening on carbon and Global and Planetary water cycle in the African Sahel-Sudano-Guinean Region. Global Planet. Change **202**, 10352 (2021)
3. Zhou, S.X., Dursma, R.A., Medlyn, B.E.: How should we model plant responses to drought? An analysis of stomatal and non-stomatal responses to water stress. Agric. For. Meteorol. **182**, 204–214 (2014)
4. Qin, X.J., Hong, J.T., Ma, X.X., Wang, X.D.: Global patterns in above-ground net primary production and precipitation-use efficiency in grasslands. J. Mt. Sci. **15**(8), 1682–1692 (2018)
5. Rigden, A.J., Salvucci, G.D.: Stomatal response to humidity and CO_2 implicated in the recent decline in US evaporation. Glob. Change Biol. **23**, 1140–1151 (2017)
6. Li, M.Q., Zhou, L., Wang, X.Y., et al.: Stem and leaf photosynthesis of seven desert woody species and its influencing factors. Chin. J. Appl. Ecol. **34**(10), 2637–2643 (2023)
7. Zhao, F.B., et al.: Predicting the climate change impacts on water-carbon coupling cycles for a loess hilly-gully watershed. J. Hydrol. **581**, 124388 (2020)
8. El Masri, B., et al.: Linking remotely sensed carbon and water use efficiencies within situ soil properties. Remote Sensing **13**(13), 2593 (2021)

9. Zhang, T., Peng, J., Liang, W., et al.: Spatial-temporal patterns of water use efficiency and climate controls in China's Loess Plateau during 2000–2010. Sci. Total Environ. **565**(15), 105–122 (2016)
10. Lin, Z.Q., et al.: Spatiotemporal variation and attribution of carbon and water use efficiency in the Yellow River Basin based on remote sensing data. Acta Ecological Sinica **44**(1), 377–391 (2024)
11. Liu, X.F., Hu, B.Y., Ren, Z.Y.: Spatiotemporal variation of water use efficiency and its driving forces on the loess plateau during 2000–2014. Scientia Agricultura Sinica **51**(2), 302–314 (2018)
12. Du, L.T., et al.: Effects of anthropogenic revegetation on the water and carbon cycles of a desert steppe ecosystem. Agric. For. Meteorol. **300**, 108339 (2021)

Variation Characteristics and Control Factors of Groundwater Level in Typical Over-Exploitation Areas of North China: A Case Study of Xiong'an New Area

Jingsi Zhu[1], Zhe Wang[2,3], and Tianyu Hou[4](✉)

[1] Hydrology Bureau of Haihe River Water Conservancy Commission, MWR, Tianjin, China
[2] State Key Laboratory of Hydraulic Engineering Intelligent Construction and Operation, Tianjin University, Tianjin 300072, China
[3] Haihe River Water Conservancy Commission, The Ministry of Water Resources, Tianjin 300170, China
[4] Tianjin Meteorological Information Center, Tianjin 300170, China
617874764@qq.com

Abstract. Xiong'an New Area is a state-level new administrative region established in 2017, and groundwater resources are an important source of water supply. Based on the comprehensive treatment of groundwater over-exploitation in North China, the groundwater resource status, groundwater dynamic change characteristics, and multi-factor influencing mechanism in Xiong'an New Area were studied under the influence of multiple water source allocation, ecological water replenishment in rivers and lakes, and restricted groundwater exploitation. Research provided a scientific basis for water resource allocation and ecological recovery in the new district. Results showed that: (1) From 2000 to 2021, the groundwater level in Xiong'an New Area showed a trend of first decreasing and then rising. Since water was transferred through the South-North Water Diversion Project in 2014, the shallow groundwater level has risen by 0.11m per year on average; (2) According to the random forest machine learning analysis, the groundwater level in Xiong' an New Area was mainly affected by the amount of groundwater extracted (contribution rate 55%), precipitation (contribution rate 42%), and ecological water replenishment; (3) Numerical simulation method was used to analyze the change of groundwater level without considering the influence of precipitation. The rising groundwater level from 2019 to 2021 was mainly affected by precipitation, and the current average groundwater extraction was slightly higher than the number of recoverable resources. Transregional water resources such as the South-North Water Diversion Project must be fully considered to ensure the safety of urban water supply and sustainable use of groundwater resources.

Keywords: Xiong'an New Area · Groundwater dynamic · treatment of groundwater over-exploitation

C. Weng (Ed.): WRE 2024, LNCE 634, pp. 110–122, 2025.
https://doi.org/10.1007/978-981-96-5095-8_10

1 Introduction

Xiong'an New Area is a state-level new administrative region located in the groundwater over-exploitation area of the North China Plain. Local surface water resources are scarce, and groundwater resources are important water supply sources in Xiong'an New Area. For the security of the water supply, it is very important to study the status of groundwater resources and the dynamic change of groundwater level in Xiong'an New Area.

Many scholars have studied the groundwater evolution and the contribution rate of control factors at different regional scales. In the field of evolution research, over-exploitation, and recharge evaluation of groundwater resources in North China, Chen Chen et al. systematically analyzed the dynamic law and control factors of groundwater resources in China and various water resources divisions in the past 60 years, as well as the variation characteristics of groundwater resources supply and discharge structure [1]. Shi et al. studied the evolution of groundwater and its depression cone in the North China Plain since the 1950s [2]. Liu et al. innovatively proposed the ecological critical water level and land subsidence volume method to evaluate the over-exploitation status in the Haihe Plain [3]. Cao et al. proposed a suitable evaluation method for groundwater recharge in the Daqing River Basin based on numerical simulation and GIS [4]. In analyzing the contribution of influencing factors, traditional statistical analysis methods such as the analytic hierarchy process and regression analysis are mostly used. As a machine learning method, the random forest model is increasingly used in the hydrology field. Wang et al. simulated the runoff of Yongding River Basin based on the RF model and quantitatively analyzed the contribution rate of different factors to runoff change [5]. Existing research results need to be deeper in the quantitative study of the dynamic evolution law and control factors of the long series of groundwater in Xiong'an New Area. Particularly, there were few influencial analysis of over-exploitation control measures such as restricted mining and ecological water replenishment on groundwater, which could not meet the precise management of water resources.

Based on the statistical analysis results of national and provincial groundwater monitoring datum, water resources investigation and evaluation, water supply and consumption bulletin of water resources, this paper analyzed the groundwater resource status, groundwater dynamic change characteristics, and multi-factor influencing mechanism in Xiong'an New Area from 2000 to 2021. The contribution rate of each influencing factor was analyzed by the random forest (RF) machine learning method, and the GMS (Groundwater Modeling System) groundwater numerical model in the study area was constructed to analyze the control factors of water level change since the over-exploitation treatment in North China. The research results of this paper can provide a basis for groundwater over-exploitation control, sustainable utilization, and scientific management of groundwater resources in Xiong'an New Area.

2 Overview of the Study Area

2.1 Regional Basic Situation

Xiong'an New Area is located in Baoding, Hebei Province. The planned area is about 1770 km^2, located in the middle reaches of the Daqing River Basin. Northwest of the area is higher than the southeast, and the terrain changes gently. The distribution of annual precipitation is uneven, mainly from June to September. It is located in the Daqing River system in the Haihe River Basin. Baiyangdian wetland is located in the southeast region. Total area is 366 km^2. Tang River and eight other rivers enter the wetland from the southwest and the north (Fig. 1).

2.2 Regional Hydro-Geological Condition

The groundwater in Xiong'an New Area is mainly quaternary loose rock pore water. The groundwater system of Xiong'an New Area belongs to the Daqinghe Plain groundwater system of the Haihe River basin. It is generally divided into four main aquifer groups from I to IV vertically. Shallow groundwater is submersible and weak confined groundwater. The deep groundwater is confined groundwater. The discharge capacity of deep groundwater is better than that of shallow groundwater [6].

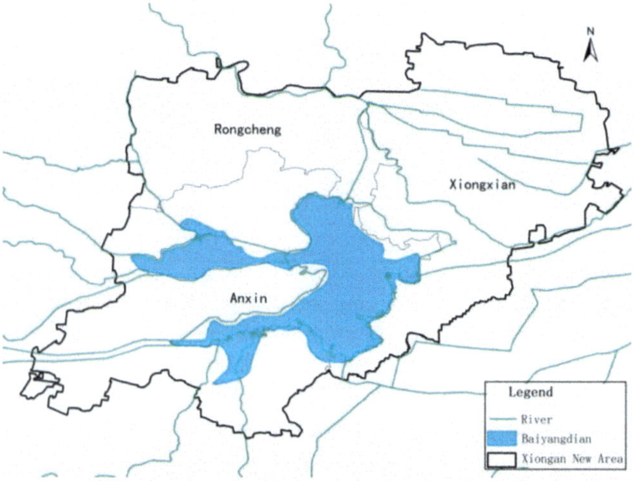

Fig. 1. Geographical location and water system map of Xiong'an New Area.

2.3 The Change of Water Resources and the Treatment of Over-Exploitation

The water resources in Xiong'an New Area are mainly groundwater. The proportion of local surface water resources in the total water resources is less than 5%. Preliminary results of the 3rd water resource investigation and evaluation showed that [7], the average annual total water resources in Xiong'an New Area from 2001 to 2016 was 146 million m^3. Local self-produced surface water resources were only 0.007 billion m^3, and groundwater resources were 142 million m^3. The amount of groundwater that can be mined was 156 million m^3. The total water resources of Xiong'an New Area were 310 million m^3 in 2021, among which groundwater resources are 293 million m^3. Water Resources Bulletin showed that [8], the total water supply in 2021 was 745 million m^3. Surface water supply accounted for nearly 80% of the total water supply, of which external water transfer accounts for more than 60% of the total water supply. It is mainly used for ecology, agriculture, industry and life, of which ecological water replenishment accounts for nearly 80% of the total water consumption. The amount of water transferred from outside to Baiyangdian for ecological replenishment increased significantly since 2018. The annual water supply is 670 million m^3. The water supply of Xiong'an New Area in 2021 is shown in Table 1.

The annual variation of groundwater resources in Xiong'an New Area is significant, and the average annual groundwater resources are 142 million m^3. Groundwater resources ranged from 298 million m^3at most (2012) to 52 million m^3at least (2006) between 2005 and 2021, a difference of 4.7 times. The latest round of national over-exploitation zone delineation results (the evaluation period is 2011–2020) showed that the area of shallow groundwater over-exploitation in Xiong'an New Area is 1566 km^2, and the area of deep confined water overproduction is 700 km^2. Over-exploitation from 2014 to 2021 was reduced by 71 million m^3.

Table 1. Water supply in Xiong'an New Area (2021).

Water supply (10^8 m^3)				Water consumption (10^8 m^3)	
Surface water	5.87	Local surface water	1.14	Agricultural water	1.24
		Interbasin water transfer	4.73	Industrial water	0.07
Underground water	1.33	Quantity of shallow groundwater extraction	1.27	Domestic water	0.33
		Production of deep confined water	0.06	Ecological water replenishment	5.80
Other water sources (reclaimed water)	0.25	/	/	/	/
Total water supply	7.45	/	/	Total water consumption	7.45

3 Data and Methods

3.1 Groundwater Level Monitoring Data

The groundwater level monitoring datum used in this paper is based on the daily monitoring datum of the automatic stations of the National Groundwater Monitoring Project and the provincial artificial stations. Forty-six national groundwater automatic monitoring stations were deployed in Xiong'an New Area (As of the end of 2021). 34 shallow and 12 deep groundwater monitor stations have been monitored since 2017. In addition, nine provincial artificial groundwater monitoring stations monitored shallow groundwater and stopped monitoring since 2020 (Fig. 2). Dynamic analysis of shallow groundwater levels is based on artificial stations that were monitored from 2000 to 2020. Analysis of deep confined water levels is based on automatic stations, which monitored from 2017 to 2021.

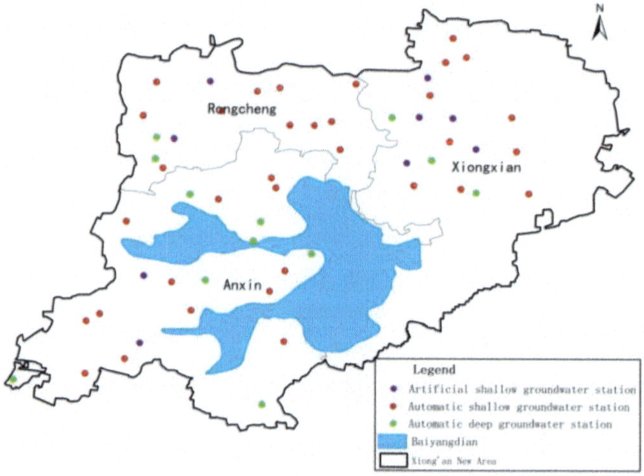

Fig. 2. Diagram of the distribution of groundwater stations in Xiong'an New Area.

3.2 Research Method

This paper analyzed the dynamic change of stratified groundwater level in the Xiong'an New Area from 2000 to 2021. Groundwater level contour and groundwater flow field were drawn using the GIS Kriging spatial interpolation method to analyze the spatial distribution characteristics of the groundwater level. The random forest (RF) machine learning method was used to analyze the contribution rates of precipitation, mining, ecological water replenishment, and other influencing factors to identify the influencing factors of groundwater level change. A numerical groundwater flow model was established using GMS. The model was used to simulate and analyze the groundwater change influenced by the treatment of over-exploitation started from 2019.

Random forest machine learning

Random forest is a highly flexible machine learning method for regression and classification of big data. Random forest is composed of a series of decision trees. The main parameters of the model are the number of decision tree classifiers and the number of randomly selected predictors when tree nodes are divided. After training and debugging by the model, the number of decision tree classifiers and tree nodes was set to 50 and 4, respectively.

GMS numerical simulation

The grid size of the study area is 2km × 2km, and the model is generalized into three layers in the vertical direction. Groundwater in the first and second aquifer groups exchange closely, which was the first layer of the model. The third and fourth aquifer groups were the second and third layers of the model, respectively. The west side of the model is the inflow boundary. The east side is the general head boundary, judged as the outflow boundary, and the bottom boundary of the model is generalized as the impermeable boundary. Source and sink data include precipitation, lateral recharge, agricultural irrigation infiltration, evaporation, and deep and shallow exploitation. The model identification verification period is from June 2010 to December 2019, and the simulation prediction period is from January 2020 to December 2021. One stress period per month. Model parameters are: permeability coefficient of phreatic aquifer is 30m/d, specific yield is 0.1, and permeability coefficient of confined water aquifer is 60m/d.

4 Dynamic Characteristics of Groundwater

4.1 Groundwater Flow Field

The shallow groundwater level in Xiong'an New Area generally flows from Baiyangdian to the surrounding area. The groundwater level in the Baiyangdian area is relatively high (Fig. 3). Deep confined water generally flows from north to south, and the local area is inconsistent with the overall flow direction. The water level in the central area of Anxin County in the south of the New Area is relatively high (Fig. 4).

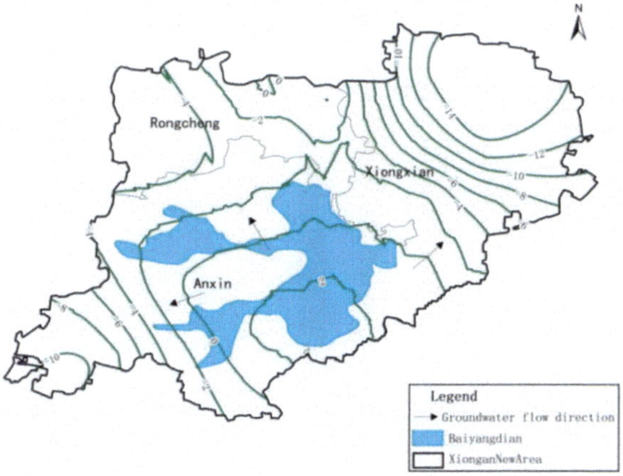

Fig. 3. Diagram of the contour map of shallow groundwater level by the end of 2021.

4.2 Dynamic Characteristics During the Year

The annual dynamic change of shallow groundwater level in the Xiong'an New Area is obvious. There is less precipitation from January to March, and agricultural exploitation

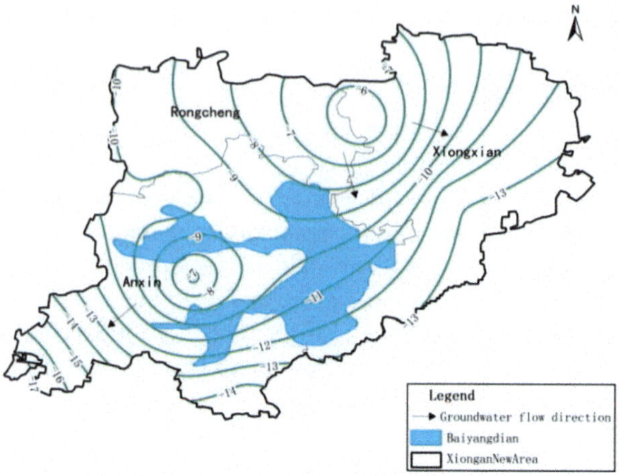

Fig. 4. Diagram of the contour map of deep confined water level by the end of 2021.

is also small. Groundwater level rised fluctuatedly, and the minimum burial depth generally occurs in February or March. Water demand for agricultural irrigation increases, and precipitation is insufficient during April and July. Therefore, the groundwater level is declining. The maximum burial depth generally occurs in June or July, with June in the majority. Precipitation is abundant during the flood season of August and September. Most precipitation recharges the groundwater, and the groundwater level rises again. The groundwater level is relatively stable due to small agricultural irrigation during October and December (Fig. 5).

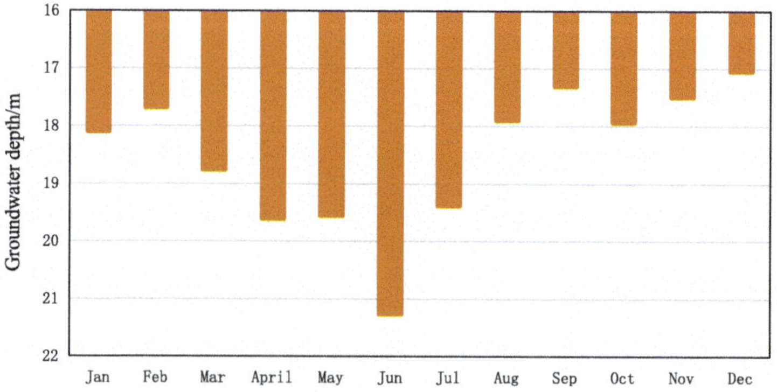

Fig. 5. Diagram of the Annual variation of shallow groundwater level in Xiong'an New Area.

In the past five years (2017–2021), the water level of deep confined water has generally recovered, and the average annual rise was 0.64m. By the end of 2019, the buried depth of the deep confined water level reached a maximum of 25.96m in the past five years. Changes in shallow groundwater level depth, precipitation, river and lake recharge, and exploitation are shown in Fig. 6. Changes in buried depth and mining volume of deep confined water level are shown in Fig. 7.

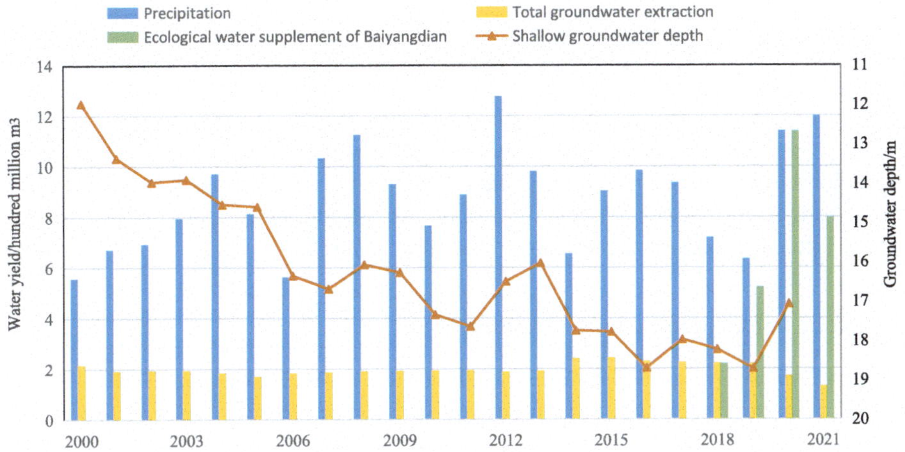

Fig. 6. Diagram of the shallow groundwater level depth, precipitation, water replenishment, and exploitation in Xiong'an New Area.

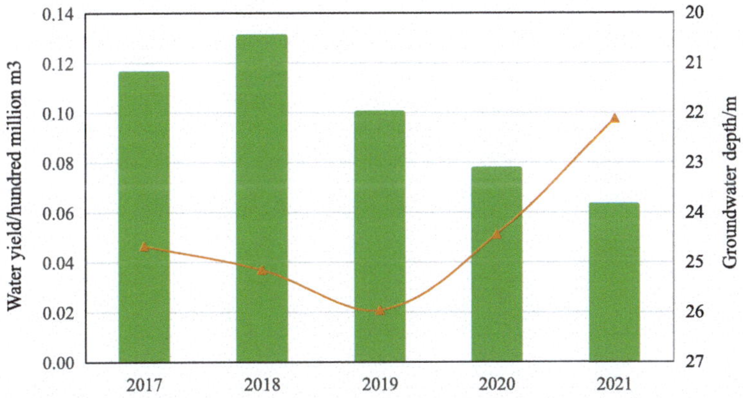

Fig. 7. Diagram of the deep confined water level depth and exploitation in Xiong'an New Area.

5 Analysis of Influencing Factors

The multi-year dynamic changes of shallow groundwater level in Xiong'an New Area were mainly affected by precipitation and artificial mining due to agricultural irrigation. On the contrary, the deep groundwater level was mainly affected by artificial mining [6]. Dynamic types of shallow groundwater in Xiong'an New Area include the precipitation-infiltration-mining type and river-lake hydrology type, mainly precipitation-infiltration-mining type. Dynamic type of deep groundwater is mainly runoff, overflow recharge, and exploitation [9].

5.1 Precipitation Change

The precipitation in Xiong'an New Area is unevenly distributed and varies greatly between years. Annual precipitation from 2000 to 2021 was 494 mm (Fig. 6).

The dynamic of shallow groundwater level buried depth is affected by precipitation. Precipitation infiltration recharge decreases in less precipitation year, leading to decreased surface runoff and seepage recharge. Thus, groundwater resources are less, and vice versa. When the precipitation anomaly was less than −20% in a dry year, the shallow groundwater level decreased by 0.5m ~ 1.8m between 2000 and 2020. When the precipitation anomaly was greater than 20%, the rise of shallow groundwater level ranged from 0.6m to 1.6m. The shallow and deep groundwater levels have obviously increased due to excessive precipitation in 2020 and 2021.

5.2 Change of Groundwater Extraction

Xiong'an New Area mainly exploited shallow groundwater. Between 2000 and 2021, the average annual groundwater exploitation in Xiong'an New Area was 199 million m^3. The average annual mining volume of shallow groundwater was 165 million m^3, which accounted for 83% of the total mining volume. The process can be roughly divided into three stages: basic stability, slight increase, and continuous decrease (Fig. 8). The mining volume was stable from 2000 to 2013. Between 2014 and 2019, it increased slightly, with an average increase of 22% compared with the previous stage. It reached a maximum of 246 million m^3 in 2015 in the past 20 years m^3. Between 2020 and 2021, it continued to decline, reaching a minimum of 133 million m^3 in 2021.

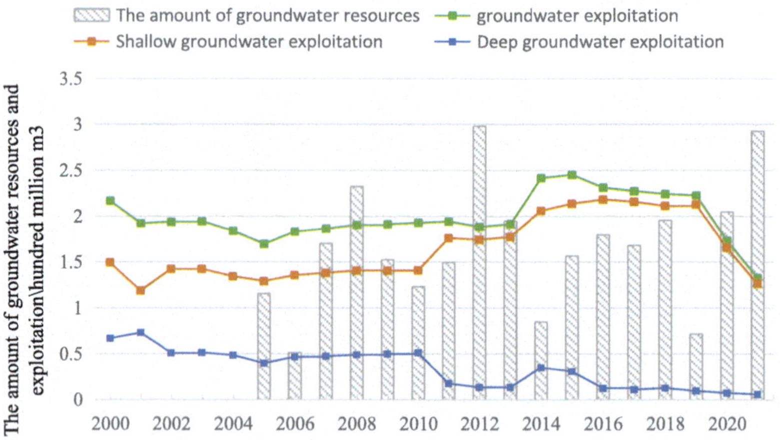

Fig. 8. Diagram of the changes in groundwater resources and groundwater extraction in Xiong'an New Area.

In recent years, with the continuous advancement of comprehensive management of over-exploitation, the control of groundwater exploitation has achieved obvious results. The shallow groundwater exploitation from the maximum of 219 million m^3 (2016)

reduced to 127 million m^3 (2021). Between 2000 and 2013, the average annual exploitation of shallow groundwater was 147 million m^3, 1% less than that in the same period of many years. The average shallow groundwater level was 0.88 m higher than that in the same many years. Between 2014 and 2020, the average annual exploitation of shallow groundwater was 207 million m^3, 26% more than that in the same period of many years, and the water level was 1.77 m lower than the average of the same period of many years. Between 2020 and 2021, the shallow mining volume was significantly reduced, and the deep groundwater was not mined.

5.3 Contribution Rate of Influencing Factors

Based on the precipitation, groundwater exploitation, river-lake ecological replenishment, and groundwater level change in the past 20 years, the period was divided into four stages (Table 2).

Table 2. The average amount of groundwater exploitation, precipitation, and ecological supplement of rivers-lakes in each stage

Phase	Precipitation (Mm)	Groundwater Exploration (10^8 M^3)	Shallow Groundwater Exploitation (10^8 M^3)	Deep Groundwater Exploitation Quantity (10^8 M^3)	Water Supplement of River and Lake (10^8 M^3)
2000–2006	409	1.91	1.37	0.54	
2007–2010	545	1.91	1.41	0.50	
2011–2013	593	1.92	1.77	0.15	
2014–2020	482	2.24	2.07	0.17	6.28

From 2000 to 2006, the exploitation of Xiong'an New Area was stable, with an average of 191 million m^3. The precipitation is 16% less than the same period in many years. Based on existing research [3], when the exploitation exceeds the capacity of regional groundwater recharge, the groundwater level continues to decline.

From 2007 to 2010, the total mining volume was stable at 191 million m^3. The precipitation was 12% more than the same period in many years. Precipitation recharge increased obviously. Although the exploitation remained stable compared to the previous stage, the correlation between groundwater level fluctuation and precipitation change was obvious. From 2007 to 2010, the groundwater level rised significantly in more precipitation year, and vice versa. Therefore, the contribution rate of precipitation to groundwater level increased significantly from 2007 to 2010 compared with 2000 to 2006.

From 2011 to 2013, the total mining volume was still stable compared with the previous stage, with an average of 192 million m^3. Amount of shallow mining increased slightly, and the amount of deep mining decreased significantly. Precipitation was 22% more than that in the same period of many years. The precipitation increased by 49% in 2012, and the groundwater level was positively correlated with the precipitation change.

From 2014 to 2020, the total mining volume increased slightly compared with the previous stage, and the precipitation was the same as that in the same period of many years. The ecological water replenishment of rivers and lakes such as Baiyangdian forms a continuous and effective supply of groundwater, and the groundwater level generally maintains a stable fluctuation. Based on existing research [4], the suitable areas for replenishment are mostly distributed along the river channel, mainly in the near-mountain rivers in the plain area of Daqing River. There are nine inflow rivers in the upper reaches of the Baiyangdian Lake, and the amount of water added to the lake is as high as 628 million m^3 from 2018 to 2020. The area is suitable for recharge. The permeability coefficient, groundwater depth, and effective water storage space are large. Good aquifer storage conditions and abundant recharge water sources formed effective recharge to groundwater resources in Xiong'an New Area.

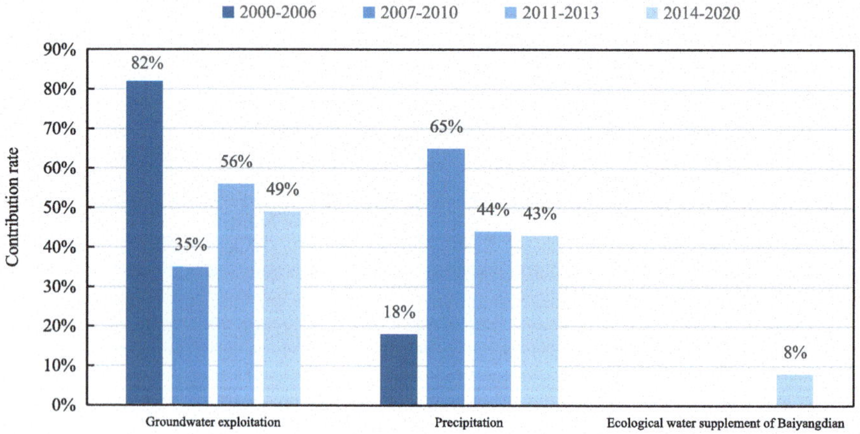

Fig. 9. Contribution rate of influencing factors of shallow groundwater level in each stage.

From the contribution rate analysis of influencing factors in each stage (Fig. 9), we can see that the contribution rate of groundwater exploitation was the largest, with an average of 55%. Second was precipitation, with an average of 42%. Ecological water replenishment of rivers and lakes has been implemented in the Xiong'an New Area since 2018. Ecological water replenishment has formed an effective supply to groundwater, and the impact on water level recovery has gradually emerged.

5.4 Groundwater Numerical Simulation Analysis

Since 2019, China has implemented comprehensive treatment of groundwater over-exploitation in North China, and the precipitation from 2019 to 2021 was significantly more than that of the past years. To further identify the influence degree of over-exploitation comprehensive treatment measures on groundwater level without precipitation, a numerical model was constructed using GMS. Groundwater level was compared under actual precipitation and normal year precipitation from 2019 to 2021.

According to the precipitation series of Xiong'an New Area from 1956 to 2021, the average annual precipitation was 478 mm. Precipitation in 2019 was 25% less than normal, and the precipitation in 2020 and 2021 were 35% and 42% more than normal, respectively.

According to the simulation analysis, the precipitation factor caused the groundwater level to rise by 1.54 m at the end of 2021 compared with the end of 2018. Based on groundwater level data of 34 shallow automatic monitor stations, the groundwater level rose by 2.44m at the end of 2021 compared with the end of 2018. After deducting the influence of precipitation, the water level at the end of 2021 rose by 0.9m compared with the end of 2018 due to the comprehensive impact of over-exploitation control measures such as reducing groundwater exploitation, river-lake replenishment, and water source replacement. The contribution rates of precipitation and over-exploitation to the shallow groundwater level were 63% and 37%, respectively. Therefore, more precipitation in the past three years influenced the groundwater level recovery. Precipitation was the main factor in the rise of water level in the past three years, which proved the reliability of attributing groundwater level change by numerical simulation method.

From the perspective of sustainable utilization of regional groundwater resources, the exploitable groundwater resources in Xiong'an New Area are 156 million m^3. Based on the statistics, the average annual groundwater exploitation from 2014 to 2021 was 213 million m^3 (133 million was exploited in 2021), which was still higher than the exploitable groundwater resources. The current exploitable groundwater resources in Xiong'an New Area are difficult to meet the water resources needs of future urban management. External water resources such as the South-north Water Diversion Project must be considered to ensure the safety of the water supply. We should control the exploitation of deep groundwater strictly, to avoid environmental geological problems such as land subsidence.

6 Conclusions

The groundwater level in Xiong'an New Area showed a trend of first decreasing and then rising. Shallow groundwater level decreased continuously from 2000 to 2014, with an average annual decrease of 0.41 m. After the implementation of over-exploitation control, the shallow groundwater level rised since 2014 with annually 0.11m.

According to the random forest machine learning analysis, the groundwater level in Xiong'an New Area was mainly affected by the amount of groundwater extracted (contribution rate 55%), precipitation (contribution rate 42%), and ecological water replenishment.

The Numerical Simulation method was used to analyze the groundwater level change without considering precipitation. Groundwater level rise from 2019 to 2021 was mainly affected by precipitation, and the current average groundwater exploitation was higher than the recoverable resources. It was necessary to continue promoting the ecological replenishment of rivers and lakes, water source replacement, and other effective management measures to maintain the over-exploitation control effects.

Acknowledgements. The financial support of the National Natural Scientific Foundation of China (NSFC) (U21A2004) and the Major Science and Technology Projects of the Ministry of Water Resources (SKS-2022041) are generally acknowledged.

References

1. Chen, F., Xu, X.Y., Yang, Y.: Analysis of the evolution trend and influencing factors of groundwater resources in China. Adv. Water Sci. **31**(6), 811–819 (2020)
2. Shi, J.S., Li, G.M., Liang, X.: Groundwater evolution mechanism and regulation in North China Plain. Acta Geol. Sin. **35**(5), 527–534 (2014)
3. Liu, R., Zhao, Y., He, X., et al.: Evaluation of cumulative recoverable over recovery of groundwater in Haihe Plain. J. Hydraul. Eng. **53**(11), 1336–1349 (2022)
4. Cao, W.G., Wen, A.X., Tian, N., et al.: Evaluation of groundwater recharge suitability in typical areas of Haihe River Basin. Adv. Water Sci. **53**(11), 227–237 (2022)
5. Wang, Y.X., Liu, X., Shen, Y.J., et al.: The applicability of the random forest model in attribution analysis of runoff change. Chin. J. Eco-Agric. **30**(5), 864–874 (2022)
6. Li, H.T., Feng, W., Wang, K.L., et al.: General situation, characteristics and exploitable potential of groundwater resources in Xiong'an New Area. Geology of China **48**(4), 1112–1126 (2021)
7. The 3th National Water Resources Investigation and Evaluation. Ministry of Water Resources
8. Haihe River Basin Water Resources Bulletin. Haihe River Water Conservancy Commission, MWR,
9. Wang H.T., Zhao B.L. Li L.F.: Analysis of groundwater dynamic characteristics and influencing factors in Xiong'an New Area. Ground Water. **41**(1), 51, 78 (2019)

A Fast and Stable Method of Water Sampling for Online Monitoring System of Water Quality in Enteromorpha Tracking Unmanned Ship

Yan Liu[1,2], Hui Su[1], Ran Ma[1,2](✉), Xiangfeng Kong[1,2], Dongzhi Chu[1,2], Fengqing Liu[1,2], and Qian Shi[1,2]

[1] Institute of Oceanographic Instrumentation, Qilu University of Technology (Shandong Academy of Sciences), Qingdao 266061, People's Republic of China
mr_sdioi@126.com

[2] Laoshan Laboratory, Qingdao, Shandong 266237, People's Republic of China

Abstract. To enable long-term real-time monitoring of the water environment during the outbreak of Enteromorpha, a drifting unmanned vessel equipped with a multifarious online water quality parameter monitoring system has been selected. Monitor changes in water quality in the area affected by Enteromorpha through real-time detection and analysis of water quality data. A pump valve is used to draw seawater into a storage tank for water sample collection, followed by detection and analysis. Because the unmanned ship needs to move in real-time during monitoring, the injection valve is fast, and impurities such as bubbles and sediment will affect the measurement of optical probes. Therefore, the water samples in the storage tank need to be quickly stabilized to minimize impact. So, this paper proposes a rapid stabilization method for water sample injection based on PSO (Particle-Swarm-Optimization) fuzzy PID(Proportional-Integral-Derivative) control. The model is built and simulated to compare the conventional PID control and the PSO fuzzy PID control algorithms. The simulation results show that the conventional PID algorithm takes 58.201 s for the water sample to reach a steady state with a stable height of 17.7 cm. In contrast, the PSO fuzzy PID algorithm achieves a steady state in 8.17 s, with a stable height of 17.96 cm. Compared with the response curves of the two, the overshoot of the PSO fuzzy PID control system is smaller, and the adjustment time is shortened by 86.01%. Reached the rapid stability of the water sample. Ensure the timeliness and reliability of the moving measurement process data.

Keywords: Enteromorpha monitoring · Rapid and stable injection method · PSO fuzzy PID control

1 Introduction

The outburst of Enteromorpha in the Yellow Sea has become a normal Marine ecological disaster in China. The Enteromorpha is non-toxic, but when it washes ashore and accumulates on the beach, it quickly decays, causing serious impacts on coastal

C. Weng (Ed.): WRE 2024, LNCE 634, pp. 123–135, 2025.
https://doi.org/10.1007/978-981-96-5095-8_11

ecosystems, tourism, and residents in coastal areas [1]. The government environmental protection department carried out many years of investigation and analysis in response to this series of issues. Then, it established the regular model of water quality changes during the moss outbreak. In India, various regression models are developed for the upper reaches of the Ganges to predict water quality status [2]. The assessment of groundwater potential areas in and around Jhansi is conducted by preparing groundwater potential area maps using Geographic Information Systems (GIS), remote sensing, and GPS surveys for field verification of groundwater resources [3]. Although these studies have made certain progress, they usually depend on field testing, facing the restrictions of human, resources, and time. And it isn't easy to carry out large-scale and long-term monitoring. As a result, long-term unmanned monitoring needs based on tracking forms are imminent. This unmanned monitoring system can automatically obtain real-time data, improving the monitoring accuracy and enabling timely detection and response to water quality changes and ecological environmental issues.

Based on this, the effects of the growth, development, and death of the Enteromorpha, an Enteromorpha intelligent tracking unmanned ship monitoring system has been developed, which integrates Enteromorpha image recognition and tracking system, meteorology, hydrology, ecology and other sensing modules in the unmanned ship, and the water quality online monitoring system is some water quality parameters used to monitor the location of the Enteromorpha.

The water quality online monitoring system uses sampling analysis to carry out water quality monitoring. The pump valve uses the pump valve to pump the water into the water storage tank. Then, the analysis under different modes is carried out. Because the unmanned ship needs to move in real-time during monitoring, the speed of the procedure is fast, and impurities such as bubbles and sediment will affect the measurement of optical probes. Therefore, the water sample in the water storage tank needs to be rapid and stable.

The traditional valve control method controls the flow rate by adjusting the valve opening. Still, various factors often affect this method, resulting in inaccurate flow control accuracy [4]. In 2018, Seslija D [5] adopted pulse width and bypass valve control. In 2019, Zhu Tianyu [6] adopted a fuzzy neural network PID control algorithm. In 2020, Cheng Qian [7] adopted an improved valve control algorithm and valve control method. In 2021, Zhao Congcong [8] adopted an intelligent control algorithm based on fuzzy reasoning. In 2022, Jiang Jiayu [9] adopted a control strategy based on a valve control parameter. The above research methods have improved the control quality of valves and applied to different fields [10]. However, there is a lack of methods to achieve rapid stability after sampling water. The PSO fuzzy PID control algorithm is a hybrid control algorithm based on PSO and fuzzy PID control algorithms [11]. It has better adaptability and robustness. It can adapt adaptive adjustment according to the real-time flow deviation and deviation change rate. Adaptively adjust the valve's opening degree to precisely control the flow rate, ensuring rapid stability in water sample monitoring.

Based on this, this paper introduces a particle swarm fuzzy PID control algorithm to achieve adaptive control of flow rates, ensuring rapid stability in water sample monitoring, addressing uncertainties and nonlinear issues, and obtaining better stability for water samples.

2 PSO Fuzzy PID Control Algorithm

2.1 PSO Algorithm

The PSO is derived from bionic research on the foraging behavior of migratory birds. Migratory birds will continue to update, move path updates, and move direction updates when looking for food, and find the best food location through information exchanges between groups. PSO merits briefness iteration, few parameters to be adjusted, and fast convergence to the best area [12, 13]. The flowchart of the PSO is shown in Fig. 1.

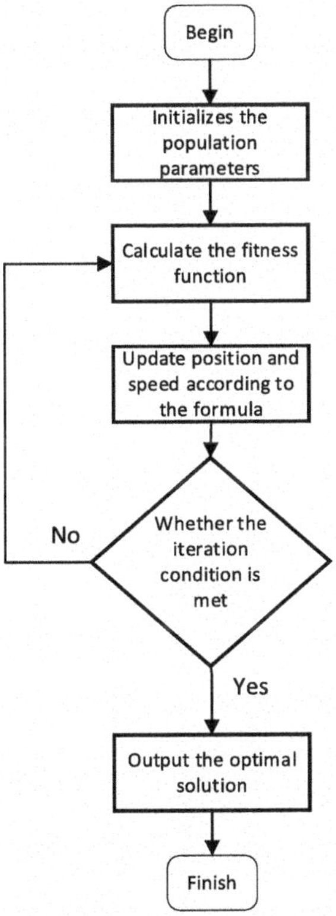

Fig. 1. Flow chart of PSO.

1) Initialized group parameters. First, determine the dimension D and particle number J of the search space, and then randomly generate the initial position X_{jd}, the initial speed V_{jd}, the best position of the individual X_{jd,p_best}, and the optimal position X_{jd,g_best}.
2) Define the adaptation function according to the required definition. According to the position of each particle, calculate the value of the fitness function.
3) Update particle location and speed. Update the particle's velocity V_{jd}^{k+1} and position X_{jd}^{k+1} based on its current position, velocity, and historical optimal position.

The next movement direction is composed of the inertial direction, the optimal direction of the individual, and the optimal direction of the group, as shown in Fig. 2.

4) After reaching the conditions of iteration N times, the output optimal solution.

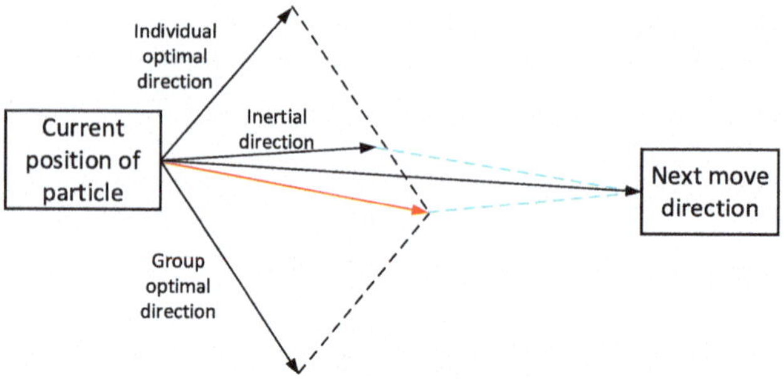

Fig. 2. Next move direction.

2.2 Fuzzy PID Controller

Fuzzy PID control is a mathematical model with nonlinear control, high stability, strong robustness, and no need for accuracy. Its working principle is to detect the position error and error deviation rate in real-time during system operation, and to calculate the corresponding fuzzy value using quantification factors, which are then matched with the established fuzzy rule table. Enter the proportional coefficient, integral coefficient, and differential coefficients obtained by the reasoning to adjust it in the PID controller to regulate it to attain control of the system [14, 15]. The specific design steps are successively fuzzification, establishing fuzzy rules, logical reasoning, and defuzzification, as shown in Fig. 3.

1) Fuzzification. Select appropriate input and output, divide the fuzzy set, use the membership function to describe the membership degree of input variables to the fuzzy set, and transform an exact value into a fuzzy value.
2) Fuzzy reasoning. Establish a fuzzy controller's control rules and establish a fuzzy rules table.
3) Defuzzification. The output of fuzzy rules is a fuzzy amount, so it needs to be converted to an accurate amount by fuzzy judgment. Common defuzzification algorithms include the Area Averaging Method, Maximum Membership Degree Method, Centroid Method, and Weighted Average Method [16].
4) Adjust the result through the PID controller.

2.3 PSO Fuzzy PID Algorithm

The fuzzy PID controller adapts its parameters during the control process, providing a fast response. By optimizing the initial control parameters using particle swarm optimization,

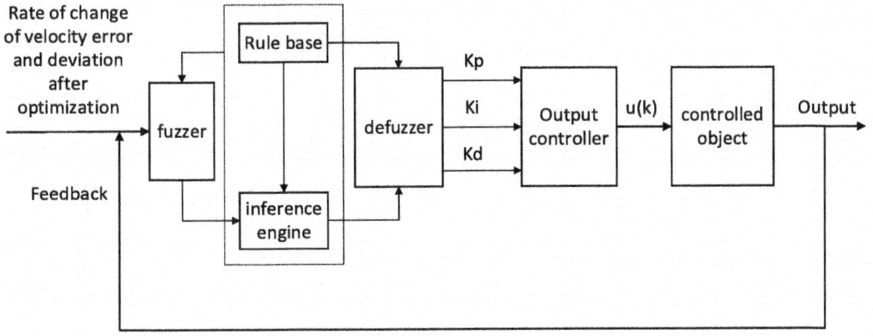

Fig. 3. Fuzzy PID controller.

the performance of the fuzzy PID controller is further enhanced, thereby improving the system's response speed and stability [12].

3 Algorithm Implementation

3.1 PSO Algorithm Implementation

1) Initialized group parameters. Select a 10-dimensional search space, 100 particles, and each particle represents a solution.
2) Define the adaptation function according to the required definition.

The fitness function is calculated using the following formula:

$$Fitness = \omega_1 * e(t) + \omega_2 * ec \tag{1}$$

Here, e(t) is the flow rate deviation, ec is the rate of change of the deviation, ω_1 and ω_2 are the inertia weight coefficients. By adjusting the values of ω_1 and ω_2 (generally selected within the range of [0, 1]), the balance between the flow rate deviation and the rate of change of deviation can be adjusted, thereby finding the optimal flow rate deviation $e_{best}(t)$ and the optimal rate of change of deviation ec_{best}.

3) Update particle location and speed.

The velocity is calculated using the following formula:

$$V_{jd}^{k+1} = \omega * V_{jd}^{k} + c_1 * r_1(X_{jd,p\_\text{best}}^{k} - X_{jd}^{k}) + c_2 * r_2(X_{jd,g\_\text{best}}^{k} - X_{jd}^{k}) \tag{2}$$

Here, ω is the inertia weight (generally selected within the range of [0, 1]); c_1 is the individual learning factor; c_2 is the group learning factor,$c_1 = c_2 = 2$; r_1 and r_2 are random numbers between [0, 1].

The position is calculated below:

$$X_{jd}^{k+1} = X_{jd}^{k} + V_{jd}^{k+1} * dt \tag{3}$$

4) After 60 iterations, the optimal solution is output. The obtained optimal flow rate deviation $e_{best}(t)$ and optimal rate of change of deviation ec_{best} are then input into the PID controller.

3.2 Fuzzy PID Control Algorithm Implementation

3.2.1 Fuzzification

Select the input variables as the optimal flow rate deviation $e_{best}(t)$ and the optimal deviation rate of change ec_{best}, and the output variables as the proportional coefficient K_p, integral coefficient K_i, and differential coefficient K_d. The domain V of the fuzzy subset is chosen as [–6, 6]. As shown in Figs. 4 and 5. Input variables and output variables are evenly divided into seven fuzzy subsets on the domain of discourse {positive large (PL), positive intermediate (PI), positive diminutive (PD), zero (ZO), negative diminutive (ND), negative intermediate (NI), negative large (NL)}. The membership function of the input variable is selected as the Gaussian distribution, and the membership function of the output variable is selected as the triangle.

The formula of the membership function of the Gaussian distribution type is

$$\mu_{gaussmf}(a; c, \sigma) = e^{-\frac{1}{2}(\frac{a-c}{\sigma})^2} \tag{4}$$

C is the center of the Gaussian belonging function. The width of the Gaussian membership function is determined by σ. $\sigma = 0.8494$.

The triangular affiliation function is calculated using the following formula:

$$\mu_{trimf}(a; x, y, z) = \begin{cases} 0 & a \le x; a \ge z \\ \frac{a-x}{y-x} & x \le a \le y \\ \frac{z-a}{z-y} & y \le a \le z \end{cases} \tag{5}$$

where x, y and z are the X-coordinates of the three tops of the triangular membership function.

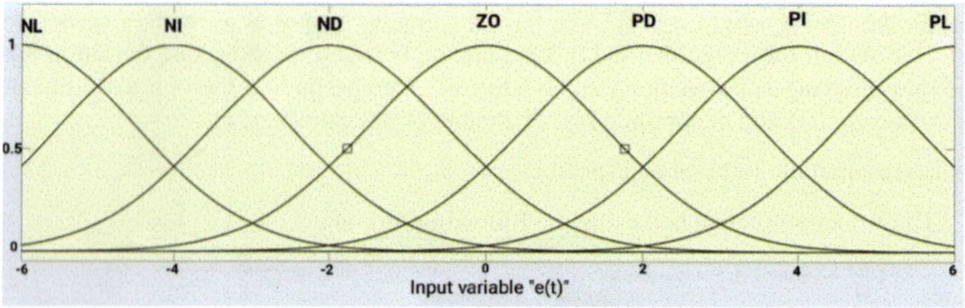

Fig. 4. Seven fuzzy subsets of Gaussian membership function partition.

Fig. 5.Seven fuzzy subsets of triangular membership function partition.

3.2.2 Fuzzy Reasoning

The control rule of fuzzy controller is established: when the flow velocity deviation $e_{best}(t)$ is large, the K_p value is large. To reduce the overshoot, the K_i value should be

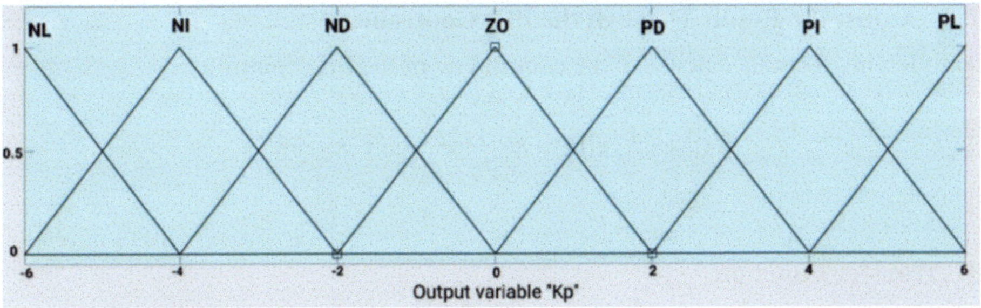

Fig. 5. Seven fuzzy subsets of triangular membership function partition.

as small as possible, and the K_d value should be small. When $e_{best}(t)$ is in the medium size, K_p takes a small value, K_i and K_d take a moderate value. When $e_{best}(t)$ is small, K_p takes a larger value, K_i takes a larger value, and K_d takes a moderate value. We get 49 fuzzy rules (Table 1).

Table 1. K_p, K_i, K_d fuzzy rule table.

K_{ec}	K_e						
	NL	NI	ND	ZO	PD	PI	PL
NL	PL/NL/PD	PL/NL/PD	PI/NL/ZO	PI/NI/ZO	PD/NI/ZO	PD/ZO/ZO	ZO/ZO/PL
NI	PL/NL/ND	PL/NL/ND	PI/NI/ND	PI/NI/ND	PD/ND/ZO	ZO/ZO/PD	ZO/ZO/PI
ND	PI/NI/NL	PL/NI/NL	PI/ND/NI	PD/ND/ND	ZO/ZO/ZO	NS/PD/PD	NI/PD/PI
ZO	PI/NI/NL	PL/ND/NI	PS/ND/ND	ZO/ZO/ND	PD/ND/ZO	NI/NI/PD	NI/PI/PI
PD	PI/ND/NL	PI/ND/NI	ZO/ZO/ND	ND/PD/ND	NI/PD/ZO	NI/NI/PD	NI/PI/PI
PI	ZO/ZO/NI	ZO/ZO/ND	PD/PD/ND	ND/PI/ND	NI/PI/ZO	PL/PD/PD	NL/PL/PD
PL	ZO/ZO/PD	ZO/ZO/ZO	PD/PD/ZO	NI/PI/ZO	NI/PL/ZO	NL/PL/PL	NL/PL/PL

3.2.3 Defuzzification

Select the area average division method for defuzzification.

The area is divided from the following format:

$$\int_{f}^{x} A(v)dv = \int_{x}^{g} A(v)dv = \frac{1}{2}\int_{f}^{g} A(v)dv \tag{6}$$

where, A(v) is the membership function of set A on the domain V; f and g are the x values of the two intersection points where the fuzzy set membership function curve intersects the horizontal coordinate.

3.2.4 Adjust the Result Through the PID Controller

The following formula calculates the control law of the PID controller:

$$U(t) = K_p e(t) + K_i \int_0^t e(t)dt + K_d ec(t) \tag{7}$$

3.3 Transfer Function

During the operation of the water quality monitoring system, a pneumatic control valve is used to draw seawater into the storage tank. The flow rate of water is two liters per minute when the opening is at its maximum. The diameter of the water tank is 15 cm, and the height is 20 cm. The water level in the water storage tank is constantly changing. When the detection unit starts to run, the valve opening should be large (80%). After reaching the target liquid level (18 cm), the opening degree should be smaller (20%). Real-time monitoring of the water level of the water storage tank is obtained from the water level changes in Fig. 6.

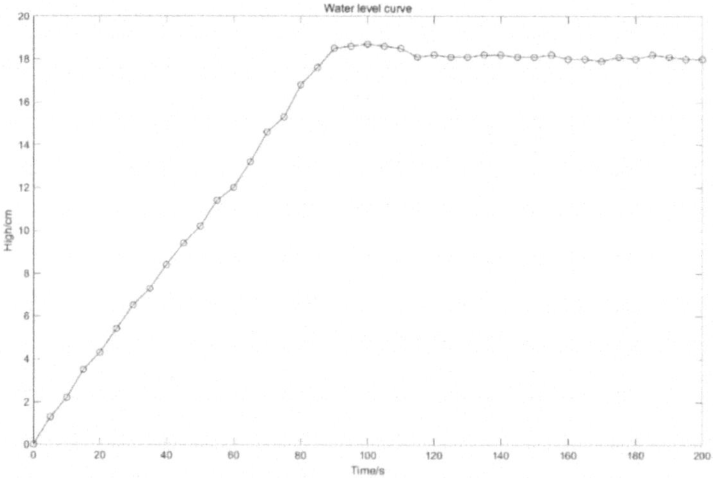

Fig. 6. The curve of water level.

Import water level changes in the program editor, and create a time sequence to pass the number of steps to pass the function in at least one order. After that, the transmission function is obtained, and the system response is drawn. Get the recognition curve in Fig. 7. The formula of the transmission function is:

$$G(s) = \frac{0.11}{100s + 0.28} \tag{8}$$

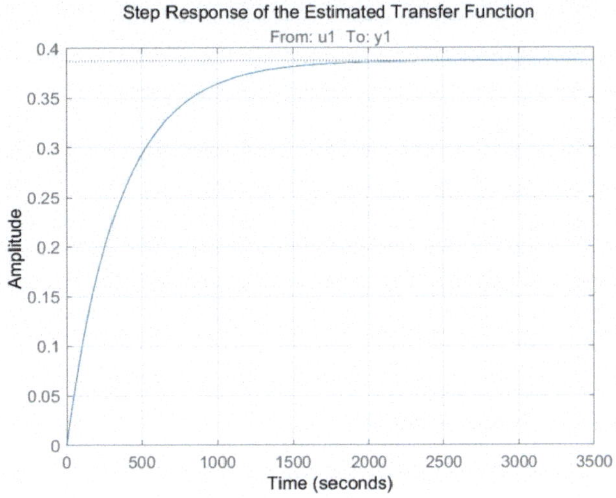

Fig. 7. System identification output curve.

4 Conclusion of PSO Fuzzy PID Simulation

4.1 Simulation Model Building

In the simulation software, set up a rapid and stable valve control simulation model for water samples, as shown in Fig. 8.

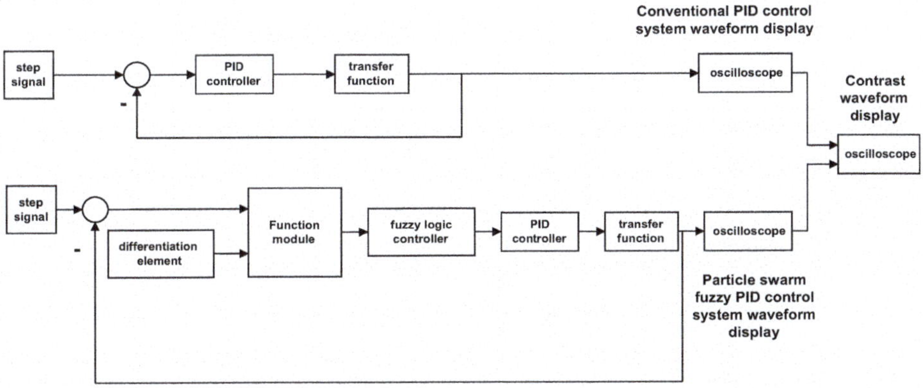

Fig. 8. PID/ PSO fuzzy PID simulation flow chart.

The traditional PID control system receives a step signal as input. After processing by the PID controller, this input signal generates a corresponding control signal. The generated control signal is then transmitted to the transfer function module of the controlled object, which describes the dynamic response characteristics of the controlled object. Finally, an oscilloscope is used to display the output waveform in real time, facilitating the observation of the system's response and steady-state error.

The PSO fuzzy PID control system also receives a step signal as input. First, the system utilizes a Function module to implement the PSO algorithm to optimize the input signal of the PID controller. This optimization process aims to enhance the controller's response performance and steady-state accuracy under varying conditions. The optimized signal is then processed by a fuzzy logic controller, which adjusts the control strategy based on uncertainty and non-linear characteristics, and is subsequently further refined by the PID controller to generate the final control signal. This control signal is transmitted to the transfer function module of the controlled object to describe the dynamic response characteristics of the controlled object. Finally, an oscilloscope is used to display the output waveform in real time, facilitating the observation of the system's response and steady-state error.

4.2 System Simulation and Analysis

4.2.1 Conventional PID Control Simulation

The results of controlling the water sample using a conventional PID algorithm are shown in Fig. 9. As can be seen from the figure, the rate of increase in water level is relatively slow, and when the water level reaches the set height, the stable height is 17.7 cm. The time to stabilize was 58.201 s.

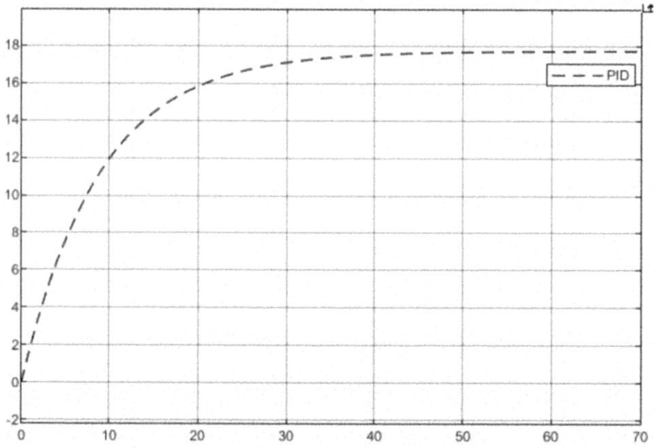

Fig. 9. PID response curve.

4.2.2 Particle Swarm Fuzzy PID Control Simulation

Optimize the initial parameters of the fuzzy PID controller by using particle group optimization algorithms. Then, the water sample is controlled. When the number of iterations of the particle swarm is 60, the optimal solution can be output, and its optimal fitness value is 0.00015 (Figs. 10 and 11).

The PSO fuzzy PID control algorithm optimizes the initial parameters e(t) and *ec* of the fuzzy PID controller, resulting in final optimized values of 0.7942 and 0.2096.

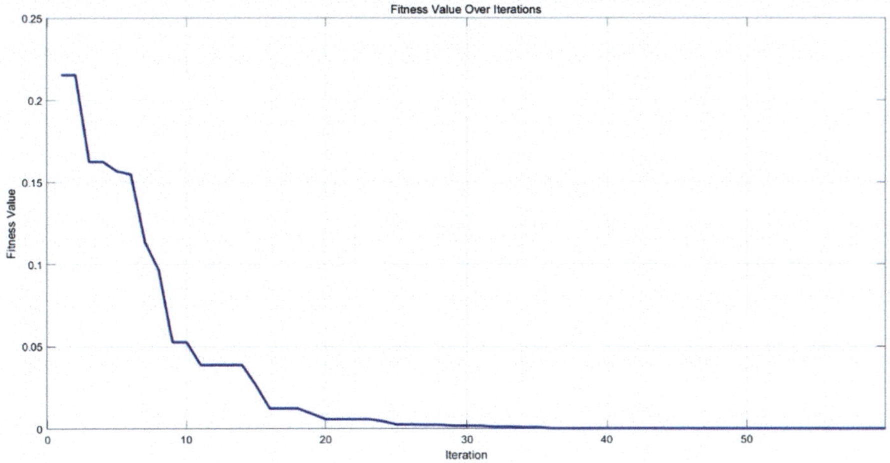

Fig. 10. Optimization results of particle swarm fitness value.

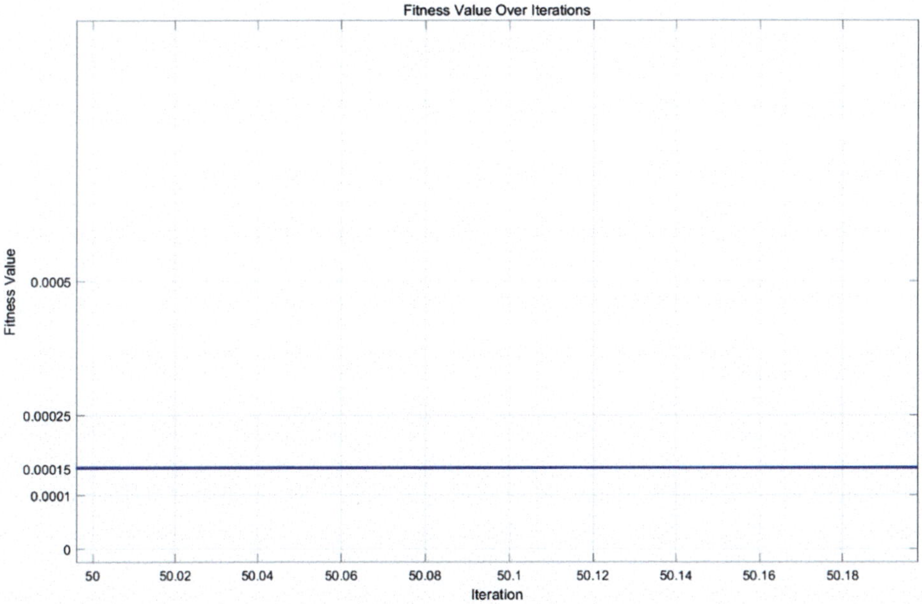

Fig. 11. Optimum fitness value.

With the final water sample height set to 18 cm, the system's response curve reached a maximum value of 17.96 cm at 8.17 s. This result indicates that the optimized control parameters can effectively improve the system's response performance. Although the set water level was not reached, the system still exhibited good dynamic characteristics (Fig. 12).

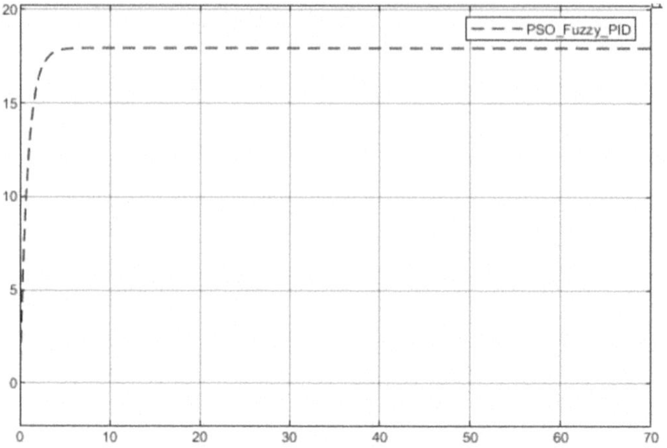

Fig. 12. Fuzzy PID response curve of particle swarm.

4.3 Results Discussion

The response curve of the simulation results shows that the time for the water sample to reach a steady state under the conventional PID algorithm is 58.201s, and the stable height of the water sample is 17.7CM. The PSO fuzzy PID algorithm takes 8.17s and 17.96CM to stabilize the water sample. By comparison, the PSO fuzzy PID control system has a smaller overshoot, and the adjustment time is shortened by 86.01%. It is proved that the control effect of PSO fuzzy PID is indeed better than that of conventional PID control, and the established PSO fuzzy PID control can realize effective suppression of some uncertain factors, has better adaptability, and can achieve better control effects in the simulation environment.

5 Conclusions

This article proposes the rapid stability of the water sample control algorithm with a partial part of the PID control algorithm to solve the detection results of the air bubbles and impurities produced by the water storage tank when the unmanned ship is running quickly. The simulation results show that the PSO fuzzy PID algorithm is able to control the valve so that the seawater in the storage tank quickly reaches the desired height of the water samples and achieves rapid stabilization of the water samples. Compared with the response curves, the PSO fuzzy PID control system has a smaller overshoot, and the adjustment time is shortened by 86.01%. This algorithm can reach the steady state faster to ensure the timeliness and reliability of the moving measurement process data.

Acknowledgment. This work was supported by the Major Innovation Project of Science and Education Integration Pilot Project of Qilu University of Technology (Shandong Academy of Sciences) (2023JBZ02), Major Innovation Project of Science and Education Integration Pilot Project of Qilu University of Technology (Shandong Academy of Sciences) (2023JBZ03), Science Education Industry Integration Pilot Major Innovation Project (2023HTZX01) of Qilu University of Technology (Shandong Academy of Sciences), Open Research Fund of Key Laboratory of Marine Ecological Monitoring and Restoration Technology, Ministry of Natural Resources, (MEMRT202301).

References

1. Yang, D., Yuan, K., Gu, X., Sun, C., Gao, L.: Influences of environmental factors on the dissipation of green tides in the Yellow Sea, China. Mar. Pollut. Bull. **189**, 114737–114737 (2023)
2. Shukla, K.A., Ojha, P.S.C., Mijic, A., Buytaert, W., Pathak, S., Garg, P.: Population growth, land use and land cover transformations, and water quality nexus in the Upper Ganga River basin. Hydrol. Earth Syst. Sci. **22**(9), 4745–4770 (2018)
3. Pandey, K.N., Shukla, K.A., Shukla, S., Pandey, M.: Assessment of underground water potential zones using modern geomatics technologies in Jhansi district, Uttar Pradesh, India. ISPRS - International Archives of the Photogrammetry. Remote. Sens. Spat. Inf. Sci. **XL-8**(1), 377–381 (2014)
4. Feng, Z., Gao, J., Huang, Q.: The fuzzy PID control method of connecting reactor-taking valve is improved[J]. China Petroleum and Chemical Standard and Quality **44**(12), 36–38 (2024)
5. Šešlija, D., Čajetinac, S., Blagojević, V., Šulc, J.: Application of pulse width modulation and by-pass valve control for increasing energy efficiency of pneumatic actuator system. Proceedings of the Institution of Mechanical Engineers, Part I: Journal of Systems and Control Engineering **232**(10), 1314–1324 (2018)
6. Zhu, T., Dong, Q., Liu, R.: Application of fuzzy network in valve opening control. Instrum. Tech. Sens. **103**(2), 94–98 (2019)
7. Cheng, Q., Liu, Z., Jiang, A., Jiang, E., Xiao, Y., Li, F., Jiang, J.: Research on control algorithm of intelligent valve positioner based on parameter self-tuning. Proceedings of the 39th China Control Conference., 259–264 (2020)
8. Zhao, C.: Design of experimental platform and research on control strategy of pneumatic regulating valve based on LabVIEW. Hefei University of Technology, Anhui (2021)
9. Jiang, J.: Performance analysis and improvement research of pneumatic valve positioning device. Hangzhou Dianzi University, Zhejiang (2022)
10. Gao, Y.: Research on modeling and control strategy of pneumatic control valve control system. Wuhan University of Science and Technology (2023)
11. Liu, M., Cheng, L., Xu, J., Zhang, X., Zhang, H.: An improved particle swarm fuzzy PID for adaptive temperature control in CFRP induction heating. Proceedings of the Institution of Mechanical Engineers, Part B: Journal of Engineering Manufacture **238**(4), 500–513 (2024)
12. Yin, Q., Yu, Y., Gong, P.: PCR temperature control system based on particle swarm optimization fuzzy PID. Journal of Changchun University of Science and Technology (Natural Science Edition) **46**(06), 89–96 (2023)
13. Oruc, R., Baklacioglu, T.: Cruise range modeling of flight strategies for transport aircraft using genetic algorithms and particle swarm optimization. Energy **294**, 130917 (2024)
14. Li, R., Wang, W., Li, X.: Control system design of variable distance seedling picking mechanism based on PSO-BP fuzzy PID[J/OL]. J. Agric. Mech. Res., 1–10 (2024)
15. Jiang, K.: Enhancing water level system based on Fuzzy PID control. J. Phys.: Conf. Ser. **2786**(1) (2024)
16. Li, H., Li, W., Hu, C., Wang, R., Luo, K.: Research on fuzzy PID signal compensation fault-tolerant control of electro-hydraulic servo system. Gas Turbine Technology **34**(02), 15–22 (2021)

Establishment of an Automatized Groundwater Monitoring Network for Assessing the Impacts of Human Activity and Meteorological Factors on Municipal Environment in the Great Hungarian Plain

Tamás Mester[1](✉), György Szabó[1], Emőke Kiss[1], and Dániel Balla[2]

[1] Department of Landscape Protection and Environmental Geography, Faculty of Science and Technology, University of Debrecen, Egyetem Tér 1, 4032 Debrecen, Hungary
{mester.tamas,szabo.gyorgy,kiss.emoke}@science.unideb.hu
[2] Department of Data Science and Visualization, Faculty of Informatics, University of Debrecen, Kassai Út 26, 4028 Debrecen, Hungary
balla.daniel@inf.unideb.hu

Abstract. Environmental data has considerable value. However, traditional sampling often lacks sufficient data for comprehensive analysis. The present study aims to demonstrate how automated monitoring devices can supplement traditional methods with high temporal resolution data. Since 2011, we have been conducting groundwater investigations in a settlement located in the Hungarian Great Plain. However, measurements for this period are only available at monthly, seasonal, or annual resolutions. Therefore, in addition to the installation of an automatized meteorological station Garni 3015 Arcus, we installed three Hungarian-developed DA-SLKTRB 122 continuous groundwater level, temperature, and electrical conductivity (EC) measuring instruments in the settlement in 2023. In the present study, data series (hourly measurements, in total: 7296) from December 2023 to October 2024 are evaluated. During this period, the precipitation amount was 408 mm. The results indicate significant fluctuation in groundwater levels (1.8 m), which correlate with the lower precipitation levels in the summer months. The EC did not show significant changes following precipitation events. The lowest EC values (1810–2200 μS/cm) were measured in well 2, while the highest values (6150–6300 μS/cm) were recorded in well 3, respectively. The water temperature in the wells reached its minimum at the end of January, with the lowest value recorded in Well 3 (9.7 °C), which had the lowest water level. The highest values were measured in September, with the maximum temperature (17.9 °C) recorded in well 2 with the highest water level. It can be stated that high temporal resolution measurements improve the assessment of water resources and complement conventional methods.

Keywords: Groundwater Monitoring · Groundwater Level · High-resolution Data · Automatized Measurements

C. Weng (Ed.): WRE 2024, LNCE 634, pp. 136–146, 2025.
https://doi.org/10.1007/978-981-96-5095-8_12

1 Introduction

Groundwater resources globally are under significant pressure due to human activities such as agriculture, industrial practices, and urban expansion [1, 2]. These activities often lead to water resource depletion and water quality degradation due to pollutants such as agricultural runoff and urban pollutants [3]. Shallow groundwater is particularly vulnerable to contamination, which can be significant around urban areas. One of the most significant challenges is the inadequate collection and treatment of urban wastewater, which is a challenge not only in the less developed countries of the world but also on a global scale [4–6].

Environmental data has considerable value, although its collection is often costly, and traditional monitoring methods may not provide sufficient data for comprehensive analysis [7]. Moreover, conventional groundwater monitoring typically relies on periodic (e.g. seasonal, annual) manual sampling and laboratory analysis, which can delay the detection of quality issues, allowing harmful contaminants to remain undetected [8]. Therefore, establishing an automatized groundwater monitoring infrastructure is crucial for understanding the effects of human activity and climate-induced variations in precipitation changes on water resources, particularly in municipal environments.

Automated monitoring increases data accuracy and lowers operational costs by reducing the need for frequent manual sampling. This efficiency is particularly valuable for villages with limited resources, as it supports timely interventions to protect water supplies. Furthermore, data from automated systems can be integrated with geospatial and predictive models, improving the ability to forecast changes in contamination and manage groundwater resources effectively over the long term [7].

Continuous monitoring systems are increasingly used to improve response times and reliability [9]. These systems allow frequent, real-time measurements, enabling responsible executives to detect changes in water quality more readily than traditional methods. The increased frequency and immediacy of these measurements make real-time monitoring valuable for protecting groundwater quality and ensuring timely corrective action to prevent adverse impacts.

Many countries, including Hungary, do not have detailed monitoring data at the municipal level. Hence, data collection at the local level is one of the major priority tasks in recent years [10]. Automated data collection devices can contribute to this effort effectively. The aim of the present study is to demonstrate how traditional monitoring can be supplemented with new automated monitoring devices. Since 2011, we have been conducting groundwater investigations in a settlement in the Hungarian Great Plain to uncover contamination resulting from human activities and assess the effects of observed changes in precipitation distribution in recent years. However, measurements for this period are only available at monthly, seasonal, or annual resolutions.

Efforts in the future will focus on expanding the network with additional water quality parameters to provide a more comprehensive understanding of the management and protection of groundwater resources in similar municipal areas.

2 Materials and Methods

The methodology adopted is presented in the flowchart of Fig. 1, while its description is presented in the following subsections.

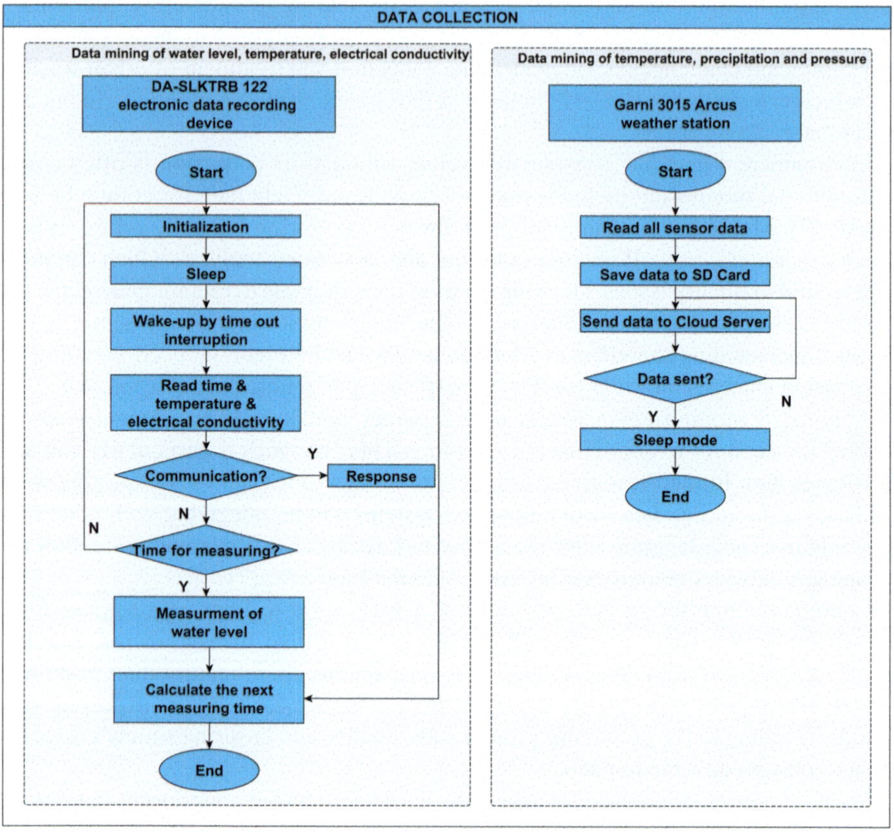

Fig. 1. The methodology of data mining.

2.1 Study Area

The municipality we examined since 2011, Báránd, is in the Hungarian Great Plain (Fig. 2). The settlement is a typical medium-sized village in Hungary with a population of 2426 [11]. The area is classified as a lowland with an altitude of 85–89 m above sea level and a relative altitude difference of 0–3 m/km^2, [12]. In recent years, 75% of the area of the sub-region has been subject to intensive agricultural activity, mainly wheat, maize and sunflower. The lower-lying areas affected by water influence are typically pastures and meadows. Due to the low altitude compared to the surrounding areas, the water table is close to the surface, at a depth of 1–2 m, and all soil types are therefore affected by water [13]. Among the soils under direct water influence, saline soils are extensive, covering 36% of the total area.

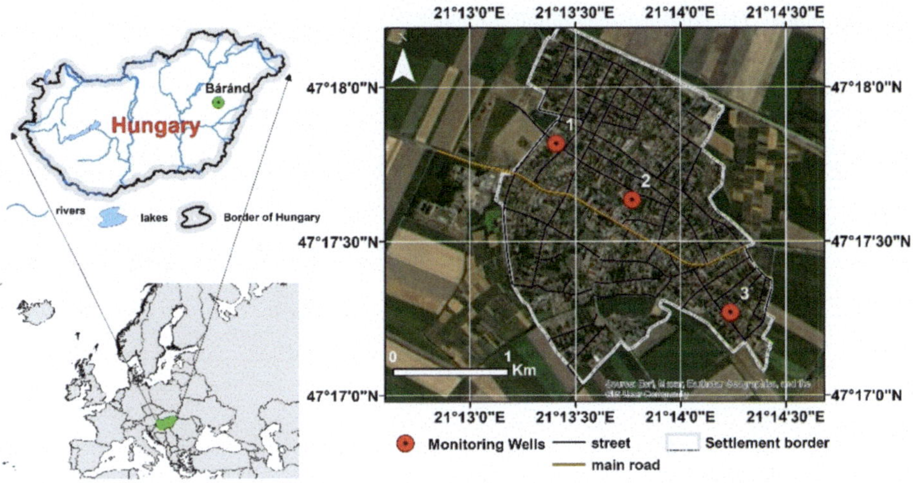

Fig. 2. Location of the study area and the monitoring wells equipped with automatized measurement devices.

For decades, municipal wastewater was stored in septic tanks in the backyards of households. Still, in line with EU efforts, a sewerage network was built in 2014 so that more than 80% of households now transport wastewater by pipeline to a treatment plant in a neighbouring town. Wastewater effluent from the septic tanks has caused severe contamination of the groundwater resources of the municipality, with significantly elevated concentrations of ammonium, nitrate and phosphate, as shown by our previous studies [14, 15]. Following the cessation of wastewater leakage, groundwater levels in the municipality have started to fall significantly, which, combined with drier summers in recent years, has already resulted in a decline of 2–4 m compared to pre-sewerage levels [16].

2.2 Establishment of an Automatized Monitoring Network

Since 2011, we have been conducting groundwater investigations in a settlement located in the Hungarian Great Plain, however, measurements for this period are only available at monthly, seasonal, or annual resolutions. Therefore, in addition to the installation of an automatized meteorological station Garni 3015 Arcus, continuous data collection devices (DA-SLKTRB 122) for water level, temperature and electrical conductivity were installed in 3 dug groundwater wells in 2023 (Fig. 2). The present study evaluated data series (hourly measurements, in total: 7296) from December 2023 to October 2024. The accuracy of the device is 0.1% for water level, 0.1 °C for temperature and 1% for specific electrical conductivity, expressed in terms of specific electrical conductivity at 25 °C.

Key information on the monitoring wells is given in Table 1. The wells are used for different purposes. The water from Well 1 is used for livestock watering during the summer, with water extraction done manually. Water from Well 2 is pumped out for irrigation, while the water from Well 3 is not utilized.

Table 1. Description of the monitoring wells and the nearby pollution sources.

Well ID	Type	Depth (m)	GW depth (m)	Use	Nearby risk factors within 100 m
1	Dug well	5.5	2–3	Manual water extraction for livestock watering	Septic tanks, agriculture, pit latrine
2	Dug well	6	2–3	Pumping for irrigation	Septic tanks
3	Dug well	6.5	4–6	Not used	Septic tanks, agriculture, pit latrine

3 Results and Discussion

3.1 Meteorological Data

Data collected every 10 min by the weather station is uploaded to an online database, which is available online (Weather Underground). The database can be displayed in graphical or tabular form daily, weekly, or monthly. For illustration purposes, we have selected a period from 01.04.2024 to 30.04.2024 (Fig. 3).

From December 2023 to October 2024, the recorded precipitation was 408 mm. This precipitation mainly fell in larger quantities during the winter months. In spring, rain events were less frequent but occurred with higher intensity. The driest period was the summer, especially August, when only 0.5 mm of rainfall. The lowest monthly average temperature (1.2 °C) was recorded in January, while the highest monthly average temperature occurred in August (26.5 °C). The low rainfall during the summer months and high temperatures led to a significant decrease in water levels in the monitoring wells.

3.2 Data Display

The water level, temperature, and EC data collected by the instrument can also be displayed separately or aggregated in a diagram in its software. Figure 4 shows all measured parameters for Well 3 during the studied period.

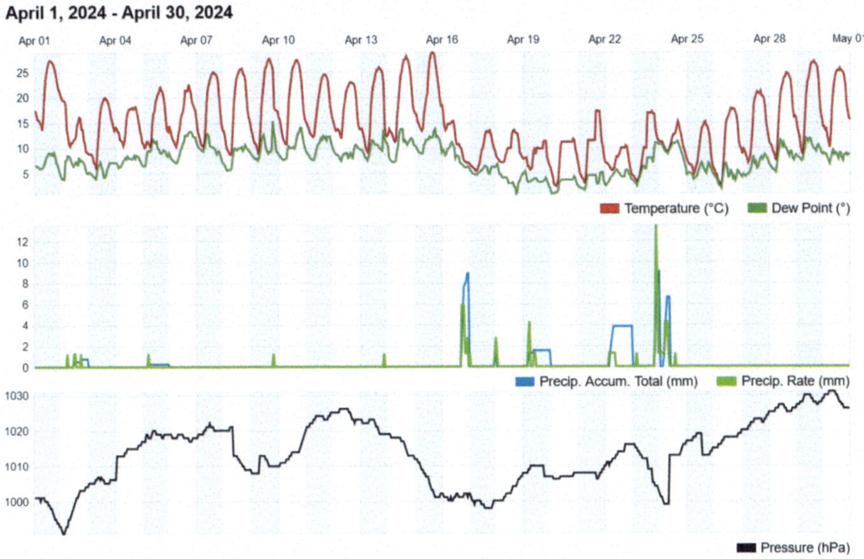

Fig. 3. Temperature, Precipitation and Pressure in the study area for a representative period from 01.04.2024 to 30.04.2024 depicted in the online database of Weather Underground.

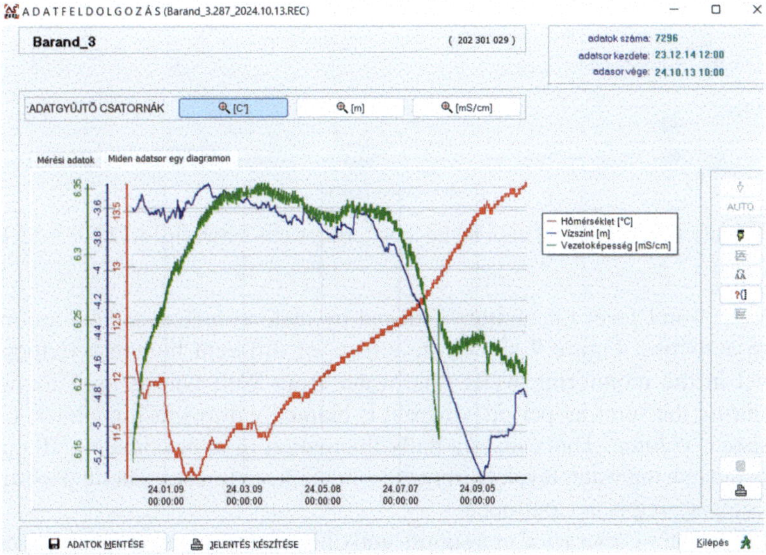

Fig. 4. Groundwater level fluctuations in Well 3 are displayed in the device software.

3.3 Water Level Fluctuations

The results show significant fluctuations in the groundwater level, reaching up to 1.8 m in the wells between winter and summer periods (Fig. 5). The highest water level was recorded in well One located in the northern part of the municipality, with a fluctuation between 1.5–3.3 m, while the lowest values were recorded in the southern part of the municipality (well 3), with a fluctuation between 3.5–5.4 m. The values are consistent with the previous finding that the local groundwater flow direction is N-S [16]. Water levels in wells respond to major rainfall events by rising for several days (Fig. 6).

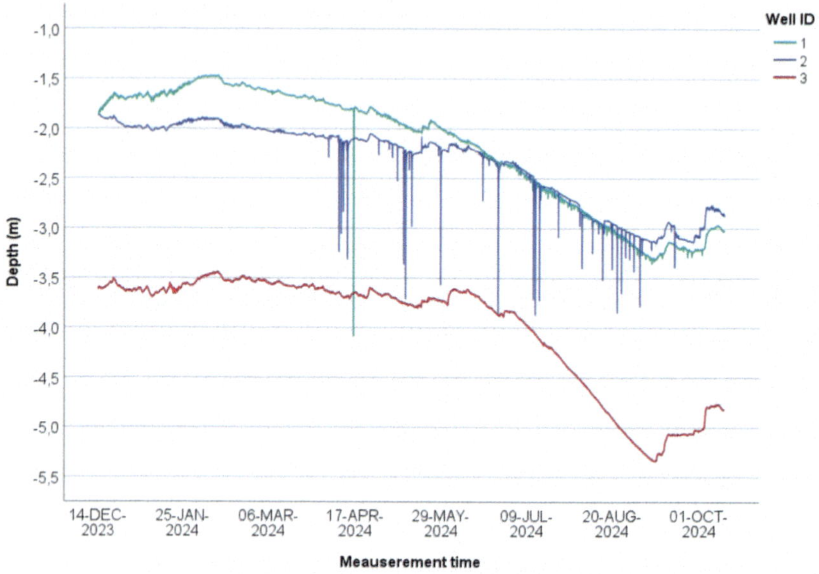

Fig. 5. Groundwater level fluctuation in the monitoring wells between 14.12.2023–13.10.2024.

However, water level fluctuations induced by meteorological factors are modified by human activities. Figure 7 shows the effects of different human activities on the water level in the monitoring wells. The water from Well One is used for watering animals during the summer period, where it is manually drawn using a bucket, always in a consistent volume. This causes a daily fluctuation of approximately 10 cm in the constantly decreasing water level. During the winter, the animals are herded to a different location, so watering is discontinued.

In Well 2, water is extracted at a significantly higher rate using mechanical pumping to irrigate the household's garden, resulting in much higher fluctuation in water levels (Fig. 7). During pumping, the water level typically drops by 0.8–1.2 m, which usually returns to normal within 4–6 h.

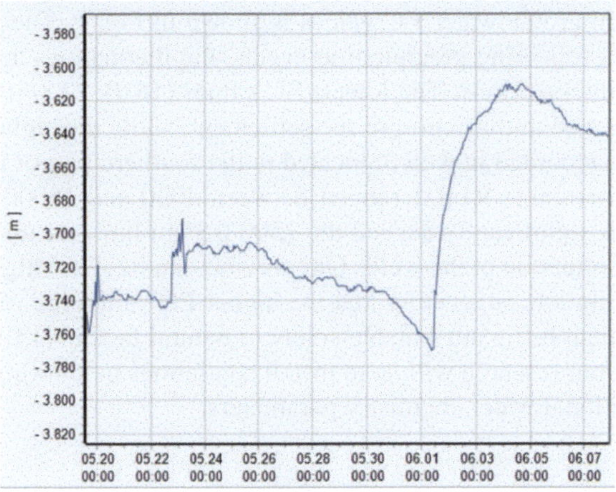

Fig. 6. Rising groundwater level due to precipitation in well 3 between 01.06.2024–05.06.2024.

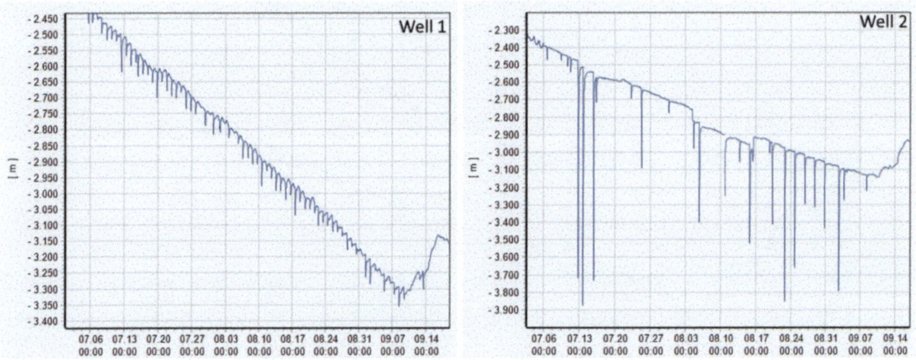

Fig. 7. Water levels affected by manual livestock watering (Well 1) and pumping for irrigation (Well 2) from 06 July to 14 September 2024.

3.4 Groundwater Temperature and Electrical Conductivity Fluctuations

The groundwater temperature in the wells reached its minimum at the end of January and the beginning of February, which can be attributed to an unusually warm February (Fig. 8). The lowest value was recorded in Well 3 (9.7 °C), which had the lowest water level. Following the winter period, a continuous increase has been observed, in parallel with the rise of external temperatures. The highest values were measured in September and October, with the maximum temperature (17.9 °C) recorded in well 2 with the highest water level. This trend persists even in October's data despite a significant decrease in air temperature. Therefore, longer datasets will be needed to determine any potential temporal lags.

The variation in groundwater EC values is shown in Fig. 9. The EC did not show significant changes following precipitation events, the fluctuations appeared primarily following human interventions. The lowest EC values (1810–2200 μS/cm) were measured in well 2, located in the centre of the settlement, while the highest values (6150–6300 μS/cm) were recorded in well 3, located in the southern part of municipality. The values in the northernmost Well 1 ranged between 4000 and 45000 μS/cm. The EC values in two wells significantly exceed the 2500 μS/cm limit for groundwater, which may indicate contamination of the wells. Our previous studies have highlighted the issue of groundwater pollution, suggesting that the higher EC values are related to this contamination rather than being attributable solely to natural factors [15, 16]. Therefore, a future objective of our research will be to install continuous monitoring devices capable of registering additional water chemistry parameters.

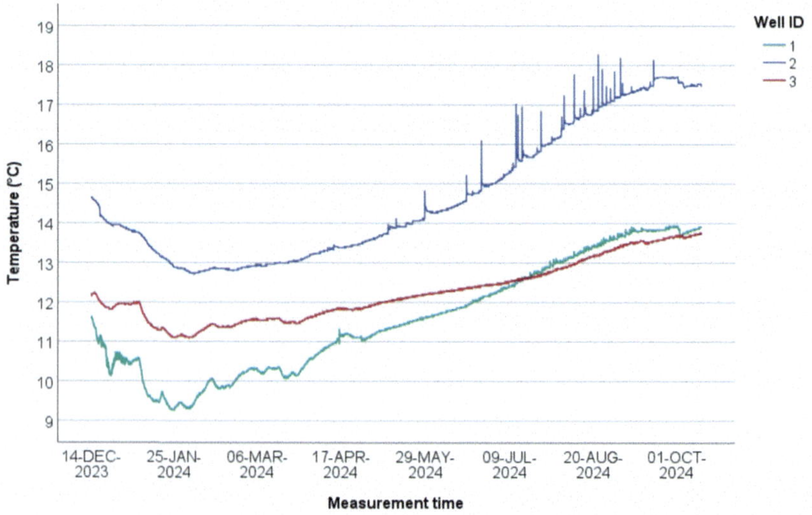

Fig. 8. Groundwater temperature fluctuation in the monitoring wells between 14.12.2023–13.10.2024.

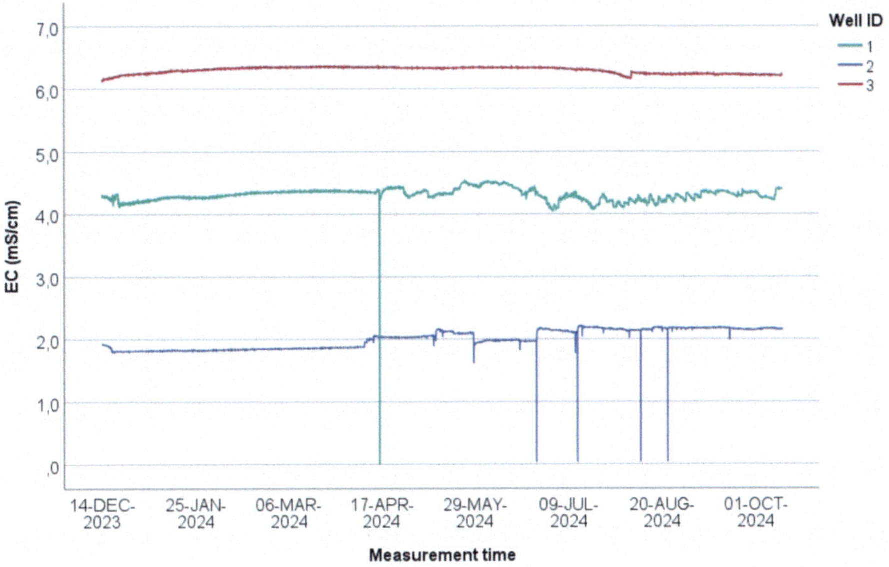

Fig. 9. Fluctuation of electrical conductivity in the monitoring wells between 14.12.2023–13.10.2024.

4 Conclusion

Based on the results, implementing an automated groundwater monitoring network in Báránd, Hungary, has provided valuable insights into the interactions between human activities, meteorological factors, and groundwater resources. The high temporal resolution data collected from December 2023 to October 2024 allowed for a detailed analysis of groundwater level fluctuations, temperature, and electrical conductivity. These measurements revealed significant seasonal and human-induced variations, underscoring the sensitivity of shallow groundwater to both natural precipitation patterns and anthropogenic pressures, such as livestock watering and irrigation.

Electrical conductivity values exceeding the contamination limit indicate contamination issues in specific wells, aligning with our earlier studies, which identified significant pollution mainly from septic tank effluent.

Automated measurement techniques provide high-resolution data that enables a more detailed understanding of groundwater dynamics, offering insights not achievable through traditional sampling methods. This continuous data collection allows for a more precise analysis of fluctuations and contaminant patterns, supporting improved resource management. Therefore, future efforts will focus on expanding this network to include additional water quality parameters, providing a more comprehensive understanding of managing and protecting groundwater resources in similar municipal environments.

Acknowledgements. Supported by the University of Debrecen Scientific Research Bridging Fund (DETKA). Supported by the Mecenatúra MEC_R_24-148879 funding scheme provided by the National Research, Development and Innovation Fund of Hungary.

References

1. Pham, T.L., Tran, T.H.Y., Tran, T.T., Ngo, X.Q., Nguyen, X.D.: Assessment of surface water quality in a drinking water supply reservoir in Vietnam: A combination of different indicators. Rendiconti Lincei. Scienze Fisiche e Naturali **33**(3), 653–662 (2022)
2. García-Torres, E., Pérez-Morales, R., González-Zamora, A., Calleros-Rincón, E.Y.: Subclinical hypothyroidism in families due to chronic consumption of nitrate-contaminated water in rural areas with intensive livestock and agricultural practices in Durango. Mexico. Water **14**(3), 282 (2022)
3. Zeydalinejad, N., Javadi, A.A., Webber, J.L.: Global perspectives on groundwater infiltration to sewer networks: A threat to urban sustainability. Water Res. **262**, 122098 (2024)
4. Famous, O.: Water caused diseases prevalence resulting from Septic Contamination of Hand-Dug Wells in Ughelli, Delta State, Nigeria. Lapai International Journal of Management and Social Sciences **16**(1), 1–17 (2024)
5. Singh, S., Raju, N. J., Mehmood, G., Gupta, S. K., Ahmed, S.: A review of the current scenario and best possible solution for fecal sludge management (FSM) in India. Groundw. Sustain. Dev. 101346 (2024)
6. Kim, M., Koh, E.H., Kim, J.: Field study and numerical modeling to assess the impact of on-site septic systems on groundwater quality of Jeju Island. South Korea. Hydrology **11**(9), 146 (2024)
7. Velasquez-Orta, S.B., Werner, D., Varia, J.C., Mgana, S.: Microbial fuel cells for inexpensive continuous in-situ monitoring of groundwater quality. Water Res. **117**, 9–17 (2017)
8. Lyons, K.J., et al.: Monitoring groundwater quality with real-time data, stable water isotopes, and microbial community analysis: A comparison with conventional methods. Sci. Total Environ. **864**, 161199 (2023)
9. Drage, J., Kennedy, G.: Building a low-cost, internet-of-things, real-time groundwater level monitoring network. Groundwater Monitoring & Remediation **40**(4), 67–73 (2020)
10. Stevanović, Z., Stevanović, A.M.: Monitoring as the key factor for sustainable use and protection of groundwater in Karst environments—an overview. Sustainability **13**(10), 5468 (2021)
11. Hungarican Central Statistical Office.: Key data on settlements according to the census in 2022. https://nepszamlalas2022.ksh.hu/. Last accessed 2024/10/26
12. Dövényi, Z., Ambrózy, P., Juhász, Á., Marosi, S., Mezősi, G., Michalkó, G., Tiner, T.: Magyarország kistájainak katasztere. Inventory of microregions in Hungary. OTKA Kutatási Jelentésekl OTKA Research Reports (2008)
13. Michéli, E., Fuchs, M., Hegymegi, P., Stefanovits, P.: Classification of the major soils of Hungary and their correlation with the World Reference Base for Soil Resources (WRB). Agrokém. Talajt. **55**(1), 19–28 (2006)
14. Mester, T., Szabó, G., Kiss, E., Balla, D.: Long-term spatiotemporal changes in nitrate contamination of municipal groundwater resources after sewerage network construction in the Hungarian Great Plain. Environ. Sci. Pollut. Res., 1–24 (2024)
15. Mester, T., Szabó, G., Sajtos, Z., Baranyai, E., Kiss, E., Balla, D.: Assessment of groundwater decontamination processes around a dismantled septic tank using GIS and statistical analysis. Water **15**(5), 884 (2023)
16. Mester, T., Szabó, G., Balla, D.: Assessment of shallow groundwater purification processes after the construction of a municipal sewerage network. Water **13**(14), 1946 (2021)

Hydrology and Hydrological Engineering

Research on Low-Temperature Outflow Mitigation Strategy for Stable Stratification of Reservoirs for Irrigation Water Temperature Requirements

Ganggang Bai[1], Lai Jiang[2], Wu Zhou[2], Youcai Tuo[1](✉), Jiayue Shi[2], Yun Deng[1], and Mingqiu Nie[3]

[1] State Key Laboratory of Hydraulics and Mountain River Engineering, Sichuan University, Chengdu 610065, China
tuoyoucai@scu.edu.cn

[2] Powerchina Huadong Engineering Corporation Limited, Hangzhou 311122, China

[3] China Meteorological Administration Hydrometeorology Key Laboratory, Hohai University, Nanjing 210024, China

Abstract. The low-temperature outflow from large hydropower projects can have adverse effects on downstream ecosystems and on crops requiring irrigation. This study, which integrates prototype observations with numerical models, examined the influence of different water withdrawal schemes on the thermal structure of reservoir water and proposed measures to mitigate the impact of low-temperature outflows. The results indicate that from May to October, the thermocline under the single-level withdrawal scheme (S1) descended by an average of 11.7 m more than that under the stratified withdrawal scheme (S2). Moreover, from May to August, the outflow temperature under S2 was consistently higher than under S1, by an average of about 3.5 °C. Specifically, during the irrigation period, the average water temperatures for S1 and S2 were 15.6 °C and 18.5 °C, respectively, resulting in a difference of 2.9 °C. Consequently, during the irrigation season, the irrigation water temperature under the S1 scheme failed to meet the minimum temperature requirements for 35.4% of the total irrigation days, whereas the S2 scheme consistently met these requirements. For the S1 scheme, measures involving the substitution of low-temperature outflow with warm water from natural rivers were proposed, with an extraction flow rate ranging from 0.005 to 0.185 m^3/s. This research provides technical support and guidance for management authorities to mitigate the impact of releasing cold water.

Keywords: Thermally Stratified · Outflow Low-temperature · Numerical Model · Irrigation · Mitigation Strategy

1 Introduction

The construction and operation of reservoirs play a crucial role in regional flood control, drainage, and irrigation water supply. However, they also result in significant alterations to the natural conditions of the water body. Particularly, the environmental issues caused

C. Weng (Ed.): WRE 2024, LNCE 634, pp. 149–161, 2025.
https://doi.org/10.1007/978-981-96-5095-8_13

by the release of cold water have attracted extensive attention from researchers [1–3]. The impoundment of reservoirs weakens the vertical interaction of the water body, altering the spatiotemporal distribution of water temperature. Vertical stratification of water temperature occurs within the reservoir. The temperature difference between the released water and natural water is significant due to different withdrawal locations [4, 5]. This not only affects the stability and health of downstream river ecosystems but also has adverse effects on the growth and development of crops in downstream irrigation areas [6–9].

To address the adverse effects of Low-temperature outflow from reservoirs on downstream areas, researchers have proposed a series of improvement measures. These include the ecological operation of reservoirs, stratified intake devices, and thermocline breaking devices. Among them, stratified intake facilities, which adjust the outflow temperature by changing the intake location, are considered an important engineering measure to mitigate the adverse effects of water temperature. This is particularly relevant for large, deep reservoirs where stratified intake is widely adopted to regulate the intake elevation and outflow temperature [10, 11]. For example, the United States Bureau of Reclamation has addressed the temperature requirements of salmon downstream of the Sacramento River by constructing multi-opening intake towers at Shasta Power. Additionally, temperature control curtains have been installed to regulate the water temperature at the fish spawning grounds downstream of the Lewiston Station on the Trinity River. Furthermore, there are other measures that can mitigate the adverse effects of Low-temperature outflow, such as extracting warmer water for replacement.

For some existing reservoirs with bottom intake structures, the outflow temperature fails to meet the ecological requirements downstream, making it urgent to evaluate the effectiveness of various measures. Currently, numerical models have been widely used in water temperature prediction and evaluation studies [12–15]. Among them, the CE-QUAL-W2 model, a two-dimensional hydrodynamic and water quality model, is proven to efficiently reproduce the thermal regimes of water bodies, especially in reservoirs, due to its advantages of user-friendly software interface, easy and flexible operation, and open-source, et al. [16–18]. Rangel-Peraza et al. utilized the CE-QUAL-W2 model to analyze the dynamic temperature changes in the Aguamilpa reservoir and observed significant thermal stratification within the reservoir, particularly in the bottom layer [16]. Wang et al. (2024a) conducted a study using the CE-QUAL-W2 model to explore the response characteristics of the thermal state of the Xiluodu reservoir in Southwest China to climate change which indicated a significant future increase in surface water temperature of the reservoir by 0.13 °C/10a to 0.34 °C/10a [17]. He et al. (2020) combined the CE-QUAL-W2 model with the HEC-RAS model to jointly analyze the cumulative effects of cascade reservoir operations in the upper reaches of the Yangtze River on downstream water temperature, which demonstrated that the cumulative effects of dams on water temperature vary by season [18].

In summary, this study selects the LYT reservoir in Southwest China as the research subject. It is a comprehensive large-scale water conservancy project mainly serving water supply and irrigation, playing a crucial role in the regional aquatic ecological environment and the water supply for residents. It is worth noting that, although the original feasibility study, environmental impact assessment, and preliminary design of the LYT reservoir proposed a double-level intake structure for the diversion tunnel, the actual construction adopted a single-level intake. Whether this change will affect the temperature structure in front of the dam and the water temperature in the downstream irrigation area, and what measures can be taken to mitigate the impact of this change, are questions worthy of our investigation.

2 Materials and Methods

2.1 Study Area and Data

The LYT reservoir (31.38° N, 108.38° E) is in the southwestern region of China, situated in the upper reaches of the Taoxi River, a second-level tributary of the Yangtze River system. The dam has a height of 103.8 m, with the annual average flow of 5.6 m^3/s. The total storage capacity of the reservoir is 10,240 million m^3, with a normal water level of 450.0 m, corresponding to a reservoir area of 3.139 km^2 and a capacity of 9,826 million m^3. Water is conveyed from the Bahou power station to the Dieshui power station via the main canal and is subsequently distributed to the irrigation area through the branch canals, as shown in Fig. 1.

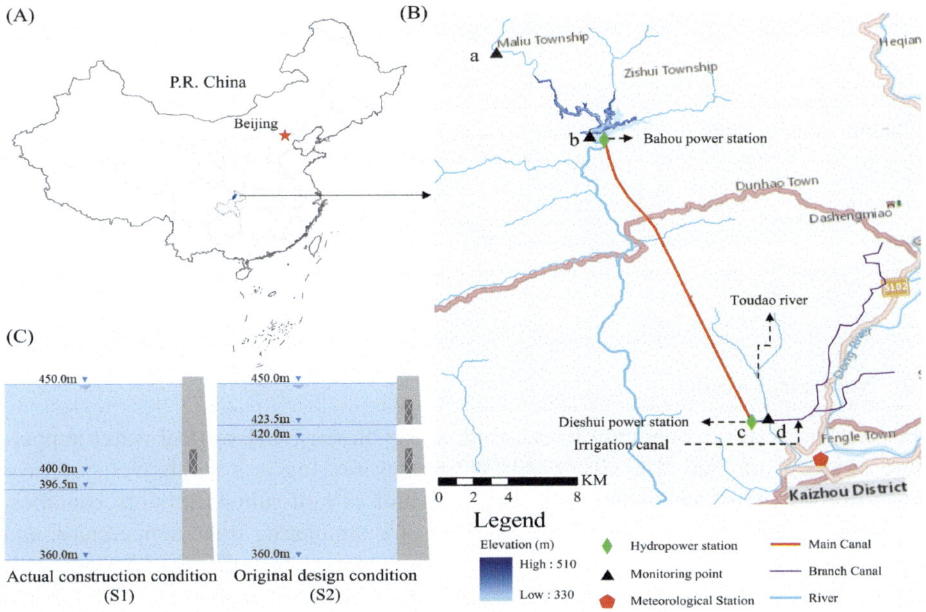

Fig. 1. Map showing the locations of the LYT reservoir (A), region description (B) and the stratified intake device (C).

For model validation, continuous water temperature observations were conducted at points a-d from November 2019 to November 2020. The operation was based on a normal water year design, and meteorological conditions were based on the multi-year

(1991–2020) monthly averages from the National Basic Monitoring Station - Kaixian Station. The inflow water temperature was inferred from the observed water temperature and meteorological relationships. The water temperature structure in the reservoir under the single-level (S1) and multi-level (S2) intake schemes was simulated. Specifically, the multi-level intake scheme used the upper intake (intake elevation: 420–423.5 m) from April to October and the lower intake (intake elevation: 396.5–400 m) for the rest of the period.

The main crop in the irrigation area is rice, and its temperature requirements at each growth stage are shown in Table 1. If the irrigation water temperature does not meet the minimum growth temperature for rice, the rice will suffer from cold water damage. Considering that the overall flow rate for agricultural irrigation is relatively small, the water can undergo sufficient heat exchange with the atmosphere after entering the paddy fields, leading to a significant temperature increase and quickly reaching a water-air thermal equilibrium state. Therefore, this study mainly verifies whether the channel water temperature meets the minimum irrigation water temperature requirements.

Table 1. Minimum Water Temperature Requirements for Rice Growth Stages in the Irrigation Area.

Growth Stage	Growth Period	Minimum Temperature Requirement (°C)
Seedling	From March 15 to April 20	12
Tillering	From May 23 to May 31	12
Jointing	From June 1 to July 10	15
Heading	From July 11 to July 21	20
Flowering	From July 22 to July 31	20
Filling	From August 1 to August 20	15

2.2 Numerical Model

For reservoir water temperature forecasting, a two-dimensional vertical water temperature model is adopted. The CE-QUAL-W2 model, serving as a width-averaged two-dimensional hydrodynamic and water quality model, is well-suited for rivers, estuaries, lakes, and reservoirs. It is particularly effective for simulating flow, temperature, and water quality in relatively narrow and deep-water bodies. Given that the LYT reservoir is a deep-water type with a maximum depth at the dam site of approximately 90 m and has a relatively narrow and elongated shape, the CE-QUAL-W2 model is employed to simulate the water temperature of the LYT reservoir. This model adheres to the principles of mass conservation, momentum conservation, and energy conservation, satisfying the Boussinesq assumption. The governing equations include the continuity equation, momentum equation, state equation, free surface equation, and heat transport equation. For more detailed information about the model, please refer to the relevant literature [19].

For river water temperature forecasting, a longitudinal one-dimensional water temperature model is adopted. The temperature variation downstream of the dam along the river and channel is predominantly longitudinal, which can be simulated by the HEC-RAS model—a one-dimensional hydrodynamic model [20]. The model calculates heat transport using a heat budget and can simulate the water temperature in the river [21].

3 Results and Analysis

3.1 Model Calibration and Validation

The two-dimensional vertical temperature model was validated using observed data of inflow and outflow temperatures in the LYT reservoir from November 2019, to December 2020. Figure 2 compares the measured and simulated water temperatures in the vertical section of the reservoir in front of the dam on April 11, May 23, August 11, and December 2, 2020, respectively. The results show that: In April, which is the warming period, the water temperature within 20 m of the surface layer of the reservoir began to stratify. From May to August, due to rising temperatures and water intake, there is an obvious stratification structure in the reservoir. In December, the temperature dropped, the surface layer of the thermocline disappeared, and the reservoir transitioned from a single thermocline structure to a vertical isothermal structure. From April to December, the measured surface water temperature varied between 17.5 °C and 31.9 °C, while the bottom water temperature remained between 12.1 °C and 13.1 °C. The simulated and measured results of the surface and bottom water temperatures and their variability trends are consistent. In 2020, the average absolute relative error (MARE) for the measured and simulated water temperatures on April 11, May 23, August 11, and December 2 were 1.4%, 2.4%, 2.9%, and 3.1%, respectively. The Nash-Sutcliffe efficiency coefficient (NSE) values 0.95, 0.97, 0.99, and 0.82, respectively, indicating good overall agreement and high model credibility.

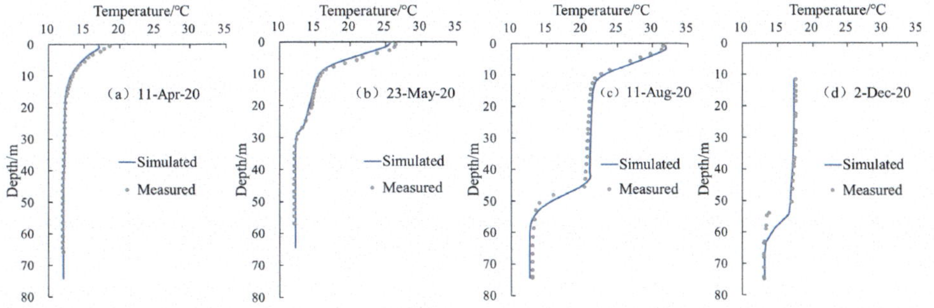

Fig. 2. Comparison of calculated and measured values of the vertical water temperature in different sections of reservoir.

Figure 3 compares the simulated and measured water temperature processes at the water supply intake during the observation period in the LYT reservoir. The comparison results show that the simulated and measured water temperature change trends are consistent, and the two water temperature change rates are in good agreement. The MARE is 2.0%, and the NSE is 0.98.

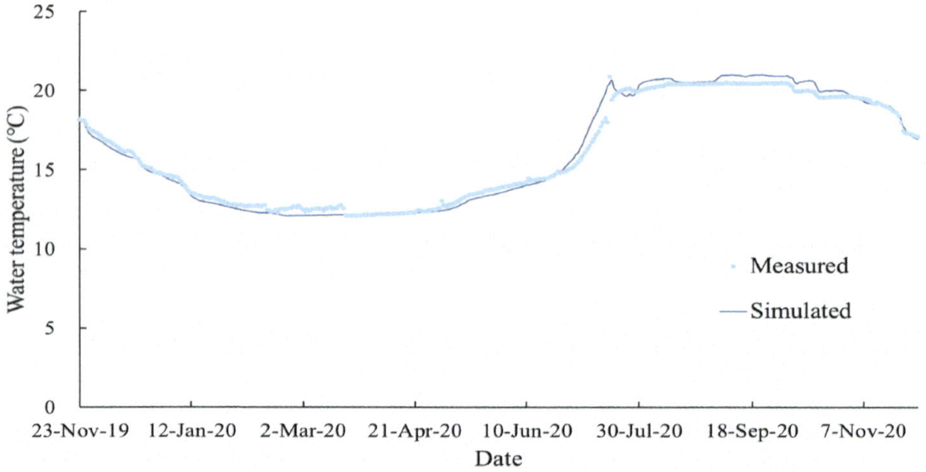

Fig. 3. Comparison between simulated and measured discharged water temperature

Overall, the established mathematical model effectively simulates the vertical water temperature structure and the outflow temperature process in the LYT Reservoir area. This verifies the model's applicability to LYT Reservoir and confirms its suitability for simulating water temperature distribution and the outflow temperature process in the reservoir.

3.2 Effect of Various Water Intake Schemes on Thermal Stratification

Figure 4 illustrates the temporal variations of water temperature (WT) in front of the dam under schemes S1 and S2. The differences in surface WT (0–1 m) are relatively small between the two conditions, ranging from 12 °C to 32 °C. For the bottom-layer WT, the values are similar, with averages of 12.3 °C and 13.0 °C for S1 and S2, respectively, differing by only 0.7 °C. In both scenarios, stratification begins to appear in the reservoir water temperature in March. From April to September, as the air temperature rises and solar radiation intensifies, the surface water temperature gradually increases. The maximum vertical temperature differences are 19.5 °C and 19.0 °C, respectively, occurring in early August. The increase in air temperature, inflow temperature, and solar radiation causes the surface temperature of the reservoir to rise. Meanwhile, the increased inflow volume and noticeable changes in water level significantly disturb the vertical water temperature in the reservoir. Additionally, the sustained stratified flow in S2 inhibits the transfer of heat to the middle and lower layers of the reservoir, leading to a significant difference in thermocline thickness compared to S1. The S2 thermocline is approximately 11.7 m thinner than S1 from May to October. In April and May, due to weaker stratification, the change in water temperature structure is not significant when altering the water intake location. From June to August, both the air temperature and inflow water temperature further increase, exacerbating the stratification phenomenon. The surface temperature reaches its maximum for the year, with the average surface temperatures differing by approximately 1.9 °C between the two measures (S1: 29.8 °C and S2: 31.7 °C). However, due to the upward movement of the thermocline, the average temperature in front of the dam in S2 is always lower than that in the S1 scenario. From September to October, the reservoir undergoes autumnal turnover, with surface temperatures decreasing by 8.12 °C (S1) and 8.8 °C (S2) due to the declining air temperatures. Subsequently, starting from December, the reservoir gradually develops a stable vertical mixing thermal structure.

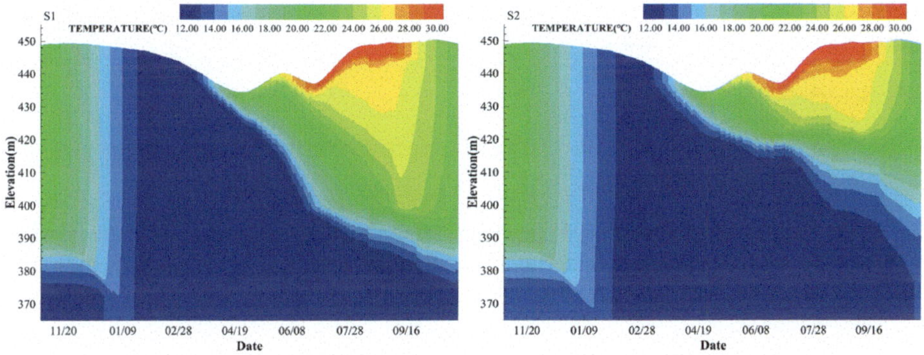

Fig. 4. Comparison of annual process of water temperature for two scenarios.

3.3 Impact Analysis of Downstream Water Temperature on Irrigation Period in the Irrigation Area

Due to the channel warming effect, the water temperature at each section shows an increasing trend along the channel. Therefore, the analysis primarily focuses on whether the water temperature at the head section of the irrigation channel meets the minimum temperature requirements, as depicted in Fig. 5.

For the S1 water intake scheme, the periods when the channel water temperature fails to meet the minimum temperature requirements for rice growth are from June 1 to June 24 and from July 11 to July 31, totaling 45 days and accounting for 35.4% of the total irrigation days. Among these, the temperature was lowest on July 11, being 2.5 °C below the minimum growth temperature requirement for rice (20 °C). For the S2 water intake scheme, the water temperature across all sections of the channel meets the minimum temperature requirements for rice growth. Compared to the S1 scheme, the average water temperature at the head of the irrigation channel is 3.5 °C higher from May to August. Specifically, during the irrigation period, the average water temperatures for S1 and S2 are 15.6 °C and 18.5 °C, respectively, with a difference of 2.9 °C. Since the LYT Reservoir is a multi-year regulating reservoir with a clear stratification phenomenon and a generally stable structure, the water temperature at the bottom of the reservoir differs significantly from the upper layer. Therefore, the single-level intake leads to relatively lower downstream water temperatures, failing to meet the minimum temperature requirements for irrigation.

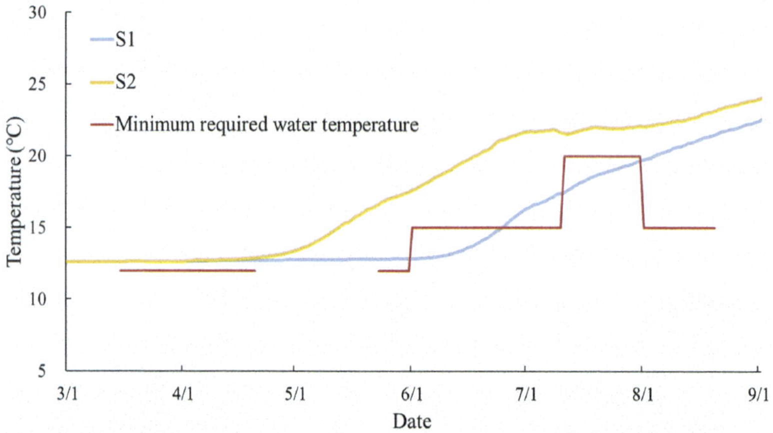

Fig. 5. Comparison of the water temperature range at the head of the channel under different water intake schemes with the minimum temperature requirements for rice growth in the irrigation area.

3.4 Evaluation of Measures to Mitigate Low-Temperature Outflow in Irrigation Channels Based on Pumping and Replacement

Considering the characteristics of the LYT project, as well as the climate and geography of the irrigation canal, measures and schemes are proposed to mitigate the impact of low-temperature outflow in the canal through pumping warm water from the Toudao (TD) River. The TD River is a natural river with average flow rates of 1.26 m^3/s and 3.95 m^3/s and average water temperatures of 24.6 °C and 25.0 °C in June and July, respectively. These temperatures are significantly higher than those at the Dieshui power station (14.7 °C and 18.5 °C). By releasing the irrigation water flow into the TD River and extracting a portion of the water from upstream of the TD River, the canal water temperature can be adjusted to meet the minimum irrigation water temperature requirements.

Based on the multi-year average daily air temperature and a design frequency of 75% for the temperature conditions, the target water temperature value for the minimum temperature requirement of rice and the water temperature process of the TD River should be considered. Combined with the predicted outflow temperature of the single-level intake scheme at the Dieshui power station, the flow process that needs to be pumped from the TD River to meet the temperature requirements during the periods when the water temperature does not meet the requirements (from June 1st to June 24 and from July 11 to July 31) is calculated, as shown in Fig. 6.

Under the multi-year average daily air temperature conditions, the flow rate required to be pumped and replaced from the TD River to meet the minimum irrigation water temperature requirement ranges from 0.005 to 0.143 m^3/s, with an average of 0.076 m^3/s. For another temperature condition, the flow rate range is 0.005 to 0.185 m^3/s, with an average of 0.090 m^3/s. It is worth noting that the maximum flow rate required for replacement under different temperature conditions is only 0.185 m^3/s, which is much smaller than the flow rate of the TD River. In conclusion, considering the 75% design frequency for the temperature, the maximum pumping design capacity of the pump should not be less than 0.185 m^3/s to meet the minimum temperature requirements for irrigation in the rice-growing area.

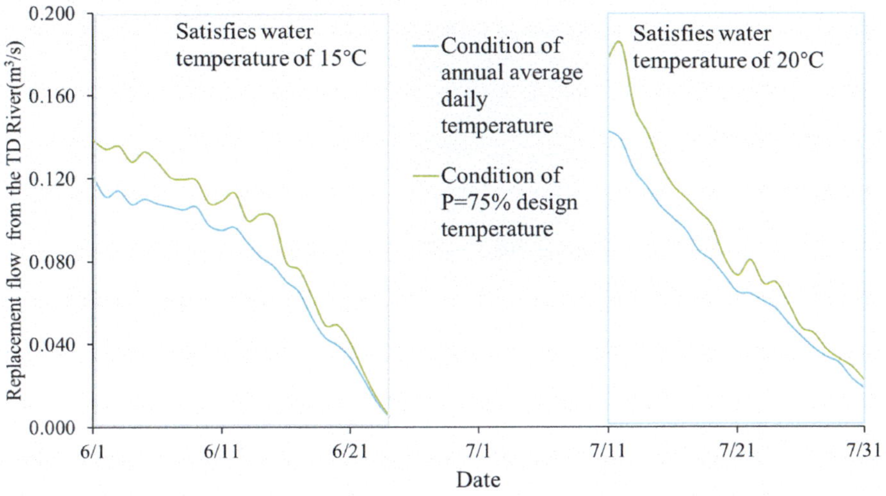

Fig. 6. Flow replacement from the TD River under different air temperature conditions.

4 Discussion

Water temperature is a critical environmental characteristic for the regional aquatic ecosystem and downstream irrigation plant growth. The thermal stratification in reservoirs leads to significant temperature impacts on both the reservoir and downstream release water, depending on the water intake schemes employed. As previously mentioned, the primary purpose of the LYT reservoir is urban water supply, supplemented by hydropower generation and irrigation. The temperature stratification in the LYT reservoir is stable, with the bottom water temperature being much lower than the surface temperature during the warming period. At the normal water storage level of 450.0 m, the water entering the water diversion pipeline of the hydropower station originates from a depth of approximately 53.5 m. In spring and summer, the water temperature for supply varies greatly from the ambient temperature due to the regional humid and hot climate characteristics. This often leads to condensation and frost formation on the exposed pipe walls of the supply pipeline, affecting the normal life of residents. Additionally, using cold water for irrigation can cause "cold damage" to crops. Therefore, it is urgent to adopt corresponding mitigation measures to improve the temperature impact of the single-level intake scheme.

Raising the intake elevation scheme can significantly increase the downstream irrigation water temperature, ensuring a 100% guarantee rate for irrigation temperature. However, considering the challenges, costs associated with its implementation, and the significant impact, this approach would not only require substantial financial, human, and material resources but also the LYT reservoir has been operational for many years with the tasks of power generation and water supply. Implementing a stratified water intake through reservoir emptying is impractical. In contrast, the strategy of water extraction and replacement from the TD River not only improves the irrigation water temperature but also has lower implementation difficulty and lower construction and operation costs. However, further evaluation and discussion are needed to determine whether the replacement will affect the water temperature of the natural water in the TD River. Here, only the worst-case scenario (P = 75% design temperature) is considered for the changes in the water temperature of the TD River before and after the implementation of the replacement measure, as shown in Fig. 7. Due to the relatively small replacement flow rate (maximum of 0.185 m^3/s) and a further decrease in the replacement flow rate with increasing water temperature, the change in water temperature before and after replacement is minimal, with an average temperature decrease of only 0.15 °C.

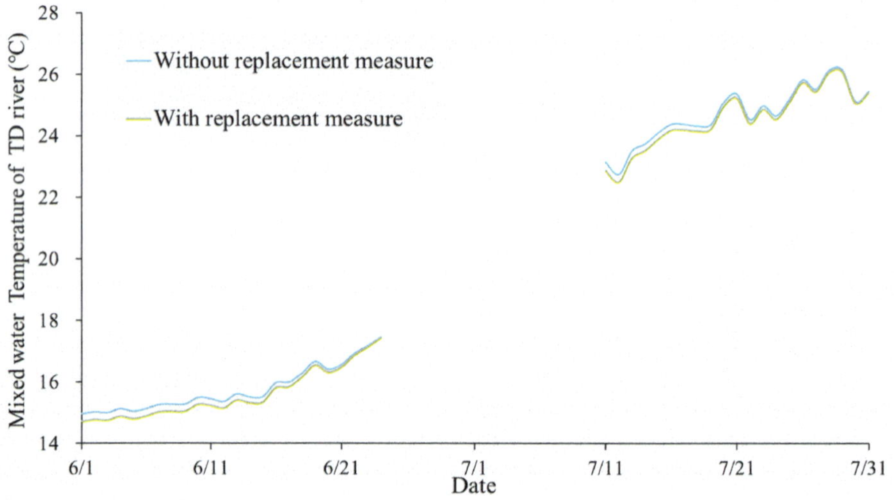

Fig. 7. The impact of the replacement measures on the water temperature of the TD River before and after implementation.

In conclusion, the water temperature of the TD River is relatively high, while the amount of water used for irrigation is relatively low. By replacing the warm water to improve the temperature of the irrigation canal, it can effectively raise the irrigation water temperature. Moreover, the interference with the natural water temperature of the TD River is minimal. It is recommended that the construction unit, under the current water intake scheme, prioritize this method to remedy the irrigation with low-temperature outflow. Additionally, pumping and replacement may affect the water quality of the water body [22], which should be considered in subsequent research and practical applications. Finally, we suggest that during the actual planning and design phase, a comprehensive assessment should be conducted on the various environmental impacts that may occur after the construction of the reservoir. This will not only save time, effort, and money but also minimize the impact on the natural environment.

5 Conclusions

The study focused on the LYT Reservoir, which was initially designed with a two-tiered water intake during the planning phase. However, only the bottom-level water intake was retained in the actual construction, which significantly alters the thermal structure of the reservoir and affects the downstream areas. The study used a numerical model to analyze the impact of different elevation intakes on the thermal status of the LYT Reservoir and the downstream irrigation channels. The results indicate that the multi-level intake (S2) shows a significant upward shift in the thermocline compared to the single-level intake (S1), with an average upward movement of 11.7 m from May to October. Furthermore, compared to the S1 scheme, the average water temperature of outflow temperature of S2 is 3.5 °C higher from May to August. Specifically, during the irrigation period, the average water temperatures for S1 and S2 are 15.6 °C and 18.5 °C, respectively, with a difference of 2.9 °C. Under the multi-level intake scheme, the outflow temperature always meets the minimum temperature requirements for crop irrigation during the growing season. In contrast, with the single-level intake, there are 45 days where the outflow temperature does not meet the minimum irrigation temperature requirement, accounting for 35.4% of the total days. Considering the current single-level intake in the LYT reservoir, the study comprehensively evaluated the feasibility, economic efficiency, and effectiveness of mitigation measures and proposed a cold water mitigation measure using water extraction from the natural river. Based on different temperature conditions, the required flow rate for water extraction from the TD River to meet the temperature requirements ranged from 0.005 to 0.185 m^3/s. This suggests that the maximum pumping capacity of the water pump should be at least 0.185 m^3/s to meet the minimum irrigation water temperature requirements. The study provides theoretical and technical support to mitigate the adverse effects of releasing cold water from the reservoir on the ecological environment and the growth of crops in the irrigation area of the downstream river. This is important for promoting the sustainable utilization of water resources.

Acknowledgments. This work is supported by the Study on the Business Model of the 'Environmental+' Pumping Project (no. KY2022-HS-02-14). We would also like to thank anonymous reviewers and editors for their helpful comments and suggestions on improving the manuscript.

References

1. Tao, Y., Wang, Y., Rhoads, B., et al.: Quantifying the impacts of the Three Gorges Reservoir on water temperature in the middle reach of the Yangtze River. J. Hydrol. **582**, 124476 (2020)
2. Zhang, D., Wang, D., Peng, Q., et al.: Prediction of the outflow temperature of large-scale hydropower using theory-guided machine learning surrogate models of a high-fidelity hydrodynamics model. J. Hydrol. **606**, 127427 (2022)
3. He, W., Zhang, X., Zhang, J., et al.: Regulating outflow temperature for multi-objective operation of cascade reservoirs: a case study. Renewable Energy **211**, 155–165 (2023)
4. Jiang, B., Wang, F., Ni, G.: Heating impact of a tropical reservoir on downstream water temperature: a case study of the Jinghong Dam on the Lancang River. Water **10**(7), 951 (2018)
5. He, W., Wang, H., Zhang, J., et al.: Diurnal variation characteristics of thermal structure in a deep reservoir and the effects of selective withdrawal. J. Environ. Manage. **333**, 117459 (2023)
6. Yang, M., Li, L., Li, J.: Prediction of water temperature in stratified reservoir and effects on downstream irrigation area: a case study of Xiahushan reservoir. Physics and Chemistry of the Earth, Parts A/B/C **53–54**, 38–42 (2012)
7. Yang, Y., Deng, Y., Tuo, Y., et al.: Study of the thermal regime of a reservoir on the Qinghai-Tibetan Plateau. China. PLOS One **15**(12), e0243198 (2020)
8. Larabi, S., Schnorbus, M.A., Zwiers, F.: A coupled streamflow and water temperature (VIC-RBM-CE-QUAL-W2) model for the Nechako Reservoir. Journal of Hydrology: Regional Studies **44**, 01237 (2022)
9. Wang, L., Xu, B., Zhang, C., et al.: Exploring the trade-offs among hydropower benefits, environmental flow, and surface water temperature in a large reservoir under deep uncertainty. J. Hydrol. **624**, 129913 (2023)
10. Deng, Y., Tuo, Y.C., Li, J., et al.: Spatial-temporal effects of temperature control device of stoplog intake for Jinping I hydropower station. Science China-Technological Sciences **54**, 83–88 (2011)
11. Zhang, L.F., Zhang, J.M., Peng, Y., et al.: Numerical simulation of flow and temperature fields in a deep stratified reservoir using water-separating curtain. Int. J. Environ. Res. Public Health **16**(24), 5143 (2019)
12. Bruce, L.C., Frassl, M.A., Arhonditsis, G.B., et al.: A multi-lake comparative analysis of the General Lake Model (GLM): stress-testing across a global observatory network. Environ. Model. Softw. **102**, 274–291 (2018)
13. Wang, Y., Zhang, N., Wang, D., et al.: Impacts of cascade reservoirs on Yangtze River water temperature: assessment and ecological implications. J. Hydrol. **590**, 125240 (2020)
14. Azadi, F., Ashofteh, P.-S., Chu, X.: Evaluation of the effects of climate change on thermal stratification of reservoirs. Sustain. Cities Soc. **66**, 102531 (2021)
15. Yang, S., Zhang, Z., Ji, Q., et al.: Study on the water temperature distribution characteristics of a mixed pumped storage power station reservoir: a case study of Jinshuitan Reservoir. Renewable Energy **202**, 1012–1020 (2023)
16. Rangel-Peraza, J.G., Obregon, O., Nelson, J., et al.: Modelling approach for characterizing thermal stratification and assessing water quality for a large tropical reservoir. Lakes Reserv. Sci. Policy Manag. Sustain. Use **17**(2), 119–129 (2012)
17. Wang, H., Deng, Y., Yang, Y., et al.: Future projections of thermal regimes and mixing characteristics in a monomictic reservoir under climate change. Sci. Total Environ. **906**, 167527 (2024)
18. He, T., Deng, Y., Tuo, Y., et al.: Impact of the dam construction on the downstream thermal conditions of the Yangtze River. Int. J. Environ. Res. Public Health **17**(8), 2973 (2020)

19. Wang, H., Deng, Y., Yang, Y., et al.: Optimization of selective withdrawal strategy in a warm monomictic reservoir based on thermal stratification. Ecol. Ind. **158**, 111294 (2024)
20. Brunner, G.W.: HEC-RAS river analysis system: hydraulic reference manual, Version 5.0. US Army Corps of Engineers–Hydrologic Eng. Center **547** (2016)
21. Drake, J., Bradford, A., Joy, D.: Application of HEC-RAS 4.0 temperature model to estimate groundwater contributions to Swan Creek, Ontario, Canada. J. Hydrol. **389**(3), 390–398 (2010)
22. Zhang, H., Wang, Y., Li, J., et al.: Whether water exchange is an effective solution for emergency mitigation of algae bloom. J. Water Process Eng. **60**, 105085 (2024)

Study of the Effect of Rainfall on the Parameters of Sandy Soils

Xinmin Yu[1], Jianhao Lin[1], and Yurong Xiao[2](✉)

[1] State GRID Fujian Economic Research Institute, No. 68 Road Chayuan, FuzhouFujian Province 350013, China
110903205@qq.com, 214030828@qq.com
[2] School of Civil Engineering and Architecture, Northeast Electric Power University, Jilin 132012, China
1456242009@qq.com

Abstract. Rainfall infiltration is an important parameter that affects the performance change of unsaturated soils. In order to study the changes in the properties of sandy soils under different rainfall intensities with different durations, the changes in pore pressure, saturation, angle of internal friction, and cohesion of soils were analyzed under three rainfall intensities of 10mm, 50mm, and 100mm with durations of 0, 3, and 7 days, respectively, by using ABAQUS finite element software. The results showed that the changes in the parameters were not significant under 10 mm rainfall; When the rainfall lasted for 3 days and 7 days, the zero pore pressure increased by 3.3% and 16.67% under 50 mm rainfall, and by 6.67% and 30% under 100 mm rainfall; In terms of saturation, there was an increase of 150 percent and 180% under 50 mm rainfall, but saturation had not yet been reached, with less change in the deeper 2–3 m. Surface soils were saturated at 0–1 m in the top layer when rainfall of 100 mm lasted for 3 days, and all soils above the water table were saturated for 7 days. The angle of internal friction and cohesion in the initial state decreased gradually with the increase of depth and showed a trend of increasing and then decreasing after the rainfall, and the depth where the peak point appeared gradually shifted downward with the increase of rainfall intensity and duration.

Keywords: Unsaturated Soil · Rainfall Infiltration · Numerical Simulation

1 Introduction

Distribution lines due to flooding, waterlogging, landslides, and other disasters can lead to downed poles, broken wires, etc. The main reason for this is that the surface soil is unsaturated, and its properties will change significantly under rainfall. The bearing capacity of foundation soils is usually studied based on the Mohr-Coulmb strength criterion. It is considered to be saturated, but in real engineering, the vast majority of soils are unsaturated. Under the influence of rainfall intensity and time, the change of unsaturated soil properties is of great significance for practical engineering and application. Current research in this area includes Li [1], who analyzed rainfall infiltration on slopes

C. Weng (Ed.): WRE 2024, LNCE 634, pp. 162–170, 2025.
https://doi.org/10.1007/978-981-96-5095-8_14

based on ABAQUS software and concluded that rainfall infiltration leads to a decrease in the strength parameter of the saturated region of the slope soil. Hou [2] discussed the effects of water table, cohesion, and angle of internal friction on foundation bearing capacity. Zhou [3] used Geo-studio finite element numerical analysis software to conclude that the higher the rainfall intensity, the faster the rainwater infiltration rate, the longer the rainfall holding time, and the increased depth of rainfall infiltration impact. Vahedifard [4] gave a theory for calculating the ultimate bearing capacity of foundations under steady seepage flow and analyzed the effect of the magnitude of infiltration flow and water table on the bearing capacity. Wang [5] and others studied the effect of soil-water characteristic curves and permeability coefficients on seepage in unsaturated soils to predict slope stability more accurately. Richards [6] studied slope stability under different rainfall types with different rainfall intensities.

This paper takes sandy soil as the research object, establishes a rainfall infiltration model based on the theory of seepage in unsaturated soil, and adopts ABAQUS finite element software to simulate the change rule of soil parameters under three different rainfall intensities and different durations, which is of great practical significance to the analysis of seepage in unsaturated soil under the condition of rainfall.

2 Soil Finite Element Simulation Establishment

The soil studied in this paper is sandy in Fujian Province, and the main soil parameters are shown in Table 1.

Table 1. Soil parameter values.

Name	Gravity (kN/m^3)	Densities ($Kg{\cdot}m^{-3}$)	Modulus of elasticity (kPa)	Cohesive force (kPa)	Angle of internal friction (°)	Poisson's ratio
Sandy soil	19.5	1900	20	20	25	0.4

ABAQUS was used to establish a three-dimensional mesh model of the soil body, assuming a uniform distribution of the soil body, which was established using C3D8RP hexahedral cells, with the side lengths set to 12 m and the water table to 3 m [7], as in Fig. 1. The meshing is shown in Fig. 2.

In terms of permeability, the depth of the soil will affect the permeability coefficient, as the deeper the layer below the ground the denser the soil is, the fewer the pores in the soil, and the greater the impediment to water seepage. Therefore, as the depth increases, the permeability coefficient of the soil body needs to be discounted to a certain extent. In ABAQUS software, the infiltration coefficient is controlled by the discount factor ks and the saturation degree by default, and their relationship curves are shown in Fig. 3.

By reviewing the literature and information [8], three rainfall intensities of light, heavy and rainstorm were set up with daily rainfall amounts of 10 mm, 50 mm, and 100 mm and durations of 0d, 3d, and 7d, respectively.

Boundary condition setting: This model is pure seepage analysis only needs to consider the pore pressure condition; the groundwater level is 3 m, so the plane where the groundwater level is located is set as the zero pore pressure surface, and the pore pressure boundary condition is set in the remaining four surfaces that increase with the depth.

Fig. 1. Foundation Dimensions.

Fig. 2. Foundation Grid Diagram.

3 Analysis of Calculation Results

3.1 Pore Pressure

The variation of pore pressure of soil with depth for different rainfall intensities and durations was obtained, as shown in Fig. 4.

In the initial state of 0d, there is negative pore pressure above the water table, and the soil is unsaturated, i.e., there is suction in the soil, and the magnitude of the pore pressure

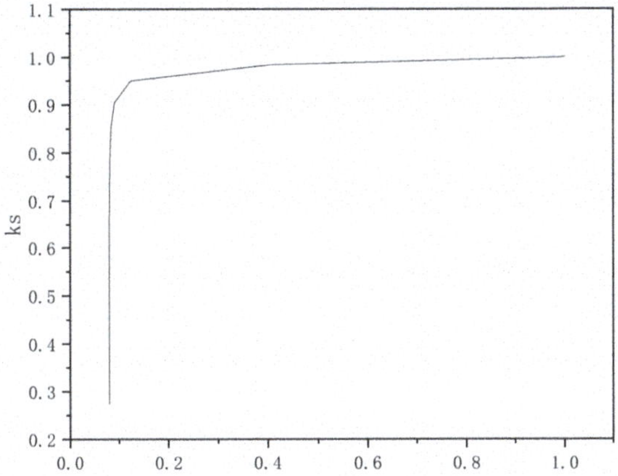

Fig. 3. Relationship between ks and saturation.

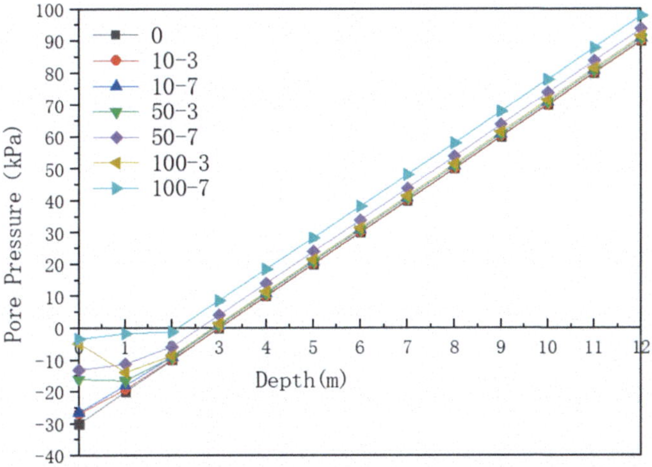

Fig. 4. Variation of pore pressure with depth.

increases linearly with the distance from the water table. There is zero pore pressure at 3 m where the water table is located, and there is positive pore pressure below the water table, which indicates that the soil is saturated. With the growth of rainfall time, water began to infiltrate from the surface to the deeper layer; the surface layer of soil was the first to approach saturation, forming a transient saturated zone [9], and the pore pressure gradually increased and tended to zero. 10 mm rainfall lasted for 0d, 3d, and 7d of the zero pore pressure did not have a big change, are located in the water table near the 3 m, which is due to the precipitation being too small, the water infiltration to the surface layer, only affect the surface layer of the pore pressure of the soil body, and the change is not big. This is because the moisture only infiltrated the surface layer, affecting the pore pressure of the surface soil, and the change is not big. The zero pore pressures for 0d, 3d, and 7d of 50 mm rainfall were 3 m, 2.9 m, and 2.5 m, respectively, and the zero pore pressure shifted upwards by 0.5 m at 7 d, indicating that the water had infiltrated the water table. The zero pore pressures for 0d, 3d and 7d of 100 mm rainfall were 3 m, 2.8 m, and 2.1 m, respectively, and the pore pressures tended to be closer to zero due to the higher rainfall, which caused the surface soil to be close to the saturated state. Close to zero value. As the intensity of rainfall increases, water can infiltrate to a greater depth at the same time. Hence, the pore pressure of the soil at the same location gradually increases as the intensity of rainfall increases at the same time, converging towards the zero value and reaching saturation more easily.

3.2 Saturation

The variation of pore pressure with depth for soils at different rainfall intensities and durations is obtained, as shown in Fig. 5.

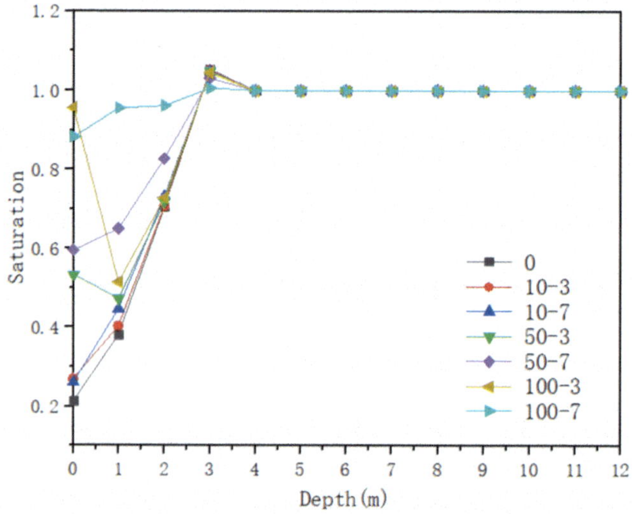

Fig. 5. Plot of saturation versus depth.

In the initial state without rainfall, due to the existence of capillary phenomenon, in a part of the soil body closer to the water table, saturation is larger; that is, the peak point at 3 m in the change diagram, the saturation of the soil body surface to the water table gradually increases and is less than 1, for the unsaturated soil. The saturation of the soil body below 3 m is stable at about 1, for the saturated soil state. The saturation of the soil did not change significantly under 10 mm of rainfall, the state of the soil saturation did not change significantly because the daily rainfall was so small that even if it lasted for 7 days, it would not be enough for infiltration to make a change in the state of the soil. Under 50mm rainfall, as water infiltrates from the top down, the top layer of soil is the first to change, with saturation reaching 0.53 and 0.6 at 3 and 7 days. There is a corresponding increase in the deeper 1 m to 3 m. Still, the change is small compared to the surface layer, which also indicates that water begins to infiltrate to a deeper depth.

There is an increase in saturation at the 2 m depth at 7 days, which suggests that water infiltration occurs only then. The saturation level increased at 2 m depth for 7 days, suggesting that water infiltrated only then. 100 mm rainfall, the saturation of the soil at all depths has greater changes, in the continuation of 3 days when the surface soil has reached saturation, saturation tends to 1; at this time, the permeability coefficient of the soil is reduced, the rate of infiltration of rainwater is slowed down [10], so the saturation of 1 m at the 1 m although there is an increase the change is not large, for 7 days, due to the sufficient amount of water, water continues to infiltrate, so the saturation of the soil at the 1 to 3 m have an increased considerably, and all tended to be saturated. As the intensity of rainfall increases, the depth of water infiltration increases at the same time due to the increase in rainfall, so the water content of the soil at the same depth increases, and the degree of saturation increases.

3.3 Angle of Internal Friction

The variation of internal friction angle with depth for different rainfall intensities and durations is plotted in Fig. 6.

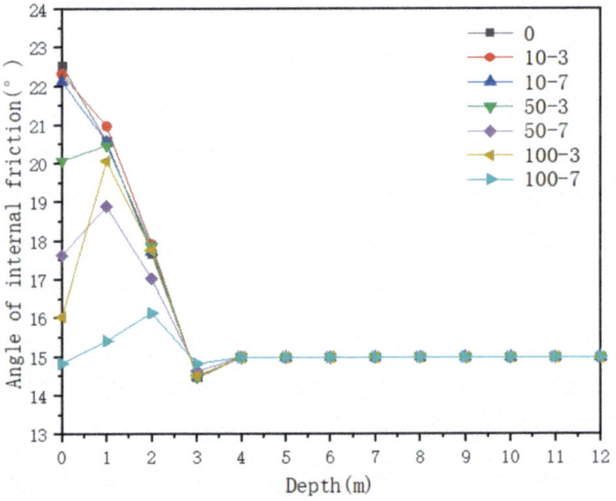

Fig. 6. Variation of the angle of internal friction with depth.

The initial state without rainfall shows a linear decrease in the angle of internal friction with increasing depth. Under 10 mm rainfall, the water content of the soil did not change significantly due to the amount of rainfall being too low, and therefore, the angle of internal friction did not change significantly, and the trend of the rainfall lasting for 3 days versus 7 days was still not very different from the initial state. Under 50 mm rainfall, water first penetrates the surface layer, and the water content of the soil surface layer increases; when with the increase of water content, there is free water present between the particles in the soil, and this water leads to an increase in the lubrication between the particles and particles [11], therefore, the angle of internal friction begins to decrease. With the growth of the rainfall time, the water content of the soil at the same depth gradually increases, and the angle of internal friction all decreases. Whereas, below 3 m, there is a levelling off trend as the soil is already saturated, and the water content does not change significantly. Compared with 50 mm rainfall, the change is more obvious at 100 mm rainfall; the angle of internal friction decreases more, but both show the trend of increasing first and then decreasing; this is due to the gradual increase of water content in the soil, the free water gradually increases, the substrate suction decreases, and the excess water plays a lubricating role in the soil to exacerbate the soil's shear until the water content in the soil increases to a certain degree, when the soil has more water. This water's effect on the internal friction of the soil is no longer very obvious, so there will be an inflection point [12].

3.4 Cohesion

The variation of cohesion with depth for different rainfall intensities and durations is plotted in Fig. 7.

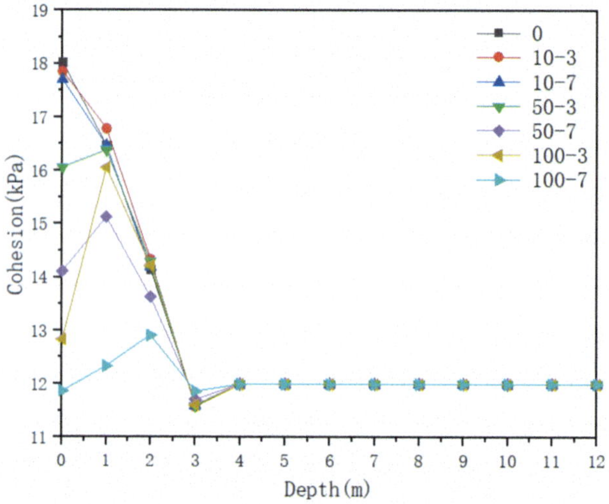

Fig. 7. Variation of cohesion with depth.

The change of cohesion is the same as the angle of internal friction, the initial state, with the increase of depth and shows a linear trend of decline, with the growth of rainfall time, the same depth out of the cohesion is decreased, and the greater the intensity of the rainfall the more decline, the overall also shows the trend of first increase and then decrease, and are smaller than the initial state.

4 Conclusions

(1) Regarding pore pressure, the pore pressure of the soil above the water table is negative in the initial state without flooding, indicating that the soil is unsaturated and has matrix suction. With the increase of rainfall time and amount, the pore pressure in the surface layer of the soil body tends to zero value, and the location of zero pore pressure gradually moves upwards.
(2) In terms of saturation, in the initial state, due to the existence of the capillary phenomenon, the saturation of a part of the soil body closer to the water table is larger. The saturation gradually decreases from bottom to top. With the increase in rainfall time and rainfall, the water content of the surface layer of the soil body grows rapidly. It is easier to reach the saturated saturation state close to the situation of the water table.
(3) About the angle of internal friction and cohesion, with the increase of rainfall time and rainfall amount, the water infiltrated deeper into the soil body from the top down. The angle of internal friction and cohesion of the soil body at the same depth decreased, although the overall tendency was to increase and then decrease, and it was always smaller than the initial state.

Acknowledgment. This research was supported by a grant (52130N23000P) from State Grid Fujian Electric Power Co., Ltd, the project title is "Research on key technologies for improving safety level of distribution line structure in Fujian Power Grid under Flood".

References

1. Li, K., Huang, Y., Jin, K.S.: Analysis of rainfall infiltration on laterite slope with finite element secondary development. China Water Transp. **2**, 156–158 (2024)
2. Hou, C.Q., Zhang, Z.Y., Sun, Z.B., et al.: Calculation of the upper limit of bearing capacity of non-homogeneous and anisotropic slope-facing foundations. Highw. Traffic Sci. Technol. **36**, 21–27+35 (2019)
3. Zhou, Y.L., He, Q., Li, M., et al.: Stability analysis of unsaturated soil slopes under different rainfall conditions considering seepage-stress coupling. J Chongqing Jiaotong Univ. (Nat. Sci. Ed.) **42**(8), 57–63 (2023)
4. Vahedifard, F., Robinson, J.D.: Unified method for estimating the ultimate bearing capacity of shallow foundations in variably saturated soils under steady flow. J. Geotech. Geoenvironmental Eng. **142**(4), 04015095 (2016)
5. Wang, L., Tang, L.B., Wang, Z.Y., et al.: Probabilistic characterization of the soil-water retention curve and hydraulic conductivity and its application to slope reliability analysis. Comput. Geotech. **121**, 103460 (2020)
6. Richards, L.A.: Capillary conduction of liquids through porous mediums. J. Appl. Phys. **1**(5), 318–333 (1931)
7. Source of data 'Collection of geological survey reports of Fujian Province'. https://wenku.baidu.com/view/b2806cf3d938376baf1ffc4ffe4733687f21fc71.html?_wkts_=1733387169393
8. Wang, J.H., Wang, Y.L., Wang, Q., Zhong, Y.L.: Research on the effect of different rainfall intensities on the stability of clayey soil slopes. J. Heilongjiang Inst. Technol. **22**(01), 104–108 (2022)

9. Cha, W.H., Yang, H., Xu, T. et al.: Analysis of seepage field and stability of binary slopes under different rainfall type cycles. Hydropower 1–7 (2024)
10. Wang, X.X.: Stability study of unsaturated soil slope under rainfall based on finite element. Western Transp. Sci. Technol. **1**, 30–33 (2024)
11. Du, Z.Y.: Stability Analysis of Unsaturated Fill Slopes Under Different Rainfall Intensities. Taiyuan University of Technology, China (2022)
12. Li, S.H., Lin, Q.X.: Experimental study on the effect of dry density and water content on the strength of unsaturated soil. Sci. Technol. Bull. **39**(2), 62–67 (2023)

Automatic Calibration and Inversion of SWMM Parameters Using HHO-BP Algorithm

Xiongdi Ma, Fan Chen, Zhikai Cai, Mengting Zheng, and Zheng Sheng(✉)

Huzhou University, Huzhou 313000, Zhejiang, China
1509914184@qq.com, 1812987854@qq.com, 1947832003@qq.com, 799763242@qq.com, 03083@zjhu.edu.cn

Abstract. In order to enhance the prediction accuracy and applicability of the Storm Water Management Model (SWMM) in urban flood management, this study conducted a case study at the East Campus of Huzhou University, systematically performing sensitivity analysis and automated calibration of the model parameters. First, the improved Morris screening method was employed to conduct sensitivity analysis on numerous parameters within the SWMM model, identifying five key sensitive parameters as the focus for subsequent calibration efforts. Subsequently, this study integrated the BP neural network with the Harris Hawk Optimization (HHO) algorithm to construct an efficient parameter calibration model. After optimizing the neural network weights and thresholds using the HHO algorithm, the model successfully established a nonlinear mapping relationship between the inputs (water depth and flow sequence) and the outputs (sensitive parameter combinations). The experimental results showed that the trained BP neural network model could accurately and efficiently deduce the optimal SWMM parameter combinations. When the deduced parameter combinations were re-entered into the SWMM model for simulation verification, the simulation results were found to have a high degree of consistency with the measured data. Nash-Sutcliffe efficiency coefficients all maintained above 0.82, significantly superior to traditional manual calibration methods. This research result verifies the feasibility and superiority of using BP neural networks in the calibration of SWMM parameters.

Keywords: SWMM · Morris · BP Neural Network · HHO

1 Introduction

The Storm Water Management Model (SWMM) is a widely used computer model for managing stormwater in urban and urbanized areas. It can simulate runoff generated by rainfall and the flow within drainage systems, aiding urban planners and engineers in evaluating various stormwater management strategies, including rainwater harvesting, infiltration, and runoff delay [1]. Since SWMM is a complex distributed model with numerous parameters, the accuracy of the model's simulation results is influenced by the precision of its parameters and input data. Therefore, to ensure the model's stability and the accuracy of its output results, precise parameter calibration is essential for achieving optimal model performance [2].

C. Weng (Ed.): WRE 2024, LNCE 634, pp. 171–184, 2025.
https://doi.org/10.1007/978-981-96-5095-8_15

Before conducting SWMM parameter calibration, sensitivity analysis is necessary. This step aims to identify which parameters are most sensitive to the model output, allowing for the prioritization of these key parameters [3]. This ensures that the parameter calibration process is more targeted and scientific, thereby improving the accuracy and efficiency of the model [4]. SWMM parameter calibration methods are primarily classified into manual and automatic calibration. Manual calibration relies on the expertise and experience of researchers. Still, it is less efficient, time-consuming, and labor-intensive and is prone to subjective influences, leading to results that may not be objective or comprehensive. This study utilized the improved Morris screening method to calculate parameter sensitivity. Based on the sensitivity analysis results, relevant parameters were calibrated using the BP neural network. The East Campus of Huzhou University, as a case study, parameter calibration and model result verification were conducted using measured rainfall data and the proposed automatic calibration method.

2 Materials and Methods

2.1 Overview of the Study Area

Huzhou University is in Huzhou City, Zhejiang Province, China. As shown in Fig. 1, the model was simplified based on various parameters of the study area, comprising 50 nodes, including 49 inspection wells and one outlet, along with 49 pipes and 74 subcatchments. The subcatchments include 10 building rooftops, 14 green spaces, 25 pavements, and 25 mixed-use areas. GIS was used to calculate the area, slope, and other information for each subcatchment, which were then input into the SWMM inp file. A schematic diagram of the model is shown in Fig. 2. Rainfall data for the study area were collected using rain gauges, and the selected rainfall events are listed in Table 1.

Table 1. Selected rainfall events.

Index	Rainfall Dates	Total Rainfall/mm	Duration of Rainfall/h	Average Rainfall Intensity/mm.h^{-1}
1	2023.6.18	13.4	3.58	3.74
2	2024.3.27	34.4	7.92	4.35
3	2023.8.28	10.2	1.5	6.8
4	2023.9.29	11	0.58	18.86
5	2023.7.22	18.8	0.75	25.05

2.2 Parameter Determination

In the process of establishing and applying the SWMM model, the initial parameter selection is a crucial step that lays the foundation for the entire parameter calibration and model optimization process. Parameters in the SWMM model can be broadly categorized into

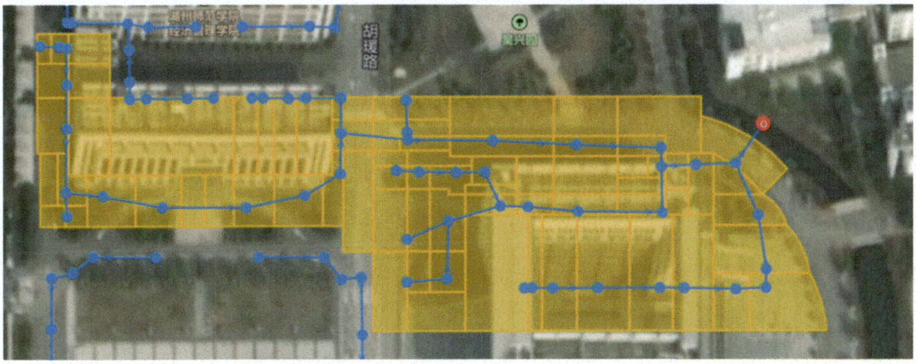

Fig. 1. Distribution of the study area.

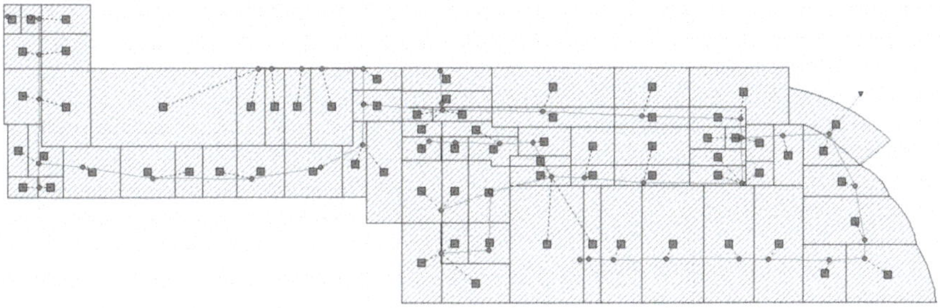

Fig. 2. Schematic diagram of the model simplification.

two types: deterministic parameters obtained through actual measurements, such as pipe diameter, pipe length, subcatchment width, and area; and empirical parameters that cannot be directly measured, such as Manning's roughness coefficient, depression storage depth, infiltration parameters, etc. [5]. These empirical parameters are determined based on empirical data and ranges of values obtained from measurements, and parameters like infiltration coefficients can vary depending on the region, requiring consultation of relevant empirical data and selected infiltration models for determination. In this study, the Horton infiltration equation was chosen as the infiltration model, and dynamic wave method was used for pipe network routing calculations. The main hydrological and hydraulic parameters selected are listed in Table 2.

2.3 Sampling Method

For the BP neural network model, having a sufficiently large number of samples is crucial for establishing an effective mapping between inputs and outputs [6]. Ensuring uniform probability distribution of input parameter combinations is essential before model execution. The learning and computation processes of the BP neural network model determine the optimality of parameter calibration results, with the representativeness of sampled data significantly impacting result accuracy.

Table 2. Main hydrological and hydraulic parameters for SWMM model calibration.

Index	Parameter Names	Physical Meaning of Parameters	Range of Values	Initial Valu	Type
1	N-Imperv	Impervious Manning's Coefficient	0.01~0.05	0.025	Fully Empirical Parameters
2	N-Perv	Pervious Manning's Coefficient	0.1~0.8	0.33	
3	S-Imperv	Impervious Depression Storage Depth	0.2~10	2.31	
4	S-Perv	Pervious Depression Storage Depth	2~10	4.65	
5	MaxRate	Maximum Infiltration Rate	20~80	40	Certain Empirical Parameters
6	MinRate	Minimum Infiltration Rate	0.5~10	5	
7	Decay	Infiltration Decay Coefficient	2~7	4	
8	DryTime	Dry Weather Time	2~14	7	
9	Manning-N	Pipe Manning's Roughness Coefficient	0.01~0.032	0.025	

Latin Hypercube Sampling (LHS) is a common statistical technique used to generate uniformly distributed and diverse sample sets in multidimensional parameter spaces [7]. It partitions the parameter space into equally spaced subintervals and randomly selects one sample point from each subinterval, ensuring uniform distribution across each parameter dimension. Compared to traditional random sampling methods, LHS more effectively explores parameter space and provides reliable estimates with fewer samples. Therefore, I employed LHS to sample independent variables within their initial ranges uniformly.

2.4 Sensitivity Analysis Methods

The Morris screening method is the most used approach for local sensitivity analysis. Its principle involves keeping all other parameters fixed while selecting one parameter to perturb within its range randomly. The analysis examines how variations in this parameter affect model outputs when included in the model calculation [8]. The local sensitivity analysis of this parameter is illustrated by Eq. (1):

$$ei = (Y^* - Y)/\Delta_i \tag{1}$$

The equation is defined as follows: Y represents the model output before parameter change. Y*represents the model output after parameter change. Δi represents the change in the parameter. ei represents the Morris coefficient, which indicates the sensitivity of the parameter.

The improved Morris screening method features fixed parameter increments [9] (in this paper, 15%), with the sensitivity evaluation metric being the average of perturbed Morris coefficients, as shown in Eq. (2). Moreover, the model performance is evaluated using the Nash-Sutcliffe coefficient [10], as shown in Eq. (3)

$$S = \sum_{i=0}^{n-1} \frac{(Y_{i+1} - Y_i)/Y_0}{(P_{i+1} - P_i)/100}/n \tag{2}$$

$$NSE = 1 - \frac{\sum_{t=1}^{n} (Q_{obs,t} - Q_{sim,t})^2}{\sum_{t=1}^{n} (Q_{obs,t} - \overline{Q_{obs,t}})^2} \tag{3}$$

In the equations: Yi and Yi + 1 represent the model output values for the i-th and (i + 1)-th runs, respectively. Y0 is the initial computed result value. Pi and Pi + 1 denote the percentage changes of the parameter values relative to the initial parameter value for the i-th and (i + 1)-th model runs. n represents the number of model runs. S represents parameter sensitivity: ($0 \le |S| < 0.05$, Not sensitive; $0.05 \le |S| < 0.2$, Moderately sensitive; $0.2 \le |S| < 1$, Sensitive; $|S| \ge 1$, Highly sensitive); NSE is the Nash-Sutcliffe Efficiency coefficient.; $Q_{obs,t}$ and Q_{sim}, t are the observed and simulated values at time ttt. $\overline{Q_{obs,t}}$ is the mean of the observed values.

2.5 Backpropagation Neural Network

The Backpropagation Neural Network (BP Neural Network) is a commonly used artificial neural network model for solving regression and classification problems [11]. It consists of an input, hidden, and output layer. Through forward propagation and backpropagation, it continuously adjusts network weights using training data to learn complex nonlinear relationships between inputs and outputs. BP Neural Networks find extensive applications in pattern recognition, prediction, and control, as they can effectively handle various complex data relationships and nonlinear mappings [12].

In the calibration of SWMM parameters, the backpropagation (BP) neural network demonstrates significant advantages [13]. Firstly, the BP neural network allows for rapid modeling and parameter optimization of complex hydrological models without requiring extensive prior knowledge. Secondly, it can handle many input parameters and nonlinear relationships, capturing complex mapping relationships between model inputs and outputs to enhance the accuracy and efficiency of parameter calibration. Additionally, the BP neural network exhibits strong generalization ability and can address challenges posed by data noise and model complexity. This ensures the model performs well in predicting performance under different environmental conditions [14]. Therefore, in this experiment, the BP neural network will be used to construct a model to calibrate SWMM parameters.

2.6 HHO-BP Neural Network Algorithm

The BP neural network is a multi-layer feedforward network that uses the computed error for backpropagation training [15]. It is characterized by excellent nonlinear learning and approximation capabilities [16]. In general research, PSO is commonly used for hyperparameter tuning [17, 18]; however, the training time and prediction accuracy of BP neural networks are also significantly influenced by the selection of initial weights and thresholds. Traditionally, initial weights and thresholds in BP neural networks are randomly assigned, potentially affecting network convergence speed and leading to longer training times or trapping in local minima [19]. Therefore, optimizing the process of determining initial weights and thresholds is crucial. Given that BP neural networks involve numerous weights and thresholds, and considering that the Harris Hawks Optimization (HHO) algorithm performs well for high-dimensional computational problems, the author adopts the HHO algorithm to optimize the BP neural network in this study.

The Harris Hawks Optimization (HHO) algorithm is an optimization algorithm proposed based on the predatory behavior of Harris's Hawks [20]. Specific algorithm model:

$$X(t+1) = \begin{cases} X_{rand}(t) - r_1 |X_{rand}(t) - 2r_2 X(t)|, q >= 0.5 \\ \Delta X - r_3[lb + r_4(ub - lb)], q < 0.5 \end{cases} \tag{4}$$

$$\Delta X = [X_{rabbit}(t) - X_{ave}(t)] \tag{5}$$

$$X_{ave} = \frac{1}{N}\sum_{i=1}^{N} x_i(t) \tag{6}$$

In the equation, q, r1, r2, r3, r4 are all random numbers within the interval [0,1]; X(t + 1) and X(t) represent the position of Harris's Hawks at the (t + 1)-th and t-th iterations, respectively; ub and lb are the upper and lower bounds of the search range; Xrand denotes a random position within the interval [0,1]; Xrabbit represents the current prey's position, Xave denotes the average position of Harris's Hawks; N represents the population size. If $q < 0.5$, it indicates that a Harris's Hawk has found prey. In this case, the hawk will choose its nesting position based on the positions of other companions and the prey. If $q \geq 0.5$, it means none of the hawks have found prey. In such cases, a Harris's Hawk will randomly choose a nesting position on a tree within the current range of the hawk group's activity.

During the prey search process, the hunting mode of Harris's Hawks is determined by how much escape energy the prey has remaining. As the prey attempts to escape, its escape energy gradually depletes, following a linear decrease based on the number of iterations. The formula is represented as:

$$E = 2E_0(1 - \frac{t}{t_{\max}}) \tag{7}$$

In the equation, E represents the escape energy of the prey, E_0 denotes the initial value of escape energy ranging from [−1, 1], reflecting variations in prey's initial escape capabilities. T_{max} indicates the maximum number of iterations. r represents the probability of prey escaping, where $r < 0.5$ signifies escape, and $r \geqq 0.5$ indicates no escape.

When $|E| \geq 1$, global search begins; when $|E| < 1$, local exploitation starts.

3 Results

3.1 Sensitivity Analysis Results

This study selected two rainfall events on June 18, 2023, and March 27, 2024. The peak flow at observation points was chosen as the target output for sensitivity analysis. The analysis aimed to assess how various parameters are sensitive to changes in water levels and quantities in the study area. Sensitivity indices S, calculated using the formula described in Eq. (2) with a perturbation value of 15%, are presented in Table 3.

The depth of depression storage in impermeable areas has the greatest impact on peak flow rates. The sensitivity index S for June 18, 2023, is approximately 1, indicating high sensitivity. Similarly, the sensitivity index for March 27, 2024, also indicates moderate sensitivity. This contrasts with the findings of Liao Rutin and colleagues in their study on the Liangshui River in Beijing [21, 22], where they identified permeable Manning coefficients and minimum infiltration rates as sensitive parameters affecting peak flow and runoff coefficients. The Manning coefficient of pipes was also identified as moderately sensitive due to the accumulation of leaves and debris, affecting the flow resistance within pipes. Furthermore, the impermeable Manning coefficient was also identified as moderately sensitive, with sensitivity indices S of 0.051 and 0.052, respectively. These findings align with those of Xiang Daifeng and others in their research in the Shenzhen River basin [13]. These comparisons highlight the need to consider local rainfall patterns and watershed conditions when assessing SWMM model parameter sensitivities. Therefore, sensitivity analyses should be tailored to specific study areas, ensuring relevance to regional conditions and backgrounds.

3.2 Model Validation Results

Before calibration, it is necessary to identify the parameters for calibration. Based on the sensitivity analysis results mentioned earlier, this study selected five relatively sensitive calibration parameters: S-Imperv, N-Imperv, N-Perv, Manning-N, and MinRate. The reference ranges for these parameters were determined based on recommendations from the SWMM model manual and typical empirical values (Table 2).

Table 3. Sensitivity Results of Two Rainfall Events.

Category	Parameter Names	2023.6.18		2024.3.27	
		Sensitivity	Sensitivity Level	Sensitivity	Sensitivity Level
Depression Storage Parameters	S-Imperv (mm)	0.960	Sensitive	0.086	Moderately Sensitive
	S-Perv (mm)	0.001	Insensitive	0.001	Insensitive
Manning's Roughness Coefficient	N-Imperv	0.051	Moderately Sensitive	0.052	Moderately Sensitive
	N-Perv	0.029	Insensitive	0.037	Insensitive
	Manning-N	0.062	Moderately Sensitive	0.076	Moderately Sensitive
Horton Infiltration Parameters	MaxRate ($mm.h^{-1}$)	0.001	Insensitive	0.000	Insensitive
	MinRate ($mm.h^{-1}$)	0.026	Insensitive	0.033	Insensitive
	Decay (h^{-1})	0.000	Insensitive	0.000	Insensitive
	DryTime (day)	0.000	Insensitive	0.000	Insensitive

The parameter calibration process is depicted in Fig. 3. First, 1000 parameter combinations were randomly generated using Latin Hypercube Sampling within the range of values for the parameters to be calibrated. These parameter combinations were used as inputs to the SWMM engine to compute corresponding water depth and flow sequences for 1000 simulations. These 1000 sets of water depth and flow sequences served as inputs, while the corresponding parameter combinations were used as outputs. The Harris Hawks Optimization (HHO) algorithm was employed to optimize weight thresholds and establish a BP neural network training model. After completing model training, observed water depth and flow sequences were input into the constructed model to deduce parameter combinations. These calibrated parameter combinations were then reintroduced into SWMM for simulation. The simulated water depth and flow sequences were analyzed with the Nash-Sutcliffe Efficiency coefficient (NSE) to assess the credibility of the calibration results against measured values.

Using the actual rainfall data from June 18, 2023, as input, the model was set to simulate for a duration of 6 h. The Horton model was chosen for infiltration, and the kinematic wave method was employed for simulation. A Python program implementing Latin Hypercube Sampling (LHS) was used to sample 1,000 parameter combinations within specified ranges. These combinations were saved to a spreadsheet file. A program was developed to automatically generate.inp files, reading data for subcatchments, nodes, outfalls, and other components from a pre-existing database of measured data. The 1,000 parameter combinations were sequentially read into the code and filled into the.inp file generation script. Upon completion, the SWMM computational engine was used to run

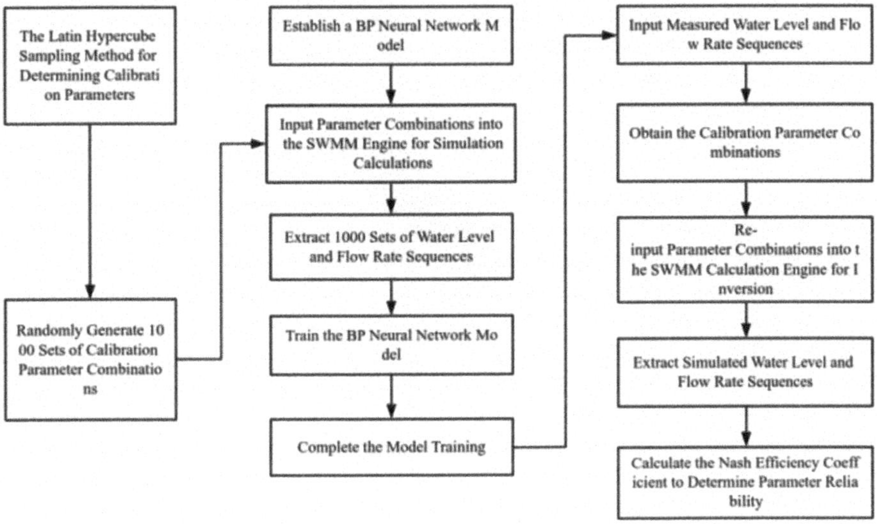

Fig. 3. BP Neural Network calibration process.

the.inp files, and generate.out and.rpt files. Water level and flow rate sequences corresponding to each parameter set were extracted from these.out and.rpt files, with data recorded at five-minute intervals. A BP neural network model was built using TensorFlow, utilizing the 1,000 sets of water level and flow rate sequences as input variables. Training data constituted 60%, with validation and test data each comprising 20%. The coefficient of determination (R^2) [20] was used to evaluate the effectiveness of BP neural network training. First, the mean of each parameter in the random combinations was calculated. Then, the sum of squared differences (SST) between parameters and their mean was obtained, along with the sum of squared residuals (SSR) between random parameter values and model outputs. The R^2 value was derived from SSR and SST; a value closer to 1 indicates better results. The model was fine-tuned to achieve an R^2 value approaching 1. An example scatter plot depicting the model training results for the maximum infiltration rate is shown in Fig. 4.

The trained BP neural network model achieved an R^2 coefficient of 0.93, indicating that the training process was successful and the mapping function relationship between the model input and output has been fully established. At this stage, the model can be used in place of the SWMM computational engine for verification analysis with actual rainfall data.

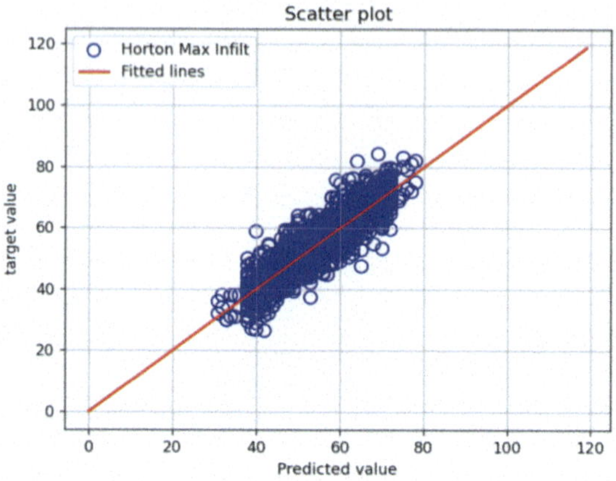

Fig. 4. Scatter plot of model training results (Using maximum infiltration rate as an example).

For this study, the rainfall on July 22, 2023, was used, and the parameter calibration results were obtained by reverse calculation through the BP neural network's function relationship. The calibration results are shown in Table 4. A comparison between the BP neural network model and manual calibration results showed that the model comprehensively identifies parameter relationships, considering overall interrelationships. This leads to higher calibration accuracy. In contrast, manual calibration primarily relies on sensitivity analysis results, making targeted adjustments only to highly sensitive parameters. This approach introduces more subjective elements and results in lower calibration accuracy.

Table 4. Comparison of SWMM Parameter Calibration Results.

Parameter Name	Manual Calibration Value	BP Neural Network Calibration Value
S-Imperv (mm)	4.5	2.5142
S-Perv (mm)	5	7.5461
N-Imperv	0.025	0.0119
N-Perv	0.55	0.3986
Manning-N	0.018	0.0114
MaxRate (mm.h^{-1})	75	55.0372
MinRate (mm.h^{-1})	7	4.8143

To further verify the accuracy and reliability of the model calibration results, the calibrated parameter combinations were reinserted into the SWMM computational engine to obtain water level and flow rate sequence data. Nash-Sutcliffe Efficiency (NSE) was calculated for these sequences to validate the results further. Typically, the closer the NSE value is to 1, the better the agreement between the simulated sequence data and the measured water level and flow rate data. An NSE value of at least 0.75 is typically required to meet practical application accuracy requirements.

First, the actual rainfall data from August 28, 2023, were selected. The calibrated parameter combinations were used in the model for this rainfall event, and the simulated water level and flow rate results were compared with the measured results. The calculated NSE value was 0.842, indicating that the model training results are good and the calibration parameters are reliable.

To avoid overfitting and the chance of occurrence of validation from a single rainfall event, additional rainfall events on September 29, 2023, and March 27, 2024, were selected for further validation. These steps were repeated, and NSE values were calculated for these events. The NSE values for all three rainfall events were found to be greater than 0.82. The fitted curves for water levels and flow rates are shown in Fig. 5. The results show that the model's predictions align well with the measured results, meeting the requirements of the calibration study.

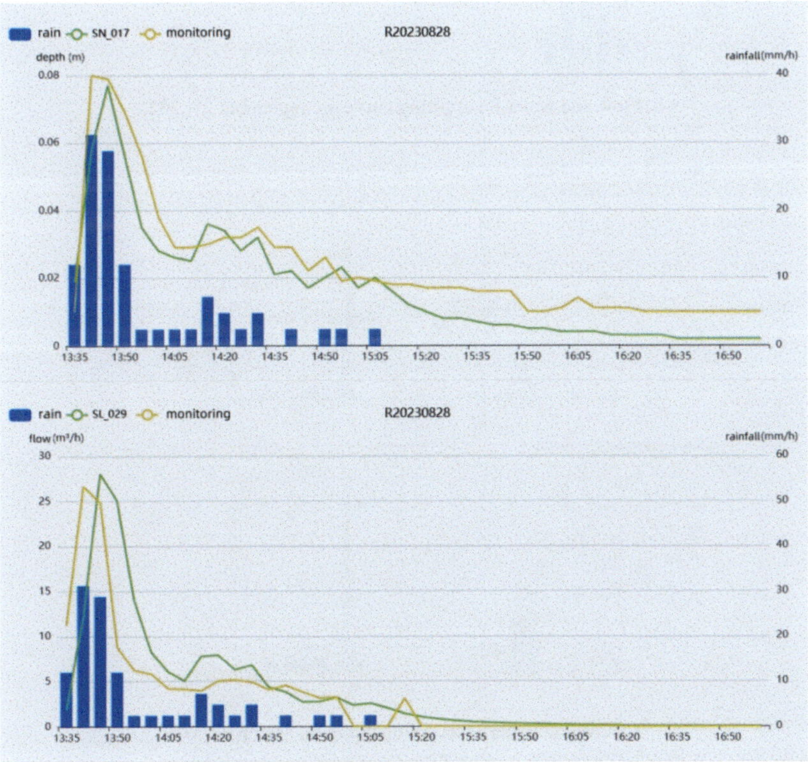

a. Flow and water level fitting curve on August 28, 2023.

Fig. 5. Fitting curve of water level and flow.

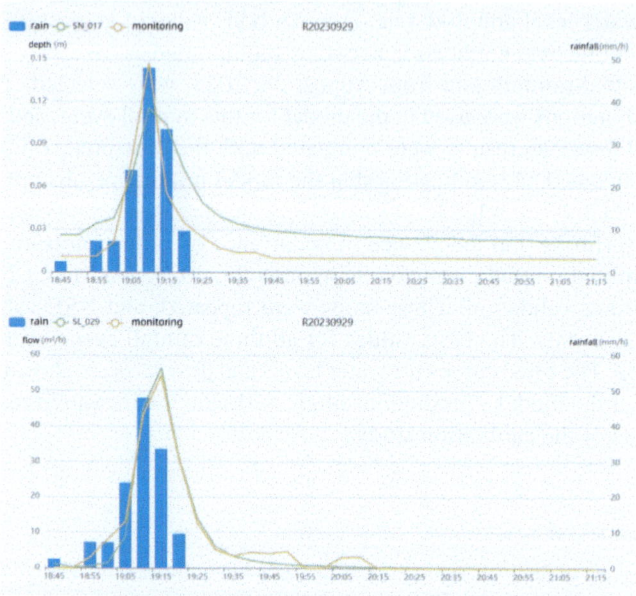

b. Flow and water level fitting curve on September 29, 2023.

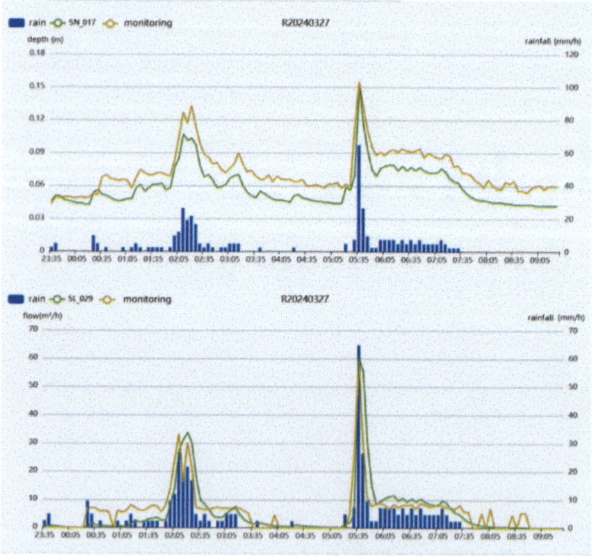

c. Flow and water level fitting curve on March 27, 2024.

Fig. 5. (*continued*)

4 Conclusions and Future Prospects

Based on the BP neural network model, this study calibrated SWMM hydrological parameters and derived the following conclusions and prospects:

1. Sensitivity analysis using the improved Morris screening method was performed on the parameters of the study area. The parameters with higher sensitivity identified were Manning's coefficient for impervious areas (N-Imperv), depression storage for impervious areas (S-Imperv), and Manning's coefficient for conduits (Manning-N).
2. Experimental results indicate that after model training is completed, parameters can be automatically based on learned mapping relationships. Traditional calibration methods often require repeating the entire process with a single rainfall event if results do not meet preset performance thresholds, which is time-consuming and inefficient. In contrast, the BP neural network model demonstrates significant advantages in calibration. If model prediction accuracy is insufficient, the training dataset can be adjusted and optimized, and the model can be retrained to improve performance effectively. This approach is more flexible and efficient, reducing the complexity and time cost of calibration.
3. Introducing the Harris Hawks Optimization (HHO) algorithm to optimize the weights and thresholds of the BP neural network effectively improves its convergence speed and stability, offering a more efficient solution for SWMM parameter calibration.
4. This study focused on calibrating the parameters with higher sensitivity. However, other parameters in the SWMM model may also affect the simulation results. Future research could consider including more parameters in the calibration scope to further enhance the comprehensiveness and accuracy of model calibration.
5. The model constructed in this study may encounter challenges fitting extreme or significantly different rainfall patterns (i.e., large discrepancies in rainfall patterns) during the training phase. However, the flexibility and scalability of the model allow for effective expansion of the training dataset. By incorporating data from previously poorly fitted rainfall events into the training set and retraining the model, its generalization ability can be enhanced, enabling adaptation to a broader range of rainfall characteristics.

References

1. Nazemi, R.A., Dolatshahi, M., Kerachian, R.: A decentralized multi-agent framework for urban flood management. Sustain. Cities Soc. 106105328 (2024)
2. Chen, T., Xia, M., Liu, Y., et al.: Study on rainwater runoff control in sponge renovation community based on SWMM. China Water Wastewater **36**(11), 103–111 (2020)
3. Kang, D., Lai, L., Qiu, F., et al.: Automatic calibration method of SWMM parameters based on machine learning theory. China Water Wastewater **40**(5), 122–129 (2024)
4. Feng, J., Duan, T., Bao, J., et al.: An improved back propagation neural network framework and its application in the automatic calibration of Storm Water Management Model for an urban river watershed. Sci. Tot. Environ. 915169886 (2024)
5. Lewis, A R.: Storm Water Management Model User's Manual. USA:US Environmental Protection Agency, pp. 163–165 (2010)

6. Zhang, J., Feng, H., Zhou, Y., et al.: Sensitivity analysis of runoff calculation parameters in SWMM. China Rural Water Hydropower (8), 1–5+13 (2022)
7. Li, C., Wang, W., Xiong, J., et al.: Sensitivity analysis for urban drainage modeling using mutual information. Entropy **16**(11), 5738–5752 (2014)
8. Peng, J., Zhao, H.C., Li, R., Xue, R.Z.: Parameter sensitivity analysis of SWMM: a case study of airport airfield area. Nat. Hazards **120**(7), 6551–6568 (2024)
9. Zhao, B.L., Wang, Z.M., Yang, H.B., et al.: Parameter optimization of SWMM model using integrated morris and GLUE methods. Water **15**(1), 149 (2022)
10. Zhou, S., Du, P., Pang, Y.: Research progress on urban stormwater management model applications. Adv. Water Resour. Hydropower Sci. Technol. **34**(06), 89–97 (2014)
11. Zhao, Y., Zhou, D., Yan, H.: An improved retrieval method of atmospheric parameter profiles based on the BP neural network. Atmos. Res. **213**, 389–397 (2018)
12. Zhai, J., Yin, Q., Ji, C.: Performance-based design of environmental parameters for offshore wind turbine foundations. Ocean Eng. **309**(P2), 118520 (2024)
13. Yuan, S., Li, D., Chen, Y., et al.: Automatic calibration method of SWMM parameters based on BP neural network algorithm. China Water Wastewater **37**(21), 125–130 (2021)
14. Gui, H., Zhang, C., Wu, Z.: Study on human-machine joint calibration of SWMM based on BP neural network. China Rural Water Hydropower **3**, 36–40 (2021)
15. Wang, M., Zhao, G., Wang, S.: Hybrid random forest models optimized by Sparrow search algorithm (SSA) and Harris hawk optimization algorithm (HHO) for slope stability prediction. Transp. Geotech. **48**, 101305 (2024)
16. Romulus, C., Subodh, C.P., Chaitanya B.P., et al.: Flood mapping based on novel ensemble modeling involving the deep learning, Harris Hawk optimization algorithm and stacking based machine learning. Appl. Water Sci. **14**(4) (2024)
17. Liu, B., Nam, V., Timon, R., et al.: A stochastic multiscale method for the prediction of the thermal conductivity of Polymer nanocomposites through hybrid machine learning algorithms. Compos. Struct. **273** (2021)
18. Liu, B., Nam, V., Zhuang, X., et al.: Stochastic integrated machine learning based multiscale approach for the prediction of the thermal conductivity in carbon nanotube reinforced polymeric composites. Compos. Sci. Technol. **224** (2022)
19. Almotairi, S., Badr, E., Salam, A. M., et al.: Three chaotic strategies for enhancing the self-adaptive harris hawk optimization algorithm for global optimization. Mathematics **11**(19) (2023)
20. Li, Y., Qian, Q.: Harris hawk optimization algorithm based on multi-population and collaborative quantumization. Control Decis. 1–9 (2024)
21. Zhu, P.: Calibration and application of SWMM model parameters based on different algorithms. Water Resour. Dev. Manag. **9**(8), 72–75+80 (2023)
22. Liu, B., Nam, V., Zhuang, X., et al.: Al-DeMat: a web-based expert system platform for computationally expensive models in materials design. Adv. Eng. Softw. **176**(C) (2022)

Assessment of Emergency Response Capacity for Urban Flooding Using Combined AHP-Fuzzy Petri Net Method

Boyu Li(✉), Rengui Jiang, Jiancang Xie, and Lingzi Wang

Xi'an University of Technology, Xi'an 710048, Shaanxi, China
18717321445@139.com, jrengui@163.com, jcxie@xaut.edu.cn

Abstract. Urban flooding has become a recurrent issue, leading to significant losses for both people and the economy. Enhancing emergency response capability is crucial for effectively managing flood events. This paper taking Xi'an city of China for an example, applying the prevention-preparation-response-recovery (PPRR) theory to structure the assessment into four criterion layers including prevention, preparation, response and recovery. Assessment indicators are selected accordingly. Analytic Hierarchy Process (AHP) method is used to determine the weights of these indicators. Fuzzy Petri Net is used to evaluate the emergency response capacity of flood disasters. The results indicate that Xi'an City's capacity is at a medium level. Key factors contributing to this rating include the arrangement of disaster prevention projects, implementation of emergency plans, the level of professional skills in disaster relief, and the disaster insurance system. The use of Fuzzy Petri Net proves to be effective in this assessment. The results offer valuable insights for improving urban flood disaster response capacity in Xi'an City, providing guidance for urban planners and policymakers to enhance flood preparedness and response strategies in rapidly urbanizing areas.

Keywords: Analytic Hierarchy Process · Emergency Response Capacity · Fuzzy Petri Net · Assessment Indicator System · Prevention-Preparation-Response-Recovery · Xi'an City

1 Introduction

Currently, urban floods occur more frequently, bringing a great threat to people's normal life and safety and the development of cities and property safety [1]. Affected by residual circulation of Super Typhoon Doksuri and subtropical high pressure, water vapor transport of Typhoon Khanun, topography and other combined effects, Beijing city and the surrounding areas of the catastrophic exceptionally suffered heavy rainstorms weather on July 29, 2023, which caused severe casualties and economic losses. As cities are characterized by high population density, intensive economic activities, high building density, etc., it will bring huge losses to the city once an urban flood disaster occurs. How to improve the emergency response capacity of urban flood disasters has become

C. Weng (Ed.): WRE 2024, LNCE 634, pp. 185–197, 2025.
https://doi.org/10.1007/978-981-96-5095-8_16

a problem that every city needs to face [1]. As an important means to improve the emergency response capacity of cities in the face of various disasters, emergency response capacity assessment not only improves the level of urban emergency response capacity and strengthens the urban emergency response capacity system, but also serves as the basis of emergency rescue work and the prerequisite for the construction of urban emergency response capacity.

Many previous studies focused on emergency response capacity assessment, which can be divided into two aspects: assessment decision-making and simulation. For example, Arrighi et al. [2] measured the reachable range of emergency response for firefighting under a transportation roadway network simulation system for sunny day and flood scenarios. Their results showed that at its worst, 22.5% of the road extension, is inaccessible; and all hospitals, fire and police stations cannot be reached, highlighting the need to take preventive action from the outset of an event that is predicted to lead to substantial inundation [2]. Zhang et al. [3] assessed the emergency response capacity and optimized the distribution of emergency response departments using simulation of the rainstorm and flooding model in the Jinshui District of Zhengzhou City, to enable the emergency response departments to achieve a more efficient and quicker emergency response capacity under the double disadvantageous factors of rainstorm and congestion in the evening rush hour [3]. Zhu [4] simulated the urban rain and flood values based on QuickFlood and Cloud-EWM to establish emergency response capacity assessment, and assessed the risk of urban rainstorm flooding disaster through the integrated risk and hard skill capacity evaluation method to provide reference for the urban flooding refinement of the emergency response management [4].

This paper takes Xi'an city for an example and uses four stages of prevention, preparation, response, and recovery from the PPRR theory as the criterion layers for the indicator system. These criterion layers are further subdivided to form the indicator layers. By analyzing the weight of each indicator in emergency response capacity through the Analytic Hierarchy Process (AHP) and Fuzzy Petri Net, we establish an AHP-Fuzzy Petri Net model. This model calculates the capacity values for each indicator to evaluate Xi'an City's preparedness for heavy rainfall and flooding. Based on this evaluation, we provide recommendations and direction for improvement to address identified shortcomings.

2 Material and Methods

2.1 Study Area

Given the frequency of urban flooding and the challenges associated with data acquisition, this paper takes Xi'an City as the study area. Xi'an City, the capital of Northwest China's Shaanxi province, is situated in the central Guanzhong Plain, characterized by higher terrain in the Southeast and lower terrain in the Northwest. Xi'an City experiences a temperate semi-arid continental monsoon climate, with hot, rainy summers and precipitation predominantly occurring from June to September. Recently, the flood season has been marked by frequent localized heavy rainfall, leading to numerous flooding events. For instance, a severe rainfall event in Xi'an city caused significant waterlogging in several road sections at Xiaozhai cross on July 24, 2016. The water depth exceeded one meter in some areas, resulting in subway disruptions and traffic congestion [5]. Map of Xi'an city of China could be found in Fig. 1.

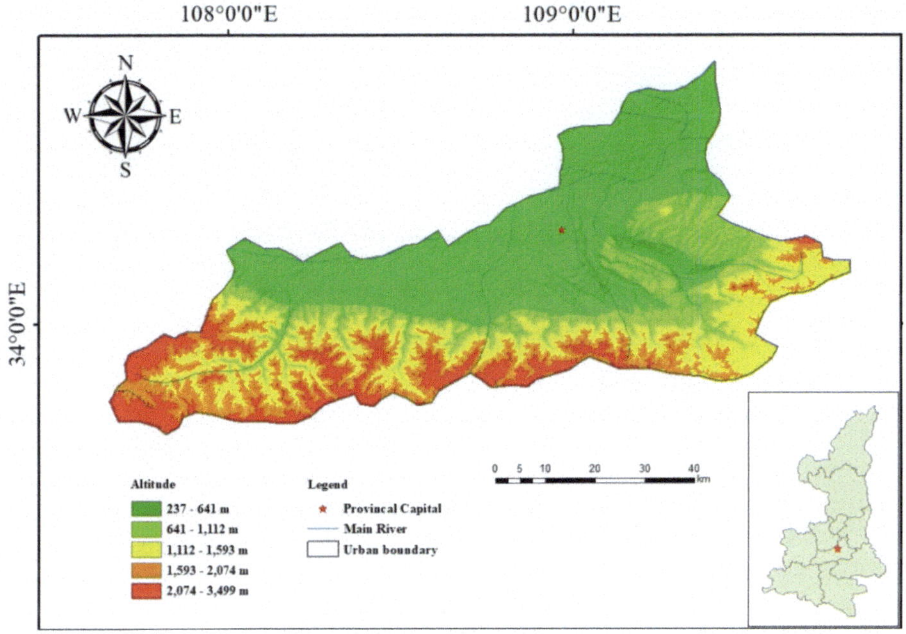

Fig. 1. Map of Xi'an city of China.

2.2 Data

The data used in this paper are obtained from the Water Statistical Yearbook and through expert scoring. To conduct this research, we invited 10 experts who are actively engaged in practice or research in the field of emergency capacity assessment to evaluate and score the indicators. These experts assess the weight of each indicator and determine its affiliation.

2.3 Assessment Indicator System

Establishment of Indicator System. By analyzing the entire process of urban flooding events, this paper uses the four stages of prevention, preparation, response, and recovery from the PPRR theory as the criterion layers for the indicator system. Through an extensive review of relevant literature, we establish an indicator library by summarizing indicators from various sources. Using the Pearson product-moment correlation coefficient method, we analyze the correlation between the indicator and target layers. Indicators with higher correlation coefficients, ease of data access, and non-redundancy

Table 1. Summary of the indicator system.

Target layer	Criterion layer	Indicator layer
A Urban flooding emergency response capacity assessment	*B1* Prevention	*C1* Arrangement of monitoring and warning facilities (density of distribution of monitoring and early warning facilities)
		C2 Arrangement of disaster prevention works (distribution and efficiency of flood storage, diversion, and drainage)
		C3 Emergency simulation exercises (number of emergency simulation drills for urban flooding scale)
	B2 Preparation	*C4* Risk information processing and dissemination (length of time for information processing and dissemination)
		C5 Disaster relief equipment reserves (distribution density of flood control sandbags, pumps, etc.)
		C6 Implementation of the emergency plans (efficiency and effectiveness of the use of the emergency plan)
	B3 Response	*C7* Level of medical assistance (number of doctors per 1,000 population)
		C8 Level of professional skills in disaster response (number of holders of national emergency responder certificates)
		C9 Emergency response speed level (average motor vehicle travel speed on urban main roads during rainfall)
	B4 Recovery	*C10* Catastrophe insurance system (percentage of disaster prevention and loss prevention in catastrophe insurance system)
		C11 Post-disaster recovery and reconstruction capacity (compliance rate of urban water supply quality)
		C12 Replenishment of emergency supplies (special funds for emergency relief supplies reserve)

are selected. Ultimately, twelve indicators are chosen to form the indicator layer of this study. The indicator system is summarized in Table 1.

Calculation of Capacity Value. Based on the affiliation degree of each indicator, we jointly use the AHP and Fuzzy Petri Net to establish an assessment model for urban flood emergency response capacity. This model is applied to evaluate the emergency response capacity for urban flooding in Xi'an City by deriving the capacity values for each indicator.

Analytic Hierarchy Process. In order to ensure the reliability of the assessment results for urban flood emergency response capacity, this paper utilizes AHP to determine the weights of the indicators [6]. AHP is a systematic and effective decision-making methodology for qualitative and quantitative analysis. AHP decomposes elements related to decision-making into multiple levels, such as objectives, guidelines, and programs. Wang et al. [7] utilized AHP to develop a community emergency management capacity evaluation model from the perspective of disaster resilience, employing hierarchical analysis and the distance between superiority and inferiority solutions [7]. Koc et al. (2021) employed AHP in conjunction with a Monte Carlo simulation to assess 35 urban flood risk factors. Their findings indicated that urban infrastructure, land use, and institutional capacity were key criteria for reducing urban flood risks [8]. Wang et al. [9] applied AHP to ascertain the weights of indicators within a two-level fuzzy comprehensive evaluation model, creating a hazard scenario and employing two level fuzzy comprehensive evaluation for hazard assessment. The method they proposed was deemed reasonable and practical [9]. Construct the judgment matrix:

$$A = \begin{bmatrix} \frac{C_1}{C_1} & \frac{C_1}{C_2} & \cdots & \frac{C_1}{C_n} \\ \frac{C_2}{C_1} & \frac{C_2}{C_2} & \cdots & \frac{C_2}{C_n} \\ \vdots & \vdots & \ddots & \vdots \\ \frac{C_n}{C_1} & \frac{C_n}{C_2} & \cdots & \frac{C_n}{C_n} \end{bmatrix} \tag{1}$$

After constructing the judgment matrix, find the largest eigenvalue λ and its corresponding eigenvector v for the judgment matrix and extrapolate the weight vector from the feature vector ω. Then, do a consistency test for the judgment matrix. As shown in Eqs. (2) and (3).

$$CI = (\lambda_{\max} - n)/(n - 1) \tag{2}$$

$$CR = CI/RI \tag{3}$$

Fuzzy Petri Net. Fuzzy Petri Net is an evolution of traditional Petri Net, integrating strict mathematical theory with visualizable modeling processes. Fuzzy Petri Net has found extensive application in assessment research. For example, Yu et al. (2019) employed Fuzzy Petri Net to evaluate the slope safety of existing buildings. This approach not only provided results on the slope safety of these buildings but also identified the primary risk indicators that most significantly impact slope stability, as well as the influence of variable factors on slope safety [10]. Additionally, Li et al. [11] proposed a method for assessing government public panic emergency management capacity using

Fuzzy Petri Net. By analyzing factors influencing public panic across four evolutionary stages—gestation, outbreak, diffusion and fading, they developed a hierarchical evaluation index system for assessing emergency management capabilities. Their work demonstrated the effectiveness and advantages of models and rules constructed with Fuzzy Petri Net, showcasing their superiority over traditional assessment methods [11].

Define the Fuzzy Petri Net for urban flood emergency response capacity assessment as a seven-tuple for:

$$ECFPN = (P, T, IN, OUT, U, \lambda, M) \tag{4}$$

Assuming that *A*, *B*, *C* and *D* are matrices of order $n \times m$, the following operator is used in the reasoning process:

(1) addition operator $\oplus$:

$$C = A \oplus B \Leftrightarrow c_{ij} = \max(a_{ij}, b_{ij}) \tag{5}$$

(2) direct multiplication operator $\otimes$:

$$D = A \otimes B \Leftrightarrow d_{ij} = a_{ij} \otimes b_{ij} \tag{6}$$

in which, i=1, 2, 3,... n, j=1, 2, 3,... m

P and *T* are the set of contingency capacity influencing factors and the set of contingency behaviors, respectively, $|P|=n$, $|T|=m$, and the matrix form is $IN = \{\ \omega ij\}$, $OUT = \{\ \gamma ji\}$, $U = (\ u1, u2, u3, \ldots, um)$, then the inference steps are as follows:

(1) Initialize the input matrix *IN*, the output matrix *OUT*, the confidence level *U* and the contingent capacity judgment matrix *Q*;
(2) Calculate the confidence level of the factors influencing the initial emergency response capacity by making k=0;

$$I^T M(k) \tag{7}$$

(3) Calculate the credibility of the occurrence of the emergent behavior tj;

$$U \otimes O \tag{8}$$

(4) After the contingency behavior has occurred, the credibility of the change in the contingency capacity influencing factors and the next round of states are calculated:

Degree of credibility:

$$U \otimes O[I^T M(k)] \tag{9}$$

next round status:

$$M(k+1) = M(k) \oplus (U \otimes O)[I^T M(k)] \tag{10}$$

(5) If $M(k+1) \neq M(\ k)$, let $k = k+1$ and fold back to step (2) and loop through the calculation of the subsequent steps. Until $M(\ k+1)=M(\ k)$, the loop ends;

(6) According to $S = M(\ k)QT$, calculate the capacity value of emergency response capacity.

$$S = M(k)Q^T \tag{11}$$

The fuzzy mathematical method of affiliation assignment is used to establish the assessment set by categorizing the competency levels into five levels: high, relatively high, moderate, relatively low, low. Table 2 shows the evaluation criteria for emergency response capability levels. The emergency response capability evaluation matrix is Q.

$$Q^T = \{10, 8, 6, 4, 2\} \tag{12}$$

Table 2. Criteria for evaluating the level of emergency response capacity.

Capacity level	Capacity value
High	8~10
Relatively high	6~8
Moderate	4~6
Relatively low	2~4
Low	0~2

3 Results

3.1 Weights of Assessment Indicators

The determined indicator system and its weights are shown in Table 3.

3.2 Determination of Affiliation Degree

By distributing questionnaires to 10 experts and judging each of the indicators listed in the previous section according to the classification of the assessment level, the results of the experts' feedback on the affiliation degree are shown in Table 4.

3.3 Calculation of Capacity Value

As shown in Fig. 2, urban flood emergency response capacity assessment indicator system is converted to Fuzzy Petri Net model. The first column p (p1~p12) in Fuzzy Petri Net model diagram for urban flood emergency response capacity assessment corresponds to the indicator layer, the second column p (p13~p16) corresponds to the criterion layer, and the third column p (p17) corresponds to the target layer.

Table 3. Distribution of the indicator system and weights.

Target level	Criterion layer	Criterion layer weight	Indicator layer	Indicator layer weight	Total weight
A	B1	0.565	C1	0.659	0.372
			C2	0.185	0.104
			C3	0.156	0.088
	B2	0.262	C4	0.694	0.182
			C5	0.132	0.035
			C6	0.174	0.046
	B3	0.117	C7	0.072	0.008
			C8	0.279	0.033
			C9	0.649	0.076
	B4	0.056	C10	0.078	0.004
			C11	0.487	0.027
			C12	0.435	0.025

Table 4. Affiliation degree table of emergency response capacity assessment indicators.

Target layer	Criterion layer	Indicator layer	Assessment set				
			High	Relatively high	Moderate	Relatively low	Low
A	B1	C1	0	0.4	0.2	0.4	0
		C2	0	0.4	0.2	0.3	0.1
		C3	0.1	0.2	0.4	0.2	0.1
	B2	C4	0.1	0.2	0.2	0.5	0.1
		C5	0.1	0.5	0.3	0.1	0
		C6	0	0.4	0.2	0.3	0.1
	B3	C7	0.1	0.5	0.4	0	0
		C8	0	0.4	0.3	0.1	0.2
		C9	0.1	0.4	0.2	0.2	0.1
	B4	C10	0	0.1	0.6	0.1	0.2
		C11	0.3	0.6	0.1	0	0
		C12	0.1	0.3	0.4	0.1	0

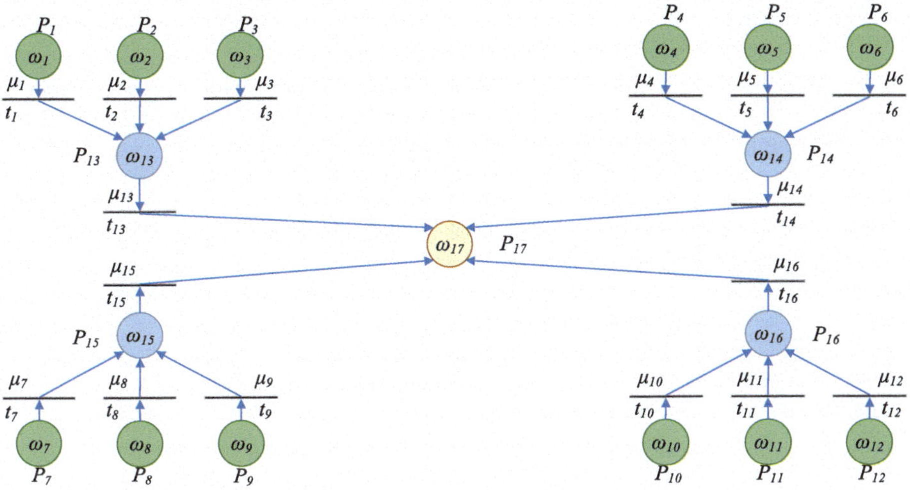

Fig. 2. Fuzzy Petri Net model for urban flood emergency response capacity assessment.

The MATLAB R2018b software is used to follow the above inference steps, and the following results are obtained.

$$M(3) = M(2) = \begin{pmatrix} 0 & 0.4 & 0.2 & 0.4 & 0 \\ 0 & 0.4 & 0.2 & 0.3 & 0.1 \\ 0.1 & 0.2 & 0.4 & 0.2 & 0.1 \\ 0.1 & 0.2 & 0.2 & 0.5 & 0.1 \\ 0.1 & 0.5 & 0.3 & 0.1 & 0 \\ 0 & 0.4 & 0.2 & 0.3 & 0.1 \\ 0.1 & 0.5 & 0.4 & 0 & 0 \\ 0 & 0.4 & 0.3 & 0.1 & 0.2 \\ 0.1 & 0.4 & 0.2 & 0.2 & 0.1 \\ 0 & 0.1 & 0.6 & 0.1 & 0.2 \\ 0.3 & 0.6 & 0.1 & 0 & 0 \\ 0.1 & 0.3 & 0.4 & 0.1 & 0.1 \\ 0.014 & 0.332 & 0.208 & 0.315 & 0.031 \\ 0.074 & 0.247 & 0.192 & 0.371 & 0.078 \\ 0.065 & 0.366 & 0.218 & 0.142 & 0.109 \\ 0.171 & 0.387 & 0.243 & 0.046 & 0.053 \\ 0.042 & 0.301 & 0.197 & 0.280 & 0.051 \end{pmatrix} \tag{13}$$

$$\begin{aligned} S &= M(3)Q^T \\ &= \big(6\ 5.8\ 6\ 6\ 7.2\ 5.8\ 7.4\ 5.8\ 6.4 \\ &\quad 5.2\ 8.4\ 6.4\ 5.367\ 5.511\ 5.674\ 6.552\ 5.232\big) \end{aligned} \tag{14}$$

4 Discussion

The results of the above MATLAB calculations are organized into three panels: indicator layer (Table 5), criterion layer (Table 6) and target layer (Table 7). The capacity values of the criterion layer and indicator layer are sorted from high to low.

Table 5. Ranking of capacity value for indicators in the indicator layer (highest to lowest).

Indicator layer	Capacity value
C11 Post-disaster recovery and reconstruction capacity	8.4
C7 Level of medical assistance	7.4
C5 Disaster relief equipment reserves	7.2
C9 Emergency response speed level	6.4
C12 Replenishment of emergency supplies	6.4
C1 Arrangement of monitoring and warning facilities	6.0
C3 Emergency simulation exercises	6.0
C4 Risk information processing and dissemination	6.0
C2 Arrangement of disaster prevention works	5.8
C6 Implementation of the emergency plans	5.8
C8 Level of professional·skills in disaster response	5.8
C10 Catastrophe insurance system	5.2

From Table 5, we can find the capacity values of seven indicators at the indicator layer, namely, C1-Arrangement of monitoring and warning facilities, C3-Emergency simulation exercises, C4-Risk information processing and dissemination, C2-Arrangement of disaster prevention works, C6-Implementation of emergency plans, C8-Level of professional skills in disaster response, range from 4 to 6, indicating that the capacity of these indicators is of Medium level. The capacity values of 5 indicators, namely, C7-Level of medical assistance, C5-Disaster relief equipment reserves, C9-Emergency response speed level, C12-Replenishment of emergency supplies, and C11-Post-disaster recovery and reconstruction capacity, are greater than 6, indicating that the capacity value of these indicators is Relatively High.

From Table 6, we could find that the capacity values of 3 criteria namely *B1*-Prevention, *B2*-Preparation and *B3*-Response in criterion layer are between 4 and 6, indicating that the capacity level of these criteria is Medium. The capacity value of *B4* recovery in the criterion layer is greater than 6, indicating that the capacity level of *B4*-Recovery is Relatively High.

The capacity value of *A*-Urban flood emergency response capacity assessment in the target layer is between 4 and 6, with a Medium capacity level. In order to better reflect the common weaknesses of each indicator layer and each criterion layer, radar chart for capacity value of indicator layer, criterion layer and target layer is drawn in Fig. 3.

Table 6. Ranking of capacity value for criteria in the criterion layer (highest to lowest).

Criterion layer	Capacity value
B4 Recovery	6.552
B3 Response	5.674
B2 preparation	5.511
B1 Prevention	5.367

Table 7. Capacity value for target layer.

Target layer	Capacity value
A Urban flooding emergency response capacity assessment	5.232

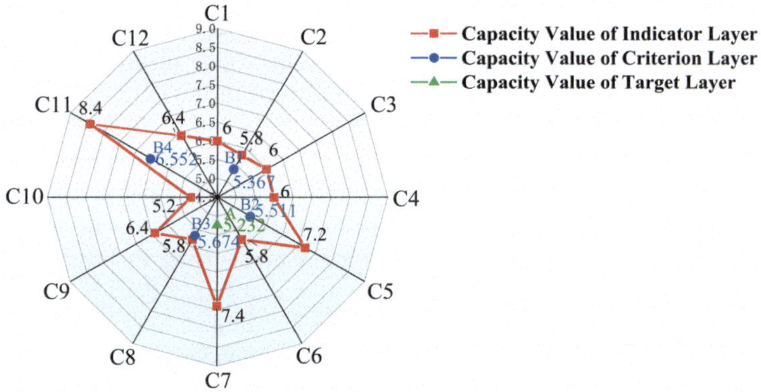

Fig. 3. Radar chart for capacity value of indicator layer, criterion layer and target layer.

Xi'an City should prioritize both the prevention and preparation stages of urban flooding. This effort can begin with engineering measures, such as increasing investment in the construction and renovation of flood storage, diversion, and drainage systems to enhance their efficiency. These improvements aim to address the root causes of urban flooding and mitigate its impacts on the city. Additionally, non-engineering measures should be implemented, including the establishment of urban rainstorm and flood prediction systems, as well as the development of effective warning systems and response plans. These actions are crucial for reducing casualties and economic losses associated with urban flooding.

Moreover, Xi'an City should focus on conducting large-scale emergency simulation exercises for urban flooding. Currently, such exercises are infrequent, leading to a general lack of preparedness among residents. For instance, people may not know how to avoid driving through water-filled tunnels during a flood.

The significance of monitoring, forecasting, and warning cannot be overstated, as it holds the highest weight in the indicator layer. Forecasting and warning systems for rainstorms face significant challenges, with reliability issues in predicting sudden rainstorms. Achieving breakthroughs in rainstorm forecasting, with accurate predictions of location, timing, and intensity, would greatly enhance the city's ability to prepare for, respond to, and recover from urban floods. This would also significantly reduce economic losses and casualty throughout the city.

5 Conclusions

With the increasing frequency of extreme weather events in recent years, Xi'an City has been experiencing severe urban flooding hazards increasingly. Current research has not evaluated the emergency response capacity of Xi'an City concerning urban flooding.

The urban flood emergency response capacity of Xi'an city is assessed as medium. Several factors contribute to this medium level, including the arrangement of monitoring and warning facilities, emergency simulation exercises, risk information processing and dissemination, disaster prevention infrastructure, implementation of emergency plans, professional skills in disaster response, and the catastrophe insurance system. These factors align with the medium capacity level observed in the corresponding criterion layer, indicating that the use of Fuzzy Petri Net for assessing urban flood emergency response capacity provides a clear and effective means of simulation and analysis.

Fuzzy Petri Net, which excels in processing information in parallel and accurately depicting the fuzzy states of uncertain indicators, is applied to assess urban flood emergency response capacity. The results of this assessment using Fuzzy Petri Net align well with the situation in Xi'an City, demonstrating that Fuzzy Petri Net is effective in evaluating emergency response capacity for urban flooding.

However, the AHP weight calculation and capability value results are mainly based on expert judgment, which introduces the potential for subjectivity to affect the final assessment results. Future research will explore integrating real-time data from IoT devices into the Fuzzy Petri Net model and incorporating quantitative metrics alongside the qualitative ones to further enhance data reliability and result credibility.

References

1. Xie, J.C., Li, B., Chai, L., Jiang, R.J.: Some realizations of dealing with the urban flood disaster problems. J. Xi'an Univ. Technol. **31**(1), 25–33 (2015)
2. Arrighi, C., Pregnolato, M., Dawson, R.J., Castelli, F.: Preparedness against mobility disruption by floods. Sci. Total Environ. **654**, 1010–1022 (2019)
3. Zhang, J.P., Li, Y.D., Zuo, Q.T., Zhang, C.Y.: Evaluation and optimization of urban flood emergency response capacity. J. North China Univ. Water Resour. Electr. Power (Nat. Sci. Ed.) **45**(4), 8–15 (2024)
4. Zhu, Y.X.: Urban rainstorm waterlogging risk and emergency response capability evaluation based on QuickFlood and Cloud-EWM. Mod. Electron. Tech. **46**(16), 97–104 (2023)
5. Li, W., Jiang, R.J., Wu, H., Xie, J.C., Zhao, Y., Li, F.W.: An integrated urban flooding risk analysis framework leveraging machine learning models: a case study of Xi'an, China. Int. J. Disaster Risk Reduct. **112**(104770), 1–22 (2024)

6. Yuan, X.M., Sang, L.H., Shen, F.X., Tian, F.C.: FAHP-based flood risk assessment on Beijing-Tianjing-Hebei region. Water Resour. Hydropower Eng. **49**(10), 37–45 (2018)
7. Wang, K., Wang, Z., Deng, J., Feng, Y.Y., Li, Q.F.: Study on the evaluation of emergency management capacity of resilient communities by the AHP-TOPSIS method. Int. J. Environ. Res. Public Health **19**(16201), 1–14 (2022)
8. Koc, K., Isik, Z.: Assessment of urban flood risk factors using Monte Carlo analytical hierarchy process. Nat. Hazard Rev. **22**(4), 1–14 (2021)
9. Wang, X., Wang, S.Y., Qi, J.S.: Open-channel landslide hazard assessment based on AHP and fuzzy comprehensive evaluation. Water Supply **20**(8), 3687–3696 (2020)
10. Yu, D.D., Yan, X.: Study on safety evaluation of existing slope based on fuzzy Petri net and its application. Yangtze River **50**(5), 145–150 (2019)
11. Li, C., Zhao, Y.M., Gao, L., Ni, Y., Liu, X.X.: Government's public panic emergency capacity assessment and response strategies under sudden epidemics: a fuzzy Petri net-based approach. Heliyon **10**(9), 1–17 (2024)

Design Water Surface Profile Calculation of Leshan City

Xing Zhou[1,2](✉), Song Sha[1,2], Qinghui He[1,2], Liping Li[3], and Ning Zhao[1,2]

[1] Changjiang Survey, Planning, Design and Research Co., Ltd., Wuhan 430010, China
261436652@qq.com, zhaoning@cjwsjy.com.cn
[2] Hubei Key Laboratory of Basin Water Security, Wuhan 430010, China
[3] Hydrology Bureau, Yangtze River Water Resources Commission, Wuhan 430010, China
lilp@cjh.com.cn

Abstract. The design of flood water surface profile calculation is the core work of river regulation projects. In this paper, the central urban area of Leshan was taken as the research object. The Wutongqiao station downstream of the confluence of the Min River, Dadu River, and Qingyi River was selected as the flood control section. The flood composition of the three rivers was calculated using the typical year flood composition method. The roughness was determined using the measured flood process and flood trace investigation. The outer line of different water surface profiles was taken as the result. The results show that the typical year combination method uses the actual flood composition as a sample for flood allocation, which is simple and easy to implement. Multiple typical years can more comprehensively reflect the characteristics of regional flood composition, which is suitable for the calculation of the design flood water surface line in areas with complex flood composition.

Keywords: Water Surface Profile · Flood · Leshan · The Typical Year Method

1 Introduction

The design flood water surface profile calculation is the core work for river regulation projects, which directly affects the embankment elevation. The higher or lower water surface profile will affect the economy and safety of the project [1–3]. Therefore, the calculation of the design of the water surface line must be scientifically rigorous. Leshan City is in the southwestern part of the Sichuan Basin and the southern part of the Chengdu Plain. It is one of the central cities of the Chengdu Plain, a national historical and cultural city, and an important tourist city. The center area of Leshan City is located at the confluence of the Qingyi River, Dadu River, and Min River, with a high risk of flooding. Due to the unique geomorphology and climate characteristics of each river, the three rivers have different flood characteristics, resulting in a complex composition of floods in Leshan urban area, bringing great challenges to the design flood water surface profile calculation work. Previous studies have only focused on individual rivers, ignoring the impact of other rivers. Therefore, this study seeks a new method for calculating the water surface profile for Leshan City considering the mutual influence of multiple rivers.

C. Weng (Ed.): WRE 2024, LNCE 634, pp. 198–205, 2025.
https://doi.org/10.1007/978-981-96-5095-8_17

2 Study Area

This study was conducted in the center city of Leshan as shown in Fig. 1. Min River is a first-class tributary of the Yangtze River with a whole basin area of 135, 387 km^2, of which the area above the mouth of Dadu River confluence is 33, 887 km^2, the total length of the main river channel is 735 km, the average specific drop of 4.84‰. Dadu River is the largest tributary of Min River, which joins Min River at Xiaogongzui in the center of Leshan City, with an area of 90, 273 km^2, and a trunk channel length of 1062 km, an average specific drop of 1.20‰. Qingyi River is a first-class tributary on the left bank of Dadu River, which injects into Dadu River at Caoxiedu in the northern part of Leshan city, with a watershed area of 12, 842 km^2, a length of the main channel of 289 km, and an average specific drop of 2.18‰.

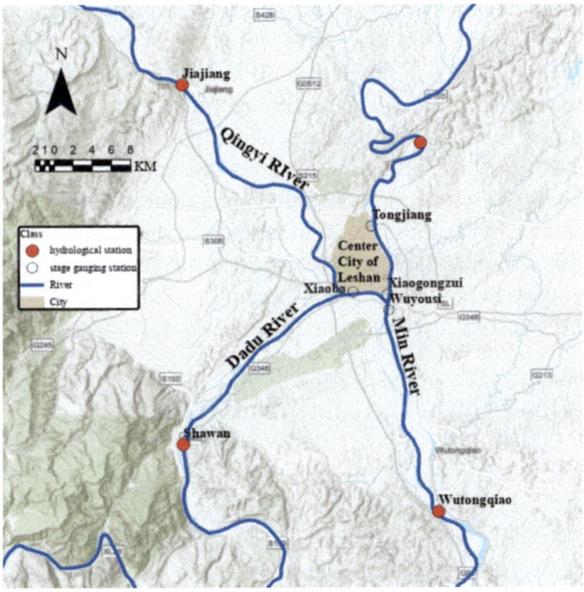

Fig. 1. Study area.

3 Design Flood

The design flood was calculated for each of four stations, namely, Penshan, Wutongqiao, Shawan, and Jiajiang, which were in the mainstream of the Min River, Dadu River, and Qingyi River, using a discrete flood series composed of long term measured flood series and historical surveyed floods. The design flood was calculated for each station by the P-III frequency analysis and the result shows in Table 1.

The result indicates that the floods of the three rivers are at different frequencies. For example, the sum of the 2% frequency flood under Pensha, Shawan, and Jiajiang

Table 1. Design flood results of representative stations.

River	Station	Design Flood/ $m^3 \cdot s^{-1}$				
		1%	2%	5%	10%	20%
Min River	Penshan	14900	13500	11600	10100	8440
	Wutongqiao	47500	43600	34300	28600	23000
Daduhe River	Shawan	10900	10100	9070	8220	7310
Qingyi River	Jiajiang	19100	17400	15100	13300	11300

is significantly less than the same frequency flood of Wutongqiao. Thus, the traditional equivalent frequency regional composition method may underestimate the flood of one river while overrating the other rivers. Thus, we use the typical year method of designing flood regional composition in this paper by choosing Wutongqiao station as a flood control section. Five top-ranked floods of 1955, 1959, 1966, 1981 and 2020 were selected as typical years shown in Table 2. The incoming water from Wutongqiao is allocated according to the flood composition of the selected five typical years. The sum of the measured flood peak flow under the three stations in each typical year is different from Wutongqiao due to the difference in controlled area, coupled with the slight difference in the time of flood peak evolution to Wutongqiao station. In order to guarantee the water balance, the design peak flow of Wutongqiao is taken as the total amount, which is distributed according to the proportion of Shawan, Penshan, and Jiejiang stations under typical years, and the results are shown in Table 3 under 2% frequency.

4 Water Surface Profile Calculation Method

The design flood water surface profile was calculated using the software MIKE 11 hydrodynamic module (HD) developed by the Danish Institute for Water Resources and Water Environment (DHI), which is widely used in the field of hydrology and hydraulic engineering and has a high degree of accuracy and stability [4–6]. The MIKE 11 HD uses the 6-point Abbott-Ionescu implicit finite difference scheme for the computation of unsteady flows in rivers by solving the one-dimensional Saint-Venant equations as below:

$$\frac{\partial A}{\partial t} + \frac{\partial Q}{\partial x} = q \tag{1}$$

$$\frac{\partial Q}{\partial t} + \frac{\partial (\alpha \frac{Q^2}{A})}{\partial x} + gA\frac{\partial h}{\partial x} + \frac{gQ|Q|}{C^2 AR} = 0 \tag{2}$$

where, Q means discharge, A means flow area, q means lateral inflow, Z means water level, C means Chezy resistance coefficient, R means hydraulic or resistance radius, and A means momentum distribution coefficient.

Table 2. Flood regional composition of typical years.

Typical Year	Term	Wutongqiao	Pensha (1)	Jiajiang (2)	Shawan (3)	Sum of 1~3
1966	Peak Flood ($m^3 \cdot s^{-1}$)	25200	7530	12800	6200	26530
	Proportions of Wutongqiao (%)	100	29.88	50.79	24.60	105.28
1959	Peak Flood ($m^3 \cdot s^{-1}$)	27500	10900	12000	4650	27550
	Proportions of Wutongqiao (%)	100	42.41	46.69	18.09	107.2
1981	Peak Flood ($m^3 \cdot s^{-1}$)	26300	8730	7960	5900	22590
	Proportions of Wutongqiao (%)	100	33.19	30.27	22.43	85.89
1955	Peak Flood ($m^3 \cdot s^{-1}$)	35300	4850	16000	8280	29130
	Proportions of Wutongqiao (%)	100	13.74	45.33	23.46	82.52
2020	Peak Flood ($m^3 \cdot s^{-1}$)	43800	13500	19700	10700	43900
	Proportions of Wutongqiao (%)	100	30.82	44.98	24.43	100.23

4.1 Model Range and Boundary Conditions

The calculation range of the model includes the Min River, Dadu River, and Qingyi River, and the hydrological station is used as the control node for the model's calculation range selection. The calculation range of the Min River is Pengshan Station ~ Wutongqiao Station, which is about 159 km, the calculation range of the mainstream of the Qingyi River is Jiajiang Station ~ estuary, which is about 32 km long, and the calculation range of the mainstream of the Dadu River is Shawan Station ~ estuary, which is about 55 km long. The cross-sectional data of the river channel measured in 2018 and 2019 are used for

Table 3. Design flood regional composition of typical year method.

Typical Year	Term	Wutongqiao	Pensha (1)	Jiajiang (2)	Shawan (3)
1955	Peak Flood ($m^3 \cdot s^{-1}$)	43600	7259	**23949**	**12391**
	Frequency (%)	2	35.43	0.12	0.21
1959	Peak Flood ($m^3 \cdot s^{-1}$)	43600	17248	18992	7360
	Frequency (%)	2	0.31	1.03	17.53
1966	Peak Flood ($m^3 \cdot s^{-1}$)	43600	12374	21037	10189
	Frequency (%)	2	3.79	0.43	1.63
1981	Peak Flood ($m^3 \cdot s^{-1}$)	43600	16851	15365	11388
	Frequency (%)	2	0.39	4.50	0.54
2020	Peak Flood ($m^3 \cdot s^{-1}$)	43600	13407	19563	10625
	Frequency (%)	2	2.28	0.81	1.10

the cross-sectional data. The cross-sectional data of the urban river section are measured in 2023, and the average cross-section spacing is about 500 m.

The upper boundaries of the Min River, Dadu River, and Qingyi River are designed or measured flood of Pengshan Station, Shawan Station, and Jiajiang Station, respectively, and the water level ~ flow relationship curve of Wutongqiao Station is used for the lower boundary conditions.

4.2 Model Calibration

Roughness is the most critical parameter for the calculation of a one-dimensional hydrodynamic model, reasonable selection of roughness is the key to a reliable water surface profile. The measured hydrological data of the "8.18" flood in 2020 and the flood trace survey data were used to calibrate the model.

The results show that the mathematical model adapts to the flow characteristics and water surface ratio of the Min River, Dadu River, and Qingyi River. The calculated water level process corresponds well to the measured peaks and valleys for each water level station, and the flood elements such as flood rise time, peak occurrence time, ebb time and maximum water level are consistent as shown in Fig. 2. The highest water level calculated by the model is generally consistent with the flood marks surveyed by 818 in 2020, with an error between −0.056~0.123 m as shown in Fig. 3. The model calibration result shows that the simulation accuracy of the model meets the needs and can be used to calculate the design water surface lines of the Min River, Dadu River, and Qingyi River in Leshan City.

The calibrated roughness of the Dadu River, Qingyi River, and Min River is 0.042~0.056, 0.039, 0.028~0.039, respectively.

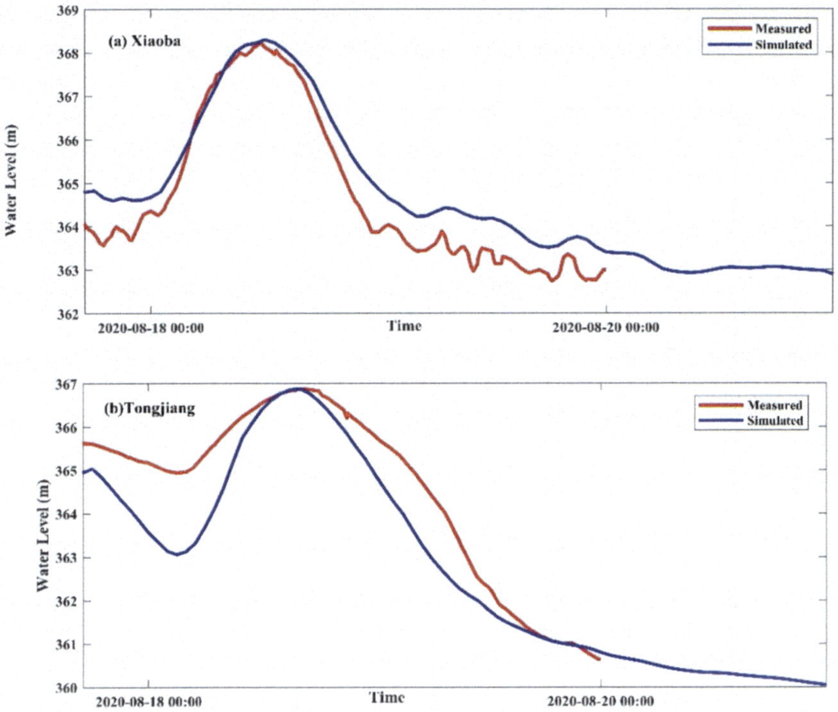

Fig. 2. Model calibration results at (a) Xiaoba (b) Tongjiang Water level station.

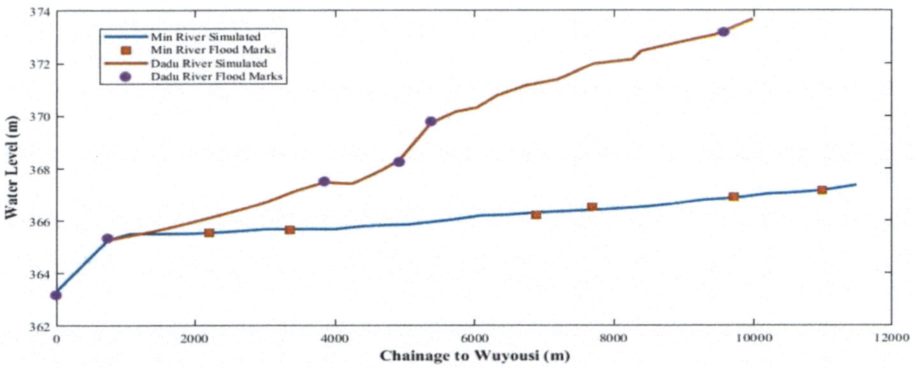

Fig. 3. Comparison between simulated flood water level and flood marks.

4.3 Results and Discussion

The calibrated model calculated 2% frequency flood water profile of 1955, 1959, 1966, 1981 and 2020 as shown in Fig. 4. The results of different typical years were significantly different. The typical year corresponding to the minimum water level profile of the Min River section above the Dadu River estuary is 1955, with the water surface line is between

366.72 m and 367.41 m, while the typical year corresponding to the maximum is 1959, with the water surface line between 367.03 m and 369.97 m. The regularity of the water profile on the side of the Qingyi-Dadu River is reversed, with the highest water profile corresponding to the 1955 typical year, while the lowest water profile occurred in the 1959 typical year. The upper envelope was selected as the final design water level profile for safety reasons.

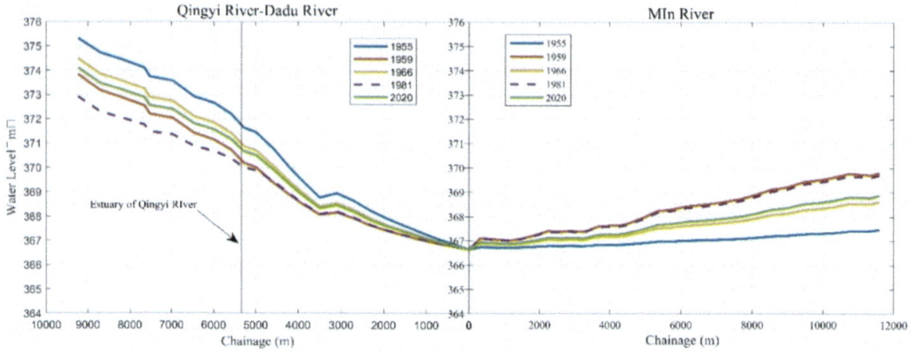

Fig. 4. Calculated design water surface profile of different typical year.

It was not difficult to discover that the highest water profile corresponded to the highest water flow allocated in the typical year. Such results reveal the importance of the impact of the design flood on water levels. The typical year combination method uses the actual flood composition as a sample for flood allocation, which is simple and easy to implement, and multiple typical years can more comprehensively reflect the characteristics of regional flood composition, which is suitable for the calculation of the water surface line in areas with complex flood composition such as Leshan.

5 Conclusions

The design of flood water surface profile calculation is the core work of river regulation projects. In this paper, the central urban area of Leshan was taken as the research object. The main conclusions are as follows:

(1) The measured hydrological data of the "8.18" flood in 2020 and the flood trace survey data were used to calibrate the model, and the calibrated roughness is 0.028~0.056.
(2) The highest water profile corresponded to the highest water flow allocated in which typical year, and the upper envelope was selected as the final design water level profile for safety reasons.
(3) Multiple typical years can more comprehensively reflect the characteristics of regional flood composition, but the selection criteria for typical years and the quantity will affect the results.
(4) The applicability and selection criteria of the typical year method will be explored in future research with more case studies.

References

1. Naik, B., Kaushik, V., Kumar, M.: Water surface profile in converging compound channel using gene expression programming. Water Supply **22**(5), 5221–5236 (2022)
2. Kaushik, V., Kumar, M.: Water surface profile prediction in non-prismatic compound channel using support vector machine (SVM). AI Civil Eng. **2**(1), 6 (2023)
3. Vatankhah, A.R.: New solution method for water surface profile along a side weir in a circular channel. J. Irrig. Drain. Eng. **138**(10), 948–954 (2012)
4. Luo, P., Luo, M., Li, F., Qi, X., Huo, A., Wang, Z., et al.: Urban flood numerical simulation: research, methods, and future perspectives. Environ. Model. Softw. **156**, 105478 (2022)
5. Panda, R.K., Pramanik, N., Bala, B.: Simulation of river stage using artificial neural network and MIKE 11 hydrodynamic model. Comput. Geosci. **36**(6), 735–745 (2010)
6. Nkwunonwo, U.C., Whitworth, M., Baily, B.: A review of the current status of flood modelling for urban flood risk management in the developing countries. Sci. Afr. **7**, e00269 (2020)

Dynamic Calculation Method and Application of Ecological Flow with Adaptive Target Guarantee Rate

Pengxin Deng(✉), Wei Li, Changjiang Xu, Jianping Bing, and Dong Wang

Bureau of Hydrology, Changjiang Water Resources Commission, Wuhan 430010, China
dpx_work@163.com

Abstract. To effectively safeguard the ecological flow of rivers and lakes, a dynamic calculation method tailored to the target guarantee rate has been developed. In alignment with current ecological flow management requirements, the target ecological flow for each month at key tributary control sections in the middle and lower reaches of the Hanjiang River has been analyzed and determined. Research indicates that the fundamental ecological flow rates of various sections within the study area, calculated using traditional hydrological methods, range from 0.492 to 7.71 m^3/s, comprising 2.56% to 14.1% of the average annual flow. Due to significant variations in water inflow in the basin, the calculation results obtained through various methods differ greatly, posing challenges in ecological environment. By considering the dynamic calculation method of ecological flow that ensures the target guarantee rate, the monthly target values of ecological flow in each section have been determined to range from 1.04 to 25.2 m^3/s, exhibiting an overall pattern of "higher during the wet season and lower during the dry season," which aligns with the inflow characteristics of the watershed. The analysis of daily average satisfaction indicates that the method produces highly satisfactory results. Under the condition of fulfilling the current management requirements, it can enhance the annual ecological water volume of the river by 16 to 317 million cubic meters, thereby better safeguarding the ecological water needs within the riverbed. The findings of this study can serve as a valuable reference for water resource protection, scheduling, and supervision in the Hanjiang River Basin.

Keywords: Target Assurance Rate · Ecological Flow · Dynamic Computation · Method Theory · Hanjiang River Basin

1 Introduction

Safeguarding the ecological flow of rivers and lakes serves as a fundamental and prerequisite basis for revitalizing their ecological environments. Ecological flow can be categorized into basic and target ecological flow, depending on the management objectives. Basic ecological flow refers to the minimum quantity that must be maintained within the river channel to ensure that the ecological environment function, corresponding to the established ecological environment and lakes, remains [1]. Target ecological flow refers

C. Weng (Ed.): WRE 2024, LNCE 634, pp. 206–219, 2025.
https://doi.org/10.1007/978-981-96-5095-8_18

to the amount of water that needs to be retained within the river channel to ensure the normal functioning of the ecological environment by the designated ecological environment protection objectives for rivers and lakes. It serves as a control value for limiting the consumption of river and lake water by the economy and society outside the river channel [1, 2]. Both basic and target ecological flows must fulfill specific protection objectives. Hence, the determination of ecological flow must be aligned with the goals of river and lake conservation as well as water resource management. The "Calculation Specification for Ecological Water Demand of Rivers and Lakes" (SL/T712-2021), promulgated in China in 2021 [3], explicitly stipulates that the assurance rate of the basic ecological flow for rivers and lakes must not fall below 90%. Concurrently, pertinent policies on water resource management and river and lake governance and protection stipulate that the compliance rate of the basic ecological flow for key rivers and lakes nationwide must be no less than 90% [4]. A 90% guarantee rate or satisfaction level of ecological flow has emerged as a crucial regulatory objective for river and lake management and conservation at the present stage.

China's current approach to managing ecological flow in rivers and lakes primarily focuses on basic ecological flow, gradually shifting towards the implementation of target ecological flow. Despite extensive academic research on target ecological flow, its practical application often falls short of expectations due to the influence of regional hydrological and meteorological factors and policies governing river and lake management and protection. At the current stage, the academic community predominantly determines ecological flow using a single numerical value. Common methods include hydrological methods, hydraulic methods, ecological habitat methods, and holistic analysis methods [5]. Among these, the hydrological method is the most widely used both domestically and internationally due to its relatively simple data requirements. Representative methods within this category include the Tennant method, the method based on the driest month in the past decade, the Q_p method, the frequency curve method, the 7Q10 method, and the RVA method [6–9]. However, for rivers with significant interannual variations in runoff, relying on a single numerical value to determine ecological flow for management purposes is inadequate for adapting to the changing natural hydrological conditions of the basin. This approach struggles to meet the management requirement of achieving an ecological flow compliance rate of over 90% in both normal and dry years. In some wet years, the excessive concentration of water during the flood season results in the annual compliance rate failing to meet the standard. Therefore, it is essential to allocate ecological flows appropriately throughout the year, considering the variations in water availability. Additionally, it is important to develop a dynamic calculation method tailored to the target assurance rate.

The Middle and Lower reaches of the Hanjiang River basin (MLHRB) serve as a vital economic corridor in Hubei Province and constitute a significant component of the Hanjiang River Ecological Economic Belt (HREEB). This region primarily encompasses essential tributaries, including the Nanhe River (NR), Manhe River (MR), Tangbai River (TBR), and Hanbei River (HR). The discharge of ecological flow in these basins is critical for promoting the high-quality development of the HREEB. However, due to substantial variations in precipitation and runoff patterns both within and across years, the historical satisfaction level of ecological flow has generally been low, failing to meet current water resource management demands. This shortfall has impeded the high-quality development of the Han River Ecological Economic Belt and has attracted widespread social attention. In order to enhance the protection and sustainable development of water resources in the Hanjiang River Basin (HRB), this paper proposes a dynamic calculation method for ecological flow tailored to specific assurance rates, building upon existing academic research. The study focuses on key control sections, including the NR, MR, TBR, and HR, located in the MLHRB. By incorporating current management requirements, dynamic adaptive calculations of target ecological flow are performed, aiming to provide technical support for the effective establishment and maintenance of a system that safeguards the ecological flow of rivers and lakes.

2 Materials and Methods

2.1 General Situation

The MLHRB traverses the administrative regions of Hubei and Henan provinces, situated between 111°E and 115°E longitude, and 30°N and 33°N latitude. The main stream extends from the Danjiangkou Dam (DJKD) to Longwangmiao in Wuhan, passing through several cities, including Danjiangkou (DJK), Laohekou (LHK), Xiangyang (XY), Zhongxiang (ZX), Jingmen (JM), Qianjiang (QJ), Xiantao (XT), Hanchuan (HC), and Wuhan (WH). Along its course, significant tributaries such as the NR MR, TBR, and HR converge with the main river [10]. The geographical location of the study area is illustrated in Fig. 1.

The MLHRB are influenced by a subtropical monsoon climate, with an annual average temperature ranging from 15 to 17 °C and annual precipitation between 700 and 1300 mm [10–12]. Due to variations in precipitation, the runoff of major tributaries, including the NR, MR, TBR, and HR, exhibits significant interannual fluctuations, with substantial differences between wet and dry seasons; the extreme ratio can reach as high as 4.41 to 14.0. Runoff distribution throughout the year is uneven, primarily concentrated from May to October, which accounts for 70 to 80% of the annual runoff. In contrast, runoff during the dry season, from December to February of the following year, constitutes only about 7% of the annual total [13, 14].

2.2 Basic Information

Based on the existing ecological flow management requirements, hydrological gauges, the distribution of water conservancy projects, and the conditions of hydrological data, sections such as Sandaohe (SDH) in the Manhe River (MR), Guotan (GT) and Xindian (XDP) in the Tangbai River (TBR), and Tianmen (TM) in the Hanbei River (HR), were selected as the focus of this research. The daily average flow series measured at the SDH, GT, XDP, and TM hydrological stations from 2003 to 2021 (data extracted from the published Hydrological Yearbook of the HRB) were collected, along with the natural monthly runoff series for each section from 1956 to 2016 (derived from the existing results of the third water resources survey and evaluation in Hubei Province). The data series utilized have undergone quality control, including analyses of consistency, reliability, and representativeness. The data quality is reliable and meets the requirements of this study. The distribution of hydrological stations in the study area is illustrated in Fig. 1.

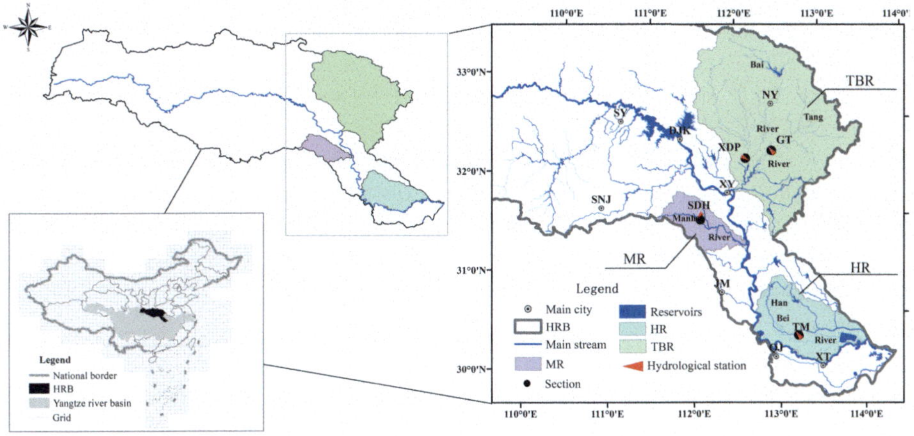

Fig. 1. Distribution of research area and hydrological gauges.

2.3 Methods

Basic Ecological Flow Calculation Method. The basic ecological flow for the study section is calculated using four widely recognized hydrological methods: the Q_p method, the Tennant method, the average flow method of the driest month over the past ten years, and the frequency curve method [15, 16]. The Q_p method and the frequency curve method define ecological flow as the flow corresponding to a 90% frequency, while the Tennant method designates 10% of the average annual flow as the ecological flow. For detailed descriptions of these methods, please refer to the reference of SL/T712-2021.

Dynamic Calculation Method of Ecological Flow with Adaptive Target Guarantee Rate. This article proposes a dynamic calculation method for ecological flow that adapts to the target assurance rate. While ensuring that the ecological flow guarantee target is met, it innovatively enables the dynamic adjustment of ecological flow throughout the year. The specific method is outlined as follows.

(1) *Determine the Intra-Annual Distribution Variance Coefficient.*

Assuming that the target ecological flow guarantee rate for the control section A is P, the quantile method is employed to determine the corresponding flow values for the monthly average flows of each month. The calculation formula is as follows:

$$Q_P(j) = \underset{i=1}{\overset{i=N}{\text{Percentile}}}\big(Q_y(i,j), P\big), j = 1, 2, \cdots, 12 \tag{1}$$

where, $Q_p(j)$ denotes the flow corresponding to the j-th month at the target assurance rate of P. **Percentile** () represents the percentile function. while, P signifies the target assurance rate. $Q_y(i,j)$ indicates the natural monthly average flow series at a cross-section, with i referring to the year of the series, j indicating the month, and N representing the overall duration of the series in years.

Based on the measured daily average flow series of Section A, the quantile method is used to calculate the flow values corresponding to the monthly daily average flow under the P. The calculation formula is:

$$Q_{dp}(j) = \underset{i=1,\,k=1}{\overset{i=N,k=M}{\text{Percentile}}}\big(Q_d(i,j,k),P\big), j = 1,2,\cdots,12 \tag{2}$$

where, $Q_{dp}(j)$ represents the daily average flow corresponding to the P in the j-th month. $Q_d(i, j, k)$ denotes the measured daily average flow series at the cross-section, where k denotes the number of days, and M denotes the total number of days in the series year.

The intra-annual distribution difference coefficient (K_1) is introduced to characterize the intra-annual distribution variation of runoff. The calculation formula is as follows:

$$K_1(j) = Q_{dp}(j)/Q_P(j),\ j = 1,2,\cdots,12 \tag{3}$$

where, $K_1(j)$ represents the intra-annual distribution difference coefficient for the j-th month.

(2) *Determine the Marginal Distribution of Monthly Flow.*

In order to capture the interannual variations in runoff, probability analysis theory is employed to ascertain the marginal distribution of monthly runoff. The selection of marginal distribution models encompasses Pearson III distribution (P-III), exponential distribution (EXP), extreme value distribution (GEV), and log-normal distribution (LOGN), with the maximum likelihood method utilized for precise parameter estimation [17–21]. The quality of model parameter estimation is comprehensively evaluated using indicators such as the Kolmogorov-Smirnov (K-S) method, the least squares method (Ordinary Least Squares, OLS), and the Akaike Information Criterion (AIC) [20, 21]. In this study, the distribution that corresponds to the minimum value of the comprehensive evaluation target, C, is considered the optimal distribution model. The formula for calculating the evaluation index is as follows:

$$\begin{cases} D_{k-s}(j) = \max\limits_{i=1}^{i=N}(|Q(i,j) - Q_1(i,j)|) \\ OLS(j) = \sqrt{\dfrac{1}{N}\sum\limits_{i=1}^{n}(Q(i,j) - Q_1(i,j))^2} \\ AIC(j) = n\log\left(\dfrac{1}{N}\sum\limits_{i=1}^{n}(Q(i,j) - Q_1(i,j))^2\right) + 2K \\ C(j) = D_{k-s}(j) + OLS(j) + AIC(j) \end{cases} \tag{4}$$

where, $Q(i, j)$ represents the actual measurement series; $Q_1(i, j)$ denotes the fitted series; C stands for the comprehensive evaluation target value; N signifies the series' annual length; K indicates the number of parameters in the marginal distribution; and j designates the month.

(3) *Determine the Marginal Distribution of the Target Annual Flow*

Calculate the monthly flow corresponding to the target assurance rate based on the annual average natural monthly flow of the cross-section and derive the target annual flow series. The formula is as follows:

$$Q_{yp}(i) = \underset{j=1}{\overset{j=12}{\text{Percentile}}}\big(Q_y(i,j), P\big), j = 1, 2, \cdots, 12;\ i = 1, 2, \cdots, N \tag{5}$$

where, $Q_{yp}(i)$ represents the monthly average flow corresponding to the P in the i-th year; $Q_y(i, j)$ represents the natural monthly average flow series of the section.

The edge distribution is determined using the same calculation method as the monthly average flow edge distribution based on the target annual flow series.

(4) *Determine the Two-Dimensional Joint Distribution of Monthly and Annual Flow*

After determining the monthly flow edge distribution, the Copula function is introduced to construct a two-dimensional joint probability distribution for each monthly average flow and the target annual flow. This approach results in a library of 12 joint probability distribution models covering the months from January to December [22, 23]. The Copula function serves as a multidimensional joint distribution function defined on the [0,1] domain, enabling the creation of a combined model that links multiple arbitrary forms of edge probability distributions. For instance, in the two-dimensional case, let $F_X(x) = u$ and $F_Y(y) = v$, representing the cumulative distribution functions of the random variables x and y, respectively. The expression for the Copula function is as follows:

$$F(x, y) = C_\theta(F_X(x), F_Y(y)) = C_\theta(u, v) \tag{6}$$

where, $F(x, y)$ denotes the joint distribution of random variables x and y; C_θ represents the Copula function; and $F_X(x)$ and $F_Y(y)$ are the marginal distribution functions of random variables x and y.

Based on existing research findings in academia, we comprehensively select easily constructible single-parameter copula functions, such as Clayton, Frank, and Gumbel, for two-dimensional joint distribution analysis.

The accuracy assessment of the Copula function also utilizes indicators such as K-S, OLS, and AIC to comprehensively evaluate the distribution that corresponds to the minimum value of C, and identify it as the optimal distribution.

(5) *Determine the Interannual Variation Coefficient*

According to China's current ecological flow management requirements, the prerequisite for meeting the monthly flow standard is that the annual flow must meet the target assurance rate. In other words, if the monthly flow assurance rate meets the standard, the annual flow assurance rate will also inevitably meet the standard. Therefore, it can be statistically understood as the conditional probability of the joint distribution. Assuming the marginal distribution of the monthly average flow is represented as $F_X[Q_y(i,j)] = u(j)$, and the marginal distribution of the target annual flow is represented as $F_Y[Q_{yp}(i)] = v$. Further define the interannual variation coefficient (K_2), representing the conditional probability of achieving the target annual flow rate [22, 23]. The expression is as follows:

$$K_2(j) = \frac{1 - u(j)}{1 - u(j) - v + C\left[u(j), v\right]} \tag{7}$$

where, $K_2(j)$ represents the interannual variation coefficient of the monthly average flow in the j-th month; $u(j)$ denotes the marginal distribution of the monthly average flow for the j-th month, while v signifies the marginal distribution of the target annual flow.

(6) *Adaptive Calculation of Target Ecological Flow*

The frequency curve method is employed to calculate the benchmark value of target ecological flow for each month. This method serves as a fundamental approach for hydrological analysis and calculations, involving the construction of a hydrological frequency curve based on the monthly average flow series. The monthly average flow corresponding to the target probability of P is designated as the benchmark value of target ecological flow for that month, denoted as $Q_c(j)$. By integrating the intra-annual distribution of runoff with the inter-annual variation coefficient of water abundance, the benchmark value of the target ecological flow is adjusted to derive the monthly adaptive value of the target ecological flow. The expression is as follows:

$$Q_{tp}(j) = K_1(j) \times K_2(j) \times Q_c(j), j = 1, 2, \cdots, 12 \tag{8}$$

where, $Q_{tp}(j)$ stands for the target ecological flow in the j-th month under the collaborative adaptation target assurance rate of P.

Evaluation of Satisfaction with Ecological Flow Targets. From the perspectives of both monthly and annual assessments, the frequency method is employed to evaluate the satisfaction level of ecological flow. The calculation formula is as follows.

$$PR_i = A_i / B_i \times 100 \tag{9}$$

where, PR_i represents the satisfaction rate achieved in meeting the ecological flow target for the control section, %; A_i denotes the number of days during the evaluation period where the daily average flow is equal to or exceeds the ecological flow; B_i signifies the total number of days of samples included in the evaluation of ecological flow satisfaction during the evaluation period.

3 Result Analysis and Discussion

3.1 Basic Ecological Flow and Its Satisfaction Degree

Based on conventional calculation methods, including the Q_p method (Approach 1), the Tennant method (Approach 2), the average flow method for the driest month over the past 10 years (Approach 3), and the frequency curve method (Approach 4), the natural monthly flow series from 1956 to 2016 was utilized to determine the basic ecological flow for sections such as the MR, TBR, and HR (see Table 1). According to the literature of Xu et al., 2024, the approved basic ecological flow results for these sections range from 1.16 to 4.9 m^3/s, which accounts for 6.36% to 12.3% of the annual average flow. In contrast, the basic ecological flow calculated by conventional methods varies from 0.492 to 7.71 m^3/s, representing 2.56% to 14.1% of the multi-year average flow. The results from each method exhibit considerable variation, highlighting a significant discrepancy between the approved results and the calculated outcomes.

To better reflect the impact of socio-economic development, water engineering regulations, storage, and off-river water extraction in the basin since the twenty-first century, the measured daily flow over the past 21 years (2003–2023) was selected to analyze the satisfaction degree with the ecological flow at the sections, as shown in Table 1. Except the SDH, which exhibited a low degree of satisfaction, most others recorded satisfaction levels exceeding 90% in most years, thereby meeting current water resource management requirements. The daily satisfaction degree for the SDH ranged from 89.0% to 91.2%. The results obtained using Approach 2 indicate that the degree of daily satisfaction falls below 90%. In contrast, the approved results represent comprehensive values derived from multiple methods which consider the requirements of water resource management. Overall, the degree of daily satisfaction in each section consistently remained above 90%.

Considering the current objectives for ecological flow control, determining the appropriate ecological flow target value through a combination of multiple methods has become a challenging issue for research. Some scholars argue that for rivers with significant variations in annual and interannual water levels, the Approach 1 or Approach 3 may be more suitable for China's current water resource management needs [5, 6]. However, the results obtained from different methods can vary significantly due to the differing water inflows across various basins. As society and the economy continue to develop, any changes in ecological protection objectives may impact the applicability of these methods.

3.2 Dynamic Adaptation Analysis of Ecological Flow

The quantile method was employed to calculate the flow values corresponding to the monthly average flow for each section based on a target assurance rate of 90%. Using the natural monthly average flow series (Q_y) for each section, the monthly average flow corresponding to the target guarantee rate is calculated annually, resulting in the target annual flow series (Q_{yp}). Utilizing both Q_y and Q_{yp}, four distribution models, named P-III, EXP, GEV, and LOGN, are constructed. Parameter estimation is conducted using the maximum likelihood method, and the optimal distribution is selected through a comprehensive evaluation objective (C). Based on the marginal distributions, single-parameter copula functions including Clayton, Frank, and Gumbel, were selected to establish two-dimensional joint distribution models for monthly average flows and target annual flows, spanning from January to December. The corresponding copula functions were constructed for fitting purposes by utilizing parameter estimation. Subsequently, the joint conditional probability was employed to calculate the co-occurrence probability

Table 1. Basic ecological flow and its guarantee rate for each section.

River-Section	Station	Annual average Flow (m^3/s)	Approach 1		Approach 2		Approach 3		Approach 4		Approved results	
			Flow (m^3/s)	Satisfaction rate (%)	Flow (m^3/s)	Satisfaction rate (%)	Flow (m^3/s)	Satisfaction rate (%)	Flow (m^3/s)	Satisfaction rate (%)	Flow (m^3/s)	Satisfaction rate (%)
TBR-XDP	XDP	77.1	2.02	100.0	7.71	97.4	2.70	100.0	5.92	98.9	4.90	99.6
TBR-GT	GT	50.8	2.34	98.7	5.08	91.6	1.30	99.2	4.23	95.2	4.50	93.7
MR-SDH	SDH	12.7	0.492	91.2	1.27	89.0	0.950	90.4	0.786	90.8	1.16	90.2
HR-TM	TM	29.6	1.51	99.0	2.96	96.1	2.45	97.6	3.43	95.1	3.67	94.4

values of the monthly average flow and the target annual flow under the 90% condition, serving as the interannual variation coefficient.

The frequency curve method is utilized to calculate the benchmark values for target ecological flow for each month, incorporating both the annual distribution coefficient and the interannual variation coefficient of runoff. These benchmark values are subsequently adjusted to derive the monthly target ecological flow rates, as detailed in Table 2. According to the table, the monthly target ecological flow rates range from 4.38 to 13.8 m^3/s for the XDP and from 4.55 to 9.20 m^3/s for the GT. Similarly, the monthly target ecological flow rates for the SDH and TM sections vary between 1.04 and 3.27 m^3/s, and 2.32 and 20.6 m^3/s, respectively. Overall, the monthly target ecological flow rates demonstrate a consistent pattern of being higher during the wet season and lower during the dry season, which corresponds to the characteristics of water inflow in the basin.

3.3 Analysis and Discussion

Utilize the measured daily average flow data spanning the past 21 years (2003–2023) to calculate the satisfaction rate of daily flow for each time period. Subsequently, these findings will be compared with the approved ecological flow results, as illustrated in Figs. 2 and 3.

From the perspective of annual average satisfaction rates over the years, the results calculated using this method generally demonstrate high satisfaction. Specifically, the new store section has achieved an average daily satisfaction rate of over 90% in all years except for 2014. Conversely, SDH fell below the 90% target guarantee rate only in 2013–2014, 2019, and 2022. In contrast, the satisfaction rates at GT and TM are relatively lower, with GT falling below 90% in 2013–2016 and 2019–2022, and TM in 2004, 2011–2013, 2019, and 2022. Compared to previously approved results, the new store section has an additional year, 2014, that did not meet the standards, while GT has 2020, and both SDH and TM have 2019 and 2022. However, the compliance situation remains largely consistent in other years. Overall, the degree of satisfaction achieved through this method is comparable to existing achievements and is sufficient to meet the current requirements for ecological flow management.

From the perspective of annual and monthly statistical analysis, the daily average satisfaction rates calculated by this method remain relatively high. Specifically, the daily average satisfaction rate for the XDP and GT are slightly lower in August and June, reaching 89.6% and 87.8%, respectively. Similarly, the daily average satisfaction rate for the TM falls below 90% only in January and December, specifically at 89.1% and 88.7%. However, the daily average satisfaction rate for all other sections exceeds 90%, thereby fulfilling the requirements of ecological flow management objectives. Compared to previously approved results, this method effectively addresses the issue of inadequate ecological flow satisfaction in the TM during January and February while maintaining an overall consistent level of satisfaction.

Based on the comprehensive analysis, the calculated results of this method are essentially comparable to the previously approved outcomes in terms of daily average satisfaction. However, it's worth noting that the target ecological flow calculated this time is not a static numerical value. Instead, it has been dynamically adjusted monthly, with specific target ecological flows tailored for each individual month.

Table 2. Target ecological flow of each section.

Section	The target ecological flow calculated by this method (m^3/s)											Approved results (m^3/s)
	Jan	Feb	Mar	Apr	May	Jun	Jul	Aug	Sept	Oct	Nov	Dec	
XDP	7.94	6.41	4.72	5.13	6.87	4.38	11.2	13.8	10.6	9.69	9.82	9.28	4.90
GT	4.80	4.55	4.58	4.83	5.87	5.86	8.00	9.20	5.65	5.56	5.49	4.97	4.50
SDH	1.22	1.04	1.16	1.39	1.40	1.62	1.62	3.27	1.47	1.36	1.40	1.37	1.16
TM	2.76	2.32	2.98	13.6	15.4	19.3	19.1	25.1	18.1	20.6	5.43	3.01	3.67

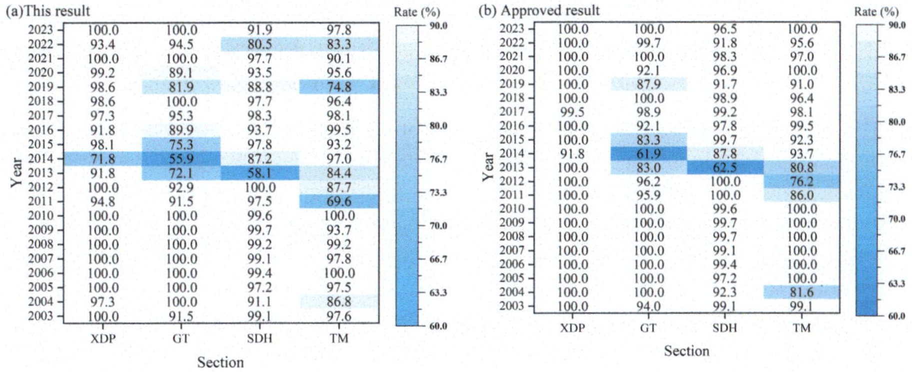

Fig. 2. Heat map of daily average satisfaction of target ecological flow at each section (over the years).

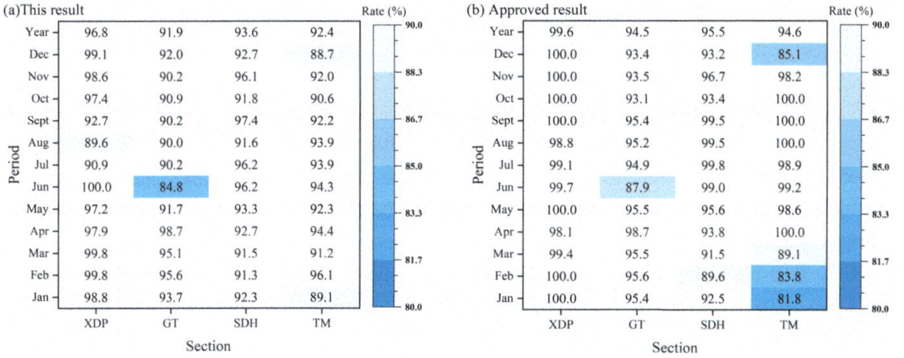

Fig. 3. Heat map of daily average satisfaction of target ecological flow at each section (year and month scale).

The advantage of this approach lies in its ability to maximize the ecological water volume within river channels while adhering to ecological flow control targets. According to statistical data, the annual target ecological water volumes for sections such as XDP, GT, SDH, and TM, calculated using this method, are 263 million m^3, 183 million m^3, 48 million m^3, and 390 million m^3, respectively. In contrast, based on the approved basic ecological flow results, the respective annual target ecological water volumes for these sections are 138 million m^3, 144 million m^3, 48 million m^3, and 73 million m^3. Notably, this approach can enhance the annual ecological water volume by 16 to 317 million m^3, thereby better safeguarding the ecological water needs within the river channels.

4 Conclusions

Utilizing key control sections in the MR, TBT, and HR in the middle and lower reaches of the Hanjiang River as focal points for our research, a dynamic calculation method tailored to the target guarantee rate has been developed, considering the variation of runoff between seasons and years. In accordance with current ecological flow management requirements, we have performed dynamic adaptation calculations for the monthly target ecological flow. The key findings of our research are summarized as follows:

(1) The approved basic ecological flow rates for various sections of the research area range from 1.16 to 4.9 m^3/s, representing 6.36% to 12.3% of the annual average flow rate. In contrast, the basic ecological flow calculated using conventional hydrological methods varies between 0.492 and 7.71 m^3/s, accounting for 2.5% to 14.1% of the multi-year average flow. Notably, there are significant disparities among the results obtained from different methods, which also deviate somewhat from the approved outcomes. Furthermore, due to fluctuations in water availability within the basin, the daily satisfaction levels of some methods' calculations are relatively low, posing challenges in meeting the requirements for ecological flow management.
(2) Taking into account the annual and interannual variations in natural inflow within the watershed, as well as the objectives for ecological flow management, an innovative method for calculating ecological flow dynamics has been proposed. This method is tailored to the target guarantee rate and incorporates a two-dimensional joint distribution model. The monthly target values for ecological flow in various sections, including the MR, TBR, and HR, range from 1.04 to 25.2 m^3/s. These values exhibit an overall pattern of being higher during wet seasons and lower during dry seasons, which aligns with the inflow characteristics of the watershed.
(3) Based on the statistical analysis of the daily average satisfaction regarding ecological flow across various years, months, and periods, it is evident that the method employed in this study generally demonstrates a high level of satisfaction. Notably, while maintaining an overall satisfaction level comparable to previously approved results, this method effectively addresses the issue of inadequate ecological flow satisfaction in the TM during January and February. Furthermore, it significantly enhances the annual ecological water volume in sections such as XDP, GT, SDH, and TM, increasing it by 16 to 317 million m^3.

Acknowledgements. The authors appreciate the Bureau of Hydrology, Changjiang Water Resources Commission for the support of data. This research work is financially supported by National Natural Science Foundation of China (Grant No. 52309004), Hubei Provincial Natural Science Foundation of China (Grant No.2024AFB054), the National Key R&D Program of China (Item Nos. 2022YFC3002701, 2023YFC3206001) and the Key R&D Program of Hubei Province (Item Nos. 2023BCB115).

References

1. Wu, Y.B., Xue, L.Q., Liu, Y.H., et al.: Uncertainty assessment of extreme flood estimation in the Dongting Lake basin, China. Hydrol. Res. **50**(4) (2019)
2. Tanaka, T., Tachikawa, Y., Ichikawa, Y., et al.: Impact assessment of upstream flooding on extreme flood frequency analysis by incorporating a flood-inundation model for flood risk assessment. J. Hydrol. **554** (2017)
3. SL/T712-2021. Calculation Specification for Ecological Water Demand of Rivers and Lakes. China Water & Power Press, Beijing

4. Wang, J.H., Hu, P., Liu, Q.T., et al.: Current status of ecological flow in key rivers and lakes in China and regulation suggestions. China Water Resour. **1**, 14–18 (2022)
5. Xu, C.J., Bing, J.P., Deng, P.X.: Phased control on target ecological flow for important tributaries in middle and lower reaches of Hangjiang River. Yangtze River **55**(05), 84–89 (2024)
6. Liu, Y.Y., Zhu, J.F., Zhao, J.S.: Development history and frontiers of river environmental flow research. J. Hydroelectr. Eng. **35**(12), 23–34 (2016)
7. Verma, K.R., Murthy, S., Tiwary, K.R.: Assessment of environmental flows for various sub-watersheds of Damodar River basin using different hydrological methods. Int. J. Waste Resour. **5**(4), 1–6 (2015)
8. Wang, S., Liu, X., Wang, X., et al.: Measuring hydrologic regime alterations and hydrodynamic characteristics in the Xijiang River Basin by the IHA-RVA method. J. Phys. Conf. Ser. **2865**(1), 012003–012003 (2024)
9. Karimi, S., Salarijazi, M., Ghorbani, K., et al.: Comparative assessment of environmental flow using hydrological methods of low flow indexes, Smakhtin, Tennant and flow duration curve. Acta Geophys. **69**(1), 1–9 (2021)
10. Deng, P.X., Zhang, M.Y., Hu, Q.F., et al.: Pattern of spatio-temporal variability of extreme precipitation and flood-waterlogging process in Hanjiang River basin. Atmos. Res. **276** (2022)
11. Deng, P.X., Bing, J.P., Jia, J.W., et al.: Pattern of spatio-temporal variability of flood season precipitation in Hanjiang River Basin between 1956 and 2016. Resour. Environ. Yangtze Basin **27**(09), 2132–2141 (2018)
12. Zhang, S., Lin, L., Wang, Z., et al.: Spatio-temporal variation of water quality in middle-lower Hanjiang River. J. Changjiang River Sci. Res. **38**(08), 47–53 (2021)
13. Zhang, O. Y., Xiong, M.: Analysis of the minimum flow variation and influencing factors in the middle and lower reaches of the Hanjiang River over the past 60 years. Yangtze River **48**(S2): 89–92+98 (2017)
14. Zhang, O.Y., Xiong, M.: Ecological flow calculation method based on measured flow results. Yangtze River **48**(S2), 61–64 (2017)
15. Zheng, Z.H., Zhang, Z.H., Huang, Q., et al.: Improvement and application on ecological water requirement tennant method. Adv. Eng. Sci. **42**(2): 34–39+27 (2010)
16. Jiang, C., Liong, L.H., Xu, C.Y., et al.: A river network-based hierarchical model for deriving flood frequency distributions and its application to the Upper Yangtze Basin. Water Resour. Res. **57**(8) (2021)
17. Jia, Y.F., Song, S.B., Ge, L.T.: Trimmed L-moments of the pearson type III distribution for flood frequency analysis. Water Resour. Manag. **37**(3), 1321–1340 (2023)
18. Rulfová, Z., Buishand, A., Roth, M., et al.: A two-component generalized extreme value distribution for precipitation frequency analysis. J. Hydrol. **534** (2016)
19. Shiau, J.T.: Effects of gamma-distribution variations on SPI-based stationary and nonstationary drought analyses. Water Resour. Manag. **34**(6), 2081–2096 (2020)
20. Busababodhin, P., Seo, A.Y., Park, J., et al.: LH-moment estimation of Wakeby distribution with hydrological applications. Stoch Environ Res Risk Assess **30**(6) (2016)
21. Edouard, S., Emmanuel, I.: Modelling extreme rainfall events in Kigali city using generalized Pareto distribution. Meteorol Appl **29**(4) (2022)
22. Bing, J., Deng, P., Zhang, X., et al.: Flood coincidence analysis of Poyang Lake and Yangtze River: risk and influencing factors. Stoch. Environ. Res. Risk Assessment **32**(4), 879–891 (2018)
23. Shahid, L., Zina, S., Ouarda, T.B., et al.: Copula-based joint modelling of extreme river temperature and low flow characteristics in the risk assessment of aquatic life. Weather Clim. Extrem. **41** (2023)

Research on Multi-objective Optimization Scheduling in the Lower Cascade Reservoir of Lancang River

Long Ji, Xide Wu, and Lingling Liu(✉)

China Waterborne Transport Research Institute, Beijing 100088, China
{jilong,wuxide,liull}@wti.ac.cn

Abstract. With the function level of Lancang-Mekong River navigation increased, the competitive water use between navigation and power generating is more and more prominent. It is important to balance the water used for power generation and navigation to ensure the benefits of generating electricity and increase the reliability of navigation water use. This study builds a multi-objective optimization scheduling model, which takes the maximum benefits of generating electricity and navigable flow assurance as the objective function. Through long-term adjust calculation, the generating capacity, the benefits of generating electricity of cascade reservoirs, the discharge, and the reliability of the hydro-power station are analyzed, respectively. The results show that the total generating capacity of the six cascaded hydro-power stations is equal to the annual output of reservoirs set in the early planning stage. The time reliability of monthly output discharge for Jinghong hydro-power station above 1000 m^3/s is 100%, which is higher than that of 2019 to 2022(45.8%). It demonstrates that the optimal operation of cascade reservoirs can significantly increase the discharge reliability in the period of 800~1000 m^3/s without sacrificing the benefits of generating electricity; it can provide extremely favorable conditions to improve the navigation level below the Jinghong hydro-power station waterway.

Keywords: Reservoir Scheduling · Optimal Operation · Navigable Discharge

1 Introduction

Lancang River rises northeast of Tanggula Mountain in Qinghai Province, China, and its drainage basin spans Qinghai Province, Xizang Autonomous Region, and Yunnan Province. It flows out of China's border at the Nanla River estuary in Dai Autonomous Prefecture of Xishuangbanna, Yunnan Province. After going abroad, it is called the Mekong River [1]. The Mekong River flows through Laos, Myanmar, Thailand, and Cambodia. Finally, it enters Vietnam, forming a delta water network in the southwest of Saigon, Vietnam, flowing into the South China Sea. The Lancang-Mekong River is the only international river in Asia that spans six countries and is known as the "Eastern Danube".

C. Weng (Ed.): WRE 2024, LNCE 634, pp. 220–234, 2025.
https://doi.org/10.1007/978-981-96-5095-8_19

The basin area of the Lancang-Mekong River is about 8×10^5 km^2. The total length from headwaters to the estuary is about 4880 km, including 2130 km in China, 31 km in China Myanmar border river, 234 km in Laos Myanmar border river, 976 km in Laos Thailand border river, 777 km in Laos, 502 km in Cambodia, and 230 km in Vietnam. The average gradient of the Lancang-Mekong River is 1.03‰, the river's average annual runoff is 475 billion m^3 at the estuary. It has abundant water resources and abundant water energy reserves, and two reservoirs and eight-level development plans are planned for the middle-lower reaches of the Lancang River, from top to bottom, the Gongguo qiao, Xiaowan, Manwan, Dachaoshan, Nuozhadu, Jinghong, Ganlanba, Mengsong hydro-power station [2, 3]. The distributions of hydro-power stations in the Lancang River's middle-lower reaches are shown in Fig. 1. All other hydro-power stations except the Ganlanba and Mengsong hydro-power stations have been built. The reservoirs have functions such as shipping, tourism, flood control, irrigated farmland, environmental protection, and power generation.

Fig. 1. Distribution map of hydro-power stations in the Lancang River's middle-lower reaches.

The joint optimization and scheduling of cascade reservoirs in a watershed need to consider the competition of flood control safety, power generation, shipping, water for industrial, agricultural, and domestic supply, environmental protection, and scheduling requirements that cannot be shared; it is a typical large-scale, strongly coupled, multi-objective, and multi constrained complex nonlinear optimization problem [4]. In reservoir scheduling reservoir scheduling is receiving increasing attention due to the random nature of natural runoff, the multi-stage nature of scheduling decision-making processes, the complexity of reservoir structures, and the multi-objective nature of applications [5]. With the construction of cascade development in river basins, cascade reservoirs' joint optimization and scheduling technology has become one of the key research directions of domestic and foreign experts and scholars. With the development of mathematical theory and computer technology, significant progress has been made in optimization theory, optimization methods, and application in the optimization scheduling problem of cascade reservoirs [6]. The dynamic programming method was developed by the American mathematician R.E. Bellman's in the 1950s, it is used to solve multi-stage decision problems; it has become one of the fundamental theories of modern optimal control [7]. Genetic algorithms, which originated from the study of natural and artificial adaptive systems in the 1960s, were initially proposed by Professor Holland from the University of Michigan in the USA [8]. Ant algorithm is a stochastic optimization method that was born in the 1990s. Huang used AA to solve the short-term optimization scheduling of Taiwan's hydro-power system [9]. The particle swarm optimization algorithm originated from the movement characteristics of bird populations and was first developed by J. Kennedy and R. C. Eberhart in 1995 [10]. In addition, there are artificial neural network methods, and simulated annealing algorithms have also been widely applied [11]. According to the relevant planning, the following waterways of Jinghong hydro-power station are planned as 4-class waterways, which can be navigable by 500-ton ships. In order to meet the needs of upgrading the waterway grade below the Jinghong hydro-power station and fully leverage the advantages of water resources, this article takes the construction and operation of cascaded hydro-power stations in the middle-lower reaches of the Lancang River as the background, based on constructing a multi-objective optimization scheduling model for cascade reservoirs, the objective function is to maximize the generating capacity efficiency and navigation flow guarantee rate of the cascade reservoirs, through long-term regulation calculations, analyze the changes in power generation, power generation efficiency, and station discharge and guarantee rate of cascaded hydro-power stations under optimized scheduling methods.

2 A Multi-objective Optimization Scheduling Model for Cascade Reservoirs

The optimization and scheduling of cascaded hydro-power stations comprehensively coordinate multiple water needs, such as power generation, flood control safety, environmental protection, and shipping, to maximize the generating capacity efficiency and navigation flow guarantee rate of cascaded hydro-power stations. A mathematical description of the objective function and constraint conditions is established by constructing an optimization scheduling model for cascaded hydro-power stations.

2.1 Decision Variables

Both the discharge flow and abandoned water flow of hydroelectric power plants can be used for downstream shipping water, which means that the flow rate of shipping water is equal to the flow rate of water used for power generation plus the flow rate of abandoned water. Therefore, the decision variables for optimizing the scheduling of cascaded hydro-power stations are the generating capacity water flow rate ($Q_{i,t}$) and the abandoned water flow rate ($SUR_{i,t}$) of each hydro-power station. Among them, i represents the level reservoir, and t represents the scheduling period.

2.2 Mathematical Description

Objective Function. The optimization scheduling of the cascaded hydro-power stations aims to maximize the comprehensive efficiency of generating power and the assurance rate of navigation flow below the Jinghong hydro-power station. See the following formula for details.

$$f(x) = \max(\lambda_1 Z_{ELC} + \lambda_2 Z_{NAV}) \tag{1}$$

$$\begin{aligned} Z_{ELC} &= \sum_{i=1}^{N}\sum_{t=1}^{T} E_{i,t} = \sum_{i=1}^{N}\sum_{t=1}^{T} P_{i,t} \times \Delta t \times \omega_t \\ &= \sum_{i=1}^{N}\sum_{t=1}^{T} [A \times Q_{i,t} \times H_{i,t}] \times \Delta t \times \omega_t \end{aligned} \tag{2}$$

$$Z_{NAV} = \sum_{t=1}^{T} f(t)/T \tag{3}$$

$$Z_{NAV} = \sum_{t=1}^{T} f(t) \tag{4}$$

$$f(t) = \begin{cases} 1, \ if\,(Q_{jh,t} \geq QNAV_t) \\ 0, \ if\,(Q_{jh,t} < QNAV_t) \end{cases} \tag{5}$$

$$Q_{jh,t} = Q_{6,t} + SUR_{6,t} \tag{6}$$

The meanings of the symbols in the formula are as follows:

Z_{ELC}: Total power generation efficiency of cascaded hydro-power stations;

Z_{NAV}: Guarantee rate of water flow for shipping;

ω_t: Efficiency coefficient of reservoir power generation in the *i* period, feed-in tariff; during the wet season (June to October), the price drops by 18%; during the dry season (December to April of the following year), the price will increase by 20%; the basic electricity price during the normal period (May and November) is 0.27 ¥/kW.h. The benefits of generating electricity are calculated based on this.

λ_1: The weight coefficient for power generation;
λ_2: The weight coefficient for shipping water assurance rate;
$Q_{jh,t}$: Jinghong hydro-power station's discharge flow, includes the water flow for power
generation and the abandoned water flow;
$Q_{6,t}$: The water flow for power generation during the t period of Jinghong hydro-power station, the sixth cascade in the Lancang River's middle-lower reaches;
$SUR_{6,t}$: Abandoned water flow during the t period of Jinghong hydro-power station, the sixth cascade in the Lancang River's middle-lower reaches;
$QNAV_t$: Water demand for the navigation water flow of class IV waterway below Jinghong hydro-power station, this calculation takes 1000 m^3/s;
$E_{i,t}$: The i-th level reservoir's generating capacity during period t, unit kW·h;
N: The cascaded hydro-power stations' total number;
T: The total number of calculated periods;
$P_{i,t}$: Generating capacity output of the i-th level reservoir during the period t, $P_{i,t} = A \times Q_{i,t} \times H_{i,t}$, unit kW;
Δt: The amount of calculated periods;
A: Comprehensive output coefficient, falling in a range between 7.5 and 8.5;
$Q_{i,t}$: The discharge flow of power generation water from the i-th reservoir during the period t, m^3/s;
$H_{i,t}$: The average water head during the calculation period Δt, which is equal to subtracting the tailwater level from the water level in front of the dam during the calculation period, $H_{i,t} = Z_{i,t} - ZWEI_{i,t}$;
$Z_{i,t}$: The water level in front of the dam during the calculation period Δt, obtained through the relationship between water level and storage capacity;
$ZWEI_{i,t}$: The tailwater level during the calculation period Δt, obtained through the relationship between the tailwater level and the flow rate.

Restraint Condition. Reservoir water balance constraint:

$$V_{i,t} = V_{i,t-1} + Q_{i-1,t} \times D_t + SUR_{i-1,t} \times D_t + I_{i,t} \times D_t - Q_{i,t} \times D_t - SUR_{i,t} \times D_t \tag{7}$$

Water level constraint:

$$ZMIN_{i,t} \leq Z_{i,t} \leq ZMAX_{i,t} \tag{8}$$

Hydro-power station output constraint:

$$NF_i \leq P_{i,t} \leq NT_i \tag{9}$$

Constraints on Reservoir Discharge Volume:

$$QMIN_{i,t} \leq Q_{i,t} + SUR_{i,t} \leq QMAX_{i,t} \tag{10}$$

Constraints on the initial operating water level of reservoir scheduling:

$$Z_{i,0} \leq ZR_{i,0} \tag{11}$$

The meanings of the symbols in the formula are as follows:

$V_{i,t}$: The *i*-th level reservoir's storage capacity during period *t*;
$V_{i,t-1}$: The *i*-th level reservoir's storage capacity during period *t*-1;
$Q_{i-1,t}$: The $(I-1)$-th level reservoir's discharge flow of generating electricity water from during period *t*;
$SUR_{i-1,t}$: The $(I-1)$-th level reservoir's abandoned water flow during period *t*;
$I_{i,t}$: The interval inflow between *i*-1 level and *i*-level hydro-power stations for the *i*-th level reservoir during period *t*;
$Q_{i,t}$: The *i*-th level reservoir's discharge flow for generating electricity during period *t*;
$SUR_{i,t}$: The *i*-th level reservoir's abandoned water flow during period *t*;
D_t: Conversion coefficient between flow rate and water volume during period *t*;
$ZMIN_{i,t}$: The minimum water level allowed for the *i*-th level reservoir during period *t*;
$ZMAX_{i,t}$: The maximum water level allowed for the *i*-th level reservoir during period *t*;
NF_i: The *i*-th level reservoir's guaranteed output of;
NT_i: The *i*-th level reservoir's maximum output;
$ZR_{i,0}$: The *i*-th level reservoir's operating water level during the zero calculation period.

3 Reservoir Scheduling Calculation Process

In multi-objective optimization scheduling, the dimension of decision variables increases linearly with the number of reservoirs; the complex hydraulic and electrical constraints divide the optimization space into discontinuous segments [12]. According to the characteristics of cascade reservoirs in the lower reaches of the Lancang River, the reservoir sets three scheduling rules: dead water level zone scheduling, scheduling between the dead water level and normal storage level, and scheduling above the normal storage level zone. The relationship between water level and storage capacity of hydro-power station is shown in Fig. 2.

3.1 The Dead Water Level Zone Scheduling

If the water level at the inception phase is lower than the dead water level, calculate the output based on the assumption that the outflow equals the inflow. If the calculated output is less than the installed capacity, the outflow of power generation from the reservoir is equal to the inflow, and there is no abandoned water; if it is greater than the installed capacity, the water level and outflow flow rate will be recalculated based on the installed capacity. If the water level at the last phase of the current period is lower than the normal storage level, it will generate power according to the installed capacity without any abandoned water; if it is higher than the normal storage level, it will generate power according to the installed capacity, and excess water will be discarded. See Fig. 3 for details.

3.2 The Scheduling Between the Dead Water Level and the Normal Storage Level

If the water level at the inception phase is between the dead and normal storage levels, lower the water level directly to the dead water level and calculate the output. If the

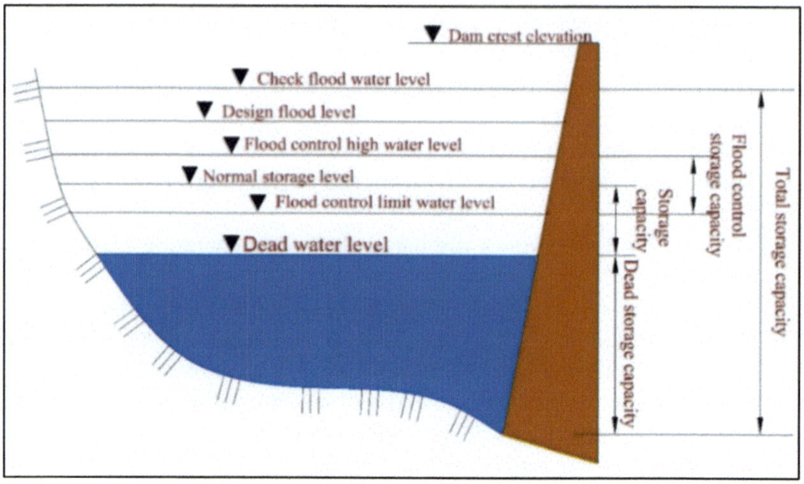

Fig. 2. Schematic diagram of the hydro-power station's water level and storage capacity.

calculated output is less than the installed capacity, the water level at the last phase of the period will directly decrease to the dead water level without abandoning water. If the calculated output is greater than the installed capacity, recalculate the water level and outflow flow based on the installed capacity. If the water level at the last phase of the calculated period is lower than the normal storage level, it will generate power according to the installed capacity without any abandoned water; if it is higher than the normal storage level, it will generate power according to the installed capacity, and excess water will be discarded. See Fig. 4 for details.

3.3 The Scheduling Above Normal Water Storage Level Zone

If the water level at the inception phase of the period is higher than the normal storage level, calculate the output based on the installed capacity. If the water level at the last phase of the calculation period is lower than the normal storage level, it will generate power according to the installed capacity without any abandoned water. If the water level is higher than the normal storage level, it will generate power according to the installed capacity, and excess water will be discarded. See Fig. 5 for details.

4 Analysis of Scheduling Results

4.1 Discharge of Jinghong Hydro-Power Station

Monthly Average Discharge. From 2019 to 2022, the maximum monthly average discharge value was 3256.8 m^3/s, which appeared in 2019; the minimum value was 303.7 m^3/s, which appeared in 2020; and the average value was 1292.7 m^3/s. The monthly average discharge statistics of Jinghong Hydro-power Station are shown in Table 1.

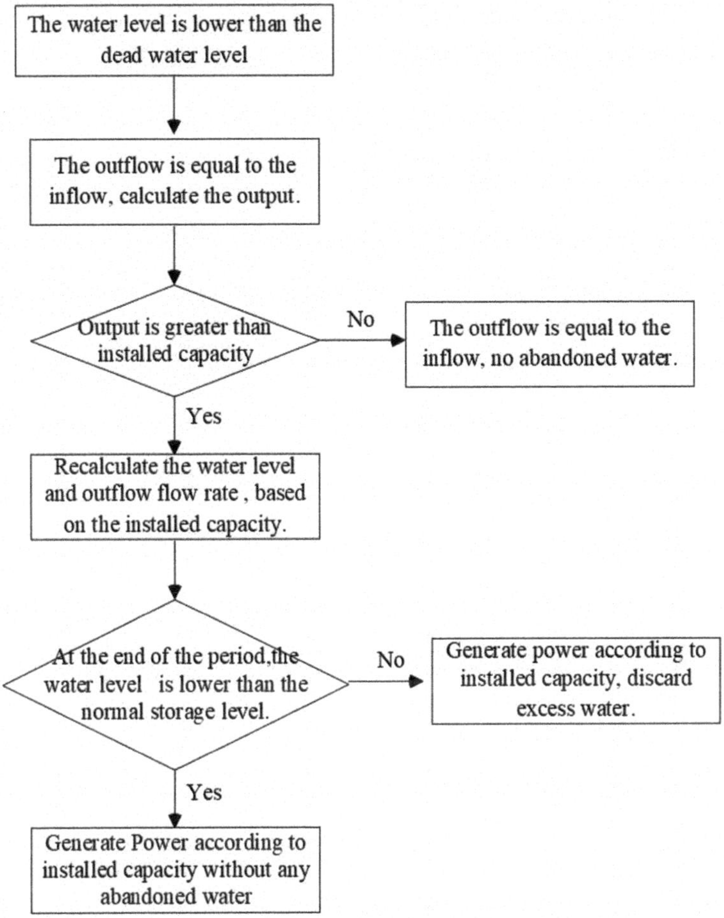

Fig. 3. Diagram of the dead water level zone scheduling.

Guarantee Rate of Monthly Average Discharge. From 2019 to 2022, the monthly guarantee rate for the discharge of Jinghong hydro-power station that exceeded 504 m^3/s was 89.6%; the monthly guarantee rate for the discharge that exceeded 800 m^3/s was 68.8%; the monthly guarantee rate for the discharge that exceeded 1000 m^3/s was 45.8%. The guarantee rate of average monthly average outflow of Jinghong Hydro-power Station is shown in Table 2.

4.2 Optimize Scheduling Plan and Boundary Condition Settings

The Hydrological Time Series Used in the Present Study Spans the Past 30 Years. Among the six cascade reservoirs from Gongguoqiao to Jinghong in the Lancang River's middle-lower reaches, only Nuozhadu is responsible for downstream flood control tasks; Xiaowan Reservoir cooperates with Nuozhadu Reservoir to undertake certain flood control tasks downstream. Therefore, during the main flood season (June 1st to September

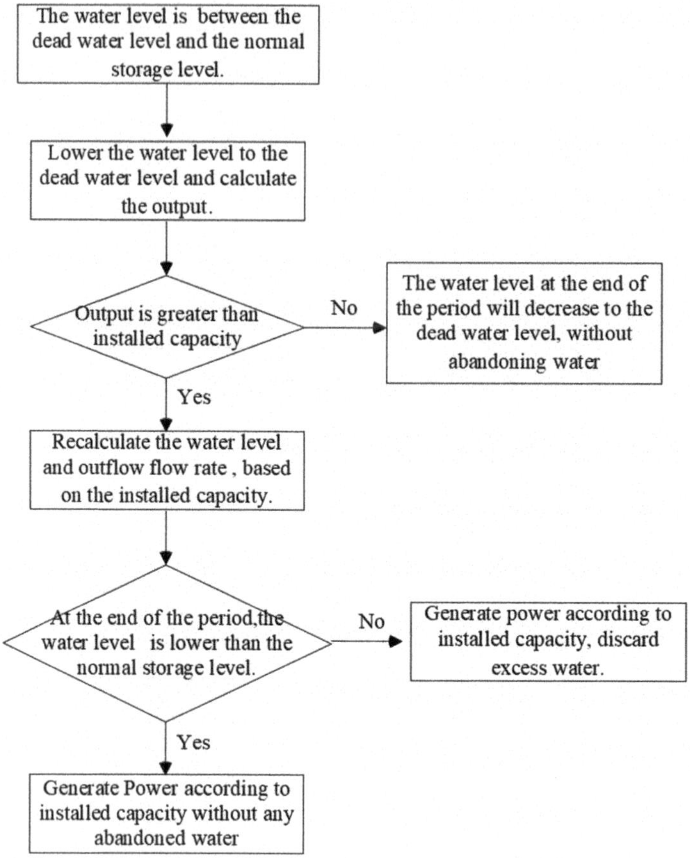

Fig. 4. Diagram of the scheduling between the dead water level and the normal storage level.

21st), to ensure the flood control tasks of the urban area and farmland embankments in Jinghong City, a flood control and storage capacity operation mode is adopted, the reservoir water level is adjusted from the flood control water level (804 m), with a flood control storage capacity of 2.002 billion m^3. The Nuozhadu Reservoir must be lowered below the flood control level by the end of May. When the reservoir's water level is higher than the flood control limit water level but lower than downstream farmland flood control requirements, the reservoir's maximum discharge flow cannot exceed 9206 m^3/s; when the water level of the reservoir is higher than 807.2 m but lower than the control water level of 810.69 m required for downstream urban flood control, the maximum discharge flow of the reservoir shall be controlled at no more than 16728 m^3/s; When the reservoir water level exceeds 810.69 m, the flood control of the reservoir shifts to meeting the safety of its own dam body, discharging according to the inflow, and the reservoir water level operates between 804 m and the flood control high water level (810.69 m). During the non-flood season, power generation scheduling is the main function, considering the water demand for shipping below the Jinghong hydro-power station.

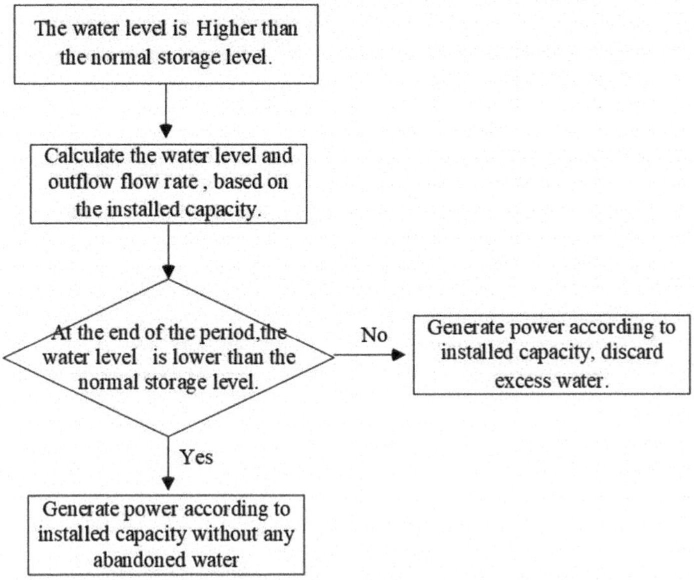

Fig. 5. Diagram of the scheduling above normal storage level zone.

Table 1. Statistics of monthly average discharge of Jinghong Hydro-power Station.

Year	2019	2020	2021	2022	2019~2022
Maximum value (m^3/s)	3256.8	2399.2	2505.4	2407.0	3256.8
Minimum value (m^3/s)	502.9	303.7	739.3	517.7	303.7
Average value (m^3/s)	1355.4	1341.4	1404.5	1069.5	1292.7

Table 2. Guarantee rate of monthly average outflow of Jinghong Hydro-power Station.

Name		2019	2020	2021	2022	2019~2022
>504 m^3/s	Amount of months	10	9	12	12	43
	Guarantee rate(%)	90.9	75.0	100.0	100.0	89.6
>800 m^3/s	Amount of months	8	8	11	6	33
	Guarantee rate(%)	72.7	67.7	91.7	50.0	68.8
>1000 m^3/s	Amount of months	5	7	6	4	22
	Guarantee rate(%)	45.5	58.3	50.0	33.3	45.8

Unlike Nuozhadu, Jinghong and other reservoirs only undertake their dam flood control safety. The reservoirs' water levels must be lowered below the flood control level by the end of May. In the flood season (June 1st to September 21st), the water level must satisfy the request of flood control and operate according to the water demand of power generation and shipping. Operate during non-flood season based on the demand for power generation and shipping water.

4.3 Scheduling Results and Discussion

Power Generation. The six hubs' average annual power generation, including Jinghong hydro-power station, is 3.88 × 10^9 kW·h, 1.897 × 10^{10} kW·h, 8.51 × 10^9 kW·h, 7.73 × 10^9 kW·h, 23.42 × 10^9 kW·h, and 7.32 × 10^9 kW·h, respectively. The average total annual power generation of the six reservoirs is 6.983 × 10^{10} kW·h, an increase of 1.64% compared to the initial average annual power generation of 68.7 billion kW·h. The comparison of power generation of different reservoirs is shown in Table 3.

Table 3. Comparison of power generation between different reservoirs. (10^9kW·h).

Name	Gongguoqiao	Xiaowan	Manwan	Dachaoshan	Nuozhadu	Jihong	Summation
Optimal scheduling	3.88	18.97	8.51	7.73	23.42	7.32	69.83
Initial design	3.94	18.89	7.6	6.71	23.9	7.62	68.7

Benefits of Generating Electricity. The six hubs' average annual benefits of generating electricity, including Jinghong hydro-power station, are 0.963 billion yuan, 5.178 billion yuan, 2.372 billion yuan, 2.147 billion yuan, 6.427 billion yuan, and 2 billion yuan, respectively, the average annual total benefit of generating electricity is 19.087 billion yuan. The benefits of generating electricity between different reservoirs are shown in Table 4.

Table 4. Benefits of generating electricity between different reservoirs.

Name	Gongguoqiao	Xiaowan	Manwan	Dachaoshan	Nuozhadu	Jinghong	Summation
Benefits of generating electricity (Billion Yuan)	0.963	5.178	2.372	2.147	6.427	2.0	19.087
Proportion (%)	5.1%	27.1%	12.4%	11.2%	33.7%	10.5%	100%

Discharge Flow Process. Under optimized scheduling, the annual distribution of runoff from Xiaowan to Jinghong is shown in Table 5. According to Table 5, except for Gongguoqiao Reservoir, the runoff of the other five reservoirs during the flood season (June to September) is less than the natural runoff, with a variation rate of −58% to −42% and an average variation rate of −49%; the non-flood season (October to May of the following year) runoff is greater than the natural runoff, with a change rate of +54% to +71%, and an average change rate of +67%. The interannual distribution changes of runoff in cascade reservoirs are shown in Table 5.

The annual average discharge flow and natural flow process of Jinghong hydro-power station in consecutive wet and dry years are shown in Figs. 6 and 7. As shown in the figure, under the optimized scheduling plan, after being regulated by the five reservoirs upstream of Jinghong, the flow during the flood season decreased and during the non-flood season increased compared to the natural runoff process. The monthly average discharge of Jinghong hydro-power station is more than 1000 m^3/s in continuous wet and dry years, and the flow changes much more smoothly after optimized dispatching than in natural conditions. The flow conditions are more suitable for navigation, which is conducive to improving the level of navigation class IV waterway below Jinghong hydro-power station.

5 Conclusions

The scheduling results show that the total power generation of six cascaded hydro-power stations, including the Jinghong hydro-power station, is equivalent to the average annual power generation during the initial design stage of the reservoir. The optimized scheduling of cascade reservoirs would change rivers' radial yearly flow distribution process, reduce the amount of discharge during flood season, and increase the amount of discharge during non-flood season. Compared with the natural flow process, the average river runoff during the flood season decreased by 49%, while the average river runoff increased by 67% in the non-flood season. The guaranteed rate of monthly average discharge flow greater than 1000 m^3/s is 100% for Jinghong hydro-power station, which is significantly higher than the monthly average discharge flow guarantee rate of Jinghong hydro-power station, which is that of 2019 to 2022 (45.8%); the optimized scheduling of cascade reservoirs can significantly improve the period guarantee rate of 800–1000 m^3/s, and the flow changes are more stable, which provides highly favorable conditions for further improving the navigation level of the waterway below Jinghong hydro-power station.

Table 5. Table of interannual distribution changes of runoff in cascade reservoirs. (m^3/s)

Hydro-power station		Flood season (June to September)			Non-flood season (October to May of the following year)			
Hydro-power station	Natural runoff	Optimal scheduling	Runoff change	Change rate (%)	Natural runoff	Optimal scheduling	Runoff change	Change rate (%)
Xiaowan	2232.41	1299.47	−932.94	−42	706.49	1174.70	468.21	66
Manwan	2256.74	1296.84	−959.90	−43	717.21	1100.97	383.76	54
Dachaoshan	2519.60	1166.07	−1353.53	−54	811.06	1425.24	614.18	76
Nuozhadu	3209.25	1704.13	−1505.12	−47	1058.46	1811.61	753.15	71
Jinghong	4409.11	1844.76	−2564.35	−58	1090.64	1822.32	731.68	67

Note Negative values of runoff variation in the above table indicate that the flow rate of the optimized scheduling scheme is lower than the natural runoff rate. In contrast, positive values indicate that the flow rate of the optimized scheduling scheme is greater than the natural runoff rate

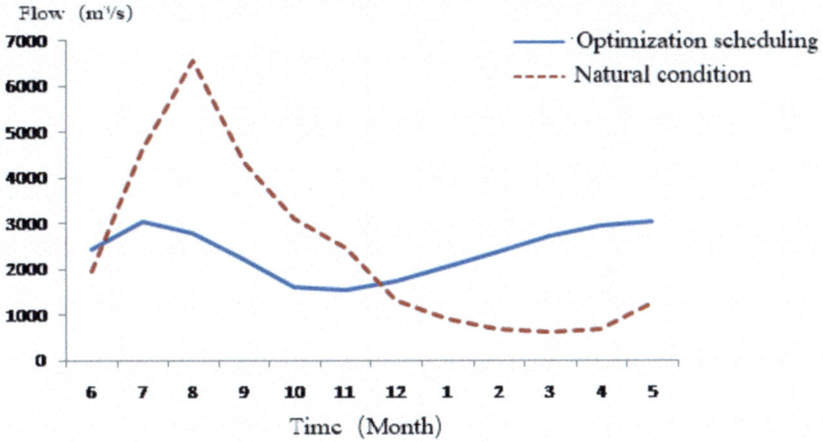

Fig. 6. Monthly average discharge and natural flow process of Jinghong hydro-power station in wet year.

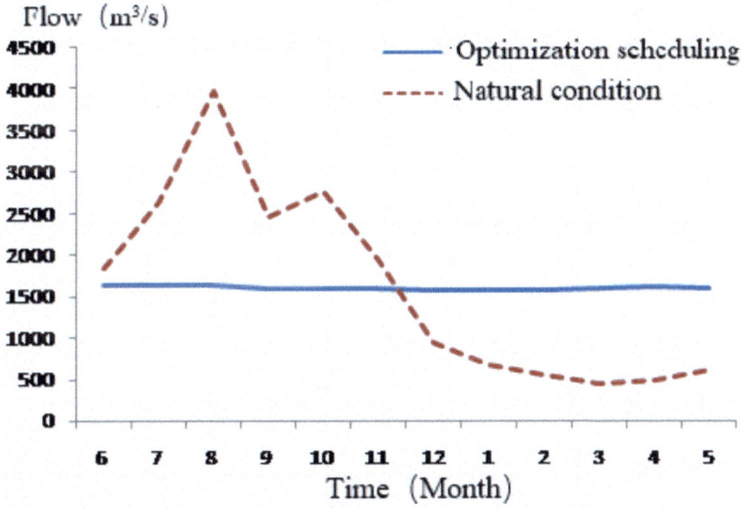

Fig. 7. Monthly average discharge and Natural flow process of Jinghong hydro-power station in dry year.

References

1. Chen, Q.: Basic data assembly of Lancang River and Mekong River Basin. Yunnan Science and Technology Press (2000)
2. Liu, J.T.: The review of Lancang River hydro-power development and Jinghong hydro-power station engineering design. Hydroelectr. Generation **34**(4), 28–31 (2008)
3. Han, D.W.: Accelerate the Lancang River hydro-power cascade scrolling development. In: The 2004 National Hydro-power Plant Academic Annual Conference and the Founding

Conference of the National Technical Cooperation Network for Large and Medium-Sized Hydro-Power Plants (2004)
4. Wang, H., Wang, X., Lei, X.H.: The development and prospect of key techniques in the cascade reservoir operation. J. Hydraul. Eng. **50**(1), 25–37 (2019)
5. Zhu, L.N.: Extensive Reservoir Water Resources Optimization Scheduling Research—Pubugou Reservoir as an Example. Sichuan Normal University (2005)
6. Zhou, C.Y., Liu, Y.: Multi-objective reservoir resources optimization scheduling decision support system and application. Comput. Eng. Appl. **43**(16), 193–195 (2007)
7. Bellman, R.E., Dreyfus, S.E.: Applied Dynamic Programming. Princeton University Press (1962)
8. Zhou, M., Sun, S.D.: The Principle and Application of Genetic Algorithm. National Defence Industry Press, Beijing (1999)
9. Huang, S.J.: Enhancement of hydroelectric generation scheduling using ant colony system-based optimization approaches. IEEE Trans. Power Syst. **16**(3), 296–301 (2001)
10. Kennedy, J., Eberhart, R.C.: Particle swarm optimization. In: Proceedings of the IEEE International Conference on Neural Networks, IEEE Service Center, Piscataway, NJ, vol. IV, pp. 1942–1948 (1995)
11. Hu, T.S., Wan, Y.H., Feng, S.Y.: Research on the artificial neural network methodology for multi-reservoir operating rules. Adv. Water Sci. **6**(1), 53–60 (1995)
12. Lu, P., Zhou, P.C., Han, B., Zhou, K.B.: Multi-objective optimal scheduling method for cascade reservoirs and its application practice. Yunnan Water Power **40**(9), 175–182 (2024)

Effects of Kernel Size and Input Channel on the Performance of CNN and Hybrid Models in Dam Water Level Forecasting

Kai Wen Ng[1], Yuk Feng Huang[1(✉)], Chai Hoon Koo[1], Kai Lun Chong[2], Ahmed El-Shafie[3], and Ali Najah Ahmed[4]

[1] Department of Civil Engineering, Lee Kong Chian Faculty of Engineering and Science, Universiti Tunku Abdul Rahman, Jalan Sg. Long, Bandar Sg. Long, Cheras, 43000 Kajang, Selangor, Malaysia
ngkaiwen7@1utar.my, {huangyf,kooch}@utar.edu.my

[2] Faculty of Engineering and Quantity Surveying, INTI International University (INTI-IU), Nilai, Negeri Sembilan, Malaysia
kailun.chong@newinti.edu.my

[3] National Water and Energy Center, United Arab Emirates University, P.O. Box 15551, Al Ain, United Arab Emirates
anasser@uaeu.ac.ae

[4] School of Engineering and Technology, Department of Engineering, Sunway University, 47500 Bandar Sunway, Petaling Jaya, Malaysia
mahfoodh@sunway.edu.my

Abstract. Deep learning applications for hydrological forecasting have become increasingly common over the past two decades. Among various deep learning models, the hybridization of the Convolutional Neural Networks (CNNs) with the Recurrent Neural Networks (RNNs), particularly the Long Short-Term Memory (LSTM) and Gated Recurrent Unit (GRU) networks, has gained popularity due to their capability of capturing both spatial and temporal features. This study investigates how different kernel sizes and input channels have an impact on the performance of the CNN, CNN-LSTM and CNN-GRU in dam water level forecasting at the Klang Gates Dam, Malaysia. The results indicated that the increase in kernel size has no impact on the performance of the CNN because the output layer was a fully connected feed forward layer. However, the LSTM and GRU layers, which process the output of the CNN layer sequentially, performed better when the kernel size was smaller due to the finer detail extracted from the CNN layer. Due to the different nature of input features, the input channels of the CNN layer were split into water level data and multivariate data input channels. Both hybrid models showed that decreasing the kernel size of the water level data input channel and increasing the kernel size of the multivariate data input channel, have improved the model performance. With the splitting of the input channels of the models, the performance of the CNN, CNN-LSTM and CNN-GRU models, with their respective best kernel size configurations, showed improvements of 13.57%, 14.39% and 15.87%, respectively. Hence, the splitting of input channels with more model configurations are recommended for further studies.

Keywords: LSTM · GRU · CNN · Kernel Size · Split Input Channels

C. Weng (Ed.): WRE 2024, LNCE 634, pp. 235–245, 2025.
https://doi.org/10.1007/978-981-96-5095-8_20

1 Introduction

In comparison to statistical models and process based hydrological models, machine learning and deep learning models can utilize different combinations of input into the model due to the strong nonlinear modeling capabilities [1]. Hence, over the past two decades, researchers have increasingly applied machine learning and deep learning for streamflow and water level forecasting [2]. Among these models, the Recurrent Neural Networks (RNNs), such as the Long Short-Term Memory (LSTM) and Gated Recurrent Unit (GRU), have exhibited superior performance across numerous studies due to their innate ability to capture temporal dependencies. For instance, Fan et al. [3] demonstrated that the Feed Forward Neural Network (FFNN) yielded inferior performance compared to the LSTM in a 1-day ahead runoff simulation. Similarly, Asaad et al. [4] found that the LSTM outperformed the FFNN and the Adaptive Neuro-Fuzzy Inference System (ANFIS) in long-term streamflow forecasting with the highest R^2 and lowest Root Mean Square Error (RMSE) values of 0.71 and 0.51, respectively. Likewise, Latif et al. [5] reported that the LSTM outperformed the Boosted Regression Tree (BRT) in reservoir inflow forecasting over 5 performance metrics. Convolutional Neural Networks (CNNs) are widely utilized to capture spatial features from multivariate time series. Shu et al. [6] demonstrated this by applying a two-dimensional CNN (2D-CNN) for rainfall and streamflow forecasting using 66 climate variables as input. The 2D-CNN showed superior performance over the FFNN and Extreme Learning Machine (ELM). Further on, the performance of the CNN and LSTM in forecasting streamflow at the Chau Doc and Can Tho stations in the Mekong River was compared by Van et al. [7]. The study demonstrated that CNN had a higher accuracy at Chau Doc with a correlation coefficient value of 0.9994 while the LSTM performed better at the Can Tho station with a correlation coefficient value 0.9872.

To further enhance the capability of RNNs in capturing both multivariate and temporal features, the CNNs with RNNs were combined as a hybrid model. For instance, Ghimire et al. [8] demonstrated that the CNN-LSTM achieved the best performance with a correlation coefficient above 0.949 during forecasting streamflow at the Brisbane River and Teewah Creek compared to their stand-alone CNN and LSTM. Meanwhile, Xu et al. [9] utilized the CNN-GRU for streamflow forecasting at the Yangtze River with streamflow data, meteorological data, monthly large-scale climate indexes and climate model hindcasts as multivariate inputs. The study illustrated that the CNN-GRU effectively utilized the multivariate data and achieved a KGE score above 0.6 for the 6-month ahead streamflow forecasting. Similarly, streamflow, precipitation and min-max temperature data with daily, weekly, and monthly time scales were applied as input data for multiple deep learning models to forecast streamflow at the Awash and Tiber River Basin [10]. The results indicated that both CNN-GRU (0.83) and CNN-LSTM (0.81) attained a coefficient of determination score higher than the GRU (0.77) and the LSTM (0.76). Further on, Fang et al. [11] combined the CNN, Transformer and LSTM for a 1-month ahead streamflow forecasting at the Shule River basin. The study showed that the CNN-transformer-LSTM outperformed the Transformer with a Nash-Sutcliffe Efficiency (NSE) value over 0.964.

Despite the advancements showcased in these studies, there remains a dearth of research investigating the performance of CNNs and their hybrid models under different kernel sizes and separated input channels. Consequently, the first objective of this study was to explore the impact of kernel size on the performance of CNNs and their hybrid models at water level forecasting at the Klang Gates Dam. Secondly, it was checking out the performance of adopting a CNN hybrid model that has split input channels with different kernel sizes for the water level data and multivariate data.

2 Methodology

2.1 Research Area

The study area at the Klang Gates Dam is in the Gombak district of Selangor, Malaysia (see Fig. 1). Since 1959, the dam has a capacity of 28,150 million liters and a catchment area of 74.76 km^2. It serves as a crucial water infrastructure, providing water resources and flood mitigation for the Klang Valley [12].

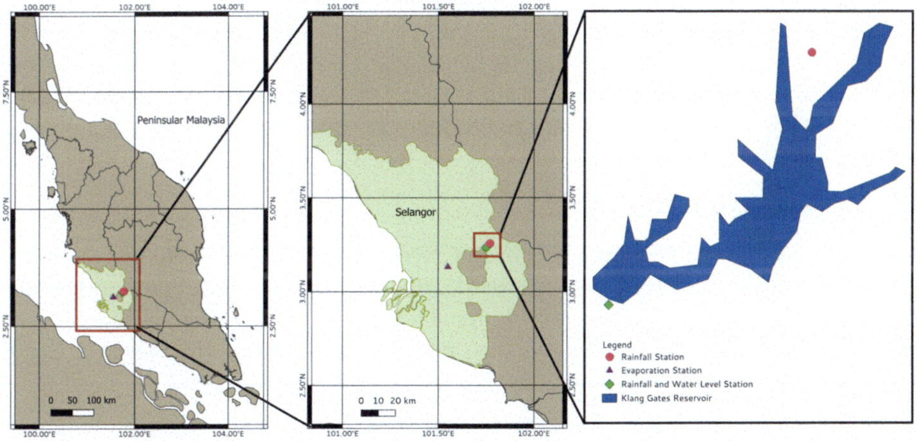

Fig. 1. Locations of water level, evaporation and rainfall stations of Klang Gates Dam.

2.2 Data Preparation

This study utilized water level (WL) data from the Klang Gates Dam from January 1, 2013, to January 31, 2023, as both the model input and target data. For local climate conditions, rainfall data from two stations (Station 3217002 and Station 3217004) and evaporation data from Station 48647 were incorporated. Additionally, global climate conditions were captured using climate indices such as the Oceanic Niño Index, Dipole Mode Index and Southern Oscillation Index. Given the role of the dam as a water supply facility, water demand data was also included in the analysis. The sources of these data are listed in Table 1.

Table 1. Data Types and Sources.

Data Types	Frequencies	Sources
Water Level (Station 3217435)	Daily	Department of Irrigation and Drainage Malaysia
Rainfall (Stations 3217002 and 3217004)	Daily	
Evaporation (Station 48647)	Daily	Malaysian Meteorological Department
Oceanic Niño Index	Monthly	National Oceanic and Atmospheric Administration
Southern Oscillation Index	Monthly	
Dipole Mode Index	Monthly	
Water Demand	Monthly	LUAS Selangor

To create the input and target dataset for model training, all the data were first normalized using the min-max normalization. Next, it is necessary to determine the appropriate number of days for input time lags. This study used the Auto-Correlation Function (ACF) to determine the input time lags for the models, which is a common practice in hydrological forecasting. Based on the ACF, a WL correlation above 0.95 was found for the past 20 days. Thus, the dataset is reshaped and sliced into input and target data with 20 days of time lag. Lastly, the models were trained and evaluated based on the K-fold cross-validation (4 folds) to evaluate the model under different conditions.

2.3 CNN and Hybrid Models

Convolutional Neural Networks (CNNs). CNNs are specialized neural networks that automatically learn features from structured data, such as images and time series. The core component of a CNN is the convolutional layer, which uses filters to convolve the input data. This convolution process detects patterns, such as edges in images or trends in time series and produces feature maps for the upcoming layers.

Long Short-Term Memory (LSTM). LSTM is a type of RNN designed to model sequential data like time series or language. They use memory cells (C_t) and three gating mechanisms: input (i_t), forget (f_t) and output gates (o_t), to maintain and regulate information over long periods.

$$i_t = \sigma\left(x_t U^i + h_{t-1} W^i\right) \tag{1}$$

$$f_t = \sigma\left(x_t U^f + h_{t-1} W^f\right) \tag{2}$$

$$o_t = \sigma\left(x_t U^o + h_{t-1} W^o\right) \tag{3}$$

$$\widetilde{C}_t = \tanh\left(x_t U^g + h_{t-1} W^g\right) \tag{4}$$

$$C_t = \sigma\left(f_t * C_{t-1} + i_t * \widetilde{C}_t\right) \tag{5}$$

$$h_t = \tanh(C_t) * o_t \tag{6}$$

Equations 1–6 show the computation process of LSTM. Where x_t and h_t are input and output values at time t. h_{t-1} is the output value at time $t-1$, while U and W are input weights and output weights for each gate. Lastly, C_t is the cell value in time t.

Gated Recurrent Units (GRUs). Like LSTM, GRU is a type of RNN designed to model sequential data, such as time series or natural language. GRU addresses the vanishing gradient problem in RNNs by utilizing gating mechanisms. Differently, GRU only has two gates: the update gate (z_t) and the reset gate (r_t).

$$z_t = \sigma\left(x_t U^z + h_{t-1} W^z\right) \tag{7}$$

$$r_t = \sigma\left(x_t U^r + h_{t-1} W^r\right) \tag{8}$$

$$\widetilde{h_t} = \tanh\left(x_t U^h + r_t W^h h_{t-1}\right) \tag{9}$$

$$h_t = (1 - z_t) * h_{t-1} + z_t * \widetilde{h_t} \tag{10}$$

Equations 7–10 show the computation process of GRU. Where x_t and h_t are input and output values at time t. h_{t-1} is the output value at time $t-1$, while U and W are input weights and output weights for each gate. Lastly, h_t is the candidate activation vector in time t.

CNN-RNN Hybrid Models. Combining CNNs with RNNs leverages the strengths of both architectures to extract spatial and temporal features from data (see Fig. 2a). CNNs excel at capturing multivariate features by convolving multiple filters across input data. In contrast, RNNs specialize in extracting temporal features by processing sequences over time. In this study, a CNN layer with a split input channel is applied to improve the performance of the conventional CNN-RNN approach. The model structure is shown in Fig. 2b, where WL data had its input channel while the other multivariate data were assigned to the second input channel. This forces the CNN layer to extract features from multivariate features without the dependency of WL and allows the use of different kernel sizes for each channel.

2.4 Model Hyperparameters

All CNN layers for CNN, CNN-LSTM and CNN-GRU in this study have 20 filters. Similarly, all GRU and LSTM layers in this study consist of 20 neurons. The number of CNN, GRU, and LSTM layers in this study was set to 1. Additionally, only at the last time step was the output of the GRU and LSTM layers passed to the Feed Forward layer for WL forecasting. For the CNN, CNN-LSTM, and CNN-GRU with split input channels, the CNN layers had 10 filters each for the WL data input channel and the multivariate data input channel, respectively, to ensure a fair comparison.

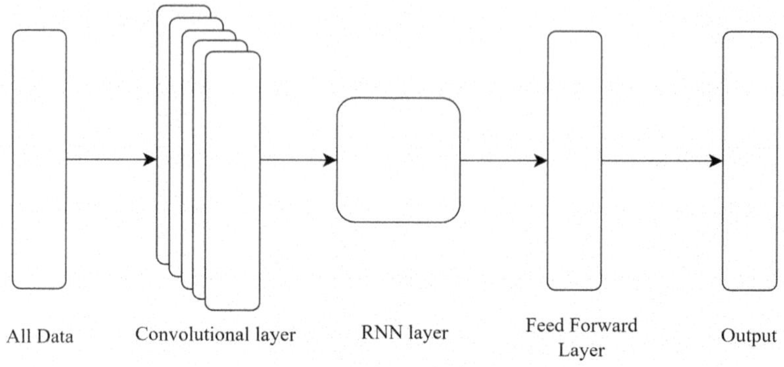

(a) CNN-RNN model structure

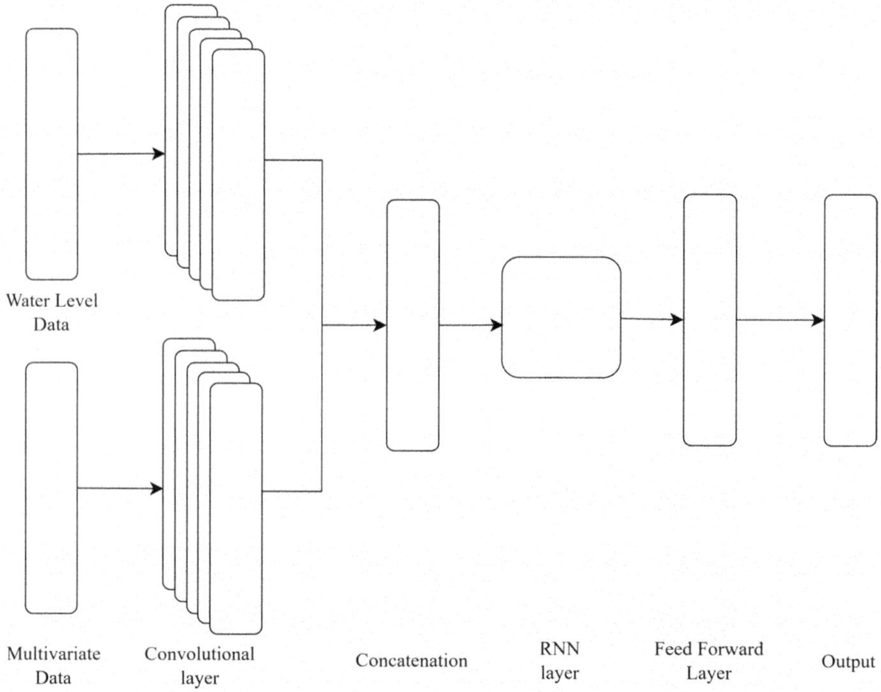

(b) CNN-RNN model structure with split input channel

Fig. 2. Structure of CNN-RNN.

2.5 Performance Evaluation

The performance of the WL forecasting was evaluated using four key metrics: Normalized Root Mean Square Error (NRMSE), Nash-Sutcliffe Efficiency (NSE) and Mean Absolute Percentage Error (MAPE). These metrics provide insights into the accuracy and reliability of the forecasting model. The Combined Accuracy (CA) metric was calculated as the average of these three individual metrics to assess the overall performance comprehensively.

3 Results and Discussion

3.1 Effects of Kernel Size

Kernel size is a crucial aspect of the CNN architecture. A larger kernel size allows the CNNs to access a broader range of features, while a smaller one focuses on capturing more localized features and abnormalities in small areas [13, 14]. Observing the boxplot as shown in Fig. 3, no obvious performance difference was observed in the CNN with the change in kernel size from 1 to 10. However, for CNN-LSTM and CNN-GRU, an increase in mean CA value with the increase in kernel size was observed. The difference between these two findings can be attributed to the different output layers. CNN uses an FFNN layer that connects all feature maps directly to the output, mitigating the effects of kernel size. In the CNN-LSTM and the CNN-GRU, the LSTM layer and the GRU layer process the CNN feature map sequentially. The course feature map made it harder to extract features from each time step, causing a slight drop in performance.

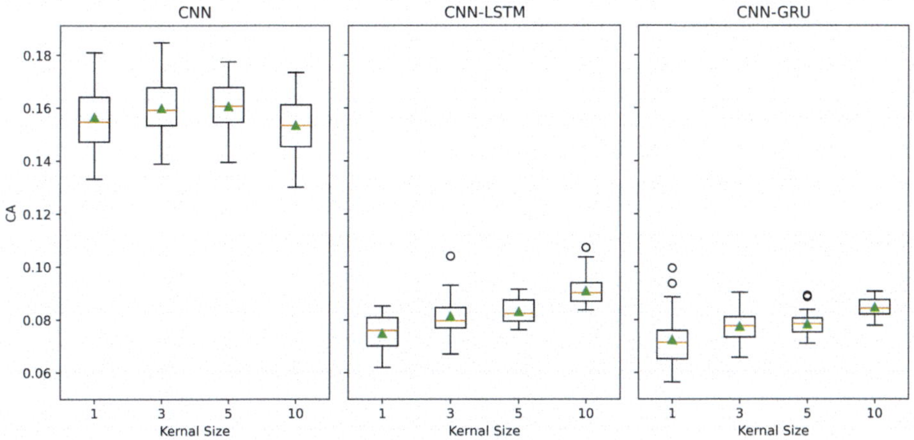

Fig. 3. Effects of CNN kernel size.

3.2 Effects of Input Channel Splitting

The process of convolution aggregates features from specific time steps based on kernel size, potentially diminishing certain features when passed through a single input channel. Therefore, input features with separate channels were adopted, one for the WL data and another for the multivariate data. This separation aims to compel the CNN layer to extract features solely from the multivariate data for WL forecasting, particularly in the absence of WL at the initial stages. The results of splitting CNN input channels are shown in Fig. 4, where different kernel combinations are displayed as (x, y) with x as the kernel size for the WL data input channel and y as the kernel size for the multivariate data input channel. From the results, the CNN did not show any performance improvements or degradation trend with the change in kernel sizes. On the other hand, both CNN-LSTM and CNN-GRU showed slight performance degradation with the increase of WL data input channel kernel size from 1 to 10. In contrast, a slight improvement was observed with the increase of multivariate data input channel kernel size from 1 to 10. This suggests that the WL data benefits from smaller kernels to capture finer details, whereas larger kernels are more suitable for multivariate data. Furthermore, it can be observed that changes in the kernel size of the WL data input channel resulted in a larger performance difference than changes in the kernel size of the multivariate data input channel. This indicates that the main reason for the improved performance of the CNN-LSTM and CNN-GRU with reduced kernel size in Sect. 3.1 is due to the availability of a finer filter for WL data, which implies that the kernel size for the WL data input channel has the most significant impact on the performance of the models.

3.3 Performance Comparison Under Best Model Configurations

Lastly, the performance of the CNN, CNN-LSTM and CNN-GRU were compared based on their best model configuration with the lowest mean CA values. By observing the boxplot results in Fig. 5, it is evident that employing separated input channels for the CNN, CNN-LSTM and CNN-GRU yielded superior and more stable performance compared to their respective baseline models when utilizing the best configurations of CNN. Furthermore, based on the mean performance metrics shown in Table 2, the CNN, CNN-LSTM and CNN-GRU with split input channels had 13.57%, 14.39% and 15.87% lower mean CA values, respectively, when compared to the base form. This result also concurs with that of the study by Chencho et al. [15], which showed that the 1D-CNN with different kernel sizes and input channels could extract different features better for structural damage classification. These results indicate that this method can be further extended for forecasting water levels, streamflow or rainfall using precipitation or temperature images of varying resolutions as input [16]. Different CNN input channels with various kernel sizes can be employed for each resolution of images to enhance forecasting accuracy.

4 Conclusions

In conclusion, the results obtained from the study suggest that changes in kernel size only have no impact on the performance of the CNN. In comparison, the decrease in kernel size showed slight performance improvements in the CNN-LSTM and CNN-GRU. Moreover, the small kernel size for the WL data input channel and the large kernel size for the multivariate data input channel resulted in the best performance for CNN-LSTM and CNN-GRU. Comparing the best configurations of all models, the CNN, CNN-LSTM and CNN-GRU with split input channels always outperform their base form. This approach facilitates the customization of kernel sizes for individual features, improving the feature extraction process. However, this study comes with its

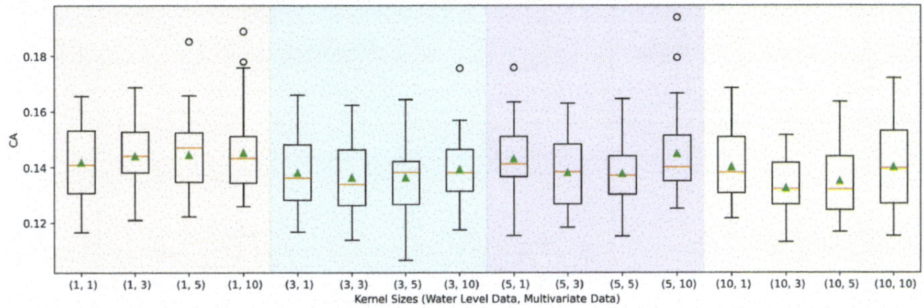

(a) CNN

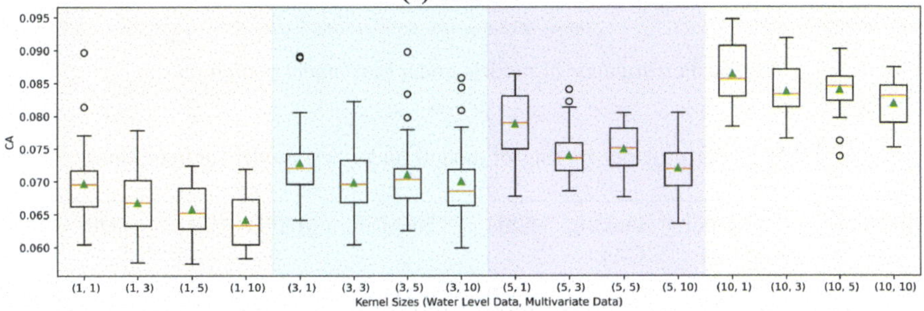

(b) CNN-LSTM

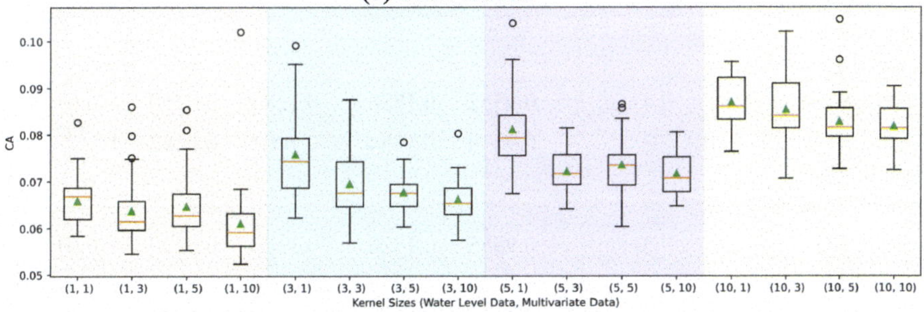

(c) CNN-GRU

Fig. 4. Performance of models with split input channels under various kernel sizes.

limitations. Firstly, the model configuration and combinations explored in this study are limited. Future research should explore additional CNN variants, such as dilated CNN, pooling layers, and different activation functions. Furthermore, more sophisticated fusion methods, such as hybrid fusion methods, which fuse features at different levels of models, should be investigated. These advancements are crucial for further enhancing the predictive capabilities of hybrid models in hydrological forecasting applications.

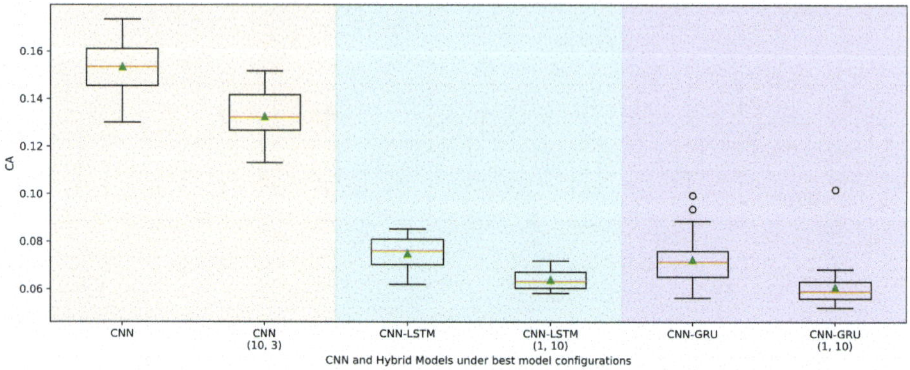

Fig. 5. Performance of models under best model configurations.

Table 2. Performance metrics of models under best model configurations.

Models	Kernel Sizes	Ranking (CA)	NSE	NRMSE	MAPE	CA	Diff CA (%)
CNN	10	6	0.8553	0.3136	0.2099	0.1535	−0.0208 (−13.57%)
CNN (splitting data input channel)	(10, 3)	5	0.8887	0.2846	0.1994	0.1326	
CNN-LSTM	1	4	0.9552	0.1788	0.1283	0.0750	−0.0108 (−14.39%)
CNN-LSTM (splitting data input channel)	(1, 10)	2	0.9638	0.1554	0.1001	0.0642	
CNN-GRU	1	3	0.9555	0.1720	0.1163	0.0725	−0.0115 (−15.87%)
CNN-GRU (splitting data input channel)	(1, 10)	1	0.9650	0.1471	0.0939	0.0610	

Acknowledgments. The research was supported by the Ministry of Higher Education (MoHE), through the Fundamental Research Grant Scheme (FRGS/1/2022/TK06/UTAR/02/14). We would like to thank Universiti Tunku Abdul Rahman through the UTAR Research Scholarship Scheme to make this research possible. The authors would like to express their gratitude and appreciation to the National Oceanic and Atmospheric Administration, Department of Irrigation and Drainage Malaysia, Malaysian Meteorological Department and LUAS Malaysia for providing the data.

References

1. Akbarian, M., Saghafian, B., Golian, S.: Monthly streamflow forecasting by machine learning methods using dynamic weather prediction model outputs over Iran. J. Hydrol. **620**, 129480 (2023)
2. Ng, K.W., Huang, Y.F., Koo, C.H., Chong, K.L., El-Shafie, A., Najah Ahmed, A.: A review of hybrid deep learning applications for streamflow forecasting. J. Hydrol. **625**, 130141 (2023)
3. Fan, H., Jiang, M., Xu, L., Zhu, H., Cheng, J., Jiang, J.: Comparison of long short term memory networks and the hydrological model in runoff simulation. Water **12**(1) (2020)
4. Asaad, M.N., Eryürük, Ş., Eryürük, K.: Forecasting of streamflow and comparison of artificial intelligence methods: a case study for Meram stream in Konya, Turkey. Sustainability **14**(10) (2022)
5. Latif, S.D., Ahmed, A.N.: Streamflow prediction utilizing deep learning and machine learning algorithms for sustainable water supply management. Water Resour. Manage **37**(8), 3227–3241 (2023)
6. Shu, X., Ding, W., Peng, Y., Wang, Z., Wu, J., Li, M.: Monthly streamflow forecasting using convolutional neural network. Water Resour. Manage **35**(15), 5089–5104 (2021)
7. Van, S.P., Le, H.M., Thanh, D.V., Dang, T.D., Loc, H.H., Anh, D.T.: Deep learning convolutional neural network in rainfall–runoff modelling. J. Hydroinf. **22**(3), 541–561 (2020)
8. Ghimire, S., Yaseen, Z.M., Farooque, A.A., Deo, R.C., Zhang, J., Tao, X.: Streamflow prediction using an integrated methodology based on convolutional neural network and long short-term memory networks. Sci. Rep. **11**(1), 17497 (2021)
9. Xu, W., Chen, J., Zhang, X.J., Xiong, L., Chen, H.: A framework of integrating heterogeneous data sources for monthly streamflow prediction using a state-of-the-art deep learning model. J. Hydro. **614** (2022)
10. Wegayehu, E.B., Muluneh, F.B.: Multivariate streamflow simulation using hybrid deep learning models. Comput. Intell. Neurosci. 1–16 (2021)
11. Fang, J., Yang, L., Wen, X., Li, W., Yu, H., Zhou, T.: A deep learning-based hybrid approach for multi-time-ahead streamflow prediction in an arid region of Northwest China. Hydrol. Res. **55**(2), 180–204 (2024)
12. Latif, S.D., Marhain, S., Hossain, M.S., Ahmed, A.N., Sherif, M., Sefelnasr, A., El-Shafie, A.: Optimizing the operation release policy using charged system search algorithm: a case study of Klang Gates Dam, Malaysia. Sustainability **13**(11) (2021)
13. Goodfellow, I., Bengio, Y., Courville, A.: Deep Learning. MIT Press (2016)
14. Yang, X., Bi, F., Cheng, J., Tang, D., Shen, P., Bi X.: A multiple attention convolutional neural networks for diesel engine fault diagnosis. Sensors **24** (2024) https://doi.org/10.3390/s24092708
15. Chencho, J. L., Hao, H., Li, L.: Structural damage classification of large-scale bridges using convolutional neural networks and time domain responses. J. Perform. Constr. Facil. **38**(4) (2024)
16. Simanjuntak, F., Jamaluddin, I., Lin, T.H., Siahaan, H.A.W., Chen, Y.N.: Rainfall forecast using machine learning with high spatiotemporal satellite imagery every 10 minutes. Remote Sens. **14**(23), 5950 (2022)

Water Intake and Sediment Control in a Run-of-River Hydropower Station

Yong Jin[1], Dangwei Wang[2](✉), Rui Qian[3], Anjun Deng[2], and Hongling Shi[2]

[1] Powerchina Resources Ltd., Beijing, China
[2] State Key Laboratory of Basin Water Cycle Simulation and Regulation, China Institute of Water Resources and Hydropower Research, Beijing, China
wangdw17@126.com, {denganj,shihl}@iwhr.com
[3] POWERCHINA International Group Ltd., Beijing, China
qianrui@powerchina-intl.com

Abstract. When the hydropower station adopts the low-dam method for water intake, it is difficult to reduce sediment concentration in water flows into the hydropower station. It is necessary to develop a systematic technology to ensure that water intake and sediment management are handled safely within the power station. We established a numerical model for reservoir sedimentation and proposed a new method for calculating sediment removal efficiency to carry out a comprehensive study on sediment problems in the upper Marsyangdi-A Hydropower Station. Based on the study, some key indices were determined, including the flow rate for reservoir desilting, the threshold of sediment concentration to avoid damage to turbines, and the sediment removal efficiency of the sediment settling tank. The measured data after the operation of the power station showed that the results of sediment analysis and calculation are in good agreement with the observations. The water intake and sediment control measures adopted by the power station improved the sediment discharge efficiency of the reservoir and met the requirements for the continuous operation of the power station.

Keywords: Mountain River · Low Dams for Water Intake · Water Intake · Sediment Control

1 Introduction

Hydropower stations typically need to build dams to create reservoirs, which are used to generate electricity. In addition to storing water, these reservoirs also capture a large amount of sediment. According to statistics from 2001, the average annual sediment that accumulates in large reservoirs worldwide accounts for 0.5% to 1% of the total remaining reservoir capacity [1], and sediment not only occupies a large amount of storage space enhancing the frequency of flood release, backwater flooding, evaporation loss but also causes the damage of turbines [2–4]. Therefore, constructing reservoirs in sediment-laden rivers generally involves designing reservoir a capacity sufficient to accommodate the local sedimentation. This reduces sediment accumulation through reservoir outflow and ensures the long-term operation of reservoirs and power stations [5].

C. Weng (Ed.): WRE 2024, LNCE 634, pp. 246–257, 2025.
https://doi.org/10.1007/978-981-96-5095-8_21

Due to development limitations, runoff power stations have been constructed along sediment laden river in recent years. However, these power stations are easily filled with silt in the reservoirs, and the impact that this has on the normal operation of the power station cannot be ignored. The existing research results are mostly concerned with the silt accumulation and treatment measures of large reservoirs [6–8], and there are relatively few studies focused on the silt accumulation process and its causes within the dammed areas of low-head hydropower stations along sediment laden rivers.

Because of the small storage capacity of hydropower stations with low dams, in conjunction with their reduced ability to regulate water and sediment, the difficulty of these systems to control water and sediment, and the poor sediment treatment in many power stations [9], sediment problems have become a key factor restricting the continuous operation of power stations.

In this study, the effect of water intake and sediment control measures in the upper Marsyangdi-A hydropower station is analyzed. At the beginning of the operation of the hydropower station, a mathematical model was used to simulate and predict the possible sediment issues that would impact the power station. This enables the determination of crucial factors such as the timing of sediment discharge from the reservoir, the threshold of the sediment concentration within the diversion channel, and an estimation of the sediment settling efficiency using a settling tank, which is based on the non-uniform non-equilibrium sediment transport theory. These analyses provide direct technical support for the effective treatment of the water intake and sediment control problems in the power station.

2 Material and Methods

2.1 Study Area

Located in the upper reaches of the Marsyangdi River in Nepal, the Upper Marsyangdi-A Hydropower Station is found in the southern foothills of the Himalayas, and the average reverbed slope of the river section where the reservoir is located is 1.25%, which is a typical mountainous river according to the type of river channel [10]. The basin area above the dam site is 2740 km^2, the annual average flow at the dam site is 97.3 m^3/s, the annual average runoff is 3.07 billion m^3, the annual average sediment concentration of the suspended load is 3.24 kg/m^3, the measured maximum sediment concentration is 27.32 kg/m^3, and the median particle size of the suspended load is approximately 0.118 mm.

According to the measured water and sediment data, the ratio of bed load to suspended load at the entrance of the reservoir is estimated to be 30%, and the total amount of sediment entering the reservoir every year is 12.92 million tons. The height of the dam of the upper Marsyangdi-A Hydropower Station is 18 m, the maximum water depth in front of the dam is 14.25 m, the depth of the backwater is small, the reservoir capacity is approximately 700,000 m^3, the ratio of reservoir capacity to river sediment volume is approximately 0.1, and the reservoir has little ability to regulate runoff and sediment. The power station utilizes a lateral water intake method, the water intake is located in front of the dam, and the water diversion system is arranged on the left bank and consists of a diversion canal, a sediment settling tank (that empties into a funnel), a buried culvert, a diversion tunnel and other buildings (Fig. 1).

Fig. 1. The layout of Marsyangdi-A Hydropower Station.

2.2 Numerical Model for Sediment Deposition Prediction

A numerical model is established by using two-dimensional shallow water equations (SWEs) to simulate the sediment transport process in the reservoir, and the details of the model are described in the literature [11]. Due to the lack of hydrological data in this region, the roughness rate is calculated according to Eq. (1) [12]:

$$n_{tot} = 7.69S^{0.28}\left(\frac{h}{d_{90}}\right)^{-0.21}\sqrt{0.009^2 + n_B^2} \tag{1}$$

where n_{tot} is the total resistance of flow, S is the bed slope, h is the water depth, d_{90} is the sediment diameter, which is larger than 90%, and n_B is the flow resistance caused by protrusions in the riverbed.

The Zhang Ruijin formula was used to determine the sediment carrying capacity of flowing water [13],

$$S_* = K\left(\frac{U^3}{gh\omega}\right)^m \tag{2}$$

where, S_* is sediment carrying capacity, K and m are coefficients, U is flow velocity, g is gravity, h is water depth, ω is settling velocity of sediment particles.

The bedload sediment transport rate was calculated following the Mayer-Peter formula [13].

$$g_{*k(dk)} = \frac{\left[\left(\frac{d_{90}}{19n}\right)^{3/2}\gamma hJ - 0.047(\gamma_s - \gamma)d\right]^{3/2}}{0.125\rho^{1/2}\left(\frac{\rho_s-\rho}{\rho_s}\right)g} \tag{3}$$

where, $g_{*k(dk)}$ is bedload transport rate, n is Mannig coefficient, d_{90} is the particle size corresponding to a proportion of 90% on the grading curve, d is the sediment size, J is water surface gradient, γ and γ_s are the unit weight of water and sediment, ρ is ρ_s are density of water and sediment, g is gravity.

2.3 Methods for Calculating Sediment Removal Rate

Sediment with a particle size greater than 0.25 mm will cause obvious abrasion to the water turbines in the upper Marsyangdi-A Hydropower Station, thus, it is critical that particles of harmful sizes are prevented from entering the turbine. The state of water in the sediment discharge funnel is complex, and it is difficult to accurately simulate the sediment transport process in the funnel. At present, there is a lack of methods to quickly and accurately calculate the removal rate of grouped sediment. Based on the theory of non-uniform non-equilibrium sediment transport, the method for calculating the sediment exclusion rate of different particle size groups is obtained:

$$E_k = \left(1 - \frac{C_* \beta_{*k}}{C_s P_k}\right) \tag{4}$$

where E_k is the exclusion rate of different water flow groups; C_* is the sediment carrying capacity of water, kg/m^3; β_{*k} is the gradation of sediment carrying capacity; C_s is the suspended sediment concentration, kg/m^3; and P_k is the proportion of suspended sediment in groups.

The comprehensive sediment discharge efficiency of the sediment funnel [14] can be calculated as follows:

$$E_t = \left(1 - \frac{Q_s}{Q_s'}\right) \tag{5}$$

where E_t is the comprehensive sediment removal efficiency; $Q_S{}'$ is the sediment transport rate at the inlet of the sediment funnel, kg/s; and Q_S is the sediment transport rate at the inlet of the culvert that leads to the power generation system and diversion channel, kg/s.

3 Sediment Problems and Countermeasures

The sediment problems encountered by the upper Marsyangdi-A Hydropower Station mainly involve sediment accumulation in the reservoir, which leads to a large amount of sediment entering the diversion channel. The sediment accumulation within the diversion channel leads to insufficient diversion flow, and the high sediment concentration through the units leads to the rapid erosion of the units. To address these 3 problems and ensure the safety of water intake and sediment control in power stations, it is crucial to solve the key problems of establishing indicators for reservoir sediment discharge, determine the critical sediment concentration for water intake in diversion channels and enhance the sediment discharge efficiency of the sediment settling tank.

3.1 Sediment Deposition

Figure 2 shows the change process along a longitudinal section of the reservoir for the upper Marsyangdi-A Hydropower Station. As seen from the Fig. 2, since the reservoir storage capacity is relatively small compared with the incoming sediment volume, the sediment is deposited in the form of a cone; therefore, sediment can be quickly transported to the front of the dam, and the reservoir becomes saturated with sediment within one year and is in a state of equilibrium. After the reservoir becomes saturated, the sedimentation height in front of the dam is approximately 900 m, and the maximum sedimentation volume in the reservoir is approximately 1.11 million tons. The sedimentation elevation predicted by the model is basically consistent with the observation results after reservoir operation, indicating that the model can accurately reflect the sedimentation form in the reservoir.

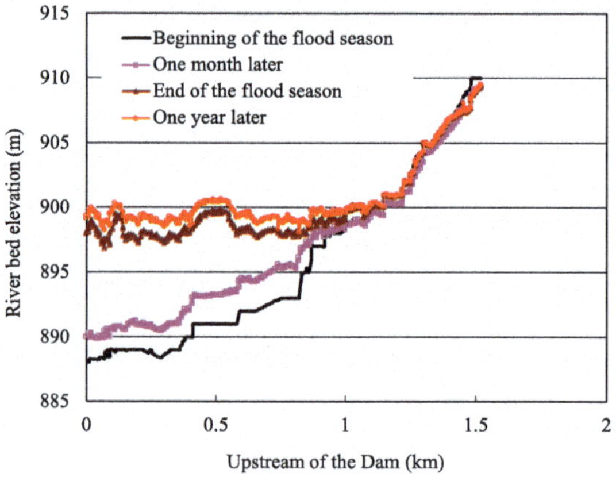

Fig. 2. Predicted deposition process.

The reservoir of the upper Marsyangdi-A hydropower station has a small storage capacity. After the reservoir quickly becomes saturated with sediment, the sediment basically loses its ability to settle the reservoir can no longer regulate runoff. The sediment is easily carried to the front of the dam by the water and large quantities enter the diversion channel, resulting in deposition within the diversion channel. The main hazard posed by sediment deposition within the diversion canal is that it leads to insufficient diversion flow and affects the power generation efficiency.

Figure 3 shows the change in the deposition and discharge capacity in the diversion canal calculated by the model after water at different sediment concentrations entered the diversion canal. It can be seen from the Fig. 3 that the sediment deposition rate in the diversion canal is proportional to the sediment concentration entering the diversion canal. With the increase in the sedimentation amount, the flow velocity in the diversion canal will gradually increase, the sediment transport capacity of the water will increase, the sediment volume within the diversion canal will not increase, and sediment scouring and deposition will reach equilibrium. The greater the sediment concentration of the water, the shorter the time needed for the diversion channel to reach equilibrium. When

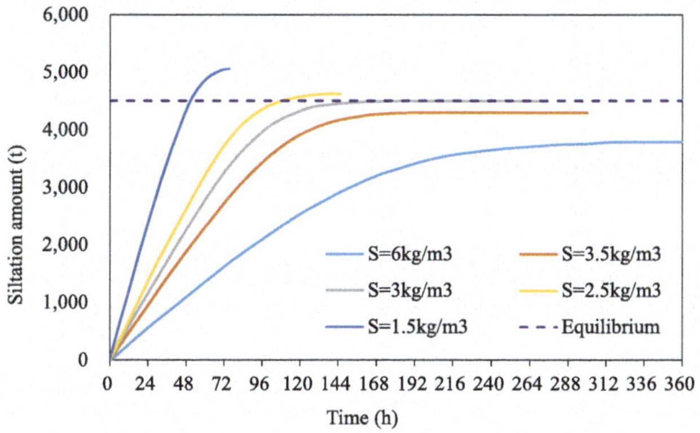

Fig. 3. The relationship between the sedimentation volume of the diversion channel and the sediment concentration (the dotted line in the figure is the maximum sedimentation volume that can meet the water intake flow).

the sediment concentration is less than 3.0 kg/m^3, despite sediment deposition within the diversion canal, it can still ensure the full output flow into the power generation system after the diversion canal reaches equilibrium. Additionally, sediment concentrations less than 3.0 kg/m^3 may lead to insufficient rates of water flow needed for power generation, and the greater the sediment concentration, the greater the sediment volume within the diversion channel when the equilibrium state is maintained.

3.2 Water Intake and Sediment Control Measures

When the reservoir reaches equilibrium with the sediment and the depth of the saturated water in front of the dam is less than 3 m, the sediment easily enters the diversion channel. According to the results of the mathematical model, when the deposition in front of the dam of the upper Marsyangdi-A Power Station is close to 900 m, it is necessary to remove sediment from the reservoir.

Especially when the sediment concentration of the inflow is greater than 3 kg/m^3, the reservoir should choose the appropriate time to stop operation to avoid insufficient discharge capacity of the diversion canal due to a high concentration of sediment, which affects the subsequent generation efficiency of the hydropower station. Stopping the operation of the dam and avoiding high sediment concentrations can be carried out in conjunction with the sediment discharge of the reservoir.

Through the calculation of the mathematical model, it is found that the sediment discharge ratio of the upper Marsyangdi-A reservoir is closely related to the flow-sediment coefficient (Fig. 4). When the flow-sediment coefficient of the reservoir is less than 0.01 kg/m^6, the sediment discharge ratio of the reservoir is approximately 100%, that is, the sediment load in the reservoir is in equilibrium with incoming sediment. When the flow-sediment coefficient of the reservoir is less than this critical value, the reservoir scouring efficiency is higher.

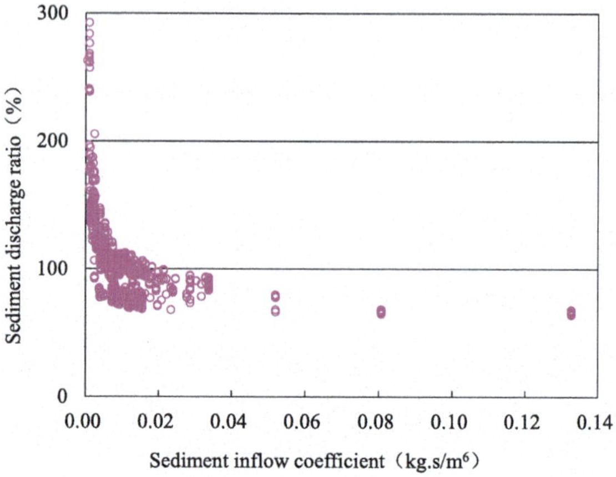

Fig. 4. Relationship between sediment discharge and flow-sediment coefficient.

Therefore, when the hydropower station stops operation to avoid high sediment concentrations during the flood season, a time with a large flow rate and a low flow-sediment coefficient should be chosen to carry out sediment discharge. At the end of the flood, the general flow is large, the sediment concentration is low, and the flow-sediment coefficient is not large, which is a favorable time to carry out sediment discharge.

According to the model prediction results, it is observed that adopting the open discharge method for hydropower stations, when the flow rate exceeds 100 m^3/s and the flow-sediment coefficient is less than 0.006 kg/m^6, leads to a higher sediment discharge efficiency for the reservoir. After approximately 6 h, the sediment discharge ratio of the reservoir stabilizes at around 1. This reduced sediment deposition in the reservoir by about 180,000 tons, allowing for the scouring and restoration of the river bed elevation at the front of the dam prior to sediment deposition (Fig. 5).

The actual sediment discharge effect of the reservoir shows that a single sediment discharge can reduce the sediment deposition in the reservoir by approximately 130,000 m^3, and the impact range of the discharge sediment is mainly concentrated within 600 m upstream of the dam. According to the dry bulk density of 1.3 t/m^3, sediment discharge removes approximately 170,000 tons of sediment, the height of the maximum water depth line in front of the dam is basically restored to the elevation of the river bottom before sedimentation, the thickness of the remaining sediment in front of the dam after sediment discharge is proportional to the flow-sediment coefficient of the river reach during sediment discharge (Fig. 6), and the actual sediment removal effect is basically consistent with the prediction result of the mathematical model.

Restoring the reservoir capacity by sediment discharge can cause some of the sediment to settle within the reservoir, especially when the sediment concentration is high, and the effect that the sediment has on the reservoir is more obvious. The actual observation results indicate that the average sediment concentration entering the diversion canal was reduced by approximately 10% after the sediment discharge (Fig. 7).

When the operation of the power station is halted to avoid high sediment concentrations, the reservoir lowers the water level to discharge sediment, while the diversion

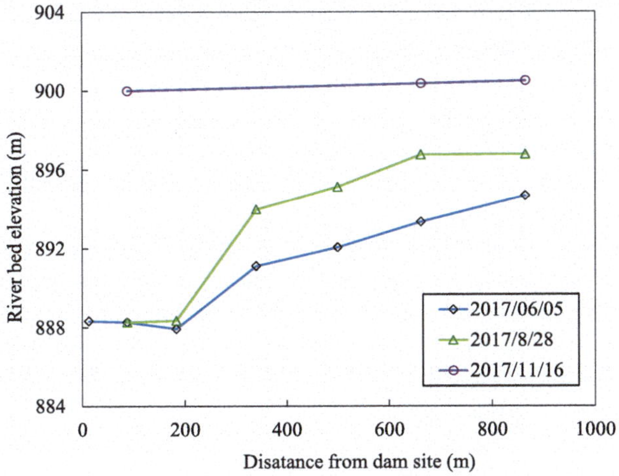

Fig. 5. Longitudinal profiles of riverbed.

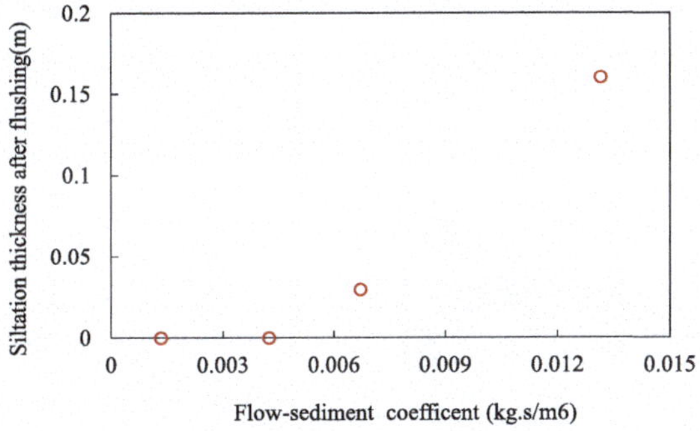

Fig. 6. Relationship between thickness of deposition upstream of the sluice gate and flow-sediment coefficient.

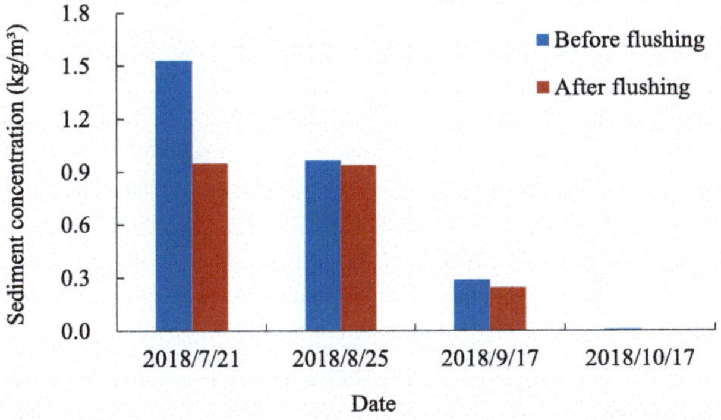

Fig. 7. Changes in sediment concentrations before and after sediment flushing.

canal also discharges sediment through hydraulic scouring. The scouring of the diversion canal of the upper Marsyangdi-A Hydropower Station makes full use of the water in the underground culvert to power water intake and electricity generation. After the power station shuts down, the sluice gate between the underground culvert and the diversion canal is closed, and all other sluice gates are opened. With the decrease in the water table in the reservoir, the water will be discharged to the river channel through the intake sluice gate and sediment funnel.

After the water in the diversion channel is discharged, the water intake gate of the culvert is opened, and water flowing along the elevation gradient between the culvert and the channel bottom is used to scour away the sediment deposited on the side of the diversion canal close to the culvert. The direction that the water flows during this process is the opposite of the direction of the water during the operation of the hydropower station and is called reverse scouring (Fig. 8a).

After the reverse scouring is over, the water intake gate for the diversion canal and the gate of the culvert are closed. When the water level of the reservoir is higher than the elevation of the diversion channel bottom by 0.3 m, the gate of the diversion canal is opened, and the water in the reservoir is used to scour the sediment in the upper section of the diversion canal. This process is called forward scouring (Fig. 8b). This practice has proven that the sediment that accumulated in the diversion channel can be basically removed by forward and reverse scouring.

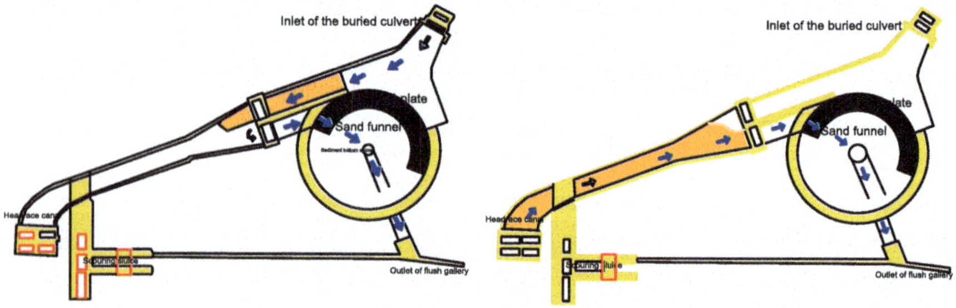

(a) Reverse scouring (b) Forward scouring

Fig. 8. Schematic diagram of diversion canal scouring.

Under the condition of the same volume of sediment, the water consumption rate is much lower than that of other types of sedimentation tanks, and it is more suitable for arid and water-deficient areas and this method has been widely used in Xinjiang Province of China since the 1980s. The sediment discharge funnel has a higher sediment removal efficiency, and the comprehensive sediment removal efficiency can reach more than 80% [15, 16].

Since the sediment settling tank and funnel is less commonly used in overseas projects, the sediment discharge effect of the sediment settling funnel is observed after the operation of the power station, and the sediment concentration change along the three key points of the inlet of the diversion canal, the outlet of the sediment settling funnel and the inlet of the culvert of the power generation diversion are recorded (Fig. 9). The measured average sediment concentration of the first diversion canal was 0.89 kg/m^3, the average sediment concentration of the sediment settling tank was 3.01 kg/m^3, and the average sediment concentration of the inlet of the power generation buried culvert was 0.70 kg/m^3. Although the sediment concentration of the sediment settling funnel increased significantly, the average water consumption rate was only 3.83%, the proportion of sediment discharged from the funnel to the amount of sediment was not high, and the sediment concentration of the inlet of the culvert was 21.3% lower than that of the inlet of the diversion canal.

According to Eq. (5), the comprehensive sediment discharge efficiency of the sediment discharge funnel of the upper Marsyangdi-A Hydropower Station is calculated to be 24.0%, which is significantly lower than that in Xinjiang Province of China, mainly because the sediment discharge of the reservoir prevents the bedload of the river from entering the diversion channel and the sediment discharge funnel. The measured median particle size of the diversion channel is between 0.02 mm and 0.125 mm, and the size of sediment entering the diversion canal in the flood season is larger, with a maximum diameter of approximately 1 mm.

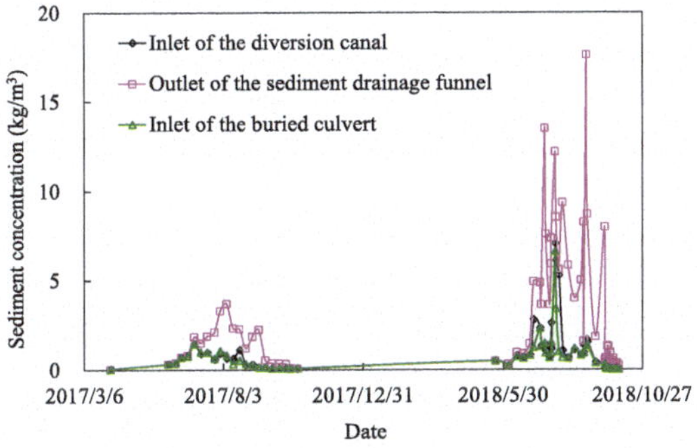

Fig. 9. The process of measured sediment concentration.

Table 1. Sediment removal rate of measured grouped sediment.

Particle size (mm)	0.001~0.002	0.002~0.005	0.005~0.01	0.01~0.02	0.02~0.05	0.05~0.1	0.1~0.25	0.25~0.5	>0.5
Sediment removal rate (%)	4.40	4.83	4.21	3.10	2.93	7.3	28.1	38.1	53.0

Table 1 shows the measured sediment removal rates of each group. Calculated from the above formula, when E_k is greater than 0, it is the theoretical removal rate of grouped sediment. Taking the typical measured sediment grading of the upper Marsyangdi-A Hydropower Station as an example, the grouped sediment removal rate calculated by Eq. (4) (Fig. 10). The grouped sediment removal rate obtained by the theoretical formula is generally proportional to the sediment particle size. The larger the sediment particle size is, the greater the reduction after passing through the sediment discharge funnel. The sediment removal rate of the sediment discharge funnel is proportional to the sediment concentration in the inflow. When the incoming sediment concentration is less than 3 kg/m^3, the sediment removal rate of less than 0.25 mm is nearly 0, and when the incoming sediment concentration increases to 10 kg/m^3, the sediment removal rate of 0.25 mm particle size is approximately 60%. This means that sediment discharge efficiency is proportional to sediment diameter and sediment concentration. The measured sediment removal rate is shown in Table 1. The sediment settling tank of the upper Marsyangdi-A Hydropower Station basically cannot exclude sediment particles with sizes below 0.1 mm, and when the particle size is greater than 0.1 mm, the sediment removal efficiency is greatly improved, and the sediment removal rate is proportional to the sediment particle size.

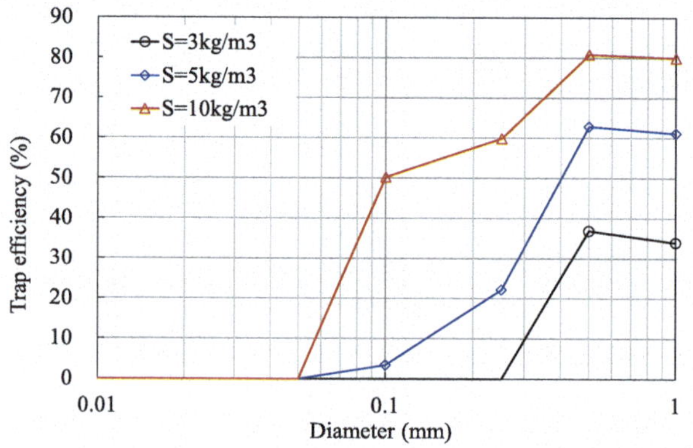

Fig. 10. The relationship between grouped sediment discharge rate and sediment concentration.

4 Conclusions

The results calculated from the sediment mathematical model for the upper Marsyangdi-A Hydropower Station agreed well with the actual operation results, which provide a reliable basis for water intake and sediment control of the hydropower station. The water flow and sediment conditions have a greater impact on the sediment discharge efficiency of reservoirs, and timely implementation of reservoir sediment discharge can reduce the sediment concentration of the suspended load entering the diversion canal and avoid bedload entering the diversion canal. This method can be done in conjunction with shutting down the power station to avoid high concentrations of sediment from entering the turbines and to discharge sediment from the reservoirs and diversion canals. It shows that hydraulic energy can be used to remove sedimentary particles within a short period. The calculation method of grouped sediment discharge efficiency is proposed, which theoretically clarifies the flow and sediment conditions applicable to the sediment discharge funnel and reveals the reasons why the removal efficiency of the sediment discharge funnel of the upper Marsyangdi-A hydropower station is not high. Finally, further research is needed to investigate more data on sediment removal efficiency from the sediment discharge funnel to verify and improve the method proposed in this paper.

Acknowledgements. This study was financially supported by the Program of the National Key Research and Development Plan (No. 2023YFC3208602), the National Natural Science Foundation of China (No. 52279079, No. U22A20237).

References

1. Xie, J.M., Wu, B.S., Liu, X.Y.: Review of reservoir sedimentation management. J. Sedim. Res. **3**, 71–80 (2013)
2. Thapa, B.S., Dahlhaug, O.G., Thapa, B.: Sediment erosion in hydro turbines and its effect on the flow around guide vanes of Francis turbine. Renew. Sustain. Energy Rev. **49**, 1100–1113 (2015)
3. Chaudhuri, D.: Empirical approaches in prediction of reservoir sediment distribution—an experience of 57 reservoirs in the USA and India. Int. J. Sediment Res. **32**, 260–276 (2017)
4. Sangal, S., Singhal, M.K., Saini, R P.: Hydro-abrasive erosion in hydro turbines: a review. Int. J. Green Energy **15**(1–5), 1–22 (2018)
5. Morris, G.L., Fan, J.H.: Reservoir Sedimentation Handbook: Design and Management of Dams, Reservoirs, and Watersheds for Sustainable Use. McGraw-Hill, New York (1998)
6. Han, Q.W., Yang, X.Q.: Review of reservoir sedimentation in China. J. China Inst. Water Resour. Hydropower Res. **1**(3), 169–178 (2003)
7. Wang, Z.Y., Hu, C.H.: Strategies for managing reservoir sedimentation. Int. J. Sedim. Res. **24**(4), 369–384 (2009)
8. Hu, C.H.: Development and consideration of engineering sediment subject from the sanmenxia reservoir to the three gorges reservoir in China. J. Sedim. Res. **2**, 1–10 (2019)
9. Xia, Y.M., Meng, G.Z.: Some sediment problems in hydropower stations in Yunan Province, China. Yunnan Hydropower **4**, 48–50 (1993)
10. Lewin, J.: British River. George Allen and Unwin, London (1981)
11. Dou, S.T., Wang, D.W., Yu, M.H., Liang, Y.J.: Numerical simulation of levee breach by overtopping in a flume with 180° bend. Nat. Hazard. **14**, 11–20 (2014)
12. Wang, D.W., Chen, J.G., Fu, X.D.: Quantification of flow resistance in mountain rivers based on resistance partitioning. In: Proceedings of 35th World Congress of the International-Association-for-Hydro-Environment-Engineering-and-Research, pp. 10339–10345, Chengdu, China (2013)
13. Wang, D.W., Lu, W.: Simulation of Flow and SEDIMENT Transport in Levee Breach. China Water Resources and Hydropower Press, Beijing (2019)
14. Li, L., Mu, Z.W., Zhou, Z.: Calculation of interception efficiency of sand funnel. Adv. Sci. Technol. Water Resour. **27**(4), 50–54 (2007)
15. Zhou, Z.: Introduction of sand funnel with strongly swirling flows. J. Xinjiang Agric. Univ. **1**, 97–98 (1987)
16. Deng, H.R.: Hydrologic measurement and analysis of suspended sediment in the sand funnel of Kashi cascade I Hydropower Station. Sichuan Water Conserv. **36**(4), 51–54 (2015)

Analysis of Erosion and Sedimentation Due to Riverbank Reinforcement Structures

Siti Murniningsih(✉)

Water Resources, Civil and Environmental Engineering Department, University of Indonesia, Jakarta, Indonesia
titik_winarno@yahoo.com

Abstract. Riverbank areas are prone to landslides, where one of the causes is an erosion of river wall soil by river flow. Pesanggrahan River, one of 13 large rivers that cross the city of Jakarta, is also threatened by landslides on its banks and slopes in several places. To protect the riverbanks from landslides, riverbank reinforcement has been built at several points in the Pesanggrahan River. However, it is necessary to analyze the impact of bank reinforcement which will affect the stability of the river channel, especially in the downstream part of the protected river channel. The flow profile was modeled using HEC-RAS software for conditions before and after the construction of bank reinforcement. Analysis of river channel stability emphasizes changes in roughness values due to changes in riverbank material, which originally came in the form of soil with wild floras to impermeable structures. The analysis results show that riverbank protection causes instability of the river channel in the downstream part due to changes in roughness value and flow shear stress along the river channel. If the value of flow shear stress (τ_0) exceeds the critical shear stress (τ_c), erosion and deposition will occur. Novelty in this study, especially in the Pesanggrahan River, is the relationship between the construction of riverbank reinforcement and channel instability as a result of changes in roughness values (n), which affect flow shear stress value, causing soil particle movement or bed channel erosion and deposition at the downstream of riverbank reinforcement structure.

Keywords: Erosion-Sedimentation · HEC-RAS Software · Landslides · Pesanggrahan River · Riverbank Reinforcement

1 Introduction

Rivers are one of the important components in the hydrological and hydraulic processes on the face of the earth. Water that has the nature of occupying space naturally moves from high places to low points; the low points filled with water will then form a flow that is usually called a river. Rivers with many components, such as flow patterns, banks, and their main channels, are units that can influence each other; these components stretch and become a unit called a river corridor, as shown in Fig. 1.

C. Weng (Ed.): WRE 2024, LNCE 634, pp. 258–271, 2025.
https://doi.org/10.1007/978-981-96-5095-8_22

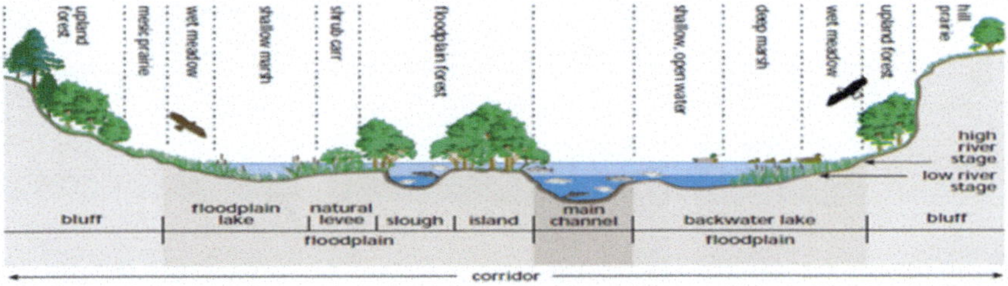

Fig. 1. River Corridor (*Source* The Federal Interagency Stream Restoration Working Group, 2001).

The river flow can be divided into main channels and floodplains. The main channel is where water flows, and the floodplain is the riverbank or channel that will be filled only when the river discharge is higher than the main channel storage volume [1]. Natural riverbanks are usually still in the form of soil with wild floras, which are easily eroded by the stream flow of the river, causing the widening of the river and generally reducing the stability of the riverbank area. Hence, to increase the stability of the river bank, the development of reinforcement structures for landslide prevention has been widely done in Indonesia, especially if the river passed on densely populated areas [2].

Pesanggrahan River, one of 13 (thirteen) major rivers that cross the city of Jakarta, is also threatened by landslides on its banks, so in several locations along the Pesanggrahan River, riverbank reinforcements have been built. However, changes in the area and shape of the cross-section due to the construction of riverbank reinforcements will also result in changes in hydraulic conditions and erosion and sedimentation of the river channel downstream of the structure [3]. Because of the change in the natural riverbank (Fig. 2) to an impermeable structure such as masonry, concrete walls, or concrete retaining walls (Fig. 3), the roughness value (n) of the river channel will change and furthermore affect the velocity (v), flow shear stress (τ_0) and sediment transport.

According to this condition, research is needed to determine whether riverbank reinforcement construction will affect the stability of the river channel as a whole.

Fig. 2. Natural River Plant (Pesanggrahan River, 2019).

Fig. 3. Riverbank Protection (Pesanggrahan River, 2022).

2 Research Methodology

The research location is the Pesanggrahan River in South Jakarta, which is approximately 9.4 km in length, with the upstream point being RPTRA Bintaro Jl. Garuda Bawah, RT.8/RW.12, Bintaro, Pesanggrahan District, South Jakarta City, and ends at the downstream point located near the Lemigas SPBG, Kebayoran Lama District, South Jakarta City (Fig. 4). Riverbank protection to prevent landslides and maintain the cross-sectional area of the channel is seen in several places along the research route, especially at 3 (three) observation locations, upstream, middle and downstream. In the upstream location, a reinforced riverbank is built as a retaining wall, while concrete piles are built in the middle and downstream locations.

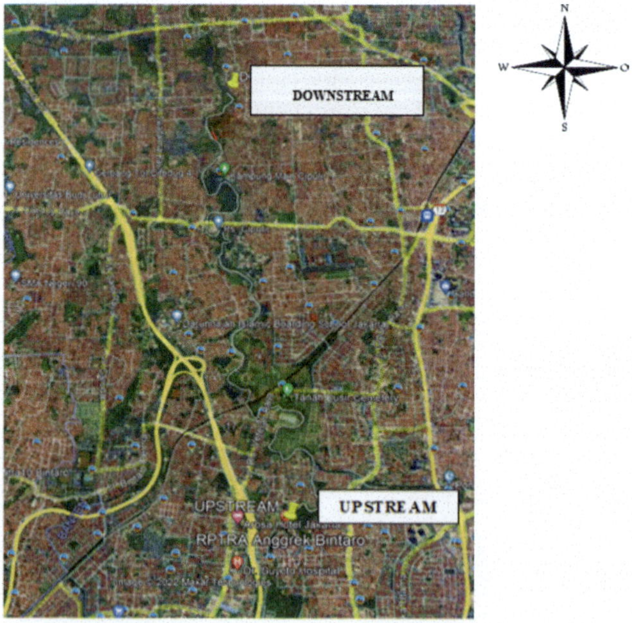

Fig. 4. Pesanggrahan River Research Path (*Source* Google Earth 2021).

2.1 Research Variable

In this study, two types of variables were used: independent variables and dependent variables. The independent variables are rainfall data, river cross-section geometry due to the construction of riverbank reinforcement, Manning's roughness coefficient (n), and soil grain diameter (D) at the research location. The dependent variable is a variable that is influenced by the independent variables, such as flow discharge (Q), average flow velocity (v), longitudinal slope of the channel (S), and the amount of sediment transport. The basic flow chart of the research is shown in Fig. 5 as follows:

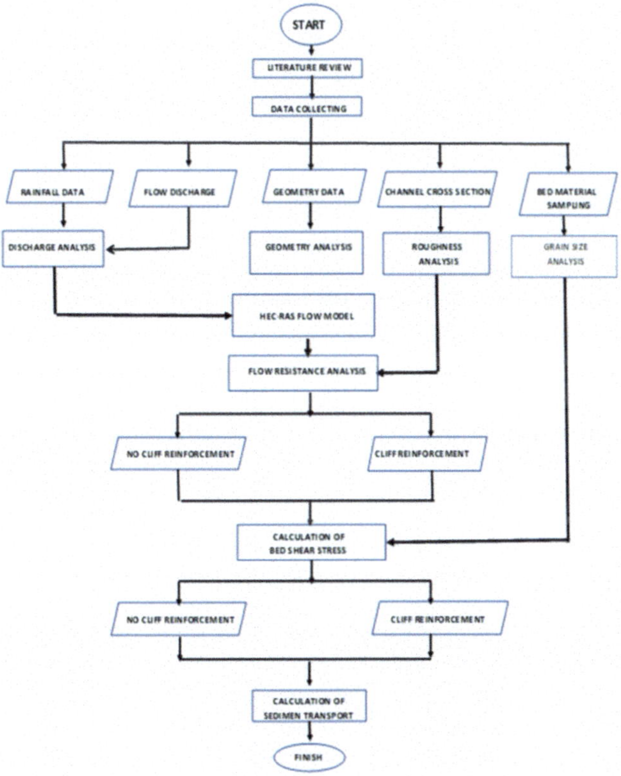

Fig. 5. Basic Flow Chart of the research.

2.2 Data Collection

Primary data used in this study include river bed material sampling at the research location and river cross-section geometry measured on the upstream and downstream sides of the three riverbank reinforcement locations. Meanwhile, secondary data that will be used in this study were obtained from the Ciliwung-Cisadane River Basin Center, i.e., rainfall data, river cross-section geometry, riverbank reinforcement materials, and river cross-section discharge at the three points reviewed in the years before and after the riverbank reinforcement was built.

2.3 HEC-RAS Flow Model

The simulation using the Hydrologic Engineering Center - River Analysis System (HEC-RAS) software aims to model the water surface profile at each location of the riverbank reinforcement structures and the sediment transport process to illustrate the river stability [4]. The first step for simulating the water surface profile and sediment transport is to create a longitudinal profile of the river, the second step is entering cross sectional geometry data at each location, and the third step is to input the flow data.

2.4 Analysis of the Channel Stability

The riverbank reinforcement built at 3 (three) locations along the Pesanggrahan River has its characteristics, ranging from the shape to the type of material between the right and left walls, as well as the bottom of the channel. For example, at the upstream and downstream locations, riverbank protection is only built on one side. Therefore, the other side is still in the form of soil and natural plants. In order to determine the effect of bank reinforcement on river channel stability, several things were done, such as Manning roughness (n) analysis, flow shear stress (τ_0) analysis, and sediment yield (Qs) analysis, especially in the downstream area of the riverbank reinforcement structure.

3 Results

3.1 Pesanggrahan River Watershed and Rainfall Analysis

Watershed area data was obtained from delineation using Arc-Gis software with the input of a DEM map of the required location and the upstream point of the research location as the Point of Origin. From the watershed delineation results, the watershed area of the Pesanggrahan River was 133.21 km^2 (Fig. 6a).

The rainfall data used in this study are rainfall data from 3 (three) rainfall stations (Fig. 6b) that affect the Pesanggrahan River discharge based on its watershed, i.e. Bogor Rainfall Station (1), the South Tangerang Rainfall Station (2) and FTUI Depok Rainfall Station (3). Using the Thiessen polygon, the area of influence was 44.55% for the Bogor Rainfall Station, 30.17% for the South Tangerang Rainfall Station, and 25.17% for the FTUI Depok Rainfall Station.

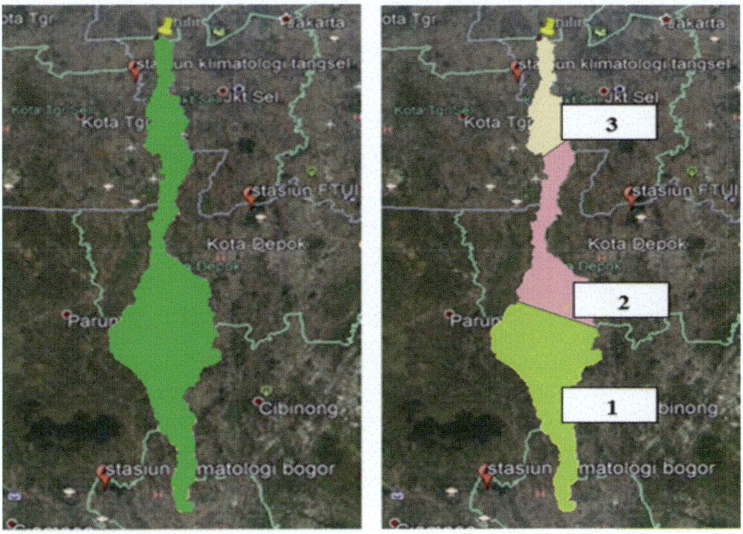

Fig. 6. **a.** Pesanggrahan River Basin (Data Processing with ArcGIS); **b.** Area of the Rain Station (Data Processing with Polygon Thiessen).

3.2 Flow Discharge

The rational method is generally used to calculate the flow discharge, which is a function of the Pesanggrahan watershed area (A), weighted flow coefficient (C), and rainfall intensity (i). The delineated Pesanggrahan watershed area is 133.210 km^2, while the weighted flow coefficient (C) value obtained from the digitization of the Pesanggrahan watershed land use map in 2017 is 0.774. From the calculation results of the distribution of rainfall for a 10-year return period using the Gumbell method, a value of 98.450 mm was obtained. Furthermore, after obtaining this value by using the IDF curve, the concentration time (Tc) value was obtained as 1636.875 and the rainfall intensity (i) is 3.369, which were then used to calculate the discharge at the point of origin. Hence, the flow discharge of 107.839 m^3/sec was obtained at the point of origin.

The results of this calculation, when compared with the discharge obtained from the Ciliwung - Cisadane River Basin Center (BBWSCC) precisely at the Sawangan Flow Discharge Station located 8 km upstream from the research location are as follows: The calculated discharge is 107.839 m^3/second, while the maximum daily average discharge is 101.800 m^3/second, and the maximum extreme discharge is 142.330 m^3/second [5].

3.3 River Cross-Section Simulation

Pesanggrahan River, which was used as the research location, was observed along 9.4 km, with 94 stations starting from station 0 in the downstream and station 94 located in the upstream. The distance between each station was 100 m (Fig. 7), and riverbank reinforcement was built at station 1 (downstream), station 46 (middle stream), and station 93 (upstream), respectively.

STATION	CROSS SECTION
0 (DOWN STREAM POINT)	NATURAL
1	PROTECTION
2	NATURAL
5	NATURAL
1	NATURAL
20	NATURAL
31	NATURAL
44	NATURAL
46	PROTECTION
48	NATURAL
49	NATURAL
58	NATURAL
71	NATURAL
84	NATURAL
91	NATURAL
93	PROTECTION
94 (UPSTREAM POINT)	NATURAL

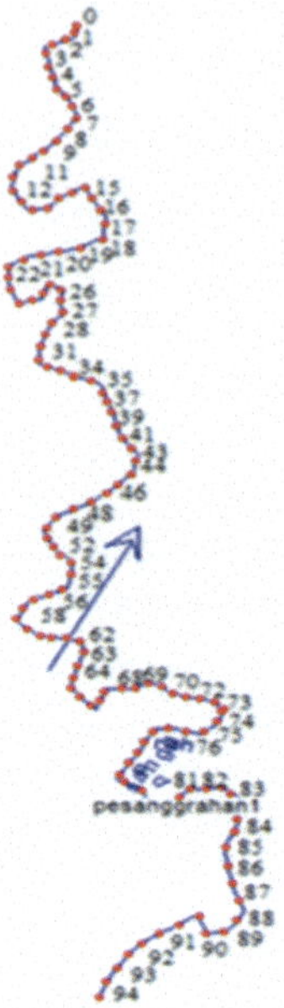

Fig. 7. Research Location along 9,4 km with 94 Station. (Google Earth Imagery).

The next step in obtaining the flow profile at each cross-section using the HEC-RAS software requires input of data including discharge, river flow geometry data, and Manning cross-section roughness data (n). Flow simulation result at Sta 1 (Fig. 8), Sta 46 (Fig. 9) and Sta 93 (Fig. 10), indicate the natural flow profile and flow profile after riverbank reinforcement construction.

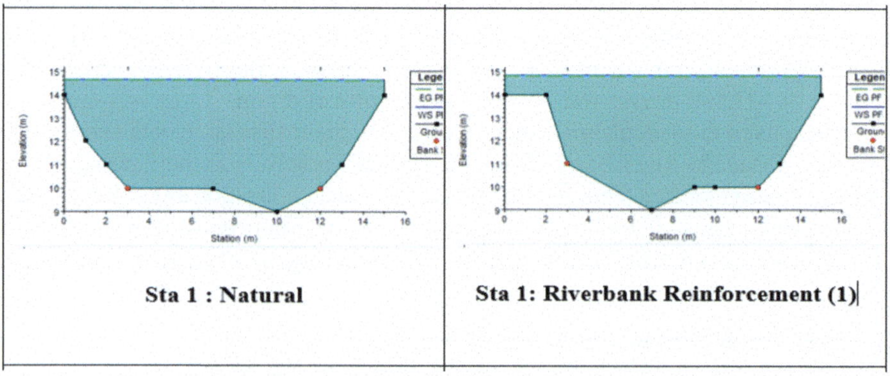

Fig. 8. Flow Simulation Result at Sta 1 (downstream).

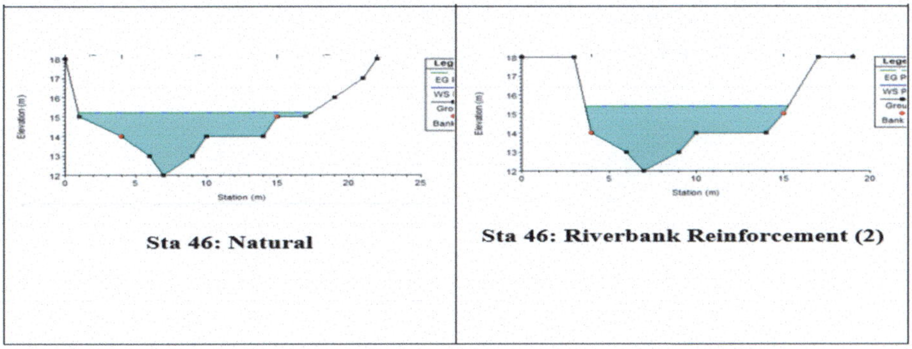

Fig. 9. Flow Simulation Result at Sta 46 (middle stream).

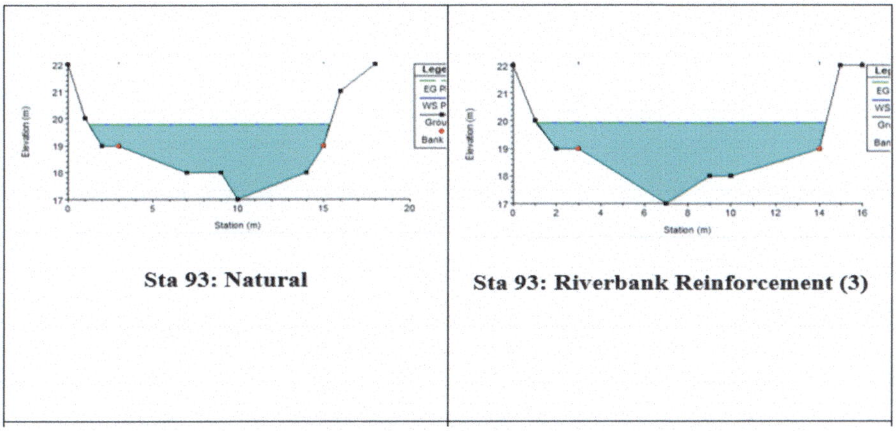

Fig. 10. Flow Simulation Result at Sta 93 (upstream).

3.4 Riverbank Reinforcement Material

The riverbank reinforcement built at Station 1, Station 46 and Station 93, uses different structures and materials. Based on the results of observations of riverbank protection materials at Sta 93 made on the right side of the river, with a cast stone wall, the roughness value (n) is 0.013; at the middle location, Sta 46 is made on the left and right sides of the river, with a concrete sheet pile wall type, then the n value for the left and right river wall is 0.012, while the downstream location, Sta 1, is made on the left side of the river, with a cast concrete wall type, then the n value for the right wall is 0.012. Therefore, the roughness values (n) on the left wall, channel bed and right wall at Sta 93 were 0.013; 0.020; 0.020, for Sta 46 was 0.012; 0.020; 0.012 and for Sta 1 was 0.013; 0.020; 0.020.

Because the roughness values n varies depending on the location of the bank reinforcement, the composite roughness n_c must be calculated using Eq. (1) [6, 7]. Figure 11 is an illustration of n composites and calculation results of n composite for Sta 1, Sta 46, Sta 93 are shown in Table 1.

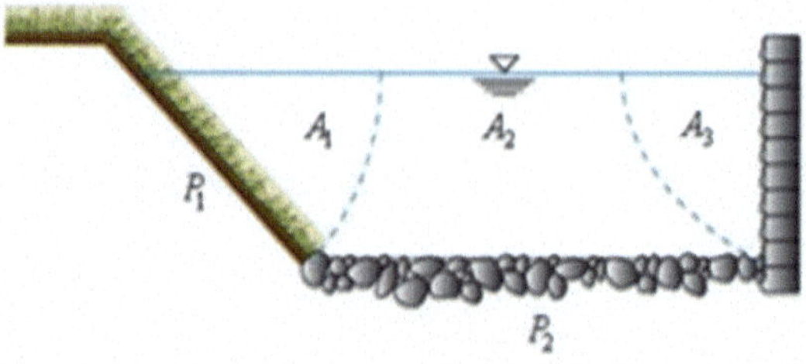

Fig. 11. Illustration of n composite (*Source* Williams & Julien, 1989).

Equation (1) for n composite (n_c) is as follow:

$$n_c = \left[\frac{\sum_1^N P_i n_i^{3/2}}{P} \right]^{2/3} \quad (1)$$

where, n_c is n composite, P_i is wet perimeter length, and n_i is Manning roughness number.

Table 1. n composite for Sta 1, Sta 46, Sta 93.

Station	Cross Section	Natural					Riverbank Reinforcement				
		n	P	ΣP	ΣP*n	n_c	n	p	Σ p	p*n	n_c
Sta 1	Right Wall	0,025	7	26	0,02767	**0.0229**	0.0200	5	19	0.0141	**0.0180**
	Bottom	0,02	11		0,03111		0.0200	9		0.0254	
	Left wall	0,025	8		0,03162		0.0120	9		0.0065	
Sta 46	Right Wall	0,025	5	21	0,01976	**0.0224**	0.0120	1	13	0.0013	**0.0183**
	Bottom	0,02	11		0,03111		0.0200	10		0.0282	
	Left wall	0,025	5		0,01976		0.0120	2		0.0026	
Sta 93	Right Wall	0,025	4	17	0,01581	**0.0218**	0.0130	1	10	0.0014	**0.0193**
	Bottom	0,02	11		0,03111		0.0200	7		0.0197	
	Left wall	0,025	2		0,00791		0.0200	2		0.0056	

3.5 Soil Investigation

The riverbank reinforcement at station 1 and 93 was only built on one side, and the other wall was still soil and vegetated soil, while at station 46, the riverbank reinforcement was built on the left and right sides.

The distribution of soil composition as result from laboratory testing and soil classification according USDA Systems [8] was obtained for Sta 93, Sta 46 and Sta 1 as at the Table 2, reflected that the results of soil investigation are conformity with the theory, for instance at the downstream point, the average diameter of the soil in the river will be smaller, and the distribution of fine grains will increase.

Table 2. Soil Composition Distribution at Sta 1, Sta 46 and Sta 93.

Station	Cross Section	Soil Type	Type of Soil Percentage (%)	Soil Classification
Sta 93 (Upstream)	Right Side	clay	19	Silty Loam
		silt	55	
		sand	26	
	Left aside	clay	18	Loam
		silt	40	
		sand	42	
Sta 46 (Middle)	Right Side	clay	22	Loam
		silt	48	
		sand	30	
	Left Side	clay	18	Silty Loam
		silt	51	
		sand	31	
Sta 1 (Downstream	Right Side	clay	28	Silty Loam
		silt	47	
		sand	25	
	Left Side	clay	33	Clay Loam
		silt	37	
		sand	30	

Using Shields Graphic [9], as shown in Fig. 12, the D-50 grain size at each Sta 93, Sta 46 and Sta 1 is used to determine the critical shear stress (τ_c), as shown in Table 3.

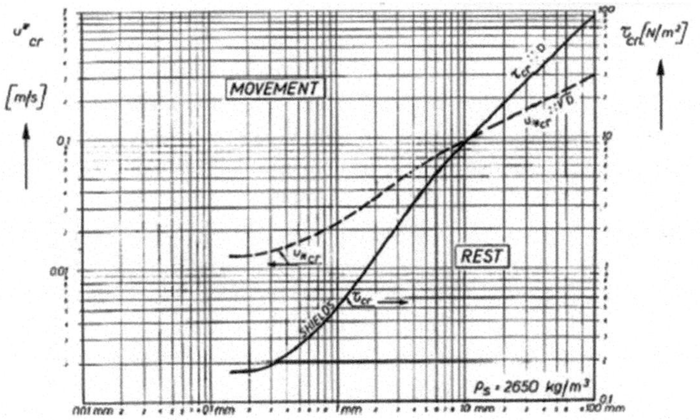

Fig. 12. Shields Graph (Shields, 1936).

Table 3. Critical Shear Stress (τ_{cr}) of Grain Diameter D_{50} at Sta 1, Sta 46 and Sta 93.

Station	Grain Diameter D_{50} (mm)		Critical Shear Stress (τ_c) (τ_{cr} N/m^2)	
	Left Wall	Right Wall	Left Wall	Right Wall
93	0.02934	0.02934	0.1880	0.1880
46	0.02023	0.01984	0.1880	0.1880
1	0.01127	0.01519	0.1880	0.1880

4 Discussion

4.1 Stability Analysis

The shear stress value on each river wall and channel bed. The calculation is done using the shear stress Eq. (2) as follows:

$$\tau_0 = \alpha\,\rho\,g\,R\,S_f \tag{2}$$

where τ_0 is flow shear stress, coefficient α is 0.97 for bottom channel, coefficient α is 0.75 for slide slope, R is hydraulic radius, and S_f is energy slope [9, 10].

From the calculation results using Eq. (2), the flow shear stress value (τ_0) will be obtained at the cross-section, Sta 93, Sta 46, and Sta 1, where the riverbank reinforcement has been built. River instability or changes in the shape of the cross-section can be caused if the flow shear stress value (τ_0) is greater than the critical shear stress (τ_c). As a result, there is a possibility that the river wall and bed channel will be eroded, and soil particles will be carried downstream. Consequently, erosion and sedimentation will occur. Analysis of the stability of the cross section based on shear stress can be seen in the Table 4:

4.2 Sediment Transport Analysis

Changes in sediment transport are one of the impacts of the construction of riverbank reinforcement structures, resulting in changes in cross-sectional area (A), wetted perimeter (P) and Manning channel roughness (n) which will result in changes in flow velocity (v_{avg}) and flow shear stress (τ_0). Based on the sediment transport equation, which is a function of flow velocity and flow shear stress, these changes will affect the magnitude of sediment transport at the cross-section being reviewed [11].

The Ackers-White equation [12] is used to calculate sediment transport at 3 (three) locations, i.e. Sta 93, Sta 46, and Sta 1, where the riverbank reinforcement was built. The formula it can calculate sediment transport with a sediment grain gradation between 0.02–4.94 mm, where the sediment grain gradation of the Pesanggrahan River is in that range. The Ackers-White Eq. (3) as follows:

$$Q_s = Q.X$$

$$X = \frac{s.D}{H}\left(\frac{V}{U*}\right)^n G_{gr};\ G_{gr} = C\left(\frac{Fgr}{A} - 1\right)^m;\ D_{gr} = D\left[\frac{g(s-1)}{A}\right]^{1/3};$$

Table 4. Cross-section Stability Analysis.

Station	Stability Analysis	Cross Section					
		Bottom Channel		Left Wall		Right Wall	
		Natural	Cliff Reinforcement	Natural	Cliff Reinforcement	Natural	Cliff Reinforcement
0	τ_0	0.2310	0.2925	0.1786	0.2262	0.1786	0.2262
	τ_c	0.1880	0.1880	0.1880	0.1880	0.1880	0.1880
	Result	y	y	n	y	n	y
1	τ_0	0.2392	0.2294	0.1850	0.1774	0.1850	0.1774
	τ_c	0.1880	0.1880	0.1880	0.1880	0.1880	0.1880
	Result	y	y	n	n	n	n
44	τ_0	0.2412	0.4042	0.1865	0.3125	0.1865	0.3125
	τ_c	0.1880	0.1880	0.1880	0.1880	0.1880	0.1880
	Result	y	y	n	y	n	y
46	τ_0	0.2963	0.2278	0.2291	0.1761	0.2291	0.1761
	τ_c	0.1880	0.1880	0.1880	0.1880	0.1880	0.1880
	Result	y	y	y	n	y	n
91	τ_0	0.3480	0.5615	0.2690	0.4341	0.2690	0.4341
	τ_c	0.1880	0.1880	0.1880	0.1880	0.1880	0.1880
	Result	y	y	y	y	y	y
93	τ_0	0.4046	0.2403	0.3128	0.1858	0.3128	0.1858
	τ_c	0.1880	0.1880	0.1880	0.1880	0.1880	0.1880
	Result	y	y	y	n	y	n
		(Ctrl)					

Note y = erosion; n = no erosion.

$$F_{gr} = \left[\frac{v0}{\sqrt{g.D.(s-1)}.\sqrt{32.log\left(\frac{\alpha H}{D}\right)}}\right]\left[\frac{U*}{V}\sqrt{32}log\left(\frac{\alpha H}{D}\right)\right]^{n} \tag{3}$$

where, D is grain diameter, υ is viscosity, A,C,n,m are constant, a depends on D_{gr} value, H is flow depth, F_{gr} is mobility function, v is average velocity, U^* is shear velocity.

From the calculation results, sediment transport on the cross-section built with bank reinforcement has increased in volume every month (Qs). This is caused by changes in the wet cross-sectional area (A) and wetted perimeter (P) due to riverbank reinforcement structures, and a decrease in Manning's roughness value (n) due to changes in river bank cover from soil with wild floras into impermeable structure.

These conditions will then result in greater average flow velocity (v), flow shear velocity (U^*), and flow shear stress (τ_0), thus increasing the ability to transport bed channel material. Consequently, erosion and sedimentation processes will occur [13, 14].

5 Conclusion

The results of this study can be concluded that riverbank reinforcement will affect the stability of the river channel as a whole due to changes in the value of n Manning roughness where the value of n is a function of the river cross section material, wet cross-sectional area A and wetted perimeter P as in formula (1). Changes in the Manning roughness value n will affect the flow velocity v, hydraulic radius R and energy slope S, where the hydraulic radius R is a function of the wet cross-sectional area A and wetted perimeter P [15].

The river channel consists of grains or soil particles that have critical shear stress where the relationship between the critical shear stress of the grains τ_c and the grain size (D) is shown by the Shields Diagram. If the critical shear stress of the grains τ_c is smaller than the flow shear stress τ_0, the soil grains will be transported downstream. This condition is occurring at the research location in the upstream (Sta 93), middle stream (Sta 46), and downstream (Sta 1) areas where the sediment transport per month (Q_s) increases after the riverbank reinforcement was built (Table 5), consequently, erosion and sedimentation processes occur.

Table 5. Calculation of Sediment Transport.

Cross Section	Sta	H (m)	A (m^2)	P (m)	R (m)	U^* (m/s)	Q (m^3/s)	V (m/s)	F_{gr}	G_{gr}	X (kg/m^3)	q_s (kg/s)	Q_s (ton/mth)
Natural	1	5.00	27.12	26.20	1.0351	0.092	107.89	3.9782	2.14	2429.75	6.66345E-07	0.000	0.2
	46	6,00	24.00	18.11	1.3252	0.091	107.89	4.4954	2.19	87084.83	2.87742E-05	0.003	8.0
	93	6.20	23,13	15.98	1.4474	0.091	107.89	4.6645	2.21	26067173.66	0.007539086	0.813	21.1
Riverbank Reinf	1	5.00	29.08	26.10	1.1141	0.098	107.89	3.7101	2.32	9775.06	2.4283E-05	0.000	0.7
	46	4.50	26.20	24.80	1.0564	0.090	107.89	4.1179	2.09	132874.58	5.45179E-05	0.006	15.2
	93	5.20	24.01	27.00	0.8892	0.093	107.89	4.4935	2.01	6466812.61	0.002106681	0.227	58.9

Thus, there is a strong correlation between the change in riverbank cover into an impermeable bank structure and the stability of the river channel as a whole, although each structure certainly has different characteristics and impacts. Equation (1), (2), and (3) show the correlation between changes in flow hydraulics due to riverbank reinforcement and the stability of the river channel.

Watershed managers should conduct flow simulations before deciding to build riverbank reinforcement to avoid negative impacts on the stability of the river channel as a whole [16, 17].

References

1. Stream Corridor Restoration: The Federal Interagency Stream Restoration Working Group. United State: USDA (2001)
2. Channel-Forming Discharge: US Army Corps of Engineer. Vicksburg US (2000)
3. Mosselman, E., et al.: A review of techniques available for delimiting the erodible river corridor: a sustainable approach to managing bank erosion. River Res. Appl. Wiley InterScience **21**, 773–789 (2005)
4. HEC-RAS River Analysis System.: US Army Corps of Engineer. Institute for Water Resources, Davis (2016)
5. Pengelolaan Sumber Daya Air Balai Bsar Wilayah Sungai Ciliwung-Cisadane (BBWSCC): Kementrian Pekerjaan Umum dan Perumahan Rakyat, Jakarta (2015)
6. Chow, V.T., Maidment, D.R., Mays, L.W.: Applied Hydrology. McGraw-Hill Inc, New York (1988)
7. McCuen, R.H.: Hydrologic Analysis and Design. Prentice-Hall, Inc., New Jersey (1988)
8. Craig, R.F.: Craig's: Soil Mechanics. Spon Press, New York (2004)
9. Breusers, H.: Sediment Transport 1. Delft: International Course in Hydraulic Engineering (1980)

10. Henderson, F.M.: Open Channel Flow. Macmillan Publishing Co., New York (1966)
11. Potter, M.C., Wiggert, D.C., Ramadani, B.H.: Mechanics of Fluid, 4th ed. Stamford (2011)
12. Wallingford, H.R.: Sediment Transport: The Ackers and White Theory Updated. Department of the Environment (1990)
13. Garde, R.: River Morphology. New Delhi: New Age International (P) Ltd. (2006)
14. Graf, W.H.: Fluvial Hydraulics. Wiley, West Sussex (1998)
15. Bekic, D., Kuspilic, N., Gilja, G.: Stability assessment of the river channel Bedform. In: 8th International Conference on Hydro-science and Engineering. Zagreb (2008)
16. Brookes: Channelized Rivers: Perspectives for Environmental Management. Earth Surface Processes and Landforms 326 (1988)
17. Paschalis, K. et al.: Estimating Stream Bank and Bed Erosion and Deposition with Innovative and Traditional Methods, Land (2024)

Hydrological Analysis of a Highland Reservoir Using GIS

Jieling Wu(✉)

Suwa University of Science, Chino 391-0292, Japan
wu_jieling@rs.sus.ac.jp

Abstract. In Japan, various reservoirs have been built since ancient times to relieve water shortages in agriculture and for irrigation. While many of these reservoirs continue to serve local communities, changing land use patterns and population decline have left many unmanaged, posing significant risks, especially during the rainy season. Highland reservoirs, in particular, can cause rapid downstream flooding and mudslides in the event of a collapse. Due to their age, relevant authorities often lack comprehensive hydrological data for these reservoirs. This study uses GIS to analyze elevation data of a highland reservoir to assess its hydrological conditions, contributing to disaster management and flood prevention efforts by providing critical insights into potential hazards. The results highlight the importance of such reservoirs in both flood risk mitigation and sustainable regional development.

Keywords: Hydrological Analysis · Highland Reservoir · GIS (Geographic Information System) · Elevation data

1 Introduction

1.1 Research Background

The numerous upland reservoirs in Japan's mountainous and hilly regions have long played an important role as water resource for agricultural use. These reservoirs support local agricultural development, provide valuable water resources in times of drought, and function as essential infrastructure for local communities, for example, by regulating floods. According to statistics, the number is more than 150,000 units. The history of upland reservoirs dates to the Heian period (794–1185), with a particularly large number of reservoirs being built during the Edo period (1603–1868), and they continue to be used as the basis for agricultural activities to this day [1].

However, the dilapidation of many old facilities that are not obvious and rights relations have become more complex. These reservoirs face many hydrological challenges due to their topographical characteristics and climatic conditions. In recent years changes in rainfall patterns due to climate change and the increasing frequency of extreme weather events have created an urgent need to review the reservoirs' water storage capacity and flood control functions. Assessing the seismic resistance of reservoirs against

C. Weng (Ed.): WRE 2024, LNCE 634, pp. 272–282, 2025.
https://doi.org/10.1007/978-981-96-5095-8_23

aging and seismic risks is also recognized as an important issue. The division of roles of administrative bodies (national, prefectural, and municipal) is unclear about reservoir disaster prevention. The respective responsibilities of owners, managers and farmers who use agricultural water are unclear. 96,000 reservoirs have information on location, owner/manager, dimensions/structure, etc., in the Database of Reservoirs. There is concern that user-based management organizations will weaken due to farmers leaving farming and the aging of the population and that routine maintenance and management will not be carried out properly [2].

1.2 Research Objectives

Constraints on the observation period of hydrological observation data and information related to water withdrawal. Basically, in reservoirs, unlike dams, there is no long-term hydrological observation data or data on water withdrawal necessary for the prediction of water storage volumes [3]. Municipalities with many reservoirs in their area need to evaluate and categorize these reservoirs and study the policies for dealing with the characteristics of each reservoir, and the methods used for evaluation need to be quantitative and simple, using data accessible to municipal staff and managers [4].

Elevation data (e.g., Digital Elevation Model (DEM)) are very important in hydrological analyses because the height of the terrain directly affects water flow paths, catchment patterns, and hydrological processes. Topographic data is the foundation of hydrological analysis; GIS can efficiently manage and analyze topographic data such as DEM, which can provide basic data support for subsequent watershed delineation, slope calculation, and analysis of catchment paths, etc. GIS provides powerful spatial data processing and analysis functions for hydrological analysis, and is an important tool for hydrological research. GIS can be used to better understand and simulate the hydrological processes in a watershed.

This study focuses on the hydrological characteristics of upland reservoirs and uses GIS to analyze readily available topographical data to understand their hydrological status. The value of reservoirs to local communities is reaffirmed and a simple analysis method is proposed. The results of this research are intended to contribute to the hydrological analysis of upland reservoirs and, ultimately, to the sustainable development of the region.

2 Methodology

2.1 Research Object

Lake Irikasa (latitude 35°55′57.8″ N, longitude 138°11′ 36.3″ E) in Fujimi Town (with a population of 13,860), Nagano Prefecture, which is located near Mount Irikasa [5]. It is next to the abandoned holiday property, which used to be a fishing and recreational mecca 30 years ago, but is now deserted and not allowed to be entered. The author first photographed the overall condition with a drone (Fig. 1).

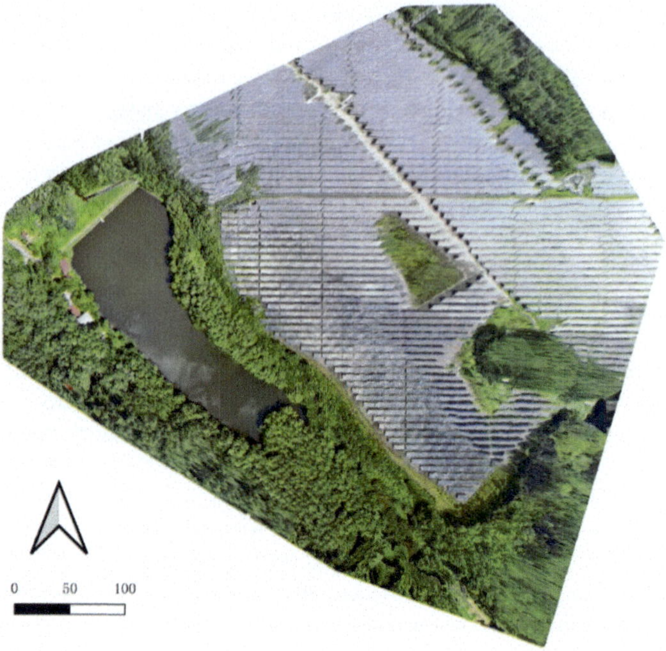

Fig. 1. Orthophoto from drone.

2.2 Software and System

The study used the software QGIS version 3.36.3 [6] running on the Windows 11 Professional operating system.

2.3 Research Data

Elevation data and administrative area data can be downloaded free of charge from Conservation GIS Consortium Japan (CGIS) [7], which offers publicly available GIS resources to support environmental conservation efforts by NPOs, governmental bodies, and educational institutions.

The elevation data utilized in this study were obtained from the Geospatial Information Authority of Japan's (GSI) 10-m mesh Digital Elevation Model (DEM), a spatial dataset that provides detailed terrain information [8]. While GSI updates the DEM irregularly, data revisions are made following significant topographical changes, technological advancements, or natural disaster recovery efforts. These updates may include using advanced surveying technologies such as LiDAR or aerial photography.

2.4 Data Processing

Elevation Data Processing. Elevation data and administrative area data of Nagano Prefecture, Japan were downloaded from CGIS. In order to increase the speed of calculation, the administrative area data of three municipalities related to the lake, namely, Chino City (with a population of 55,198) [9], Haramura (with a population of 7,734) [10], and Fujimi Town (Fig. 2), were intercepted first. Then the elevation data of the three municipalities were intercepted again.

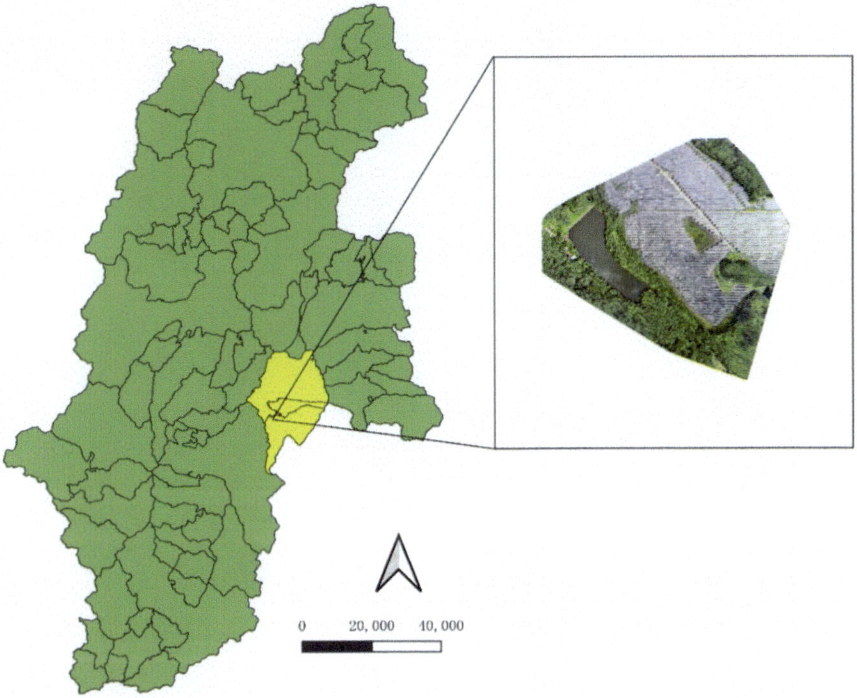

Fig. 2. Nagano Prefecture. Chino City, Haramura, and Fujimi Town is in yellow.

Filled Sinks. Depressions or "sinks" in the DEM data, which can obstruct natural water flow and affect hydrological accuracy, were addressed through a critical preprocessing step known as sink-filling. This step ensures the continuity of water flow by eliminating localized depressions, enabling correct flow direction analysis and accurate catchment delineation. Filling sinks is essential to: (1) Maintain continuous flow paths during hydrological simulations. (2) Accurately delineate watershed boundaries by preventing water from being trapped in local depressions. (3) Enhance the efficiency of model calculations by reducing computational errors caused by unfilled sinks.

The SAGA terrain analysis module in QGIS was used to perform the sink-filling operation. The procedure involved navigating to Terrain Analysis > Hydrology > Filled sinks (Fig. 3).

Channel Network and Drainage Basins. The channel network and catchment areas were derived using the processed DEM with filled sinks to further understand the water flow and drainage patterns. This step was crucial for outlining the river directions and defining the catchment areas associated with Lake Irikasa. The specific operation in

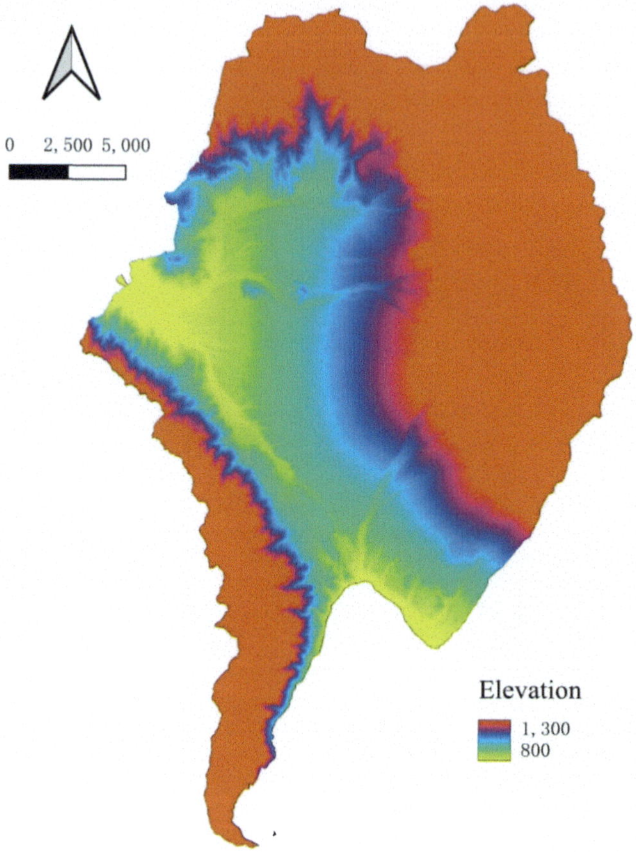

Fig. 3. The filled dem data of Chino City, Haramura, and Fujimi Town.

QGIS involved Terrain Analysis > Channels > Channel network and drainage basins (Fig. 4).

Catchment Extraction. A catchment area is the area where surface water ends up in a particular outlet (e.g., a river, lake, or reservoir) through a confluence. The delineation of a catchment helps to understand the flow paths within the area and how water is pooled on the surface. This is essential for analyzing hydrological cycle processes, predicting water availability and developing water management strategies.

Find the downstream coordinates on the satellite map and use them to extract the catchment area of the lake (Fig. 5). The raster data of this catchment is converted to vector data for more overlay analysis. The relevant QGIS steps were Terrain Analysis > Hydrology > Upslope area, followed by Raster > Conversion > Polygonise (raster to vector) (Fig. 5). Subsequently, the elevation data for the delineated catchment area were isolated for additional analysis (Fig. 6).

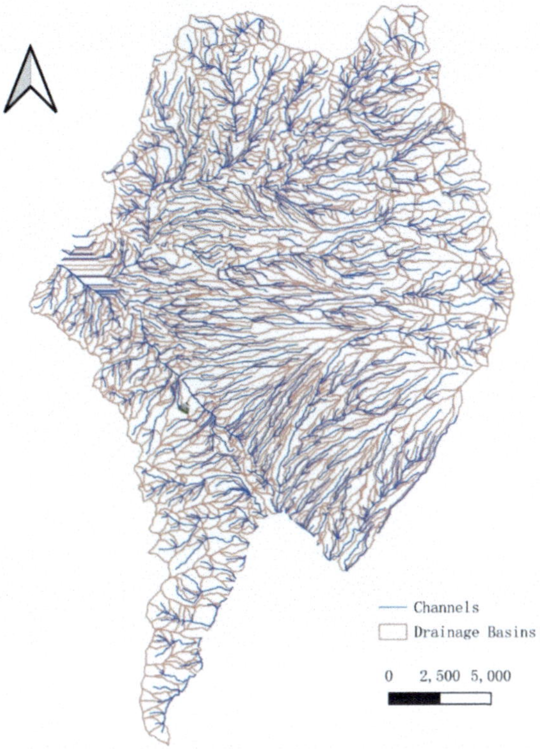

Fig. 4. Channel Network and Drainage Basins.

Overlay analysis. In order to cross-verify the extracted catchment boundaries and hydrological features, the vector data were overlaid onto satellite maps from Google Maps and OpenStreetMap for comparison and validation.

2.5 Data Verification

In order to verify the correctness of the GIS hydrological analysis method, the same method was verified by using Sasahara Reservoir (latitude 36°01′44.9″ N, longitude 138°16′35.0″ E) in Chino City, Nagano Prefecture, which is a highland reservoir located in Mount Yatsugatake range as a sample. It is examined by overlaying elevation maps, watershed maps and Google Maps.

3 Results and Discussions

As seen from Figs. 6 and 7, the catchment area is dendritic, with high south and low north situation and a maximum elevation difference of 500 m. Combined with Fig. 8, the southern end of the starting point is a high mountain forest, with a certain soil and water retention capacity. Lake Irikasa is in the middle of the catchment area, storing water

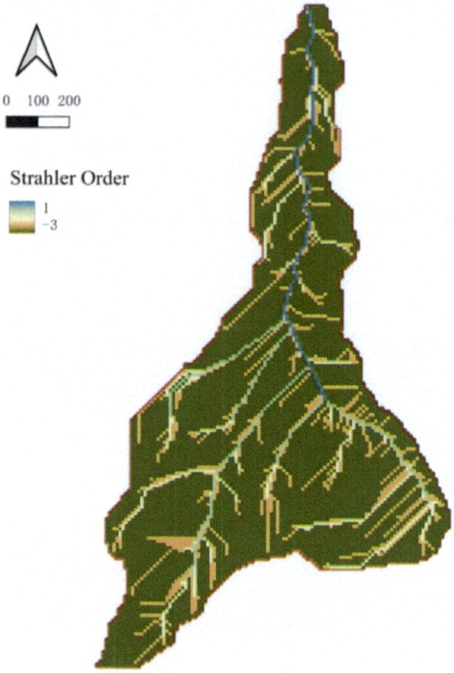

Fig. 5. Catchment area of Lake Irikasa.

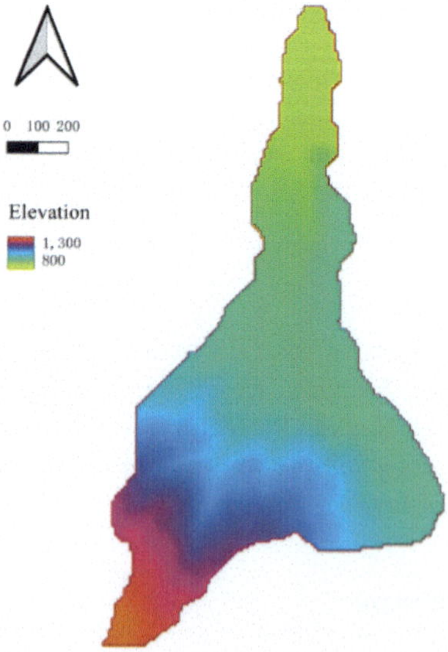

Fig. 6. The elevation data of the Lake Irikasa catchment area.

and, at the same time, relieving the velocity of high-water flow. Although the reservoir Lake Irikasa is in Fujimi Town, the catchment area straddles Fujimi Town and Chino City, and when there is heavy rainfall, the amount of water flowing downstream from the higher elevation increases dramatically. There is also a Daze River downstream, which is a serious water hazard for the residents of Chino City and Haramura downstream.

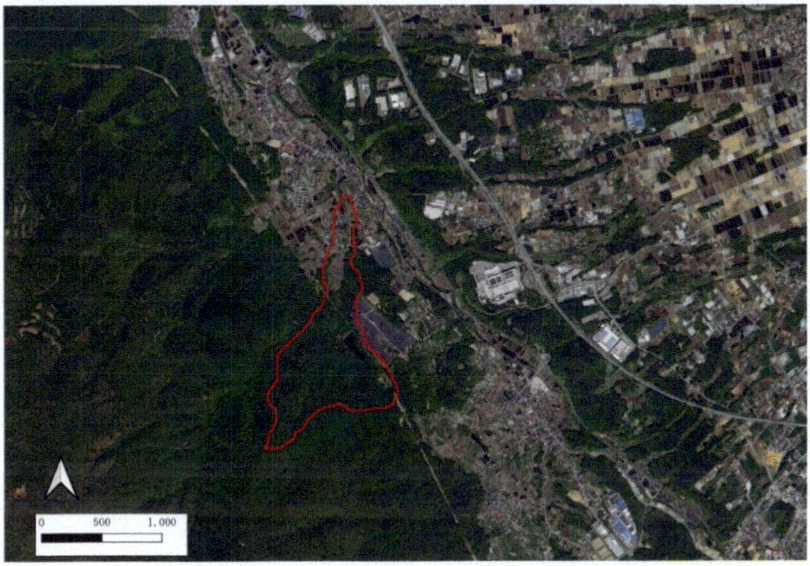

Fig. 7. Lake Irikasa Catchment area in Google map.

As seen in the elevation and watershed overlay of Sasahara Reservoir in Fig. 9, this catchment has an elevation difference of 300 m and a narrow dendritic watershed. As seen in Fig. 10 with Google Maps, abundant forest vegetation mitigates the flow of water. Sasahara Reservoir is also located in the center of the catchment, and this impoundment on the hillside plays a very important role in irrigating the farmlands. However, there are many farmlands and houses downstream, and because the impoundment is located in the Mount Yatsugatake range, surrounded by eight large mountains, there is a risk of overflow from the associated agricultural watercourses during snowmelt and rainfall in the spring.

4 Conclusions

This study performed a hydrological analysis of Lake Irikasa using GIS and validated the method with Sasahara Reservoir, focusing on the reservoir's catchment area. The catchment boundaries were delineated, and critical drainage patterns were identified. The results reveal that while highland reservoirs like Lake Irikasa pose potential flood hazards during heavy rainfall, they also serve as important structures for water storage and flood regulation. The analysis highlighted the risk to downstream areas, particularly

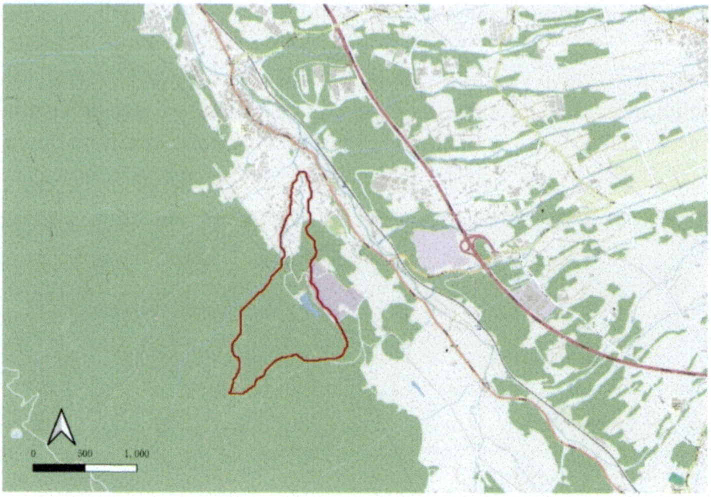

Fig. 8. Lake Irikasa Catchment area in Open Street Map.

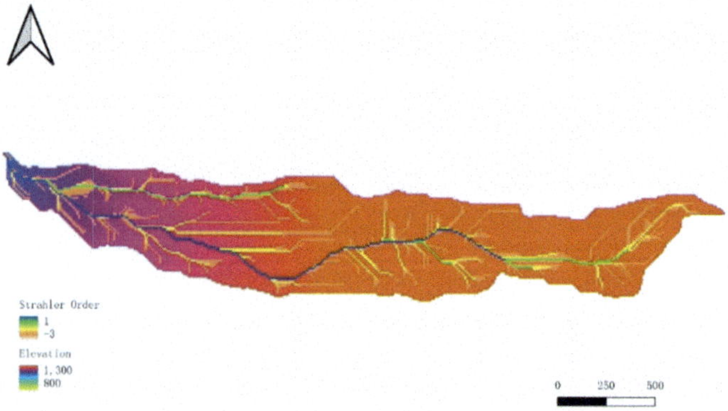

Fig. 9. Sasahara Reservoir catchment in elevation and watershed overlay.

during periods of extreme weather, and demonstrated the reservoir's capacity to buffer rapid water flow from elevated terrain.

The methodology outlined, which includes catchment extraction, terrain analysis, and overlay of hydrological and satellite data, provides a straightforward yet effective approach for understanding the hydrological conditions of unmanaged reservoirs. It offers local authorities a means to assess flood risks and plan appropriate disaster prevention measures.

While this study utilized readily available elevation and administrative data, it is important to acknowledge the potential limitations in these datasets. The 10-m mesh DEM, while detailed, may still need to be revised, particularly in densely vegetated or steep terrain areas, which could influence the precision of hydrological analyses. Additionally, the absence of long-term hydrological data, such as historical flow rates and

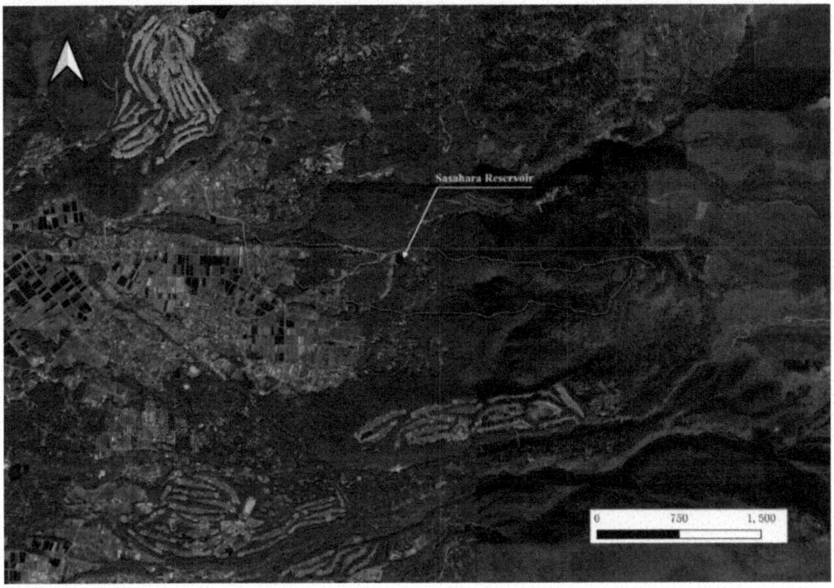

Fig. 10. Sasahara Reservoir catchment area in Google Maps.

precipitation patterns, limits a complete understanding of the reservoir's behavior under varying seasonal and weather conditions. Future research should address these limitations by incorporating long-term hydrological data to provide a more comprehensive analysis of reservoir dynamics and improve the accuracy of flood risk assessments.

In future research, integrating drone-based remote sensing to estimate the volume of water stored in the reservoir and the outflow during heavy rainfall will provide a more dynamic assessment of potential flood risks. Additionally, combining this data with rainfall models will enable the development of a comprehensive disaster prevention strategy that can be applied to similar reservoirs in mountainous regions. This will be critical for safeguarding communities and ensuring sustainable water management.

References

1. Ministry of Agriculture, Forestry and Fisheries, Japan Homepage (2024): Overview of reservoirs, https://www.maff.go.jp/j/nousin/bousai/bousai_saigai/b_tameike/. Last accessed 12 Aug 2024
2. Rural Development Bureau, Ministry of Agriculture, Forestry and Fisheries, Japan: Outline of the Act on the Management and Conservation of Agricultural Reservoirs, https://www.maff.go.jp/j/nousin/bousai/bousai_saigai/b_tameike/attach/pdf/kanrihozenhou-11.pdf. Last accessed 12 Aug 2024
3. Lee, S.J., Yoshizako, H., Kojima, H.: Predicting agricultural reservoir storage using deep learning. J. AI Data Sci. **3**(J2), 693–703 (2022)
4. Tsujioka, Y., et al.: Integrated evaluation index of reservoirs in consideration of disaster prevention and maintenance management. J. Jpn. Soc. Hydrol. Water Resour. **35**(2), 122–133 (2022)

5. Fujimi Town Homepage.: https://www.town.fujimi.lg.jp/site/toukei/jinkou.html. Last accessed 12 Aug 2024
6. QGIS Homepag.: A Free and Open Source Geographic Information System http://www.qgis.org/en/site/. Last accessed 12 Aug 2024
7. CGIS Homepage.: https://cgisj.jp/. Last accessed 12 Aug 2024
8. GSI base map information download service.: http://fgd.gsi.go.jp/download/. Last accessed 12 Aug 2024
9. Chino City Homepage.: https://www.city.chino.lg.jp/soshiki/kikakuzaisei/1159.html/. Last accessed 12 Aug 2024
10. Haramura Homepage.: https://www.vill.hara.lg.jp/sonsei/toke/jinkou//. Last accessed 12 Aug 2024

Hydrogeological Modeling in the Archaeological Site Muyu Muyu—Chincheros—Apurímac

Rubén Esaú Mogrovejo Gutiérrez[1](✉), Douglas Donal Sarango Julca[2], Efraín Jhiaromir Contreras Gutiérrez[2], Daniel Minaya Gutiérrez[3], and Luis Fernando Miranda Gutiérrez[3]

[1] Universidad Peruana de Ciencias Aplicadas—UPC, Lima, Perú
pciprmog@upc.edu.pe
[2] Universidad Nacional Mayor de San Marcos, Lima, Perú
{dsarangoj,efrain.contreras}@unmsm.edu.pe
[3] Universidad Nacional José María Arguedas, Andahuaylas, Perú
{dminaya,lmiranda}@unajma.edu.pe

Abstract. Peru is known for its archaeological wealth, ranging from the world-famous ruins of Machu Picchu to numerous lesser-known but equally significant sites. One such site is the Muyu Muyu archaeological center, located in the district of Uranmarca, Apurímac. Despite its historical and cultural value, Muyu Muyu has not received the attention and care it deserves, leading to conservation issues that threaten its integrity. Among the main challenges this archaeological site faces is the slope inclination caused by soil pressure due to the saturation of the surrounding terrain. The accumulation of water in the soil can lead to an increase in pore pressure, reducing slope stability and increasing the risk of landslides. Solving this problem requires a multidisciplinary approach combining civil engineering, soil mechanics, and hydrogeology. Therefore, the installation of subdrains has been proposed, where 56.7 L per second (l/s) were obtained at the Muyu Muyu site. For this reason, the installation of crown channels is recommended as a preventive and corrective measure to mitigate the negative effects of soil saturation. The subdrains allowed for proper groundwater drainage, thus reducing pore pressure, and improving slope stability. In turn, the crown channels will collect and divert surface water, preventing it from infiltrating the soil and contributing to its saturation.

Keywords: Groundwater · Hydrogeology · Modeling

1 Introduction

In the Apurímac region of Peru, slope stability in archaeological areas like Muyu Muyu is a significant challenge due to soil saturation, which can lead to landslides and the degradation of ancient structures. Previous studies have addressed the relationship between soil saturation and slope stability, highlighting the importance of understanding how hydrological conditions and soil properties affect its strength and cohesion. For example, Jiang et al. (2019) [1] and Li et al. (2023) [2] investigated how saturation and

C. Weng (Ed.): WRE 2024, LNCE 634, pp. 283–297, 2025.
https://doi.org/10.1007/978-981-96-5095-8_24

wetting-drying cycles affect soil stability, providing valuable insights for the design of engineering interventions in archaeological sites.

Regarding proposed solutions, the installation of subdrains and crown channels has been identified as an effective measure to mitigate the effects of soil saturation. This technique relies on redirecting groundwater and surface water away from vulnerable areas, thus reducing hydrostatic pressure and improving slope stability. According to Montrasio and Valentino (2008) [3], the implementation of proper drainage systems can significantly reduce the risk of shallow landslides on rain-affected slopes.

Furthermore, soil loss assessment through models like the (R)USLE (Revised Universal Soil Loss Equation) and the use of GIS (Geographic Information Systems) can help quantify soil erosion and plan more effective conservation strategies. Alexakis et al. (2019) demonstrated how the integration of remote sensors, artificial neural networks, and field spectroscopy can provide accurate estimates of key soil parameters regarding erosion, which applies to the management of archaeological sites [4].

A significant study in the region relates to landslides and terrain stability at the Kuelap Archaeological Complex in Amazonas, Peru. Kuelap, a pre-Inca fortress, has faced significant landslide issues due to its geographic location and climatic conditions. Recent research has indicated that investment in infrastructure, such as the installation of drainage systems and slope stabilization, has been crucial in protecting this archaeological heritage [5].

The management of landslides at archaeological sites like Kuelap has involved the implementation of slope stabilization techniques and underground drainage systems to control soil saturation and minimize the risk of structural collapses. These methods have proven effective not only in preserving cultural heritage but also in promoting sustainable tourism, attracting visitors, and generating income for local communities [6].

Another significant example is slope stabilization in Machu Picchu, where advanced geotechnical technologies have been used to mitigate risks associated with erosion and landslides. The drainage systems and erosion control techniques implemented in Machu Picchu have been the subject of extensive studies and have served as a model for other archaeological sites in Peru and around the world [7].

The Apurímac region in Peru is of great cultural and economic importance. The presence of numerous archaeological sites, agricultural lands, and rural communities makes geotechnical studies essential for the preservation and sustainable development of the region. Landslides and soil instability not only threaten archaeological heritage but also put local infrastructure and livelihoods at risk.

Firstly, Apurímac is home to important archaeological sites such as Sondor, Muyu Muyu, Inka Raqay, and Waqrapukara, which represent invaluable intangible heritage for Peruvian culture and humanity. The conservation of these sites is crucial, as they not only hold historical and cultural value but also attract tourism, generating income for local communities. Landslides and erosion can cause irreparable damage to these structures. Moreover, agriculture is a fundamental economic activity in Apurímac. Landslides can destroy croplands, resulting in significant economic losses for local farmers and generating food insecurity, as well as losses in tourism income if the sites suffer deterioration due to rainfall and mass movements [8]. Investment in geotechnical studies and the implementation of erosion control and soil stabilization measures are essential to protect agricultural lands and tourist sites, ensuring the safety and well-being of rural communities. Figure 1 shows the exact location of the cultural site Muyu Muyu, which is the site developed in this research work [9].

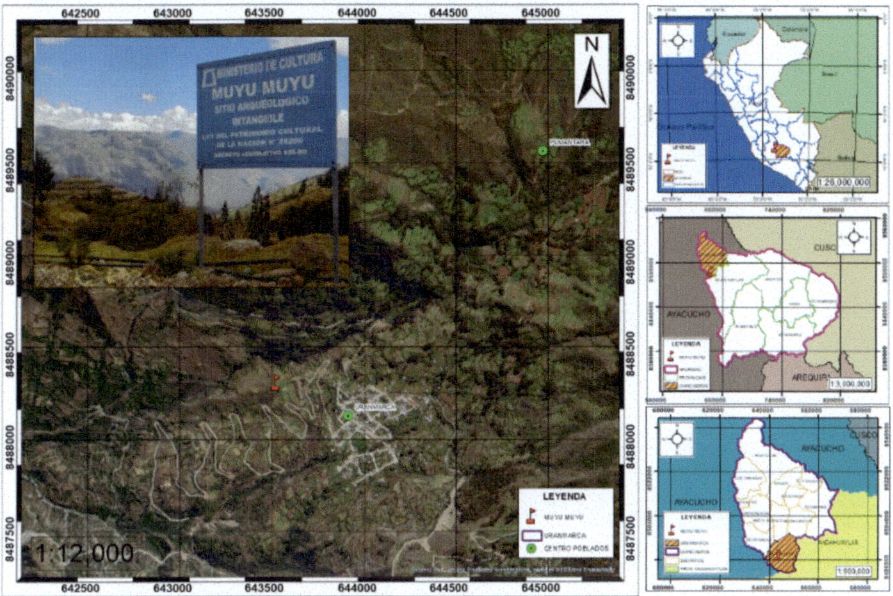

Fig. 1. Location map of the Muyu Muyu Archaeological Center. *Source* Own elaboration.

2 Materials and Methods

2.1 Geological Assessment

The study area covers the [10] quadrangles of Huncapi, Chincheros, Querobamba, and Chaviña, as indicated in Bulletin No. 70. This area is characterized by the presence of the Ferrobamba Formation, whose lithology consists of a sequence of dark to black limestones, beige calcareous sandstones, and shales ranging from black to dark green, as shown in Fig. 2. Due to weathering, the limestones exhibit tones ranging from yellowish-brown to occasionally reddish.

2.2 Geotechnical Assessment

During the visit to the slope at the Muyu Muyu archaeological site, disturbed soil samples were taken from strata E-1 (by conducting a test pit 100 m from the archaeological center due to its intangible nature) and two rock samples (E-2 and E-3) for analysis in the Soil Mechanics Laboratory (Fig. 3).

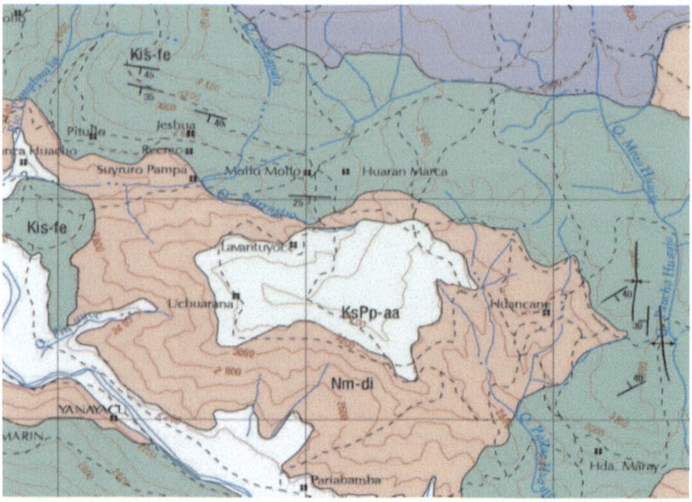

Fig. 2. Geological formation of the study area. *Source* INGEMMET.

Fig. 3. View of the test pit (E-1). *Source* Own elaboration.

The analysis of stratum E-1 reveals that it contains 68.52% fine soil, 26.31% sand, and 5.17% gravel, with a plasticity index of 2.26%, as shown in Fig. 4. According to the USCS classification, this soil is categorized as low-plasticity inorganic silt with sand (ML), and according to the AASHTO classification, it is classified as silty clayey material (A-4).

Additionally, strata E-2 and E-3 consist of rocks from the 'Ferrobamba' formation, known as limestone, with different levels of permeability. The results of the soil parameters are shown in Table 1.

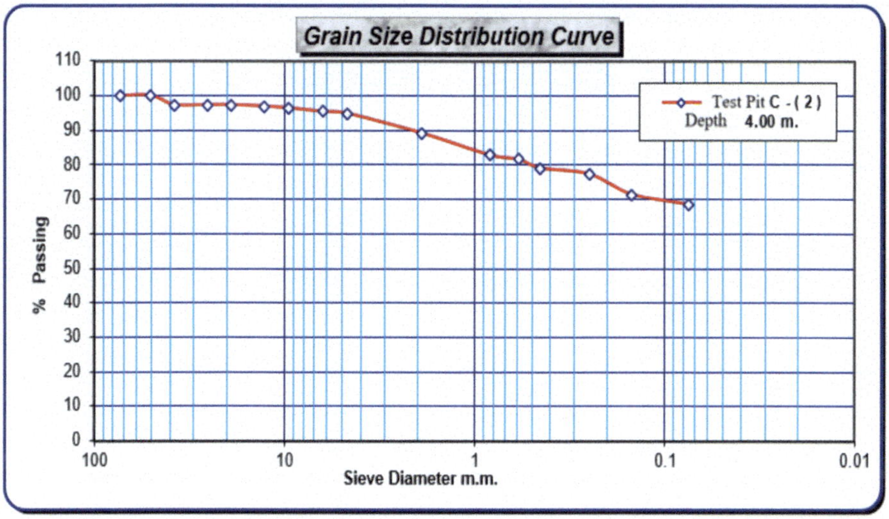

Fig. 4. Grain size distribution curve of stratum E-1. Own elaboration.

Table 1. Soil parameters.

Item	Properties	Test Results
1.0	% passing the sieve N°4	94.83
2.0	% passing the sieve N°10	89.09
3.0	% passing the sieve N°40	79.12
4.0	% passing the sieve N°100	71.48
5.0	% passing the sieve N°200	68.52
6.0	Coefficient of uniformity	It does not exist
7.0	Coefficient of curvature	It does not exist
8.0	Group Index	0
9.0	Liquid Limit (%)	32.07
10.0	Plastic Limit (%)	29.81
11.0	Plasticity Index (%)	2.26
12.0	Natural Moisture (%)	20.62
Classification AASHTO		A-4
Classification SUCS		ML
LOW-PLASTICITY INORGANIC SILT WITH SAND		

Table 2. Complete Monthly Precipitation Data. *Source* ANA and MINAGRI.

YEAR	JAN	FEB.	MAR.	APR.	MAY.	JUN.	JUL.	AUG.	SEP.	OCT.	NOV.	DEC.	MAX
1965	91.3	233.1	155.1	70.5	9.2	4.1	0.0	13.0	81.5	57.3	73.4	198.2	233.1
1966	98.9	182.1	141.8	34.1	82.8	0.0	0.0	14.5	56.0	194.2	133.8	152.6	194.2
1967	163.5	155.7	152.2	65.9	20.8	3.9	3.9	13.7	39.7	88.4	86.7	150.5	163.5
1968	242.7	243.8	212.9	29.1	10.4	13.0	5.0	10.4	27.8	148.5	112.3	131.1	243.8
1969	176.0	117.0	153.0	41.0	0.0	7.0	1.0	15.6	38.0	81.0	122.0	97.0	176.0
1970	273.0	147.0	150.0	115.0	11.0	2.0	0.0	0.0	53.0	41.0	71.0	175.0	273.0
1971	206.0	205.0	171.0	67.0	3.0	4.0	1.0	11.0	7.0	40.0	66.0	166.0	206.0
1972	184.3	117.7	221.8	104.0	22.0	1.0	24.2	13.5	22.2	46.6	37.5	134.7	221.8
1973	227.7	180.0	146.6	207.3	105.5	3.7	11.2	18.7	8.5	40.6	66.7	164.7	227.7
1974	193.7	198.7	196.7	72.1	16.1	7.2	0.4	20.8	17.8	187.2	110.8	162.5	198.7
1975	188.8	295.2	223.1	103.8	17.9	0.4	0.6	4.3	52.7	64.8	94.2	164.0	295.2
1976	170.5	138.1	134.5	43.9	22.3	0.8	7.4	5.9	83.9	79.6	96.3	119.4	170.5
1977	164.3	139.7	164.3	56.7	38.9	0.7	0.7	6.9	16.7	59.0	137.8	99.9	164.3
1978	336.7	164.2	158.2	27.6	4.0	0.7	0.2	4.7	74.0	62.3	93.9	187.1	336.7
1979	70.7	110.1	148.5	45.1	1.6	1.4	0.8	13.5	39.8	56.8	61.4	205.9	205.9
1980	69.5	253.5	168.3	14.7	10.6	0.9	1.0	21.1	28.5	58.9	79.0	176.8	253.5
1981	160.1	177.9	147.2	25.1	0.5	8.3	4.2	0.1	17.9	84.0	110.4	159.7	177.9
1982	80.1	152.4	148.2	34.5	0.6	8.1	0.1	3.4	68.5	42.9	66.8	177.5	177.5
1983	187.3	220.1	159.1	56.5	8.8	3.0	1.1	0.3	62.5	26.0	51.9	115.1	220.1
1984	325.9	131.2	157.6	70.7	0.1	4.4	0.9	21.8	65.8	22.4	40.9	173.0	325.9
1985	48.7	147.5	155.0	100.7	18.7	2.7	7.6	12.6	33.6	56.7	85.6	171.2	171.2
1986	155.6	220.2	184.6	61.4	14.8	0.2	0.5	17.7	33.5	56.1	48.8	170.1	220.2
1987	153.0	253.7	158.9	122.1	32.0	0.9	0.1	21.5	16.1	81.6	49.6	161.2	253.7
1988	162.1	195.2	170.9	81.2	8.7	0.9	0.4	22.0	59.2	61.4	68.1	171.4	195.2
1989	146.3	149.7	194.9	67.6	13.3	1.3	0.9	6.5	33.3	78.2	68.2	160.4	194.9
1990	242.6	166.9	139.1	88.7	48.5	8.3	0.3	1.8	44.8	129.3	146.1	134.2	242.6
1991	221.9	96.0	138.1	39.8	39.9	13.2	0.0	0.4	33.7	41.7	54.7	206.5	221.9
1992	76.4	133.8	175.9	93.3	14.9	13.8	0.9	11.0	6.6	85.2	163.8	103.4	175.9
1993	258.3	187.1	158.5	95.4	7.1	0.1	0.0	20.7	87.6	60.3	71.3	177.5	258.3
1994	119.8	253.4	219.2	42.9	0.6	0.1	0.3	22.0	50.2	140.8	231.5	152.1	253.4
1995	337.2	219.9	140.8	35.1	6.6	0.5	6.8	13.1	38.3	34.9	65.0	149.4	337.2
1996	226.3	140.7	183.7	71.7	0.1	2.2	0.1	18.1	31.6	75.5	86.0	184.6	226.3
1997	129.2	160.5	168.5	51.1	72.2	0.6	0.9	13.4	45.4	41.9	92.5	156.2	168.5
1998	162.8	192.8	158.8	37.8	1.3	0.8	0.7	20.2	20.6	58.4	86.5	141.9	192.8
1999	159.1	139.8	144.7	47.6	9.7	9.0	0.5	19.3	72.9	76.8	91.4	129.2	159.1
2000	170.5	166.4	175.9	29.5	9.1	2.6	2.6	12.1	14.4	97.2	79.3	128.9	175.9
2001	258.0	261.2	206.7	35.9	18.2	0.5	0.5	4.6	41.5	85.3	91.0	155.3	261.2
2002	115.0	139.3	145.4	123.1	19.9	2.4	2.0	1.7	9.8	74.2	189.1	115.4	189.1
2003	203.9	91.2	146.8	99.6	4.1	1.5	0.7	10.8	40.7	61.8	67.6	164.4	203.9
2004	129.4	163.0	164.4	110.9	61.4	1.9	1.0	0.0	15.8	72.5	115.9	103.7	164.4
2005	217.7	184.4	191.3	49.0	2.4	1.3	0.1	20.4	63.5	56.1	78.2	146.5	217.7
2006	244.3	167.4	174.8	86.2	0.9	3.1	0.8	20.4	48.2	73.4	97.3	124.9	244.3
2007	194.0	169.8	156.4	29.4	15.0	0.6	5.2	0.5	16.7	73.3	92.4	139.2	194.0
2008	260.5	160.5	154.6	29.5	38.6	6.8	0.3	9.5	22.3	96.9	75.8	148.4	260.5
2009	266.6	116.7	197.5	67.0	14.3	0.3	1.8	18.0	7.0	73.0	82.4	174.8	266.6
2010	161.2	209.5	201.1	74.2	25.1	0.4	0.9	8.5	29.1	76.9	64.1	207.0	209.5
2011	179.3	177.3	169.7	102.5	10.7	3.6	4.2	6.8	11.8	75.2	116.6	116.7	179.3
2012	265.7	126.1	156.0	109.9	0.8	6.1	5.9	5.6	16.1	84.4	67.7	95.9	265.7
AVERAGE	184.9	174.0	167.5	68.1	18.6	3.3	2.3	11.5	37.6	74.2	90.4	152.1	220.2
STAND. DESV	69.6	46.6	23.7	36.2	22.6	3.7	4.1	7.3	22.5	35.4	37.2	29.4	45.8
MÁX	337.2	295.2	223.1	207.3	105.5	13.8	24.2	22.0	87.6	194.2	231.5	207.0	337.2
MÍN	48.7	91.2	134.5	14.7	0.0	0.0	0.0	0.0	6.6	22.4	37.5	95.9	159.1
Nº Data	48	48	48	48	48	48	48	48	48	48	48	48	48

2.3 Hydrological Analysis

Using data provided by ANA and MINAGRI (Table 2) from the meteorological station named "Chincheros," with coordinates 13°31′1″ (South Latitude), 73°43′1″ (West Longitude), at 2862 m above sea level, data from the project "Evaluation of Water Resources in the Headwaters of the Sub-basins in the provinces of 'Andahuaylas' and Chincheros' was extracted for this research.

The meteorological data includes 48 years of maximum 24-h precipitation, with an average of 220.2 mm, a standard deviation of 45.8, and a skewness coefficient of −1.18.

We applied the Kolmogorov-Smirnov Goodness of Fit Test to the meteorological data using Normal (0.0799), Gumbel (0.08), LogNormal (0.36), Pearson III (0.36), and LogPearson III (0.4) distributions. The Normal distribution provided the best fit, as shown in Fig. 5.

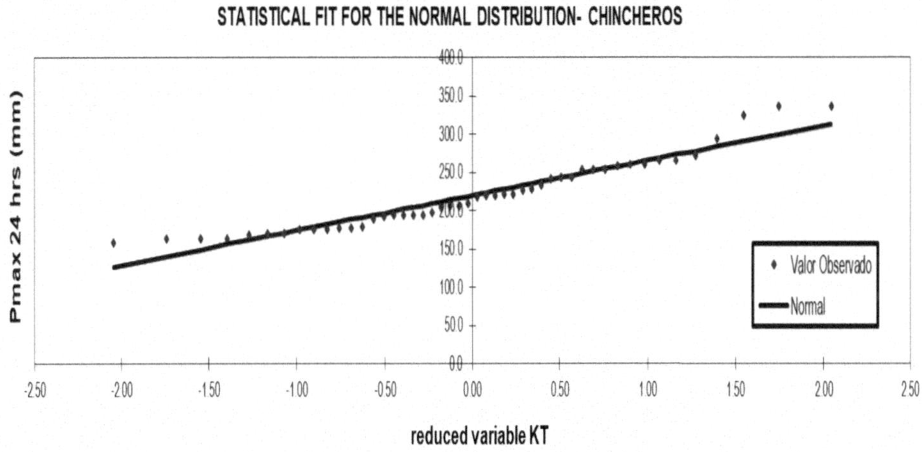

Fig. 5. Normal Distribution Fit. Own elaboration.

Based on the Normal distribution results, we proceeded to develop the design intensities, generating the IDF equation using the Bell Omate method.

$$I = \frac{(10^K * T^m)}{t^n} \tag{1}$$

where: I: Intensity (mm/h); T: Return Period; t: Duration and k = 2.62, m = 0.18, n = 0.53: Coefficients. These coefficients are obtained by creating graphs of intensity, duration, and return period from Table 2.

We will generate Fig. 6 by plotting the intensity, duration, and frequency values from Table 3 for return periods of 20, 50, 100, 200, and 500 years.

According to Table 3, we will use the design intensity of 281.49 mm/h for infiltration and 47.84 mm/h for infiltration (input parameters for the software).

2.4 Infiltration Analysis

The infiltration analysis based on precipitation analysis is conducted using the Soil Conservation Service (SCS) method, where average infiltration and runoff are obtained. This is an important parameter to input into the numerical model. This study reports infiltration data of 47.84 mm/h for a 100-year return period and a 1-h interval (Fig. 7).

2.5 Topographic Survey

Through the coordination of the photogrammetry area of the Geomatics Laboratory at the National University José María Arguedas, a topographic survey was conducted at the Muyu Muyu Archaeological Center. A total of 142 UTM 18 S points were obtained.

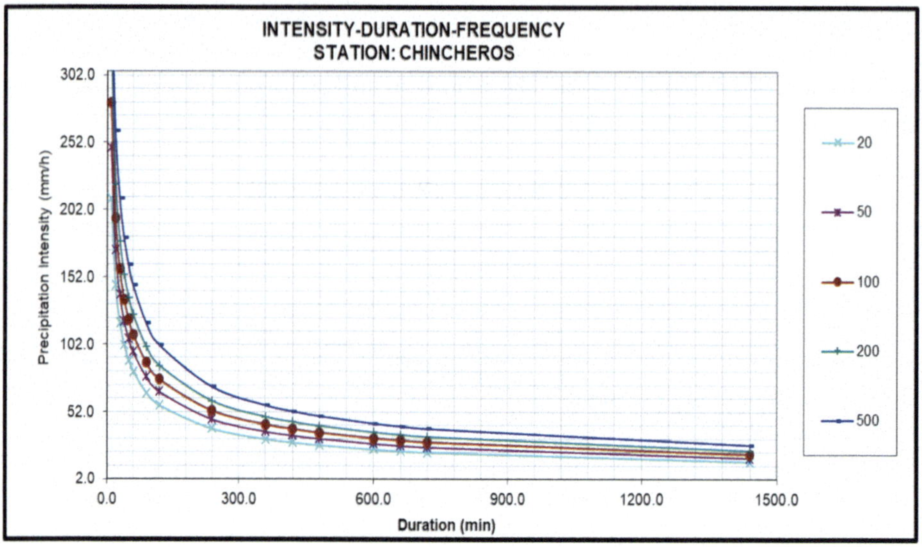

Fig. 6. Design IDF curve.

2.6 Methodology

The conceptual hydrogeological model was developed at the Muyu Muyu archaeological site by clearly defining the study area and establishing an appropriate grid to calculate infiltration levels in the three identified strata. Figure 8 shows the three soil strata of the study area.

The grid consists of 42,148 nodes, covering an area of 10.66 hectares. The strata consist of low-plasticity silt with sand, and limestone rocks from the Ferrobamba Formation.

This study uses FEFLOW software to develop a 3D Hydrogeological model, where the Kriging method is employed to interpolate topographic points three - dimensionally. This method allows for estimating variable values in unsampled locations using the information from the samples, specifically designed to provide the best unbiased linear estimator with the lowest possible variance. Figure 9 shows the discretization of the numerical model into finite volumes, generating a mesh, a necessary process for calculating flow rates across the three soil strata.

Table 3. Design intensities using the Bell Omate Method.

DESIGN INTENSITY (mm/h)										
DURATION		RETURN PERIOD (years)								
Hr	min	2	5	10	20	25	50	100	200	500
0.17	10	139.15	164.12	185.94	210.66	219.30	248.46	281.49	318.92	376.13
0.33	20.00	96.57	113.89	129.04	146.19	152.19	172.42	195.35	221.32	261.03
0.50	30.00	77.99	91.98	104.21	118.06	122.91	139.25	157.76	178.74	210.80
0.67	40.00	67.02	79.04	89.55	101.45	105.61	119.66	135.57	153.59	181.15
0.83	50.00	59.58	70.27	79.61	90.20	93.90	106.38	120.52	136.55	161.05
1.00	60.00	54.12	63.83	72.32	81.93	85.29	96.63	109.48	124.04	146.29
1.50	90.00	43.71	51.55	58.40	66.17	68.88	78.04	88.42	100.17	118.15
2.00	120.0	37.56	44.30	50.19	56.86	59.19	67.06	75.98	86.08	101.52
4.00	240.0	26.07	30.74	34.83	39.46	41.08	46.54	52.73	59.74	70.46
6.00	360.0	21.05	24.83	28.13	31.87	33.17	37.58	42.58	48.24	56.90
7.00	420.0	19.41	22.89	25.93	29.38	30.59	34.65	39.26	44.48	52.46
8.00	480.0	18.09	21.33	24.17	27.38	28.51	32.30	36.59	41.46	48.89
10.00	600.0	16.08	18.97	21.49	24.35	25.34	28.71	32.53	36.86	43.47
11.00	660.0	15.29	18.04	20.44	23.15	24.10	27.31	30.94	35.05	41.34
12.00	720.0	14.61	17.23	19.52	22.12	23.02	26.08	29.55	33.48	39.49
24.00	1440	10.14	11.96	13.55	15.35	15.98	18.10	20.51	23.23	27.40

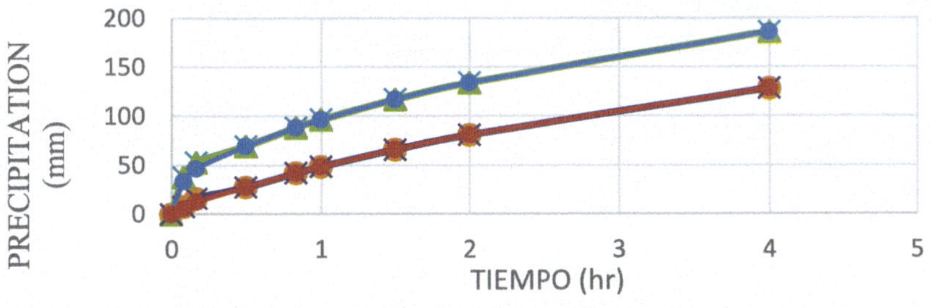

Fig. 7. Precipitation and Effective Precipitation Graph for 100-year Return Period.

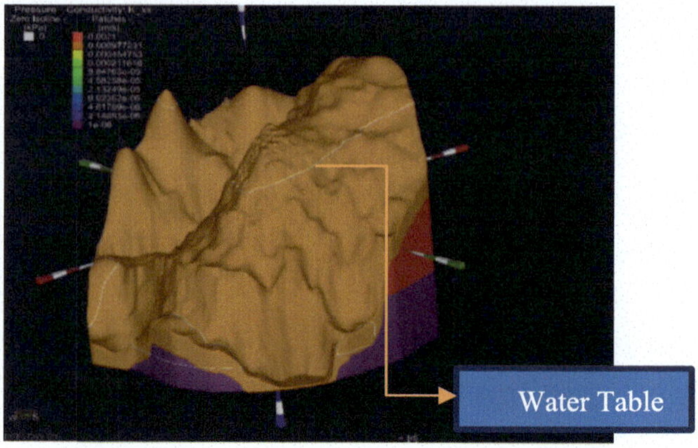

Fig. 8. Hydrogeological Model. Own elaboration.

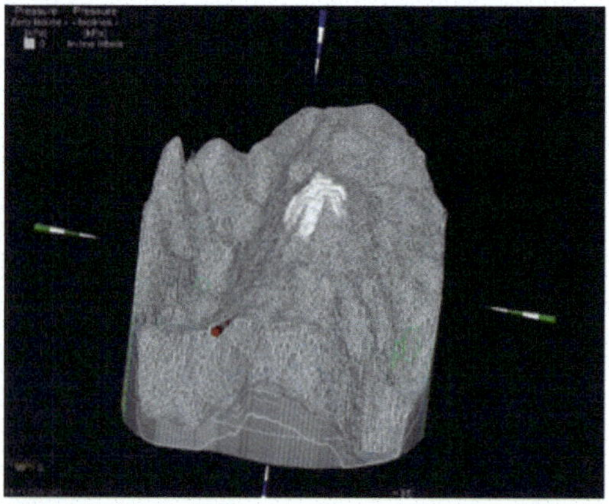

Fig. 9. Grid of the study area. Own elaboration.

3 Results

A numerical model was developed to determine the water table level along the Muyu Muyu fortress. This research focuses on the filtration analysis to visualize the lowering of the water table, obtaining a result of 56.7 L per second at the Muyu Muyu ruins. As shown in the cross-section in Fig. 10, the water table lowers near the base of the fortress, which could lead to future landslides and result in the loss of invaluable cultural heritage.

Figure 11 shows the direction vectors generated by the intense rainfall in the area, causing the saturation of the fortress. This result was obtained by inputting the parameters shown in Table 4.

Figure 12 shows the distribution of observation wells along the archaeological site to proceed with the calibration of the hydrogeological model. Additionally, Fig. 13 displays the interface of the FEPEST software where the numerical model calibration is performed.

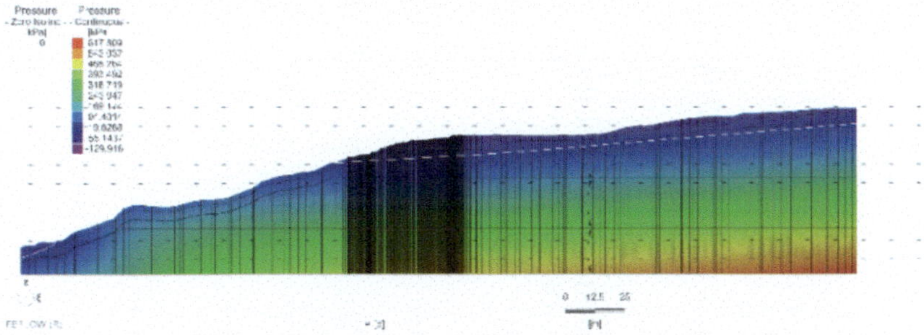

Fig. 10. Water table cross-section. Own elaboration.

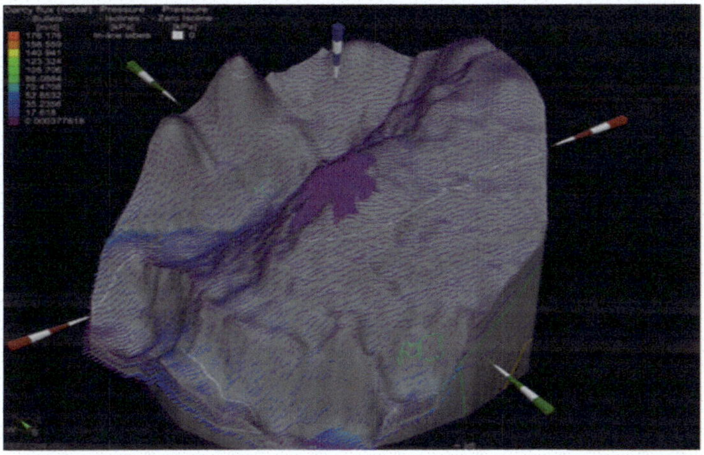

Fig. 11. Filtration vectors in the study domain. Own elaboration.

Table 4. Design Intensities Using the Bell Omate Method

STRATA	PERMEABILITY (m/s)	PERMEABILITY KXX = KYY/10 = KZZ (m/s)	PRECIPITATION INTENSITY (mm/h)	INFILTRATION INTENSITY (mm/h)	HYDRAULIC LOAD (NORTE) (msnm)	HYDRAULIC HEAD (SUR) (msnm)
E-1: ML	0.001	0.0001				
E-2: ROCK 1	0.0021	0.00021	281.47	42.22	3033	2977
E-3: ROCK 2	1.00E-06	1.00E-07				
DRAINS	0.05	0.005				

Source Own Elaboration

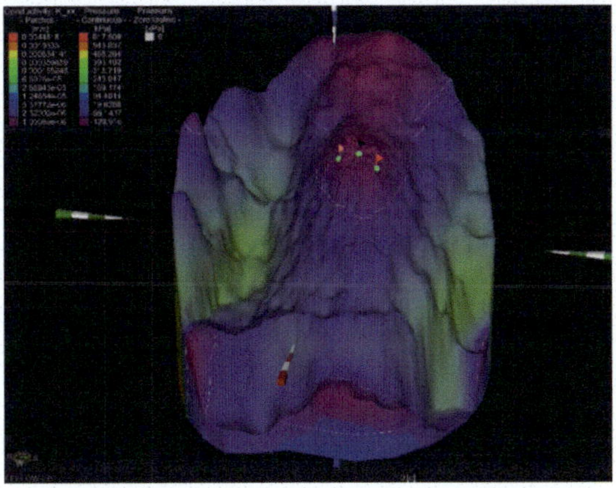

Fig. 12. Installation of observation wells.

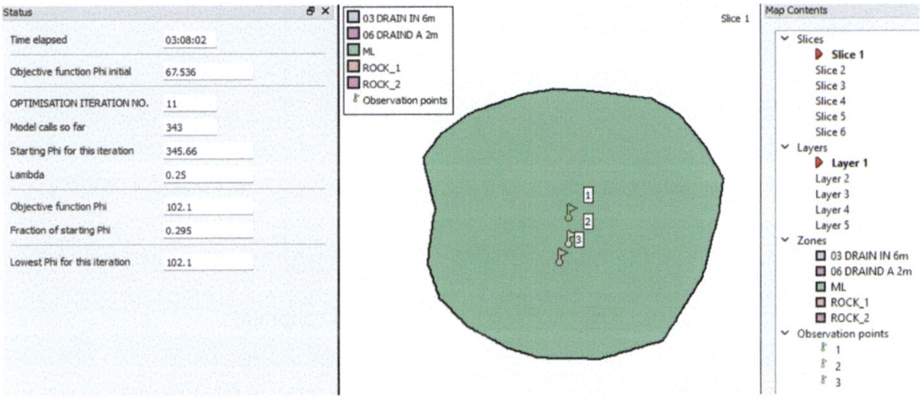

Fig. 13. Delimitation of the study area in FEPEST software for model calibration.

4 Conclusion

The hydrogeological modeling conducted at the Muyu Muyu Archaeological Center revealed significant filtration of 56.7 L per second (l/s) in the study area, for permeabilities of Kxx = 0.001 m/s (ML), Kxx = 0.0021 m/s (Rock 1), Kxx = 1.00E-06 m/s (Rock 2) and Kxx = 0.05 m/s (drains), likewise; Kyy = kzz = 0.0001 m/s (ML), Kyy = kzz = 0.00021 m/s (Rock 1), Kyy = kzz = 1.00E-07 m/s (Rock 2) and Kyy = kzz = 0.005 m/s (drains). This substantial groundwater flow highlights the high permeability of the subsurface, particularly in areas with fractured soils and rocks. The magnitude of this filtration could have a significant impact on the stability and conservation of the archaeological structures, increasing the risk of erosion and material deterioration. These findings emphasize the urgent need to implement groundwater management and control measures to protect and preserve the archaeological site. Crown channels should be designed at the archaeological site.

Through the calibration of the hydrogeological model, an RMS value of 1.70 was obtained, indicating that our hydrogeological modeling is acceptable. With these results, we were able to estimate the filtration rates with the installation of subdrains using FEPEST 7.0 software, as shown in Fig. 14.

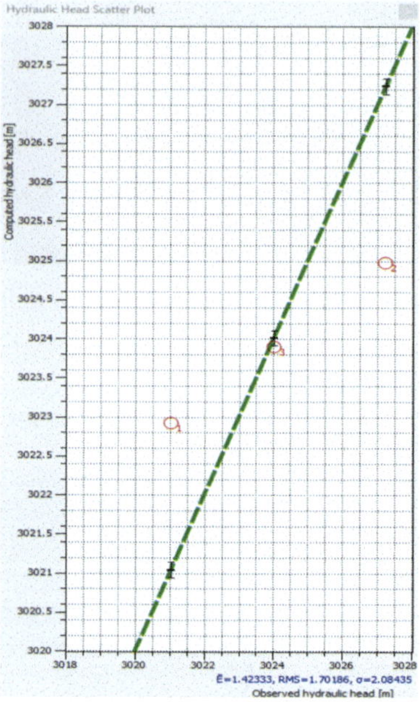

Fig. 14. Favorable results of the model calibration.

The hydrogeological modeling at the Muyu Muyu Archaeological Center provided a detailed understanding of the groundwater dynamics in the area. This included the identification of the main aquifers and the evaluation of their hydraulic characteristics.

Critical areas where groundwater seepage may affect the stability of the archaeological structures were identified. The most vulnerable zones are located near water sources and in areas with fractured soils and rocks.

The use of advanced tools such as FEFLOW proved to be essential for conducting a precise and detailed analysis. The ability to simulate different scenarios allowed for the evaluation of the impact of various interventions and management strategies.

Acknowledgment. This research was supported by the Research Directorate of the Peruvian University of Applied Sciences through the UPC-EXPOST-2024-2 incentive, the Soil Mechanics, Aggregates, and Concrete Laboratory of GEOINCO APURIMAC S.A.C., and the Geomatics and Photogrammetry Laboratory of the National University José María Arguedas, which provided essential topographic equipment. The contributions and support from these institutions were crucial in facilitating the completion of this work.

References

1. Jiang, Q.Q., Liu, L.L., Jiao, Y.Y.: Strength properties and microstructure characteristics of slip zone soil subjected to wetting-drying cycles. Rock Soil Mech. **40**(3), 1005–1012 (2019)
2. Li, Z., Zhou, P.: Research progress of coarse-grained slip zone soil in China. Nat. Hazards **118**, 1–29 (2023)
3. Montrasio, L., Valentino, R.: A model for triggering mechanisms of shallow landslides. Nat. Hazards Earth Syst. Sci. **8**, 1149–1159 (2008)
4. Alexakis, D.D., Tapoglou, E., Vozinaki, A.E.K., Tsanis, I.K.: Integrated use of satellite remote sensing, artificial neural networks, field spectroscopy, and GIS in estimating crucial soil parameters in terms of soil erosion. Remote Sens. **11**(1106) (2019)
5. Empirical Economics: The effect of infrastructure investment on tourism demand: a synthetic control approach for the case of Kuelap, Peru. Empirical Economics (2021)
6. Laidback Trip: Kuelap Travel Guide: Visiting Pre-Incan Ruins in Northern Peru. Laidback Trip (2023)
7. SouthAmerica.travel: Kuelap Fortress—An Archaeological Wonder of Peru. SouthAmerica.travel (2023)
8. INGEMMET: Geological hazard assessment due to landslides in the Yanama community, Apurímac district, Andahuaylas province, Apurímac region. Technical Report No. A7044, pp. 1–25 (n.d.)
9. Mogrovejo, A., Contreras, E.: Flow analysis of slope stability for Landslide 1 - Hill Huayhuaca—Andahuaylas, Peru. In: IEEE Xplore (2021)
10. Marocco, R.: Geology of the quadrants Andahuaylas, Abancay, and Cotabambas. Institute of Geology and Mining (1975)

Study on Wind and Wave Extreme Parameters Around Hainan Island

Wanchun Yu[1], Rongwei Pan[2], Mengke Wang[2], and Zhifeng Wang[2](✉)

[1] State Nuclear Electric Power Planning Design & Research Institute Co. Ltd., Beijing, China
[2] College of Engineering, Ocean University of China, Qingdao, China
panrongwei@stu.ouc.edu.cn, {wmk,wzf1984}@ouc.edu.cn

Abstract. Hainan Island is in an area affected by tropical marine monsoon. Typhoons are frequent, often accompanied by strong winds, extreme waves, and rainstorms. The driving wind field is a reconstructed wind field, which is a combination of the Jelesnianski typhoon model and the ERA5 wind field, to drive the SWAN model and to obtain hourly wind and wave data for 20 years (2003–2022) of Hainan Island. Through comparison, this study found that the P-III distribution is more appropriate for calculating wind and wave extremes in this sea area. It was found that the maximum wind speeds and wave heights mainly occur on the northeast side of Hainan Island and the southwest side of the Xisha Islands. The maximum wind speed with a 50-year return period reaches 55 m/s, and the maximum wave height with a 50-year return period reaches 13 m.

Keywords: Hainan Island · Typhoon · Extreme Conditions · Wave Characteristics

1 Introduction

Hainan Island is in the northwestern part of the South China Sea and is surrounded by a vast sea area with rich marine resources. But it is frequently affected by Þtyphoons, often accompanied by extreme weather events such as strong winds, typhoon waves and heavy rain. For example, Super Typhoon Maki in 1973 and Super Typhoon Wilson in 2014 brought serious climate impacts to Hainan Island and the surrounding areas, causing huge property losses and casualties. Studying the extreme wind and wave characteristics will help scientifically and rationally assess and utilize wind and wave energy resources. And for ocean engineering and energy resources development, it is necessary to consider the impact of the extreme weather that may be brought about by typhoons in order to design and arrange the facilities in a more rational way, and to improve their disaster-resistant capability and stability. To make the simulated wind field closer to the real situation, the driving wind field combines the Jelesnianski typhoon model and the Fifth Generation of European Reanalysis (ERA5) wind field. Shashank et al. [1] derived the optimal wind model suitable for the Bay of Bengal area by comparing the reconstructed wind fields. Pan et al. [2] demonstrated that the reconstructed wind field of the Holland typhoon model and the Cross-Calibrated Multi-Platform (CCMP) wind field can significantly improve

C. Weng (Ed.): WRE 2024, LNCE 634, pp. 298–315, 2025.
https://doi.org/10.1007/978-981-96-5095-8_25

the peak wind speed and thus more accurately calculate the waves during typhoons. The use of reconstructed wind fields is a prerequisite for accurate wave calculations. The simplicity of the SWAN model and its continuous improvement have made it an important part of wave simulations worldwide. Islek et al. [3] calculated 40 years of waves in the Black Sea using the SWAN model and analyzed the wave characteristics in the region using various statistical methods. Wu et al. [4] simulated extreme waves in the South China Sea (SCS) using the coupled atmosphere-wave model, and the statistical distribution of typhoon waves was simulated using the Pearson type III distribution (P-III). Yin [5] chose SWAN to simulate waves in the SCS, and derived the extreme information of waves in the SCS with an accuracy of 0.05°.

In this study, the South China Sea (105–125°E, 5–25°N) is selected as the computational area, with Hainan Island as the center and the sea area within 500 km as the key study area. After comparing with the measured data to verify the reliability, the reconstructed wind field drives the SWAN model for wave computation, and hour-by-hour continuous wind and wave data were obtained for 20 years (2003–2022). Finally, the wind and wave characteristics under extreme environments were calculated and analyzed based on the available wind and wave data. Figure 1 (left) shows the tropical cyclone paths passing through the computational region from 2003 to 2022, which can be seen as the Hainan Island is frequently affected by typhoons. Figure 1 (right) shows the specific locations of 12 characteristic points and measurement stations within 500 km, centered on the O-point (109.6°E, 19.1°N) of Hainan Island.

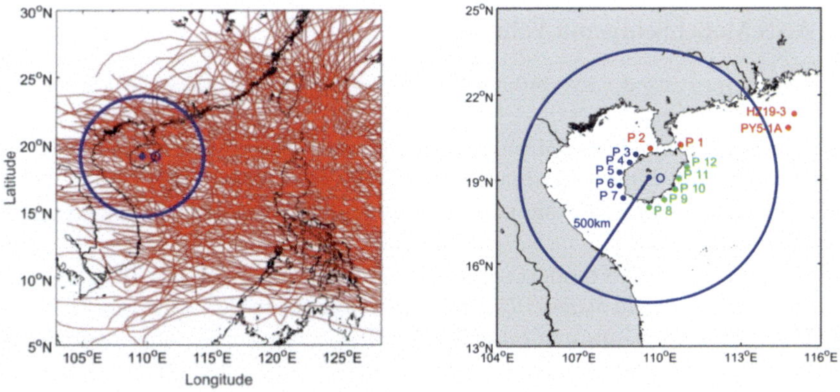

Fig. 1. The tropical cyclone paths passing through the computational region in 2003 to 2022 (left) and specific locations of 12 characteristic points and measurement stations in Hainan Island (right).

2 Data and Method

2.1 Driving Wind Field

Lin et al. [6] compared the quality of three reanalyzed wind fields in different Chinese seas from the point of view of wind speed and direction. The results showed that the ERA5 wind field performed best in the northern part of the SCS, and the ERA5 reanalyzed data have a high temporal resolution. In this study, the ERA5 reanalysis data at 10 m height is used as the background wind field, which has a temporal resolution of 1 h and a spatial resolution of 0.25° × 0.25°, the calculation area is taken as 5°N–25°N and 105°E–125°E.

Tropical cyclones have been the meteorological disaster with the greatest impact on Hainan Island in recent decades, and extreme events are closely related to the number of tropical cyclones [7]. The average wave height intensity in this sea area is the highest in August, and a typhoon is the main weather system that produces the above mentioned large-value waves, and the waves above 12m are the ones that occur during the typhoon [8]. Therefore, it is crucial to study the extreme wind and wave environments in the sea around Hainan Island to reproduce the wind field during typhoon transit. However, the ERA5 wind field cannot reproduce the exact maximum wind speeds during the typhoon, but it can provide relatively accurate wind field data away from the typhoon centre. Therefore, the Jelesnianski [9, 10] model is used in this study to correct the wind field during typhoons.

2.2 SWAN Model Setup and Validation

Currently, the most used wave models in various regions of the world are the WAM, WAVEWATCHIII, and SWAN models. The SWAN models are more comprehensive than the other two models in terms of physical factors such as wave refraction, wave wrap, wave breaking, and whitecap dissipation terms. The simulation results of SWAN model in the Yellow Sea and the SCS are more representative [11]. However, the computation time of the model is longer.

In this study, the SWAN model (version 41.45) was used to simulate 20 years of waves in the sea around Hainan Island. The bathymetry data are derived from the GEBCO data. The specific data information and SWAN structured grid setup parameters are shown in Table 1. The calculated results were verified with actual observations, and the specific locations of the observation stations HZ19–3 (115.02°E, 21.35°N) and PY5-1A (114.80°E, 20.87°N) are shown in Fig. 1 (right). The error statistics are shown in Table 2, and the comparison plots are shown in Figs. 2, 3, 4, 5, 6 and 7. It can be found that the four error indicators of significant wave height (SWH) and mean wave period (MWP) are relatively low, and the mean relative error of wind speed is a little bit worse compared with the other three error indicators. However, considering the instability of the wind speed measurements and the good fitting of the scatterplot, the study concludes that the accuracy of the model calculations compares favorably with the measured data and can be used for the subsequent characterization of wind and waves in the extreme environment.

Table 1. Major configuration and schemes in the SWAN model.

Domain	Initial parameter setting	Value
SWAN model	Computational range	105°E-125°E, 5°N-25°N
	Computational resolution	0.1° × 0.1°
	Time step	15 min
	Output step	1 h
	Grid size	201 × 201
Bathymetric data (GEBCO)	Range	105°E-125°E, 5°N-25°N
	Resolution	0.1° × 0.1°
Wind field data	Range	105°E-125°E, 5°N-25°N
	Resolution	0.1° × 0.1°
	Time Resolution	1 h

Table 2. Error analysis table of simulation results and measured data.

Observatory	Element	Mean Absolute Error (MAE)	Mean Absolute Percentage Error (MAPE)	Root Mean Square Error (RMSE)	Correlation coefficient (R)
HZ19–3	Wind Speed (m/s)	1.40	31.44	1.89	0.73
	SWH (m)	0.14	14.21	0.20	0.95
	MWP (s)	0.65	12.76	0.82	0.75
PY5-1A	Wind Speed (m/s)	1.19	22.79	1.55	0.92
	SWH (m)	0.21	14.54	0.29	0.97
	WMP (s)	0.58	11.03	0.75	0.87

2.3 Extremal Distribution Theory

In engineering construction design, it is often necessary to calculate the maximum wind speed and wave height for different return periods, which is an important reference for the safe and stable operation of engineering projects. The study compares three common extremal distribution functions: the Weibull distribution, the Gumbel distribution, and the P-III distribution. The following is a brief introduction.

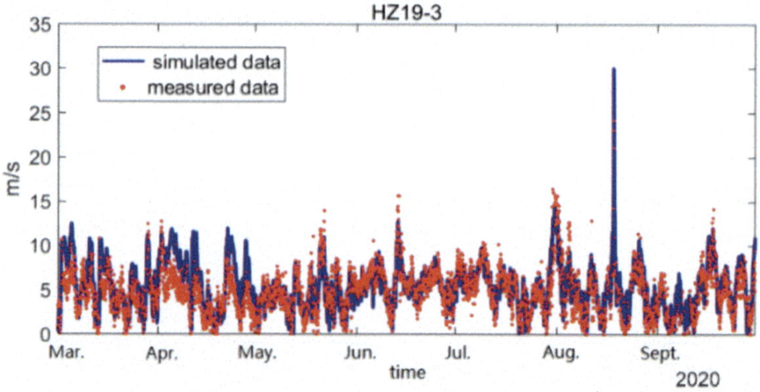

Fig. 2. Comparison of measured wind speed and simulation results for validation (HZ19–3).

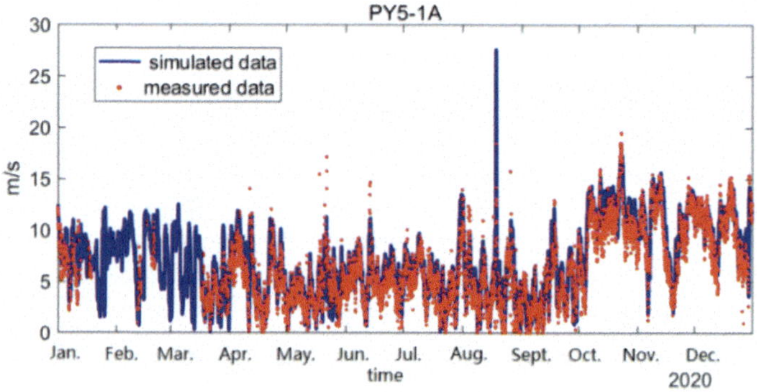

Fig. 3. Comparison of measured wind speed and simulation results for validation (PY5-1A).

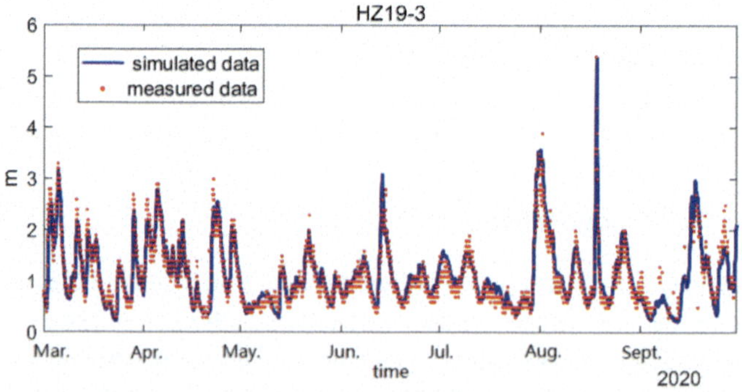

Fig. 4. Comparison of measured significant wave height and simulation results for validation (HZ19–3).

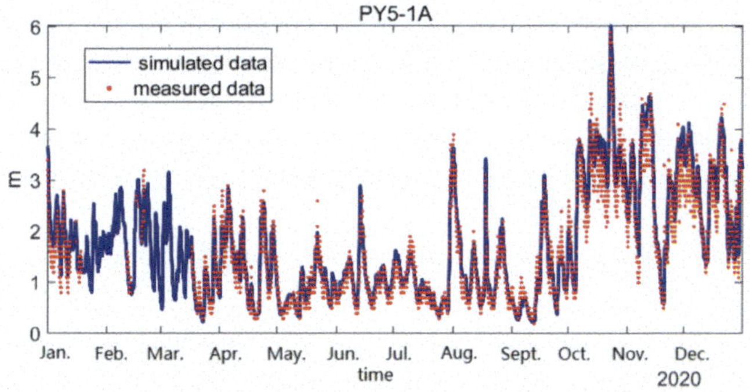

Fig. 5. Comparison of measured significant wave height and simulation results for validation (PY5-1A).

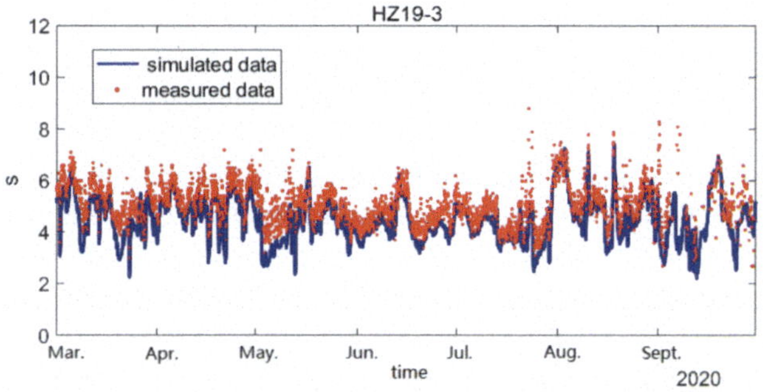

Fig. 6. Comparison of measured mean wave period and simulation results for validation (HZ19–3).

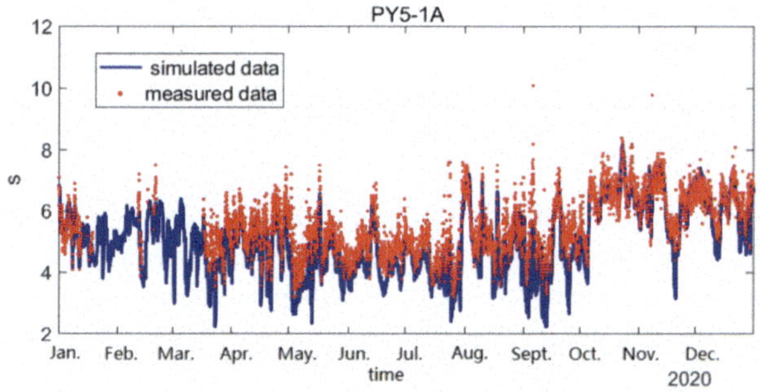

Fig. 7. Comparison of measured mean wave period and simulation results for validation (PY5-1A).

The Weibull distribution was initially applied mainly in the study of materials, and its applicability has become increasingly widespread. In general, there are mainly two-parameter and three-parameter methods. The two-parameter is widely used in hydrological studies, and the following equation shows the two-parameter probability model [12]:

$$f(x, \alpha, \beta) = \frac{\alpha}{\beta} x^{\alpha - 1} \exp\left[-\frac{x^{\alpha}}{\beta} \right] (\alpha > 0, \beta > 0, x \geq 0) \tag{1}$$

where, x is the random variable; α is the shape parameter; and β is the scale parameter.

The Gumbel distribution was published in the 1940s, named after the researcher Gumbel E J, who applied the theoretical model to the field of hydrology and had a wide impact. Its probability density function is distributed as follows [13]:

$$f(x, \beta, \mu) = \frac{1}{\beta} exp\left[-\frac{x - \mu}{\beta} \right] - exp\left(-\frac{x - \mu}{\beta} \right), x > \mu, \beta > 0 \tag{2}$$

There, x is the random variable; β is the scale parameter; and μ is the position parameter.

The P-III distribution is a very common imputation method in hydrological research. Its curve fitting effect can be changed by adjusting the deviation coefficient and departure coefficient, so the fitting effect performs well in many cases. Generally, as the number of samples increases, the better the fit is, and in the case of insufficient sample size, a relatively large error may occur. Its probability density function is distributed as follows [14]:

$$f(x, \alpha, \beta, a_0) = \frac{\beta^{\alpha}}{\Gamma(\alpha)} (x - a_0)^{\alpha - 1} e^{-\beta(x - a_0)}, x \geq a, \alpha > 0 \tag{3}$$

where, x is a random variable; α is a shape parameter; β is a scale parameter; and a_0 represents a position parameter.

2.4 Goodness-Of-Fit Theory (GFT)

In the article, three fitting tests are selected in order to select a more suitable extremal distribution for the wind field near Hainan. The metrics include Kolmogorov-Smirnov (KS) test, RMSE, and the improved Akaike information criterion (AIC) [15, 16]. The KS test statistic can be defined as Eq. (4).

$$\hat{D}_n = \max|F_n(x_i) - F(x_i)| \tag{4}$$

where, $F_n(x)$ is the cumulative probability distribution obtained according to the selected extreme value distribution model. $F(x)$ is the cumulative probability distribution of the original wind speed extremes. If the statistic $\hat{D}_n$ is smaller than the predetermined value D_n determined by the significant level, then it is considered that the fitted distribution model is in accordance with the original data.

The RMSE value is also an important indicator for evaluating the goodness of fit of the data. The RMSE value can be estimated according to Eq. (5).

$$RMSE = \sqrt{\frac{1}{n}\sum_{i=1}^{n}[F_n(x_i) - F(x_i)]^2} \tag{5}$$

where, n is the sample capacity, $F(\mathrm{x})$ and $F_{\mathrm{n}}(\mathrm{x})$ are the original extreme wind speed and the value obtained by the model, respectively.

The AIC criterion was proposed in the 1970s as an important metric for evaluating fitted distribution models. However, a large bias may occur when the sample size is insufficient. Therefore, this paper adopts the improved AICC criterion, which takes the sample length n into account in the assessment, and its mathematical relationship is as follows:

$$AICC = -2L(X|\theta) + 2k + \frac{2k(k+1)}{n-k-1} \tag{6}$$

where, k is the number of parameters identified by the cumulative probability model.

In the evaluation, if the statistical value of the KS is smaller and less than the preset value of the α, the model is better fitted. Similarly, if the values of RMSE and AICC are smaller, the extremal distribution is better fitted for calculating wind speeds under different return periods.

3 Extreme Wind and Wave Characteristics

3.1 Selection of Extremal Distribution of Wind

This section evaluates the imputation of the three distributions using the three goodness-of-fit tests described above. The results of the 12 characteristic points are recorded in Table 3. By comparison, it can be found that the estimator $\hat{D}_n$ of the 12 characteristic points is smaller than the value of Dn. Among them, the $\hat{D}_n$ of both Gumbel distribution and P-III type distribution is smaller than that of Weibull's, which indicates that these two distributions are relatively better fitted. Secondly, the AICC values of the three distributions are roughly the same, and for different characteristic points, the AICC of the three fitted distributions slightly have their own advantages. However, the RMSE of the P-III distribution has the smallest fitting errors for all 12 points. Finally, the P-III distribution model performs the best and was used to calculate extreme wind speeds for different return periods in this study area.

Table 3 shows that the P1 point and P12 point of the 50-year return period maximum wind speed of more than 40 m/s. These two points are located on the northeast side of Hainan Island, which suggests that the wind is relatively strong here. Figure 8 shows the P-III distribution fitting diagrams of wind speed, which shows that the fit is of high quality and the model is reliable.

Table 3. Comparison of wind speed fitting effects at characteristic points and calculation of winds with different return periods around Hainan Island.

Wind	Extremal Distribution	$\hat{D}_n$	Dn (0.05)	RMSE	AICC	Wind speed for different return periods (m/s)			
						50	25	10	5
P1	Gumbel	0.1797	0.2941	0.0718	137.81	44.87	40.08	33.62	28.51
	Weibull	0.2055		0.1006	146.10	42.30	39.23	34.46	29.99
	P-III	0.1305		0.0484	152.60	42.23	37.85	31.91	27.21
P2	Gumbel	0.1508	0.2941	0.0653	123.15	32.61	29.71	25.81	22.73
	Weibull	0.1901		0.0805	127.84	29.92	28.43	26.03	23.69
	P-III	0.1536		0.0576	151.36	34.05	30.82	26.39	22.86
P3	Gumbel	0.1222	0.2941	0.0626	116.20	29.22	26.91	23.79	21.33
	Weibull	0.137		0.064	118.70	26.52	25.46	23.75	22.05
	P-III	0.1103		0.033	120.31	31.14	28.64	25.11	22.17
P4	Gumbel	0.1549	0.2941	0.0669	114.06	28.04	25.92	23.05	20.78
	Weibull	0.1351		0.0536	115.32	25.41	24.47	22.94	21.42
	P-III	0.1051		0.0346	118.34	28.81	26.93	24.17	21.75
P5	Gumbel	0.1081	0.2941	0.0476	107.90	28.31	26.43	23.91	21.91
	Weibull	0.131		0.0532	111.42	26.13	25.31	23.96	22.59
	P-III	0.1022		0.0296	111.63	28.70	27.01	24.58	22.50
P6	Gumbel	0.1537	0.2941	0.0636	115.60	31.02	28.55	25.21	22.57
	Weibull	0.1952		0.0926	122.43	28.66	27.43	25.46	23.50
	P-III	0.1143		0.0465	116.93	32.38	29.59	25.78	22.73
P7	Gumbel	0.1631	0.2941	0.0592	119.03	31.27	28.64	25.10	22.30
	Weibull	0.2148		0.0801	124.37	28.77	27.45	25.32	23.23
	P-III	0.1234		0.0478	123.40	32.89	29.83	25.66	22.36
P8	Gumbel	0.1216	0.2941	0.044	110.61	29.61	27.50	24.64	22.38
	Weibull	0.1592		0.0714	116.80	27.48	26.48	24.85	23.21
	P-III	0.1043		0.0323	113.28	30.59	28.34	25.23	22.68
P9	Gumbel	0.2452	0.2941	0.1179	98.53	24.60	23.26	21.46	20.04
	Weibull	0.1747		0.0799	96.22	22.58	22.08	21.26	20.41
	P-III	0.1783		0.0765	101.50	25.32	24.15	22.42	20.87
P10	Gumbel	0.092	0.2941	0.0282	114.46	31.20	28.78	25.51	22.92
	Weibull	0.1479		0.0748	122.90	29.40	28.11	26.03	23.98

(continued)

Table 3. (*continued*)

Wind	Extremal Distribution	$\hat{D}_n$	Dn (0.05)	RMSE	AICC	Wind speed for different return periods (m/s)			
						50	25	10	5
	P-III	0.0876		0.027	116.32	30.82	28.53	25.36	22.79
P11	Gumbel	0.1113	0.2941	0.0422	129.80	38.74	35.03	30.04	26.08
	Weibull	0.1648		0.0773	137.79	36.40	34.21	30.75	27.44
	P-III	0.116		0.0392	130.60	37.99	34.48	29.65	25.76
P12	Gumbel	0.1668	0.2941	0.0705	129.26	37.25	33.83	29.22	25.56
	Weibull	0.1873		0.0784	134.16	34.09	32.30	29.43	26.66
	P-III	0.1399		0.0488	128.91	41.34	36.70	30.49	25.69

3.2 Extreme Wind for Different Return Periods

In this section, the maximum wind speeds were selected for each year (2003–2022) and the wind speeds of different return periods (50, 25, 10, and 5 years) at 13431 grid points, and the distributions are shown in Fig. 9.

As can be seen from Fig. 9, the maximum wind speed in the study area reaches 55 m/s for a 50-year return period, the wind speed of a 25-year return period reaches 45 m/s, the wind speed of a 10-year return period reaches more than 34 m/s, and the wind speed of 5-year return period reaches 32 m/s. Strong winds in the sea area are mainly distributed in two areas, the first is the northeast side of Hainan Island, where points P1 and P12 are located in the vicinity, so the influence of extreme wind speed should be considered in the process of engineering construction and utilization of wind resources. The second is the southwest side of the Xisha Islands, due to its farther offshore and larger water depth, while the design problems caused by strong wind conditions should be considered. These two places are located on the east side of Hainan Island, Hainan far east of the sea, by the northeast monsoon climate and the Taiwan Strait, Luzon Strait topography "narrow tube effect", the wind speed increases significantly. The Beibu Gulf is a semi-enclosed bay, and due to the topography, wind speeds at the entrance are higher than inside the bay.

3.3 Selection of Extremal Distribution of Wave

As for wind speed, this section selected the extremal distribution suitable for waves in the study area. Table 4 shows the results of the three extremal distributions and their evaluation of them. $\hat{D}_n$-values indicates that all of them can calculate the extreme wave height in the case of $\alpha = 0.05$ and the results are good. Among the 12 characteristic points, the $\hat{D}_n$-values of the P-III and Gumbel are smaller than those of Weibull, and P-III distribution is the smallest for nine of the points. For the AICC, the values of the three distributions are similar, and the Gumbel distribution is slightly better than the other two, but the RMSE of the P-III distribution is smaller than the other two. In general, the P-III distribution is selected to analyze extreme wave characteristics.

According to the projection results of the P-III distribution in Table 4, the extreme wave height at P12 is the largest, and the extreme wave heights at the east side of Hainan Island (P8-P12) are larger than those at the west side (P3-P7). Figure 10 shows the P-III distribution fitting diagrams, which shows that the curve fits well with the extreme wave height data.

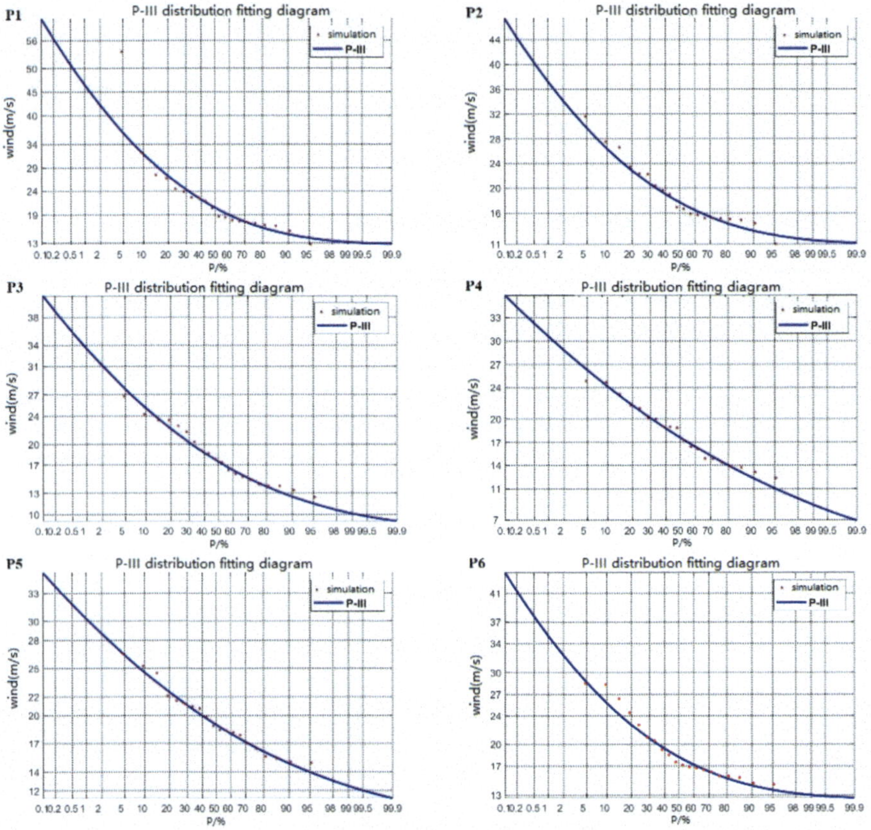

Fig. 8. P-III distribution fitting diagrams of wind speed at 12 characteristic points around Hainan Island.

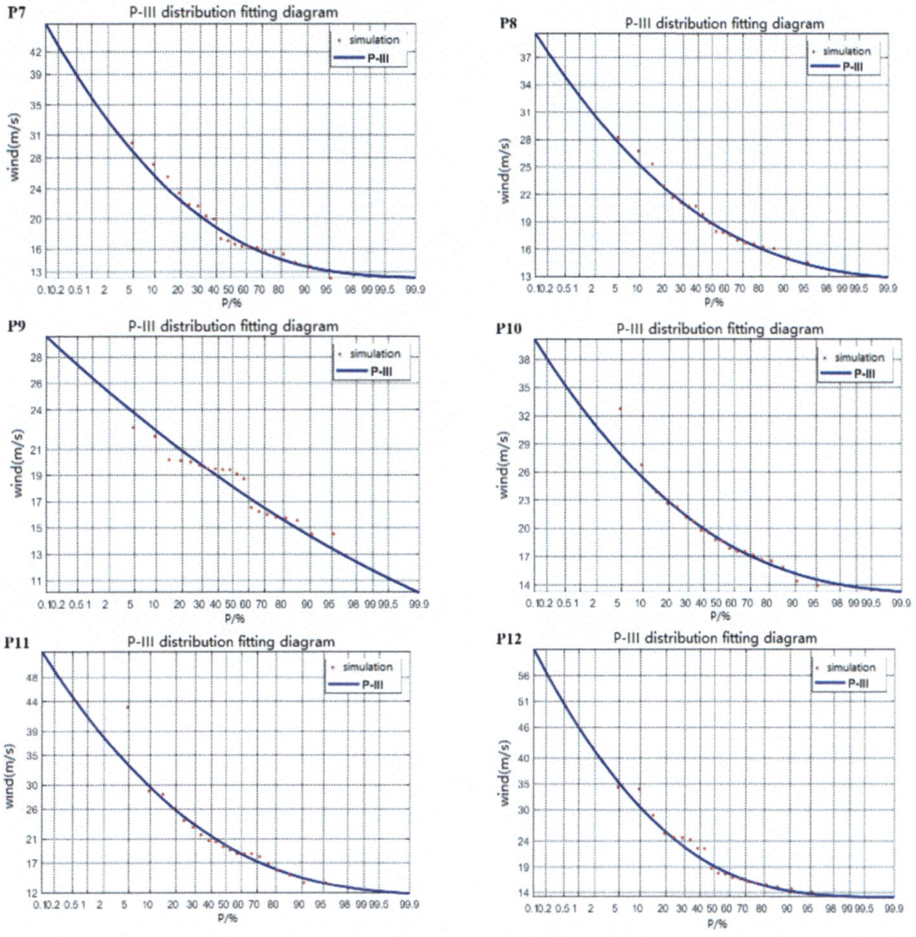

Fig. 8. (*continued*)

3.4 Extreme Wave for Different Return Periods

In this section, based on 20-year wave height data of 13431 SWAN grid points, the extreme wave heights of each grid point are projected for different return periods using the P-III distribution, and the wave height distribution is shown in Fig. 11. The wave height distribution pattern shows that the farther offshore, the larger the wave height. Within the study area, the maximum wave heights for the 5-, 10-, 25-, and 50-year return periods reached 9 m, 10 m, 12 m, and 13 m. The high wave heights were mainly distributed off the northeastern part of Hainan Island at the O-point, the southwestern part of Xisha Islands, and the entrance of the Beibu Gulf. The reason for this may be due to the southeast of Hainan Island facing the open South China Sea, the depth of water from the shelf shallow water rapidly down to 1000 m, and farther offshore, so by the influence of deep-sea waves and topography, the region wave height is large. In addition to the

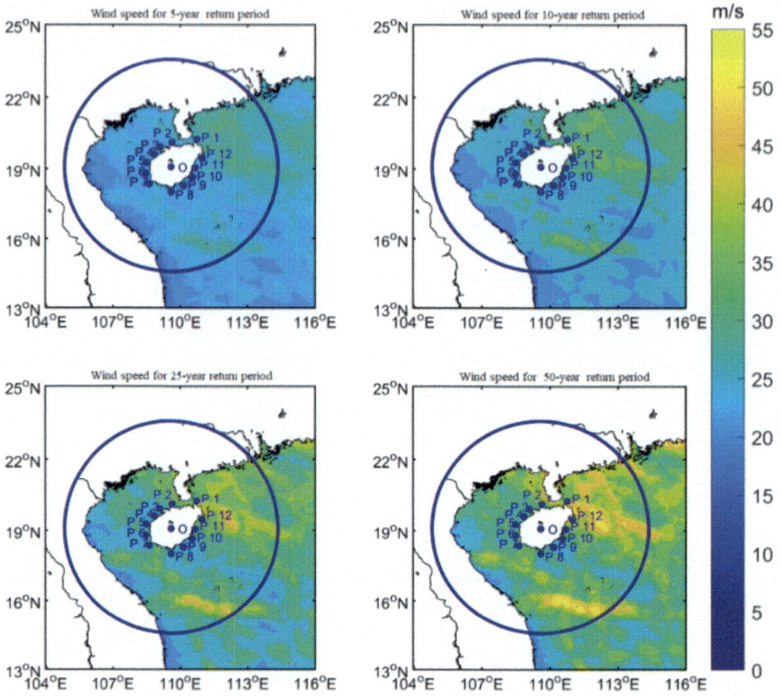

Fig. 9. Wind speed for 5-year, 10-year, 25-year, 50-year return periods around Hainan Island (unit: m/s).

monsoon climate, in winter, the eastern part of Hainan Island is significantly affected by the South China Sea from east to west waves. The Beibu Gulf in the western part of Hainan Island is semi-enclosed, and the surrounding topography forms a natural barrier against the propagation of deep-sea waves, so the wave height in the Gulf is relatively small.

4 Summary

Based on the reanalysis wind field and Jelesnianski model, the reconstructed wind field is constructed and used to drive the wave model SWAN. 20 years of wind and wave data in the computational area (5–25°N, 105–125°E) are obtained, and the simulated data are verified to be reliable by comparing with the measured data, Based on the 20-year data, the P-III distribution is used to calculate the extreme wind and wave characteristics around Hainan Island.

Wind and wave characteristics are analyzed under the extreme environment. The maximum wind speed and wave height mainly occur in the northeast side of Hainan Island and the southwest side of the Xisha Islands. The maximum wind speed for the 5-, 10-, 25-, and 50-year return periods reached 32 m/s, 34 m/s, 45m/s, and 55 m/s. And the maximum wave heights for the 5-, 10-, 25-, and 50-year return periods reach 9 m, 10 m, 12 m, and 13 m, respectively.

Table 4. Comparison of wave height fitting effects at characteristic points and calculation of waves with different return periods around Hainan Island.

Wave	Extremal Distribution	$\hat{D}_n$	*Dn* (0.05)	RMSE	AICC	Wave height for different return periods (m/s)			
						50	25	10	5
P1	Gumbel	0.1258	0.2941	0.0604	68.20	8.06	7.37	6.44	5.70
	Weibull	0.1848		0.0716	70.03	7.27	6.94	6.42	5.90
	P-III	0.1038		0.0397	71.99	8.60	7.85	6.81	5.93
P2	Gumbel	0.1517	0.2941	0.0629	40.46	4.31	3.97	3.50	3.14
	Weibull	0.1913		0.0772	42.55	3.91	3.75	3.49	3.24
	P-III	0.1199		0.0451	43.65	4.59	4.21	3.68	3.23
P3	Gumbel	0.1365	0.2941	0.0634	45.85	5.14	4.72	4.17	3.73
	Weibull	0.1546		0.0834	49.91	4.67	4.48	4.17	3.86
	P-III	0.1241		0.0398	47.90	5.54	5.04	4.35	3.80
P4	Gumbel	0.104	0.2941	0.0475	42.11	4.89	4.52	4.01	3.62
	Weibull	0.179		0.0768	46.43	4.48	4.31	4.02	3.74
	P-III	0.0854		0.0325	44.40	5.18	4.75	4.15	3.68
P5	Gumbel	0.1335	0.2941	0.0681	48.70	5.97	5.53	4.94	4.47
	Weibull	0.1858		0.0785	53.00	5.47	5.27	4.95	4.62
	P-III	0.1268		0.0425	52.19	6.41	5.89	5.17	4.59
P6	Gumbel	0.1483	0.2941	0.0694	34.31	4.54	4.26	3.88	3.57
	Weibull	0.1663		0.0604	34.70	4.15	4.04	3.85	3.66
	P-III	0.0952		0.0383	38.07	4.64	4.40	4.03	3.71
P7	Gumbel	0.1137	0.2941	0.0467	46.97	5.44	5.01	4.45	4.00
	Weibull	0.1669		0.0655	51.57	5.01	4.80	4.47	4.15
	P-III	0.1113		0.0401	50.27	5.53	5.12	4.56	4.08
P8	Gumbel	0.1417	0.2941	0.0578	66.31	8.58	7.87	6.92	6.16
	Weibull	0.1574		0.0699	72.97	8.03	7.65	7.04	6.44
	P-III	0.1279		0.0538	69.99	8.52	7.90	7.01	6.27
P9	Gumbel	0.1083	0.2941	0.0445	63.82	8.36	7.73	6.88	6.20
	Weibull	0.1396		0.043	68.47	7.78	7.47	6.96	6.45
	P-III	0.1111		0.0311	67.51	8.10	7.64	6.95	6.34
P10	Gumbel	0.1601	0.2941	0.0569	64.37	8.69	8.05	7.20	6.52
	Weibull	0.1691		0.0624	68.47	8.05	7.75	7.25	6.76

(*continued*)

Table 4. *(continued)*

Wave	Extremal Distribution	$\hat{D}_n$	Dn (0.05)	RMSE	AICC	Wave height for different return periods (m/s)			
						50	25	10	5
	P-III	0.1254		0.0479	67.68	8.45	7.97	7.26	6.65
P11	Gumbel	0.1132	0.2941	0.0452	67.17	8.84	8.15	7.21	6.47
	Weibull	0.1908		0.0681	71.68	8.16	7.82	7.27	6.73
	P-III	0.1322		0.0431	70.09	8.80	8.18	7.30	6.57
P12	Gumbel	0.0907	0.2941	0.0391	71.07	9.37	8.59	7.54	6.71
	Weibull	0.1416		0.0584	75.87	8.62	8.23	7.60	6.98
	P-III	0.0942		0.0321	73.93	9.61	8.82	7.73	6.82

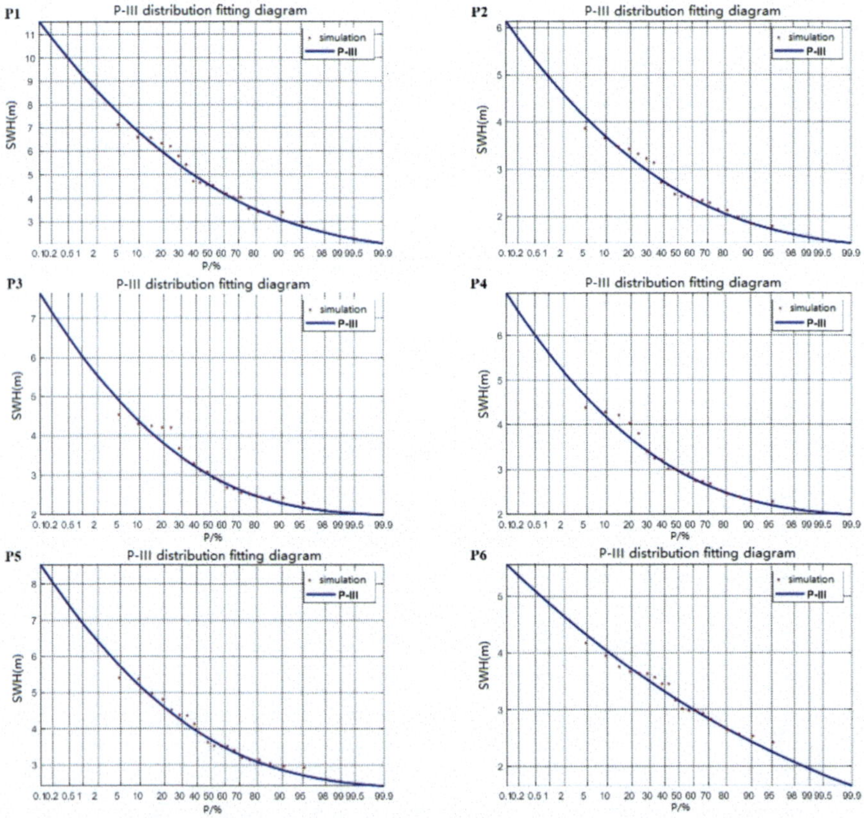

Fig. 10. P-III distribution fitting diagrams of wave height at 12 characteristic points around Hainan Island.

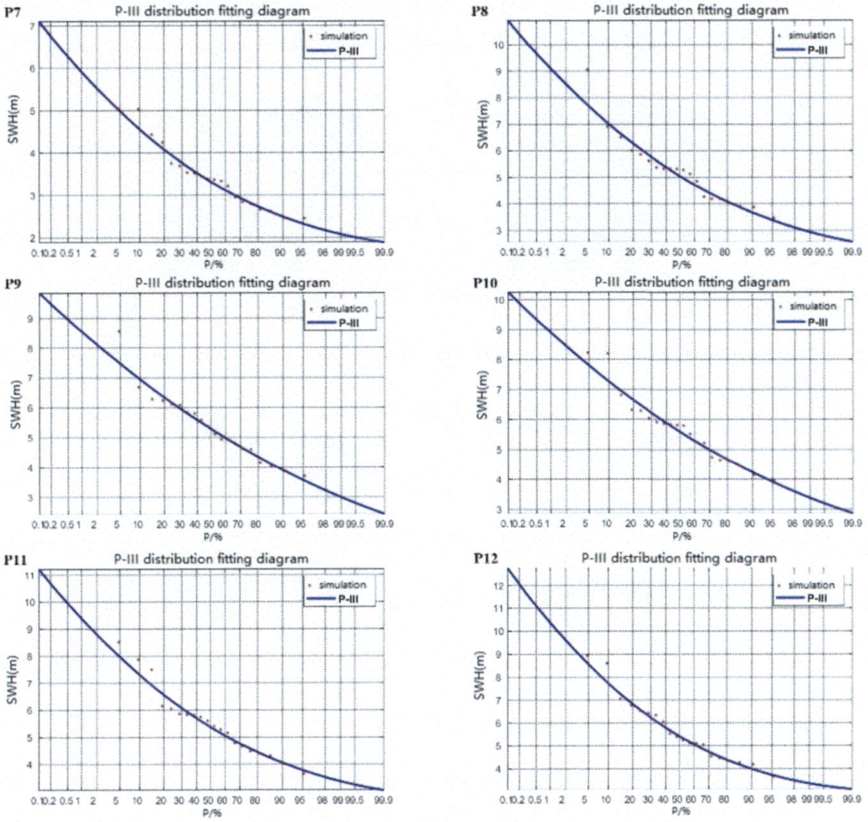

Fig. 10. *(continued)*

This study contributes to the scientific and rational assessment and utilization of wind and wave energy resources. And it provides reference value for the rational design and arrangement of facilities for ocean engineering to improve their disaster resistance and stability. However, limited by the computational capacity and cost, longer time and higher precision data computation can be carried out in order to make a more accurate analysis.

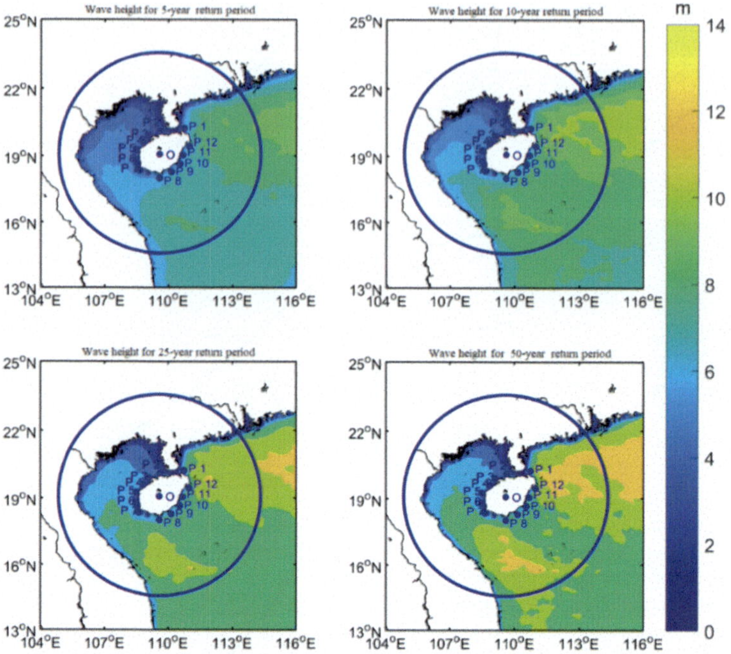

Fig.11. Wave height for 5-year, 10-year, 25-year, and 50-year return periods around Hainan Island (unit: m).

Acknowledgements. This study was funded by the National Natural Science Foundation of China (No. 42377457). We would like to thank Marine Big Data Center of Institute for Advanced Ocean Study of Ocean University of China for data conversion and archiving, and the professional and technical services provided by Yujie Dong.

References

1. Shashank, V.G., Sriram, V., Sannasiraj, S.A.: Improvements in wind field hindcast for storm surge predictions in the Bay of Bengal: a case study for the tropical cyclone Varadah. Appl. Ocean Res. **127**, 103324 (2022)
2. Pan, Y., Chen, Y.P., Li, J.X., et al.: Improvement of wind field hindcasts for tropical cyclones. Water Sci. Eng. **9**, 58–66 (2016)
3. Islek, F., Yuksel, Y., Sahin, C., et al.: Long-term analysis of extreme wave characteristics based on the SWAN hindcasts over the Black Sea using two different wind fields. Dyn. Atmos. Oceans **94**, 101165 (2021)
4. Wu, Z.Y., Chen, J., Jiang, C.B., et al.: Simulation of extreme waves using coupled atmosphere-wave modeling system over the South China Sea. Ocean Eng. **221**, 108531 (2021)
5. Yin, H.Q.: Typhoon wave forecast in the South Sea by SWAN. Dalian University of Technology (in Chinese) (2014)
6. Lin, Y.F., Liu, Y.F., Wang, X.H., et al.: Comparative study of three kinds of sea wind field data for the wind energy utilization in the China adjacent sea. Trans. Ocean. Limnol. **45**(06), 1–11 (in Chinese) (2023)

7. Wang, J.C., Liu, J.C., Wang, Y., et al.: Spatiotemporal variations and extreme value analysis of significant wave height in the South China Sea based on 71-year long ERA5 wave reanalysis. Appl. Ocean Res. **113**, 102750 (2021)
8. Wang, L.Q., Xia, Y.Q., Liang, B.C., et al.: Basic research on the characteristics of hazardous waves in the South China Sea. Haiyang Xuebao **41**(3), 23–34 (in Chinese) (2018)
9. Jelesnianski, C.P.: A numerical calculation of storm tides induced by a tropical storm impinging ON a continental shelf. Mon. Weather Rev. **93**(6), 343–358 (1965)
10. Jelesnianski, C.P.: Numerical computations of storm surges without bottom stress. Mon. Weather Rev. **94**(6), 379–394 (1966)
11. Mei, C.J., Zhao, D.L., Shi, J.: The analysis of Yellow Sea wave with WAVEWATCH and SWAN models. Mar. Forecast. **25**(02), 92–98 (in Chinese) (2008)
12. Muraleedharan, G., Rao, A.D., Rao, et al.: Modified Weibull distribution for maximum and significant wave height simulation and prediction. Coast. Eng. **54**(08), 630–638 (2007)
13. Wang, X.N., Chen, X.F.: Calculation of return period for storm surge at several stations along the coast of China. Mar. For. Serv. **1**, 18–25 (1984)
14. Isaacson, M.S.Q., MacKenzie, N.G.: Long-term distributions of ocean waves: a review. J. Waterw. Port, Coast. Ocean Div. **107**(2), 93–109 (1981)
15. Zhang, X.Y., Cheng, W.H.: Statistical inference for binomial-generalized paretocompound extreme value distribution model. Acta Math. Appl. Sin. **35**(03), 560–572 (2012)
16. Brewer, J.M., Butler, A., Cooksley. L.S.: The relative performance of AIC, AICsubC/sub and BIC in the presence of unobserved heterogeneity. Methods Ecol. Evol. **7**(6), 679–692 (2016)

Research on Automatic Identification of Atmospheric Fronts Based on Target Detection Network

Shaoyu Hou[1,2,3], Wenxia Yang[1,2,3(✉)], Chaoyong Tu[4], and Xinya Ding[5]

[1] China Meteorological Administration Xiong'an Atmospheric Boundary Layer Key Laboratory, Xiong'an, China
ywx418@163.com

[2] Hebei Key Laboratory of Meteorology and Ecological Environment, Shijiazhuang, China

[3] Weather Modification Center of Hebei Province, Shijiazhuang, China

[4] Yunnan R&D Institute of Natural Disaster On Chengdu, University of Information Technology, Kunming, China

[5] College of Meteorological and Ocean, National University of Defense Technology, Changsha, China

Abstract. Severe weather phenomena are often associated with atmospheric fronts, making their identification crucial for weather analysis and forecasting. However, current machine learning methods for identifying atmospheric fronts face challenges, including a mismatch between the input data and the network, which increases training difficulty and affects accuracy. Additionally, these methods only provide category information of grid points without specifying the front they belong to, causing ambiguity when similar fronts are close together. In order to address these deficiencies, this paper proposes a Cage R-CNN architecture based on a target detection network for automatically identifying atmospheric fronts. The proposed method introduces a fusion module before the R-CNN network, which reduces conflicts between multi-meteorological factors by adjusting their weights, enhancing the identification of weak frontal features. Experimental results demonstrate that adding the fusion module to the R-CNN network improves overall performance by about 6%, making the target detection network more effective for front identification tasks. This advancement enhances the generalizability of machine learning applications to other specific tasks.

Keywords: Atmospheric Front · Automatic Identification · Cage · Target Detection Network · Cage R-CNN Network

1 Introduction

An atmospheric front is the interface or transition zone between two air masses of different densities. It forms when air masses with different thermodynamic properties meet, usually in low-pressure troughs. The significant baroclinicity near the front often leads to severe weather changes. For example, areas near frontal systems are often accompanied

C. Weng (Ed.): WRE 2024, LNCE 634, pp. 316–325, 2025.
https://doi.org/10.1007/978-981-96-5095-8_26

by heavy rainfall and storms, frequently serving as the center of thunderstorms. Statistics show that frontal precipitation accounts for 51% of global extreme precipitation [1], with up to 90% of extreme precipitation events related to mid-latitude fronts [2, 3]. Therefore, accurate identification of fronts is beneficial for weather analysis and forecasting and has significant implications for socio-economic life and even military operations.

Traditional front identification methods mainly involve numerical front analysis (NFA), which relies on selected parameters and manually set functions and thresholds to diagnose fronts. These methods include the gradient method (G_τ), thermal parameter method (TFP) [4], wind shift method (WV) [5], and F diagnostic method (F). The method calculates the gradient of the parameter τ directly, determining front results through a threshold, making it the simplest method but with the lowest accuracy. The TFP method further derives the method, defining the front as the directional derivative of the thermodynamic quantity τ along its horizontal gradient direction. However, like the TFP method, it requires additional derivatives to determine the warm edge of the front formation area as a front, and substituting potential temperature with other parameters degrades the identification performance [6]. Recently, a method based on the dynamic state index (DSI) of non-adiabatic processes has been proposed to diagnose fronts, but it still suffers from the limitations of NFA methods. It relies on a single physical model and threshold to identify fronts, easily missing fronts with weak meteorological factor diagnostic values due to special weather conditions and unique terrain or misidentifying non-front meteorological factor features with high diagnostic values as front features.

Machine learning, particularly convolutional neural networks (CNNs), has seen significant progress in atmospheric front identification [7, 8]. However, methods like those of McGovern and Lagerquist [9], Biard et al. [10] face challenges due to feature conflicts among multi-meteorological inputs, leading to inaccurate identification, especially under special weather conditions. This paper proposes an automatic front identification method using a target detection network with a fusion module that effectively resolves these feature conflicts, improving accuracy. The paper is structured as follows: Sect. 2 presents the ablation and comparative experimental results, Sect. 3 discusses the network model and experiments, and Sect. 4 details the methodology.

2 Methods

2.1 Dataset and Preprocessing

In this study, the fifth-generation global atmospheric reanalysis dataset produced by the European Centre for Medium-Range Weather Forecasts (ECMWF), known as ERA5, is used as the input for the network. Surface bulletins from the Weather Prediction Center (WPC) of the United States are used to create the training labels for the model. This section briefly describes the ERA5 reanalysis data and the meteorological factors used as network inputs. Additionally, it details the format of the frontal data obtained from the WPC surface bulletins and the preprocessing methods applied to this data.

ERA5 Reanalysis Data. The ERA5 reanalysis dataset, provided by the European Centre for medium-range Weather Forecasts (ECMWF), offers comprehensive global atmospheric data, including temperature (T), dew point temperature (Td), specific humidity (Q), and mean sea level pressure (Pmsl) at 850 hPa, with a spatial resolution of 0.25° × 0.25° and temporal resolution of 3 h, spanning from 1979 to the present. For training, validation, and testing of the model, the dataset is partitioned as follows: The training set includes data from 0000 UTC January 1 2010 to 2100 UTC December 28 2011 (96, 014 fronts) and from 0000 UTC January 1 2015 to 2100 UTC December 28 2018 (242, 584 fronts); the validation set covers 0000 UTC January 1 2012 to 2100 UTC December 28 2013 (112, 523 fronts); and the testing set is from 0000 UTC January 1 2014 to 2100 UTC December 28 2014 (49, 843 fronts).

WPC Fronts and Their Preprocessing. Manually marked fronts from the Weather Prediction Center (WPC), represented by discrete latitude and longitude pairs, were used, ensuring consistency with ERA5 reanalysis data. The data underwent three preprocessing steps: Bilinear interpolation of the WPC front data into a gridded format, expansion of front grid points by marking adjacent non-front points as the front, and creation of bounding boxes for the expanded front areas. The label expansion process is depicted in Figs. 1 and 2 illustrates the overall preprocessing workflow.

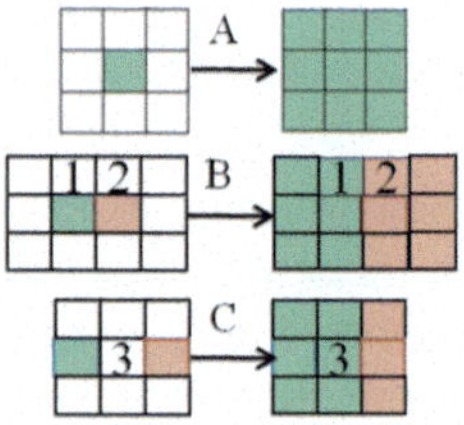

Fig. 1. WPC Frontal extension method. Green indicates cold fronts and red indicates warm fronts.

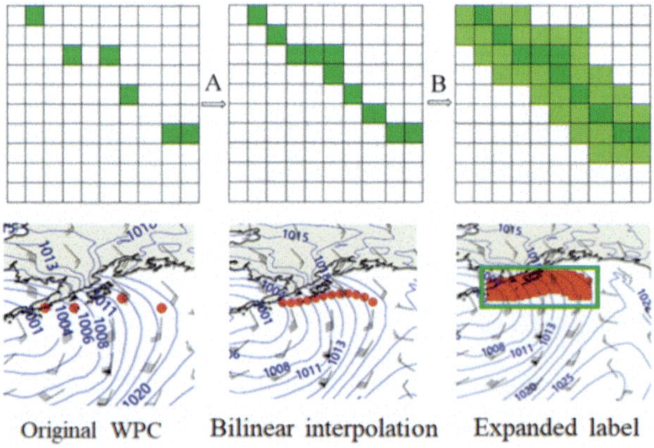

Fig. 2. Schematic and visualization of the WPC front preprocessing. The top panel illustrates the preprocessing steps, while the bottom panel shows the corresponding visualization.

2.2 Network Architecture

Given the current machine learning methods applied to automatic atmospheric front identification, there is a lack of consideration for the compatibility between network inputs and the network itself, leading to training difficulties. Additionally, a network architecture that integrates a fusion module into a target detection network is proposed to avoid the attribution issues present in semantic segmentation networks. The structure of this architecture is shown in Fig. 3, with the purple dashed box representing the fusion module and the green dashed box representing the R-CNN.

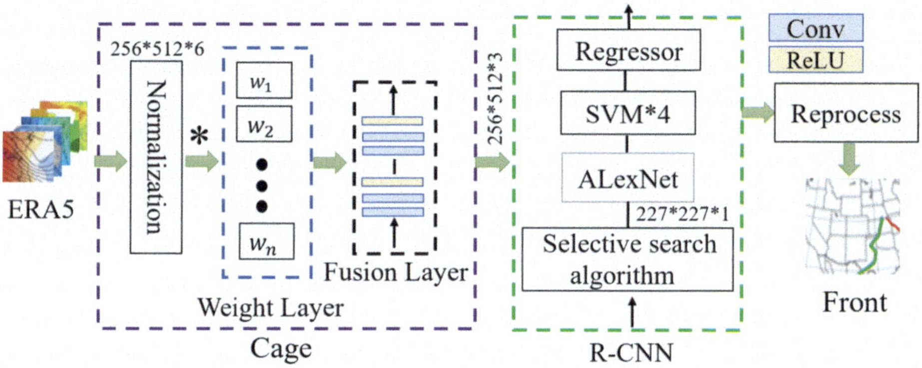

Fig. 3. Schematic diagram of the network architecture.

Cage. A fusion module was designed to improve the network's handling of multi-meteorological factor data for automatic atmospheric front identification. It consists of a normalization layer, weight layer, and fusion layer, shown in the purple dashed box in Fig. 3. The normalization layer standardizes input data from six meteorological factors using min-max normalization to ensure consistent value ranges, speed up convergence, and prevent issues like gradient vanishing or explosion. The calculation formula for min-max normalization is as follows:

$$x' = \frac{x - x_{min}}{x_{max} - x_{min}} \tag{1}$$

where x is the original data value, x_{min} and x_{max} are the minimum and maximum values of the dataset, and x' is the normalized value.

The weight layer of the fusion module assigns variable weights to grid points in multi-meteorological data, adjusting during training to reduce misidentification, such as distinguishing non-frontal temperature features. For example, grid points in coastal regions with large temperature differences may receive lower weights, while areas with weaker differences are given higher weights. The weight values for the six types of meteorological factors at each grid point are adjusted according to the following formula:

$$w_{ij}^{m} = e^{\lambda_{ij}^{m}} \Big/ \sum\nolimits_{n=1}^{6} e^{\lambda_{ij}^{n}} \tag{2}$$

where m represents the six types of meteorological factors, ij represents the grid point position in the recognition area, w_{ij}^{m} represents the weight value of the six types of meteorological factors, λ_{ij}^{m} and λ_{ij}^{n} represent the data values of the six types of meteorological factors.

The fusion layer improves training and recognition accuracy by performing multiple convolutions on the six meteorological factors to extract comprehensive features. It consists of two sets of convolutional layers with ReLU activation, transforming the 6-channel input into a 3-channel output to match the target detection network's input, enhancing segmentation performance.

R-CNN. The R-CNN network consists of four components: the Selective Search (SS) algorithm, AlexNet, an SVM classifier [11], and a regressor. SS generates 2000 candidate boxes based on meteorological features. AlexNet extracts 4096-dimensional feature vectors from these boxes. The fully connected layers of AlexNet are used without the Softmax layer, and the boxes are resized to 227 × 227 pixels for input. SVM classifiers predict the probability of each front type, and Non-Maximum Suppression (NMS) removes overlapping boxes. Finally, regressors refine the remaining boxes, outputting the highest-scoring bounding box for each category. The translation (t_x, t_y) and scale (t_w, t_h) adjustments for the regressor-corrected candidate box positions are calculated using the following formulas:

$$t_x = \frac{G_x - P_x}{P_w},\ t_y = \frac{G_y - P_y}{P_h} \tag{3}$$

$$t_w = \log\left(\frac{G_w}{P_w}\right),\ t_h = \log\left(\frac{G_h}{P_h}\right) \tag{4}$$

where G represents the ground truth, and P denotes the CNN features corresponding to the candidate box, the objective function is expressed as follows:

$$d_*(P) = w_*^T \Phi_5(P) \tag{5}$$

where $\Phi_5(P)$ represents the predicted feature vector, w_*^T are the parameters to be learned, and $d_*(P)$ is the obtained predicted value.

The network's loss function, derived by comparing the predicted values and ground truth, is defined as follows:

$$Loss = \sum_i^N \left(t_*^i - \hat{w}_*^T \phi_5(P^i)\right)^2 \tag{6}$$

$$W_* = \arg_{w_*} \min \sum_i^N (t_*^i - \hat{w}_*^T \varphi_5(P^i))^2 + \lambda||\hat{w}_*||^2 \tag{7}$$

3 Results

Ablation experiments with the fusion module were conducted on the ERA5 dataset and WPC frontal data to evaluate performance using multiple metrics and visualization analysis. The experiments were run on an Intel Xeon W-2102 CPU, 32GB RAM, and Nvidia 3080 GPU, using Python and Pytorch. The network was trained with a batch size of 32, adaptive learning rate optimization, and hyper-parameter optimization via Bayesian Optimization, which involved iterating through model training and cross-validation to refine hyper-parameters.

3.1 Ablation Experiment Results

Ablation experiments were conducted on the R-CNN [12] network with and without the fusion module, using POD, CSI, SR, MPA, and F1 Score as evaluation metrics. The results (Fig. 4) showed that adding the fusion module improved the average of the five metrics by about 6%, demonstrating its effectiveness in the target detection network.

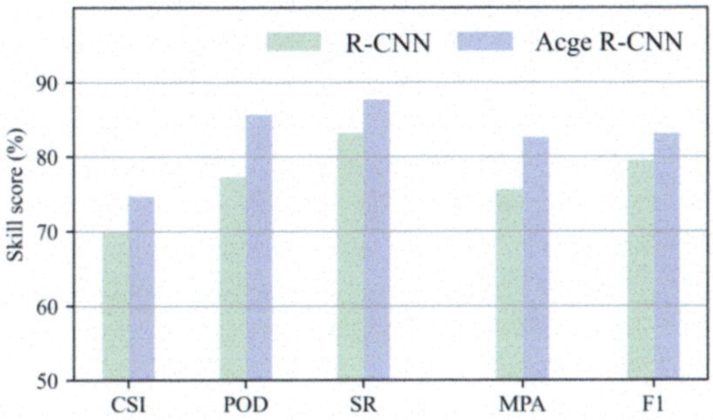

Fig. 4. MCA results of ablation experiments.

Ablation experiments on the R-CNN network were conducted to visualize the impact of the fusion module at 15:00 UTC on September 10 2009, as shown in Fig. 5. A comparison of Fig. 5a and b reveals that the fusion module improved identification by correctly identifying the front within the black dashed box and filtering out the misleading front in the blue dashed box. This demonstrates the module's effectiveness in reducing feature conflicts and enhancing weak frontal features while suppressing strong non-frontal features caused by special weather conditions or geographical factors.

Comparison experiments were conducted on CNN, FCN, UNet, and R-CNN to verify the superiority of R-CNN. Figure 6 shows the loss curves of the four networks, with R-CNN achieving the fastest convergence, completing training in about 30 epochs. This demonstrates that R-CNN outperforms the other networks in recognition accuracy and convergence speed.

The results of the multi-indicator scores are shown in Fig. 7. From it, we can see that the R-CNN network has the highest score in each metric, which indicates the superior performance of the R-CNN network.

3.2 Comparison Results with Multiple Methods

The effectiveness of the proposed method was evaluated by comparing it with several front identification methods using POD, CSI, and SR metrics [13]. The results, shown in Table 1, highlight the performance of the proposed method relative to others.

From Table 1, it is evident that the G_τ and TFP methods score quite low across multiple metrics. This is because these methods rely on a single function and cannot fit the complex patterns of frontal systems. Additionally, the determination of thresholds

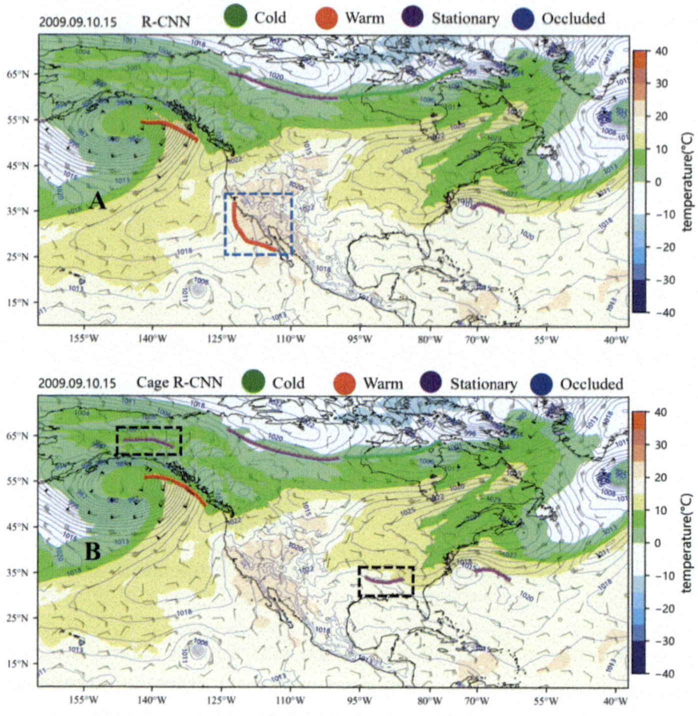

Fig. 5. Visualization of the ablation experiment to identify fronts.

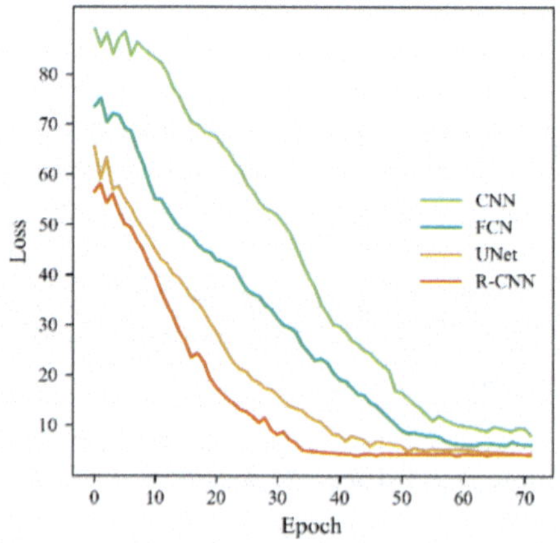

Fig. 6. Comparison of network loss during training of four network methods.

is subjective and challenging, often leading to misclassification of non-front values as fronts under special weather conditions and geographical locations, causing a high rate of false positives.

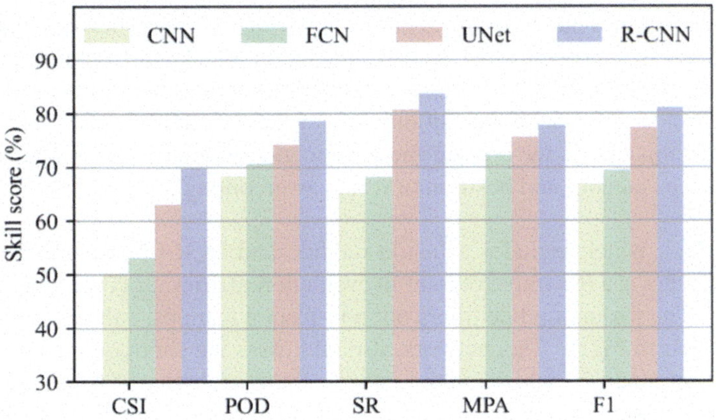

Fig. 7. Comparison of the scores of the four types of networks on multiple metrics.

Table 1. Comparison of Skill Scores across Multiple Methods.

Method	CSI (%)	POD (%)	SR (%)	MPA (%)	F_1 (%)
G_τ	44.1	65.6	57.3	58.9	62.3
TFP	48	69.6	61.8	61.8	63.5
F-diagnosis	50.7	71.2	63.7	63.4	66.7
CNN	52	73.3	65	67	70.9
FCN	63	75.2	79.5	72.2	75.8
Unet	66.9	77.3	83.2	75.6	79.5
Cage R-CNN	75.7	84.7	87.7	82.6	83.1

The F diagnostic method, which integrates thermal and wind parameters, shows significant improvement over the single-parameter methods G_τ and TFP. This improvement is primarily due to considering the systematic nature of multiple meteorological factors.

Machine learning methods outperform traditional front analysis by incorporating multi-meteorological inputs, enhancing identification accuracy. However, semantic segmentation networks struggle with feature conflicts, limiting performance. The proposed method improves this by using a fusion module to resolve input-network mismatches and a target detection network (R-CNN) instead of segmentation. This approach directly identifies both the category and position of fronts, overcoming attribution issues and sample imbalance present in semantic segmentation networks.

4 Discussion

Current machine learning methods for front identification face challenges such as mismatched inputs, weak frontal features, and feature conflicts, particularly under special weather conditions [14]. To address this, we propose a fusion module that resolves feature conflicts and integrates meteorological data, improving training efficiency and accuracy. This is particularly important when weak frontal features and non-frontal meteorological factors interfere with model performance. Semantic segmentation networks like CNN, FCN, and UNet struggle with attribution issues when multiple fronts of the same type are nearby. In this study, we replace these networks with an R-CNN network, which outputs both the category and position of fronts, overcoming attribution issues and label imbalance.

Our method, which includes a fusion module and an R-CNN-based target detection network, shows a 6% improvement in evaluation metrics, as confirmed by ablation experiments. The fusion module filters out non-frontal features, enhancing front identification, as seen in Fig. 6. This network also addresses the shortcomings of traditional numerical front analysis, which cannot capture the complexity of frontal systems and suffers from threshold-based misclassifications. Comparative analysis shows that the R-CNN model outperforms other machine learning methods, as it avoids the attribution and imbalance issues inherent in semantic segmentation networks. The R-CNN's ability to output both categorical and positional information makes front identification more accurate and efficient. R-CNN's ability to process grid points without ambiguity is also a significant advantage.

However, R-CNN networks are computationally intensive, requiring substantial feature extraction and classification resources. While this results in longer training and detection times compared to semantic segmentation networks, using high-performance hardware (e.g., GPUs or TPUs) and optimization techniques such as distributed training can significantly speed up the process.

Future work will focus on replacing the R-CNN with a more efficient Fast R-CNN network and refining the fusion module to improve performance. This will enhance both the speed and accuracy of front identification and offer a robust framework for applying machine learning to similar tasks.

Acknowledgements. This work is supported by the Class B Pilot Project of the Chinese Academy of Sciences titled "Physical and Chemical Exploration of the Atmosphere in Extreme Meteorological Environments," the Hebei Meteorological Bureau's scientific research project 23ky08, and the Research Fund of Chengdu University of Information Technology under Grant KYTZ376016.

References

1. Parfitt, R.L., Czaja, A., Seo, H.: A simple diagnostic for the detection of atmospheric fronts. Geophys. Res. Lett. **44**(9), 4351–4358 (2017)
2. Catto, J.L., Pfahl, S.: The importance of fronts for extreme precipitation. J. Geophys. Res. Atmos. **118**, 10791–10801 (2013)
3. Dowdy, A.J., Catto, J.L.: Extreme weather caused by concurrent cyclone, front and thunderstorm occurrences. Sci. Rep. **7**, 40359 (2017)
4. Renard, R., Clarke, L.: Experiments in numerical objective frontal analysis. Mon. Wea. Rev. **93**, 547–556 (1965)
5. Simmonds, I., Keay, K., Bye, J.: Identification and climatology of Southern Hemisphere mobile fronts in a modern reanalysis. J. Climate **25**, 1945–1962 (2012)

6. Schemm, S., Rudeva, I., Simmonds, I.: Extratropical fronts in the lower troposphere—global perspectives obtained from two automated methods. Quart. J. Roy. Meteor. Soc. **141**, 1686–1698 (2015)
7. Fukushima, K.: Neocognitron: a self-organizing neural network model for a mechanism of pattern recognition unaffected by shift in position. Biol. Cybern. **36**, 193–202 (1980)
8. Fukushima, K., Miyake, S.: Neocognitron: a new algorithm for pattern recognition tolerant of deformations and shifts in position. Pattern Recogn. **15**, 455–469 (1982)
9. McGovern, A., Lagerquist, R., Gagne, D.J.: Making the black box more transparent: understanding the physical implications of machine learning. Bull. Am. Meteor. Soc. **100**(11), 2175–2199 (2019)
10. Biard, J.C., Kunkel, K.E.: Automated detection of weather fronts using a deep learning neural network. Adv. Stat. Climatol., Meteorol. Ocean. **5**(2), 147–160 (2019)
11. Krizhevsky, A., Sutskever, I., Hinton, G.: Imagenet classification with deep convolutional neural networks. Adv. Neural Inf. Process. Syst. **25**(2) (2012)
12. Uijlings, J.R., Sande, K.E., Gevers, T.: Selective search for object recognition. Int.: J. Comput. Vis. **2013**(2) (2013)
13. Niebler, S., Miltenberger, A., Schmidt, B.: Automated detection and classification of synoptic-scale fronts from atmospheric data grids. Weather. Clim. Dyn. **3**(1), 113–137 (2022)
14. Klanar, G., Zdear, A., Krishnan, M.: Robot navigation based on potential field and gradient obtained by bilinear interpolation and a grid-based search. Sensors **22**, (2022)

Mix Proportion Optimization of Hydraulic Asphalt Concrete Using the Orthogonal Experimental Design Method

Chunyan Wang[1,2](✉), Rengui Jiang[1], Liangshu Yu[1], and Changhong Yang[1]

[1] Xi'an University of Technology, Xi'an Shaanxi, China
416896247@qq.com
[2] Xi'an Thermal Power Research Institute Co. Ltd., Xi'an Shaanxi, China

Abstract. Hydraulic asphalt concrete is one of the important materials for interior core walls or face panel construction, and it has been widely used in hydraulic engineering, such as reservoirs, dams, or hydropower stations. The mix proportion of hydraulic asphalt concrete is vital for the construction quality of hydraulic asphalt concrete, and the construction quality of asphalt concrete core wall can be ensured by optimizing the content of hydraulic asphalt, the amount of filler, and the aggregate grading index. Because of the traditional single factor analysis method, the selection of the mixed proportion of hydraulic asphalt concrete is time-consuming and laborious, and the results are not satisfactory. The orthogonal experimental design method is applied to the mix proportion optimization of hydraulic asphalt concrete, and the results are processed using range and variance analysis methods to analyze the influences of different oil-stone ratios, filler dosage, and grading index on the optimization efficiency of the mix proportion of hydraulic asphalt concrete. The optimal mix proportion of hydraulic asphalt concrete was obtained using the oil-stone ratio, filler dosage, and grading index as independent variables, and the porosity and separation of asphalt concrete as dependent variables. The results indicate that the orthogonal experimental design method is helpful in improving the optimization efficiency of the mix proportion of hydraulic asphalt concrete.

Keywords: Hydraulic Asphalt Concrete · Orthogonal Experimental Design · Mix Proportion · Range Analysis · Variance Analysis

1 Introduction

The pumped storage power stations have developed rapidly in recent years in China. The pumped storage power station pumps water to the upper reservoir when the electric energy is in the low load, and then releases water to the lower reservoir to generate electricity in the peak load [1]. Hydraulic asphalt concrete has excellent anti-seepage and deformation properties, which have been extensively used in hydraulic engineering, such as impervious interior core walls or face panels in pumped storage power station project [2]. With the wide application of the orthogonal experimental design (OED) method in

C. Weng (Ed.): WRE 2024, LNCE 634, pp. 326–337, 2025.
https://doi.org/10.1007/978-981-96-5095-8_27

the hydraulic asphalt concrete experiment, the advantages of the OED method are more and more prominent. In the selection test of asphalt concrete mix ratio, the workload is larger to obtain an ideal asphalt concrete mix ratio. Most of the traditional asphalt concrete mix ratio is determined by the single factor analysis method, which is time-consuming and laborious, but the results are not often ideal. Asphalt, coarse aggregate, fine aggregate, and filler have been determined in advance before the mix ratio design of hydraulic asphalt concrete, so the key to the selection of mix ratio lies in the selection of oil-stone ratio, filler content, and aggregate grading index, which are considered as independent variables, and the influence of independent variable changes on porosity, separation, and other indicators.

The OED method scientifically arranges and analyzes multi-factor tests using an orthogonal table. It is one of the most used methods to analyze multi-factor and multi-level experiment designs. This method uses a set of normalized tables named "orthogonal table" to arrange the experiment and uses the principle of mathematical statistics to process and analyze the test results scientifically. The main advantage is that it can select a few representative test schemes among many test schemes and select the best scheme through the results of different schemes. At the same time, it can obtain more information than the test results with in-depth analysis.

The OED method is commonly used in designing highway asphalt concrete ratios but is less frequently applied in tests for hydraulic asphalt concrete ratio selection. For example, Chen et al. (2022) used the orthogonal test method to the mix proportion design of high-content fly ash concrete. The influence law of the main variables in the mix proportion on the mechanical properties of concrete was obtained [3]. Wang et al. (2022) conducted an orthogonal experimental analysis of double-sided direct shear composite concrete test blocks, and the results show that the increase of additional normal force can significantly improve the bond shear strength [4]. Wang et al. (2021) analyzed the factors affecting the low-temperature performance of porous asphalt mixture through the OED method. The results showed that the modifier content was the most important factor [5]. Dong et al. (2022) used the OED method to determine the representative combination of prefabricated base structures in the construction process of asphalt pavement [6]. This paper uses the OED method to design and analyze the effects of asphalt content, filler content, and grading index on the porosity and other performance indexes of hydraulic asphalt concrete in selecting hydraulic asphalt concrete mix. By applying the OED method in selecting hydraulic asphalt concrete mix, better test results can be obtained with fewer test times, and the experiment efficiency can be improved. Therefore, it is of great importance and basic theoretical significance to investigate the optimization method of hydraulic asphalt concrete mix ratio.

2 Material and Methods

2.1 Mix Proportion Optimization Test

The selection of hydraulic asphalt concrete mix proportion is based on different materials, different levels of distribution index, different filler content, and oil-stone ratio to form a variety of different mix proportions, among which, oil-stone ratio refers to the ratio of asphalt quality in asphalt mixture and mineral quality in asphalt mixture, after density, porosity and segregation and other performance tests, select the mix proportion to meet the engineering requirements [7]. The density and porosity index can reflect the anti-seepage performance of asphalt concrete. The degree of segregation, the analysis method used to determine the density of asphalt concrete, is assessed by pouring a hot asphalt mixture into a φ100 mm × 100 mm test mold. This is done using self-weight compaction without the application of vibration. After the specimen is cooled to below 15 °C, the specimen is then cut into upper and lower parts. The density of two parts was measured at 20 °C. If the ratio of the density of the lower half compared with the upper half is less than 1.05, it can be considered homogeneous. The performance of hydraulic asphalt concrete is primarily influenced by the quality of raw materials, mineral grading, filler dosage, and oil-stone ratio.

Mineral grading parameters include maximum aggregate particle size D_{max}, grading index r or coarse and fine aggregate rate, and filler content F. The maximum density grading theory and grading formula were used to determine the mineral grading [8]. The equation is shown as follows:

$$P_i = F + (100 - F)\frac{d_i^r - d_{0.075}^r}{D_{\max}^r - d_{0.075}^r} \tag{1}$$

where, P_i represents the passing rate of sieve holes %. F indicates filler content with particle size less than 0.075 mm, %. D_{max} is the maximum particle size of aggregate, mm. d_i indicates a certain sieve hole size, mm. $d_{0.075}$ indicates the maximum size of the filler particles. r is the gradation index.

According to previous engineering and experimental research experiences, the maximum particle size $D_{max} = 19$ mm was selected. The mineral grading index r generally ranges from 0.30 to 0.36. Three gradation index factors, including 0.31, 0.34, and 0.37, were selected as research objects [9].

The mixed proportion primary experiment used porosity and separation tests. The porosity and separation values were the average values of specimens. The porosity calculation method is shown as follows:

$$P = (1 - \frac{\rho_0}{\rho}) \times 100\% \tag{2}$$

where P represents the porosity of asphalt concrete, %. ρ_0 is the density of asphalt concrete, g/cm^3. ρ indicates the maximum density of asphalt concrete, g/cm^3. The smaller the porosity value, the better the anti-seepage performance.

The separation degree value can be calculated using the density of the lower half of the specimen divided by the density of the top half of the specimen.

According to the specification requirements, a separation degree value below 1.05 is deemed qualified. The closer this value is to 1, the better it is.

2.2 Material Inspection

The experiment was carried out according to the *Test code for hydraulic asphalt concrete*, and the technical specifications were carried out according to the *Design specification of asphalt concrete facings and cores for embankment dams*. Raw material inspection includes coarse aggregate, fine aggregate, filler, and asphalt.

Coarse Aggregate. The quality inspection results of coarse aggregate are shown in Table 1.

Table 1. Performance results of the coarse aggregate of asphalt concrete.

Technical Index	Apparent Density (G/Cm^3)	Water Absorption (%)	Adhesion to Asphalt (Grade)	Durability (%)	Crushing Value (%)	Acid-Basic
Specification	≥ 2.6	≤ 2	≥ 4	≤ 12	≤ 30	–
Test value	2.761	0.766	5$^-$	0.43	16.41	alkali

Fine Aggregate. There are two kinds of fine aggregate. One is artificial aggregate, processed from coarse aggregate by crushing and screening in the laboratory into artificial sand with a particle size of 2.36 ~ 0.075 mm, and the other is river sand. Both artificial sand and river sand are tested for quality in the experiment. The quality inspection results of fine aggregate are shown in Table 2.

Table 2. Performance results of the fine aggregate of asphalt concrete.

Technical Index	Apparent Density (g/cm^3)	Water Absorption (%)	Adhesion to Asphalt (grade)	Durability (%)	Crushing Value (%)
Specification requirements	≥ 2.55	≤ 2	≤ 15	≥ 6	≤ 2
Artificial sand detection value	2.753	0.766	5.026	9.0	0
River sand detection value	2.683	0.300	1.320	5$^-$	14.8

Filler of Asphalt Concrete. The filler test results are shown in Table 3.

Table 3. Performance results of filler aggregate of asphalt concrete.

Technical Index	Density (G/Cm^3)	Moisture Content (%)	Hydrophilic	Coefficient Particle Composition Passing Rate (%)			
				0.6	0.3	0.15	0.075
Specification Requirements	≥ 2.5	≤ 0.5	≤ 1	100	–	> 90	> 85
Test Value	2.717	0.24	0.766	100	99.9	99.8	99.3

Asphalt. Karamay No. 70 asphalt was used in this experiment, and the results comply with specifications. The testing of coarse aggregate, fine aggregate, filler, and asphalt raw materials indicated that their performances meet the technical requirements of hydraulic asphalt concrete. Therefore, these materials can be used in the construction of hydraulic asphalt concrete panels.

2.3 Orthogonal Experimental Design Method

The OED method is a commonly used method to investigate multiple levels of factors. It uses a set of normalized tables named orthogonal tables for experiments and uses mathematical statistics methods to analyze the results. Its main advantage is that it can select representative schemes among many test schemes and finally select the best scheme. This method has been widely used in the highway asphalt concrete ratio design. However, it is rarely used in the ratio design for hydraulic asphalt concrete. The OED method uses mathematical statistics and the orthogonality principle to select appropriate representative points from large experimental data, utilizing the orthogonal table to arrange the experiment effectively [10]. The principle of the orthogonal table is balanced dispersion and neat comparability, characterized by the arrangement of the least number of tests and can reflect the law of changes in objective things. L represents the orthogonal table, $L_n(t^m)$ represents the specific orthogonal table form, where n represents the number of tests, m represents the maximum number of factors that can be arranged, and t represents the number of levels of factors. The flowchart of the OED method is shown in Fig. 1.

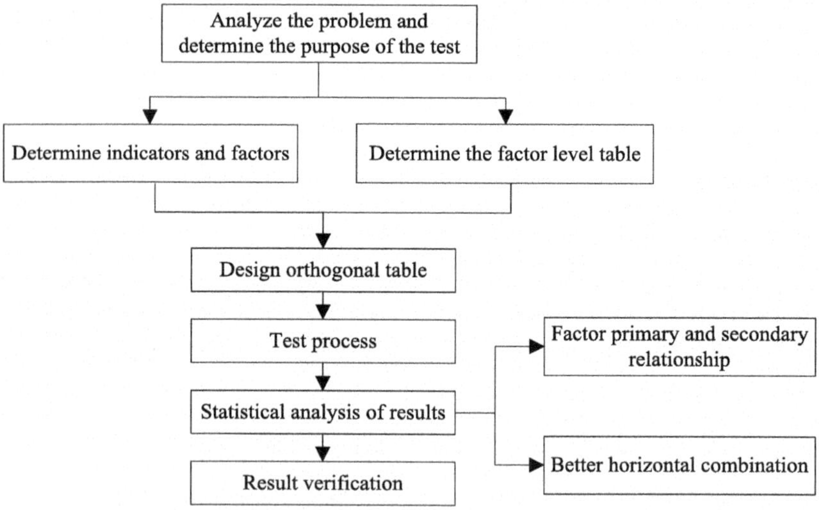

Fig. 1. Flowchart of the OED method.

2.4 Range Analysis Method

Range analysis is one of the most used methods for the study of orthogonal test results, which can determine the optimal and optimal level combination of factors, determine the primary and secondary order of influencing factors, and draw the trend chart of factor levels and indicators. It has the characteristics of a simple calculation process, an intuitive image, simple and easy to understand [11]. The range analysis method includes two steps of calculation and judgment. The advantages of the range analysis method include being intuitive, simple in calculation, and convenient in application. The flowchart of the range analysis method is shown in Fig. 2.

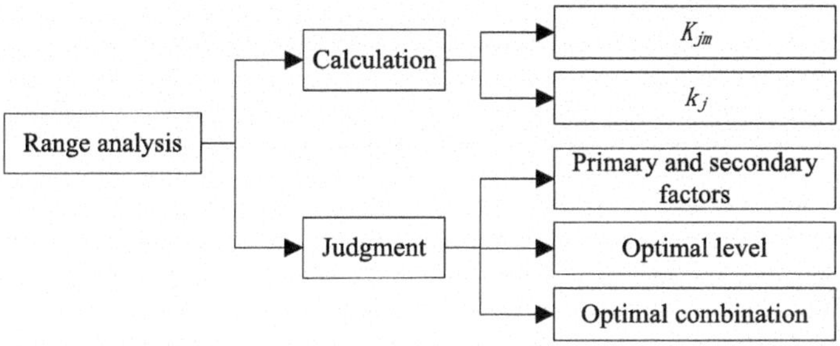

Fig. 2. Flowchart of the range analysis method.

$L_n(t^m)$ is used to design the calculation scheme. Assume the result of the calculation scheme i be y_i, $i = 1, 2, 3, \ldots, n$. Calculate the sum of the calculation results in y_i at the i level in column j of the orthogonal table respectively, denoted as K_{1j}, K_{2j}, K_{3j}, etc., $i = 1, 2, 3, \ldots t$, $j = 1, 2, 3 \ldots, m$. Range refers to the difference between the maximum and minimum values of K_{1j}, K_{2j}, and $K_{3j}, \ldots$, represented as R_j. Three levels are taken as examples to explain the meanings of K_{1j}, $K_{2j,}$ and K_{3j}. According to the nature of the orthogonal table, in the corresponding tests of K_{1j}, $K_{2j,}$ and K_{3j}, the occurrence times of other factors are the same except for the factor where column j is located. Therefore, it can be understood that other factors have roughly the same influences on K_{1j}, $K_{2j,}$ and K_{3j}, and the differences among K_{1j}, $K_{2j,}$ and K_{3j} can be attributed to the factor in column j taking on three different level values. Therefore, the values of K_{1j}, $K_{2j,}$ and K_{3j} can generally reflect the influence of the three levels of the factor on the index. The size of the corresponding range R_j represents the degree of influence of the factors on the indicator. The greater the range, the greater the influence of the factor level change on the experimental results. The factor with the largest range is also the most important, and the factor with the smaller range indicates that the factor has little impact on the index when it changes within the selected range.

2.5 Variance Analysis Method

Variance analysis is an important method to analyze experiment data in experimental research. It is mainly used as a statistical method to examine whether the mean values of multiple normal populations are equal under the condition of the same variance. The basic concept is to break down the sum of squares of data variation into the sum of variation squares from factors and random error. The F test can determine whether the effect of the factor is significant. Variance analysis can be divided into single-factor analysis of variance, double-factor analysis of variance, and multiple-factor analysis of variance according to the number of factors that affect the experimental index.

Ln(t^m) is used to design the calculation scheme. Set the result of calculation scheme i for y_i, $i = 1, 2, 3, \ldots n$, and $y_1, y_2, \ldots y_n$ is independent of each other and follows the normal distribution of homoscedasticity σ^2, which can be represented as $y_i \sim N(\mu_i, \sigma^2)$. The variance analysis is performed on y_i, that is, the hypothesis H_0: $\mu_1 = \mu_2 = \mu_3 = \ldots\ldots = \mu_n$ for the significance test.

As can be seen from the hypothesis test, the key is to construct the test statistics of the hypothesis. Then, the distribution of the statistics under the condition of H_0 can be obtained, and the negative domain of H_0 can be constructed. By judging whether the given sample falls into the negative domain, the hypothesis can be judged whether it is valid. The variance analysis method is one of the important and commonly used experimental data analyses, often used to explore the influences of various factors on experimental performance. In the experiment, the performance and yield indicators are known as experimental indicators. The conditions that affect the experiment are called factors, and the different states of factors are called levels.

3 Results and Discussion

3.1 Influencing Factors on Optimization Efficiency of Grade Grading of Hydraulic Asphalt Concrete

The quantities of asphalt and filler greatly influence the grading optimization efficiency of hydraulic asphalt concrete. According to the engineering experience, the actual situation of raw materials, and requirements of design specifications, the mix ratio of asphalt concrete with a fine aggregate of 50% artificial sand and 50% river sand is selected. According to the orthogonal table design requirements, this experiment conducted a total of nine groups of tests, among which there are three main factors affecting the optimization efficiency of grading of hydraulic asphalt concrete: grading index, filler content, oil-stone ratio, and referring to the test specification and the summary of previous practical engineering experience, select three commonly used different levels for each factor. Therefore, the orthonormal table L9(3^3) is adopted, including:

Factor *A* is the gradation index: 0.31, 0.34, 0.37. Factor *B* is the filler content: 12%, 14%, 16%. Factor *C* is the oil-stone ratio: 10%, 11%, 12%.

According to the number of factors and the number of levels to choose the appropriate orthogonal table, combined with the specific conditions of this experiment, the fine aggregate of selecting the proportion of mixture artificial sand 50%, river sand 50%, using the orthonormal table L9 (3^3), The design orthogonal table is shown in Table 4..

3.2 Range Analysis Results of Hydraulic Asphalt Concrete Orthogonal Test

In order to optimize the mix proportion of the hydraulic asphalt concrete, it is necessary to carry out the mix proportion optimization test of hydraulic asphalt concrete, and two indexes of porosity and separation need to be checked [12]. The smaller the porosity, the better, while the larger the separation degree, the closer to 1, the better. The following is

Table 4. Standard orthogonal design.

Factors	Gradation Index	Filler Content (%)	Oil-stone Ratio (%)
Test number	A	B	C
1	0.31	12	10
2	0.31	14	11
3	0.31	16	12
4	0.34	12	11
5	0.34	14	12
6	0.34	16	10
7	0.37	12	12
8	0.37	14	10
9	0.37	16	11

a comprehensive balance method to analyze the test data. The analysis results are shown in the following table.

The average porosity of three different combinations at the same level under each factor is calculated, and the minimum value k_1 of the average porosity at three different levels under each factor is 1.426, 1.458, and 1.231, respectively. In other words, the corresponding optimal scheme is as follows: grading index 0.37, filler content 16%, oil-stone ratio 12%. In the same way, taking the average value of small porosity, the second-best scheme can be obtained as follows: grading index 0.34, filler content 12%, oil-stone ratio 10%. The range analysis result is $0.512 > 0.387 > 0.259$. Therefore, the degree of influence of factors on porosity is ordered as follows: oil-stone ratio > grading index > filler content.

Similarly, the results of the single-factor separation range analysis are as follows. The optimal scheme is a gradation index of 0.31, filler content of 14%, and oil-stone ratio of 10%. The next best scheme is a gradation index of 0.31, filler content of 12%, and oil-stone ratio of 10%. The factors affecting the degree of separation are oil-stone ratio > gradation index > filler content.

The above optimal scheme is obtained under the given factors and level conditions. If the given level is not limited, it is possible to get a better experimental scheme, so when the selected factors and levels are inappropriate, the obtained optimal scheme may not meet the test purpose, not the real sense of the optimal scheme, then the selected factors and levels should be adjusted appropriately. It can find a new and better scheme. At this time, the factor level can be taken as the horizontal coordinate, and the average value of its test index *ki* can be taken as the longitudinal coordinate to draw the trend chart of the influencing factors and indicators (see Figs. 3 and 4).

From Figs. 3 and 4, we can see the influence of the oil-stone ratio on each index. For the porosity and separation value, the range of the oil-stone ratio is the largest. Therefore, the oil-stone ratio significantly influences porosity and separation value, with 10% being optimal.

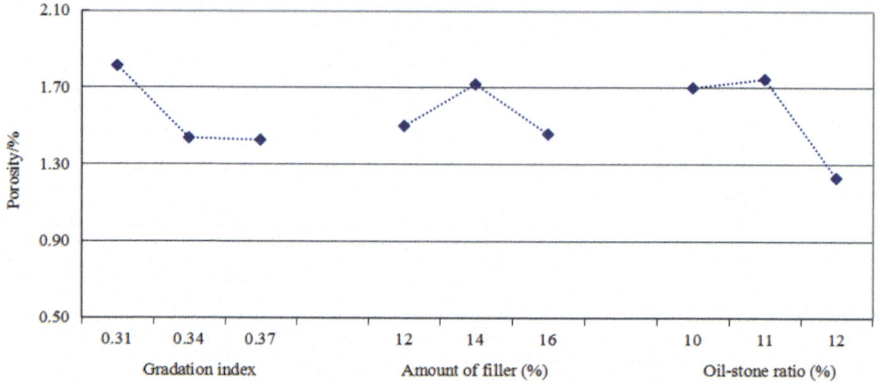

Fig. 3. Trends of porosity and influencing factors.

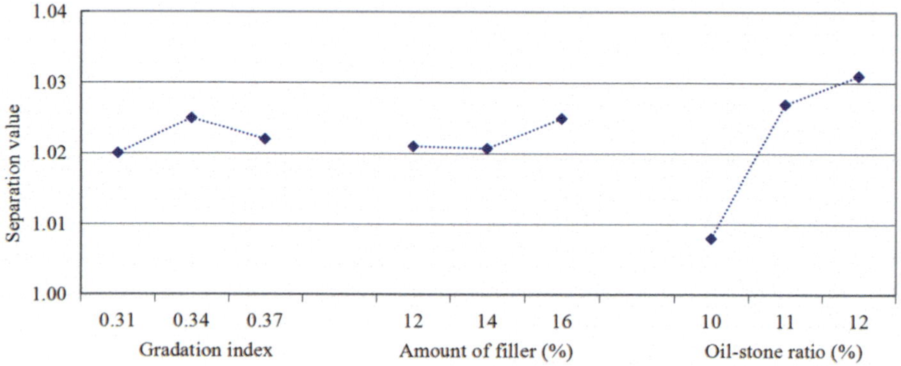

Fig. 4. Trends of separation and influencing factors.

The influence of filler content on each index. For the porosity and separation of the two indicators, filler content is not the largest difference. Therefore, it is not the biggest influencing factor and is a less important factor. Regarding porosity, the filler content is 12%, 16% is better, and 14% is better for separation.

The effect of the gradation index on each index. For the two indexes of porosity and separation, the gradation index is not the biggest difference. Therefore, it is not the biggest influencing factor but a lesser factor. In terms of porosity, it is better to use 0.34 and 0.37 as the packing order index and 0.31 and 0.37 for separation.

A better proportion of the mixture scheme is obtained through the comprehensive analysis of various factors related to the index, specifically a gradation index of 0.37, filler content of 14%, and an oil-stone ratio of 10%.

3.3 Variance Analysis Results of OED for Hydraulic Asphalt Concrete

The range analysis method of OED has the advantages of being simple and intuitive, with a small amount of calculation. However, the range analysis method cannot estimate the size of the error or accurately estimate the importance of the influence of various factors on the test results. This is particularly true for tests with many levels and interactions, making the range analysis method cumbersome to use. At this time, the variance analysis method can make up for these shortcomings of the intuitive analysis method.

Variance is the mean of the sum of squares of a certain deviation. Its size reflects the degree of dispersion of the data and is an important measure of the stability of the test conditions. Different variances have different meanings, and there is a certain relationship between different variances, reflecting the statistical law between the data. If the variance belongs to the category of experimental error, it can be separated from the variance caused by experimental factors and their interaction from the total variance formed by the influence of conditional factors and experimental factors, and two types of variances can be compared, the influence of each experimental factor on the test index can be understood. It provides a reference for controlling various test factors and improving test conditions. The significance of influencing factors was examined using the F-test, and the results are shown in Table 5.

Table 5. Results of variance analysis.

Sources of Variance		The Sum of Partial Variance	Degrees of Freedom	The Sum of Mean Squares	F-value	Significance level $\alpha\alpha$
Variance analysis of porosity	Factor A	0.2910	2	0.1455	8.5969	0.1
	Factor B	0.1157	2	0.0579	3.4191	0.1
	Factor C	0.4835	2	0.2417	14.2830	0.05
	Error	0.0338	2	0.0169	—	—
	Total	0.9240	8	$F_{0.1}(2,2) = 9$; $F_{0.05}(2,2) = 19$		
Variance analysis of the separation degree	Factor A	0.000034	2	0.000017	1.9868	0.1
	Factor B	0.000030	2	0.000015	1.7500	0.1
	Factor C	0.000963	2	0.000481	57.0132	0.05
	Error	0.000017	2	0.000008	—	—
	Total	0.001043	8	$F_{0.1}(2,2) = 9$; $F_{0.05}(2,2) = 19$		

It can be seen from Table 5. That when the significance level α is taken as 0.1, the horizontal changes in the gradation index (factor A) and the filler content (factor B) have no significant effects on the porosity and separation value. In contrast, changes in the oil-stone ratio (factor C) have a significant impact on both porosity and separation test results. The confidence level for this conclusion is 90% (calculated as 100(1-α)%). The possibility of making an error is 10% (calculated as 100α%).

4 Conclusions

The orthogonal experiment design was carried out to optimize the mix ratio of hydraulic asphalt concrete based on the actual hydraulic asphalt concrete project. Two methods, range analysis and variance analysis, were used to analyze the experimental results. These analyses helped identify the optimal mix ratio scheme, which can provide a scientific basis for the selection of the optimal ratio of hydraulic asphalt concrete and improve the optimization efficiency of the mix ratio of hydraulic asphalt concrete.

(1) By selecting three main factors that affect the optimization efficiency of grading of hydraulic asphalt concrete, including the grading index, filling content, and oil-stone ratio, the orthogonal table is constructed, and the values of each factor and index are analyzed. The results of range analysis show that the optimal mix scheme is a grading index of 0.37, filling content of 16%, and oil-stone ratio of 12% from the perspective of porosity. The optimal mix ratio scheme has a grading index of 0.31, filler content of 14%, and oil-stone ratio of 10% from the perspective of separation degree after comprehensively analyzing the above two indexes. The optimal mix scheme is a grading index of 0.37, filler content of 14%, and oil-stone ratio of 10%.
(2) To offset the deficiency of range analysis, the F-test method was used to test the significance of each factor. The results of variance analysis show that when the significance level α is 0.1, the gradation index and the filler content have no significant influence on the porosity and separation value. The horizontal variation of the oil-stone ratio significantly affects the porosity and separation values.

The results indicate that the application of the OED method can improve the efficiency of the mix selection of hydraulic asphalt concrete, which is of great significance in improving the construction quality and shortening the construction period. We can refer to the optimal mix ratio scheme obtained by orthogonal test design in this study to improve the optimization efficiency of the mix ratio in subsequent projects. In our further work, the influence of the optimal mix scheme on the performance of hydraulic asphalt concrete based on the OED needs to be further improved.

References

1. Zhou, E.L., Ning, Z.H., Huang, K.K., Guo, S.L., Xu, C.Y., Chang, F.J.: Sustainable energy integration: Enhancing the complementary operation of pumped-storage power and hydropower systems. Renew. Sustain. Energy Rev. **210**, 115175 (2025)
2. Li, Y., Li, Y.L., Liu, Y.H., Li, W.M., She, L.: Optimization of some properties of hydraulic asphalt concrete mix using RSM and NSGA-II. Constr. Build. Mater. **438**, 137317 (2024)
3. Chen, W.Y., Fan, X., Li, S., Yan, P.Y.: Mix proportion design of high-volume fly ash concrete using orthogonal test method. Concrete **8**, 1–5 (2022)
4. Wang, J.M., Xiao, Z.Q., Fan, Y.T., Wang, N.J., Wang, W.Z., Liu, J.Z.: Orthogonal experiment analysis on the multiple factors influence of the interfacial bonding performance of the composite concrete. Mater. Rep. **36**(2), 87–92 (2022)
5. Wang, J., Ng, P.L., Gong, Y.H., Han, S., Du, J.S.: Experimental study of low temperature performance of porous asphalt mixture. Appl. Sci. **11**(9), 4029 (2021)

6. Dong, Z.J., Guo, Z.S., Zhang, H.Q., Li, J.G.: Optimal design of prefabricated base joint for asphalt pavement based on finite element method and field deflection test. Constr. Build. Mater. **345**, 128301 (2022)
7. Dong, J., Yu, C.Y., Liu, Y.H., Qi, Y.J., Chen, L.L.: Study on applicability of recycled aggregate in hydraulic asphalt concrete. Water Resour. Power **41**(2), 168–172 (2023)
8. Dong, J., Ding, Y.X., Liu, Y.H., Ning, Z.Y., Qi, Y.J.: Tensile test and discrete element study on recycled aggregate hydraulic asphalt concrete. Constr. Build. Mater. **450**, 138538 (2024)
9. Yang, Z.H., He, J.X., Li, Z.H., Yang, W., Lu, J.N.: Scheme optimization of mix proportion for asphalt concrete based on PPR-TOPSIS analysis. Adv. Sci. Technol. Water **43**(2), 82–88 (2023)
10. Xu, X.Q., Gu, Y.W., Huang, W., Chen, D.K., Zhang, C., Yang, X.: Structural optimization of steel-epoxy asphalt pavement based on orthogonal design and GA-BP Algorithm. Crystals **11**(4), 417 (2021)
11. Liu, F.P., Xu, J., Tan, S.C., Gong, A.M., Li, H.M.: Orthogonal experiments and neural networks analysis of concrete performance. Water **14**(16), 2520 (2022)
12. Yang, S.H., Zhou, D., Wang, Y.P., Li, P.: Comparing impact of multi-factor planning layouts in residential areas on summer thermal comfort based on orthogonal design of experiment. Build. Environ. **182**, 107145 (2020)

Water Pollution and Treatment Technology

Comparison and Analysis of China Domestic and International Standards on Fire-Resistant Oil Spill Containment Boom

Yuxiang Yuan[1], Jianping Luo[2](✉), Guodong Qian[1], Dewen Zhang[2], Chaoyu Shen[2], and Bailong An[3]

[1] CNOOC EnerTech-Safety & Environmental Protection Co, Beijing, China
{yuanyx16,qiangd}@cnooc.com.cn
[2] China Waterborne Transport Research Institute, Beijing, China
{luojp,zhangdw,shenchaoyu}@wti.ac.cn
[3] Qingdao Huahai Environmental Protection Industrial Co., Ltd, Qingdao, China

Abstract. Fire-resistant oil spill containment boom is widely and effectively used to contain burning oil spills. Standards and specifications worldwide related to the oil spill containment boom have been summarized. By comparing and studying the differences between these standards, the shortage of fire-resistant oil spill containment boom in China is analyzed, and reasonable suggestions are put forward for the development of high-performance fire-resistant oil spill containment boom.

Keywords: Fire-resistant Oil Spill Containment Boom · Oil Spill Containment · Standards and Specifications · Fire Resistance Test

1 Introduction

In recent years, with the rapid development of offshore oil development and the transportation industry, severe oil spill accidents have occured from time to time, causing huge financial losses and extremely bad ecological damage. After the occurrence of offshore oil spill accidents, the oil containment boom is commonly used [1]. In order to effectively contain the burning oil spill, it is necessary to adopt the fire-resistant oil spill containment boom. At the same time, fire-resistant oil spill containment booms can also be used to gather the oil spill and facilitate the disposal of the oil spill by burning [2]. The performance of a fire-resistant oil spill containment boom directly determines the emergency response effect of the oil spill. Therefore, based on the summary and comparison of relevant standards at home and abroad related to the fire-resistant oil spill containment boom, this paper analyzes the shortcomings of the existing fire-resistant oil spill containment boom in China. In addition, the direction of attention and efforts to develop high-performance fire-resistant oil spill containment boom has been discussed.

C. Weng (Ed.): WRE 2024, LNCE 634, pp. 341–349, 2025.
https://doi.org/10.1007/978-981-96-5095-8_28

2 Standards and Specifications of Oil Spill Containment Boom

Relevant standards and specifications covering the design, manufacturing, and installation of oil containment booms have been formulated at home and abroad to improve the standardization of oil containment boom products and ensure that the design and classification of oil containment booms meet the needs of use in different environments.

Existing domestic and foreign standards related to oil containment booms and their main technical content has been summarized in Table 1.

Table 1. Summary of standards and specifications regarding oil containment booms.

Origin of Standards	Standard Title	Main Content
Chinese Standards	GB/T 34621—2017	This standard defines the naming and model composition, structure, technical requirements, test methods, inspection rules, marking, packaging, transportation, and storage of oil containment booms
Chinese Standards	GBT 36148.1—2018 (ISO 17325–1:2014, IDT)	This standard specifies the basic design requirements, general functions, marking and logo, minimum information that manufacturers must provide on the design, dimensions, and materials of oil containment booms
Chinese Standards	GBT 36148.2—2018 (ISO 17325–2:2014, IDT)	As supplement to GB/T 36148.1, this standard specifies detailed strength and performance requirements of oil containment booms and corresponding test methods. It does not involve safety precautions related to the application
Chinese Standards	GB/T 29132—2018 (ISO 16446:2013, IDT)	This standard specifies a common method of connecting booms with different types of connectors using standard adapters
ASTM Standards	ASTM F1093–99 (2023)	This standard, which originated from 1991 and was latest revised and re-approved in 2023, defines static laboratory tests of booms under tensile loading

(*continued*)

Table 1. (*continued*)

Origin of Standards	Standard Title	Main Content
ASTM Standards	ASTM F1523–94 (2023)	This standard guides the selection of oil containment booms and other accessories
ASTM Standards	ASTM F2084/F2084M-01(2018)	This standard guides the evaluation of the effectiveness of full-scale containment booms in a controlled test facility
ASTM Standards	ASTM F2438–04 (2022)	This specification covers design criteria, geometry, material characteristics, and desirable features for oil containment boom slide connections
ASTM Standards	ASTM F2682–07 (2018)	This guide describes the method for determining the buoyancy to weight (B/W) ratio of oil containment booms
ASTM Standards	ASTM F962–04 (2023)	This specification covers design criteria, geometry, material characteristics, and desirable features for Z-connections of oil containment boom
ASTM Standards	ASTM F1657/F1657M-96 (2018)	This practice provides a standard for joining oil spill containment boom connectors in emergencies. However, this connection method may adversely affect the total tensile strength of the connected booms
ASTM Standards	ASTM F2152–07 (2021)	As one important standard regarding fire-resistance booms, they cover a set of criteria to evaluate the performance, material characteristics, and essential features of fire-resistant oil containment booms
ISO Standards	ISO 17325–1:2014 Part 1	This part of IS0 17325 specifies the basic design requirements, general function, designations, and marking of oil booms
ISO Standards	ISO 17325–2:2014 Part 2	This standard specifies the strength and performance requirements of oil booms and associated test methods

(*continued*)

Table 1. (*continued*)

Origin of Standards	Standard Title	Main Content
ISO Standards	ISO 17325–3:2018 Part 3	This document specifies the design requirements for end connectors
ISO Standards	ISO 17325–4:2018 Part 4	This document specifies basic requirements for the design, geometry, and application of boom accessories
ISO Standards	ISO 16446:2013	This standard specifies a method for joining oil containment booms with dissimilar connectors using a standard adapter with prescribed mating specifications

3 Comparison and Analysis of Standards at Home and Abroad

Based on the summary in Table 1, at present, the most outstanding standards on fire-resistant oil spill containment booms are Chinese Standards GB/T 34621–2017 Oil Boom and ASTM F2152–07 (Re-approved 2021) Standard Guide for in-situ Burning of Spilled Oil: fire-resistant Boom. Table 2 summarized a detail comparison.

Based on the comparison of standard content, it is found that both standards specify the performance requirements and test methods for fire-resistant oil spill containment booms. GB/T 34621–2017, compared to ASTM F2152–07 (Re-approved 2021), fails to provide a discussion about corrosion resistance, extreme temperature properties and hazardous waste. These properties can be further investigated by Chinese researchers.

GB/T 34621–2017, compared to ASTM F2152–07 (Re-approved 2021), fails to define the operability testing method.

ASTM F2152–07 (Re-approved 2021) illustrates the containment test method, while GB/T 34621–2017 does not discuss it. For a fire-resistant oil boom, an oil containment test shall be conducted as per Guide F2084/F2084M. Three burns will be conducted, each lasting one hour. Before the first burn and after each burn, the containment parameter will be checked.

GB/T 34621–2017 defines the final inspection content and type test requirement for fire-resistant oil booms, which provides a good quality control method for workshop fabrication. In contrast, ASTM F2152–07 (Re-approved 2021) only requests certain documentation.

GB/T 34621–2017 Oil Boom stipulates that the water part of the fire containment boom should meet the requirements of high temperature resistance above 1000 °C for 30 min. However, ASTM F2152–07 (Re-approved 2021) does not mention the specific resistance temperature. Still, it takes the heat flux as the assessment characteristic and stipulates that the boom should be burned three times with an average heat flux of 100 kW/m^2 (peak value of 150 kW/m^2). Each burn will last for one hour, then cool down

Table 2. Comparison of technical requirements of fire-resistant oil spill containment boom.

Key Indicators	GB/T 34621—2017	ASTM F2152–07(Re-approved 2021)
Operability Characteristics	There is no definition of operability characteristics. However, the dimensions and strength parameters have been listed, and the connector parameters have been illustrated	The Operability Characteristics can be divided into physical dimensions, strength parameters, corrosion resistance, extreme temperature properties, fabric tests, hazardous waste, and end connectors
Oil Containment Characteristic	The fire-resistant boom shall be equal to the performance oil containment with conventional boom	There is no oil leakage between sections of the oil boom
Fire-Resistance Characteristics	The above part of the Boom should meet the requirements of high temperature resistance above 1000°C	The fire-resistant boom shall contain oil and survive in heat fluxes equivalent to an in-situ burn of diesel with a minimum diameter of 4.5 m for three one-hour burn cycles, with a minimum one-hour cool-down between cycles
Operability Testing	No definition	At least 150m of fire-resistant boom shall be towed in a U-configuration and in a straight line to confirm its towing characteristics and its short-term durability. The gap ratio will be checked, and no significant damage is allowed
Oil Containment Testing	No definition	Oil containment test shall be conducted as per Guide F2084/F2084M Three burns will be done, and each will last for one hour. Before the first burn and after each burn, the containment parameter will be checked
Final inspection and Type Testing	Final inspection and type testing have been defined	No definition

(*continued*)

Table 2. (*continued*)

Key Indicators	GB/T 34621—2017	ASTM F2152–07(Re-approved 2021)
Fire Resistance Testing Environment	The water depth should not at least 1.5 times the draft of the oil containment boom, and the test water should be calm. There should be no safety hazard around the water area that may cause burning or explosion	No definition
Fire Resistance Testing Samples	A section of boom. Or one part of the boom containing at least five floating bodies and not less than 6m in length, provided that the two ends of this part of the boom be equipped with the same joints as those of a section of the boom	A minimum burn diameter of 4.5 m
Fire Resistance Testing Oil and Film Thickness	1. Diesel 0 is adopted 2. When ignition difficulty occurs, less than 10% gasoline can be added 3. The thickness of the test oil is not less than 20mm	1. The thickness of the oil layer shall be the lesser of 200 mm or 50% of the boom's draft 2. Prior to and during a burn exposure, the boom shall be capable of containing a layer of low viscosity oil of 200 mm in thickness without significant losses 3. After the second or third of the three burns, the boom shall contain a layer of diesel of 200 mm in calm water. The losses from the boom due to leakage shall be no greater than 10%, measured by volume, over 10 min from the start of the test

(*continued*)

Table 2. (*continued*)

Key Indicators	GB/T 34621—2017	ASTM F2152–07(Re-approved 2021)
Fire Resistance Testing Methods	1. Fix the test samples in the test location and connect the two ends of the boom sample with the joint so that the fire-resistance test section is enclosed in a ring, polygon, or triangle. Test oil no less than 20 mm will be injected and ignited, and the burning will be maintained for 30 min. Then, the boom will be visually checked after the test oil is fully burned and self-extinguished. The metal parts should not be significantly deformed, and the underwater skirt body and flexible section should be intact. As to the parts with fire retardant paint or cloth, if the main structure of the fire containment boom is not damaged, the burn or peel-off of fire resistance paint or cloth can be accepted 2. The water-cooled layer of the boom does not have structural damage 3. The boom freeboard and floating weight ratio after combustion cooling shall be at least 50% of that before combustion. The tensile strength shall meet the requirements of this standard	1. The fire-resistant boom shall survive direct fire exposure with an average heat flux of 100kW/m^2 or greater (heat flux measurements averaged over the one hour burn period), with peaks of 100k W /m^2 to 150 kW/m^2, while exposed to wave action. The boom shall be exposed to a series of three one hour burns with a minimum one-hour cool-down period between burns 2. Heat flux is very difficult to measure accurately. Heat flux gauges respond quickly to changes in the fire, and substantial fluctuation is normal for these measurements. Readings will vary significantly with altitude and direction of a sensor, and with the wind speed and direction 3. Given the difficulties in accurately measuring heat flux, the recommended test is a burn of diesel with a minimum burn diameter of 4.5 m. This test meets the requirements for this guide without the need to conduct heat flux measurements 4. During the heat exposure and intervening cool-down periods (except the leak test), the boom shall be exposed to wave action. A wave height of at least 15 cm and a duration of less than 3 s is recommended. The wave height shall be the minimum of 15 cm or the minimum freeboard following the burn for the appropriate boom category, as shown in this standard

for one hour. The definition of ASTM F2152–07 (Re-approved 2021) seems more practical. However, the test environment is not easy to build precisely by different researchers, and the measurement of heat flux is difficult.

In addition, both standards indicate that diesel oil is used as the test oil to conduct the combustion test. To investigate the performance parameters of oil containment booms, it is essential to carry out simulated offshore oil burning testing. The method of building a test platform and the adoption of the proper test procedure are the key issues. The collected data shows that most experimental studies on oil spill combustion adopted crude or diesel oil as the testing oil. For example, as mentioned in the article Interface Insulation Systems for Enhancing in Situ Burni [3], Alaska North Slope crude oil is used for large-scale simulated combustion tests in the research process. However, the combustion of crude or diesel oil will inevitably cause serious environmental pollution. At present, with the implementation of the relevant policies of "carbon peak, carbon neutrality," China's environmental protection requirements are increasingly strict, and it is difficult to use crude oil or diesel oil directly to carry out relevant experimental research.

Certain results have been achieved in the research of fire-resistant oil spill containment booms worldwide [4]. However, the fire-resistant capability of product boom produced by some foreign manufacturers has reached a relatively ideal level. For example, fire booms produced by ELASTEC Company [5] can resist the high temperature of 1100°C and maintain effectiveness for 11 h and 21 min, and were successfully used in the Deepwater Horizon oil spill in the Gulf of Mexico in 2010 [6]. As a comparison, the domestic fire-resistant oil spill containment booms are not resistant to impact and friction and are easy to disintegrate after a long combustion time. Considering the high price and longtime procurement cycle of foreign products, improving the technical performance of China's fire-resistant oil spill containment booms is imperative.

As a reasonable target, the development of high-performance, fire-resistant oil spill containment booms should match or even exceed the performance parameters of existing foreign oil booms. To be specific, when it comes to fire resistance, it is advisable to develop fire-resistant oil spill containment booms that can effectively contain operations for 12 h or more under high temperature conditions of 1000°C.

Since it is impossible to conduct combustion tests using crude oil or diesel in China, alternative testing mediums need to be considered. Recently, propane spray guns were used in some small combustion tests at home and abroad with satisfactory results. The researchers conducted the propane mixed air combustion test to verify its effect. The results showed that the combustion temperature of the propane spray gun mixed with air can reach 1000°C, and the average heat flux in the combustion area can reach 128kW/m^2, as shown in Fig. 1, which meets the combustion test requirements of relevant domestic and foreign standards. In addition, the test platform built and the test plan formulated must also meet relevant policy requirements such as environmental protection and safety.

Meanwhile, it's important to formulate or revise relevant standards with technological progress, new product development, and the introduction of advanced foreign technology. Latest and advanced standards can reflect the latest technical requirements and business practices. They will benefit different countries and regions to coordinate better and unify the normative requirements of products and services.

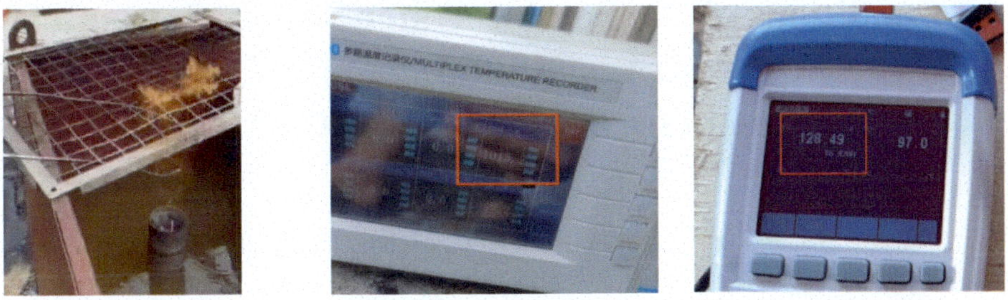

Combustion test | Temperature measurement | Heat flux measurement

Fig. 1. Propane and air mixture combustion test.

4 Summary

Fire-resistant oil spill containment booms provide important support and guarantee for dealing with offshore oil spill fire accidents. The improvement of their technical performance is helpful to improve China's offshore oil spill emergency response capability. By comparing the relevant technical standards of fire-resistant oil spill containment booms at home and abroad, this paper analyzed the shortage and advantages of standards in China, and the shortcomings of oil booms in China. In addition, this paper also guided the further development of standards for fire-resistant oil spill containment booms.

References

1. Zhang, M.R.: A brief analysis of on-site disposal methods for marine oil spill accidents. China Pet. Chem. Stand. Qual. **44**(15), 148–150 (2024)
2. Man, C.Z., Liu, H.: Technology of emergency management for marine oil spill. Environ. Prot. Oil & Gas Fields **22**(6), 50–52 (2012)
3. James, L.M., Stephen, G.P., Ian. A.B.: Interface insulation systems for enhancing in-situ burning. Bureau of Safety and Environmental Enforcement (BSEE) Report
4. Li, Y., Sun, S.W., Yuan, Y.X., et al.: Applicability analysis and development prospect of fire barrier and skimmer in offshore oil blowout disposal. Technol. Superv. Pet. Ind. **40**(1), 47–52 (2024)
5. ELASTEC Homepage: https://www.elastec.com/products/floating-boom-barriers/fire-resistant-oil-boom/american-fireboom/, last accessed 2024/11/29
6. Peng, B, Wang, Y.N., Liao, S.M., Cai, S.Y.: Review on emergency disposal technology of marine oil spill. Guangdong Chem. Ind. **46**(07), 161–162+184 (2019)

Conditions of the Potable Water Supply System of the Administrative Building at the University of the Armed Forces ESPE Using the Geophone

David Carrera-Villacrés[1,2](✉), David Asipuela[1], Melany Cárdenas[1], Vielka Carrillo[1], Hans Guzmán[1], and Sebastián Toapanta[1]

[1] Universidad de Las Fuerzas Armadas ESPE. Departamento de Ciencias de La Tierra y La Construcción. Av. Gral. Rumiñahui S/N, Sangolquí, Ecuador
{dvcarrera,daasipuela,mmcardenas5,vvcarrillo,higuzman1, sstoapanta}@espe.edu.ec

[2] Facultad de Ingeniería en Geología, Universidad Central del Ecuador, Petróleos y Ambiental (FIGEMPA). Av. Universitaria, Minas, Quito, Ecuador

Abstract. The detection of unidentified water leaks leads to loss of potable water, high repair and maintenance costs, and structural damage associated with water treatment and supply. The implementation of geophones for leak detection becomes crucial at the University of the Armed Forces ESPE, especially in the administrative building, since previous investigations have confirmed the presence of water leaks at night using IoT flowmeters. In response to this problem, zoning of the area has been carried out using the geophone, a tool that has proven to be effective in the accurate identification of leaks. The presence of large leaks was identified on the upper floors, corroborating the 116.14% increase in non-visible leaks in comparison with previous investigations.

Keywords: Leaks · Geophone · Detection · Plumbing Systems

1 Introduction

Water, an essential pillar for life on the planet, has garnered constant attention to preserve and sustainably manage this vital resource [1]. Access to safe drinking water and basic sanitation is fundamental for human health, well-being, and sustainable development, as well as poverty eradication, economic growth, and environmental sustainability [2]. Water conservation is a shared responsibility that necessitates collaboration among governments, businesses, and civil society [3–6].

The Administration Building faces a serious water loss problem during nighttime hours when only security personnel are present [7, 8]. Previous investigations have shown unusually high-water consumption during this period, suggesting leaks in the pipes [8–10].

These leaks not only involve wasted resources but also potential structural damage and additional costs [11, 12]. The lack of direct monitoring makes it difficult to detect

C. Weng (Ed.): WRE 2024, LNCE 634, pp. 350–360, 2025.
https://doi.org/10.1007/978-981-96-5095-8_29

and correct leaks [13]. Autonomous monitoring measures and an action plan are urgently needed to address this problem and minimize water losses and associated costs [8, 14].

Leaks in water systems are caused by various factors, such as old and poor-quality piping, lack of corrosion control, poor valve maintenance, and mechanical damage. They are classified into visible leaks, detected by humidity, and non-visible leaks, identified by the frequency emitted by the water [11, 15]. Studies reveal that the amount of water wasted by leaks varies widely between countries, in some cases as high as 70% in distribution networks [16, 17]. Early detection of leaks not only improves services for consumers but also increases the efficiency and extent of those services [18]. Technological advances and experience can reduce losses to below 10%, assessing costs and cost-effectiveness of corrective actions [19].

Water leaks, constituting more than half of the national supply, are caused by old distribution networks, lack of maintenance, and poor-quality materials [20, 21]. In Ecuador, these leaks represent 48.4% of the potable water supply, equivalent to 696.2 million cubic meters of water, with an estimated annual cost of USD 320 million [22]. Additionally, within the University of the Armed Forces ESPE, monthly leaks represent a total cost of USD 2,713.27, resulting in significant economic loss for the institution [10, 14].

Geophones, a non-destructive technique used in this project, are crucial for identifying and locating leaks in subway pipes in ESPE's administrative building [23]. This tool significantly contributes to efficient water management by enabling early detection and repair of leaks, conserving water resources, and reducing economic losses [24, 25]. Leak detection is based on the identification of sounds generated by water flow and vibrations in the pipes, providing accurate and effective results to maintain hydraulic infrastructure in optimal conditions [26, 27].

In Ecuador, geophones have gained popularity in recent decades due to their efficiency and cost-effectiveness in detecting water leaks [28]. The Empresa Pública Metropolitana de Agua Potable y Saneamiento de Quito has confirmed that the use of geophones has significantly reduced the time needed to locate leaks, highlighting their importance in the efficient management of water resources at local and national levels [29, 30].

Non-visible leak detection also plays an important role in promoting sustainability, aligning with the Smart University objectives and the Sustainable Development Goals (SDGs) established by the United Nations [4]. Timely leak detection contributes directly to UN SDG 6, ensuring clean water and sanitation, and SDG 11, seeking sustainable cities and communities by improving urban infrastructure and reducing water losses [31, 32].

The objective of the research was based on creating a map of non-visible water leaks for effective detection of losses in the distribution and supply system of the administrative building of the UFA "ESPE" based on previous research to create a comparison of the percentage increase of water losses in the last academic periods, through geophonation, to provide information to reduce economic losses and promote efficiency in the consumption of water resources.

2 Study Area

The Administrative building of the University of the Armed Forces ESPE is in Sangolquí on General Rumiñahui Av and Quito Av (Fig. 1). It is a contemporary construction of eight floors that began in 2008 and ended in 2014. Inside it are the offices of the university authorities and various administrative and support units. The building has a capacity for 500 people; however, on average 168 people are currently in it [33].

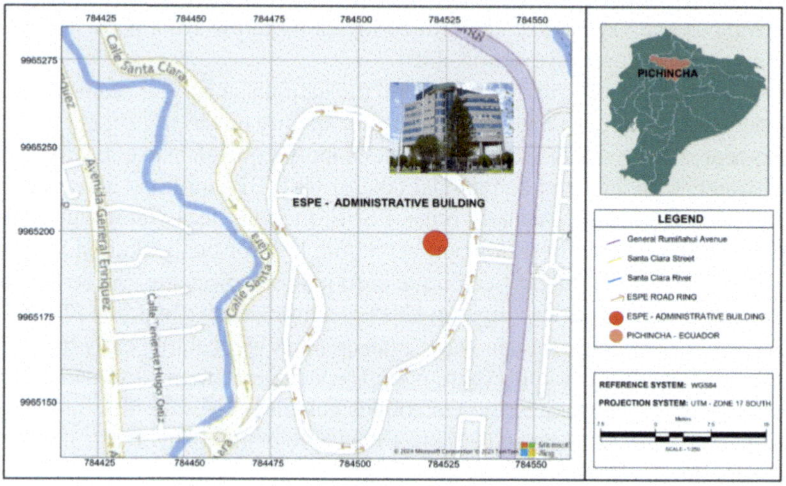

Fig. 1. Location of the ESPE Administrative Building, with coordinates 784527.00 m E, 9965195.00 m S. (Retrieved from: GoogleEarth y AutoCAD)

A study carried out in 2021 through the installation of the IoT flowmeter in the Administration Building it recorded a consumption of 128.16 L per hour (L/h) (Fig. 2) this study was conducted 24 h a day for the entire year, with notable consumption peaks during nighttime hours [8].

Subsequently, in the year 2022, a consumption of 145.76 (L/h) (Fig. 2) was determined during the night hours through the analysis of the flow meter. The losses observed in this period due to leaks were 14.45%, according to the comparison between the flow of potable water measured in the administrative building and the flow of wastewater [10].

In July 2023, another analysis of potable water consumption in the same building was performed, revealing a flow of 130.25 (L/h) (Fig. 2) during the hours recorded with high consumption at 06:00, 05:00, 23:00 and 02:00 h, where the personnel found during those hours are only security personnel. It is observed that the percentage of leaks in the Administrative Building is 31.21% [14].

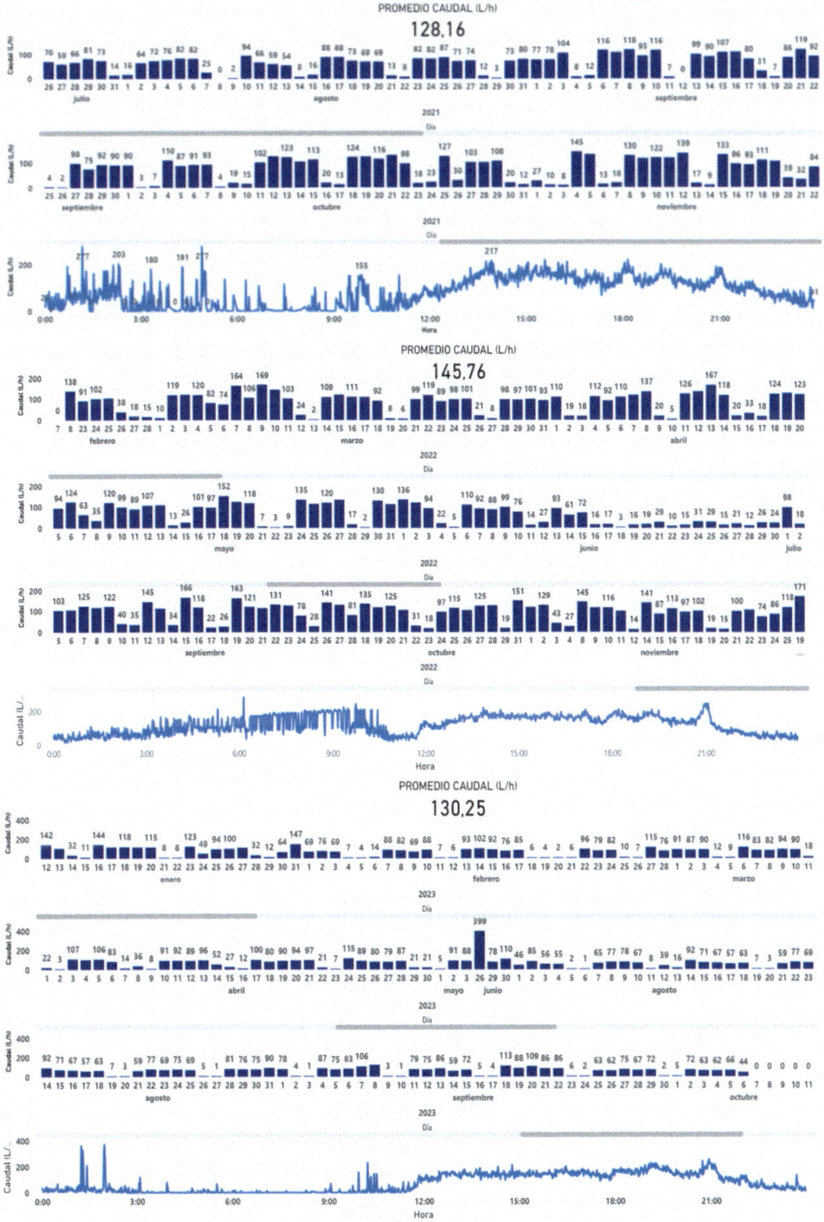

Fig. 2. Water consumption in Flow unit year (2021, 2022, 2023) in the ESPE administrative building. (*Source* [8])

3 Materials and Methods

The AquaTest T10 geophone is a water leak detection tool that uses the electro-acoustic method, is portable and ergonomic, and does not require an additional receiver [34]. It has an SDR radio module that allows one select the listening frequency among 8 preset levels. This helps to improve the reception of leak sounds and reduce the influence of ambient noise [35–37].

3.1 Locating Leaks Using the AquaTest T10 Geophone

Detection of leaks in the potable plumbing systems was carried out through a three-stage process. The first phase consisted of a review of the plans provided by the ESPE Physical Development unit [38]. In the second phase, a visual inspection of the distribution networks was carried out to identify visible leaks, while the third stage focused on locating non-visible leaks using the AquaTest T10 geophone, divided into pre-location and actual location [39–41].

During the pre-location, factors such as the time of geophone sounding were considered, as well as avoiding the geophone sounding during the working day to minimize noise interference [42, 43]. To locate leaks, tools such as the geophone bell and the scanning rod were used, allowing precise identification of the exact location of the leak by listening at close range on the surface of the wall, following the route of the pipe indicated in the plumbing system plans [44, 45].

3.2 Sound Classification

The process was carried out using frequency analysis, which is divided into two stages: the first consists of avoiding the sound generated by environmental noises that could have interfered with the classification, and the second phase involves the filtering used for the leakage measurement, using the AquaTest10 geophone [34, 46]. Filter level 2 was chosen, which affects individual sound perception in a range from low to medium. Since the inspection was performed in sections on all floors, it was not considered necessary to use higher filter levels [47]. The following (Fig. 3) shows the frequency level of each filter:

Filter level	Filter band display [low ← frequency → high]
1	■■■■□□□□
2	□■■■■□□□
3	□□■■■■□□
4	□□□■■■■□
5	■■■■■■■□
6	■■■■■■■■
7	□■■■■■■■
8	□□□■■■■■

Fig. 3. AquaTest10 Geophone Filter Levels. (*Source* SEWERIN Manual, 2024)

3.3 Leakage Representation

The representation of sound intensity at different points of a pipeline was achieved by using a heat map scale for noise intensities. In order to visualize this information, a logarithmic scale (Fig. 4) was used in the creation of a heat map, assigning a specific color to each noise intensity interval [48]. This facilitates the clear and didactic interpretation of the different noise levels present in the analyzed area [49, 50]. Typical sound values for PVC pipes range from 20 Hz to 500 Hz, depending on factors such as pipe diameter, water pressure, and installation conditions [51, 52].

The elaboration of the heat map was done using specific software, in this case, SolidWorks, under student license, allowing the drawing of the piping and hydraulic fittings. To represent the color scale on the pipe, the Blender program, which specializes in lighting model rendering, and is freely available to the public. Table 1 shows examples of sound values for different conditions [53–56].

Table 1. Dominant frequencies according to the flow condition.

Condition	Dominant frequencies
Low water flow	20–100
Normal water flow	100–250
High water flow	250–500
Turbulent water flow	>500

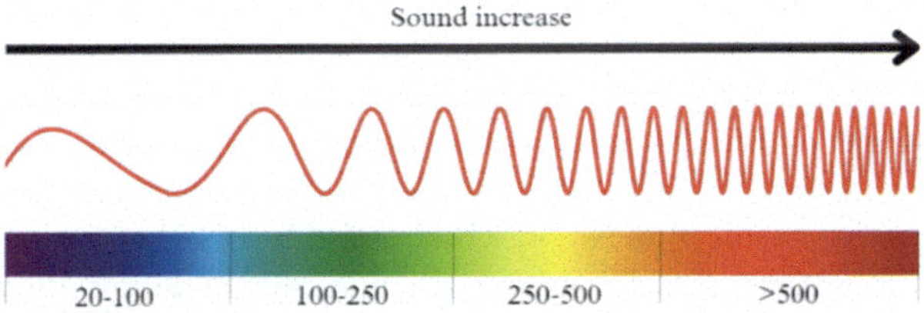

Fig. 4. Logarithmic scale of sound intensities.

4 Results and Discussion

4.1 Isometric Heat Maps Drinking Water Columns

Heat maps are shown for each drinking plumbing system or each conduit (Fig. 5), where the logarithmic scale indicates the areas with the highest probability of leakage according to the level of sound generated [57].

The heat map analysis revealed the distribution of sound levels in the column of drinking water pipe No. 1, which comes from the pumping system; high sound levels were detected in the upper level, where the rectory and the cafeteria are located, at the ceiling level. High sound levels were also observed at the water outlet along this column. However, unlike the other duct mentioned, no high sound levels were detected along the column. These results are consistent with previous investigations that recorded high nighttime consumption in the Administration Building during periods of low activity.

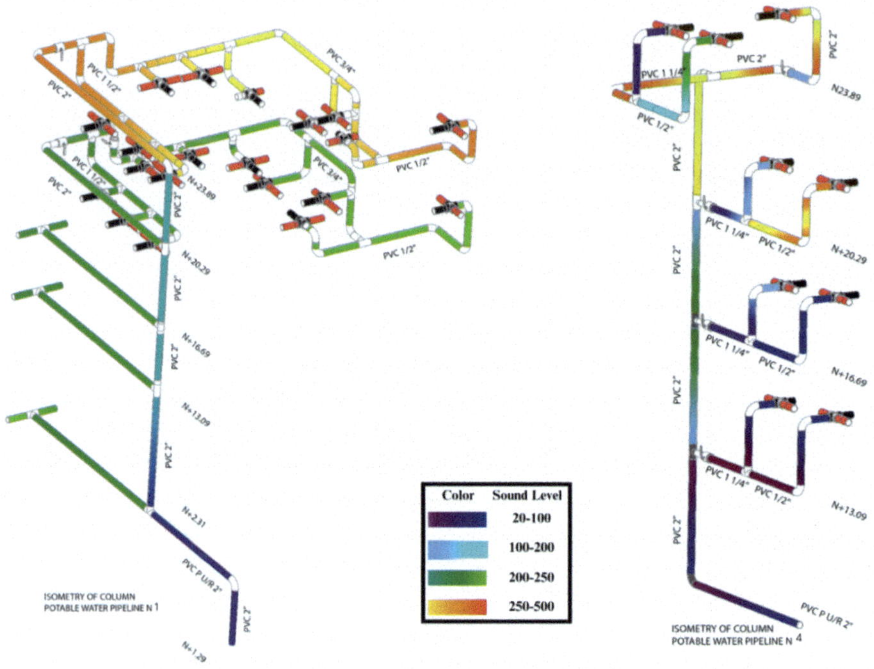

Fig. 5. Isometric heat maps drinking water columns No. 1 and No. 4.

As for the drinking water pipe No. 4, which is fed from the pumping system, in this analysis, it was observed that, at a height of 20.29 m, specifically in the men's restroom at ceiling level, a notable sound intensity was recorded, indicating that the main water loss originates in that location. In addition, it was identified that in the last two upper levels of the building, sound intensity values were reflected, highlighting especially on the top floor at ground level, where high sound levels were detected.

The geophonic survey was effective in determining the location and magnitude of water leaks in the Administrative Building.

The increase in leakage in the administrative building from 14.45% in 2022 to 31.21% in 2023 represents an increase of 116.14% compared to the previous year, which has a negative impact on water resources and costs associated with the increase in potable water leakage [8].

In addition, it causes damage to infrastructure, such as roof deterioration due to moisture from leaks. Water leaks result in excessive waste of water resources, increasing water and associated repair costs [10].

5 Conclusions

The study confirms the existence of non-visible water leaks in the water distribution and supply system of the UFA "ESPE" administrative building using the geophonation technique. The accurate detection of these leaks on the upper floors of the building supports the effectiveness of geophones as a tool to optimize water management and reduce economic losses.

The results are consistent with previous research from three preceding years, which have confirmed the presence of potential non-visible leaks in the drinking plumbing systems during nighttime hours using IoT flow meters. The location of the possible leaks was determined through geophoning. It is important to conduct periodic evaluations with geophoning to prevent damage to the infrastructure and optimize water consumption.

References

1. Fezai, R., Bouzrara, K., Mansouri, M., Nounou, H., Nounou, M., Puig, V.: A novel leak detection approach in water distribution networks. In: *2021 18th International Multi-Conference on Systems, Signals & Devices (SSD)*, pp. 607–612. IEEE (Mar. 2021)
2. Negm, A., Ma, X., Aggidis, G.: Review of leakage detection in water distribution networks. IOP Conf. Ser. Earth Environ. Sci. **1136**(1), 012052 (2023)
3. World Health Organization, Drinking-water. Accessed: Dec. 25, 2024. [Online]. Available: https://www.who.int/news-room/fact-sheets/detail/drinking-water
4. United Nations, Goal 6: Ensure access to water and sanitation for all. Accessed: Dec. 25, 2024. [Online]. Available: https://www.un.org/sustainabledevelopment/water-and-sanitation/
5. UFA ESPE: Plan Director Institutional 2022–2027, Quito - Ecuador (2022)
6. Carrera Villacrés, D.V.: ESPE-DCTC-2023–032-M, Quito (2023)
7. University of the Armed Forces Espe, Report on the results obtained in the implementation of strategies to promote environmental sustainability, Presented by The 2020 Integrated Security Unit., Sangolquí, 2020. Accessed: Dec. 26, 2024. [Online]. Available: https://sin.espe.edu.ec/wp-content/uploads/2021/11/INFORME-DE-LA-SOSTENIBILIDAD-AMBIENTAL-2020.pdf
8. Carrera, D., Montaño, F., Quintuña, M.: Dynamic and statistical modeling to find critical points in space and time through Iot sensors of water resources in the smatriz field of the University of the Armed Forces ESPE. University of the Armed Forces ESPE, Quito (2023)
9. Carrera, D.: ESPE-DCTC-2023–032–3, Sangolquí (Dec. 2023)
10. Carrera, D., Garcés, R.: Diagnosis and proposal for improvement of drinking water and sewerage at the University of the Armed Forces ESPE Matrix implemented IoT smart flowmeters. University of the Armed Forces ESPE, Sangolquí (2022)
11. Fuentes, M., Palma, N., Rodríquez, V.: Estimation and location of leaks in a drinking water pipeline network using genetic algorithms. Ingeniería, investigación y tecnología, vol. 12 (Jun. 2011)
12. CONAGUA: Drinking Water, Sewerage and Sanitation Handbook. Tlalpan. Accessed: Dec. 26, 2024. [Online]. Available: https://files.conagua.gob.mx/conagua/mapas/SGAPDS-1-15-Libro4.pdf
13. ARCA, Public Consultation Process No. ARCA-AR-2024-CP-003 of the Institutional Regulatory Plan 2025 of the Water Regulation and Control Agency. Accessed: Dec. 26, 2024. [Online]. Available: https://www.regulacionagua.gob.ec/proceso-de-consulta-publica-nro-arca-ar-2024-cp-003-del-plan-regulatorio-institucional-2025-de-la-agencia-de-regulacion-y-control-del-agua/

14. Carrera Villacrés, D.V., Garcés, R.A., Tierra Criollo, A. R., Duran, R., Raura, G.: Drinking water and sewerage at the universidad de las fuerzas armadas ESPE and implementation of an internet of things flowmeter, pp. 73–84 (2022). https://doi.org/10.1007/978-3-031-18272-3_6
15. Piao, C., Kim, S.H., Lee, J.K., Choi, W.G., Kim, Y.Y.: Non-invasive ultrasonic inspection of sludge accumulation in a pipe. Ultrasonics, vol. 119, p. 106602, Feb. 2022 (2020)
16. Fonseca J., Rodríguez, J.: Detection and location of leaks in the drinking water networks of the Acuarrayan San Francisco, Cundinamarc aqueduct of the Acuarrayan San Francisco, Cundinamarc aqueduct using electroacoustics and pressure analysis using electroacoustics and pressure analysis. Accessed: Dec. 26, (2024). [Online]. Available: https://ciencia.lasalle.edu.co/ing_civil/131?utm_source=ciencia.lasalle.edu.co%2Fing_civil%2F131&utm_medium=PDF&utm_campaign=PDFCoverPages
17. Wang, F., Lin, W., Liu, Z., Qiu, X.: Pipeline leak detection and location based on model-free isolation of abnormal acoustic signals. Energies (Basel) **12**(16), 3172 (Aug. 2019)
18. Republic of Ecuador, Statistics of Environmental Economic Information in Autonomous Decentralized Municipalities. Accessed: Dec. 26, 2024 (2024). [Online]. Available: https://www.ecuadorencifras.gob.ec/documentos/web-inec/Encuestas_Ambientales/Municipios_2021/Agua_potable_alcantarrillado_2021/PRESENTACION%20APA%202021_V7%20(Rev.%20Dicos).pdf
19. World Water: Water: a shared responsibility; the United Nations world water development report 2, executive summary, 2006. Accessed: Dec. 25, 2024. [Online]. Available: https://unesdoc.unesco.org/ark:/48223/pf0000144409
20. Villanueva, H.: Smart metering system for household water consumption using IoT and Cloud Computing. National University of San Marcos (2021). Accessed: Dec. 26, 2024. [Online]. Available: https://hdl.handle.net/20.500.12672/17259
21. Izadi, A., Yazdandoost, F., Ranjbar, R.: Asset-based assessment of resiliency in water distribution networks. Water Resour. Manag. **34**(4), 1407–1422 (Mar. 2020)
22. EPMAPS: Update of the DMQ distribution system. Accessed: Dec. 26, 2024. [Online]. Available: https://www.aguaquito.gob.ec/sistema-de-distribucion/
23. Dighade, M., Kadu, M., Pande, A.: Challenges in water loss management of water distribution systems in developing countries. Int. J. Innov. Res. Sci. Eng. Technol. **3**(6), (Jun. 2014)
24. Almeida, F.C.L., Brennan, M.J., de Lima, F.K., Iwanaga, M.K., Scussel, O.: Using a geophone as an actuator to estimate the velocity of leak noise propagation in buried water pipes. Appl. Acoust. **184**, 108251, (Dec. 2021)
25. Chiarini, B., Agostino, A., D'Marzano, E., Regoli, A.: The perception of air pollution and noise in urban environments: a subjective indicator across European countries. J. Environ. Manag. **263**, 110272, (Jun. 2020)
26. Tornyeviadzi, H.M., Seidu, R.: Leakage detection in water distribution networks via 1D CNN deep autoencoder for multivariate SCADA data. Eng. Appl. Artif. Intell. **122**, 106062 (Jun. 2023)
27. Ladino-Moreno, E.O., García-Ubaque, C.A., García-Vaca, M.C.: Leakage estimation in pressurized pipes for drinking water systems using artificial neural networks and Epanet. Rev. Científica **43**(1), 2–19 (2021)
28. Terán, C., Arquello, J.: Document methodological statistics of environmental economic information in autonomous decentralized municipalities. Republic of Ecuador. Accessed: Dec. 26, 2024. [Online]. Available: https://www.ecuadorencifras.gob.ec/documentos/web-inec/Encuestas_Ambientales/Municipios_2022/Agua_potable_alcantarillado/Documento_Metodologico_APA_2022_vF.pdf
29. De Santis, D., Shams, A.: Numerical modeling of flow induced vibration of nuclear fuel rods. Nucl. Eng. Des. **320**, 44–56 (2017)

30. Market Research Intellect: Intelligent leak detection: transforming water management in manufacturing and construction, Packaging and construction. Accessed: Dec. 26, 2024. [Online]. Available: https://www.marketresearchintellect.com/es/blog/smart-leak-detection-transforming-water-management-in-manufacturing-and-construction/
31. United Nations: Goal 11: Make cities inclusive, safe, resilient and sustainable. Accessed: Dec. 26, 2024. [Online]. Available: https://www.un.org/sustainabledevelopment/cities/
32. EMAC: Environmental education. Accessed: Dec. 26, 2024. [Online]. Available: https://emac.gob.ec/educacion-ambiental/
33. University of the Armed Forces Espe, Resolution ESPE-HCU-RES-2021–106. Sangolquí (2021). Accessed: Dec. 26, 2024. [Online]. Available: University of the Armed Forces Espe
34. SEWERIN: Operating Instructions. Accessed: Dec. 26, (2024). [Online]. Available: https://www.sewerin.com/fileadmin/redakteure/BEA/en/bea_at-t10_en.pdf
35. Lobos, C., Ignacio, J., Campos, N.: Detection and localization of leaks in drinking water distribution networks in a large Chilean city using a neural network classification algorithm, University of Chile (2022). Accessed: Dec. 26, 2024. [Online]. Available: https://repositorio.uchile.cl/handle/2250/185782
36. Rodríguez, N., Delgado, J., Aguilera, G.: Locating leaks in interscience networks, *Ciencia* **47**, (2022)
37. Innovaequipos Desing Zerox Devs: Water Leak Detection Geophones. Accessed: Dec. 26, 2024. [Online]. Available: https://innovaequiposecuador.com/geofonos-detectores-de-fugas-de-agua/
38. Fuchs, H.V.: Generation and control of noise in water supply installations Part 3: rating and abating procedures. Appl. Acoust. **39**(3), 165–190 (1993)
39. Muggleton, J.M., Brennan, M.J., Gao, Y.: Determining the location of buried plastic water pipes from measurements of ground surface vibration. J. Appl. Geophy. **75**(1), 54–61 (2011)
40. Fonseca Pérez, J.D., Rodríguez Sastoque, J. C.: Detección y localización de fugas en las redes de agua potable del acueducto acuarrayan (san francisco, cundinamarca) mediante electroacústica y análisis de presiones. Universidad de La Salle, Bogotá (2017). Accessed: Jan. 27, 2024. [Online]. Available: https://ciencia.lasalle.edu.co/cgi/viewcontent.cgi?article=1130&context=ing_civil
41. Cai, Y., Santos, R.B., Givigi, S.N., Fileti, A.M.F.: A pipeline leak classification and location estimation system with convolutional neural networks. IEEE Syst. J. **14**(3), 3072–3081 (2020)
42. Pérez, D., García, A., López, A.: Water stress: problems and solutions. a analysis with a legal approach. Iberoamerican University Puebla, Puebla (2022). Accessed: Dec. 26, 2024. [Online]. Available: https://hdl.handle.net/20.500.11777/5598
43. Deoreo, W., Hodgins, M., Aquacraft: Residential end uses of water, version 2, Water use & efficiency. Accessed: Dec. 26, 2024. [Online]. Available: https://www.waterrf.org/research/projects/residential-end-uses-water-version-2
44. World Bank Group: After a decade of prosperity, a less favorable external context demands that Ecuador ensure fiscal sustainability, strengthen the foundations of dollarization, promote new engines of growth and protect the social gains achieved. Accessed: Dec. 26, 2024. [Online]. Available: https://www.worldbank.org/en/country/ecuador
45. UNESCO: The United Nations world water development report 2019: leaving no one behind. Paris (2019)
46. Argo, T.F., Guild, M.D., Wilson, P.S., Schröter, M., Radin, C., Swinney, H.L.: Sound speed in water-saturated glass beads as a function of frequency and porosity. J. Acoust. Soc. Am. **129**(4), EL101–EL107 (Apr. 2011)
47. Yu, Y., Safari, A., Niu, X., Drinkwater, B., Horoshenkov, K.V.: Acoustic and ultrasonic techniques for defect detection and condition monitoring in water and sewerage pipes: a review. Appl. Acoust. **183**, 108282 (Dec. 2021). https://doi.org/10.1016/j.apacoust.2021.108282

48. Caesarendra, W., Kosasih, B., Tieu, A.K., Zhu, H., Moodie, C.A.S., Zhu, Q.: Acoustic emission-based condition monitoring methods: Review and application for low-speed slew bearing. Mech. Syst. Signal Process **72–73**, 134–159 (May 2016)
49. Gómez, F.: Application of BIM technology for the calculation and design of facilities in an industrial building. Master Degree, University of Sevilla, Sevilla (2020). Accessed: Dec. 26, 2024. [Online]. Available: https://idus.us.es/items/1f93e314-4871-4f69-8e1b-c27c33bad2dc
50. Fereidooni, Z., Tahayori, H., Bahadori-Jahromi, A.: A hybrid model-based method for leak detection in large scale water distribution networks. J. Ambient. Intell. Humaniz. Comput. **12**(2), 1613–1629 (2021)
51. Lan, Z., Cai, M.: Dynamic traffic noise maps based on noise monitoring and traffic speed data. Transp. Res. D Transp. Environ. **94**, 1361–9209 (2021)
52. Lewis, R.C., Gershon, R.R.M., Neitzel, R.L.: Estimation of permanent noise-induced hearing loss in an urban setting. Environ. Sci. Technol. **47**, 6393–6399 (2013)
53. Domke, H.W.: Acoustic Regelung von Sanitärinstallationen (2020)
54. Rindel, J.H.: Noise control in water supply systems (2021)
55. Muggleton, J., Brennan, M., Gao, Y.: Vibration radiation from buried plastic water pipes. Dynamics of Civil Structures (2020)
56. Fenini, L.: Numerical modelling of flow-induced (2019)
57. Tsai, K.T., Lin, M.D., Chen, Y.H.: Noise mapping in urban environments: a Taiwan study. Appl. Acoust.coust. **70**(7), 964–972 (2009)

Research on Prediction of Marine Dissolved Oxygen Concentration Based on Modal Decomposition

Yan Liu[1,2], Yupeng Zhao[1], Fengqing Liu[1,2(✉)], Xiangfeng Kong[1,2], Xuan Cao[1,2], Ran Ma[1,2], and Yang Wang[1,2]

[1] Institute of Oceanographic Instrumentation, Qilu University of Technology (Shandong Academy of Sciences), Qingdao, Shandong, China
lfqemail@163.com, kongxiangfeng@qlu.edu.cn, cx@sdioi.com, mr_sdioi@126.com, yangwangphy@139.com
[2] Laoshan Laboratory, Qingdao, Shandong, China

Abstract. Marine Dissolved Oxygen Concentration (MDOC) is one of the key factors affecting marine aquaculture and marine ecosystems. Aiming at the problems of complexity and diversity of factors affecting MDOC and the difficulty of prediction, this paper proposes a research method of marine dissolved oxygen concentration prediction based on modal decomposition. Based on the MDOC data monitored by self-research sensors, this study selects the best data for subsequent analysis by pre-processing the raw data, analyzing the results of temporal resampling and the initial decomposition of Complementary Ensemble Empirical Mode Decomposition with Adaptive Noise (CEEMDAN). An adaptive Variable-division Modal Decomposition (VMD) decomposition method is proposed to decompose the high-frequency decomposition terms of the initial decomposition in a secondary decomposition to fully explore the internal features of the data. Prediction is performed by constructing a Convolutional Neural Networks (CNN)-Bidirectional Long Short-Term Memory (BiLSTM)-Attention Mechanism (attention). The experimental results show that the method proposed in this paper can fully extract many features from the data and achieve high prediction accuracy and effect, providing a new research idea for the prediction of MDOC.

Keywords: Marine Ecological Environment · Marine Dissolved Oxygen Concentration Forecast · Mode Decomposition · Neural Network

1 Introduction

As marine ecological natural disasters have occured frequently in recent years, real-time monitoring of marine ecological environmental parameters is crucial for preventing disasters such as red tide and seagrass. Marine Dissolved Oxygen Concentration (MDOC), as a key indicator, directly affects the survival of marine organisms, especially during natural disaster outbreaks [1]. Therefore, constructing a timely and accurate MDOC prediction model can not only effectively respond to marine disasters and provide real-time reference data for the marine aquaculture industry, but also provide a reference for related research [2].

C. Weng (Ed.): WRE 2024, LNCE 634, pp. 361–376, 2025.
https://doi.org/10.1007/978-981-96-5095-8_30

Dissolved Oxygen (DO) prediction methods soon will mainly include a combination of various types of models, neural networks and artificial intelligence. DO data are usually collected continuously over a long period by sensors [3], and machine learning models are constructed based on them for prediction, such as plain Bayes [4], support vector machine [5], correlation vector machine [6], Light GBM [7], etc. However, these traditional methods make it difficult to mine the long-term patterns of data, compared to which neural networks are more advantageous. For example, Roger Rozario used a Dolphin Glow Worm Optimized Radial Basis Function Neural Network (DGO-RBFNN) to predict the dissolved oxygen content in shrimp ponds [8]. Huang et al. proposed a novel interval prediction method based on a deep auto regression recurrent neural network (DeepAR) for predicting dissolved oxygen content in aquaculture systems [9]. Long Short-Term Memory (LSTM) neural networks are outstanding in time series prediction [10]. For example, Cao et al. effectively improved the accuracy and robustness of DO prediction by optimizing the parameters of the LSTM model and combining K-means clustering and improved particle swarm optimization [11]. With the development of artificial intelligence, it has become a trend to synthesize multiple methods to improve prediction accuracy [1]. Hu et al. used a hybrid Convolutional Neural Networks (CNN-LSTM) model, combined with meteorological data and high-frequency water level, water temperature, and specific conductivity (CTD) data measured within the water sheet, to predict DO concentration. Comparison with LSTM showed that the hybrid model exhibited better prediction stability [12]. Huan et al. combined the gradient boosting decision tree (GBDT) with the long short-term memory network (LSTM). They proposed a prediction model of DO, which further improved the accuracy of DO prediction in aquaculture [13]. Although these methods can capture the long-term patterns of data, they are still insufficient in dealing with nonlinear and complex time series, so the periodicity, noise and randomness in the data need to be further explored to improve the prediction accuracy of the model [14].

Modal decomposition methods, as decomposition strategies adapted to nonlinear data, have been widely used in several fields in recent years. For example, Le et al. used Empirical Modal Decomposition (EMD) to decompose battery data into intrinsic modal functions (IMFs) and residual sequences (Res), and used a hybrid model to make predictions [15]. However, EMD has some issues, among which modal aliasing and endpoint effects are the two most common limitations. Researchers have proposed various improvement methods based on the EMD method to overcome these limitations. For example, Jin et al. used an improved EEMD decomposition strategy to effectively decompose the monthly mean sea level (MSL) time series, and combined it with a machine learning model to achieve effective prediction [16]. Sareen et al. combined Complementary Ensemble Empirical Mode Decomposition with Adaptive Noise (CEEMDAN) with adaptive noise and BiLSTM deep learning algorithm to achieve accurate wave height prediction, solving the problem of wave energy prediction in power systems [17]. Similar studies have been conducted in DO prediction. Schäfer et al. used EMD to decompose river water quality data (including DO data, etc.), dividing the data into two parts: trend and fluctuation. Then, they used hyper statistical methods to analyze the fluctuation part and explore the phenomenon that the fluctuation of river water quality time series follows hyper statistics, in order to reveal the simple local distribution hidden behind complex time series [18]. Kim et al. combined Variational Modal Decomposition (VMD) with Bidirectional Long Short-Term Memory (BiLSTM) to make long-term and short-term predictions of DO data in Poyang Lake [19].

The marine environment is more complex than ordinary waters, and the nonlinear complexity of MDOC is relatively higher. Based on the above research, this paper proposes a quadratic decomposition strategy that uses CEEMDAN and adaptive VMD decomposition method, and builds a hybrid prediction model with CNN, BiLSTM and Attention Mechanism (attention) for predicting the MODC. First, CEEMDAN has the qualities of adaptive noise and decomposition completeness, which can effectively improve the EMD decomposition results and accelerate the computation process. Adaptive noise and decomposition completeness can effectively improve the EMD decomposition results and accelerate the computation process. The Sample Entropy (SE) is used to analyze each decomposition term, and an adaptive VMD method is used to decompose the complex decomposition term twice to extract the information and features inside the data fully. Then, local spatial features in the relevant data are extracted by CNN, BiLSTM is used for bidirectional learning of temporal correlation information between the input data, and attention is introduced to enhance the key features of the model to predict MDOC more accurately. In addition, this paper combines the described model with LSTM, BiLSTM, hybrid models (CNN-BiLSTM, CNN-BiLSTM-attention), and other models. The results show that the model proposed in this paper has better results than other models with sufficient data samples.

2 Related Methods

2.1 Ceemdan

CEEMDAN is an improved method based on EMD while incorporating the idea of Ensemble Empirical Mode Decomposition (EEMD) by adding Gaussian noise to the data and superimposing and averaging multiple times to cancel out the noise [20]. Compared with EMD and EEMD, CEEMDAN has the property of completeness that the original signal can be obtained by summing the decomposed components, which to some extent overcomes the defects of EEMD and Complementary Ensemble Empirical Mode Decomposition (CEEMD) in completeness the defects of EEMD and CEEMD. In addition, CEEMDAN has faster computation speed and better modal decomposition results, which are also superior to other modal decompositions. The following figure shows a simple flowchart of CEEMDAN decomposition (see Fig. 1).

2.2 Sample Entropy

SE is a metric used to measure the complexity of a time series, the core idea is to measure the complexity of similar patterns in a time series based on information theory. It aims to assess the degree of disorder or complexity of a time series. SE quantifies this complexity by calculating the frequency of similar patterns in a sequence, and its main process is as follows:

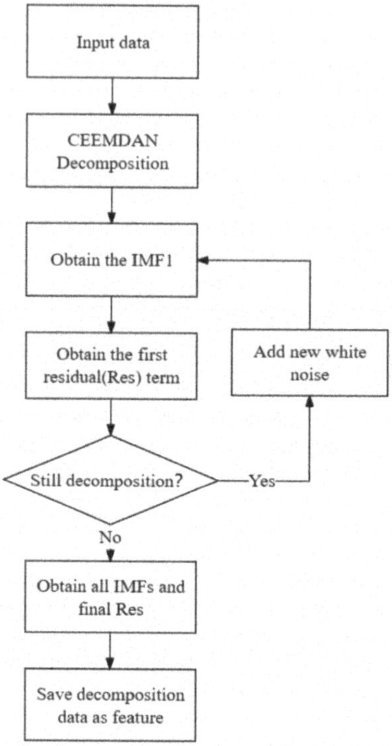

Fig. 1. Simplified flowchart of CEEMDAN decomposition process.

- Define the embedding vector and similarity:
  For the time series $x_1, x_2, \ldots x_N$ and embedding dimension m, define the m-dimensional embedding vector and calculate the distance of each pair of (X_i, X_j) of the time series are shown as Eqs. 1 and 2.
- Calculating similarity ratios:
  For the given tolerance r, compute the similarity ratio of the m-dimensional embedding vector pairs and compute the similarity ratio of the m +1-dimensional embedding vector pairs are shown as Eqs. 3 and 4.
- Calculate the SE:

$$X_i = [x_i, x_{i+1}, \ldots x_{i+m-1}] \tag{1}$$

$$d(X_i, X_j) = \max_{k=1,\ldots,m} |x_{i+k-1} - x_{j+k-1}| \tag{2}$$

$$C(m, r) = \frac{\text{Number of pairs } (X_i, X_j) \text{ with } d(X_i, X_j) \le r}{\text{Total number of pairs } (X_i, X_j)} \tag{3}$$

$$C(m+1, r) = \frac{\text{Number of pairs } (X_i, X_j) \text{ with } d(X_i, X_j) \le r}{\text{Total number of pairs } (X_i, X_j)} \tag{4}$$

$$\text{SE}(m, r, N) = -\ln\left(\frac{C(m+1, r)}{C(m, r)}\right) \tag{5}$$

The final SE(m, r, N) is calculated by Eq. 5.

where $C(m, r)$ is the proportion of similarity within tolerance r for an embedding dimension of m. $C(m + 1, r)$ is the proportion of similarity within tolerance r for an embedding dimension of $m + 1$.

SE reflects the complexity between samples by taking a logarithmic form. Higher SE values indicate greater complexity or irregularity, while lower SE values indicate lower complexity or more regular patterns. In this paper, the SE index is mainly used to discriminate the decomposition terms, to seek to decompose the data into inherent frequency terms and noise terms, and to filter the high-frequency terms of the initial decomposition to judge the decomposition terms that need to be decomposed twice. SE plays a good screening role in this paper.

2.3 Improved Adaptive VMD Decomposition

VMD improves the stability and accuracy of decomposition by optimizing the bandwidth and center frequency of each modal function to decompose a complex signal into intrinsic modal functions with different frequency characteristics. Compared with traditional decomposition methods, VMD enhances the stability and accuracy of decomposition by optimizing the objective function to control the bandwidth and center frequency of the modes. However, VMD requires several parameters to be set manually when using it, including the penalty term parameter alpha, which controls the decomposition stringency, the time step tau, the modal number K, the parameter DC, which controls the DC component of the signal, the initialization mode init, and the convergence criterion tol. The selection of these parameters requires a lot of preliminary work, among which the modal number K is the most critical parameter, which directly affects the decomposition finesse, quality, computational complexity and reliability of the subsequent analysis.

In this paper, an adaptive VMD decomposition method incorporating spectral analysis is proposed for automatic estimation of the modal number of the signal K. Spectral analysis is used to identify the major frequency components in the signal and estimate the modal number accordingly, thus making the VMD decomposition more adaptive and accurate. The specific steps of spectral analysis are as follows:

- Fast Fourier Transform (FFT):
  Convert the time series to the frequency domain is shown as Eq. 6.
- Spectral energy calculations:
  The spectral energy distribution is calculated by Eq. 7.
- Cumulative energy calculation:
  The cumulative distribution of spectral energy to estimate the modal number is calculated by Eq. 8. When the accumulated energy reaches a certain threshold, the modal number K is determined.
- VMD decomposition:

$$Y(f) = FFT(x(t)) \tag{6}$$

$$\text{spectrum} = \frac{2}{N}|Y(f)| \tag{7}$$

$$\text{energy} = \frac{\sum_{i=1}^{k} \text{spectrum}[i]}{\sum_{i=1}^{\frac{N}{2}} \text{spectrum}[i]} \tag{8}$$

$$\min_{u_k,\omega_k}\left\{\sum_{k=1}^{K} \partial_t\left[\left(\delta(t) + \frac{j}{\pi t}\right) * u_k(t)\right]e_2^{-j\omega_k t2} + \alpha\sum_{k=1}^{K} \partial_t u_k(t)_2^2\right\} \tag{9}$$

VMD decomposition using adaptively estimated modal number K: Shown as Eq. 9, decomposes the signal $x(t)$ into K modal functions $u_k(t)$ to satisfy the relational optimization objective.

where $x(t)$ is the time series and $Y(f)$ is the frequency domain signal, N is the signal length and $|Y(f)|$ is the amplitude of the signal in the frequency domain, ω_k is the center frequency of each mode, ∂_t denotes the time derivative, and $*$ denotes the convolution operation.

2.4 CNN-BiLSTM-Attention

In the study of time series data, CNN extracts the local features of the data through the convolutional layer, and the pooling layer, reducing the dimensionality in the time step dimension to improve the computational efficiency. Traditional RNNs are prone to gradient vanishing and explosion problems when dealing with long time series, while LSTM effectively controls the information flow by introducing memory cells and gating mechanisms to capture dependencies in long time series. BiLSTM is further improved by introducing a bidirectional LSTM layer to better capture pre- and post-contextual information and improve prediction accuracy. The attention helps the model to focus on important time points by dynamically assigning weights to improve the understanding and prediction of complex temporal dependencies.

In this paper, we generate a CNN-BiLSTM-attention based prediction model for MDOC data by combining the above methods (see Fig. 2). The model consists of an input layer, a CNN layer (convolutional and pooling layers), a BiLSTM layer (forward LSTM layer and backward LSTM layer), an attention layer and an output layer. Among them, the input layer takes the decomposed features and the corresponding data as input; the convolutional layer in the CNN layer performs feature extraction, the pooling layer performs dimensionality reduction of the feature data, and the reshaping layer reshapes the pooled feature map into a shape adapted to the input of the subsequent model; the BiLSTM layer takes the data after the dimensionality reduction of the extracted features in the CNN layer and performs time sequence data to obtain the output data; attention layer calculates the attention value of the feature state of the data at different moments to the predicted value, and obtains the correlation between each data of the time series and the predicted value; output layer carries out the output calculation of the data outputted from the attention layer to obtain the predicted value.

2.5 Evaluation Metrics

In this paper, mean-square error (MSE), Root Mean Squared Error (RMSE), Mean Absolute Error (MAE), Mean Absolute Percentage Error (MAPE) and Coefficient of

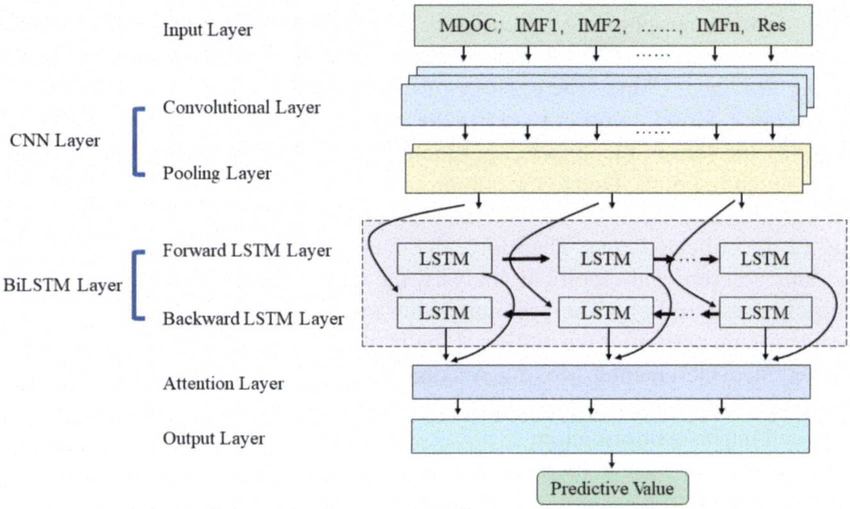

Fig. 2. Structure of CNN-BiLSTM-Attention Model.

Determination (R^2) are selected as the evaluation indexes for model prediction evaluation. They between the predicted and actual values of the said model and the comparison model are measured and compared. The indicators can objectively reflect the process of comparing the actual data with the predicted data. The calculation process for each is shown as Eqs. 10 to 14.

$$\mathrm{MSE} = \frac{1}{n}\sum_{i=1}^{n}(y_i - \hat{y}_i) \in [0, +\infty] \tag{10}$$

$$\mathrm{RMSE} = \sqrt{\frac{1}{n}\sum_{i=1}^{n}(y_i - \hat{y}_i)^2} \in [0, +\infty] \tag{11}$$

$$\mathrm{MAE} = \frac{1}{n}\sum_{i=1}^{n}|y_i - \hat{y}_i| \in [0, +\infty] \tag{12}$$

$$\mathrm{MAPE} = \frac{100\%}{n}\sum_{i=1}^{n}\left|\frac{y_i - \hat{y}_i}{y_i}\right| \in [0, +\infty] \tag{13}$$

$$R^2 = 1 - \frac{\sum_{i=1}^{n}(y_i - y_i)^2}{\sum_{i=1}^{n}(y_i - y_i)^2} \in [0, 1] \tag{14}$$

3 Experimental Process

3.1 Experimental Data Description and Preprocessing

Two sets of data were used in this paper: Data. 1 from August 10 to September 19, 2023, and Data. 2 from October 27 to December 27, 2023, with a sampling frequency of 30s. The data used in this paper came from an MDOC sensor independently developed by the

Institute of Oceanographic Instrumentation, Qilu University of Technology (Shandong Academy of Sciences), collected in the nearshore waters of the Jiaozhou Bay, Qingdao, Shandong Province, China. The ecology of the sea area was relatively stable for most of the sampling period. In order to verify the objectivity and reasonableness of the data collected by the sensor, this paper experimentally analyzes and compares the seawater near the sampling points. Due to the complexity of the marine environment, the sensors are susceptible to noise, missing data, and anomalies, which lead to a decrease in the quality of the collected data. Therefore, the data were first pre-processed. According to the data structure and empirical analysis, the triple standard deviation method was used to eliminate outliers. The proximity interpolation method was used to fill in the gaps and missing parts after the outliers. Finally, the filled data were smoothed using the Time-Weighted Incremental Moving Average (TWIMA) method. As a result of these preprocessing operations, the data quality was improved while retaining its original features and intrinsic information.

3.2 Data Sampling Analysis and Initial Decomposition of CEEMDAN

CEEMDAN decomposition experiments are conducted on the preprocessed Data. 1 combined with the characteristics of time series data. It is found that the data can be regarded as short-term or small samples, but its sampling frequency is sufficient to reflect the detailed changes, which could reflect the regularity and periodicity of the MDOCs to a certain extent. However, excessive data in the short term may introduce noise and anomalies, and although the decomposition method can effectively separate the data, the dense data may generate too many decomposition terms, which affects the subsequent research. For this reason, in this paper, we refer to the data types used in the literature [1–18], and down sample Data. 1 using the time resampling method. Under the premise of maintaining data regularity and integrity as much as possible, the analysis efficiency is improved by condensing the amount of data. The datasets after down sampling at different frequencies show (see Fig. 3) that the data are gradually sparse in the visualization as the resampling interval increases, but their macroscopic rule of change is not destroyed. However, resampling intervals that are too large will lose local details and affect data integrity (see Fig. 3(e) and (f)). The analysis shows that reasonable time resampling can reduce data complexity, retain major trends, and save computational costs.

By conducting 100 CEEMDAN decomposition experiments on the above data (see Fig. 4), it is found that (a) exhibits obvious randomness and nonlinearity in IMF1-IMF8, indicating that the excess data may not fully reflect the situation. As the resampling frequency decreases, the decomposition term decreases, and in the interval from (b) to (d), the decomposition term is concentrated in the middle, which can capture the stationary pattern and trend better. And under longer resampling (see Fig. 4(e) and (f)), the high-frequency term is nearly lost and the Res term shows a flatter trend, which may reflect the seasonal long-term trend. By comprehensive comparison, this paper chooses (d) of Fig. 3 as the best initial decomposition data for subsequent analysis.

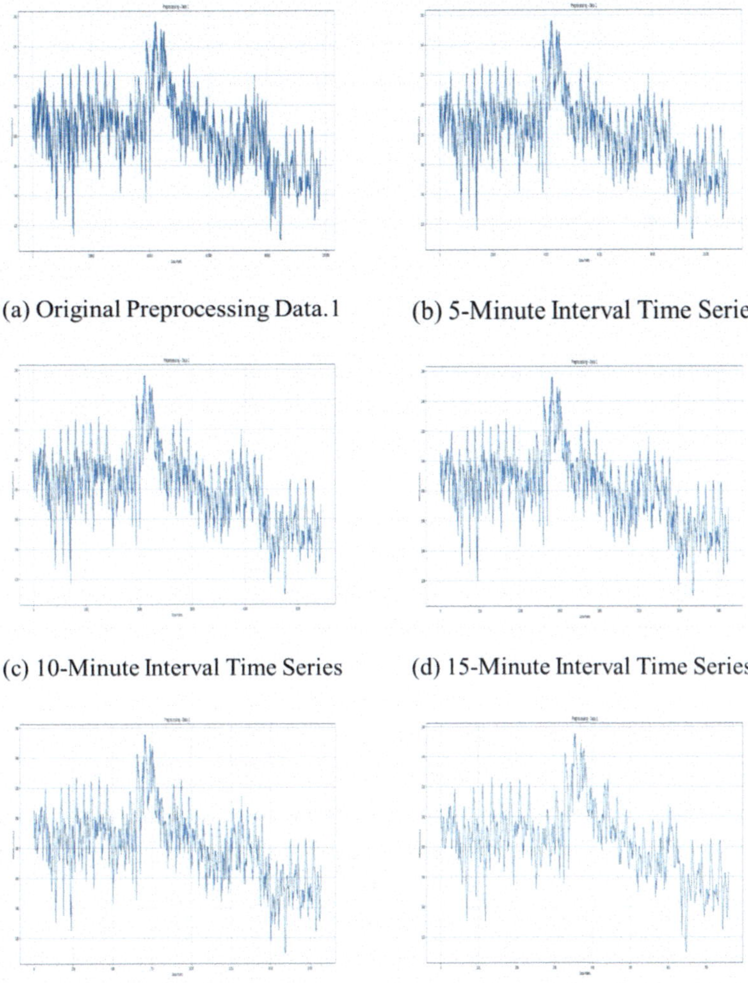

(a) Original Preprocessing Data. 1 (b) 5-Minute Interval Time Series

(c) 10-Minute Interval Time Series (d) 15-Minute Interval Time Series

(e) 30-Minute Interval Time Series (f) 60-Minute Interval Time Series

Fig. 3. Comparison between original preprocessing Data. 1 and compression of interval data.

3.3 Sample Entropy Calculation

According to Sect. 3.2, combined with the results after CEEMDAN decomposition, the SE of each decomposition term is calculated (see Fig. 5), and the SE of the difference between the neighboring decomposition terms is shown in Table. 1.

From Fig. 5(a), the SE value of each decomposition term shows a decreasing trend, which indicates that the fluctuation frequency of each decomposition term is gradually reduced, and the complexity and randomness are also gradually reduced. The SE value of IMF1 is the largest, and its SE difference is also the largest (see Fig. 5(b)), which indicates that there is less similarity between IMF1 and IMF2, and the feature information and

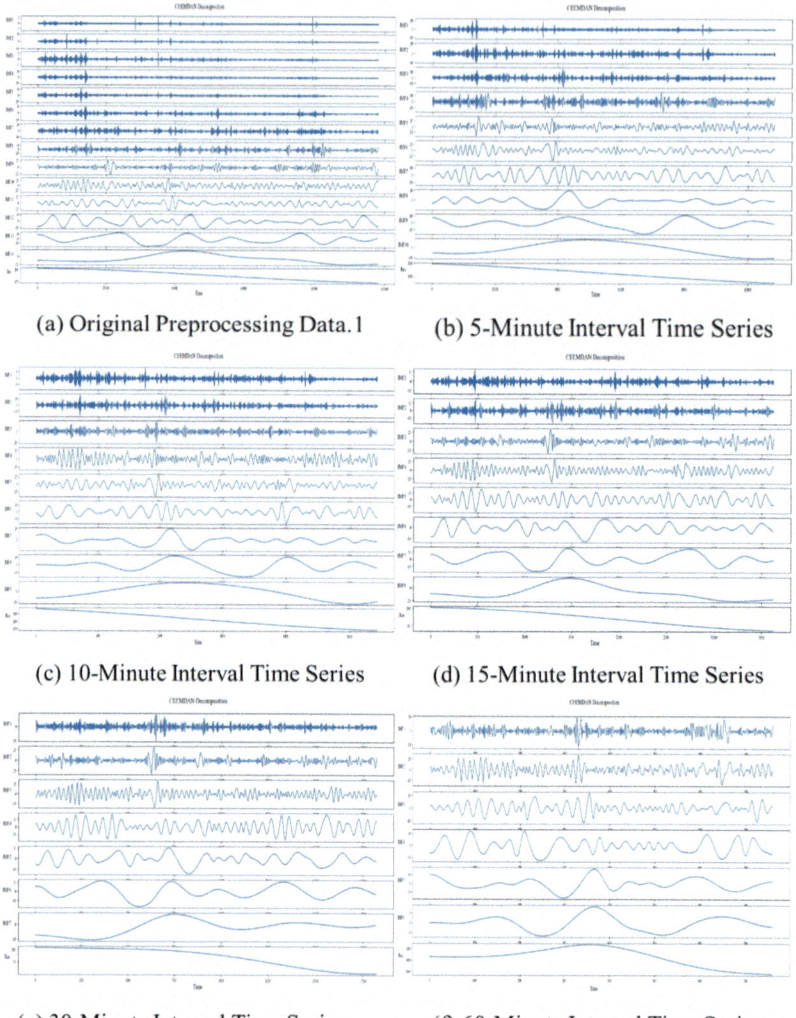

(a) Original Preprocessing Data.1

(b) 5-Minute Interval Time Series

(c) 10-Minute Interval Time Series

(d) 15-Minute Interval Time Series

(e) 30-Minute Interval Time Series

(f) 60-Minute Interval Time Series

Fig. 4. CEEMDAN decomposition comparison.

change details of both are more different. The SE values of IMF2-IMF4 are large, and the entropy difference is also large, indicating that the feature information and change details contained in IMF2-IMF4 have some influence on the prediction, and can be used as the characterization factor for the subsequent prediction inputs. -IMF8 has a smaller SE value and lower complexity, but IMF5 is larger than the remaining decomposition terms of magnitude, indicating that the feature information contained in IMF5 is less like the remaining decomposition terms and can be used as a feature factor alone. IMF6-IMF8 have certain similarities; in order to reduce the computational cost, it is reasonable to superimpose these three decomposition terms as a new mixed decomposition term as a new feature factor. The contribution of the Res term in the SE is very small, but it may affect the effect of the subsequent prediction in the weak bias, so it can also be used as a feature factor.

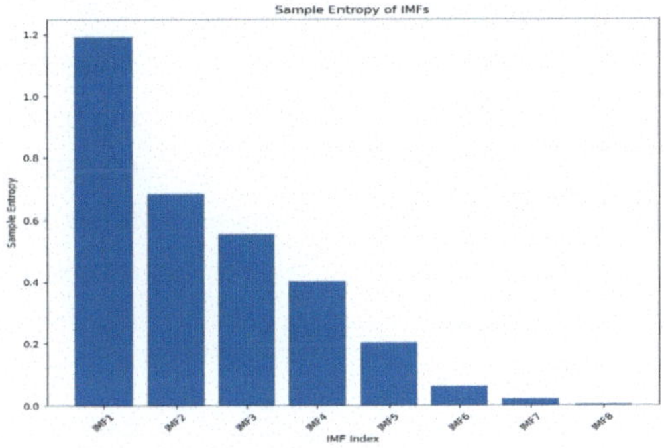

(a)

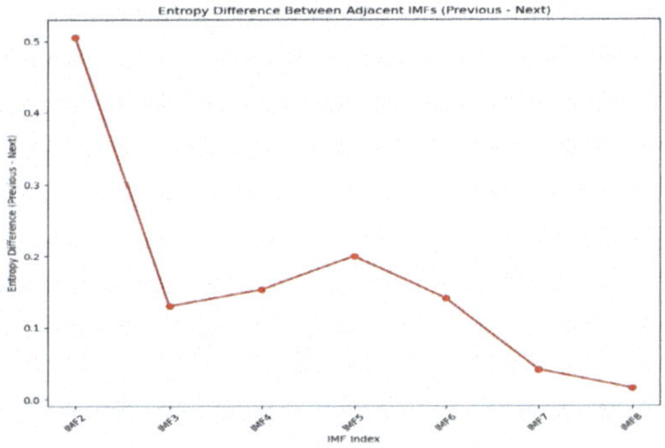

(b)

Fig. 5. The bar chart of SE for each decomposition term and the line chart of the difference in SE between adjacent decomposition terms.

3.4 Adaptive VMD Quadratic Decomposition

According to the conclusion of 3.3, the IMF1 is decomposed using adaptive VMD to further analyze the characteristics of the decomposition term after the decomposition of IMF1. The relevant parameters of the adaptive VMD model described in this paper are: alpha = 2000, tau = 0, DC = 1, init = 1, tol = 1e-7. As can be seen in Fig. 6, the adaptive VMD model selects $K = 10$ as the optimal modal number after relevant calculations and carries out the relevant decomposition. The SE and entropy difference of each decomposition term after secondary decomposition are visualized (see Fig. 7). Figure 7 visualizes the SE and entropy difference of each decomposition term after quadratic decomposition. After adaptive VMD decomposition, the complexity of IMF1

Table 1. SE and differences of Data. 1.

IMF/Res	SE	Entropy Diff
IMF1	1.190230624	–
IMF2	0.68444313	0.505787494
IMF3	0.554141854	0.130301276
IMF4	0.401214396	0.152927458
IMF5	0.201932619	0.199281777
IMF6	0.061089064	0.140843556
IMF7	0.019549376	0.041539688
IMF8	0.003730256	0.01581912
Res	0.000418068	0.003312188

is effectively reduced. The SE of each decomposition term is in the range of 0.2–0.5. The absolute value of the entropy difference is mostly in the range of 0.2, which indicates that the feature information and change details contained in each quadratic decomposition term have certain influence on the prediction, and can be used as the feature factor of the subsequent prediction inputs.

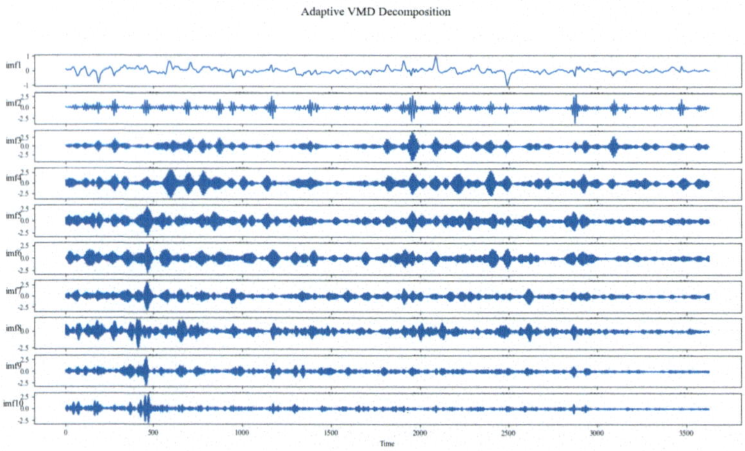

Fig. 6. IMF1 decomposition diagram based on adaptive VMD.

3.5 CNN-BiLSTM-Attention Model Parameter Settings

Data. 1 is used as the sample experimental dataset, and combined with each feature factor described in 3.3 and 3.4, respectively, as the features combined with Data. 1 data for prediction. The predicted data will be summed up to generate the final prediction Data. 2 will be used as the dataset for verifying the performance of the model.

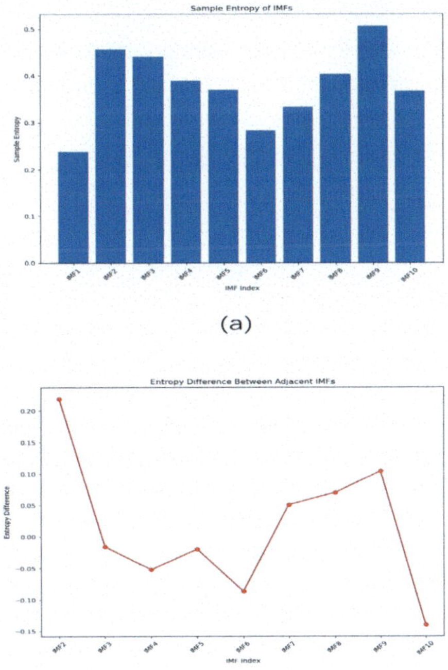

(a)

(b)

Fig. 7. The histogram of SE after secondary decomposition of IMF1 and the line graph of SE difference between adjacent decomposition terms.

After many experimental effect comparisons, this paper chooses to build the CNN-BiLSTM-attention model with 3 CNN layers, 2 BiLSTM layers, and 32 nodes of attention. After comparing several optimizers, the Adam optimizer was able to help the model the most. The model uses 16 as the batch size and divides the raw data into 75% as the training set and 25% as a test set. The programming environment used in this paper is Python 3.9, Tensorflow 2.15.0 and Keras 2.15.0. The parameters related to the compared models are consistent with the corresponding parameters of the models described in this paper.

3.6 Prediction Results and Discussion

To verify the effectiveness of the method proposed in this paper, two sets of comparison experiments are set up for analysis and evaluation. Data. 1 is mainly used to train the model, and Data. 2 is used to test the effectiveness of the method. The two groups of prediction line graphs (see Fig. 8) and evaluation indexes (see Table 2) are shown below.

In Data. 1, compared with other models, the method proposed in this paper achieves the best values in all evaluation metrics, indicating that the method can better capture data patterns and make accurate predictions. Figure 8(b) demonstrates the prediction effect in Data. 2, with an R^2 of 95.60%, which is also capable of predicting the case of a larger amount of data with a higher accuracy. However, the prediction is still somewhat affected in local intervals with more frequent changes, indicating that the model mechanism prioritizes accurate modeling of most data points. However, there is still room for improvement in capturing extreme values or outliers in the data.

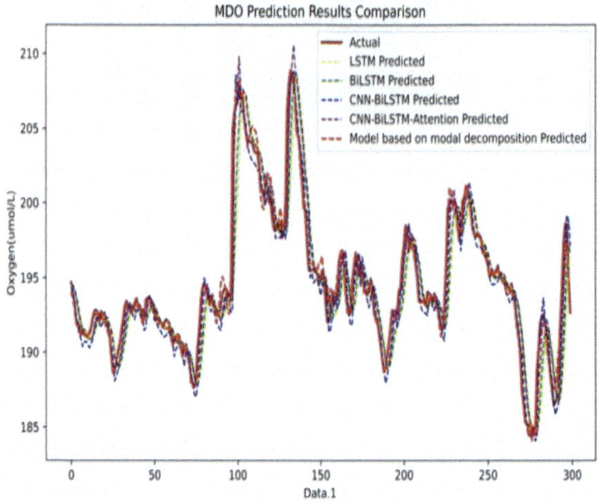

(a)

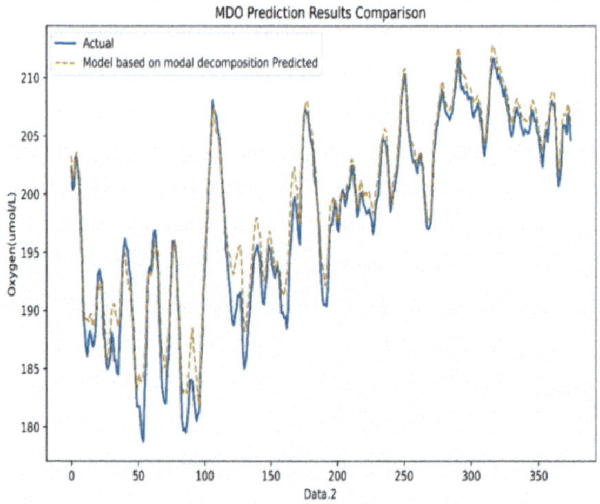

(b)

Fig. 8. Prediction results between Data. 1 and Data. 2.

Table 2. Comparison of predictions from different models by Data. 1.

Models\ Evaluation Metrics	MSE	RMSE	MAE	MAPE	R^2
LSTM	5.70	2.39	1.59	0.81%	74.44%
BiLSTM	3.77	1.94	1.27	0.68%	83.10%
CNN-BiLSTM	2.78	1.67	1.13	0.58%	87.52%
CNN-BiLSTM-attention	1.81	1.35	0.87	0.44%	91.86%
The model described in this paper	0.39	0.62	0.43	0.22%	98.26%

4 Conclusion

Compared with the existing methods, this paper proposes a research method for seawater dissolved oxygen prediction based on modal decomposition. The main research conclusions are as follows:

1. A secondary decomposition strategy proposed in this paper can effectively decompose the data and extract the noteworthy decomposition terms as feature factors to assist the model to make more accurate predictions. This provides a reference for applying data decomposition methods in marine complex data.
2. The experimental analysis shows that the model described in this paper is more accurate in prediction than the traditional model. This indicates that the model can further capture and learn the characteristics of the data to achieve higher accuracy prediction based on the premise of effective data decomposition.

Acknowledgements. This work was supported by the Major Innovation Project of Science and Education Integration Pilot Project of Qilu University of Technology (Shandong Academy of Sciences)(2023JBZ02), Major Innovation Project of Science and Education Integration Pilot Project of Qilu University of Technology (Shandong Academy of Sciences)(2023JBZ03), Science Education Industry Integration Pilot Major Innovation Project (2023HTZX01) of Qilu University of Technology (Shandong Academy of Sciences), Open Research Fund of Key Laboratory of Marine Ecological Monitoring and Restoration Technology, Ministry of Natural Resources, (MEMRT202301).

References

1. Wang, L., Jiang, Y., Qi, H.: Marine dissolved oxygen prediction with tree tuned deep neural network. IEEE Access **8**, 182431–182440 (2020)
2. Wu, J., et al.: Multi-step ahead dissolved oxygen concentration prediction based on knowledge guided ensemble learning and explainable artificial intelligence. J. Hydrol.Hydrol. **636**, 131297 (2024)
3. Park, S., et al.: An antifouling redox sensor with a flexible carbon fiber electrode for machine learning-based dissolved oxygen prediction in severely eutrophic waters. Water **15**(13), 2467 (2023)

4. Sun, J., Li, D., Fan, D.: A novel dissolved oxygen prediction model based on enhanced semi-naive Bayes for ocean ranches in northeast China. PeerJ Comput. Sci. **7**, e591 (2021)
5. Li, Y., et al.: Dissolved oxygen prediction model for the Yangtze River estuary basin using IPSO-LSSVM. Water **15**(12), 2206 (2023)
6. Sun, X., et al.: Multivariate seawater quality prediction based on PCA-RVM supported by edge computing towards smart ocean. IEEE Access **8**, 54506–54513 (2020)
7. Liang, W., et al.: Spatiotemporal-aware machine learning approaches for dissolved oxygen prediction in coastal waters. Sci. Total. Environ. **905**, 167138 (2023)
8. Roger Rozario, A.P.: Prediction of dissolved oxygen in shrimp pond using dolphin glow worm optimization based radial basis function neural network. 2024 IEEE 5th World AI IoT Congress, AIIoT 2024, 286–291 (2024)
9. Huang, J., et al.: Dissolved oxygen content interval prediction based on auto regression recurrent neural network. J. Ambient. Intell. Humaniz. Comput.Intell. Humaniz. Comput. **14**(6), 7255–7264 (2023)
10. Khabusi, S.P., Huang, Y.P.: A deep learning approach to predict dissolved oxygen in aquaculture. 2022 International Conference on Advanced Robotics and Intelligent Systems (ARIS), Taipei, Taiwan, pp. 1–6 (2022)
11. Cao, S.Q., Zhou, L.X., Zhang, Z.: Prediction model of dissolved oxygen in aquaculture based on improved long short-term memory neural network. Trans. Chin. Soc. Agric. Eng. (Trans. CSAE) **37**(14), 235–242 (2021)
12. Hu, Y., et al.: Prediction of riverine daily minimum dissolved oxygen concentrations using hybrid deep learning and routine hydrometeorological data. Sci. Total. Environ. **918**, 170383–170393 (2024)
13. Huan, J., et al.: Prediction of dissolved oxygen in aquaculture based on gradient boosting decision tree and long short-term memory network: a study of Chang Zhou fishery demonstration base, China. Comput. Electron. Agric.. Electron. Agric. **175**, 105530–105539 (2020)
14. Wang, Z., et al.: A deep learning interpretable model for river dissolved oxygen multi-step and interval prediction based on multi-source data fusion. J. Hydrol.Hydrol. **629**, 130637–130647 (2024)
15. Le Thi Minh, L., et al.: Forecasting state-of-health of battery based on combined EMD-CNN-BiLSTM-SM Method. 2023 Asia Meeting on Environment and Electrical Engineering (EEE-AM), Hanoi, Vietnam, 01–05 (2023)
16. Jin, H., Zhong, et al.: Using EEMD mode decomposition in combination with machine learning models to improve the accuracy of monthly sea level predictions in the coastal area of China. Dyn. Atmos. Ocean. **102**, 101370–101386 (2023)
17. Sareen, K., et al.: An integrated decomposition algorithm based bidirectional LSTM neural network approach for predicting ocean wave height and ocean wave energy. Ocean Eng. **281**, 114852–114868 (2023)
18. Schäfer, B., et al.: Fluctuations of water quality time series in rivers follow superstatistics. iScience **24**(8), 102881–102881 (2021)
19. Kim, J., et al.: A novel hybrid water quality forecast model based on real-time data decomposition and error correction. Process. Saf. Environ. Prot.Saf. Environ. Prot. **162**, 553–565 (2022)
20. Torres, M.E., et al.: A complete ensemble empirical mode decomposition with adaptive noise. 2011 IEEE International Conference on Acoustics, Speech and Signal Processing (ICASSP), Prague, Czech Republic, pp. 4144–4147 (2011)

Oil and Grease Removal in Fast-Food Wastewater Using Electrocoagulation Designed Reactor: A Bench-Scale Approach

Maria Christine B. Latosa[1,2](✉) and Kathlia A. De Castro-Cruz[2]

[1] Department of Science and Technology—Industrial Technology Development Institute, Taguig City, Philippines
mcblatosa@itdi.dost.gov.ph
[2] Mapua University, Manila, Philippines

Abstract. The rapid growth of the fast-food industry leads to substantial wastewater production, particularly rich in oil and grease, posing significant environmental and public health risks. Conventional wastewater treatment methods often face limitations, such as excessive sludge generation and high operational costs. This study investigates electrocoagulation (EC) as a sustainable method for treating fast-food wastewater, focusing on oil and grease removal. A bench-scale EC reactor was developed and optimized using a full factorial design with 81 experimental runs to assess key parameters like electrode type (aluminum, iron, stainless), voltage, current, and rotational speed. The optimized EC process was applied to actual wastewater from a fast-food establishment in Pasig City, Metro Manila, confirming its effectiveness and compliance with environmental standards. Results showed that stainless-steel electrodes, operated at 30 V, 1 A, and 100 rpm, achieved a remarkable 98.8% reduction in oil and grease concentrations. EC offers a viable and cost-effective solution for wastewater management in the fast-food industry, providing significant advantages over traditional methods. Future research should explore scalability and alternative electrode materials to enhance the applicability of EC technology.

Keywords: Electrocoagulation · Fast-Food Industry Wastewater · Oil and Grease Removal

1 Introduction

The fast-food industry is rapidly expanding and generates substantial wastewater, which poses significant environmental challenges. Effective wastewater management is crucial for mitigating these impacts, ensuring compliance with regulatory standards, and protecting public health [1]. Proper treatment and disposal of wastewater are essential to prevent the contamination of soils and groundwater, preserve biodiversity, and reduce risks from harmful pathogens and chemicals [2]. This integrated approach promotes sustainability and enhances the industry's reputation, highlighting the importance of responsible waste management practices.

C. Weng (Ed.): WRE 2024, LNCE 634, pp. 377–386, 2025.
https://doi.org/10.1007/978-981-96-5095-8_31

Fast food establishments produce wastewater with high levels of oil and grease, leading to clogs and unpleasant odors in drainage systems [3]. Such accumulations can cause sewer overflows, diminish system capacity, and potentially rupture sewer pipes, ultimately resulting in environmental degradation and public health risks [4, 5]. Although most fast-food chains employ pre-treatment systems to segregate solid waste and oil, these systems often fall short. The rapid decomposition of solid waste during peak operational hours can lead to pollutant accumulation at the effluent surface, undermining effective oil and grease removal. Current treatment techniques in the fast-food sector, including activated sludge processes, chemical treatments, anaerobic digestion, and membrane filtration, each have advantages but also face limitations, such as high sludge production, membrane fouling, and elevated operational costs [6]. These constraints highlight the urgent need for advanced treatment methods.

Electrocoagulation (EC) has emerged as a promising alternative due to its sustainability and adaptability in treating diverse wastewater streams [7]. This process employs metal electrodes as anodes and cathodes, connected to an external direct current (DC) power source and immersed in the wastewater [8]. The electrochemical process facilitates the dissolution of sacrificial electrodes, generating metallic hydroxides that destabilize pollutants, allowing them to agglomerate into larger masses that can be easily separated [9, 10]. Unlike traditional chemical coagulation, which often requires slow operation and produces significant sludge, EC offers precise control over coagulant dosage through modulation of electrical current, eliminating the need for additional chemical additives [11]. Despite its potential, existing wastewater treatment facilities often struggle with inefficiencies in pollutant removal.

To address these challenges, this study developed a bench-scale electrocoagulation reactor for the treatment of fast-food wastewater in a batch mode operation for thirty minutes. Specifically, optimum operating conditions with the highest reduction of oil and grease contaminants in simulated wastewater were investigated. A full factorial design comprising eighty-one experimental runs was used to evaluate key parameters such as electrode type, voltage, current, and rotational speed. A comprehensive statistical analysis of these parameters was conducted to identify the optimum parameters, which were subsequently applied to the treatment of oil and grease contaminants in wastewater collected from a fast-food establishment in Pasig City, Metro Manila.

2 Methodology

2.1 Materials and Wastewater

The three reactor electrodes used in this study are made from stainless, aluminum, and iron. These materials were selected for their excellent electrical conductivity, which is essential for efficient current flow during the electrocoagulation process. These rod-shaped electrodes were fabricated at a local metal shop located at Sta. Cruz, Manila. The simulated wastewater used in the optimization study was prepared by mixing waste palm oil and dishwashing detergent with 10 L of tap water. This mixture was designed to replicate the characteristics of sewage generated by fast food establishments, which contain fats, oils, and surfactants.

Actual wastewater was used to assess the optimum condition of the EC reactor that had been obtained. The wastewater was collected from the under-sink chamber of the fast-food restaurant in Pasig City, Metro Manila. This chamber is directly connected to the pipe of the kitchen sink, serving as a primary point of collection for the wastewater generated during various kitchen activities. In this specific area, washing of cooking utensils and food preparation takes place. During collection, the sample was placed in a 10-L plastic container and transported to the Mapua University laboratory for treatment using the EC-designed reactor. Subsequently, the wastewater was brought to the Department of Science and Technology—Industrial Technology Development Institute (DOST-ITDI) for oil and grease analysis.

2.2 Experimental Set-Up and Reactor Design

The EC process of wastewater was performed in a bench-scale batch mode using a rectangular 10-L reactor chamber (see Fig. 1). The chamber comprises an acrylic sheet to ensure clear visibility of the reactions occurring.

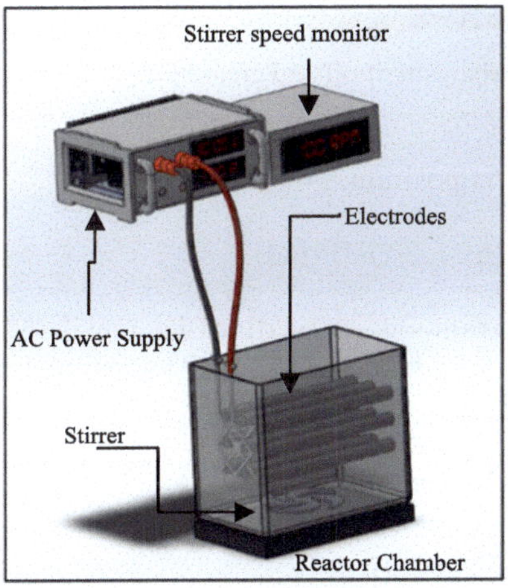

Parameters	Description
Reactor Chamber	
Material:	Acrylic Sheet
Shape:	Rectangular
Volume:	10 Liters
Operation:	Batch Mode
Power Supply	
Voltage:	12V, 24V, 30V
Current:	1A, 2A, 3A
Electrode	
Material:	Stainless Steel, Aluminum, Iron
Shape:	Cylindrical rods
Configuration:	Monopolar
Arrangement:	Parallel, equally spaced
Stirrer	
Rotational Speed:	50 rpm, 100 rpm, 150 rpm

Fig. 1. Electrocoagulation Set-up.

The cylindrical rod-shaped electrodes (aluminum, iron, or stainless) are arranged in parallel monopolar configuration and connected to an alternating current (AC) power supply. The power supply features adjustable settings for voltage (12V, 24V, and 30V) and current (1A, 2A, and 3A). The stirrer operating at different rotational speeds (50 rpm, 100 rpm, and 150 rpm) was also incorporated to facilitate uniform mixing throughout the EC process. The speed of the stirrer, measured in revolutions per minute, is displayed on the speed monitor.

2.3 Mechanism of Oil and Grease Removal Through Electrocoagulation

EC is an effective technique for removing oil and grease from wastewater through chemical reactions induced by an electric current. The EC removal mechanism (see Fig. 2) begins with electrolysis, where water (H_2O) is converted into hydrogen gas (H_2) and hydroxide ions (OH^-).

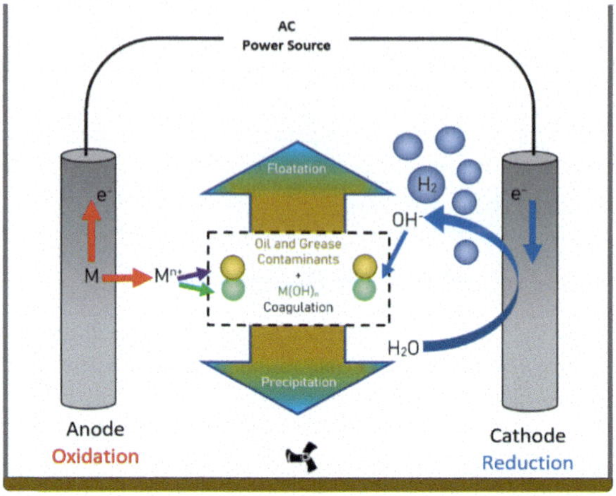

Fig. 2. Modified EC removal mechanism of oil and grease [12].

This reaction alters the solution's ionic composition.

$$2H_2O + 2e^- \rightarrow H_2(g) + 2(OH)^-(aq) \quad (1)$$

Simultaneously, the anode undergoes oxidation, converting the metal (M) into metal cations (M^{n+}). These cations react with hydroxide ions (OH^-) generated during electrolysis to form metal hydroxide coagulants in the solution.

$$M \rightarrow M^{n+} + ne \quad (2)$$

$$M^{n+} + nOH \rightarrow M(OH)n \quad (3)$$

Due to their polar moment and oxidizing capacity, metal hydroxyl ions that act as coagulants form complexes with oil and grease to form larger flocs [13]. The removal mechanism may involve charge neutralization, sweep coagulation, or adsorption [14]. These flocs are separated from the water via sedimentation or flotation. The generated H_2 gas promotes flotation, allowing the flocs to rise to the surface of the liquid. This action facilitates the removal of lighter contaminants, while heavier particles may settle to the bottom as a precipitate. [15].

2.4 Optimization of Bench-Scale EC Reactor

A full factorial approach comprising eighty-one (81) experimental runs was performed for the optimization study of the simulated wastewater. In the experimental design presented in Table 1, several independent variables were examined to assess their impact on the dependent variable, which is the percent reduction of oil and grease contaminants. Three types of electrode materials—aluminum, iron, and stainless—were evaluated for their unique properties and potential impact on the effectiveness of the electrocoagulation process. In addition, three voltage levels (15 V, 20 V, and 30 V) and current levels (1 A, 2 A, and 3 A) were assessed. Both voltage and current are critical determinants of the electrochemical reactions and coagulant generation within the reactor.

Table 1. Design of experiments for simulated wastewater.

Independent variables	Levels	Dependent variable/Response
Electrode Material	Aluminum	Oil and Grease (Percent Reduction)
	Iron	
	Stainless	
Voltage	15 V	
	20 V	
	30 V	
Current	1 A	
	2 A	
	3 A	
Rotational Speed	50 rpm	
	100 rpm	
	150 rpm	

In addition, rotational speeds of 50 rpm, 100 rpm, and 150 rpm were examined to enhance mixing and improve contact between contaminants and coagulants, thereby influencing the overall removal efficiency. By systematically varying these parameters, the study aims to identify the optimum conditions for maximizing the reduction of oil and grease in the simulated wastewater. Each simulated wastewater sample was treated for 30 min in a batch mode process with a 5-min lag time before starting each experimental run.

After each run, approximately 2 L of representative water sample was collected for analysis. The untreated and treated wastewater were placed in a 1 L glass bottle, stored at 4 °C, and analyzed for oil and grease content using the Hexane Extractable Gravimetric Method (EPA Method 1664). A 350 mL aliquot of the sample underwent a sequential extraction process with 20 mL of analytical-reagent grade n-hexane in a separatory funnel. After the extraction, the two phases were allowed to separate, and the hexane layer was collected. This layer was then filtered, and the solvent was evaporated, leaving behind a residue that was weighed to determine the oil and grease content of the sample. The data from eighty-one (81) experimental runs were statistically analyzed and evaluated using Response Surface Modeling (RSM) to determine the optimum operating conditions for the bench-scale EC reactor.

2.5 Application of Optimized EC Conditions to Fast-Food Wastewater

The optimized operating conditions of the bench-scale EC reactor were applied to actual wastewater collected from a fast-food establishment. Approximately ten (10) liters of wastewater samples were treated in the EC reactor using a batch process for 30 min. To assess the effectiveness of the EC treatment, three replicate runs were conducted under the optimized conditions, including the optimum voltage and current applied to the more efficient electrode type while maintaining the optimum stirrer rotational speed. Following each treatment, approximately 4 L of representative samples of the treated wastewater were collected to analyze oil and grease content.

3 Results and Discussion

3.1 Removal Efficiency of Oil and Grease in Simulated Wastewater

The optimum operating conditions for the electrocoagulation (EC) reactor were determined by evaluating its efficiency in reducing oil and grease contaminants in simulated wastewater samples. The percentage reduction of these contaminants was calculated for eighty-one samples treated in varying operating conditions. A statistical analysis was performed to identify the optimum conditions for the reactor to reduce oil and grease content.

Various parameters were systematically varied and analyzed across three replicates employing a full factorial statistical design. This methodology enabled a thorough assessment of the performance of the bench-scale EC reactor to identify the operating conditions with the highest removal efficiency of oil and grease in simulated wastewater.

3.2 Analysis of Variance and Findings

An analysis of variance (ANOVA) was performed to validate the factorial models and assess the impact of various parameters on the electrocoagulation (EC) treatment process for fast food wastewater. The Model F-value of 5.09 indicates statistical significance, with only a 0.01% probability that such a large F-value occurred by chance. P-values below 0.0500 confirmed that parameters such as voltage, electrode type, and their interactions (voltage-current, voltage-electrode, and current-electrode) significantly affected the effectiveness of the EC reactor in reducing oil and grease contaminants.

The significance of voltage and electrode interactions aligns with findings from other studies, which demonstrate that varying electrode configurations and voltage levels can effectively reduce contaminant concentrations in wastewater [16]. Conversely, the P-value for stirrer speed exceeded 0.100, indicating a lack of statistical significance. In contrast, other research has reported that an optimal stirring speed significantly enhances contaminant removal efficiency in wastewater treatment through electrocoagulation [17]. Thus, in this study, the stirrer speed did not significantly impact the removal of oil and grease from fast food wastewater within the parameters tested.

The predicted R^2 value of 0.2591 showed reasonable alignment with the adjusted R^2 value of 0.4171, suggesting that the model adequately fits the data. The high signal-to-noise ratio of 10.46 further reinforces the reliability of the model's predictions, indicating that the variations in the response parameter were primarily driven by relevant factors rather than random fluctuations.

3.3 Response Surface Modeling

Through a three-dimensional response surface modeling, the matrix of the effects of voltage and current on the reduction efficiency of oil and grease was determined. These demonstrate the interaction effects between the two factors with the most significant impact. Such representations are commonly employed to identify optimal process parameter conditions, particularly when the optimal values tend to be located within the same region.

The data from the electrocoagulation treatment through a three-dimensional response surface modeling of oil and grease reveal an irregular surface. The highest percent reduction at the peak occurred at a current of 1 A and a voltage of 30 V. This indicates that these specific operating conditions yield the most favorable reduction efficiency for oil and grease in electrocoagulation.

Through statistical analysis, the optimum operating conditions of the EC reactor, such as a voltage of 30 V, a current of 1 A, and an electrode rotational speed of 100 rpm at a 30-min reaction time while utilizing stainless-steel electrodes for the EC treatment, were determined.

While numerous studies suggest that aluminum electrodes may achieve higher efficiencies, this research demonstrates that stainless-steel electrodes are more effective in removing oil and grease from simulated wastewater [18]. Additionally, some studies have found no significant differences in performance between aluminum and stainless-steel electrodes [19]. Moreover, stainless offers important advantages over aluminum and iron, such as excellent corrosion resistance, high strength, and versatility.

In conjunction with an electrode material, stirring is crucial for enhancing ion mass transfer and promoting effective coagulation. Our results indicate that a rotational speed of 100 rpm maximizes efficiency while preventing excessive turbulence, which can disrupt particle collisions and hinder aggregation [20]. Previous studies have suggested that intense stirring can accelerate the breakage or aging of flocs, reducing the final floc size at equilibrium. Additionally, other studies have reported that optimum removal efficiency is achieved at a stirring speed of 100 rpm, further supporting our results [21].

A significant correlation between voltage and reduction efficiency is also observed in the electrocoagulation process of simulated wastewater. Our findings indicate that at higher voltages, the reduction percentages approach 99%. This observation is consistent with research studies reporting higher removal efficiencies at elevated voltages [22].

Conversely, lowering the voltage may decrease efficiency, highlighting the importance of maintaining optimal operating parameters. Notably, a substantial interaction exists between voltage and current, implying that the influence of one parameter depends on the other. In this study, the highest reduction efficiencies are observed when the voltage is maintained at elevated levels, while the current is kept at lower settings. This is supported by the study of Pasciucco et al. in 2024 [23], which concluded that lower electrical current represents the best compromise, given the reduced energy consumption. At higher voltages, both the size and growth rate of the generated flocs increases, thereby enhancing the overall efficiency of the process.

3.4 Treatment of Fast-Food Wastewater

The findings of this study demonstrate the effectiveness of EC in treating wastewater from a fast-food establishment, particularly under optimized operating conditions. Wastewater collected from Pasig City was subjected to treatment in an EC reactor utilizing stainless-steel electrodes at 30 V, 1 A, and a stirrer rotational speed of 100 rpm. Table 2 summarizes the oil and grease concentrations (expressed in mg/L) for each 30-min replicate run, including the average values and percent reductions achieved.

Initial measurements indicated an oil and grease concentration of 18.95 mg/L. Following 30 min of treatment, concentrations were recorded at 0.12 mg/L, 0.35 mg/L, and 0.22 mg/L across the replicate runs, resulting in an average concentration of 0.23 mg/L. This significant reduction results in an average percent decrease of approximately 98.8%, demonstrating the effectiveness of the EC process in achieving nearly complete contaminant removal. These results are consistent with the research of Fotovat et al. in 2023 [24] on the efficacy of electrocoagulation for oily wastewater treatment. They reported similar reductions, achieving a removal efficiency of 99%, which slightly decreased to 98% when energy consumption was optimized, resulting in a 37% reduction in operating costs. Additionally, Ji et al. [25] demonstrated effective treatment of restaurant wastewater, achieving over 95% removal efficiency under optimal conditions. The considerable decline in oil and grease concentration is attributed to the destabilization and subsequent separation of these contaminants from the wastewater, emphasizing the high efficiency of this treatment method [26].

Table 2. Oil and Grease Analysis Results.

Replicate runs	Oil and Grease (mg/L)		
	0 min	30 min	% Reduction
1	18.95	0.12	99.37
2		0.35	98.15
3		0.22	98.79
Average		0.23	98.79
Standard Deviation		0.12	0.61

Moreover, the results comply with DENR Administrative Order 2016–08, which sets the standard limit for oil and grease in the food industry in the Philippines at 5 mg/L [27]. The achieved concentration of 0.12 mg/L, along with a standard deviation of 0.61% across the three 30-min treatments, indicates consistent performance and reliability of the electrocoagulation process. The ability to significantly reduce oil and grease content using electrocoagulation supports its adoption as a sustainable treatment solution [7]. As environmental regulations become increasingly stringent, industries may benefit from integrating electrocoagulation into their wastewater management strategies to enhance compliance and reduce treatment costs.

The EC process mechanism corroborates the effectiveness of oil and grease removal based on the maximum voltage and low current values considered in this study. It is also complimented by the stainless-steel electrodes at moderate stirring speed. Thus, the effectiveness of EC may vary based on differences in wastewater compositions and electrode materials.

4 Conclusion

The study demonstrates that electrocoagulation is a highly effective method for treating fast-food wastewater, achieving substantial reductions in oil and grease concentrations under optimized operating conditions. These conditions include the use of stainless-steel electrodes, a voltage of 30V, a current of 1A, and a stirring speed of 100 rpm, leading to a significant 98.8% reduction in oil and grease levels. These results confirm electrocoagulation's potential as a viable alternative to conventional treatment methods, offering benefits such as minimized chemical usage, reduced operational costs, and improved environmental sustainability.

Moreover, the achieved oil and grease concentration reduction meets regulatory standards, positioning electrocoagulation as a valuable solution for wastewater management in the fast-food industry. Future research should focus on scaling the process and exploring alternative electrode materials to enhance treatment efficacy. Overall, electrocoagulation represents a promising solution for managing wastewater with high concentrations of oil and grease, particularly in the fast-food sector.

References

1. Etsuyankpa, M.B., et al.: An overview of wastewater characteristics, treatment, and disposal: a review. J. Appl. Sci. Environ. Manag. **28**(5), 1553–1572 (2024)
2. Singh, B.J., Chakraborty, A., Sehgal, R.: A systematic review of industrial wastewater management: evaluating challenges and enablers. J. Environ. Manage. **348**, 119230 (2023)
3. Ahmad, I., Abdullah, N. a. C., Iwamoto, K., Yuzir, A., Ahmad, M.D., Rachmadona, N., Al-Dailami, A., Show, P.L., Khoo, K.S.: Micro and macro analysis of restaurant wastewater containing fat, oil, grease (FOG): an approach based on prevention, control, and sustainable management. Chemosphere **325**, 138236 (2023)
4. Williams, J., Clarkson, C., Mant, C., Drinkwater, A., May, E.: Fat, oil and grease deposits in sewers: characterisation of deposits and formation mechanisms. Water Res. **46**(19), 6319–6328 (2012)
5. De Medeiros, A.D.M., Da Silva, C.J.G., De Amorim, J.D.P., Durval, I.J.B., De Santana Costa, A.F., Sarubbo, L.A.: Oily wastewater treatment: methods, challenges, and trends. Processes **10**(4), 743(2022)
6. Ahmad, I., Abdullah, N.a.C., Iwamoto, K., Yuzir, A., Mohamad, S.E., Show, P.L., Cheah, W.Y., Khoo, K.S.: The role of restaurant wastewater for producing bioenergy towards a circular economy: a review on composition, environmental impacts, and sustainable integrated management. Environ. Res. **214**, 113854 (2022)
7. Tahreen, A., Jami, M.S., Ali, F.: Role of electrocoagulation in wastewater treatment: a developmental review. J. Water Process. Eng. **37**, 101440 (2020)
8. Rusdianasari, J., Taqwa, A., Wijarnako, Y.: Effectiveness of electrocoagulation method in processing integrated wastewater using aluminum and stainless-steel electrodes. J. Phys. Conf. Ser. **1167**, 012040 (2019)

9. Emamjomeh, M.M., Sivakumar, M.: Review of pollutants removed by electrocoagulation and electrocoagulation/flotation processes. J. Environ. Manage. **90**(5), 1663–1679 (2009)
10. Nasrullah, M., Singh, L., Krishnan, S., Sakinah, A.M.M., Zularisam, A.W.: Electrode design for electrochemical cell to treat palm oil mill effluent by electrocoagulation process. Environ. Technol. Innov. **9**, 323–341 (2018)
11. Cañizares, P., Martínez, F., Jiménez, C., Sáez, C., Rodrigo, M.A.: Technical and economic comparison of conventional and electrochemical coagulation processes. J. Chem. Technol. Biotechnol. **84**(5), 702–710 (2009)
12. Priya, T., Mishra, B.K., Prasad, M.N.V.: Physico-chemical techniques for the removal of disinfection by-products precursors from water. In Elsevier eBooks (pp. 23–58) (2020)
13. Collana, J.T.M., Ormeño, M.A., Meza, C.J., Cuadros, G.M., Venegas, L.a.C., Dextre, C.a.A., Taranco, O.J.R., Carhuaricra, C.A., Bravo, P.D., Pisfil, J.a.M.: Processes coupled to electrocoagulation for the treatment of distillery wastewater. Sustainability **16**(15), 6383 (2024)
14. Safwat, S.M.: Treatment of real printing wastewater using electrocoagulation process with titanium and zinc electrodes. J. Water Process. Eng. **34**, 101137 (2020)
15. Jasim, M.A., et al.: Studying the effect of reactor design on the electrocoagulation treatment performance of oily wastewater. Heliyon **9**(7), e17794 (2023)
16. Oktiawan, W., et al.: Effect of electrode configuration and voltage variations on electrocoagulation process in surfactant removal from laundry wastewater. IOP Conf. Ser. Earth Environ. Sci. **896**(1), 012049 (2021)
17. Bayar, S., Yıldız, Y.Ş, Yılmaz, A.E., İrdemez, Ş: The effect of stirring speed and current density on removal efficiency of poultry slaughterhouse wastewater by electrocoagulation method. Desalination **280**(1–3), 103–107 (2011)
18. Zailani, L.W.M., Zin, N.S.M.: Application of electrocoagulation in various wastewater and leachate treatment-a review. IOP Conf. Ser. Earth Environ. Sci. **140**, 012052 (2018)
19. Karhu, M., Kuokkanen, V., Kuokkanen, T., Rämö, J.: Bench scale electrocoagulation studies of bio-oil-in-water and synthetic oil-in-water emulsions. Sep. Purif. Technol. **96**, 296–305 (2012)
20. Yu, J., Xu, H., Yang, X., Sun, H., Jin, Z., Wang, D.: Floc formation and growth during coagulation removing humic acid: effect of stirring condition. Sep. Purif. Technol. **302**, 122084 (2022)
21. Sitorus, I.S., Astono, W., Iswanto, B.: Batch leachate treatment using stirred electrocoagulation reactor with variation of residence time and stirring rate. IOP Conf. Ser. Earth Environ. Sci. **106**, 012095 (2018)
22. Saad, M.S., Balasubramaniam, L., Wirzal, M.D.H., Halim, N.S.A., Bilad, M.R., Nordin, N.a.H.M., Putra, Z.A., Ramli, F.N.: Integrated Membrane–Electrocoagulation system for removal of celestine blue dyes in wastewater. Membranes **10**(8), 184 (2020)
23. Pasciucco, E., Pasciucco, F., Iannelli, R., Pecorini, I.: A Fenton-based approach at neutral and un-conditioned pH for recalcitrant COD removal in tannery wastewater: experimental test and sludge characterization. Sci. Total. Environ. **926**, 172070 (2024)
24. Fotovat, F., Hosseini, M.: Treatment of oily wastewater by electrocoagulation: simultaneous optimization of oil removal efficiency and specific energy consumption. J. Water Process. Eng. **55**, 104221 (2023)
25. Ji, M., Jiang, X., Wang, F.: A mechanistic approach and response surface optimization of the removal of oil and grease from restaurant wastewater by electrocoagulation and electroflotation. Desalin. Water Treat.. Water Treat. **55**(8), 2044–2052 (2014)
26. An, C., Huang, G., Yao, Y., Zhao, S.: Emerging usage of electrocoagulation technology for oil removal from wastewater: a review. Sci. Total. Environ. **579**, 537–556 (2017)
27. Department of Environment and Natural Resources Administrative Order No. 2016–08. Water Quality Guidelines and General Effluent Standards of 2016

Microplastics in Institutional Wastewater Treatment Plant: Abundance, Characterization, Removal

Yash Choudhary[1], Ayushi Chaudhary[1], Anirban Dutta[2], and Amit Kumar[1(✉)]

[1] Department of Civil Engineering, Malaviya National Institute of Technology Jaipur, Jaipur, India
2020RCE9072@mnit.ac.in, amitrathi.ucf@gmail.com

[2] Department of Physics, Malaviya National Institute of Technology Jaipur, Jaipur, India
anirban.phy@mnit.ac.in

Abstract. This study examined the fate of MPs in an institutional WWTP situated on an academic campus in Jaipur, India. The research assessed MP presence at various stages of the treatment process, including the inlet (IN), removal efficiency post-settler (SET), and outlet (OUT), as well as their accumulation in sludge. The quantity, colour, and size of MPs in the gathered samples were examined. The polymer composition was analyzed using the Fourier Transform Infrared Microscope System (FT-IR). The study found that the number of detected MPs was 53.11 ± 4.89 MPs/L at the IN, 20.88 ± 3.26 MPs/L after the SET, and 3.64 ± 0.71 MPs/L at the OUT, resulting in a total removal efficiency of 88.6%. However, considering the total volume of wastewater treated daily in the plant and the sludge being disposed of in the open, the potential daily release of MPs to the environment was estimated to be millions of MPs per day. The results emphasize that, given the potential reuse of treated wastewater and indiscriminate disposal of WWTP sludge in open areas, these plants could be a notable source of MPs to agroecosystems.

Keywords: Institutional WWTP · Microplastics · Urban pollution

1 Introduction

With the advancement of the global economy, plastic materials have become essential in modern life. Synthetic polymers has surged dramatically in recent years, with worldwide plastic production rising from 234 million tons in 2000 to 460 million tons by 2019 [1]. Plastics remain in the environment for long periods because they break down very slowly, resulting in a swift buildup over time. Currently, the prevalence of microplastics (MPs) and nano-plastics (NPs) has garnered substantial attention from researchers worldwide, given their serious environmental impacts [1–3]. MPs refer to synthetic particles with sizes ranging from 5 mm down to 1 μm along their longest axis. They consist of various polymeric materials, including polyethylene (PE), polystyrene (PS), polypropylene (PP), polyethylene terephthalate (PET), and polyvinyl chloride (PVC) [4]. Over time, these MPs have persisted in the environment, dispersing into aquatic ecosystems, terrestrial

C. Weng (Ed.): WRE 2024, LNCE 634, pp. 387–397, 2025.
https://doi.org/10.1007/978-981-96-5095-8_32

areas, living organisms, food sources, and even the atmosphere. With degradation periods spanning hundreds of years, MPs are readily ingested and can accumulate in both living organisms and aquatic food webs [5–7].

MPs are classified into primary and secondary types based on their sources. Primary MPs are produced by industries to use in cosmetics, personal care products, dermal exfoliators, and other applications. In contrast, secondary MPs result from the breakdown of larger plastic items, such as fishing gear, food packaging, plastic bottles, synthetic textiles, car tires, paints, and cosmetics [8].

WWTPs are specifically designed to remove organic substances and nutrients from wastewater. However, they cannot effectively remove other pollutants such as Per- and Polyfluoroalkyl Substances (PFAS), pesticides, and microplastics. As a result, these contaminants can escape into the water system, posing a threat to aquatic ecosystems [5, 9]. Since the fact that sewage sludge has application for fertilizer in agriculture, this aspect could also pose a threat leading to terrestrial pollution [10]. Although the toxicological effects of MPs are still mostly unidentified, MPs are considered an evolving pollutant, including concerns about probable effect on human health. Virgin MPs have already been shown to negatively impact several organisms [11]. In addition to MPs' inherent effects, they can act as a conduit for transmitting other dangerous substances [12–14].

Several works have been done on the detection of MPs in marine environments [15–17] also in freshwater [18, 19] in India. Every year, tons of plastic trash are dumped into various surface water sources. MPs are largely distributed by wastewater treatment facilities that release their effluent primarily into surface water bodies [20]. This study aims to assess the prevalence, and physical and chemical properties of MPs in institutional WWTPs, by sampling at various stages of treatment of wastewater and in activated sludge.

2 Methodology

2.1 Wastewater and Sludge Sampling

The wastewater treatment plant (WWTP) is situated at MNIT Jaipur, located in Western India (Fig. 1). The facility is designed to perform both secondary and tertiary treatment processes, including dual media filtration, and activated carbon filtration. The plant has an average influent flow rate of approximately 1×10^6 liters daily.

In order to evaluate the presence, of MPs in the WWTP, sampling was done in March-April of the year 2022 at four points of the treatment plant: inlet (IN), settling tank (SET), and outlet (OUT) and sludge [5].

In addition, since MPs can settle, the recycled activated sludge was sampled. The sampling was done in the morning from 10 am to 12 pm thrice a week to consider the uncertainties. The sample sizes for inlet, secondary treated and outlet were 10 L, 10 L and 30 L. Meanwhile, for sludge the collected sample was 50 ml [21]. For each sampling point, there were three samples collected. At all sampling points, wastewater sample collection was done using steel containers. The samples were then sieved with sieves (ISO 3310–1:2000) of the pore sizes 5 mm and 63 μm [22, 23]. The sample for activated sludge was collected using a glass beaker (50 ml in volume).

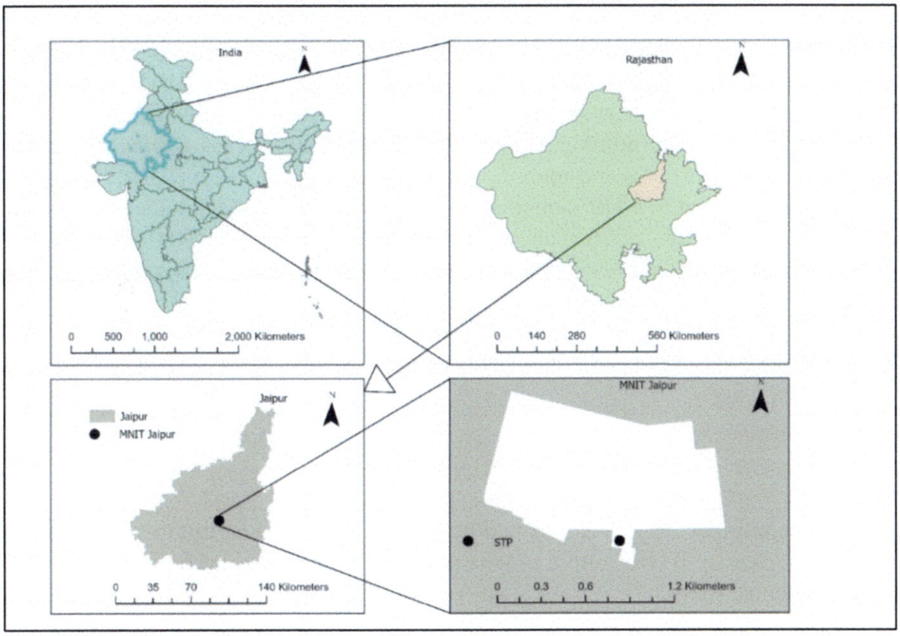

Fig. 1. Location of study area and sampling site.

2.2 Analysis of MPs

After screening, the samples were digested using hydrogen peroxide (30% H_2O_2) and an aqueous ferrous solution (0.05M Fe (II)) [24, 25]. Subsequently, the glass beaker containing the mixture is positioned on a hot magnetic stirrer plate, which is utilized for both heating and stirring the sample [26].

Following the digestion process, density separation was conducted to differentiate particles based on their densities. This method effectively isolates MPs from inorganic particles, such as sand, especially since organic matter has been minimized during digestion. A 5M sodium chloride solution is commonly used for density separation because it is cost-effective and safe for handling [5, 26]. Following density separation, the sample undergoes vacuum filtration, wherein the density-separated components pass through a filter paper. This process collects all particles onto the filter paper, which was then utilized for MP identification analysis.

After the filtration process, filter paper (Whatman Glass Microfiber filter having pore size 1.2-micron) of wastewater and sludge were visually inspected under a stereomicroscope (Optika SZM-LED 2 SN455779) [27]. All suspected plastic particle, and particles whose nature was uncertain were collected, moved to a fresh filter paper, and taken for further analysis that permitted the separation of plastic from non-plastic components.

Visual inspection of particles before their chemical classification is essential in separating particles from filter paper. The assessment of the original method on synthetic samples showed a recovery, ranging from 77 percent to 97 percent, according to particle size [28]. MPs were classified according to their color and size (measured using image analysis software Mag Vision), classifying MPs into three size classes (5–2.5 mm, 2.5–1 mm, 1 mm–0.5 microns).

The suspected microplastic particles gathered on filter papers underwent chemical identification through a Fourier Transform Infrared Microscope System (FT-IR Spectrum 2 by Perkin Elmer, serial number PODL 13070924). This process verified the plastic composition of the particles and identified their specific polymer types. Each particle's spectrum was recorded in FTIR's attenuated total reflectance (ATR) mode, with spectra captured over an average of 16 scans across a wavelength range of 400 to 4100 cm^{-1}. In order to classify the polymers, the obtained spectra were compared with reference spectra using the essential FTIR software [7].

Matching of the spectrum of the sample and the standard spectrum was taken with a matching $\geq$ 60% [29]. The total abundance of MPs on each filter paper was computed as follows (Eq. (1)):

$$\text{Total Number of MPs} = \frac{a}{b} * n \tag{1}$$

where *a*—the number of MPs confirmed by FTIR, items; *b*—the number of suspected particles analyzed by FTIR, items; *n*—the number of suspected particles examined by stereomicroscope, items.

3 Results and Discussion

3.1 MPs Abundance in Sewage and Recycled Activated Sludge

The microplastic (MP) concentration was quantified, yielding an average (mean $\pm$ standard deviation) of 53.11 $\pm$ 4.89 MPs/L in the influent, 20.88 $\pm$ 3.26 MPs/L following the sedimentation stage, and 3.64 $\pm$ 0.71 MPs/L in the effluent. These findings indicate a progressive reduction in MP levels throughout the wastewater treatment process. Specific data on MP removal efficiencies and the residual MP count in the treated effluent are provided in Table 1.

The cumulative removal efficiency of microplastics (MPs) across the entire treatment process, from influent to effluent, was found to be 88.59%. The highest reduction (60.68%) occurred between the influent (IN) and the settling stage (SET). Given that this wastewater treatment plant (WWTP) processes approximately 1×10^6 L per day, the daily MP load at the inlet is estimated to be around 5.311×10^7 particles, with an effluent discharge of approximately 1.82×10^6 MP particles per day. The number of MPs given as mean value $\pm$ SD, and the number of observed MPs was 7945.49 $\pm$ 495.81 MPs/L of the sludge. Given that the investigated WWTP generates around 3.3×10^4 L of sludge daily, then approximately 262,201,170 MPs are collected every day in the sludge.

3.2 MPs Characteristics

The sizes of identified MPs for different sampling points are shown in Fig. 2. The detected MPs are classified into three size ranges: 500–1000, 1000–2500, and 2500–5000 μm.

Table 1. Average daily and yearly MP release rates at each sampling point

Sampling Point	MP/L	MPs/Day	Million MPs/year
IN	53.11 ± 4.89	5311×10^4	19385.15
SET	20.88 ± 3.26	2088×10^4	7621.2
OUT	3.64 ± 0.50	182×10^4	664.3

From Inlet to Outlet, by seeing the size distribution it can be observed that the bigger sized MPs in the sewage treatment process dropped while the particles that are smaller in size continued to rise. In addition, the WWTP can remove MPs with bigger sizes than those with smaller sizes. The two reasons for this incidence are: (1) During treatment, bigger MPs are readily fragmented and absorbed by suspended particulates; (2) smaller MPs can remain in suspension without being removed, especially fibers.

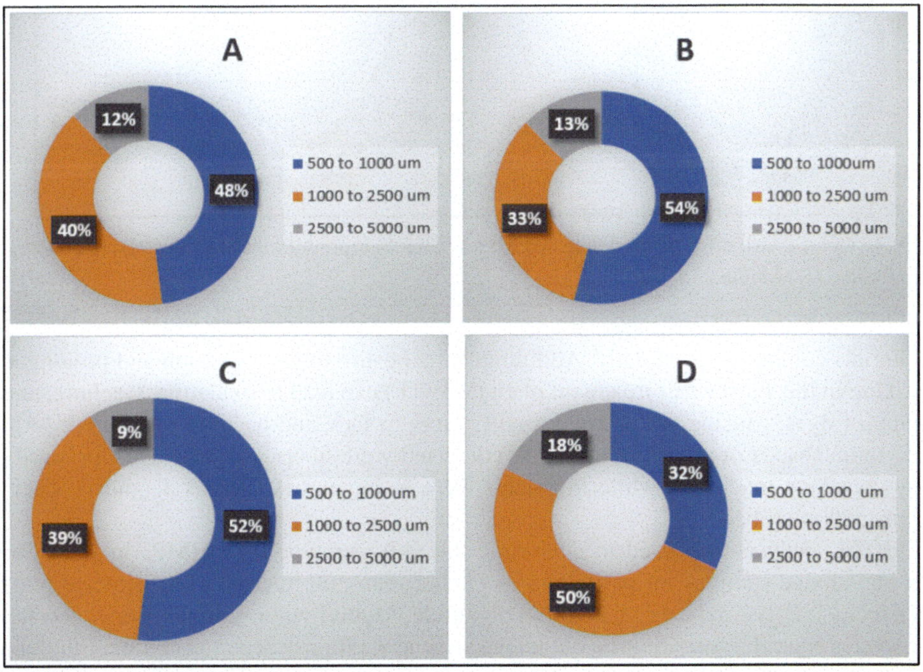

Fig. 2. Size distribution of MPs at each Sampling Point (A) Inlet, (B) Secondary, (C) Outlet, (D) Sludge.

The numerical distribution of MPs by color is presented in Fig. 3. Six primary colors of MPs were observed in the samples after the settling tank, with the majority being transparent, followed by green, red, white, yellow, and blue. Overall, transparent MPs were the most prevalent, while the abundance of white, yellow, and black MPs was comparatively low. The transparent MPs likely correspond to thin plastic films and packaging

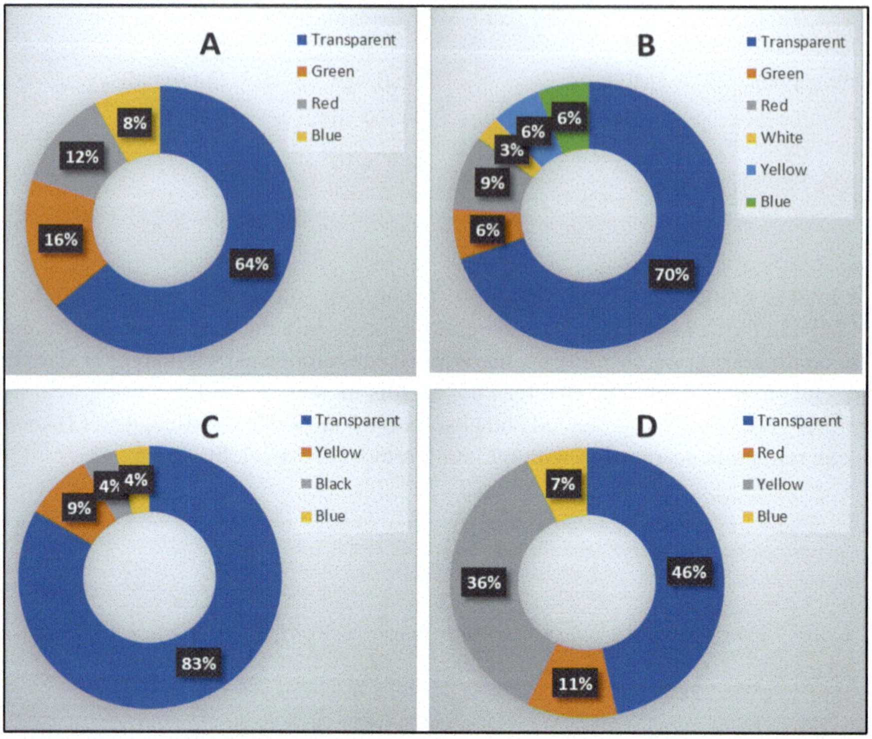

Fig. 3. Color distribution of total detected MPs at each Sampling Point (A) Inlet, (B) Secondary, (C) Outlet, (D) Sludge.

materials commonly used daily. Additionally, exposure to UV radiation and prolonged soaking in the wastewater treatment plant (WWTP) can lead to oxidation, leaching, and aging of dyes, causing the original bright colors to fade and become transparent [30]. The main shapes of microplastics (MPs) detected were fibers and fragments, primarily from washing machines, while fragments were from sources such as degraded plastic items and microbeads.

FTIR analysis was used to determine the polymer type of the MPs in wastewater and sludge samples for particle sizes greater than 500 microns. The detected plastics include Polyvinyl formal, Nylon, Polyamide, Methyl vinyl, Ethylene vinyl acetate, Polyester epoxide, Acrylonitrile butadiene styrene, Cellulose acetate butyrate, Ethylene ethyl acrylate, Polyether urethane, Polyvinyl acetate, Polyvinyl chloride, Polypropylene, Polyvinyl stearate, Polyvinyl alcohol, and Polyurethane in wastewater. Wastewater and sewage sludge both contain similar types of MPs. Figure 4 shows one of the FTIR spectra of a sample done using FTIR.

The primary finding of this study demonstrates that millions of microplastics (MPs) originating from wastewater treatment plant (WWTP) effluents and sewage sludge can

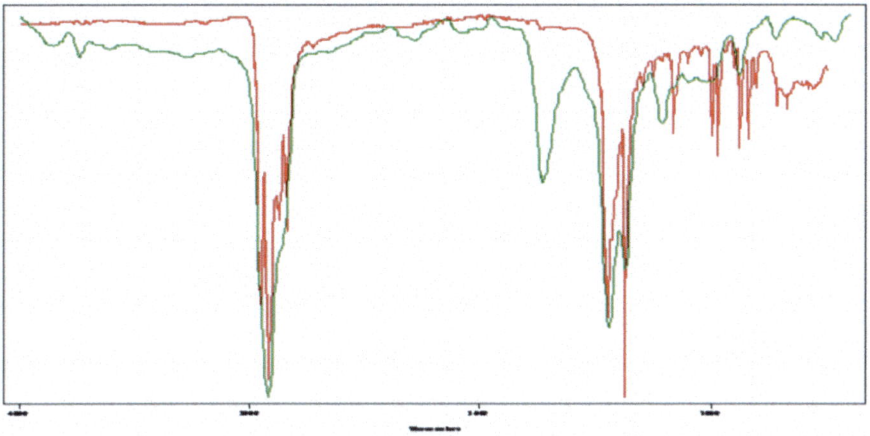

Fig. 4. FTIR spectrum.

be introduced into terrestrial and aquatic ecosystems through irrigation and soil fertilization. These observations align with recent studies from other WWTPs globally, highlighting the role of WWTPs as accumulation points for MPs generated by anthropogenic activities, thereby contributing to environmental contamination [31].

Plastic fragments in water are believed to originate from the breakdown of different plastic products, including packaging materials, personal care items, and household cleaners. Washing machine wastewater, often carried through sewage systems, commonly contributes plastic fibers to water environments. MPs can be generated from both primary and secondary sources. Primary sources involve plastic particles intentionally added to personal care products or industrial materials, such as the plastic beads used in drilling fluids for oil and gas extraction [30]. In the present study, most of the MPs recorded in the wastewater effluent and sludge were likely to be the secondary MPs, as they are fragments that look like broken parts and fibres from textiles. Researchers have observed that as particles become smaller, the concentration of MPs rises. One major risk posed by MPs is their high capacity to attract and hold a range of contaminants, such as persistent organic pollutants (POPs) like polychlorinated biphenyls (PCBs), dioxins, metals, and harmful microorganisms [32]. Due to increased surface area, MPs' adsorption capacity might rise as their size decreases [2].

The average concentration of microplastics detected in the inlet, measured at 53.11 $\pm$ 4.89 MPs/L, was significantly lower than the levels recently observed at the inlet of other sewage treatment plants in India (1860 $\pm$ 265 MPs/L; [33]) and was more than other eastern Indian MBBR-based hybrid growth STP system (64.3 $\pm$ 4.89–47.66 $\pm$ 4.71 n/L in sewage [34]. The difference may be attributed to the source of wastewater for these plants. The treatment plant in this study is an institutional one getting its wastewater from academic laboratories, office toilets, hostels, and residences on the campus. Various types of polymers arise from different processing methods. For instance, microplastics in wastewater primarily originate from synthetic fibres produced in the washing of clothes, as well as from personal care products. These microplastics ultimately find their way into wastewater treatment plants by discharging wastewater from washing machines and using personal care products via household drains [35]. Different weather and seasonal conditions can also affect the variability in MP concentration [30, 34].

The removal efficiency of 88.59% observed in this study is in the same order as other Indian WWTPs ranging from 80% to 90% [5, 33, 34]. With approximately 60.68% of removal achieved during the initial treatment stages, grease removal and sedimentation processes play crucial roles in reducing the floating and settling of MPs from wastewater. Additionally, the sand filters positioned at the final stage of the WWTP contributed significantly, further decreasing the microplastic content in the treated effluent from (20.88 $\pm$ 3.26 MPs/L) to the OUT (3.64 $\pm$ 0.71 MPs/L) by approximately 80%. This is another important finding in identifying straightforward, affordable methods of reducing MP in wastewater.

Despite the advanced effectiveness of wastewater treatment plants (WWTPs) in capturing the majority of microplastics (MPs), the substantial daily volume of wastewater processed means that a significant number of MPs is still released, which remains a critical concern. In order to minimize the release of MPs into the environment through treated effluent, it is essential to enhance MP removal efficiency across all treatment stages within WWTPs.

4 Conclusion

This study investigated the presence, characteristics, and removal efficiency of microplastics (MPs) across various treatment stages in a Moving Bed Biofilm Reactor (MBBR)-based institutional wastewater treatment plant at MNIT, Jaipur. MPs concentrations were detected in the range of 53.11 $\pm$ 4.89 MPs/L in the IN, 20.88 $\pm$ 3.26 MPs/L after the SET, and 3.64 $\pm$ 0.71MPs/L in the OUT and 7945.49 $\pm$ 495.81 MPs/L (sludge). The overall MP removal efficiency of the studied WWTP was estimated to be around 88.6%. The MPs between 500 and 1000 μm were prevalent in sewage and between 1000 to 2000 μm in sludge.

In India, wastewater treatment plants utilize various technologies, including Moving Bed Biofilm Reactor (MBBR) systems, to treat wastewater from households and municipalities. Numerous urban local bodies (ULBs) have recently implemented programs to repurpose treated wastewater for horticulture, irrigation, non-contact water bodies, and cleaning activities. Given the amount of sewage that is routinely treated in India, there is a good likelihood that MPs may re-enter surface water channels and agricultural land. STPs are generally not intended to eliminate MPs, and it would not be easy to do so quickly through technological means. Furthermore, an evaluation of the intended MP removal by a STP must be conducted to determine the likely interventions for implementing cutting-edge treatment techniques. Therefore, it is crucial to limit MPs pollution by improving the current STPs or integrating new STPs with an improved treatment technique to improve the effectiveness of MPs removal. It is essential to establish a standardized protocol for microplastic (MP) monitoring and adopt a uniform unit for expressing MP abundance to facilitate the comparison of data across various sampling sites. The observed variations in MP concentrations are likely influenced by different MP detection methods, underscoring the need for consistent methodologies.

Less plastic and diverse separation techniques can be used to remove plastics based on size. For plastics to be removed and recycled effectively the right technologies must be chosen and the separation process must be designed. Specifically, due to their size, MPs require extra separation stages after the standard procedures. They can be separated through filtration utilizing a range of filter diameters, from macro to nano, and they can be eliminated using membrane filtering, sand filtration, gravity filtration, ultrafiltration, and reverse osmosis.

Future work in this area may also include determining material density to determine the settling rate of the particles. More studies are essential to draw a relationship between the seasonal changes and MP variation in the WWTP.

References

1. OECD: Global Plastics Outlook, Organisation for Economic Cooperation and Development (Feb. 2022). https://doi.org/10.1787/de747aef-en
2. Silva Filho, C.R., Velis, C.A.: United Nations' plastic pollution treaty pathway puts waste and resources management sector at the centre of massive change. Waste Manag. Res. J. a Sustain. Circ. Econ. **40**(5), 487–489 (May 2022). https://doi.org/10.1177/0734242X221094634
3. Mensah, K., Magdaleno, A., Yaparatne, S., Garcia-Segura, S., Apul, O.G.: Emerging investigator series: suspended air nanobubbles in water can shuttle polystyrene nanoplastics to the air-water interface. Environ. Sci. Nano, 3721–3728, (2024). https://doi.org/10.1039/d4en00188e
4. Wei, F., et al.: Distribution of microplastics in the sludge of wastewater treatment plants in Chengdu, China. Chemosphere **287**, 132357 (Jan. 2022). https://doi.org/10.1016/j.chemosphere.2021.132357
5. Magni, S., et al.: The fate of microplastics in an Italian Wastewater Treatment Plant. Sci. Total. Environ. **652**, 602–610 (2019). https://doi.org/10.1016/j.scitotenv.2018.10.269
6. Vaid, M., Mehra, K., Gupta, A.: Microplastics as contaminants in Indian environment: a review. Environ. Sci. Pollut. Res. **28**(48), 68025–68052 (Dec. 2021). https://doi.org/10.1007/s11356-021-16827-6
7. Pandey, D., Banerjee, T., Badola, N., Chauhan, J.S.: Evidences of microplastic in air and street dust: a case study of Varanasi City, India (Jan. 12, 2022). https://doi.org/10.21203/rs.3.rs-1151250/v1
8. Alavian Petroody, S.S., Hashemi, S.H., van Gestel, C.A.M.: Transport and accumulation of microplastics through wastewater treatment sludge processes. Chemosphere **278**, 130471 (Sep. 2021). https://doi.org/10.1016/j.chemosphere.2021.130471
9. Carr, S.A., Liu, J., Tesoro, A.G.: Transport and fate of microplastic particles in wastewater treatment plants. Water Res. **91**, 174–182 (2016). https://doi.org/10.1016/j.watres.2016.01.002
10. Nizzetto, L., Futter, M., Langaas, S.: Are agricultural soils dumps for microplastics of urban origin? Environ. Sci. Technol. **50**(20), 10777–10779 (Oct. 2016). https://doi.org/10.1021/acs.est.6b04140
11. Chae, Y., An, Y.J.: Effects of micro- and nanoplastics on aquatic ecosystems: current research trends and perspectives. Mar. Pollut. Bull. **124**(2), 624–632 (Nov. 2017). https://doi.org/10.1016/j.marpolbul.2017.01.070
12. Andrady, A.L.: Microplastics in the marine environment. Mar. Pollut. Bull. **62**(8), 1596–1605 (Aug. 2011). https://doi.org/10.1016/j.marpolbul.2011.05.030
13. Liebezeit, G., Liebezeit, E.: Synthetic particles as contaminants in German beers. Food Addit. Contam. Part A **31**(9), 1574–1578 (Sep. 2014). https://doi.org/10.1080/19440049.2014.945099

14. Van Cauwenberghe, L., Janssen, C.R.: Microplastics in bivalves cultured for human consumption. Environ. Pollut. **193**, 65–70 (Oct. 2014). https://doi.org/10.1016/j.envpol.2014.06.010
15. Sathish, N., Jeyasanta, K.I., Patterson, J.: Abundance, characteristics and surface degradation features of microplastics in beach sediments of five coastal areas in Tamil Nadu, India. Mar. Pollut. Bull. **142**, 112–118 (2019). https://doi.org/10.1016/j.marpolbul.2019.03.037
16. Jayasiri, H. B., Purushothaman, C. S., Vennila, A.: Plastic litter accumulation on high-water strandline of urban beaches in Mumbai, India, pp. 7709–7719 (2013). https://doi.org/10.1007/s10661-013-3129-z
17. Jayasiri, H.B., Purushothaman, C.S., Vennila, A.: Quantitative analysis of plastic debris on recreational beaches in Mumbai, India. Mar. Pollut. Bull. **77**(1–2), 107–112 (2013). https://doi.org/10.1016/j.marpolbul.2013.10.024
18. Sarkar, D. J., *et al.*: Microplastics removal efficiency of drinking water treatment plant with pulse clarifier. J. Hazard. Mater. **413**(Nov. 2020), 125347 (2021). https://doi.org/10.1016/j.jhazmat.2021.125347
19. Veerasingam, S., et al.: Trends in analytical chemistry microplastics in different environmental compartments in India: analytical methods, distribution, associated contaminants and research needs. Trends Anal. Chem. **133**, 116071 (2020). https://doi.org/10.1016/j.trac.2020.116071
20. Badola, N., Bahuguna, A., Sasson, Y., Chauhan, J.S.: Microplastics removal strategies: a step toward finding the solution. Front. Environ. Sci. Eng. **16**(1), 1–18 (2022). https://doi.org/10.1007/s11783-021-1441-3
21. Crawford, C.B., Quinn, B.: Plastic production, waste and legislation. In: Microplastic Pollutants, Elsevier, pp. 39–56 (2017). https://doi.org/10.1016/B978-0-12-809406-8.00003-7
22. Murphy, F., Ewins, C., Carbonnier, F., Quinn, B.: Wastewater Treatment Works (WwTW) as a source of microplastics in the aquatic environment (2016). https://doi.org/10.1021/acs.est.5b05416
23. Pivokonsky, M., Cermakova, L., Novotna, K., Peer, P., Cajthaml, T., Janda, V.: Occurrence of microplastics in raw and treated drinking water. Sci. Total. Environ. **643**, 1644–1651 (2018). https://doi.org/10.1016/j.scitotenv.2018.08.102
24. Ziajahromi, S., Neale, P.A., Rintoul, L., Leusch, F.D.L.: Wastewater treatment plants as a pathway for microplastics: development of a new approach to sample wastewater-based microplastics. Water Res. **112**, 93–99 (2017). https://doi.org/10.1016/j.watres.2017.01.042
25. Bretas Alvim, C., Bes-Piá, M.A., Mendoza-Roca, J.A.: Separation and identification of microplastics from primary and secondary effluents and activated sludge from wastewater treatment plants. Chem. Eng. J. **402**, 126293 (Dec. 2020). https://doi.org/10.1016/j.cej.2020.126293
26. Masura, J., Baker, J., Foster, G., Arthur, C.: Laboratory methods for the analysis of microplastics in the marine environment. NOAA Mar. Debris Progr. Natl., pp. 1–31 (2015). [Online]. Available: https://marinedebris.noaa.gov/sites/default/files/publications-files/noaa_microplastics_methods_manual.pdf
27. Mason, S.A., Welch, V.G., Neratko, J.: Synthetic polymer contamination in bottled water. Front. Chem. **6**, (Sept. 2018). https://doi.org/10.3389/fchem.2018.00407
28. Avio, C.G., et al.: Pollutants bioavailability and toxicological risk from microplastics to marine mussels. Environ. Pollut. **198**, 211–222 (2015). https://doi.org/10.1016/j.envpol.2014.12.021
29. Shan, W., et al.: Distribution, characteristics and daily fluctuations of microplastics throughout wastewater treatment plants with mixed domestic—industrial influents in Wuxi City, China. Front. Environ. Sci. Eng. **16**(1), 6 (Jan. 2022). https://doi.org/10.1007/s11783-021-1440-4
30. Yang, Z., et al.: Characteristics and removal efficiency of microplastics in sewage treatment plant of Xi'an City, northwest China. Sci. Total. Environ. **771**, 145377 (2021). https://doi.org/10.1016/j.scitotenv.2021.145377

31. Prata, J.C.: Microplastics in wastewater: State of the knowledge on sources, fate and solutions. Mar. Pollut. Bull. **129**(1), 262–265 (2018). https://doi.org/10.1016/j.marpolbul.2018.02.046
32. McCormick, A., Hoellein, T.J., Mason, S.A., Schluep, J., Kelly, J.J.: Microplastic is an abundant and distinct microbial habitat in an urban river. Environ. Sci. Technol. **48**(20), 11863–11871 (2014). https://doi.org/10.1021/es503610r
33. Patil, S., Bafana, A., Naoghare, P.K., Krishnamurthi, K., Sivanesan, S.: Environmental prevalence, fate, impacts, and mitigation of microplastics—a critical review on present understanding and future research scope. Environ. Sci. Pollut. Res. **28**(5), 4951–4974 (2021). https://doi.org/10.1007/s11356-020-11700-4
34. Parashar, N., Hait, S.: Occurrence and removal of microplastics in a hybrid growth sewage treatment plant from Bihar, India: a preliminary study. J. Clean. Prod. **376**(June), 134295 (2022). https://doi.org/10.1016/j.jclepro.2022.134295
35. Liu, X., Yuan, W., Di, M., Li, Z., Wang, J.: Transfer and fate of microplastics during the conventional activated sludge process in one wastewater treatment plant of China. Chem. Eng. J. **362**, 176–182 (2019). https://doi.org/10.1016/j.cej.2019.01.033

Novel Modified Corn Stalk Powder Enhancing Removal of Natural Humic Acids from Water

Feng Lin[1], Shuxin Zhang[1], Baoyu Wang[1], Jiangtao Lv[1], Rui Guo[1], Jing Li[1], Jiongxian Wei[1], and Weihua Qin[2,3](✉)

[1] School of Chemical Engineering and Technology, Guangdong Industry Polytechnic University, Guangzhou, China
2021090107@gdip.edu.cn

[2] State Key Laboratory of Pulp and Paper Engineering, School of Light Industry and Engineering, South China University of Technology, Guangzhou, China
19124369319@163.com

[3] Guangdong Baofujia Environmental Protection Technology Co. Ltd, Foshan, China

Abstract. The formation of carcinogenic by-products via the interaction between HA in water and chlorinated disinfectants underscores the crucial importance of studying HA removal methods. This study utilized abandoned corn stalks powder (CSP) as raw materials and employed an alkali-etherification method to prepare modified corn stalks powder (MCSP) to degrade humic acid (HA) in water, t. Single-factor experiments were conducted to investigate the effects of alkali concentration, alkalization time, etherification temperature, etherification time, and etherification concentration on the removal efficiency of HA. Response surface methodology was further used to optimize the preparation conditions of MCSP. The results revealed that the optimal parameters for the preparation of MCSP were: alkali concentration of 10%, alkalization time of 1.083 h, etherification temperature of 60 °C, etherification time of 2.083 h, and etherification concentration of 50%. To further verify the effectiveness of MCSP, an experiment was conducted utilizing it for the removal of HA from water, and the results showed that the removal rate of HA gradually decreased with the increase of MCSP. This phenomenon was mainly because MCSP could neutralize the charge and increase the particle size in the HA system, thus achieving the goal of removing HA. Furthermore, it demonstrated the promising application prospects of MCSP in the treatment of HA pollutants in water.

Keywords: Alkali-etherification Method · Modified Corn Stalk Powder · Humic Acid · Kinetic Model

Abbreviations

CSP	Corn stalks powder
MCSP	Modified corn stalks powder
HA	Humic acid
NOM	Natural organic matter
DOC	Dissolved organic carbon

C. Weng (Ed.): WRE 2024, LNCE 634, pp. 398–407, 2025.
https://doi.org/10.1007/978-981-96-5095-8_33

CTMAB	Cetyltrimethylammonium bromide
NaOH	Sodium hydroxide
H_2SO_4	Sulfuric acid
RSM	Response Surface Methodology
R1	Response value
BBD	Box-Behnken experimental design
d50	Median particle size
3D-EEM	Three-dimensional fluorescence spectrophotometer

1 Introduction

Access to clean water is vital for sustaining life, yet it remains a formidable challenge in developing nations, largely due to burgeoning populations and inadequate water purification techniques [1]. Natural water bodies harbor a myriad of soluble ions and compounds, many of which are deemed contaminants [2]. Notably, humic acid (HA) constitutes a significant portion of natural organic matter (NOM) and dissolved organic carbon (DOC) in both groundwater and surface water, accounting for as much as 90% of DOC [3].

HA exists in colloidal form in natural water bodies, developed through the prolonged accumulation of different biological byproducts. [4], accounting for 30% to 50% of the total organic matter in water. HA is fully solubilized at pH values above 2 and obtains a negative charge in groundwater (20 μg/L) and surface water (0.030 g/L) [5]. Although HA can be used as an organic fertilizer to improve soil properties [6], high concentrations of HA, serving as a nutrient source for biological development, can interact with pesticides and metallic contaminants, affecting the migration and transformation of metal ions and increasing their biotoxicity. In the process of chlorinating drinking water for disinfection, HA may undergo unfavorable reactions with chlorinated substances, resulting in the production of harmful by-products like carcinogenic trihalomethanes, posing a risk to human health [7]. Furthermore, when the concentration of HA in water exceeds 5 mg/L, it appears yellow to brown and can lead to various issues, such as unpleasant taste, odor, and color, as well as reduced efficiency of water treatment facilities [8]. Health issues are not only a by-product of disinfection, but consuming large amounts of HA may also be one of the causes of human blackfoot disease [9]. Therefore, it is urgent to find efficient and rapid methods to eliminate HA from wastewater while preventing the creation of secondary pollutants.

Corn stalks are an important agricultural biomass resource, with an annual production exceeding 200 million tons in China alone. Stalks are often considered waste and are mostly disposed of through stacking or open-air burning, causing severe environmental pollution [10]. Corn stalks are a natural high-molecular substance containing abundant functional groups such as hydroxyl, providing reactive sites for subsequent modification and increasing the possibility of reactions [11]. Additionally, corn cobs contain a protein colloid called zein, which is easily biodegradable and can be completely decomposed in soil within 14 days and in water within only 7 days [12]. Therefore, utilizing modified waste corn stalks for the degradation of HA not only provides a new approach to turning waste biomass into valuable resources but also reduces carbon emissions and achieves energy-saving and emission reduction [13].

Response surface methodology (RSM) serves as an optimization technique tailored for multivariate challenges, capable of modeling and examining issues influenced by various factors, ultimately enhancing the desired response values. RSM was utilized to optimize the incorporation of a renewable-based solid oxide fuel cell with a Brayton cycle and a heating production unit for a cogeneration application. The other research studies examined the optimum conditions of a coproduction scheme using a solid oxide fuel cell, an S-CO_2, and an organic Rankine cycle using RSM [14]. The Box-Behnken design (BBD) often necessitates a lower number of experimental trials compared to alternative methods within response surface methodology (RSM), such as the Central Composite Design. This is because the BBD excludes extreme level combinations of all factors (i.e., the vertices) and instead places experimental points at the midpoints of each edge of the cube, thereby reducing the total number of experiments. Due to the reduced number of experimental runs, the BBD is more cost-effective. This is particularly important for research projects with limited resources or high experimental costs. The BBD efficiently estimates first-order and second-order coefficients, which are crucial for constructing the response surface model. These coefficients aid in understanding the interactions between variables and predicting responses under different experimental conditions.

Based on this, solid waste corn straw was used as raw material, and the RSM was used to optimize the preparation conditions of modified corn straw. This was applied to the removal of HA in wastewater, and the degradation process of HA was studied to realize waste to waste. This has important implications for the development and innovation of new technologies for HA degradation and can provide valuable references for industry standards.

2 Materials and Methods

2.1 Preparation of Modified Corn Stalk Powder

After passing corn stalk powder (CSP) through a 100-mesh sieve and drying it, place it in a beaker. Add an appropriate amount of sodium hydroxide (NaOH) solution to alkalinize the CSP at room temperature, then add an appropriate amount of cetyltrimethylammonium bromide (CTMAB) solution for further etherification reaction under water bath conditions for several hours. After the alkalization-etherification reaction, rinse and filter the materials in the beaker with deionized water, repeating this process three times. Then, dry the obtained product in an oven at 105 ± 5 °C to remove free water, then grind and pass it through a 100-mesh sieve for thorough contact reaction with HA to prepare the desired product. The resulting product is the modified corn stalk powder (MCSP), which should be placed in a dryer for later use.

2.2 Response Surface Methodology (RSM) Optimization of Preparation Conditions

Using the removal rate of HA as the response value (R1), the BBD was employed to evaluate the comprehensive effects of five independent variables: alkali concentration

(A), alkalization time (B), etherification temperature (C), etherification time (D), and etherification concentration (E). These variables were selected based on single-factor experiments.

2.3 Experimental Procedure

Batch experiments were conducted to explore how various factors influenced the removal of HA. A flask with a capacity of 150 mL, filled with 100 mL of HA solution, was utilized, and an appropriate quantity of MCSP was introduced. The pH of the solution was adjusted using 1 mol/L sulfuric acid (H_2SO_4) and NaOH solutions. The flasks were placed on a shaker individually. The vibrating screen was set to a speed of 200 revolutions per minute at a temperature of 25 °C. At specified intervals, 5 mL samples were drawn from the flask, filtered through a syringe filter, and subjected to concentration analysis of HA. This analysis was performed using a UV-visible spectrophotometer, set at a wavelength of 254 nm. Each experiment was replicated three times for accuracy.

2.4 Characterization

The particle size analyzer was used to ascertain the median diameter of the HA particles (MS3000, Malvern, UK). The zeta-potential was determined using the laser diffraction instrument (Zetasizer Nano-ZS90, Malvern, UK). The analysis of HA before and after the reaction was performed using a three-dimensional fluorescence spectrophotometer (3D-EEM, F-7000, Hitachi, Japan). The scanning speed was set to 1200 nm/min. The emission spectrum (Em) ranged from 250 to 500 nm, while the excitation spectrum (Ex) extended from 200 to 500 nm. Both spectra were documented with a 5 nm sampling interval, maintaining a consistent slit width of 5 nm throughout the measurements [15].

3 Results and Discussion

3.1 Preparation and Optimization of MCSP

Single Factor Experiment. The effects of alkaline concentration, alkalization time, etherification temperature, etherification time, and etherification concentration on the removal rate of HA were investigated during the preparation process of MCSP. Figure 1 shows that with increasing alkaline concentration, the removal rate of HA initially decreased and then increased. When the alkalization concentration was 8%, after a reaction time of 60 min, the removal rate of HA reached as high as 95.5%. However, when the alkalization concentration increased to 20%, the removal rate of HA decreased to 56.2%. Similarly, the effect of alkalization time on the removal rate of HA exhibited a similar trend in Fig. 1b, with the highest removal rate observed at an alkalization time of 1 h. This indicated that alkalization conditions have a significant impact on the properties of MCSP, thereby playing an important role in the removal of HA. During the alkalization process, paired sodium and hydroxide ions form hydrated ions. At lower alkaline concentrations, these hydrated ions combine with many water molecules, resulting in their size being too large to effectively penetrate the fiber crystal region and disrupt the crystal structure. As the NaOH concentration increases, the number of water molecules available for forming hydrated ions in the liquid phase decreases. This reduces the size of these hydrated ions, allowing them to penetrate the fiber crystal region and participate in the reaction. However, when the alkaline concentration further increases, the size of these hydrated ions continues to decrease, resulting in a decrease in the efficiency of crystal region disruption. Additionally, the increased quantity of sodium ions in the solution competes with bound water, leading to a reduction in the degree of fiber swelling.

During the etherification reaction, the removal rate of HA initially increased and then decreased with increasing etherification temperature. The highest removal rate of HA, reaching 90.9%, was observed when the etherification temperature was 60 °C after a reaction time of 60 min. Additionally, etherification time and concentration effects on HA exhibited a similar trend.

RSM Establishment and Analysis. Utilizing the optimal ranges identified for factors A, B, C, D, and E through single-factor experiments, a 5-factor, 3-level Box-Behnken design within the response surface methodology was executed, focusing on the removal rate of HA as the response variable R1. A total of 46 sets of experiments were obtained. The quadratic regression equation between the removal rate of HA and the preparation conditions of MCSP was established using Design Expert 8.0 software, as illustrated by the formula below.

$$\begin{aligned} R1 = {} & 1.218 - 0.026A - 0.136B - 0.009C - 0.178D \\ & - 0.22E - 0.004AB + 0.0019BC + 0.041BD + 0.014A^2 \\ & + 0.128B^2 + 0.33D^2 + 0.00016E^2 \end{aligned} \quad (1)$$

As shown in Table 1, an analysis of variance was performed on the regression equation for the removal rate of HA. The "Prob > F" values indicate the significance of the model equation. A "Prob > F" value below 0.0500 signifies a statistically significant model term, whereas a value exceeding 0.1000 implies non-significance [16]. The quadratic model boasts an F-value of 13.21 and a P-value less than 0.0001, underlining the strong significance of this quadratic polynomial regression model. The "Lack of Fit" F-value is 2.72, indicating that the model lack of fit is not significant, which is favorable for the model. The coefficient of determination (R^2) for the quadratic model is 0.9135, indicating that 91.35% of the data variability can be explained by this response surface statistical analysis model. Additionally, the significance test of the quadratic polynomial regression equation reveals that the first-order terms A, B, C, and E all exert a notable influence on HA removal.

Furthermore, based on the results obtained from experiments and mathematical analysis using a second-order regression model, the optimal process parameters for preparing modified corn stalks were determined as follows: A-alkali concentration at 10%, B-alkalization time at 1.083 h, C-etherification temperature at 60 °C, D-etherification time at 2.083 h, and E-etherification concentration at 50%. Under these conditions, the removal experiments of HA solution using the prepared MCSP were carried out five times, resulting in an HA removal rate of 94.3%, which is close to the theoretically predicted value. This indicates that the use of the BBD method to optimize the preparation conditions of MCSP can accurately fit the relationship between the response value and the influencing factors, demonstrating a high level of credibility.

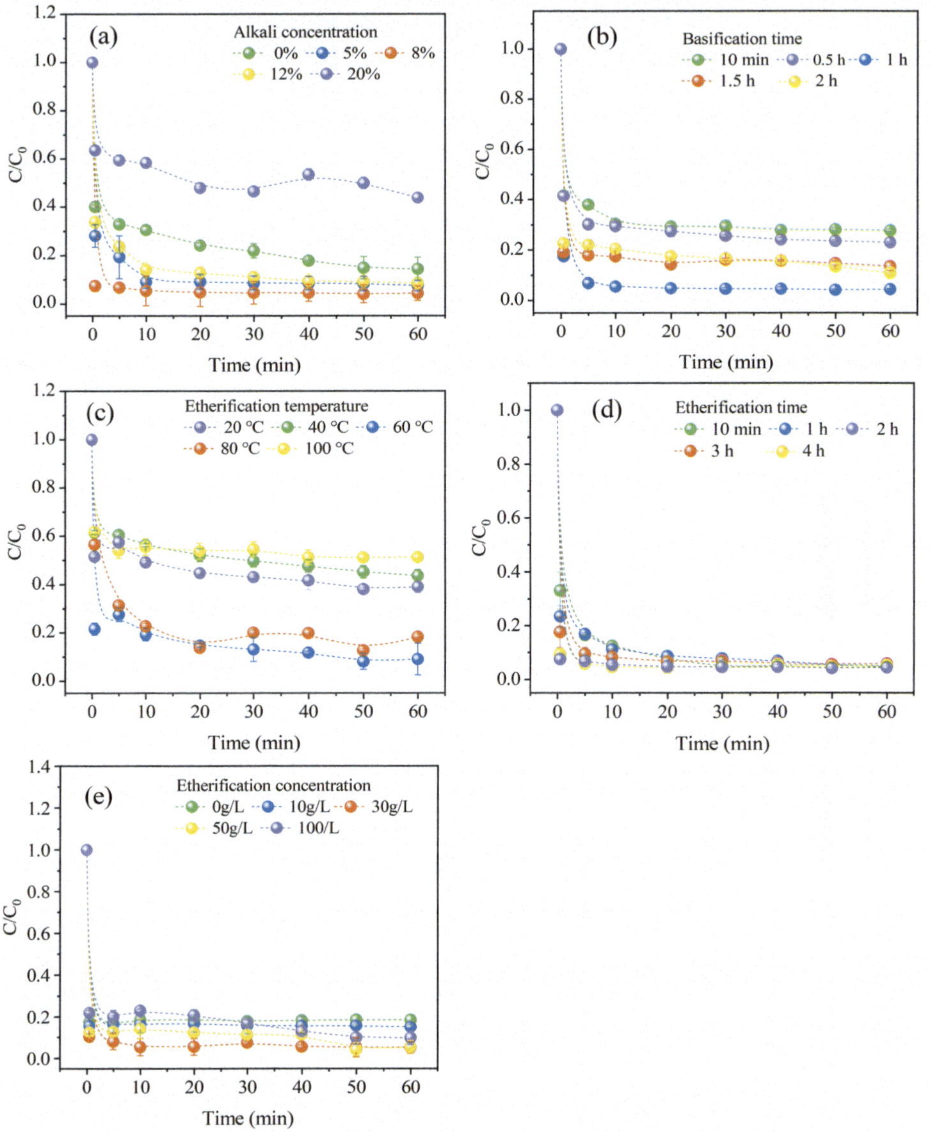

Fig. 1. The effects of preparation conditions of MCSP on the removal of HA. a: Alkali concentration; b: Alkali treatment time; c: Etherification temperature; d: Etherification time; e: Etherification concentration.

3.2 Optimization of HA Removal Conditions

As shown in Fig. 2a, as the dosage of MCSP increased, the removal efficiency of HA gradually decreased. When the dosage of MCSP reached 0.1 g/g L and the reaction time was 20 min, the removal efficiency of HA was 90.3%. With a further increase in the MCSP dosage to 0.4 g/g L, the removal efficiency of HA only increased to 93.8%.

Table 1. Response surface model analysis.

Source	Sum of squares	df	Mean squares	F-value	P-value	
Model	2.27	20	0.1137	13.21	< 0.0001	significant
A	0.0499	1	0.0499	5.80	0.0237	
B	0.0656	1	0.0656	7.62	0.0107	
C	0.0874	1	0.0874	10.15	0.0039	
D	0.0017	1	0.0017	0.2026	0.6565	
E	0.5092	1	0.5092	59.15	< 0.0001	
Residual	0.2152	25	0.0086			
Lack of Fit	0.2152	20	0.0108			
Pure Error	0.0000	5	0.0000			
Cor Total	2.49	45	$R^2 = 0.9135$			

Therefore, considering the economic cost, the optimal dosage of MCSP was chosen as 0.1 g/g L. Subsequent studies on different initial concentrations, different initial pH values, and different temperatures all used 0.1g/g L of MCSP.

Figure 2b showed the removal efficiency of HA by MCSP at different initial concentrations. As the initial concentration increased, the removal rate of HA decreased. When the initial concentrations of HA were 10 mg/L and 20 mg/L, the removal rate by MCSP was relatively fast and the removal efficiency was high. However, as the initial concentration of the pollutant increased, the production of intermediate products during the reaction also increased, thereby reducing the removal efficiency.

Figure 2c demonstrates the impact of varying initial pH levels on the removal efficiency of HA. Specifically, at a pH of 1, the removal efficiency stood at 68.4%. As the pH rose to 3 and 5, the removal efficiencies increased to 89.8% and 96.6%, respectively. However, at pH levels of 7, 9, and 11, the removal efficiencies declined, suggesting that alkaline environments impeded the reaction. Figure 2d illustrates the influence of temperature on the removal efficiency of HA by MCSP. Notably, at 20 °C, the lowest removal efficiency of HA was recorded at 91.9%.

3.3 Particle Size, Zeta Potential and 3D-EEM Analysis of HA

Smaller particles in the system could cause pore blockage during filtration, making it difficult to remove HA [17]. Therefore, reducing the number of smaller particles in the HA system was crucial for removing of HA. As shown in Fig. 3a, the median particle size (d50) in the HA system increased from 161 μm to 175 μm after the MCSP reaction. This indicated that the addition of MCSP promoted the aggregation of smaller particles in HA, leading to an increase in the particle size of HA.

As shown in Fig. 3b, the zeta potential in the HA system increased from −31.8 mV to −19.3 mV after the MCSP reaction. This indicated that the negative charges in HA were neutralized by the positive charges carried by MCSP, allowing the flocs to quickly

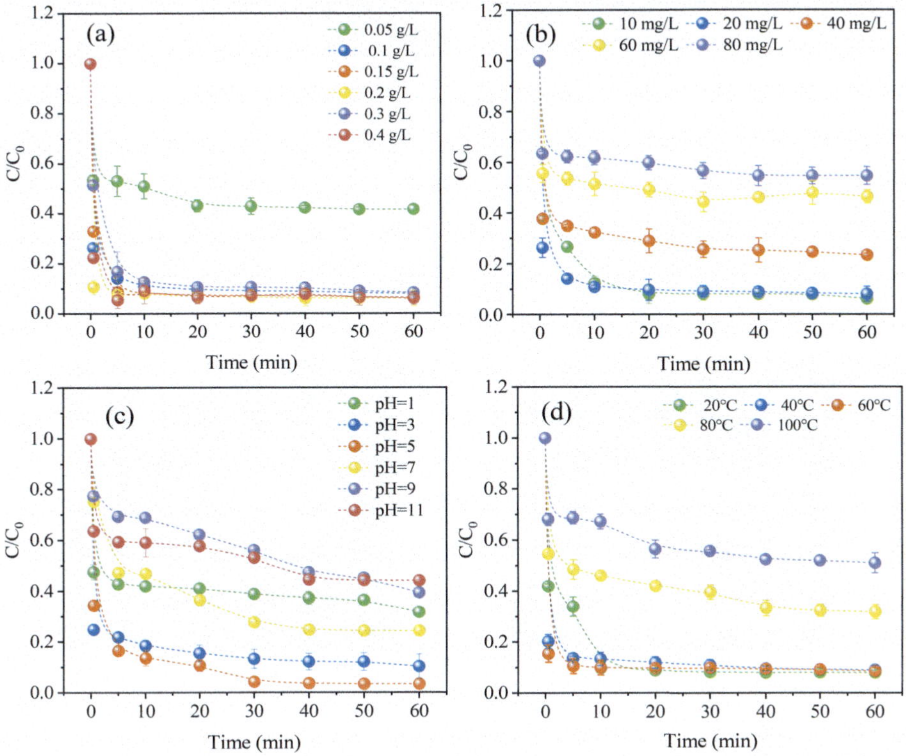

Fig. 2. Influence of MCSP on HA removal rate under different conditions. a: different dosage of MCSP; b: different concentrations of HA; c: different pH of HA; d: different temperature of HA.

aggregate and adsorb smaller particles, leading to an increase in floc structure aggregation and consistent with the change in particle size. This further demonstrated that MCSP could increase the particle size in the HA system through charge neutralization, thereby achieving the removal of HA.

Three-dimensional fluorescence spectroscopy (3D-EEM) serves as an efficient and sensitive tool for examining the constituents and architecture of DOM in water. As depicted in Fig. 3c and d, the 3D-EEM spectrum of the HA solution, prior to the MCSP reaction, predominantly featured Peaks II (representative of fulvic acid), IV (indicating aromatic humic acid), and VI (resembling polycarboxylic acid-like humic acid). Following a 60 min reaction period, the fluorescence intensity of the HA spectrum underwent a notable decrease, albeit with minimal alterations in its spectral profile.

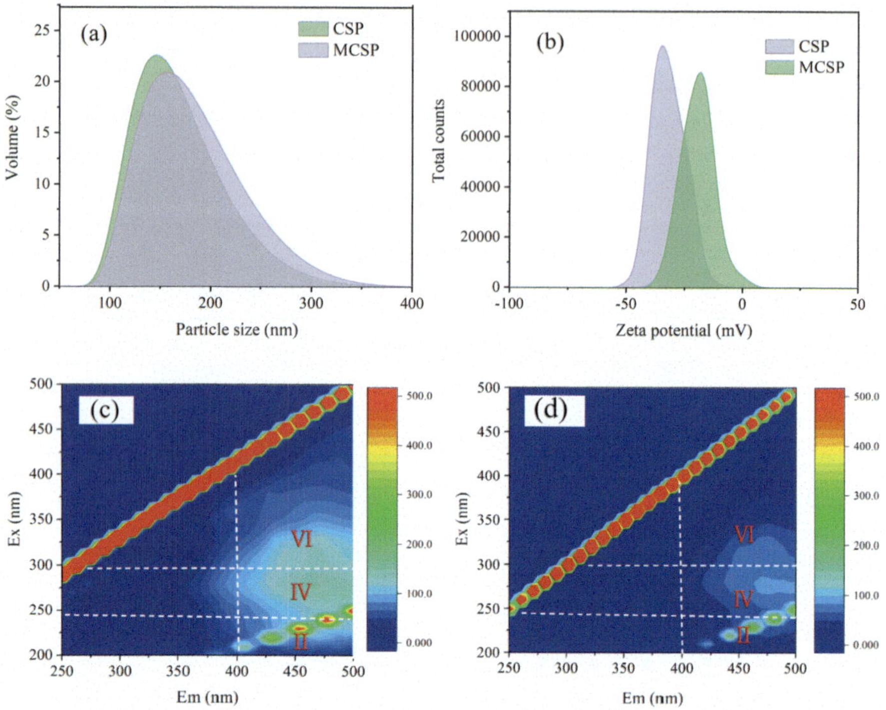

Fig. 3. The particle size, zeta potential and fluorescence spectra changes of HA before and after the reaction (a) particle size; (b) zeta potential; (c) and (d) fluorescence spectra.

4 Conclusion

The optimal preparation process parameters for MCSP were obtained using a Box-Behnken design experimental plan with five factors and three levels, with the removal rate of HA as the response value. The optimal parameters included: A—alkali concentration of 10%, B—alkalization time of 1.083 h, C—etherification temperature of 60 °C, D—etherification time of 2.083 h, and E—etherification concentration of 50%. Furthermore, the removal rate of HA from water decreased gradually with the increase of MCSP dosage and concentration. Through investigation of changes in particle size, zeta potential, and 3D-EEM before and after the removal of HA, the mechanism of HA removal by MCSP was further revealed. After MCSP reaction, the d50 in the HA system increased from 161 μm to 175 μm, and the zeta potential in the HA system increased from -31.8 mV to -19.3 mV, indicating that MCSP could increase the particle size in the HA system through charge neutralization, thereby achieving the goal of HA removal. 3D-EEM further confirmed the effective removal of HA by MCSP.

Acknowledgements. The current study was supported by the Project of Educational Commission of Guangdong Province of China (2022KQNCX172 and 2023KTSCX248), Science and Technology Program of Guangzhou, China (2023A04J1635), University's Scientific Research

Project (KJ2022–01), and Special Funds for the Cultivation of Guangdong College Students' Scientific and Technological Innovation ('Climbing Program' Special Funds) (pdjh2022a0794 and pdjh2023b0801).

References

1. Li, Y., et al.: Humic acid fertilizer improved soil properties and soil microbial diversity of continuous cropping peanut: a three-year experiment. Sci. Rep. **9**, 12014 (2019)
2. Llad'o, J., Lao-Luque, C., Sol'e-Sardans, M., Montemurro, N., P'erez S., Fuente, E., Ruiz, B.: Elimination of persistent anthropogenic pollutants by micro-mesoporous carbon xerogels. Natural organic matter on surface water and textural properties influences. J. Environ. Chem. Eng. **9**, 104885 (2021)
3. Southararajah, P., Loganathan, P., Kandasamy, J., Vigneswaran, S.: Effects of humic acid and suspended solids on the removal of heavy metals from water by adsorption onto granular activated carbon. Int. J. Environ. Res. Publ. Health **12**, 10475–10489 (2015)
4. Tang, S., Tang, J., Yuan, D., Wang, Z., Zhang, Y., Rao, Y.: Elimination of humic acid in water: comparison of UV/PDS and UV/PMS. RSC Adv. **10**, 17627–17634 (2020)
5. Susic, M.: Replenishing humic acids in agricultural soils. Agronomy **6**, 45 (2016)
6. Sarangapani, C., Lu, P., Behan, P., Bourke, P., Cullen, P.: Humic acid and trihalomethane breakdown with potential by-product formations for atmospheric air plasma water treatment. J. Ind. Eng. Chem. **59**, 350–361 (2018)
7. Khan, S., Kim, J., Sotto, A., Bruggen, B.: Humic acid fouling in a submerged photocatalytic membrane reactor with binary Ti O_2-Zr O_2 particles. J. Ind. Eng. Chem. **21**, 779–786 (2015)
8. Xie, L., et al.: Probing the intermolecular interaction mechanisms between humic acid and different substrates with implications for its adsorption and removal in water treatment. Water Res. **176**, 115766 (2020)
9. Xu, C., Sun, C., Wan, H., Tan, H., Zhao, J., Zhang, Y.: Microstructure and physical properties of poly (lactic acid)/polycaprolactone/rice straw lightweight bio- composite foams for wall insulation. Constr. Build. Mater. **354**, 129216 (2022)
10. Tian, Y., Xu, Y., Hou, S.: Effect of storage methods on properties of corn straw after storage. Trans. CSAE **31**(9), 223–229 (2015)
11. Elisângela, C., Priscila, S., Adriano, M.: Recent advances in food-packing, pharmaceutical and biomedical applications of zein and zein-based materials. Int. J. Mol. SCI **15**(12), 22438–22470 (2014)
12. Zhang, X., Li, B., Han, X., Wang, N.: Preparation, characterization and application of CS@PDA@Fe3O4 nanocomposite as a new magnetic nano-adsorber for the removal of metals and dyes in wastewater. RSC Adv. **11**(38), 23679–23685 (2021)
13. Zhang, X., Wang, Z., Can, M., Bai, H., Ta, N.: Analysis of yield and current comprehensive utilization of crop straws in China. J. China Agric. Univ. **26**(9), 30–41 (2021)
14. Yan, M., et al.: Design and evaluation of a novel heliostat-based combined cooling, heating, and power (CCHP) system: 3E analysis and multi-criteria optimization by response surface methodology (RSM). Energy **285**, 129389 (2023)
15. Chen, W., Westerhoff, P., Leenheer, J.: Fluorescence excitation-emission matrix regional integration to quantify spectra for dissolved organic matter. Environ. Sci. Technol. **37**(24), 5701–5710 (2003)
16. Almahbashi, N., et al.: Sustainable sewage sludge biosorbent activated carbon for remediation of heavy metals: optimization by response surface methodology. Case Stud. Chem. Environ. Eng. **8**, 100437 (2023)
17. Neyens, E., Baeyens, J., Dewil, R.: Advanced sludge treatment affects extracellular polymeric substances to improve activated sludge dewatering. J. Hazard. Mater. **106**(2–3), 83–92 (2004)

Towards Sustainable Water Treatment: Optimizing Coagulation Through Experimental Design

Naruethat Moolwong, Atiwat Nanphang, Pongchanun Luangpaiboon, and Kullaya Naranong(✉)

Industrial Statistics and Operational Research Unit (ISO-RU), Department of Industrial Engineering, Faculty of Engineering, Thammasat School of Engineering, Thammasat University, Pathumthani, Thailand
{naruethat.moo,atiwat.nan,kullaya.na}@dome.tu.ac.th, lpongch@engr.tu.ac.th

Abstract. This investigation investigates the optimization of coagulation processes in water treatment using experimental designs that integrate controlled and obstructing factors. The research is dedicated to enhancing the efficacy of turbidity elimination by conducting container experiments that replicate conditions pertinent to the Thailand Metropolitan Waterworks Authority. The concentrations of coagulants, PolyAluminum Chloride (PACl) and alum, and the rotational speed of the turbine are among the primary factors examined. Nephelometry Turbidity Units (NTU) are used to quantify the response variable, water turbidity. The obstructing factor of untreated water with variable initial turbidity levels mitigates the variability in water quality. The variability in water quality is mitigated by the obstructing factor of untreated water with variable initial turbidity levels. In order to guarantee the resilient performance of the process, a desirability function is implemented to regulate the mean and standard deviation of water turbidity levels. A factorial design is employed to systematically evaluate factor interactions in order to identify the optimal coagulation conditions. The results indicate that turbidity reduction is considerably improved by specific combinations of turbine rotational speed, PACl, and alum concentrations, resulting in more efficient water treatment. The new operating scenario exhibits superior performance in terms of both central tendency and dispersion effects. The initial water turbidity blocking factor significantly influences the efficacy of the coagulation process, underscoring the significance of integrating input variability into treatment optimization. This study offers practical suggestions for enhancing coagulation processes and valuable insights into the application of advanced experimental designs in water treatment, thereby promoting the sustainable and reliable management of water supplies.

Keywords: Coagulation Optimization · Turbidity Reduction · Experimental Design · Desirability Function · Water Treatment Efficiency

C. Weng (Ed.): WRE 2024, LNCE 634, pp. 408–419, 2025.
https://doi.org/10.1007/978-981-96-5095-8_34

1 Introduction

Access to pure and secure potable water is fundamental for public health and economic development. It is imperative that water treatment is conducted effectively to ensure the safety and quality of drinking water. The coagulation process is essential for water treatment, as it enables the aggregation and removal of suspended particles and impurities, thereby enhancing water clarity and reducing turbidity. However, the complexity of optimizing the coagulation process lies in the intricate interactions among treatment parameters and the variability in water quality [1].

The significance of this study is underscored by its focus on the Thailand Metropolitan Waterworks Authority, where the maintenance of consistent water quality in the presence of fluctuating raw water conditions is a critical challenge. Previous studies have often overlooked the dynamic nature of untreated water, leading to suboptimal treatment outcomes [2]. To address this issue, we employ sophisticated experimental designs that integrate both controlled and blocking factors, aiming to improve the efficacy of the coagulation process.

The research gap addressed in this study involves the lack of systematic approaches to optimize coagulation processes under varying water conditions. The primary factors examined in this study are the turbine rotational speed and the concentrations of aluminum sulfate (alum) and polyaluminum chloride (PACl), which are essential components of the coagulation process. The quantitative evaluation of water clarity is achieved by measuring the response variable, turbidity, using Nephelometry turbidity units (NTU).

This research aims to identify optimal conditions for turbidity reduction and systematically investigate the interactions between these factors. This research employs factorial design and desirability function approaches to achieve this goal. The results are anticipated to offer practical insights and recommendations for optimizing water treatment processes, thereby promoting sustainable resource utilization and enhanced water management practices [3, 4].

The contributions of this paper are threefold: Firstly, it illustrates the utilization of factorial design to optimize the coagulation process, thereby offering a systematic approach to improving the efficacy of turbidity removal. Secondly, it underscores the critical role of blocking factors, such as initial water turbidity, in enhancing the robustness of experimental outcomes, thereby addressing the challenges posed by variability in raw water quality. Thirdly, the paper provides practical suggestions that can be immediately implemented to enhance operational practices within water treatment facilities, particularly in regions with variable water sources.

The structure of the paper is as follows: Sect. 2 reviews the pertinent literature on experimental design methodologies and coagulation processes, providing a benchmark for the current study. Section 3 elaborates on the experimental apparatus, including materials, methodologies, and the factorial design used. Sections 4 and 5 present the related methods and experimental results, examining the impact of various factors and their interactions on turbidity reduction. Finally, Sect. 6 discusses the implications of the findings for water treatment practices, summarizes the main conclusion, and offers recommendations for future research.

2 Literature Reviews

Wu et al. [5] investigated the removal of molybdate ($MoO4^{2-}$) from water using Fe electrocoagulation (EC). RSM uses a Box-Behnken design that optimizes pH, current density, and electrode distance. The ideal settings were pH 5.5–7.0, current density 3.5–4.0 A/m^2, and electrode spacing 0.5–0.9 cm. The research found that lepidocrocite and magnetite flocs remove Mo (VI) via adsorption and co-precipitation. Fe EC showed an average removal effectiveness of 83.4% in continuous flow testing, suggesting it could remove Mo(VI) from water.

Chemical coagulation was used to optimize petroleum refinery wastewater treatment [6]. The research indicated that aluminum sulfate was the most efficient coagulant over ferric chloride and ferrous sulfate. CCD optimized pH, coagulant dose, and mixing speed. The best conditions were pH 9.43, coagulant dose 36.14 mg/L, and mixing speed 145.93 rpm, resulting in COD removal efficiency of 78.75%, TSS 91.21%, color 72.89%, and turbidity 98.10%. The research shows chemical coagulation treats complicated industrial effluent well. Joaquin et al. [7] tested blended natural-synthetic coagulants for sewage effluent. Watermelon and rockmelon seeds were compared to synthetic hydrated lime utilizing a central composite design to optimize pH and coagulant dose. The turbidity removal efficiency was 90%, and suspended particles removal was 91% at pH 5.5, and coagulant dose 70 mg/L. Natural-synthetic coagulants work well in large-scale sewage wastewater pretreatment, providing a sustainable water treatment method.

Sasanya and Adesogan [8] investigate a pilot-scale surface runoff quality improvement for rural Nigerian water supply. The study uses storage, filtration, groundnut coagulation, and inactivated charcoal adsorption. A four by five factorial experiment in a randomized complete block design (RCBD) with four replicates optimizes these treatments. This experimental approach permits systematic assessment of treatment combinations and retention times on water quality. After 14 days of treatment, runoff water fulfills WHO requirements for household use and agricultural irrigation, with improved turbidity, coliform count, conductivity, and nitrate levels. The research shows that treated runoff water may be a sustainable supply for water-scarce places utilizing affordable materials and methods. These studies emphasize the need of improving coagulation-flocculation and electrocoagulation processes utilizing sophisticated DOE methods like RSM and CCD. Novel materials and machine learning algorithms increase treatment procedures' efficiency, improving water quality and sustainability.

3 Coagulation Process of Thailand Metropolitan Waterworks Authority

Coagulation is a fundamental step in water treatment designed to remove suspended particles and impurities. The effectiveness of coagulation depends on various factors, including the type and concentration of coagulants and operational parameters like turbine rotational speed.

The coagulation process is conducted in several stages. Initially, in the rapid mixing stage, coagulants are quickly mixed with water to ensure even distribution, optimizing their effectiveness in destabilizing particles. Afterward, the water moves to the flocculation stage, where it is gently stirred to promote the formation of larger flocs. This gentle agitation allows the destabilized particles to collide and bond together, forming visible flocs that can be more easily removed.

After flocculation, the water undergoes sedimentation in a basin or clarifier, where the flocs settle out due to gravity, significantly reducing water turbidity. The settled flocs are collected and removed from the basin. The final step, filtration, involves passing the clarified water through filters to remove any remaining fine particles and residual flocs, ensuring that the treated water meets quality standards before further use.

Colloidal particles pose a particular challenge in potable water production due to their small size and has a negative electrical charge This causes a force of repulsion between the particles. Therefore, inducing colloidal particles to aggregate into larger flocs is crucial for effective water treatment.

The jar test is a widely used laboratory procedure that simulates the coagulation process on a smaller scale (see Fig. 1). It is designed to evaluate the performance of different coagulants and treatment conditions. In this study, the jar test includes specific parameters such as turbine rotational speed, PACl, and alum. The use of untreated water as a blocking factor helps account for variations in initial water quality, ensuring that the experimental findings are robust and applicable to real-world conditions.

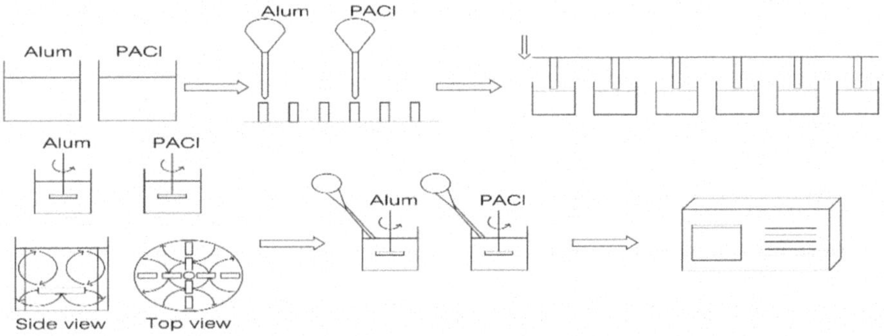

Fig. 1. Jar test setup for coagulation-flocculation optimization by the Thailand metropolitan waterworks authority.

In the jar test setup, untreated water samples are placed into a series of jars, each equipped with adjustable mechanical stirrers. Each experimental treatment is assigned specific conditions, including different turbine rotational speeds, concentrations, and the type of coagulant is PACl (an expensive option) and alum (a less expensive alternative). After adding the coagulants, the jars undergo a rapid mixing phase using the mechanical stirrers at a standardized speed and duration. Following this, the jars enter a flocculation phase, where they are gently agitated at a lower speed than during rapid mixing. After flocculation, the jars are left undisturbed to allow the flocs to settle during the sedimentation phase. This phase mimics the full-scale treatment process, where flocs settle at the bottom of the container, leaving clarified water at the top. Standardized sedimentation times ensure consistent flocculation conditions across all experiments. The turbidity of the clarified water is measured using a turbidity meter after sedimentation, serving as an indicator of each coagulant's efficacy in reducing water turbidity. Uniformity is ensured by taking measurements at the same depth in each jar.

The results from the jar test provide valuable insights into how different coagulant types and concentrations, combined with varying turbine rotational speeds, affect turbidity reduction.

4 Related Methods

4.1 Coagulation-Flocculation

In the coagulation-flocculation process, the preparation and application of coagulants like alum and polyaluminum chloride (PACl) are crucial for effective water treatment. Both coagulants destabilize colloids to facilitate coagulation, but they operate through different mechanisms. Alum primarily uses surface adsorption and charge neutralization, where alum particles adhere to colloids to neutralize their charges, and sweep coagulation, where colloidal particles are captured within aluminum hydroxide precipitates. Sweep coagulation is typically the dominant mechanism when using alum. On the other hand, PACl mainly functions through adsorption coagulation, where it rapidly hydrolyzes in water to form aluminum complexes that effectively destabilize colloids by adhering to their surfaces, requiring relatively smaller quantities for effective treatment.

The choice between alum and PACL as coagulants depends on factors like water quality, suspended particle nature, and treatment objectives. Alum is effective for high turbidity water, while PACL is suitable for situations requiring lower coagulant doses to minimize chemical residuals. Understanding these mechanisms helps optimize coagulation for efficient water treatment with minimal environmental impact. A mixture of alum and water reacts to form $2Al(OH)_3$ and $3H_2SO_4$. $2Al(OH)_3$ are crystals that allow the colloid particles to be attached because they are sticky and can wrap up particles and make the surface of particles sticky.

$$Al_2(SO_4)_3 18H_2O + 6H_2O \rightarrow 2Al(OH)_3 \downarrow +3H_2SO_4 + 18H_2\mathrm{O} \tag{1}$$

In coagulation experiments, PACl and alum solutions are prepared in advance to ensure their stability and efficacy. Untreated water samples with known turbidity levels are collected simultaneously under similar conditions to maintain consistency. The jar test apparatus is used to replicate the coagulation-flocculation process, starting with rapid mixing after adding coagulants, followed by gentle stirring during flocculation, and allowing sedimentation to occur undisturbed. Multiple replicates are included in the experimental design to ensure reliability and account for variability, closely mimicking real-world conditions. The optimal coagulant type, dosage, and treatment parameters are determined through turbidity measurements and statistical analysis, ensuring robust and reliable results.

4.2 Experimental Design

The two-level factorial design is a fundamental method for investigating the effects of multiple factors by assessing each factor at two distinct levels, typically high and low.

This design enables a thorough analysis of both the principal effects and the interaction effects. Researchers can evaluate the impact of each factor and its configurations on the response variable by conducting a systematic variation of the factors. This approach offers valuable insights into the relative importance and interactions of the factors [9].

A randomized block design is implemented to extend the results' robustness. This design divides experimental units into segments according to a source of variability that is not associated with the factors of interest, such as varying experimental conditions or time periods. The design guarantees that the treatment effects are precisely evaluated across varying conditions by randomizing the assignment of treatments within each block, thereby mitigating potential biases [10].

The experimental procedure commences with the identification of the objectives and the selection of pertinent factors for investigation. In this context, operational conditions, concentrations, and varieties of coagulants are selected for investigation. The factorial design with two levels is subsequently employed to generate a comprehensive experimental plan that encompasses all feasible combinations of these factors and their respective levels. This method enables a comprehensive evaluation of both interactions and main effects [11].

After the design phase, the investigation is conducted using six vessels as experimental modules. Each vessel is treated with different combinations of coagulant types, concentrations, and operational conditions. The treatment combinations are designated at random within each block to guarantee that the results are representative and impartial. Under each treatment condition, NTU (Nephelometric Turbidity Units) measurements are recorded from each vessel, which can be used to assess the efficacy of various treatments.

The desirability function approach is implemented to evaluate the collected data. This method converts each response variable into a desirability score that ranges from 0 to 1, with 1 denoting the most desirable outcome. The desirability function enables the simultaneous optimization of multiple response variables, such as the minimization of the mean NTU and its standard deviation, rather than relying on natural yields. Researchers can identify optimal treatment conditions more effectively by incorporating these scores into a composite measure [12].

The significance of the factors and their interactions is evaluated through statistical analyses, such as analysis of variance (ANOVA). The desirability function improves the interpretation of these results by offering a unified measure of treatment efficacy. This methodology guarantees reliable and robust outcomes by identifying the optimal parameters for each factor and accounting for variability and interactions among factors (see Fig. 2).

This integrative approach provides numerous benefits. In conjunction with randomized blocks, the factorial design enables a thorough investigation of factor effects while accounting for external variability. The desirability function is particularly advantageous in fields like environmental engineering and quality control, where it is crucial to achieve optimal performance across various variety of parameters, as it enables the simultaneous optimization of multiple response criteria.

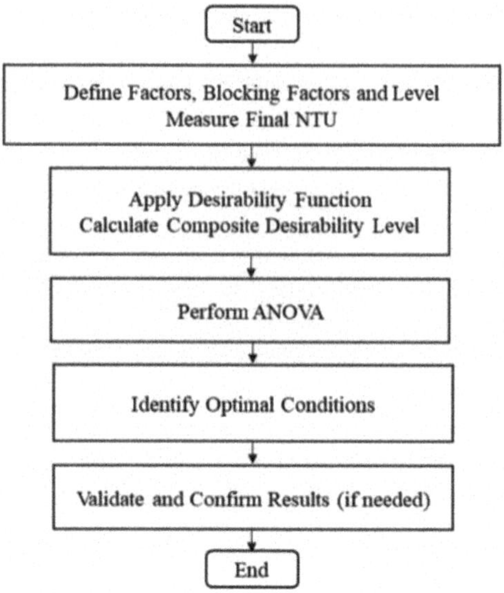

Fig. 2. Sequential procedures of the experimental design method

4.3 Desirability Function

The Desirability Function Approach (DFA) is a well-established method used for optimizing multiple, often conflicting objectives in fields like engineering and experimental design. DFA converts each objective function into a scale-free measure known as the desirability function, which ranges from zero to one, representing the level of satisfaction or alignment with the desired outcome [13]. As described by Zionts and Wallenius [14], this approach helps balance various performance criteria by quantifying how well a set of conditions meets the objectives, such as minimizing a response variable like turbidity mean for increased satisfaction. Thus, the desirability function level presents the degree of desirability or satisfaction for the corresponding objective. If the smaller-the-better (STB) scenario is the objective that must be minimized, the desirability function is defined as

$$d_i = \begin{cases} 1 & \text{for } NTU_i \le NTU_i^{MIN} \\ \left[\dfrac{NTU_i^{MAX} - NTU_i}{NTU_i^{MAX} - NTU_i^{MIN}}\right]^{\alpha_i} & \text{for } NTU_i^{MAX} \ge NTU_i \ge NTU_i^{MIN} \\ 0 & \text{for } NTU_i \ge NTU_i^{MAX} \end{cases} \quad (2)$$

where F_i, F_i^{MIN} and F_i^{MAX} are respectively the candidate solution and the minimum and the maximum acceptable values of the ith objective function. α_i is the parameter ($\alpha_i > 0$) that determines the shape of the ith desirability function (d_i).

The desirability function approach (DFA) addresses various optimization scenarios, such as the "smaller-the-better" (STB) scenario, where the goal is to minimize a response. In STB scenarios, the desirability function is structured so that higher desirability scores are associated with lower response values. Conversely, higher values would be assigned higher desirability scores for scenarios aiming to maximize a response. This adaptability allows DFA to be effectively applied to different optimization objectives, ensuring that the approach aligns with the specific goals of the analysis.

To handle multiple objectives simultaneously, DFA typically involves integrating individual desirability functions into a composite desirability score (D) using geometric means [15]. This method balances diverse objectives by geometrically aggregating the individual desirability scores into a unified overall performance measure. In practical applications, such as optimizing the mean and standard deviation of NTU in factorial and randomized block designs, DFA enables the conversion of response variables into desirability scores, which are then combined to identify optimal experimental conditions that achieve the best balance across performance criteria. Overall, DFA offers a systematic and efficient approach to solving complex multiple objective optimization problems.

5 Experimental Results

The jar test is a frequently employed method for simulating the coagulation-flocculation processes in water treatment, thereby assisting researchers and practitioners in tuning the parameters necessary for effective water purification. The jar test's predictive accuracy was compared to that of a pilot-scale system using a completely randomized design (CRD) with a significance level of 0.05 in this study. By altering critical operational parameters, including the concentrations of coagulants like alum and polyaluminum chloride (PACl), the objective was to ascertain the extent to which the jar test could accurately predict the performance of a full-scale system. Through meticulous management of these variables, our objective was to improve the reduction of turbidity, a critical factor in water quality.

Turbine rotational velocities of 100, 50, and 20 revolutions per minute (RPM) were tested in conjunction with varying concentrations of alum and PACl at 30 ppm and 14 ppm, respectively, at the experimental setup. A triplicate of each test was administered to guarantee the data's consistency and reliability. Upon the identification of optimal conditions through the container test, these conditions were subsequently implemented in a pilot-scale system. The jar test and pilot-scale system were compared using a t-test, which found no significant differences, as indicated by a p-value of 0.264. The reliability of the container test in predicting the efficacy of larger-scale water treatment operations is also confirmed by this result.

The effectiveness of the jar test as an instrument for refining coagulation-flocculation parameters is reinforced by the strong correlation between the jar test results and actual process performance. The jar test's predictive accuracy was robust through both t-test analysis and the wholly randomized design. The value of the container test in efficiently optimizing treatment conditions prior to their implementation on a larger scale is further substantiated by these findings. This validation emphasizes the jar test's significance in the effective purification of water, thereby establishing it as a practicable and reliable approach to the preliminary optimization of water treatment processes.

A general factorial design was employed to investigate the impact of various factors on coagulation efficiency further. Through this methodology, we were able to investigate the impact of three critical parameters: the concentration of coagulant, and the turbine rotational speed during the first and third mixing cycles. The initial cycle was designed to have three levels of rotational speed, the third cycle to have two levels, and the coagulant concentrations to have two. This framework was intended to facilitate the comprehension of the interactions between these factors. Table 1 provides a detailed explanation of the scenarios that were specifically designed to investigate the effects of different raw water turbidity levels (low and high) and coagulant varieties (alum and PACl). A structured approach was provided by these scenarios to investigate the impact of various combinations of these variables on turbidity reduction.

Table 1. Experimental scenarios for coagulation process optimization.

Scenario	Raw Water Turbidity	Coagulant Type	First Cycle Rotational Speed (RPM)	Third Cycle Rotational Speed (RPM)	Coagulant Concentration
1	Low	Alum	90, 100, 110	20, 25	25 ppm, 30 ppm
2	High	Alum	90, 100, 110	20, 25	25 ppm, 30 ppm
3	Low	PACl	90, 100, 110	20, 25	12 ppm, 14 ppm
4	High	PACl	90, 100, 110	20, 25	12 ppm, 14 ppm

Scenario 1 (Low Turbidity, Alum): The results from the ANOVA analysis for this scenario showed that the interaction between alum concentration and third cycle rotational speed had a significant effect on turbidity reduction, with a p-value of 0.014. The optimal conditions were identified as a third cycle speed of 20 RPM and an alum concentration of 30 ppm. Additionally, the main effects analysis revealed that the first cycle rotational speed significantly impacted the coagulation process, with an optimal speed of 110 RPM (p-value = 0.002). This combination allows for sufficient mixing to enhance particle aggregation without causing over-agitation. Alum concentration itself had a highly significant main effect (p-value = 0.000), emphasizing the importance of using a higher concentration to improve water clarity (see Fig. 3).

Scenario 2 (Turbidity, Alum): A significant main effect (p-value = 0.013) was observed in scenarios with high raw water turbidity, indicating that alum concentration was the most critical factor for effective coagulation. Concentrations of 30 ppm were more effective in reducing turbidity than those of lower concentrations. A p-value of 0.023 was also observed for interaction effects between the first and third cycle rotational speeds, indicating that the optimal turbidity reduction is achieved by combining a first cycle speed of 100 RPM with a third cycle speed of 25 RPM. This implies that the effectiveness of coagulation under high turbidity conditions is contingent upon the management of blending velocities and the concentration of the coagulant.

Scenario 3 (Low Turbidity, PACl): The concentration of polyaluminum chloride (PACl) was a critical factor in the use of PACl as the coagulant at low raw water turbidity, with a highly significant main effect (p-value = 0.001). When it comes to particle aggregation and turbidity reduction, a concentration of 14 ppm was most effective.

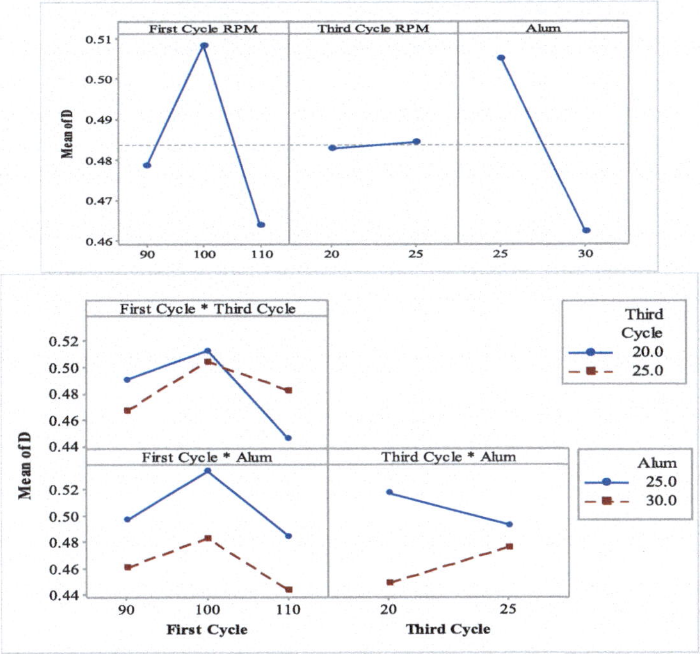

Fig. 3. Main and interaction effects based on *D* of the first scenario.

Statistically significant interactions were observed between the first cycle rotational speed and PACl concentration (p-value = 0.043) and between the third cycle speed and PACl concentration (p-value = 0.025). When a higher PACl concentration is combined with a first cycle speed of 100 RPM and appropriate agitation levels are maintained in the third cycle, these findings indicate that coagulation efficiency is significantly enhanced.

Scenario 4 (High Turbidity, PACl): The first cycle rotational speed was the most influential parameter in scenarios of high turbidity and PACl as the coagulant, demonstrating a significant main effect (p-value = 0.003). In order to ensure the effective distribution of coagulants and the aggregation of suspended particles, it was imperative to maintain an optimal speed of 100 RPM during the initial blending stage. A significant main effect was also observed for the PACl concentration (p-value = 0.021), with 14 ppm being the preferred concentration for reducing turbidity. In this scenario, the results underscore the significance of precisely regulating the first cycle speed and coagulant concentration to enhance the quality of treated water, particularly in high turbidity situations, even though interaction effects were not statistically significant.

6 Conclusion

This study focuses on optimizing the coagulation process in water treatment by utilizing sophisticated experimental designs controlled and blocking factors. The findings reveal that specific combinations of turbine rotational speed, PACl concentration, and alum dosage significantly enhance turbidity eradication. Precise control over these parameters substantially improves treatment efficiency, highlighting the importance of optimal settings. By accounting for variations in water quality, the study also highlights the alignment between jar test results and actual coagulation processes. Optimized coagulation conditions improve water quality, reduce chemical and energy consumption, and enhance turbidity reduction. These findings are highly relevant for water management professionals, translating into cost savings and more sustainable operational practices.

Moreover, integrating blocking factors into experimental designs creates a robust framework for evaluating and optimizing treatment processes. This methodological advancement could extend beyond coagulation to other aspects of water treatment, allowing facilities to adapt more effectively to operational challenges and varying water qualities. The potential for these advanced designs to enhance treatment efficacy and sustainability is underscored by the study's conclusions.

The limitations of this study include the focus on specific coagulants and turbidity levels, which may not be universally applicable across all water treatment scenarios. Future research should focus on applying these optimized conditions in full-scale treatment systems to verify their efficacy in real-world scenarios. Additionally, investigating other factors that may affect coagulation efficiency, such as alternative coagulants or varying water chemistries, could further improve treatment outcomes and provide a more comprehensive understanding of the coagulation method.

This research underscores the importance of sophisticated experimental designs in optimizing water treatment processes, contributing to more effective and sustainable water management practices by addressing water quality variability and refining coagulation parameters. The results offer practical suggestions for enhancing treatment efficacy and promoting the long-term sustainability of water management practices.

References

1. Khettaf, S., et al.: Optimization of coagulation-flocculation process in the treatment of surface water for a maximum dissolved organic matter removal using RSM approach. Water Supply **21**(6), 3042–3056 (2021)
2. Mensah-Akutteh, H., Buamah, R., Wiafe, S., Nyarko, K.B.: Optimizing coagulation–flocculation processes with aluminium coagulation using response surface methods. Appl Water Sci **12**(8), 188 (2022)
3. Sheng, D.P.W., Bilad, M.R., Shamsuddin, N.: Assessment and optimization of coagulation process in water treatment plant: a review. ASEAN J. Sci. Eng. **3**(1), 79–100 (2023)
4. Sun, Y., Zhou, S., Chiang, P.-C., Shah, K.J.: Evaluation and optimization of enhanced coagulation process: water and energy nexus. Water-Energy Nexus **2**(1), 25–36 (2019)
5. Wu, Y., et al.: Fe electrocoagulation technology for effective removal of molybdate from water: main influencing factors, response surface optimization, and mechanistic analysis. J. Environ. Chem. Eng. **12**(2), 112127 (2024)
6. Ayhan, N.N., Aldemir, A., Özgüven, A.: Treatment of petroleum refinery wastewater by chemical coagulation method: determination of optimum removal conditions using experimental design. Braz. J. Chem. Eng. **41**(1), 121–137 (2024). https://doi.org/10.1007/s43153-023-00419-3
7. Joaquin, A.A., Sivamani, S., Gnanasundaram, N.: Statistical experimental design and analysis of mixed natural-synthetic coagulants for the reduction of total suspended solids and turbidity

in sewage wastewater treatment. Biomass Convers. Biorefinery **14**(4), 4583–4590 (2024). https://doi.org/10.1007/s13399-023-03408-7

8. Sasanya, B.F., Adesogan, S.O.: A simple technique to enhance surface runoff quality for water supply in a Nigerian rural community: a pilot-scale study. Sustain. Water Resour. Manag. **9**(6), 190 (2023). https://doi.org/10.1007/s40899-023-00912-8
9. Haddaji, C., et al.: Optimization of the coagulation-flocculation process using response surface methodology for wastewater pretreatment generated by vegetable oil refineries: a path towards environmental sustainability. Environ. Nanotechnol., Monit. Manag. **22**, 100973 (2024)
10. Trinh, T.K., Kang, L.S.: Response surface methodological approach to optimize the coagulation-flocculation process in drinking water treatment. Chem. Eng. Res. Des. **89**(7), 1126–1135 (2011)
11. Khedher, M., et al.: Using the Flocculation Index to optimise coagulant dosing during drinking water treatment. J. Water Process. Eng. **51**, 103394 (2023)
12. Derringer, G., Suich, R.: Simultaneous optimisation of several response variables. J. Qual. Technol. **12**(4), 214–219 (1980). https://doi.org/10.1080/00224065.1980.11980968
13. Kim, K., Lin, D.: Simultaneous optimisation of mechanical properties of steel by maximising exponential desirability functions. J. R. Stat. Soc.-Ser. C **49**(3), 311–325 (2000). https://doi.org/10.1111/1467-9876.00194
14. Zionts, S., Wallenius, J.: An interactive programming method for solving the multiple criteria problem. Manage. Sci. **22**(6), 652–663 (1976). https://doi.org/10.1287/mnsc.22.6.652
15. Jeong, I., Kim, K.: Interactive desirability function approach to multi-response surface optimisation. Int. J. Reliab. Qual. Saf. Eng. **10**(2), 205–217 (2003)

Detection Model and Experimental Study of Dissolved Organic Carbon in Seawater Based on Ultraviolet Absorption Spectroscopy

Feng Xiong[1], Yang Wang[1,2,3,4], Haikuan Ma[1,2,3,4,5], Yanmin Zhang[1,3,4], Yan Liu[1,2,3,4], Xiangfeng Kong[1,2,3,4], Jingru Wang[1,2,3,4], Xuan Cao[2(✉)], Yunzhou Li[1,2,3,4,5], Qian Shi[1,2,3,4], and Ning Wu[1,2,3,4]

[1] Institute of Oceanographic Instrumentation, Qilu University of Technology (Shandong Academy of Sciences), Qingdao, China
jingru426@qlu.edu.cn, srdwjm@sina.com
[2] Laoshan Laboratory, Qingdao, China
cx@sdioi.com
[3] Shandong Provincial Key Laboratory of Marine Monitoring Instrument Equipment Technology, Qingdao, China
[4] National Engineering and Technological Research Center of Marine Monitoring Equipment, Qingdao, China
[5] Academician Workstation of Shandong Province, Shandong Academy of Sciences, Jinan, China

Abstract. Dissolved organic carbon (DOC) is the total carbon content of compounds dissolved in water, which is an important indicator of the marine carbon cycle and ecosystems and is of great significance to the carbon sink capacity of the oceans and the maintenance of marine ecosystems in equilibrium. The commonly used detection method for DOC in seawater is the laboratory chemical analysis method, but this detection method is complicated in operation and cumbersome in processing, and it does not have the ability to detect all-weather rapidly. To this end, this paper presents a method for detecting DOC in seawater based on ultraviolet absorption spectroscopy. A prediction model of DOC in seawater was constructed by partial least squares (PLS) and calibrated by multiple linear regression (MLR) methods, which completed the detection of DOC in 10 sets of actual unfiltered seawater samples. The results showed that using the constructed model to correct the spectral data containing impurities such as particles, significantly enhanced the correlation of measured dissolved organic carbon. Specifically, the linear correlation coefficient R^2 improved from 0.8891 to 0.9838. Furthermore, the mean absolute error (MAE) decreased from 9.775% to 3.781%, and the individual absolute errors were maintained within 9%. The PLS-MLR model achieves accurate measurement of DOC in seawater based on ultraviolet absorption spectrometry without pre-processing, and the method provides a theoretical basis for rapid detection and analysis of DOC in water bodies and the development of sensors.

Keywords: Dissolved Organic Carbon · Ultraviolet Absorption Spectroscopy · Partial Least Squares · Multiple Linear Regression

C. Weng (Ed.): WRE 2024, LNCE 634, pp. 420–432, 2025.
https://doi.org/10.1007/978-981-96-5095-8_35

1 Introduction

Dissolved Organic Carbon (DOC) is a general term for dissolved organic carbon compounds in the water column, mainly from plant residues, microbial metabolites, and decomposition of soil organic matter and other processes [1]. The study of DOC is of great significance. Firstly, it is an important indicator of the quality and ecological health of a water body, and its concentration can reflect the degree of organic pollution of a water body and its potential ecological impact [2]. In addition, as a source and sink of carbon, DOC plays a key role in the global and local carbon cycle, especially in forests, wetlands, lakes, and marine ecosystems [3]. DOC also provides a source of nutrients for aquatic organisms, especially microorganisms, affecting the health and productivity of the aquatic food chain [4]. With global climate change, the amount and composition of DOC may change, which has important implications for carbon release and storage [5]. Therefore, the study of DOC not only contributes to the protection and management of ecosystems, but also provides a scientific basis for the sustainable management of water resources.

Chemical methods are commonly used to detect DOC, including combustion, wet chemistry, and photometric methods [6]. The main advantages of these methods are their high accuracy and wide applicability to a wide range of water samples, including freshwater, seawater, and sewage, as well as the maturity of the technology and stable operational standards [7]. However, they have some drawbacks, such as a time-consuming and complex inspection process, which usually requires specialized equipment and technicians, increasing costs. Environmental and safety risks are also associated with high-temperature combustion and reagents such as strong acids used in chemical methods [8]. Ultraviolet (UV) absorption spectrometry is a relatively new detection method used in various fields. Advantages of the method include a fast response time, which makes it suitable for situations where rapid analysis is required. It is easy to operate and usually does not require complex sample pre-treatment [9]. It is highly sensitive and can detect low concentrations of DOC, has online monitoring capability, is suitable for large-scale monitoring, and does not require expensive reagents, which reduces the cost of testing and safety risks. In addition, UV absorption spectrometry is usually non-destructive and samples can be retained after measurement for further analysis [10]. Therefore, this paper proposes the use of UV absorption spectroscopy for measuring DOC in seawater.

Many researchers have used UV absorption spectroscopy to detect DOC. Sandford et al. used a UV spectrophotometric sensor for real-time, in situ detection of DOC in freshwater. A hybrid linear analytical curve fitting algorithm was used for quantitative analysis in the 230–300 nm wavelength range. Compared with high-temperature catalytic oxidation, it was demonstrated that the method has good linear results and can provide a simple and fast technical analysis tool for quantitative analysis of DOC in watersheds [11]. Zheng et al. used UV-Vis absorption spectroscopy to study the seasonal variation of DOC concentration in Niushan Island located in the East China Sea and the waters of southeastern Zhejiang Province and found that the slope of the spectrum at 280 nm was positively correlated with the DOC concentration [12]. Li et al. investigated the relationship between DOC on spectral slopes at 254 and 280 nm in surface water and wastewater treatment using a home-made miniature LED UV sensor. The results showed that the spectrum at 280 nm can complement the traditional 254 nm measurement [13].

Although these studies involved UV absorption spectroscopy for DOC detection, most used a single wavelength in the UV absorption spectrum, which does not fully utilize spectral information. Therefore, in this paper, the collected broad-spectrum data are used as the data set, and the partial least squares method establishes the detection model. For the influence of the presence of particulate carbon and other impurities in seawater on the detection, a multiple linear regression correction method is used to achieve the accurate detection of marine DOC concentration.

2 Experimental Section

2.1 Samples and Equipment

The samples used in this paper were potassium hydrogen phthalate procured by China National Pharmaceutical Chemical Reagent Co. Ltd, potassium humate and xanthate procured by Hefei BASF Biotechnology Co. Ltd, and tryptophan and tyrosine procured by Nanjing Biological Co. Ltd. The seawater samples involved were collected on two occasions from five different locations in Qingdao, China, resulting in a total of 10 samples. The coordinates for the sampling locations are: 36 35'N 120 18'E, 36 11'N 120 22'E, 36 09'N 120 47'E, 36 26'N 120 33'E, and 36 18'N 120 15'E.

The UV absorption spectrometry collection setup mainly consisted of a deuterium light source produced by Heraeus in China, model Fibre Optics DTM 6/10, a quartz cuvette measuring 40 mm × 10 mm, and a UV spectrometer from Ocean Optics in the USA, model Ocean ST-UV, featuring a detection range of 185–650 nm and a resolution of 13 nm. The equipment related to the sample configuration consisted of a water purifier produced by Germany, model EQ 7000 Milli-Q, and a vacuum pump for seawater filtration produced by JOANLAB in China, model VP-15L. A total organic carbon analyzer produced by Shimadzu in Japan, model TOC-VCPH was used to analyze the DOC content in seawater.

2.2 Measuring Device

The setup for this UV spectroscopy method of measuring dissolved organic carbon is shown in Fig. 1. A deuterium lamp is utilized as the UV light source, emitting a wide range of UV spectra. As the light beam passes through the DOC solution in the sample cell, the organic carbon in the sample absorbs UV light of a specific wavelength, weakening the light intensity. A fiber optic transmits the light from the deuterium lamp to a collimator, which ensures that the beams are parallel to each other to improve the measurement accuracy. After passing through the convergence lenses, the transmitted light is focused by a condensing lens onto a fiber optic coupler and transmitted to the UV spectrometer. The UV spectrometer analyses the received light signal and calculates the concentration of DOC in the sample by comparing the change in absorbed light intensity to a standard curve. Ultimately, a computer processes, analyses, and outputs the results of this data, providing accurate information on the measurement of dissolved organic carbon.

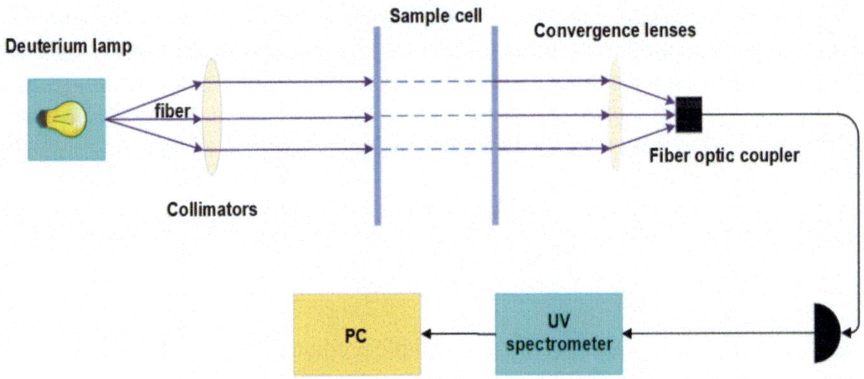

Fig. 1. Schematic diagram of the spectral acquisition process.

2.3 Partial Least Squares Detection Model

Partial Least Squares (PLS) is a powerful statistical modelling method that is particularly suitable for modelling the detection of DOC concentrations in seawater. Its core lies in dealing with the complex relationship between the spectral data (independent variable) and the DOC concentration (dependent variable) to extract potential variables from it, as well as to cope with the problem of possible multicollinearity between the independent variables [14]. The specific procedure of PLS is as follows [15]:

Firstly, data preparation was carried out by letting X be a matrix of spectral data with size [n × p], where n denotes the total number of samples and p denotes the dimension of the spectral data. Meanwhile, let the true concentration Y of DOC be a vector of size [n × 1].

Next, PLS aims to capture the covariance between the independent and dependent variables by extracting latent variables, or factors. Let T be the matrix of latent variables, which is expressed as a linear combination [16]:

$$T = XW \tag{1}$$

where, W is the weight vector describing the mapping of the spectral data in the latent space, with the formula:

$$\mathrm{W} = \left(\mathrm{X}^{\mathrm{T}}\mathrm{Y}\right)(\mathrm{Y}^{\mathrm{T}}\mathrm{Y})^{-1} \tag{2}$$

Among the extracted latent variables, PLS builds a linear regression model:

$$\mathrm{Y} = \mathrm{TB} + \mathrm{E} \tag{3}$$

where, Y is the true DOC concentration value, B represents the regression coefficient, while E stands for the error term. To guarantee the model's accuracy, PLS obtains the optimal regression coefficients by minimizing the squared error $\|\mathrm{Y\text{-}TB}\|^2$ between the predicted and actual values [17]:

$$\hat{\mathrm{B}} = (\mathrm{T}^{\mathrm{T}}\mathrm{T})^{-1}\mathrm{T}^{\mathrm{T}}\mathrm{Y} \tag{4}$$

where, $\hat{B}$ is the regression coefficient obtained by least squares estimation. The use of this least squares method is effective in capturing the features of the sample data.

In addition, to evaluate the model's accuracy and its ability to generalize, PLS uses methods such as cross-validation. The model's reliability is validated by calculating the coefficient of determination (R^2) between the actual and predicted values.

Once the model has been built and validated, Y_N predictions of DOC concentrations for new samples can be made using the latent factors input via the spectral data, with the formula:

$$Y_N = T_{new}\hat{B} \tag{5}$$

where, Y_N is the concentration of the new sample predicted by the PLS model, and T_{new} is a potential variable after the transformation of the spectral data of new samples.

Through the above steps, PLS effectively extracts key information from the spectral data to build an accurate DOC concentration detection model. The advantage of this approach is its ability to handle multicollinearity among independent variables and to perform stable modelling even when the number of samples is far smaller than the number of variables.

2.4 Multiple Linear Regression Correction Model

The presence of particulate carbon materials and other impurities in seawater may impact the detection of DOC concentrations. To enhance the precision of predictions, this paper uses a multiple linear regression correction method (MLR) to correct for these effects. The basic principle is to adjust the prediction results of the PLS model by establishing a new linear model describing the correlation between the predicted DOC concentration and the effects of impurities [18].

Firstly, Y_N is set as the predicted DOC concentration obtained from the PLS model. Z is the matrix of influence factors of impurities with the size of $[n \times m]$. Where n denotes the number of samples and m denotes the number of impurity species. Next, a multivariate linear regression model was created to explain the relationship between the predicted values and impurities. [19]:

$$Y_C = Y_N + Z\beta + E \tag{6}$$

where, Y_C is the sample concentration corrected by MLR., β represents the regression coefficient that needs to be estimated, and E signifies the error term.

Next, to optimize the parameter $\hat{\beta}$ of the regression model, we employed the least squares method to reduce the discrepancy between the predicted values and the actual values:

$$\hat{\beta} = (Z^T Z)^{-1} Z^T (Y - Y_N) \tag{7}$$

This calculation allows us to obtain the optimal correction factor $\hat{\beta}$.

After completing the parameter estimation, it is applied to the regression model to obtain the corrected DOC concentration:

$$Y_C = Y_N + Z\hat{\beta} \tag{8}$$

Through this process, the calibration model can effectively eliminate the influence of impurities on DOC prediction and improve its accuracy and reliability of prediction. Ultimately, by combining the PLS model with the multiple linear regression correction method, this paper presents an effective method for DOC concentration detection, aiming to improve the accuracy of DOC determination in seawater.

2.5 Modelling Evaluation

The prediction performance of the PLS-MLR model was evaluated using R^2, RMSE, MSE, and MAE, where $R^2$is the coefficient of determination, RMSE is the root mean square error, MSE is the mean square error, and MAE is the mean absolute error. The larger R^2, the more stable the model is, and R^2 ranges from 0 to 1; the lower the values of RMSE, MSE, and MAE, the greater the accuracy of the model. The plotting of this study and data processing were conducted with the Origin 2018 and MATLAB 2020aOrigin 2018 software platforms.

3 Results and Discussion

3.1 Measuring Device

DOC is composed of various mixtures, including humic substances, amino acids, proteins, and various other compounds. [20]. Since the concentration gradient of collected seawater does not meet the criteria for constructing a DOC detection model, this paper uses a laboratory configuration to prepare various samples, including potassium hydrogen phthalate, potassium humate, fulvic acid, tryptophan, and tyrosine. The spectral data of these samples were combined into one dataset to construct a detection model, which was then validated with actual seawater to detect DOC in seawater more accurately.

Each sample was diluted in ultrapure water, initially prepared as a 100 mg/L (milligrams of carbon per liter) solution, and then further diluted to create a concentration gradient ranging from 0 to 25 mg/L. The samples were then analyzed by UV absorption spectroscopy. Potassium hydrogen phthalate, potassium humate, fulvic acid, tryptophan and tyrosine sample solutions were measured by UV absorption spectrometry, and 10 sets of each sample were taken singly for averaging and three times to minimize errors. Additionally, shading was applied during the measurement process to prevent ambient light from affecting the experiment. Due to the high concentration of some samples in the collection of spectral data, it is easy to distort, removing part of the spectral data, a total of 109 samples. The UV spectra of each sample are shown in Fig. 2. From Fig. 2, as the concentration increases, the peak value of each sample follows. As shown in Fig. 2, the UV data exhibited considerable fluctuations between 250 and 300 nm. Consequently, this wavelength range was chosen for the dataset used to develop the DOC detection model, with each sample in the dataset comprising 151 data points.

3.2 Detection Model Building

After acquiring the spectral data, the DOC detection model was developed utilizing PLS. Firstly, the spectral data and their corresponding DOC concentration values were

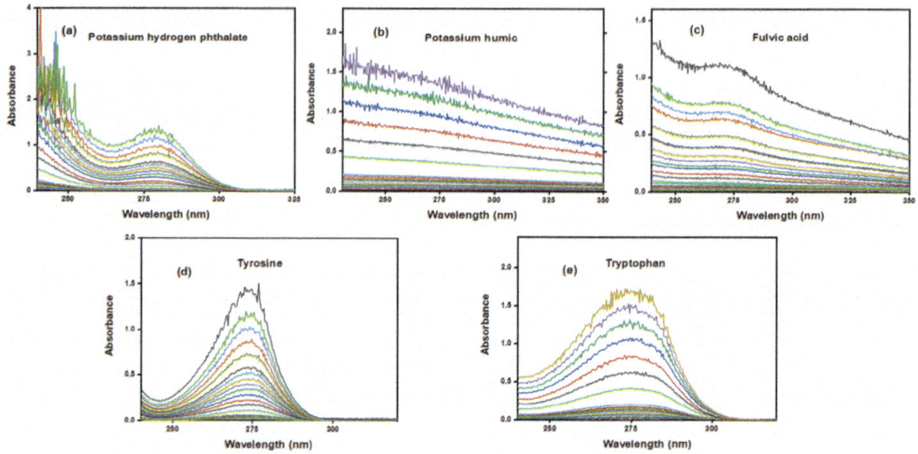

Fig. 2. Ultraviolet absorption spectra data plot. The samples were (a) Potassium hydrogen phthalate, (b) Potassium humate, (c) Fulvic acid, (d) Tyrosine, and (e) Tryptophan.

organized into a dataset and divided into a training set and a test set. The PLS then extracts latent variables to reduce the data dimensionality and maximize the covariance between the independent and dependent variables. The model is then trained using the training set to minimize the prediction error. Once the model is built, it is validated using the test set [21]. The effectiveness of PLS in DOC concentration prediction was finally concluded by comparing the predicted values with the true values through graphs and analyzing the model's degree of fit R^2, RMSE, MSE, and MAE. The results are shown in Fig. 3(a) and (b), the R^2, RMSE, MSE, MAE of the model's training and validation sets constructed with PLS are 0.93148, 0.81836, 0.66971, 0.52447 and 0.99371, 0.5786, 0.33478, 0.24669, respectively, which proves that the model can explain the relationship between the spectral data and DOC concentration better.

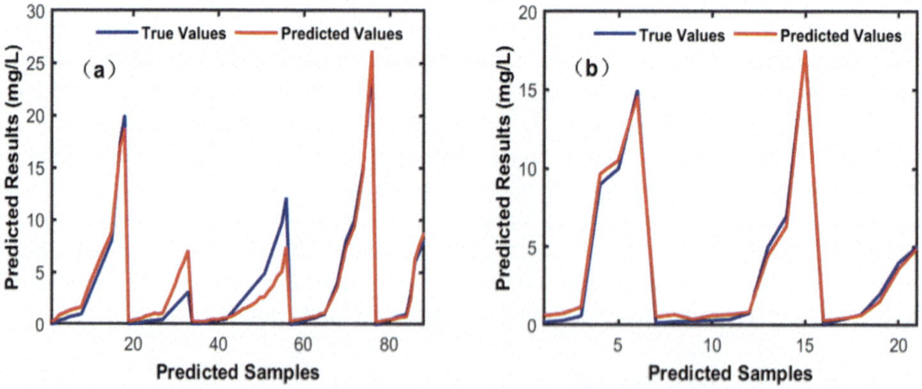

Fig. 3. (a) the results of the simulation training set and (b) the test set.

3.3 Seawater Validation

Organic carbon in seawater is known as total organic carbon, which includes DOC and particulate organic carbon, and the main difference between the two is the ability to pass through a 45 nm filter membrane [22]. To study the concentration of DOC in seawater, 10 sets of seawater samples were collected and filtered. The corresponding spectral data were then collected, while the DOC concentration for each sample was determined using a TOC analyzer. The accuracy of the model was verified after inputting the seawater spectral data into PLS detection model. The results are shown in Fig. 4(a), which compares the true and predicted values of the filtered seawater validation experiment. As can be seen from the figure, the detection error of the model is very small, indicating a strong correlation between the predicted results and the true values. In addition, Fig. 4(b) demonstrates the correlation between the predicted and true values, showing an $R^2$of 0.9671. Apart from this, RMSE, MSE, and MAE are 0.2525, 0.0637 and 0.1572, respectively, indicating that the developed PLS detection model has good accuracy and reliability in the prediction of DOC concentration in seawater.

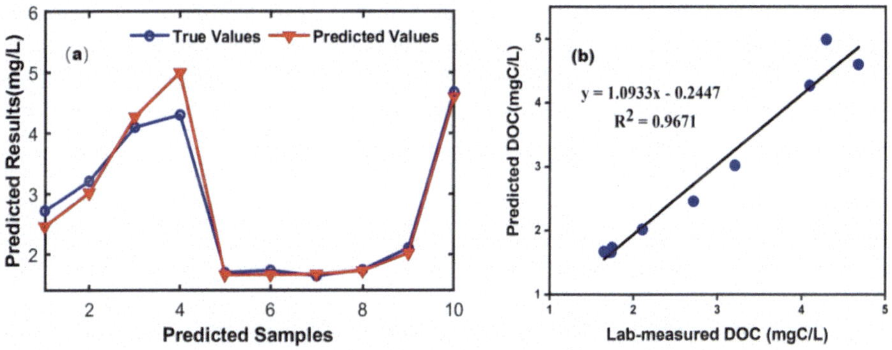

Fig. 4. (a) comparison of the true and predicted values of the filtered seawater validation experiment and (b) fitted plot of predicted DOC values for filtered seawater.

This study focuses on unfiltered seawater samples to overcome the effect of impurities, such as particulate matter and color pigments in seawater, on the prediction of DOC concentration. Spectral data were collected directly from 10 sets of collected seawater samples without filtering. Subsequently, these spectral data were fed into the PLS detection model to validate the model's precision. As shown in Fig. 5(a), the relationship between the predicted DOC concentrations and the true values is demonstrated. The blue points in the figure represent the true measurements, while the black line is the fitted line of the model. Although, the model could capture a certain trend, its R^2 value was only 0.8891, indicating that the predicted results were not satisfactory and there was a large gap between them and the true results. This phenomenon is mainly caused by the interference from particulate matter and other impurities in seawater, which contributes to a lower accuracy of the model when dealing with unfiltered samples.

To overcome the effects of impurities in seawater, this research employs a multiple linear regression correction method to improve the model's detection precision. Firstly,

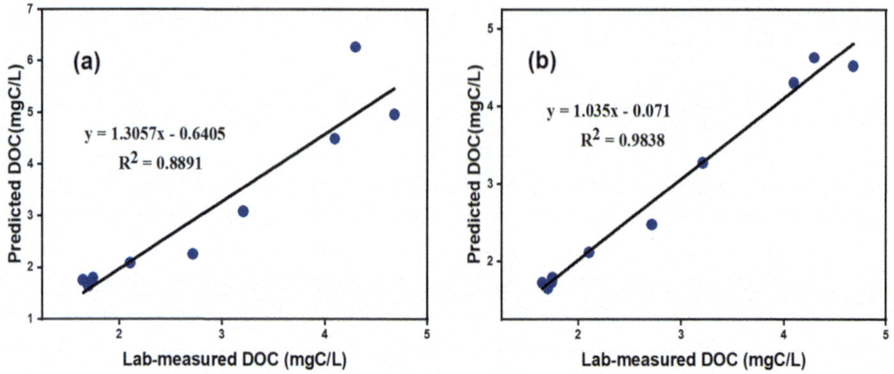

Fig. 5. (a) fitted plot of predicted DOC values for unfiltered seawater and (b) fitted plot of predicted DOC values after correction.

the PLS model was used to obtain preliminary predicted values of DOC concentration. Next, an impurity influence factor matrix was constructed and a multiple linear regression model was built to illustrate the relationship between the predicted values and impurities. The regression coefficients were then optimized using the least squares approach to reduce the error between the predicted values and the true concentrations. Finally, the correction coefficients were applied to the model to generate corrected DOC concentrations, thereby improving the precision and reliability of the predictions.

To guarantee the reliability and accuracy of the calibration model, we divided the actual seawater samples into two groups for validation. The validation results are presented in Tables 1 and 2, where the overall average absolute error is reduced from 9.775% to 3.781%, and the individual absolute errors are controlled within 9%, significantly improving accuracy. Apart from two samples with high levels of contamination, all water samples were within 5% error. As shown in Fig. 5(a) and (b), the overall R^2 value was improved from 0.8891 to 0.9838. In addition to this, the adjusted RMSE, MSE, and MAE are 0.174, 0.03054, and 0.10848, respectively, demonstrating that the corrected model could predict the DOC concentration in seawater with greater accuracy. These results further validate the effectiveness of the multiple linear regression correction method in enhancing the precision of DOC concentration prediction in seawater and provide more reliable technical support for future ocean monitoring.

Figure 6 depicts the link between the true values, the predicted values before and after correction. As can be seen from the figure, there is a significant deviation between the predicted values before correction and the true values. Especially in sample 4 and sample 9, the predicted values are significantly higher than the true values. This suggests that in the absence of correction, the model is influenced by impurities in seawater, leading to imprecise detection results. After the multiple linear regression correction, the correspondence between the predicted and actual values was markedly enhanced, particularly in sample 9, where the corrected predicted values were nearer to the true values. This suggests that the correction method successfully minimizes the impact of impurities on the detection results and improves the accuracy of the model.

Table 1. Comparison of the errors before and after calibration for the predicted concentration of the first group of seawater samples.

Water sample	True concentration (mg/L)	Concentration before calibration (mg/L)	Error (%)	Concentration after calibration (mg/L)	Error (%)
1	1.7464	1.8013	3.14	1.7879	2.38
2	4.6774	4.9665	6.18	4.5230	3.29
3	1.7374	1.7169	1.18	1.7274	0.58
4	2.7198	2.2537	17.13	2.4766	8.94
5	4.0958	4.4949	9.75	4.3039	5.08

Table 2. Comparison of the errors before and after calibration for the predicted concentration of the second group of seawater samples.

Water sample	True concentration (mg/L)	Concentration before calibration (mg/L)	Error (%)	Concentration after calibration (mg/L)	Error (%)
6	1.6484	1.7527	6.34	1.7245	4.62
7	3.2118	3.0876	3.87	3.2735	1.92
8	1.7018	1.6453	3.32	1.6518	2.94
9	4.2978	6.2734	45.98	4.6288	7.70
10	2.1074	2.0894	0.86	2.1149	0.36

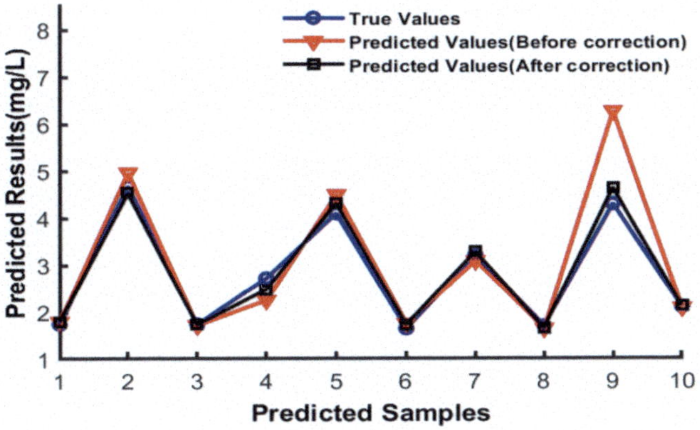

Fig. 6. Comparison of the actual concentration with the predicted concentration before and after calibration.

4 Conclusions

In this study, a method based on ultraviolet absorption spectroscopy for the detection of dissolved organic carbon in seawater was proposed. The method constructs a prediction model for dissolved organic carbon in seawater by partial least squares, and the R^2 values of the training set and test set of the prediction model are 0.93148 and 0.99371, respectively. The R^2 values of filtered and unfiltered seawater are 0.9671 and 0.8891, respectively, by using the method, which is attributed to the effect of the presence of particles and pigments in the seawater, such as impurities in the seawater, on the detection. To improve the detection accuracy, this study used the multiple linear regression correction method to correct the predicted data, and the R^2 value of DOC detection after correction was 0.9838. The overall MAE before and after correction was also reduced from 9.775% to 3.781%, and the individual absolute error was controlled within 9% after correction. Except for two samples with high levels of contamination, all water samples were within 5% error. These results indicate that the method can effectively predict DOC content in seawater. The advantage of this method is that the partial least squares algorithm can effectively extract the key information in the spectral data and make full use of it to establish a stable and accurate detection model. Meanwhile, the multiple linear regression correction method effectively conquers the interference of impurities in seawater on the results and improves the detection accuracy. The PLS-MLR model accurately measure DOC in seawater based on UV absorption spectroscopy without preprocessing. The method provides a theoretical basis for the rapid detection and analysis of DOC in water and the development of sensors. In the next step, our team will further investigate the multi-source spectral fusion technique for the detection of dissolved organic carbon.

Acknowledgements. This work was supported by the Major Innovation Project for the Science Education Industry Integration Pilot Project of Qilu University of Technology (Shandong Academy of Sciences) (Grant No. 2023HYZX01); Qingdao Natural Science Foundation (Grant No. 23–2-1–211-zyyd-jch); Major Innovation Project for the Science Education Industry Integration Pilot Project of Qilu University of Technology (Shandong Academy of Sciences) (Grant No.2023JBZ02); Major Innovation Project for the Science Education Industry Integration Pilot Project of Qilu University of Technology (Shandong Academy of Sciences) (Grant No.2023JBZ03); Key Research and Development Plan of Shandong Province (Grant No. SDZLYSTD2022); Key Research and Development Plan of Shandong Province (Grant No. 2021CXGC010705); Scientific and Technological Innovation projects for Laoshan Laboratory (Grant No. LSKJ202204603); Scientific and Technological Innovation projects for Laoshan Laboratory (Grant No. LSKJ202204700); The National Key R&D Program of China (Grant No. 2021YFB3901300).

References

1. Palviainen, M., Pumpanen, J., Mosquera, V., et al.: Extending the SUSI peatl and simulator to include dissolved organic carbon formation, transport and biodegradation-Proper water management reduces lateral carbon fluxes and improves carbon balance. Sci. Total. Environ. 950175173–175173 (2024)

2. Ramanathan, T., Ollivier, Q., Rahman, A., et al.: Long-term dissolved organic carbon changes in Woronora drinking water system in Australia. Chemosphere 364143047 (2024)
3. Prakash, G., Rajula, G., Sarma, V.: Role of phytoplankton composition on exudation of dissolved organic carbon in the Bay of Bengal. Cont. Shelf Res. 279105296–105296 (2024)
4. Huang, C., Jiang, T., Zhu, B., et al.: Assessment of refractory dissolved organic carbon in sediments of marine culture ponds: an example of two culture systems for Penaeus vannamei. Aquaculture 593741368–741368 (2024)
5. Qu, Q., Hu, X., Feng, R.: Using the potential transformation of dissolved organic matter to understand carbon emissions from inland rivers. Environ. Sci. & Technol. (2024)
6. Zhang, S., Zeng, Y., Liu, R., et al.: Impact of varying dissolved organic carbon load on sediment phosphorus release and its periodic mechanisms. Environ. Res. 259119558–119558 (2024)
7. Bristol, M.E., Behnke, I.M., Spencer, M.G.R., et al.: Eroding permafrost coastlines release biodegradable dissolved organic carbon to the Arctic Ocean. J. Geophys. Research. Biogeosciences **129**(7), e2024JG008233–e2024JG008233 (2024)
8. Qin, C., Li, L. S., Wu, Y., et al.: High sensitivity of dissolved organic carbon transport during hydrological events in a small subtropical karst catchment. Sci. Total. Environ. 946174090 (2024)
9. Hong, W.B., Bo, J.F., Zhe, L., et al.: In-situ measurement of total organic carbon concentration in seawater based on ultraviolet absorption spectrometry. Spectrosc. Spectr. Anal. **40**(8), 2484–2489 (2020)
10. Wang, X.Y., Xu, J.Y., Ji, R.D., Bian, H.Y., Feng, X.T., et al.: Detection of dihydrocoumarin in coconut juice via photoelectric detection system based on ultraviolet absorption spectrometry. Sensors **22**(21), 8267 (2022)
11. Charles, R.S., Roland, B., John, P.W.: In situ determination of dissolved organic carbon in freshwaters using a reagentless UV sensor. J. Environ. Monit. **12**(9), 1678–83 (2010)
12. Zheng, H., et al.: Seasonal variations of dissolved organic matter in the east China sea using EEM-PARAFAC and implications for carbon and nutrient cycling. Sustainability **10**(5), 1444 (2018)
13. Li, W., Cao, M., Young, T., et al.: Application of UV absorbance and fluorescence indicators to assess the formation of biodegradable dissolved organic carbon and bromate during ozonation. Water Res. 111154–162 (2017)
14. Zhu, K.J., Zhang, S.S., Yue, K.Y., et al.: Rapid and nondestructive detection of proline in serum using near-infrared spectroscopy and partial least squares. J. Anal. Methods Chem. 20224610140–4610140 (2022)
15. Zhang, Y.F., Jiang, L.L., Chen, Y., et al.: Detection of chlorpyrifos residue in apple and rice samples based on aptamer sensor: improving quantitative accuracy with partial least squares model. Microchem. J. **194**, 109352 (2023)
16. Eledum, H., Elmahgop, O.F.: The effect of corporate social responsibility on corporate performance in the food industry in Saudi Arabia: a partial least squares structural equation modeling approach. Sustainability **16**(18), 7925 (2024)
17. Philippe, K.B., Nsabimana, T., Kala, R.J., et al.: Comparative study of probabilistic and least-squares methods for developing predictive models. Open J. Appl. Sci. **14**(07), 1775–1787 (2024)
18. Tang, S.M., Li, T.T., Guo, Y., Zhu, R., Qu, H.Y.: Correction of various environmental influences on Doppler wind lidar based on multiple linear regression model. Renew. Energy **184**, 933–947 (2022)
19. Juan, S., Ken, T., Carlo, J.E., et al.: Correction: a multiple linear regression model for the prediction of summer rainfall in the northwestern Peruvian Amazon using large-scale indices. Clim. Dyn. **62**(5), 4453 (2024)

20. Wang, J.J., Pu, X., Zhang, R.N., et al.: Spatial-temporal distribution and source apportionment of riverine dissolved organic carbon in the Miyun section of the Chaobai River in Beijing, China. Water-Energy Nexus **6**, 160–166 (2023)
21. Ding, Q., Yao, M., Wu, S.H., et al.: Quantitative analysis of Al in flour products by laser-induced breakdown spectroscopy combined with partial least squares. J. Appl. Spectrosc. **89**(4), 712–718 (2022)
22. Yang, H.K., Graham, N.J.D., Wang, W.Y., et al.: Evaluating and improving the reliability of the UV-persulfate method for the determination of TOC/DOC in surface waters. Water Res. **196**, 116918 (2021)

Design of Phytoplankton Photosynthesis Intensity Measurement Device Based on Light Pulse-Induced Fluorescence

Youqiang Li[1], Yan Liu[1,2], Jingru Wang[1,2](✉), Haikuan Ma[1,2], Yang Wang[1,2], and Xiangfeng Kong[1,2]

[1] Institute of Oceanographic Instrumentation, Qilu University of Technology (Shandong Academy of Sciences), Qingdao, Shandong, China
jingru426@163.com
[2] Laoshan Laboratory, Qingdao, Shandong, China

Abstract. Measurement of chlorophyll fluorescence of phytoplankton provides real-time information on their physiological status. The determination of the physiological state of phytoplankton is of vital significance to the fields of natural disasters and water ecology research. In this study, for the detection of chlorophyll fluorescence in phytoplankton photosynthesis in seawater, a weak signal detection circuit and a high-speed data acquisition circuit were designed, and the cross-group amplification and conversion of weak photocurrent signals, as well as the design of adjustable bias second-stage amplification circuits were implemented. Actual measurements were carried out on samples from the Chlorophyta (Platymonas), Chrysophyta, Cyanobacteria, and Pyrrophyta (Prorocentrum donghaiense Lu). The results show that the designed circuit has a minimum detection current of 80 nA under excitation from a 500 kHz light source, a detection range of 80 nA-100 μA, and achieves a sampling frequency of 1 MHz. In the application of the phytoplankton chlorophyll weak fluorescence detection device, the circuit ensures the sensitivity of the system. It extends the detection range of weak signals, providing important data support for research related to chlorophyll fluorescence detection.

Keywords: Phytoplankton · Photosynthesis · Weak Signal Acquisition · Photoelectric Conversion

1 Introduction

Real-time monitoring of phytoplankton photosynthesis in marine and inland water bodies has become critical when global climate change and environmental monitoring are of increasing concern. As the primary producers in aquatic ecosystems, the health of phytoplankton directly impacts on the stability of the entire food chain and biodiversity. The chlorophyll fluorescence method [1] has been widely used in phytoplankton photosynthesis parameter measurements as a non-invasive, convenient, and reliable method [2, 3]. This technique uses chlorophyll to emit fluorescence during photosynthesis to detect and

C. Weng (Ed.): WRE 2024, LNCE 634, pp. 433–446, 2025.
https://doi.org/10.1007/978-981-96-5095-8_36

acquire of weak fluorescence signals [4, 5]. Currently, the main techniques for detecting weak chlorophyll fluorescence signals are Pulse Amplitude Modulation [6] (PAM), Fast Repetition Rate Fluorometry [7] (FRR), and Fast Fluorescence Induced Kinetics [8] (Fast Fluorescence Kinetics, FFK). Shi Chaoyi [9] et al. designed a high-frequency excitation flash system based on a pulse amplitude modulation technique to measure microsecond PSII light yield curves. Kolber Z S [10] investigated the effect of excitation mode on fluorescence parameters using single and multiple excitation lights. In 2018, Yin Gaofang [11] proposed a method to measure photosynthetic rate based on the kinetics of tunable pulsed light-induced fluorescence. However, these methods may still face several limitations in practical applications, such as the complexity and stability of the system, which need to be further optimized. Therefore, to measure chlorophyll fluorescence more accurately and lay the foundation for subsequent in-depth investigation of the mechanism of plant photosynthesis, a new chlorophyll fluorescence detection system is proposed in the present study. Since fluorescence signals possess transient response characteristics at the microsecond level, the measurement system must accurately capture and acquire weak fluorescence signals at high speed to resolve fast light response dynamics. In this paper, a system for detecting the fluorescence-induced dynamics of chlorophyll in phytoplankton was designed based on the fast fluorescence-induced dynamics technique. The system can provide the fluorescence signal state of phytoplankton in real-time, accurately, and in situ.

2 Detection Principle

Photosynthesis is a fundamental biochemical process centered on the conversion of energy and matter and can be divided into light and dark reactions (Calvin cycle). The light reaction occurs in the chloroplast thylakoid membrane and is a process of light energy utilization, oxygen release and electron transfer. Photosynthesis is essentially an electron transfer process, which is the core mechanism for converting light energy into chemical energy and is mainly carried out in the chloroplast thylakoid membrane, involving PSII, cytochrome b6f complex, PSI, and several electron carriers. Firstly, PSII chlorophyll molecules absorb photons, and light energy excites electrons to higher energy levels and passes into the electron transport chain. In PSII, photolysis of water molecules produces oxygen, hydrogen ions, and electrons, which replenish the PSII electron deficit. Electrons are transferred to PSI via the cytochrome b6f complex and carriers such as PQ and PC [12].

According to the optical characteristics of chlorophylls, chlorophylls can be divided into four types: a, b, c, and d. Chlorophylls a and b are the main pigments, with a ratio of approximately 3:1. Chlorophyll a is present in all algae and largely ensures that plants are green. The other pigments show marked variations among different algae. In photosynthesis, chlorophyll a is the main pigment. Chlorophylls b, c, d, carotenoids, and phycobiliproteins are auxiliary pigments. After receiving light energy, they are efficiently transferred to chlorophyll a. Therefore, the process by which algae produce fluorescent characteristics of chlorophyll under light source excitation is what we have detected [13].

The fast fluorescence-induced kinetics technique is used to analyze photosynthesis and study photochemical and non-photochemical processes in photosystem II (PSII) by measuring changes in chlorophyll fluorescence. The method divides fluorescence changes into two phases: a fast phase and a relaxation phase. The rapid phase, in which the fluorescence intensity rises from a minimum to a maximum in one hundred microseconds, reflects the photochemical reactions of PSII. The relaxation phase lasts hundreds of milliseconds to several seconds, during which the fluorescence signal decreases, reflecting energy dissipation, electron transfer, and PSII resilience [14].

Accurate measurement of the electron transport chain dynamics in photosynthesis, particularly the switching state of PSII reaction centers and photochemical reactions, requires fluorescence detection systems with microsecond time resolution. This involves using high-frequency modulated laser or LED light sources for excitation and photomultiplier tubes or photodiodes to capture and record fluorescence signals via high-speed data acquisition. The fluorescence-induced kinetic curve is shown in Fig. 1.

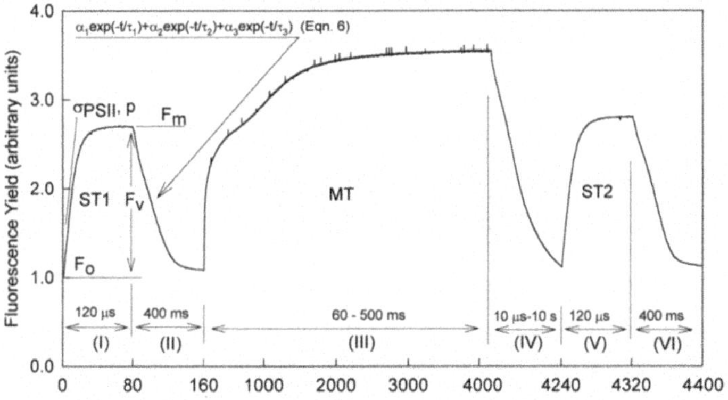

Fig. 1. Fluorescence-induced kinetic curves [15].

3 Detection System Design

Chlorophyll fluorescence detection focuses on the precise measurement of fluorescence produced by the excitation of chlorophyll and involves the design of optical and electronic modules. The key to optical design lies in the precise excitation of light sources and effective fluorescence collection, which requires selecting suitable light sources, such as LEDs or lasers, and designing filters and lens systems [16]. The electronic hardware includes core control, optoelectronic conversion, and high-speed data acquisition, which are used to control the intensity and frequency of the excitation of light sources, amplify weak fluorescence signals, and collect data in real time, The overall structure is shown in Fig. 2.

In the entire measurement process, the sample is first irradiated with a light source, which causes the sample to be excited and emit fluorescence. Subsequently, the emitted fluorescence undergoes photoelectric conversion by a photodiode and is amplified in the first stage. After the completion of the first amplification, the signal undergoes further second amplification in order to effectively enhance the signal strength. After this series of amplification, the signal will be under the precise control of the micro control unit (MCU) for the required specific signal acquisition work and then obtain the corresponding data.

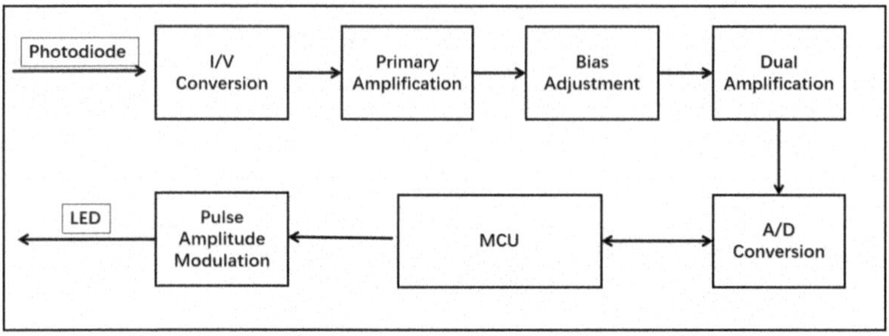

Fig. 2. Overall structure.

3.1 Excitation Optical Path Design

The optical path design is crucial for the construction of the system, and it is necessary to choose a suitable excitation light source and filter to eliminate interfering light, ensuring the detector accurately captures the fluorescence signal and improving the measurement accuracy and reliability of the system. According to the spectral characteristics of chlorophyll fluorescence, a 440 nm blue LED was selected as the excitation light source, and an orthogonal optical path structure was adopted. At the same time, filters were used to prevent interference light from entering the detector [17, 18].

Given that the duration and intensity of fluorescence emission are closely related to the intensity of the excitation light source, we were able to measure the fluorescence signal under different conditions by adjusting the voltage and frequency of the excitation light source, and the duty cycle of the driving pulse. We utilized the flexibility of the adjustable pulse generator to adjust the LEDs, achieving precise control of the LED brightness and flicker frequency by varying its output voltage, frequency, and duty cycle of the driving pulse, The excitation position of the light source is shown schematically in Fig. 3.

3.2 Photoelectric Conversion High-Voltage Drive Circuit

Chlorophyll fluorescence signals are weak, and high-performance detectors improve sensitivity and accuracy. Avalanche diodes are widely used in fluorescence detection due to their characteristics. This system uses hamamatsu brand avalanche diodes (S5344, Hamamatsu, Japan) to optimize signal reception and improve resolution [19]. Considering the avalanche diode's need for high reverse bias and low current, the system selects the TPS series switching power supply chip (TPS55340, TEXAS INSTRUMENTS, UK) for boosting purposes, which has a low-side MOSFET switching function and reduces

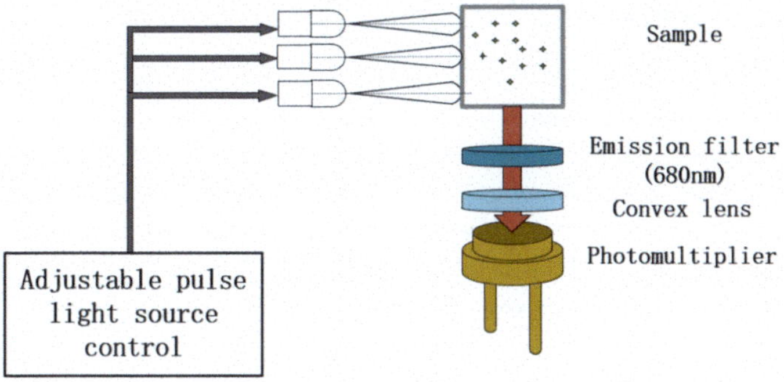

Fig. 3. Schematic diagram of the excitation position of the light source.

ripple interference by separating the digital ground from the analog ground. The switching frequency of the chip can be adjusted to adapt to different detection requirements. The driving voltage circuit diagram is indicated in Fig. 4.

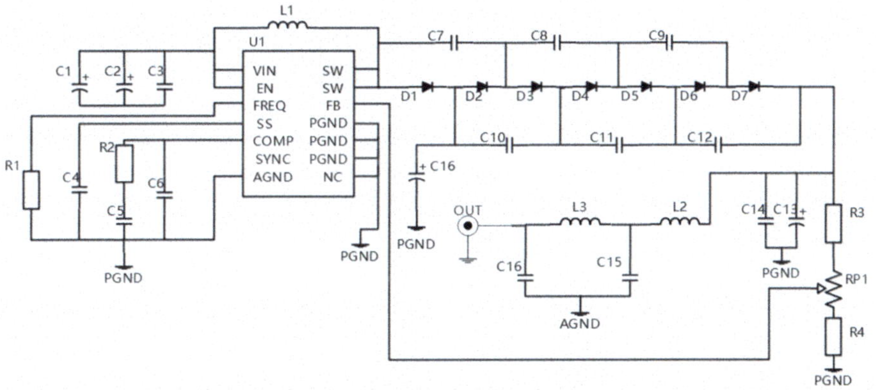

Fig. 4. Driving Voltage Circuit Diagram.

The circuit is designed to boost 12V to 150V and includes voltage regulation and feedback control mechanisms. After power-on, the chip controls the MOSFET switching and gradually raises the voltage by utilizing the principle of inductive energy storage. The output voltage is fed back to the FB (Feedback control pin) pin of the chip through the feedback network in real-time, forming closed-loop control. The control loop adjusts the MOSFET duty cycle based on the feedback to ensure a stable and accurate output voltage. When the voltage is stabilized at 150V, the chip enters steady-state mode, continuously monitoring and fine-tuning the voltage to mitigate external interference. The output capacitor is used to smooth the voltage and reduce ripple noise, ensuring a stable power supply [20].

3.3 IV Conversion Circuit

The role of the avalanche diode is to convert the optical signal into a current signal. The - 3dB bandwidth is an important parameter for characterizing the frequency characteristics of a system or signal. Specifically, it refers to the frequency range within which the power spectrum or amplitude spectrum of a signal drops by 3dB (that is, the power decreases to half of its original value, or the amplitude drops to 70.7% of its original value). Therefore, I/V conversion is required, and the signal bandwidth of the I/V conversion circuit can be calculated using a formula for calculating the $-3dB$ bandwidth of the I/V conversion circuit [21].

$$f_{-3dB} = \sqrt{\frac{G_{\mathrm{BW}}}{2\pi C_{\mathrm{T}} R_{\mathrm{F}}}} \tag{1}$$

where, G_{BW} is the op-amp gain-bandwidth product, RF is the feedback resistor, and C_{T} is the total capacitance of the inverting end of the op-amp, including the detector capacitance (C_{PD}) and input capacitance (C_{IN}). Since the $-3dB$ bandwidth of the I/V conversion circuit is inversely proportional to the value of the feedback resistor RF, an appropriate trans-impedance gain stage must be selected for I/V conversion to ensure optimal bandwidth performance. This means we need to balance gain and bandwidth to ensure that the circuit responds well over the desired frequency range.

Since chlorophyll fluorescence is generally at the nA level. The current signal obtained through the detector is amplified by the OPA chip (TPS55340, TEXAS INSTRUMENTS, UK) ($G_{\mathrm{BW}} = 275$ MHz) and fixed gain to achieve one-stage amplification, and the gain multiplier can be adjusted according to the current signal intensity to improve the detection limit. The amplification is:

$$\mathrm{Vout} = -\mathrm{Iin}^{*} R_F \tag{2}$$

where, Iin is the current of the detector after photoelectric conversion, and R_F is the feedback resistance. The amplification of the first stage of the signal processing circuit should be moderate to avoid amplifying too much interference signal and noise so as not to increase the difficulty of subsequent processing. A level of amplification after the output voltage signal is usually a millivolt level. The photoelectric conversion amplifier circuit design is shown in Fig. 5.

Considering the role of the phase compensation capacitor in the design process, the C_T in the circuit compensates for the phase lag of the output voltage, preventing circuit oscillations and providing filtering and amplification, while overcompensating for the phase zero point [22]. The impedance Z of the capacitor in parallel with the feedback resistor is typically used to represent the mutual resistance gain of the variable gain [23], i.e.:

$$A_1 = \frac{V_a}{I_{IN}} = |Z_F| = \frac{R_F}{\sqrt{1+(2\pi f C_1 R_F)^2}} \tag{3}$$

where, A_1 is the mutual resistance gain and Z_F is the impedance. From Eq. (3), the larger the value of the feedback resistor R_F, the higher the mutual resistance gain; the lower the frequency of the input signal, the higher the mutual resistance gain.

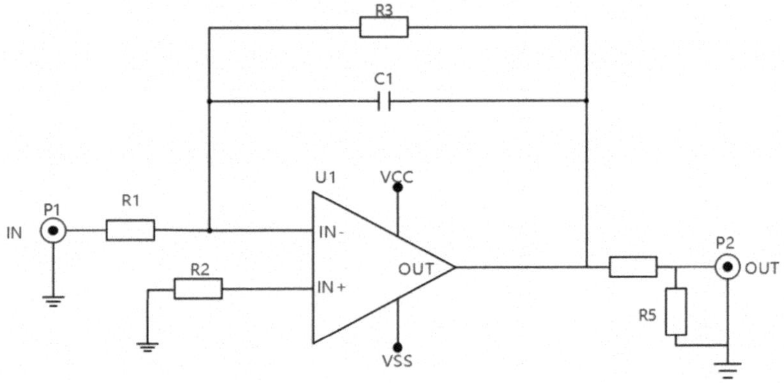

Fig. 5. Photoelectric conversion first stage amplification circuit.

3.4 Secondary Amplifier Circuit

When building a weak fluorescent signal detection system, two amplification stages are typically required. After the first amplification stage, the signal remains weak and needs further enhancement in the second stage to ensure accurate acquisition. The detector current signal outputs a negative-phase voltage after I/V conversion, which must be converted to a positive-phase voltage for acquisition. Therefore, an adjustable bias amplification circuit must be added before the second amplification stage, using the SGM chip (SGM8271, SGMICRO, China) design. The second amplification stage uses the TL chip (TL082, TEXAS INSTRUMENTS, UK) IC to achieve signal DC offset and gain adjustment, supporting a wide voltage supply range suitable for solving phase problems. The adjustable bias amplification circuit diagram is indicated in Fig. 6.

The circuit diagram mainly consists of an operational amplifier that includes biasing and amplification. The input signal is first biased by DC through resistor R_1 to ensure the linear operation of the amplifier. The operational amplifier's power supplies, V_{CC} and V_{SS}, provide bias voltages that change the phase of the signal. The signal then enters the operational amplifier U_1 through R_{P2} and R_1, with the gain set by R_3 and R_{P1}. The output signal of U_1 undergoes further amplification and feedback processing. Finally, the desired amplified signal is output through R_6. The amplification gain is:

$$G = \frac{R_{P1}+R_{13}}{R_3+R_5} \tag{4}$$

3.5 High-Speed Data Acquisition Circuit

To meet the demand for chlorophyll fluorescence signal acquisition, the high-speed signal acquisition function is implemented. The maximum sampling frequency of the analog signal is set to 1 Msps. According to the Nyquist theorem, the sampling rate of the analog-to-digital converter (ADC) needs to reach at least 2 Msps to ensure the distortion-free acquisition of the signal. The AD series chip (AD9220, Analog Devices, Inc., United States of America) is selected, which adopts successive approximation technology, integrates a digital-to-analog converter (DAC), a comparator, and a control logic

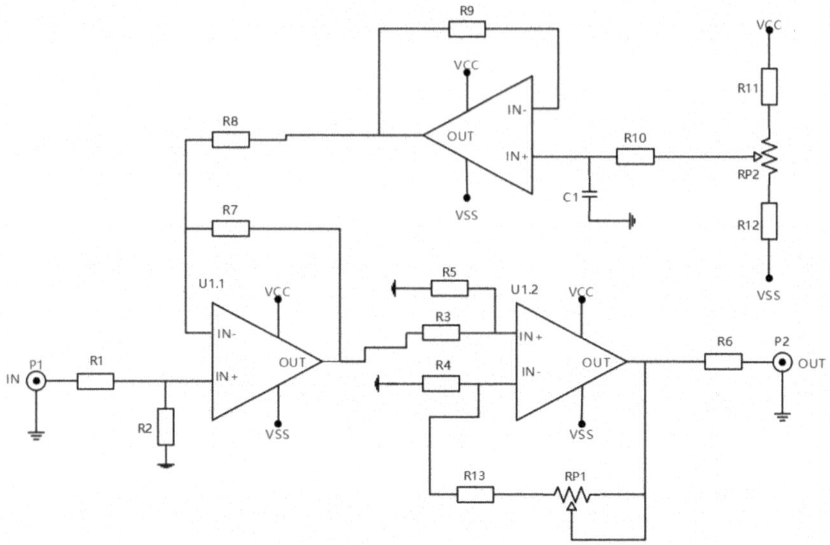

Fig. 6. Adjustable bias amplifier circuit.

circuit, and ensures the stability of the signal through the sampling hold circuit. Digital signals are output via 12 parallel lines with termination resistors to reduce interference. The chip has a 12-bit sampling accuracy, a maximum sampling rate of up to 10 Msps, a voltage acquisition range of 0 to 5 V, and a minimum resolution of 1.22 mV. The high-speed ADC acquisition circuit is shown in Fig. 7:

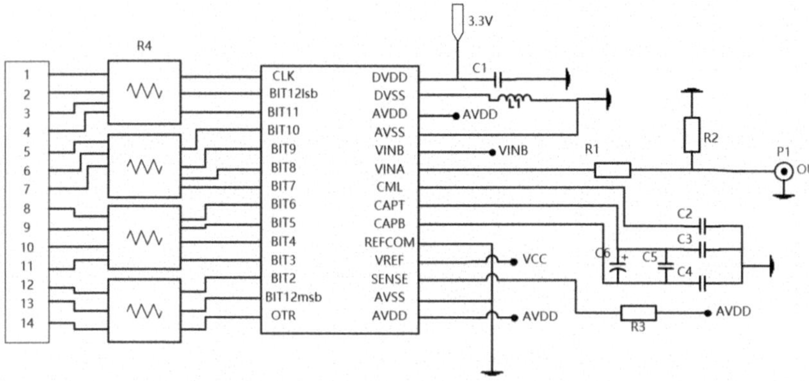

Fig. 7. AD acquisition core circuitry.

After amplification, the analog signal needs to be converted into a digital signal for acquisition. The signal is transmitted to the chip through 12 parallel lines and processed to output a digital signal. The voltage resolution of the 12-bit AD chip refers to the minimum voltage change corresponding to each code value. The calculation formula is as follows:

$$Res = \frac{V_{ref}}{2^n} \tag{5}$$

The analog-to-digital conversion formula is:

$$V_{out} = \frac{V_{in}}{V_{ref}} \times 2^n \tag{6}$$

where, Res is the voltage resolution, and V_{out} is the output voltage value of the ADC. V_{in} is the input analog voltage value, V_{ref} is the reference voltage, and n is the resolution of the ADC. According to the design, the reference voltage is set to 5V, then the voltage resolution is 1.22mV, which meets the acquisition requirements.

3.6 Core Control

In the microcontroller unit, an ARM Cortex M4 microprocessor is selected to meet the demands for precise pulse control and clock control. With high performance, low power consumption, and rich peripheral interfaces, this processor suits various application scenarios and ensures reliable data processing. Therefore, Cortex M4, as the core processing unit, can meet the current control requirements.

When the system starts, it is first initialized to ensure stability. After initializing the clock, basic and general-purpose timers are configured to generate PWM signals, where the basic timer is used as a timekeeper, and the two channels of the general-purpose timer control the LED and AD chip clock signals, respectively. The clock tree cascade architecture is used to ensure the stability of the ADC sampling clock. Through the ADC's SYSREF monitor register, the build-up and hold times of SYSREF are accurately monitored, and the SYSREF signal is adjusted. After initialization and clock configuration are completed, the pulse width of channel one is detected, and the AD chip acquisition is controlled by the timer count value. When the count value reaches half, the microcontroller reads the analog-to-digital converter data from the AD chip, with a total duration of 163 microseconds, and outputs 160 valid data. From the excitation start phase to the data acquisition, the AD chip acquires one hundred and sixty data sequentially. These data are then acquired and saved as a file through the serial assistant. The flowchart is shown in Fig. 8.

4 Experimental Tests and Data Analysis

4.1 –3dB Bandwidth Calculation

The amplifier-based I/V conversion circuit design is calculated by applying Eq. (3) in conjunction with the amplifier and the detector parameters to determine the configuration parameters of the I/V conversion circuit in the fluorescence detection circuit and its performance specifications. The results are shown in Table 1, and these parameters and metrics meet the signal bandwidth requirements.

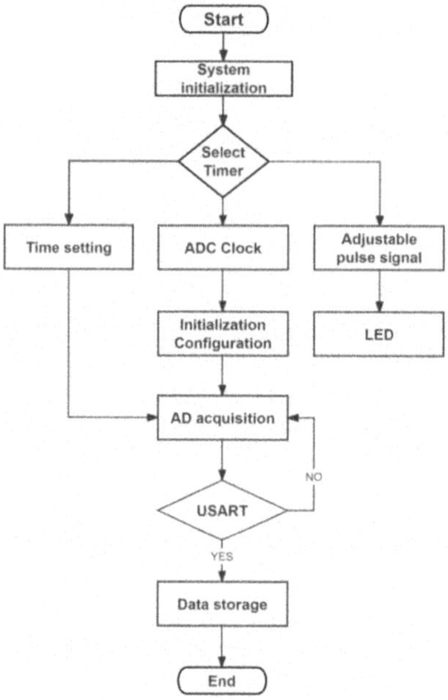

Fig. 8. MCU Control Flowchart.

Table 1. I/V conversion circuit configuration parameters and performance specifications.

Signal	Detector capacitance C_{PD}/pF	Input capacitance C_{IN}/pF	Feedback resistance R_F/kΩ	$-3dB$ bandwidth /MHz
Fluorescence	120	2.8	10	5.9

4.2 Circuit Testing

To confirm the relationship between the input and output of the weak signal detection circuit, a square wave current signal with variable frequency and amplitude was used as the signal source in this study. Since the response time of the weak fluorescent signal is at the microsecond level, the signal frequency was kept constant in the experiment, and the current output change of the photodiode was simulated by adjusting only the signal amplitude. An oscilloscope was used to monitor the corresponding changes in the output signal [22]. According to the input range of the photoelectric conversion of chlorophyll fluorescence signal, four approximate input signals are set up for simulation and real measurement. The amplification is calculated from Eqs. (2) and (4). The test results are shown in Table 2.

The designed circuit can detect weak fluorescent signals at the nanoampere level, with a minimum detection limit of 80 nA and an input current range of 80 nA to 100 μA.

Table 2. Voltage output at different amplitude currents.

Input Current	Theoretical voltage	Simulation Voltage	Measured voltage
10uA	100mV	99.33mV	99.15mV
1uA	10mV	9.92mV	9.89mV
100nA	1mV	0.95mV	0.93mV
80nA	0.8mV	0.79mV	0.72mV

4.3 Sample Measurements

In this study, a phytoplankton chlorophyll fluorescence detection device was developed based on fast fluorescence induction kinetics technology. To verify weak signal detection and acquisition, fluorescence induction experiments were conducted on four species of algae cultivated in the laboratory, namely, Chlorophyta (Platymonas), Chrysophyta, Cyanobacteria, and Pyrrophyta (Prorocentrum donghaiense Lu). The test results are shown in Fig. 9.

For the chlorophyll fluorescence induction experiments, the phytoplankton samples were placed in the dark suitable for 15 min. This process returns the plant's photosynthetic system to its initial state, with all reaction centers open. After excitation using a light source, a fluorescence yield curve was obtained. Each experiment was therefore one to reveal the state of chlorophyll fluorescence at a specific moment.

When excited by a light source, a fluorescence yield curve is obtained. At the initial stage of light exposure, all kinds of algae show significant activity during the increasing fluorescence yield. This is mainly because light causes the chlorophyll molecules of photosystem II to absorb photons, resulting in the excitation of electrons and their transfer in the electron transport chain. In the process of electron transfer in the excited state, energy is released, leading to an increase in fluorescence yield. Among them, the Chlorophyta (Platymonas) and Chrysophyta are particularly prominent, with fluorescence yield values reaching 0.8, demonstrating a high activity level.

However, the fluorescence yield gradually decreases as time passes and enters a declining phase. This is because as photosynthesis continues, the photosystem gradually reaches saturation, and the rate of electron transfer decreases. At the same time, some reaction processes may consume excited-state electrons, leading to a decrease in fluorescence yield and eventually tending to a steady state of 0.4–0.5. The relatively low fluorescence yields of Cryptophyta and Pyrrophyta (Prorocentrum donghaiense Lu) reflect the differences in photosynthetic performance among different algal species, providing a key reference for further investigation of their adaptive mechanisms.

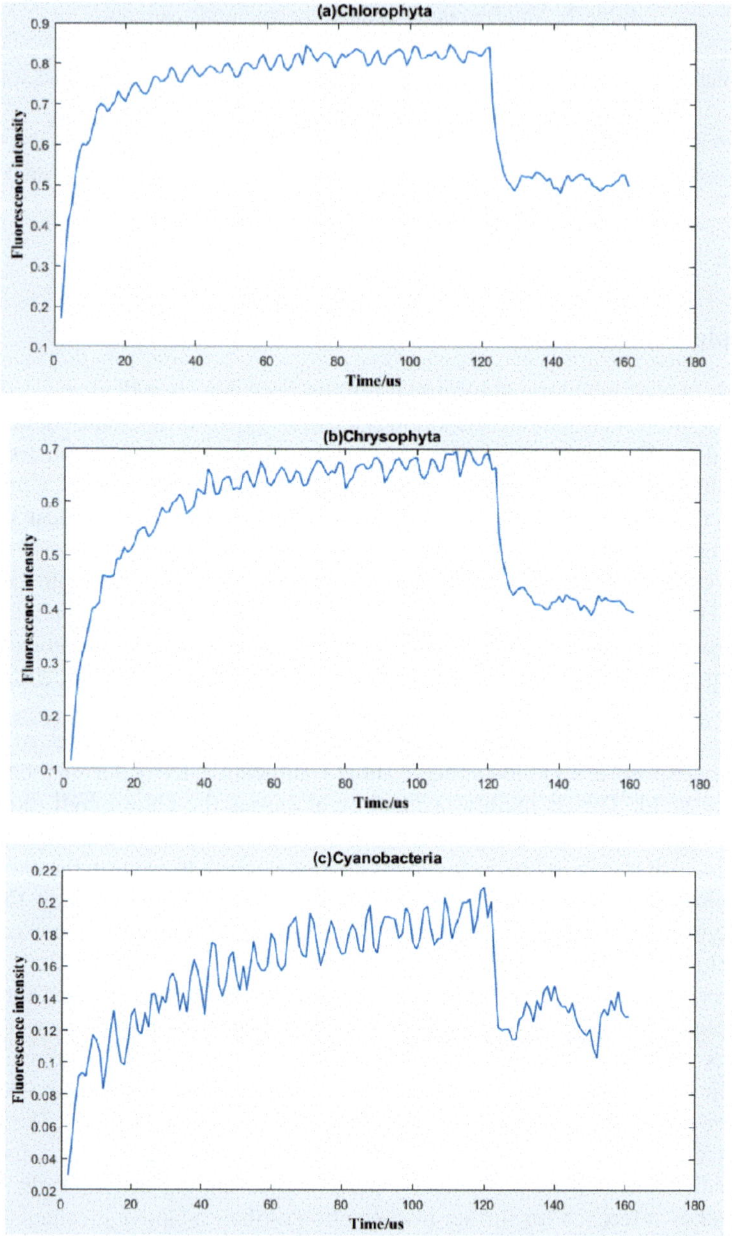

Fig. 9. Chlorophyll fluorescence yield plot for different algae species.

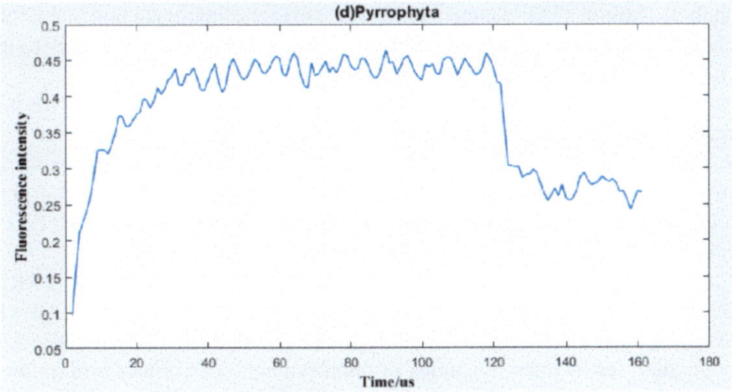

Fig. 9. (*continued*)

5 Conclusion

The detection of chlorophyll's weak fluorescence signals with high sensitivity and a wide range is valuable for practical applications. A weak fluorescence signal detection circuit was designed for the chlorophyll fluorescence signals of phytoplankton in water bodies, and experimental tests and sample measurements were carried out. The weak fluorescence signal was accurately acquired through photoelectric conversion and adjustable bias amplification by ensuring gain and bandwidth. The acquisition of chlorophyll fluorescence intensity from phytoplankton photosynthesis using this system can provide important data support for related research.

Acknowledgements. This work was supported by the Major Innovation Project of Science and Education Integration Pilot Project of Qilu University of Technology (Shandong Academy of Sciences) (2022JBZ01–02, 2023JBZ02), Major Innovation Project of Science and Education Integration Pilot Project of Qilu University of Technology (Shandong Academy of Sciences) (2023JBZ03), Science Education Industry Integration Pilot Major Innovation Project (2023HTZX01) of Qilu University of Technology (Shandong Academy of Sciences), Open Research Fund of Key Laboratory of Marine Ecological Monitoring and Restoration Technology, Ministry of Natural Resources, (MEMRT202301).

References

1. Falkowski, P.G., Kolber, Z.S.: Estimating phytoplankton photosynthesis by active fluorescence. J. Plankton Res. **14**(8), 1–10 (1992)
2. Suggett, D.J., Prasil, O., Borowitzka, M.A.: Chlorophyll a Fluorescence in Aquatic Sciences. Methods and Applications. Springer, Netherlands (2010)
3. Schreiber, U., Klughammer, C., Kolbowski, J.: Assessment of wavelength-dependent parameters of photosynthetic electron transport with a new type of multi-color PAM chlorophyll fluorometer. Photosynth. Res.: Int. J. (2012)

4. Oxborough, K., Moore, C.M., Suggett, D.J., et al.: Direct estimation of functional PSII reaction center concentration and PSII electron flux on a volume basis: a new approach to the analysis of Fast Repetition Rate fluorometry (FRRf) data. Methods (2012)
5. Shi, C.Y., Zhang, Y.J., Yin, G.F., et al.: Suppression of sunlight interference for fluorescence in situ measurement. Appl. Mech. Mater. (2014)
6. Higo, S., Maung-Saw-Htoo-Thaw, Y., Yamatogi, T., et al.: Application of a pulse-amplitude-modulation (PAM) fluorometer reveals its usefulness and robustness in the prediction of Karenia mikimotoi blooms: A case study in Sasebo Bay, Nagasaki, Japan. In: Proceedings of the Harmful Algae Conference, pp. 63–70. Springer, Heidelberg (2017)
7. Wei, Y., Zhao, X., Sun, J., et al.: Fast Repetition Rate Fluorometry (FRRF) derived phytoplankton primary productivity in the Bay of Bengal. Front. Microbiol. **10**, 1164 (2019)
8. Guha, A., Sengupta, D., Reddy, A.R.: Polyphasic chlorophyll a fluorescence kinetics and leaf protein analyses to track dynamics of photosynthetic performance in mulberry during progressive drought. J. Photochem. Photobiol. B. **119**, 71–83 (2013)
9. Shi, C., Zhang, Y., Yin, G., et al.: Measurement of algae PSII photosynthetic parameters using high-frequency excitation flashes. In: 12th Chinese Optics Letters, 080101 (2014)
10. Kolber, Z.S., Prasil, O., Falkowski, P.G.: Measurements of variable chlorophyll fluorescence using fast repetition rate techniques: defining methodology and experimental protocols. Biochim. Biophys. Acta-Bioenerg. **1367**(1–3), 88–106 (1998)
11. Yin, G., Zhao, N., Shi, C., et al.: Phytoplankton photosynthetic rate measurement using tunable pulsed light-induced fluorescence kinetics. Opt. Express **26**(6), A293–A300 (2018)
12. Malone, L.A., Proctor, M.S., Hitchcock, A., et al.: Cytochrome b6f - Orchestrator of photosynthetic electron transfer. Biochim. Biophys. Acta-Bioenerg. 148380 (2021)
13. Zhang, Y., Han, X., Yang J.: Selection of optimal spectral features for leaf chlorophyll content estimation. Scientific Reports (2024)
14. Qin, Z., Liu, J., Liu, W., et al.: Photosynthesis parameter measurement technology based on light pulse-induced fast phase and relaxation fluorescence. Acta Photonica Sin. **46**(9), 8.0930003 (2017)
15. Karsten, U., Klimant, I., Holst, G.: A new in vivo fluorimetric technique to measure growth of adhering phototrophic microorganisms. Appl. Environ. Microbiol. **62**(1), 237–243 (1996)
16. Shi, C.Y., Zhang, Y.J., Yin, G.F., Zhang, N.J.: Determining the optimal excitation condition of high-frequency flash method for algae photosynthetic parameters measurement. Acta Photonica Sin. **44**(2), 1–10 (2015)
17. Gong, Z.L., Zhu, X., Zhou, Z., et al.: Frontiers in circularly polarized luminescence: molecular design, self-assembly, nanomaterials, and applications. SCIENCE CHINA Chem. **64**, 2060–2104 (2021)
18. Zeng, L.H., Li, D.L.: Development of in situ sensors for chlorophyll concentration measurement. J. Sens. **2015**, 1–16 (2015)
19. Graeme, J.: Photodiode Amplifiers: Op Amp Solutions. McGraw-Hill, New York (1995)
20. Ren, M.Y., Tian, L., Wang, W., et al.: Design of pre-amplifiers for photoelectric detector. Appl. Mech. Mater. **380–384**, 3308–3311 (2013)
21. Gao, J.: Detection of Weak Signals. Tsinghua University Press, Beijing (2011)
22. Dong, M., Yin, G.F., Zhao, N.J., et al.: High sensitivity and large dynamic range detection technology of weak fluorescence signal and its application. Chin. J. Quantum Electron. (2018)
23. Integrated Device Technology Inc.: Patent issued for method and apparatus for current steering in high sensitivity, high linearity and large dynamic range high speed trans-impedance amplifiers (USPTO 9935591). J. Eng. (2018)

Environmental Pollution and Sustainability

Combating Climate Change Through Renewable Energy Development

Mohammed Sharif[1](✉) and Zavvar Kazim[2]

[1] Department of Civil Engineering, Jamia Millia Islamia Central University, New Delhi, India
msharif@jmi.ac.in

[2] Municipal Corporation of Delhi, New Delhi, India
zavvar.kazim@mcd.nic.in

Abstract. Renewable energy development in India is intrinsically linked to climate change mitigation efforts, primarily because renewable energy can reduce greenhouse gas emissions significantly. A key challenge is balancing economic growth with the need to reduce emissions, particularly when the demand for energy in India is continually rising. The present research aims to explore the critical role of renewable energy in mitigating climate change in India—a nation confronted with significant social and environmental challenges. In order to achieve this aim, an analysis of the temporal trends in installed renewable energy capacity over the past decades and the share of various renewable sources to the overall energy mix has been presented. Analysis indicates that the solar energy has seen the most rapid and exponential growth among renewable sectors, with installed capacity reaching approximately 85 GW. Despite a potential of around 145 GW, hydropower capacity is 45 GW. Wind energy has shown steady and consistent growth, with capacity gradually rising to about 45 GW. The findings highlight significant progress in India's renewable energy sector, though achieving the target of 420 GW by 2030 remains a challenge. Addressing issues, such as habitat disruption, biodiversity loss, and the displacement of local communities, remains a critical concern. In contrast, the environmental concerns associated with solar and wind energy development are comparatively minimal.

Keywords: Renewable · Energy · Wind · Solar · Hydropower · Climate · Change · India

1 Introduction

Climate change refers to long-term shifts in the Earth's climate, mainly attributed to anthropogenic activities. These activities include the release of greenhouse gases like carbon dioxide, methane, and nitrous oxide into the atmosphere, which significantly contribute to these changes [1, 2]. Climate change presents a serious threat to the rich biodiversity of India, impacting both ecosystems and species. The livelihoods of indigenous communities and the availability of natural resources are significantly impacted by climate change. Due to existing social inequalities, vulnerable communities in India face

C. Weng (Ed.): WRE 2024, LNCE 634, pp. 449–457, 2025.
https://doi.org/10.1007/978-981-96-5095-8_37

limited access to resources, leading to disproportionate challenges in implementing adaptive measures. To combat climate change, India has implemented various mitigation and adaptation strategies, including the National Action Plan on Climate Change (NAPCC). As a signatory of international agreements like the Paris Agreement, India participates in global initiatives to tackle climate change. Considering the adverse impacts of climate change, it is crucial for India to develop adaptation strategies.

There is a significant link between climate change mitigation and renewable energy development in India. As one of the world's fastest growing economies, India faces immense challenges and opportunities in addressing climate change. Fossil fuels currently in use are mainly noted for their high emission CO_2 and other pollutants, uneven resource distribution, high production costs, etc. In contrast, renewable energy sources (RES) are widely available, eco-friendly, and can greatly meet the demands that fossil fuels fulfill in generating electricity, heating, and fuel for transportation. Renewable energy offers a sustainable and indigenous alternative to fossil fuels, reducing India's dependency on imported coal and gas. Due to this, energy security and economic stability are improved, leading to a more resilient energy infrastructure. It is well acknowledged that investing in renewable energy development stimulates economic growth and creates jobs. The renewable energy sector, including manufacturing, installation, and maintenance, can generate millions of jobs, contributing to the socio-economic development of the country. The alignment of renewable energy development with climate change mitigation offers India a path toward sustainable and inclusive growth. Over the past decade, India has made substantial investments in expanding its renewable energy sector.

2 Motivation for Work

India has made significant commitments under the Paris Agreement to reduce the impacts of climate change. The country aims to achieve 40% of its installed power capacity from non-fossil fuel sources and reduce its emissions intensity of GDP by 33–35% by 2030 from 2005 levels. Developing renewable energy is, therefore, key to meeting these international climate commitments. Transitioning to renewable energy reduces air pollution, which is a significant health hazard in India. Cleaner energy sources contribute to improved air quality, reducing respiratory and cardiovascular diseases among the population. Additionally, renewable energy development minimizes environmental degradation associated with fossil fuel extraction and use. The creation of jobs through renewable energy development is another sector that the government of India is currently focusing on. Therefore, the aim of the present research is to assess the impact of renewable energy development in combating climate change in India.

3 Literature Review

The Sustainable Development Goals (SDGs) play a vital role in directing India's efforts to combat climate change, as they tackle the interconnected challenges of sustainable development, poverty reduction, and climate action. Many researchers have highlighted the importance of SDGs in combating climate change. The concept of marginal SDG-emissions-reduction values (MSVs), which measure the incremental impact on SDG indicators for each unit of CO_2 emissions reduced, has been introduced [3]. The study identified significant correlations between CO_2 emission reductions and various SDG targets. For instance, reducing CO_2 emissions by 1% could result in a 0.57% reduction in premature fatalities from air pollution (SDG 3) and a 0.026% decrease in mean species richness. Despite current efforts, global actions remain inadequate to achieve SDG targets [4]. The authors advocate for a more robust sustainable development strategy that includes international climate finance, fair distribution of carbon pricing revenue, and broader energy access. An investigation on how environmental improvements influence economic output in E7 and G7 nations over time, emphasizing the need to reduce pollution and shift towards green and innovative energy sources, has been carried out [5]. The study also offered policy recommendations for stakeholders to develop financial strategies that promote green investments and climate change mitigation. The relationship between CO_2 emissions, economic growth, and employment structures in 297 Chinese cities utilizing machine learning was investigated [6]. Contrary to the belief that transitioning mainly in the service sector is the most effective way to decouple growth in the economy from CO_2 emissions, their findings suggest that with technological advancements, the manufacturing and construction sectors can also achieve this decoupling while generating employment opportunities.

Several innovative ways to enhance energy efficiency have been described to enhance energy efficiency. For instance, Asonja et al. [7] explore the use of corn cob as a fuel to enhance energy efficiency. Aksoy and Ispir [8] discuss the technical and economic feasibility of various photovoltaic techniques to enhance solar power generation. Several studies in different parts of the world have evaluated the impacts of renewable energy development on air pollution mitigation. Most studies increasingly acknowledge the adverse environmental impacts of non-renewable energy sources, especially concerning air pollution and greenhouse gas emissions [9–11]. Chen et al. [12] identified a unidirectional relationship, where renewable energy contributes to reducing air pollution, whereas non-renewable energy sources result in an increase, specifically in the context of China. Studies by Ao et al. [13] and Chien et al. [14] have shown that stringent regulations effectively promote the shift to cleaner energy sources and enhance air quality. Some studies also explored the complex connections between the economy, energy consumption, and environmental issues, highlighting the need for sustainable energy strategies to mitigate adverse impacts [15–18].

4 Renewable Energy Development in India

Figure 1 shows the growth in the installed capacity of solar power in India from fiscal year (FY) 2010–11 to FY 2023–24. There is a clear exponential increase in solar power capacity over this period. The initial years (FY2010–11 to FY2014–15) show relatively modest growth, with installed capacity remaining below 5 GW. However, starting around FY2015–16, there is a significant acceleration in the growth rate, indicating a shift towards rapid expansion of solar energy infrastructure. The sharp increase in capacity from FY2015–16 onwards corresponds with government initiatives and policies aimed at boosting solar power, such as the National Solar Mission and various incentives for solar energy development. The continued rise in capacity shows that these policies have had a sustained impact over the years, promoting the scaling up of solar projects across the country. In the most recent fiscal years (FY2021–22 to FY2023–24), the installed capacity sees a substantial jump. This surge could be attributed to intensified efforts by the government to meet its renewable energy targets, including commitments made under international agreements like the Paris Agreement and the Panchamrit targets announced at the 26th Conference of Parties (COP26).

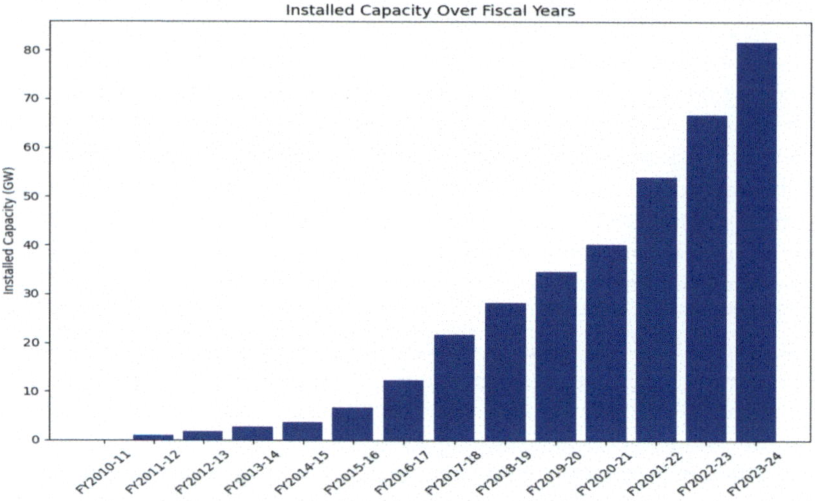

Fig. 1. Growth of solar energy capacity in India.

Figure 2 shows the growth in the installed capacity of wind power in India from fiscal year (FY) 2010–11 to FY 2023–24. A steady and consistent increase in wind power capacity over the years can be clearly seen from Fig. 2. The capacity rises gradually from FY2010–11 to FY2023–24 which indicates a sustained effort to expand wind energy infrastructure in India. Unlike solar power, which has exhibited exponential growth, the increase in wind power capacity appears to be linear. This suggests a stable yet moderate rate of development, reflecting the mature nature of the wind power sector in India. The growth in capacity from FY2010–11 to FY2015–16 is noticeable but relatively gradual, indicating that wind power growth did not experience a rapid surge during this period. In the most recent fiscal years (FY2021–22 to FY2023–24), the capacity has shown continuous growth but at a relatively stable rate. This suggests that the sector is reaching a mature phase where large-scale additions are consistent but do not show the same acceleration as in earlier years.

The growth in installed capacity of hydropower in India from FY2010–11 to FY2023–24 is shown in Fig. 3. It can be seen from Fig. 3 that the installed hydropower capacity has remained relatively stable over the years, ranging between 35 GW and 45 GW. This indicates that the hydropower sector has reached a plateau with limited growth in recent years. There was a slight increase in capacity from FY2010–11 to FY2014–15, but after this period, the growth rate appeared to stagnate, with the installed capacity

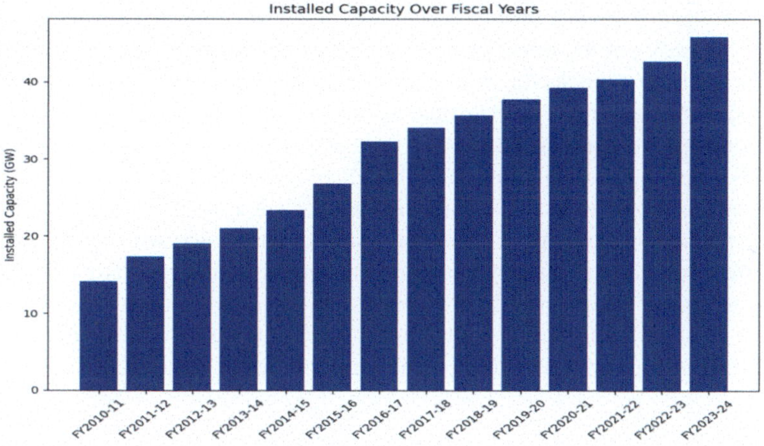

Fig. 2. Growth of wind energy capacity in India.

stabilizing around 45 GW. This stagnation could be attributed to the challenges associated with hydropower development, such as environmental concerns, displacement issues, and long gestation periods for new projects. The steady trend in hydropower capacity suggests that the sector is mature, with most economically and technically feasible sites already developed. Expanding hydropower capacity may require upgrading existing plants or developing smaller, more sustainable hydro projects.

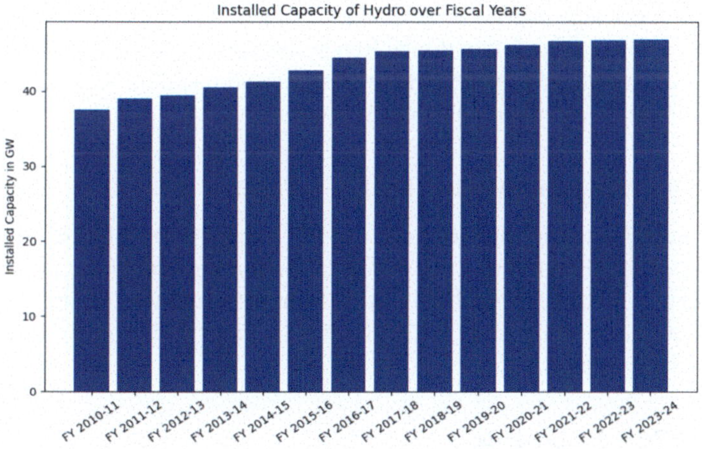

Fig. 3. Growth of hydropower development in India.

The growth rates of solar, wind, and hydropower energy in India have been influenced by a combination of government policies, economic factors, and resource availability. Solar energy has experienced the most rapid and exponential growth, primarily driven by substantial government initiatives such as the National Solar Mission, which provided financial incentives, subsidies, and favorable policy frameworks. The reduction in the cost of solar technology has further accelerated this growth. Wind energy, in contrast, has shown a steady yet moderate growth rate. This trend can be attributed to the early establishment of the wind sector, which matured faster than solar energy. However, the expansion of wind power has been limited by land availability, logistical challenges, and the need for improved grid infrastructure in high-potential states like Tamil Nadu and Gujarat. Hydropower, on the other hand, has displayed relatively slow growth. This is due to several environmental and social constraints, such as habitat disruption, biodiversity loss, and the displacement of local communities associated with large hydropower projects. Additionally, the lengthy development timelines and regulatory hurdles have hindered the rapid expansion of hydropower. Although hydropower remains a crucial component of India's renewable energy mix, recent efforts have focused on smaller, more sustainable hydro projects to minimize environmental impacts.

5 Impact of Renewable Energy on Air Pollution

The replacement of coal-fired power plants and other fossil fuel-based energy sources with renewable energy has led to considerable improvements in air quality in India. Quantification of the decrease in air pollution due to the development of renewable energy in India involves analyzing reductions in emissions of major air pollutants, such as particulate matter (PM2.5 and PM10), sulfur dioxide (SO_2), nitrogen oxides (NOX), and carbon dioxide (CO_2).

5.1 Reduction in CO_2 Emissions

The growth in renewable energy capacity has helped India significantly reduce its CO_2 emissions. According to the International Renewable Energy Agency (IRENA) [19], the expansion of renewable energy in India avoided an estimated 60 million to 70 million tons of CO_2 emissions annually in recent years.

5.2 Impact on Particulate Matter (PM)

Coal-fired power plants are a major source of PM2.5 and PM10 pollution. The transition to renewable energy has led to a reduction in the emission of these pollutants [20]. Estimates suggest that the air quality benefits from solar and wind energy installations could lead to a 5–10% decrease in PM2.5 levels in regions with a high concentration of renewable energy projects [21].

5.3 SO_2 and NO_x Reductions

Fossil fuel combustion releases large amounts of SO_2 and NO_x, contributing to respiratory illnesses and acid rain. Renewable energy projects have helped cut these emissions, with research indicating that areas with significant solar and wind deployment have seen a reduction of 10–15% in SO_2 and NO_x emissions. Cao et al. [22] analyzed the impact of the Joint Prevention and Control (JPC) policy on reducing SO_2 emissions across 116 Chinese cities between 2003 and 2017. They advocated for policies that foster regional cooperation under the JPC framework to achieve further reductions in SO_2 emissions.

6 Conclusions

The analysis of the growth in solar, wind, and hydropower energy installed capacity in India from FY2010–11 to FY2023–24 demonstrates a clear trajectory toward increasing reliance on renewable energy sources. Each renewable energy sector exhibits a unique growth pattern, reflecting the country's strategic approach to diversifying its energy mix and addressing specific challenges associated with each technology. Solar energy has shown the most rapid and exponential growth among the renewable sectors. The solar installed capacity has reached around 85 GW. The surge in solar capacity, particularly post-2015, aligns with India's policy push, such as the National Solar Mission and other incentive-based programs. Wind energy has displayed a steady and consistent growth trend, with capacity gradually increasing over the years and reaching around 45 GW. Although not as rapid as solar, the growth in wind capacity reflects a mature sector that continues to expand at a stable rate. The installed capacity of hydropower has remained largely stable at around 48 GW, indicating a saturation of economically and technically feasible projects. While hydropower still contributes significantly to the energy mix, the stagnation suggests that India may need to focus on modernizing existing plants and exploring sustainable, smaller-scale hydro projects to further enhance its hydropower potential.

Over the past decade, significant research efforts have been dedicated to renewable energy, particularly wind energy [23]. The development of renewable energy not only contributes to climate mitigation but also creates economic opportunities, such as job creation and rural electrification, which further support India's socio-economic development goals. Overall, the growth of renewable energy in India reflects a comprehensive and strategic approach that utilizes the diverse potential of various energy sources, such as solar, wind, and hydropower, to effectively address climate change. This balanced approach lowers greenhouse gas emissions, supports sustainable development, and creates a more resilient and secure energy infrastructure. By diversifying the energy portfolio and investing heavily in renewable technologies, India is making significant strides towards reducing its dependence on fossil fuels and lowering carbon emissions. However, mitigating the environmental impacts of large hydropower projects, including habitat disruption, biodiversity loss, and the displacement of local communities, continues to be a significant challenge. In contrast, there are considerable fewer environmental concerns associated with solar and wind energy development.

The analysis presented in this research indicates that the large-scale deployment of renewable energy has had a substantial impact on reducing air pollution levels, which is crucial for improving public health and protecting the environment. Cleaner air resulting from renewable energy usage can decrease in respiratory and cardiovascular illnesses, providing significant health benefits to the population, particularly in urban areas suffering from severe pollution. However, despite these positive outcomes, further research and continuous monitoring are essential to provide a more precise quantification of these benefits and to assess the long-term impacts of renewable energy development. Ongoing studies will help policymakers and stakeholders better understand how renewable energy projects influence environmental and social factors over time, ensuring that India continues on a path of sustainable and equitable growth while effectively mitigating the adverse effects of climate change. Future studies shall investigate how renewable energy development can be aligned with circular economy principles, focusing on the recycling and reusing materials in solar panels, wind turbines, and batteries.

References

1. Sharif, M., Burn, D.H.: Simulating climate change scenarios using an improved K-nearest neighbor model. J. Hydrol. **325**, 179–196 (2006)
2. Gupta, M.K., Sharif, M.: Spatio-temporal analysis of temperature projections based on representative concentration pathways for Satluj River Basin, India. Cog. Eng. **8**(1), 15 (2021)
3. Fujimori, S. et al.: Measuring the sustainable development implications of climate change mitigation. Environ. Res. Lett. **15**(8) (2020)
4. Soergel, B. et al.: A sustainable development pathway for climate action within the UN 2030 Agenda. Nat. Clim. Chang. **11**, 656–664 (2021)
5. Wu, X., Sadiq, M., Chien, F., Ngo, Q.T., Nguyen, A.T., Trinh, T.T.: Testing role of green financing on climate change mitigation: Evidences from G7 and E7 countries. Environ. Sci. Pollut. Res. **28**(47), 66736–66750 (2021)
6. Liu, Z., et al.: Climate action may reduce the risk of unemployment: An insight into the city-level interconnections among the sustainable development goals. Resour. Conserv. Recycl. **194**, 107002 (2023)
7. Asonja, A., Desnica, E., Radovanovic, L.: Energy efficiency analysis of corn cob used as a fuel. Energy Sources Part B **12**(1), 1–7 (2016). https://doi.org/10.1080/15567249.2014.881931
8. Aksoy, M.H., Ispir, M.: Techno-economic feasibility of different photovoltaic technologies. Appl. Eng. Letters, **8**(1), 1–9 (2023). https://doi.org/10.18485/aeletters.2023.8.1.1
9. Hailemariam, A., Ivanovski, K., Dzhumashev, R.: Does R&D investment in renewable energy technologies reduce greenhouse gas emissions? Appl. Energy **327**, 120056 (2022). https://doi.org/10.1016/j.apenergy.2022.120056
10. Sharifzadeh, M., Hien, R.K.T., Shah, N.: China's roadmap to low-carbon electricity and water: disentangling greenhouse gas (GHG) emissions from electricity-water nexus via renewable wind and solar power generation, and carbon capture and storage. Appl. Energy **235**, 31–42 (2019)
11. Zhang, D., Chen, X.H., Lau, C.K.M., Xu, B.: The implications of cryptocurrency energy usage on climate change. Technol. Forecast. Soc. Change **187**, 122219 (2023)
12. Chen, X.H., Tee, K., Elnahass, M., Ahmed, R.: Assessing the environmental impacts of renewable energy sources: A case study on air pollution and carbon emissions in China. J. Environ. Manage. **345**, 118525 (2023). https://doi.org/10.1016/j.jenvman.2023.118525
13. Ao, Z., Fei, R., Jiang, H., Cui, L., Zhu, Y.: How can China achieve its goal of peaking carbon emissions at minimal cost? A research perspective from shadow price and optimal allocation of carbon emissions. J. Environ. Manag. **325**, 116458 (2023). https://doi.org/10.1016/j.jenvman.2022.116458
14. Chien, F., Sadiq, M., Nawaz, M.A., Hussain, M.S., Tran, T.D., Le Thanh, T.: A step toward reducing air pollution in top Asian economies: the role of green energy, eco-innovation, and environmental taxes. J. Environ. Manag. **297**, 113420 (2021). https://doi.org/10.1016/j.jenvman.2021.113420
15. Fan, W., Hao, Y.: An empirical research on the relationship amongst renewable energy consumption, economic growth and foreign direct investment in China. Renew. Energy **146**, 598–609 (2020)
16. Pinz´on, K.: Dynamics between energy consumption and economic growth in Ecuador: a granger causality analysis. Econ. Anal. Pol. **57**, 88–101 (2018). https://doi.org/10.1016/j.eap.2017.09.004

17. Saint Akadiri, S., Alola, A.A., Akadiri, A.C., Alola, U.V.: Renewable energy consumption in EU-28 countries: Policy toward pollution mitigation and economic sustainability. Energy Pol. **132**, 803–810 (2019). https://doi.org/10.1016/j.enpol.2019.06.040
18. Yuan, X., Mu, R., Zuo, J., Wang, Q.: Economic development, energy consumption, and air pollution: A critical assessment in China. Human Ecol. Risk Assess. Int. J. **21**, 781–798 (2015). https://doi.org/10.1080/10807039.2014.932204
19. IRENA, COP28, COP29, GRA, MoEA and Government of Brazil: Delivering on the UAE Consensus: Tracking progress toward tripling renewable energy capacity and doubling energy efficiency by 2030, International Renewable Energy Agency, COP28 Presidency, COP29 Presidency, Ministry of Energy of the Republic of Azerbaijan, and Government of Brazil, Abu Dhabi (2024)
20. Khan, Y., Bounade, C.D.: Particulate matter 2.5 air pollution mitigation strategy: the role of green investment, digitalization, and renewable energy in the organization for economic co-operation and development (OECD) countries. Clean Techn. Environ. Policy. (2024). https://doi.org/10.1007/s10098-024-02864-5
21. Gi, K., et al.: A model-based analysis on energy systems transition for climate change mitigation and ambient particulate matter 2.5 concentration reduction. Mitig. Adapt. Strateg. Glob. Chang. **24**, 181–204 (2019)
22. Cao, D., Sun, Y., Chai, J., Xue, J., Sun, Q.: An assessment of China's joint prevention and control policy on sulfur dioxide emissions reduction: A spatial econometric analysis. J. Manage. Sci. Eng. **8**(4), 498–511 (2023). https://doi.org/10.1016/j.jmse.2023.03.003
23. Koondhar, M.A., Ali, M., Keerio, M.U., Junejo, A.K., Laghari, I.A., Chandio, S.: Wind energy conversion system using maximum power point tracking technique—A comprehensive survey. Appl. Eng. Lett. **6**(4), 148–156 (2021). https://doi.org/10.18485/aeletters.2021.6.4.2

A Solar-Powered Pumping System for Agricultural Irrigation: Design Methodology and Practical Application

Wuchang Wang[1(✉)], Mingyu Fu[2], Yizi Shang[1(✉)], and Ada Won[3]

[1] Institute of Water Resources and Hydropower Research, Beijing 100038, China
wwuch@qq.co, shangyz@iwhr.com

[2] Beijing Renda Machinery Equipment Co., Ltd, Beijing 100044, China

[3] Resun Solar Power Co., Ltd, Suzhou 215000, China

Abstract. Pumping irrigation is a vital method for delivering water to farmland using water pumps, ensuring an adequate water supply for crop growth, and securing grain yields. However, in remote mountainous regions and rural areas with limited access to electricity, traditional irrigation systems typically rely on fuel-powered engines to operate water pumps. This method not only results in considerable energy consumption and high operational and maintenance costs but also contributes to carbon emissions, further impacting the climate and environment. With the continuous advancements in renewable energy sources such as solar and wind power, exploring the application of these new energy technologies in agricultural irrigation is crucial for promoting energy conservation and emission reduction in the agricultural sector. The solar-powered pumping system offers a practical and feasible technological solution. This paper proposes a design methodology for a solar-powered pumping irrigation system, where a solar photovoltaic power generation system serves as the power source for the pumps, which are then integrated with the field pipelines for farmland irrigation. The system has been implemented in a demonstration farmland in the Philippines. Operational results demonstrate that it successfully provides adequate irrigation water for the crops in about 10 hectares of farmland according to design specifications, without fuel, thereby validating the system's practical feasibility and reliability.

Keywords: Pumping Irrigation · Solar-Powered Pumping System · Photovoltaic Water Pump · Agricultural Energy Conservation and Emission Reduction · "Water-Energy-Food" Nexus

1 Introduction

Amidst the backdrop of escalating global climate change and the rapid transformation of energy systems, coupled with the increasing scarcity of water resources, the interdependence of water, energy, and food—referred to as the "Water-Energy-Food" nexus—has emerged as a critical factor affecting the survival and sustainable development of humanity, as well as the stability of natural ecosystems [1, 2]. In food production, especially in

C. Weng (Ed.): WRE 2024, LNCE 634, pp. 458–471, 2025.
https://doi.org/10.1007/978-981-96-5095-8_38

arid and semi-arid regions, agricultural irrigation plays a crucial role in ensuring crop growth and food security. An efficient and well-designed irrigation system can not only enhance crop yield and quality but also optimize the use of water and energy resources [3, 4].

According to global statistics, approximately two billion people worldwide reside in water-scarce areas, predominantly in remote, rural, and mountainous regions of developing countries [5, 6]. These regions typically suffer from unreliable power supplies and primarily rely on traditional irrigation methods for pumping irrigation. However, traditional irrigation methods often involve the use of fossil fuel-powered engines to operate water pumps, which not only consumes vast amounts of energy and increases irrigation costs but also contributes to carbon emissions and environmental pollution [7, 8]. With the continuous advancement of renewable energy technologies, such as wind and solar power, exploring the research and development (R&D) and application of these new energy sources in agricultural irrigation has become a significant trend, holding great promise for promoting energy conservation and emission reduction in the agricultural sector. Solar energy stands out as a clean and renewable energy source that is abundant and universally accessible. Consequently, the development of solar-powered pumping technology presents a viable and practical solution [9, 10].

A solar-powered pumping irrigation system utilizes solar photovoltaic (PV) technology to convert solar energy into electrical power, which drives pumps for water lifting and irrigation. This system does not rely on fossil fuels and avoids environmental pollution. By integrating PV technology with agricultural irrigation practices, it offers an innovative approach to address water scarcity in areas lacking both water and electricity, thereby enhancing agricultural productivity. Moreover, with the continual improvement of PV modules, water pumps, and intelligent control systems, the efficiency of PV modules is on the rise, while their costs, along with those of related equipment (e.g., inverters, pumps, etc.), are significantly decreasing. This trend has led to a steady reduction in the overall cost of solar-powered pumping systems, making them increasingly technically and economically feasible, and thus widening their application potential [11–13].

This study is dedicated to exploring the design and application of solar-powered pumping irrigation systems in real-world agricultural settings, with the goal of providing a practical and feasible strategy to foster a positive cycle within the "Water-Energy-Food" nexus.

2 Materials and Methods

2.1 Technical Principle of Solar-Powered Pumping System

Solar-powered pumping technology harnesses solar energy through PV cell panels, converting solar radiation into electrical energy, which is then utilized to power water pumps and supply water for agricultural irrigation or human and livestock consumption. This method of water supply leverages clean and renewable energy technology and finds extensive application in areas such as agricultural irrigation, domestic water use, and industrial water supply due to its key advantages, including sustainability, environmental friendliness, and cost-effectiveness [14, 15]. A typical solar-powered pumping irrigation system comprises several components, including PV modules, controllers, inverters, electric motors, water pumps, storage tanks, pipelines, etc. [16]. The system's working principle is depicted in Fig. 1.

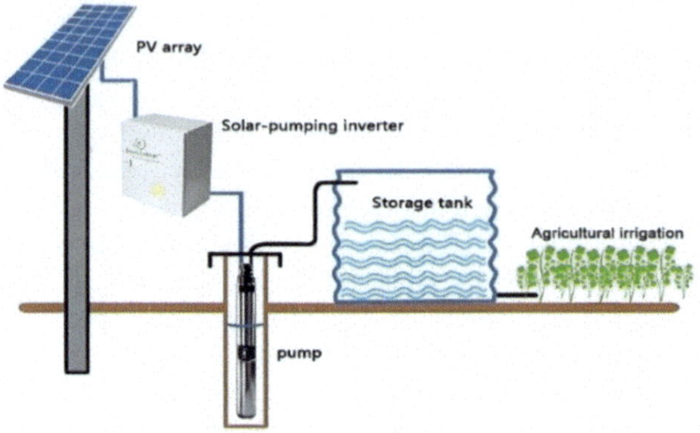

Fig. 1. Schematic diagram of a solar-powered pumping irrigation system.

The PV modules first absorb light energy and facilitate energy conversion through the photoelectric effect, converting it into electrical energy and generating direct current (DC). The PV controller, functioning as the system's intelligent control center, monitors and manages the system's operational status. It automatically detects the output voltage of the solar panels and the light intensity, assesses the availability of solar energy resources and controls the operation of the electric motor and water pump accordingly. Next, the inverter converts the DC into alternating current (AC) to power the electric motor, which then drives the water pump. In practical applications, the PV controller and inverter are typically integrated into a single unit known as a solar-pumping inverter. Finally, water from deep sources is pumped out and directed into a storage tank through the pipeline system or directly connected to the irrigation system for agricultural irrigation purposes.

2.2 System Design Methodology

When conducting application-oriented research on solar-powered pumping systems for agricultural irrigation, it is imperative to reasonably calculate and configure the parameters of each component to guarantee that the entire system achieves optimal efficiency and effectiveness during the design process [17, 18]. The primary design steps and methods are as follows:

Calculation of required pumping flow and head. Based on the water requirements of farmland crops, the irrigation quota, the irrigation area, and the irrigation method, calculate the water flow required for the pump. Then, the head required for the pump will be determined by considering the water level difference and hydraulic loss between the water source and the irrigation pipeline system or storage tank.

Selection of water pump. The performance of the water pump directly affects the system's pumping efficiency and water delivery capacity. Factors such as the depth of the water source, water quality, water delivery volume, delivery distance, durability, and reliability of the pump must be considered when selecting water pump types. Common water pump types include centrifugal pumps, deep well pumps, and screw pumps. The selection should be based on the calculated pumping flow and head, and the water source's actual conditions. Then, the water pump type, rated flow rate, head, and motor power can be determined according to the technical specifications of pumps.

Configuration of solar-pumping inverter. Factors such as input voltage, current, output power, efficiency, and stability need to be considered when selecting the inverter. In addition, a controller module for controlling start-stop and protection should also be set up to ensure the safe and stable operation of the entire system.

Configuration of PV modules. PV modules are the core components of the system, and their performance directly affects the system's power generation efficiency and stability. Factors such as the power, voltage, and conversion efficiency of the modules need to be considered when selecting PV panels. Then, the PV string design can be conducted to form a PV array combined with the inverter's input voltage range.

Design of PV mounting structure. The PV mounting structure supports PV panels, and its stability and safety directly relate to the service life and safety of the PV power generation system. Factors such as material, structure, and assembly method need to be considered when designing the mounting structure to ensure that the PV modules can achieve optimal power generation efficiency.

3 Case Study

3.1 Project Profile

The case study is drawn from a demonstration project on solar-powered pumping irrigation launched by the agricultural department of the Philippines. The demonstration farmland is situated in the suburban countryside near Manila City, encompassing an irrigation area of 10 hectares (ha). Initially, local farmers relied on traditional water supply methods for irrigation as depicted in Fig. 2, where a water pump was directly powered by diesel engines to extract water from wells with a diameter of 4-inch pipe on the farmland. The water was then channeled through earthen ditches to the fields for flood irrigation, supplementing the crops' water needs. This method not only incurred a low utilization rate of water resources but also typically led to significant fuel consumption during irrigation, resulting in high operation and maintenance costs, carbon emissions, and even environmental pollution. With the backing of the local agricultural department, plans are underway to convert the original irrigation facilities into a solar-powered pumping irrigation system. This renovation aims to enhance irrigation efficiency, reduce operation and maintenance costs, and alleviate environmental impact.

The feasibility study report for this project, furnished by the local agricultural department, outlines the relevant design parameters for the solar-powered irrigation system as follows: (a) considering the local crop irrigation quota, the water requirement for a single irrigation event is 24 cubic meters per hectare (m^3/ha), with a total of 6 to 7 irrigation cycles needed throughout the entire growth period; (b) the designed flow rate at the most disadvantageous water outlet point on the farm should not fall below 50 cubic meters per hour (m^3/h), and the water outlet pressure should not be less than 5 m (m) of the head; (c) the irrigation water utilization rate should not be lower than 70%; (d) the operational duration for each irrigation, during which the pump's water output is not less than 70% of the rated water output, is 6.5 h.

Fig. 2. Original irrigation facilities of the demonstration farmland.

3.2 Design of Field Irrigation System

In line with the design data, the original irrigation method employed on the farmland was flood irrigation, with water transported through earthen ditches across the fields. However, due to factors such as seepage, evaporation, and blockage by weeds, the irrigation efficiency and water utilization rate were notably low. Consequently, it is imperative to revamp the original irrigation method and design a new field pipeline irrigation system to supplant the existing earthen ditches for water conveyance and irrigation.

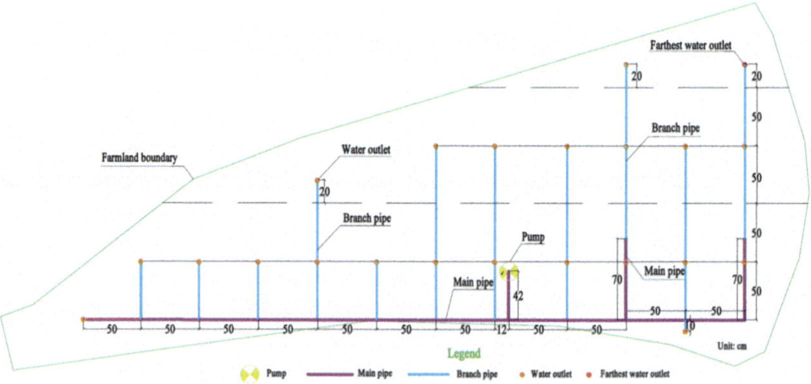

Fig. 3. Layout plan of the farmland irrigation pipelines.

The proposed pipeline system is primarily composed of main pipes, branch pipes, and outlet pipes, all constructed from polyethylene (PE) pipes. Specifically, the main pipe diameter is DN140, the branch pipe diameter is DN125, and the outlet pipe diameter is DN80. The pipeline intersections or diameter changes are connected using tees, crosses, or reducers to ensure a smooth water flow transition. Once the pipelines enter the cultivated area, they are all buried underground to prevent interference with farming activities. The buried pipe depth is set at 70 cm, with the water outlet height above ground at 20 cm. A ball valve is installed at the water outlet to manage water discharge. The layout plan for the farmland irrigation system is depicted in Fig. 3.

3.3 Calculation of Required Pumping Flow and Head

Based on the design data, the crops on the demonstration farmland have a water requirement (ET) of 24 m^3/ha, an irrigation area (A) of 10 ha, and an irrigation water utilization rate (η) of 70%. To ensure that the water outlet points at various locations receive adequate flow rate and pressure, a rotational irrigation approach is adopted. The total irrigation water consumption for the demonstration farmland is calculated as W = 342.86 m^3 based on formula (1). Considering that the irrigation duration (t) at a pump flow rate not lower than 70% of the rated flow rate is 6.5 h, the minimum water flow rate required for the pump can be calculated using formula (2) as $Q_{req} = 52.75$ m^3/h.

$$W = (ET \times A)/\eta \tag{1}$$

$$Q_{req} = W/t \tag{2}$$

where ET represents the crop water requirement (m^3/ha), A is the irrigation area (ha), η is the irrigation water utilization rate (%), W is the water consumption per irrigation event (m^3), t is the irrigation duration (h), and Q_{req} is the minimum water flow rate required for the pump (m^3/h).

As the farmland irrigation system employs pipelines for water conveyance and distribution, the frictional and local head losses of the pipeline system must be considered when calculating the required water head for the pump. The frictional head loss (h_f) and local head loss (h_j) are crucial in determining the total head loss (h_w) in the pipeline system. The coefficient of frictional resistance (λ), which depends on the roughness of the pipe walls and the flow rate, is a significant factor in calculating h_f. Additionally, the local head loss (h_j) is influenced by the presence of pipe fittings, such as elbows, valves, and tees, each contributing to the overall resistance in the system. The local resistance coefficient (ζ) quantifies the head loss caused by these fittings.

To ensure that each water outlet in the farmland has adequate water pressure, the pipeline corresponding to the most disadvantageous point i.e. the farthest water outlet, is used as the basis for calculation. Its frictional head loss and local head loss can be calculated respectively using formula (3) and formula (4).

$$h_f = \lambda \cdot \frac{l}{d} \cdot \frac{v^2}{2g} \tag{3}$$

$$h_j = \zeta \cdot \frac{v^2}{2g} \tag{4}$$

$$h_w = h_f + h_j \tag{5}$$

where h_f is the frictional head loss (m); h_j is the local head loss (m); h_w is the total head loss (m); λ is the coefficient of frictional resistance; l is the length of the pipeline (m); ζ is the coefficient of local resistance, if there are multiple local resistance coefficients along the pipeline, they should be summed together; d is the diameter of the pipeline (m); g is the acceleration due to gravity (m/s^2); and v is the average velocity of the cross-section of the pipeline (m/s).

The design drawing of the farmland irrigation pipelines provides the necessary information about the size and quantity of pipe fittings on the pipeline for the farthest water outlet. Along with the coefficients of frictional and local resistance (λ and ζ), this data is tabulated in Table 1, enabling precise calculations of the head losses in the irrigation pipeline system.

Table 1. Calculation parameters of head losses for the pipeline of the farthest water outlet.

Pipe fittings	Size	Quantity	λ	ζ
PE pipe	DN 80	0.9 m	0.022	
PE pipe	DN 125	150 m	0.018	
PE pipe	DN 140	312 m	0.016	
Steel pipe	DN 100	10 m	0.03	
PE ball valve	DN 80	1		0.1
PE elbow 90°	DN 80	1		0.6
PE reducer	DN 125 × 80	1		0.1
PE reducer	DN 140 × 125	1		0.1
PE reducing tee	DN 125 × 125 × 80	1		0.1
PE reducing tee	DN 140 × 140 × 80	1		0.1
PE elbow 90°	DN 140	1		0.1
PE equal tee	DN 140	1		1.5
Steel elbow 90°	DN 100	1		1.0

By substituting the relevant coefficients into formulas (3) and (4) for calculation, the total head loss of the pipeline system for the farthest water outlet is calculated using formula (5) as $h_w = 4.45$ m. As per the initial survey, the static water level of the well is 5 m below ground level, the maximum dynamic water level is 7 m below ground level, the water pump is installed at an elevation of 0.5 m above ground level, and the water outlet point at the farthest end is at an elevation of 0.2 m above ground level. This results in a maximum height difference of 6.8 m between the farthest water outlet and the water level surface of the well. According to the design requirement, the pressure at the farthest water outlet point should not be less than 5 m water head. Consequently, the minimum head required for the water pump can be calculated using formula (6) as $H_{req} = 16.25$ m.

$$H_{req} = \Delta z + h_w + \frac{p}{\rho g} \tag{6}$$

where H_{req} is the minimum head required for the water pump (m); Δz is the maximum height difference between the farthest water outlet and the water level of the well (m); p is the required pressure of the farthest water outlet (Pa); ρ is the density of water (kg/m^3); g is the acceleration due to gravity (m/s^2).

3.4 Pump Selection

Given the existing well's diameter of only 4 inches, installing a deep well pump into the well is not feasible due to the dimensional constraints. Consequently, selecting a self-priming centrifugal pump to be installed above ground is more appropriate. This type of pump is equipped with a self-priming feature and eliminates the need for manual priming, thereby enhancing irrigation efficiency. Referring to the technical specifications and type-spectrum chart for self-priming centrifugal pumps, as depicted in Fig. 4, where flow rate is plotted on the x-axis and head and power are on the y-axis, this chart typically illustrates the relationships among parameters such as flow rate, head, and power of various pump models. It also reflects the performance curves of a range of pumps and the optimal operation range for each pump type, facilitating the swift selection of an appropriate pump model.

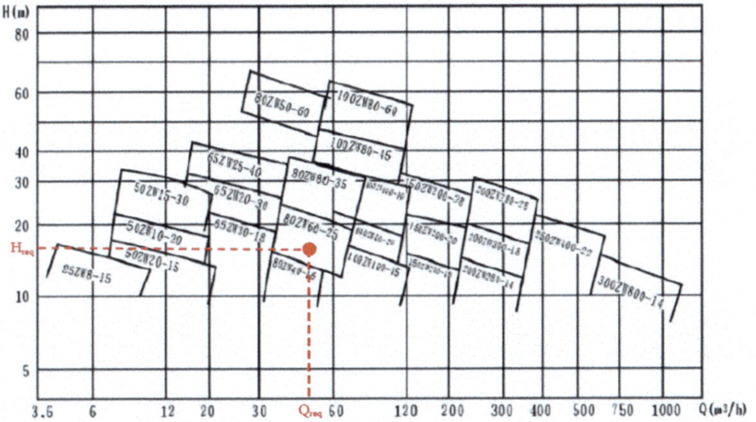

Fig. 4. Self-priming pump type-spectrum chart (Note: The figure is sourced from the technical specifications of self-priming pumps.)

Upon marking the calculated Q_{req} and H_{req} values for the water pump on the type-spectrum chart, it becomes apparent that the required working flow rate and head fall within the performance curve range of pump models between 80ZW40-16 and 80ZW60-25. In order to ensure a safety margin and enhance the reliability of the pumping output, the pump model 80ZW60-25 is selected as the most suitable option. This model features a rated flow rate of $Q = 60\ m^3/h$, a head of $H = 25$ m, a power of $P = 7.5$ kW, and operates at a motor voltage level of $V = 380$ V AC.

3.5 Configuration of Solar-Pumping Inverters

Given that PV power generation is susceptible to weather-related factors, such as solar radiation intensity and temperature, the resulting power output tends to be unstable. To ensure that the water pump operates as closely as possible to its rated working conditions, it is imperative that the inverter's rated output power exceeds the pump motor's rated power of 7.5 kW. By referring to the technical specifications of compatible solar-pumping inverters, as detailed in Table 2, an inverter with a rated power of 11 kW is chosen, balancing both economic and reliability considerations. This inverter is equipped with a Maximum Power Point Tracking (MPPT) function, which automatically adjusts the output frequency in response to the PV array's output voltage, thereby maximizing power output. The MPPT efficiency can reach up to 99.8%, and the MPPT working voltage range spans from 450V to 800V. It has the capability to start or stop based on light intensity and can be controlled either manually on-site or remotely. Additionally, the inverter is furnished with various protection functions, including undervoltage, overvoltage, lightning, overcurrent, overload, short circuit, and anti-dry running protections.

Table 2. Technical specifications of the solar-pumping inverter.

Model	Rated Power (kW)	Max DC Input (V)	MPPT Voltage (V)	Rated Output (V)	Output Frequency (Hz)	DC Power (kW)	AC Power (kW)
SP1100–5.5B	5.5	800	450–800	380V	0–50/60	8.25	5.5
SP1100–7.5B	7.5	800	450–800	380V	0–50/60	11.25	7.5
SP1100-11B	11	800	450–800	380V	0–50/60	16.5	11
SP1100-15B	15	800	450–800	380V	0–50/60	20	15

3.6 Configuration of PV Modules

In line with the technical specifications of the solar-pumping inverter, for an inverter with a rated output power of 11 kW, the optimal DC input power is determined to be 16.5 kW. This implies that the total power of the required PV modules should be 16.5 kW, with the DC input voltage for MPPT ranging from 450V to 800V. After reviewing the technical specifications of various PV modules, it is recommended to select 30 pieces of 550 Wp monocrystalline modules. These modules should be interconnected in a configuration of 15 in series and 2 in parallel to form the PV array. The selected PV module type exhibits superior electrical performance under Standard Test Conditions (STC: Irradiance 1000 W/m^2, Cell Temperature 25 °C, Air Mass AM1.5 as per EN 60904-3), i.e. Rated Power $P_{max} = 550$ Wp; Open Circuit Voltage $V_{oc} = 50.02$ V; Maximum Power Voltage $V_{mp} = 42.00$ V; Module Efficiency $\eta_0 = 21.27\%$. Consequently, under standard temperature conditions, the maximum open-circuit voltage of the PV array is 750.03 V, and the working voltage at MPPT is 630 V, satisfying the inverter's requirements for maximum input voltage and the MPPT working voltage range.

3.7 Design of PV Mounting Structure

The design of the PV mounting structure is based on the size and quantity of the PV modules, as well as the stringing arrangement of the PV array. For this project, the 550 Wp module measures 2279 × 1134 × 35 mm, with 30 panels arranged in a 2 × 15 configuration to form the PV array. Initially, the optimal tilt angle of the PV modules is determined to be 16°, considering the latitude of the demonstration farmland and NASA solar radiation data. Next, referring to local hydrological data, the distance from the lower edge of the mounting structure to the ground is set at 1.5 m to prevent submersion of the PV panels during extreme weather events. Furthermore, the mounting structure is designed to withstand strong winds and is capable of withstanding the designed maximum wind speed of 50 m/s for 3 s. The PV mounting structure, including purlins, beams, and supports, is constructed from galvanized steel and features a double-row column arrangement. Lastly, based on geological conditions, screw piles with a length of 1.6 m are used as the foundation. The design drawing of the PV mounting structure is presented in Fig. 5.

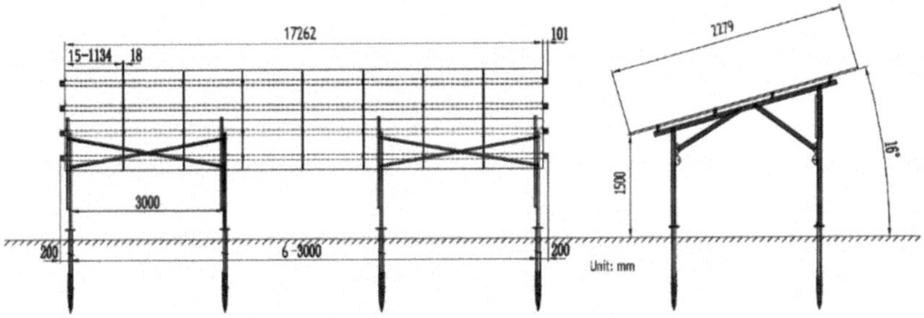

Fig. 5. Design drawing of the PV mounting structure.

4 Results and Discussion

4.1 Design and Application Results

Following the completion of the system design, the primary technical parameters for the key equipment have been established, as depicted in Table 3.

The demonstration project utilized the Engineering, Procurement, and Construction (EPC) model. In the role of contractor, all equipment procurement was conducted in China following the completion of engineering works and subsequently transported via sea to the site of the demonstration farmland in the Philippines. The installation outcomes of the primary equipment at the site are depicted in Fig. 6.

4.2 Discussion

Following the completion of construction, equipment installation, and on-site commissioning tests, a one-month trial operation was conducted to validate the performance

Table 3. Primary design parameters for the solar-powered pumping irrigation system.

No	Equipment	Items	Technical Parameter
1	PV module	Module type	Mono-crystalline
		Rated Power in Watts (P_{max})	550 Wp
		Maximum Power Voltage (V_{mp})	42 V
		Module Efficiency	21.27%
		Power of PV array	16.5 kWp
2	PV mounting structure	PV array layout	2 × 15
		Structure	Double-row columns
		Material	Galvanized steel
		Optimal tilt angle of PV modules	16°
3	Solar-pumping inverter	Rated output power	11 kW
		MPPT input voltage range	380 V AC
		Rated AC output voltage	450~800 V
		MPPT efficiency	99.8%
4	Pump	Model	Self-priming pump
		Flow rate and head	60 m^3/h, 25 m
		Mating motor power	7.5 kW
5	Farmland irrigation piping system	Main pipe	PE pipe DN 140
		Branch pipe	PE pipe DN 125
		Water outlet pipe	PE pipe DN 80

Installation of pump and inverters

Installation of Pipeline and PV array

Fig. 6. On-site installation status of the solar-powered pumping irrigation system.

and reliability of the solar-powered pumping irrigation system. The operational results demonstrated that the flow and head of the new system fully met the design specifications. It was observed that the average daily water supply exceeded 300 m^3, with no fuel consumption or manpower costs. In contrast, the original irrigation system had a rated flow rate of 80 m^3/h and a head of 20 m, with a rated power of 10 kW for the diesel engine. However, the irrigation water utilization rate for crops was relatively low, approximately 50%. This implies that if the same amount of water is used daily for crops, the diesel engine would need to operate under full load for at least 5.25 h, calculated as (300 * 70%)/50%/80 = 5.25 h. Given the diesel engine's fuel consumption rate of about 2.5 L/h, the renovation of the original irrigation system could save approximately 14 L of diesel per day.

The costs of the solar-powered pumping irrigation system primarily include expenses for PV modules, PV mounting structures, solar-pumping inverters, water pumps (including motors), irrigation pipelines, transportation, construction, installation, and maintenance. The economic benefits are mainly realized through fuel savings, reduced irrigation costs, and increased crop yields. Preliminary cost calculations indicate that the total investment for this project is approximately $16,800, fully funded by the local agricultural department as it is a government demonstration project. However, once the project is widely implemented, costs will be shared between the government department and local farmers in a predetermined ratio. According to the information from local farmers, the annual fuel and irrigation costs were approximately $3,000 when using the fuel pump irrigation system. This suggests that the equivalent economic benefit is achieved by replacing the traditional pumping irrigation system with the new system. Consequently, the static payback period for this system is within 16800/3000 = 5.6 years, with an anticipated service life exceeding 20 years, which makes it highly promising for investment potential.

5 Conclusion

This paper presents a methodology for designing a solar-powered irrigation system and demonstrates its practical application in the renovation of a traditional irrigation system at a demonstration farmland. The system design begins by calculating the required water flow rate for the pump based on the farm's crop irrigation needs. Next, the necessary head for the pump is determined by analyzing the water source level and irrigation pipeline system, leading to the selection of a water pump with a specific type, flow rate, head, and motor power. Additionally, the scale of the PV array is determined by integrating the performance parameters of solar-pumping inverters and PV modules, and the optimal tilt angle of PV mounting structures should be established by referencing the local latitude and geographical location to ensure the efficiency of the PV modules.

In terms of project implementation, the project adopts the EPC model for international collaboration, effectively controlling the project's total investment, shortening the construction period, and mitigating project risks while fostering technical exchanges and cooperation between countries. The successful application provides valuable experience for the renovation of similar farmland irrigation systems and further validates the solar-powered irrigation system's practical feasibility and advantages.

Looking forward, the solar-powered irrigation system holds the potential for broader application in farmland areas, aiming to achieve energy conservation and emission reduction goals in agricultural production and further promote the "Water-Energy-Food" nexus in a green and efficient cycle.

Acknowledgments. The authors express their gratitude to the international project team from the Electro-Mechanical Department of Beijing IWHR Technology Co., Ltd. For their invaluable contributions to data collection, figure preparation, and technical guidance throughout this study. Additionally, the authors acknowledge the support provided by the Natural Science Foundation of Beijing Municipality, grant No. JQ21029.

References

1. Purwanto, A., Sunik, J., Suryadi, F.X., Fraiture, C.D.: Water-energy-food nexus: critical review, practical applications, and prospects for future research. Sustainability, **13** (2021)
2. Baker, J.S. et al.: A hydro-economic methodology for the food-energy-water nexus: valuation and optimization of water resources. RTI Press Publication No. MR-0044-2105. RTI Press, Research Triangle Park, NC (2021)
3. Tian, H.Q., et al.: Optimizing resource use efficiencies in the food–energy–water nexus for sustainable agriculture: from conceptual model to decision support system. Curr. Opin. Environ. Sustain. **33**, 104–113 (2018)
4. Schutze, N., Kloss, S., Lennartz, F., Bakri, A.A., Schmitz, G.H.: Optimal planning and operation of irrigation systems underwater resource constraints in Oman considering climatic uncertainty. Environ. Earth Sci. **65**(5), 1511–1521 (2012)
5. Marbler, A.: Water scarcity and local economic activity: spatial spillovers and the role of irrigation. J. Environ. Econ. Manage. **124**: 1.1–1.25 (2024)
6. Rathore, L.S., Kumar, M., Hanasaki, N., Mekonnen, M.M., Raghav, P.: Water scarcity challenges across urban regions with expanding irrigation. Environ. Res. Lett. **19**(1), 1–15 (2024)
7. Sapkota, A., Haghverdi, A., Avila, C.C.E., Ying, S.C.: Irrigation and greenhouse gas emissions: a review of field-based studies. Soil Syst. **4**(2), 20 (2020)
8. Gavilanez, L.F.C., Barzola, O.C.M., Falconi Zambrano, C.J. Loqui Sanchez, A.J.: Effect of traditional irrigation on production, water efficiency and methane emission of rice crop. Bioagro, **36**(1): 15–26 (2024)
9. Ssenyimba, S., Kiggundu, N., Banadda, N.: Designing a solar and wind hybrid system for small-scale irrigation: a case study for Kalangala district in Uganda. Energy, Sustainability and Society **10**, 6 (2020)
10. Rejekiningrum, P., Apriyana, Y.: Design and implementation of solar pump irrigation systems for the optimization of irrigation and increase of productivity. IOP Conf. Ser.: Earth Environ. Sci. **622**(1): 012046 (18pp) (2021)
11. Nemouchi, W., Amrane, Y., Nemouchi, H. B. N. L.: Energy management for optimal design of PV/wind/diesel system for water pumping irrigation in semi-arid climate. Energy Convers. Manage. **304**: 1.1–1.14 (2024)
12. Mobtaker, H.G., Mostashari-Rad, F., Saber, Z., Chau, K.W., Nabavi-Pelesaraei, A.: Application of photovoltaic system to modify energy use, environmental damages and cumulative exergy demand of two irrigation systems-a case study: Barley production of Iran. Renew. Energy **160**(11), 1316–1334 (2020)

13. Franco, V.G., Natalia, D.S.R.: Photovoltaic solar energy applied to irrigation: an analysis of the financial impact in Brazil. Electr. Eng. **106**(1), 847–856 (2024)
14. Kumar, L.A., Lakshmiprasad, C.N., Ramaraj, G., Sivasurya, G.: Design, simulation of different configurations and life-cycle cost analysis of solar photovoltaic-water-pumping system for agriculture applications: use cases and implementation issues. Clean Energy **6**(2), 1100–1117 (2022)
15. Sharma, U., Singh, B.: Utility-tied solar water pumping system for domestic and agricultural applications. Journal of the Institution of Engineers (India) Series B, 101(3): 79–91 (2020)
16. Klimenta, D., Lekic, J., Radosavljevic, K.D.: A novel procedure for quick design of off-grid PV water pumping systems for irrigation. Electron. Electr. Eng. **27**(2), 64–77 (2021)
17. Liu, Z.M.: Design of photovoltaic irrigation water pump system. Doctoral dissertation, Heilongjiang Bayi Agricultural University (2019)
18. Pastoral Water Conservancy Science Research Institute of the Ministry of Water Resources: SL 540-2011 Specification for photovoltaic pumping engineering. China Water & Power Press, Beijing (2012)

Effects of Municipal Incineration Bottom Ash from Different Incineration Plants on the Engineering Properties of Low-Density Recycled Pervious Concrete

Wei-Jhu Wang[1](✉), Deng-Fong Lin[1], Lisa Y. Chen[2], and Show-Ing Shieh[3]

[1] Department of Civil Engineering, I-Shou University, Kaohsiung City 84001, Taiwan, R.O.C
isu6b001@cloud.isu.edu.tw, dflin@isu.edu.tw

[2] Department of Information Management, I-Shou University, Kaohsiung City 84001, Taiwan, R.O.C
lisachen@isu.edu.tw

[3] College of Liberal Education, Shu-Te University, Kaohsiung City 82445, Taiwan, R.O.C
sishieh@stu.edu.tw

Abstract. In this study, the Municipal Incinerator Bottom Ash (MIBA) was applied to the Low-Density Recycled Pervious Concrete (LRPC) to investigate the effects of MIBA from three different incineration plants in Taichung City, Tawan, with different substitution amounts on the engineering properties of LRPC. First, the differences in MIBA from the three different incineration plants in Taichung City, Taiwan, including Wuri Incineration Plant, Houli Incineration Plant, and Wenshan Incineration Plant, were analyzed through the tests of specific gravity, water absorption rate, abrasion test, SEM, XRD, and TCLP. Afterward, the substitution amounts of MIBA were set as 0%, 10%, 20%, and 30% for LRPC, and the effects of MIBA from the three incineration plants in Taichung City, Taiwan, with different substitution amounts were evaluated, including the compressive strength, porosity, water permeability, unit weight, and chlorine content of the LRPC. The results of the study showed that the water absorption rate and the abrasion rate of MIBA were higher than that of natural coarse particles, and the MIBA with the highest water absorption rate among the three-incineration plant was Wuri Incineration Plant and the MIBA with the best abrasion resistance was obtained from Wenshan Plant. According to the TCLP test, it was found that the amount of heavy metal dissolved in the MIBA of all three incineration plants in Taichung City meets the standard of reusing. The compressive strength and water permeability of LRPC decreased with the increase of MIBA substitution, while the use of MIBA from all incineration plants did not significantly affect the compressive strength of LRPC. It was also found that using incineration slag from the Wuri Incineration Plant would improve water permeability. Considering the engineering properties of LRPC, it was recommended that the substitution amount of different MIBA for LRPC should be at 20%.

Keywords: Low-Density Recycled Pervious Concrete (LRPC) · Municipal Incinerator Bottom Ash (MIBA) · SEM · TCLP

C. Weng (Ed.): WRE 2024, LNCE 634, pp. 472–485, 2025.
https://doi.org/10.1007/978-981-96-5095-8_39

1 Introduction

In recent years, it has become a common strategy to promote sustainable development policies by utilizing waste as recycled materials, which not only helps in the reutilization of resources but also effectively reduces waste accumulation [1]. With the advancement of technology, a large amount of solid waste is generated, which becomes a burden to the environment, such as the MIBA generated from the incineration of garbage daily. Taiwan has a small land area of 36,000 km^2 only. Therefore, the landfill sites are extremely limited. About 6.5 million tons of garbage is incinerated annually, producing about 1.2 million tons of MIBA [2]. Kundariya et al. [3] highlight that if solid waste is not properly managed, it will have adverse effects on the environment. Therefore, it is necessary to manage and utilize solid waste effectively. According to Chen et al. [4], when the MIBA is applied to the inorganic polymers, the inorganic polymers can effectively solidify the heavy metal elements in the MIBA and bring it to a stable state. MIBA can be categorized as a non-hazardous material when it reaches a stable state, as stated by Kumar et al. [5]. Nowadays, MIBA has been widely used in the construction industry as an alternative to construction material [6]. Silva et al. [7] emphasize that MIBA can be considered as an alternative material to substitute natural resources after proper treatment and has the potential to be used as a road grading particle [8]. Zhang et al. [9] used MIBA as a raw material in ceramic production. They found that the ceramics made of 50% MIBA and 50% fly ash have good corrosion resistance during kiln firing at a temperature of 1000 °C. MIBA also has the potential to be used as a thermal insulation material, as claimed by Baino et al. [10]. However, it was found by Lin et al. [11] that MIBA would lead to a decrease in the structural strength. Therefore, MIBA is difficult to use ceramic materials. Compared to the original natural environment, most surfaces in today's surroundings lack rainwater permeability, heat absorption, and air permeability, and many urban surfaces are covered with impermeable reinforced concrete structures [12]. The gradual replacement of green spaces by infrastructures led to the exacerbation of extreme climates and the urban heat island effect [13]. Okada [14] pointed out that usig porous materials can block the transfer of external heat energy, thus mitigating the urban heat island effect. Liu et al. [2] calculated the trend of reuse pathways and substitution of MIBA in Taiwan between the years from 2015 to 2019 and found that the main reuse of MIBA includes road-graded aggregate materials, base fill materials, and controlled low-strength backfill materials and the application of LRPC only accounts for a few. LRPC should consider both porosity and compressive strength at the design stage [15], as its compressive strength decreases with the increase of permeability [16]. Still, by adding a small number of fine aggregates, its compressive strength can be significantly improved [17].

According to the studies mentioned above, the problem of waste accumulation and the heat island effect are global issues. If the MIBA could be cured by cement curing technology and applied to LRPC, not only the problem of MIBA could be alleviated, and the impact of the heat island effect could be reduced at the same time. Therefore, in this study, the basic properties of MIBA from the three different incineration plants in Taichung City, Taiwan, were investigated to determine the differences in heavy metal solubilization and then applied to the LRPC. After investigating the effects of MIBA from three different incineration plants, including its compressive strength, unit weight, water permeability, and chloride ion content of LRPC, a recommended substitution amount of MIBA is proposed for application to LRPC.

2 Test Materials and Design

2.1 Basic Properties of Materials

The natural coarse aggregates used in this study were obtained from natural river sand and gravel in Taiwan. The MIBA was supplied by three waste incineration plants in Taichung City, Taiwan: the Wenshan Incineration Plant, the Houli Incineration Plant, and the Wuri Incineration Plant. Table 1 shows the basic test results of natural coarse aggregates and MIBA. The test results show that the specific gravity of MIBA is significantly lower than that of natural aggregates, and its water absorption rate is much higher than that of natural coarse aggregates. Further comparison of the nature of the MIBA produced by the three incineration plants revealed that the water absorption rates of MIBA from the Wenshan Incineration Plant, Houli Incineration Plant, and Wuri Incineration Plant were 17.02%, 14.94%, and 18.74% respectively, with the water absorption rate of MIBA from the Wuri Incineration Plant being the highest. Since the water absorption rate of the material can be used to estimate its porosity, the MIBA can also be determined as a porous material. The abrasion test results show that the abrasion rates of MIBA from Wenshan Incineration Plant, Houli Incineration Plant, and Wuri Incineration Plant were 35.37%, 41.22%, and 44.12%, respectively. The abrasion rates of these MIBA were significantly higher than that of natural coarse aggregates, indicating that the MIBA has poor resistance to abrasion. Among the three mentioned incineration plants, the MIBA obtained from the Wenshan Incineration Plant has a relatively better abrasion resistance. The test results of the CNS486 [18] standard sieve analysis after the particle size distribution are shown in Fig. 1. The results show that the MIBA from all three incineration plants contain coarse and fine particles, out of which the Wenshan Incineration Plant and Houli Incineration Plant MIBA particle size distribution is similar, and the MIBA from Wuri Incineration Plant contains a higher proportion of coarse aggregates.

Table 1. Basic properties of materials.

Type of Material	Coarse Natural Aggregate	MIBA			Standard
		Wenshan	Houli	Wuri	
Specific gravity	2.66	2.49	2.56	2.29	CNS487 [19] CNS488 [20]
Water absorption (%)	1.03	17.02	14.94	18.74	
Wear rate (%)	27.24	35.39	42.12	44.12	CNS490 [21]

2.2 Microscopic Analysis of MIBA

In microscopic analysis, the microstructure and chemical composition of MIBA were evaluated by SEM and XRD tests. The equipment used for SEM tests was HITACHI-S4700, while the equipment used for XRD tests was PANalytical. The rapid heating phase

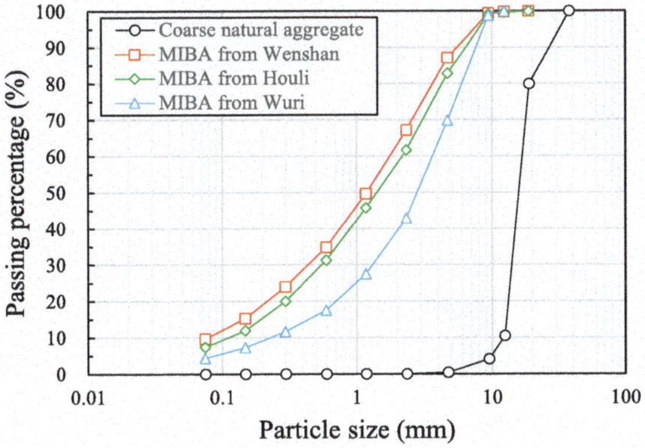

Fig. 1. Sieve analysis of materials.

during the waste incineration process and the subsequent water quenching phase both can lead to the formation of pores in the structure of the incineration bottom ash. Figure 2 shows the SEM images of MIBA from the three different incineration plants in Taichung City, magnified 10,000 times, from which it can be observed that the surface of MIBA is rough and porous, which is the main reason for its high-water absorption rate. Wang et al. [23] stated that when aggregates have a high-water absorption rate, it can easily lead to stress concentration in the structure. Figure 3 shows the XRD peaks of the MIBA, which indicates that the MIBA from the three incineration plants has strong $CaCO_3$ and SiO_2 peaks. By using the MATCH! analysis software developed by CRYSTAL IMPACT, the compound composition and the incineration bottom ash proportions were further analyzed. The compound analysis results for the bottom ash from Wenshan, Houli, and Wuri Incineration Plants are shown in Table 2. The results show that the main compounds of the MIBA from these three incineration plants are $CaCO_3$ and SiO_2, and all contain the oxides of ClO. Li et al. [24] also found that incineration bottom ash contains Cl components. Wu et al. [25] pointed out that a higher content of plastic and kitchen waste in the waste stream leads to a higher chlorine content in the incineration bottom ash. Since ClO corrodes metals, paying attention to the chloride ion content in LRPC, which contains MIBA, is necessary.

2.3 TCLP Test

He et al. [26] stated that the heavy metal components in incineration bottom ash could lead to potential environmental pollution. Quek et al. [27] indicated that even with increased rinses using rainwater and seawater, the concentration of As in the incineration bottom ash showed no downward trend. In order to avoid the environmental impacts caused by the reuse of MIBA, the analysis of heavy metal leaching of MIBA is conducted according to the test standard of NIEA R201.15C [28] in order to determine whether it meets the heavy metal leaching standard for the reuse of general utility wastes. Table 3 shows the TCLP test results of MIBA from the Wenshan Incineration Plant, Houli Incineration Plant, and Wuri Incineration Plant. The test results showed that no significant difference in heavy metal leaching from the MIBA from Wenshan, Houli, and Wuri

(a) Wenshan (b) Houli

(c) Wuri

Fig. 2. SEM test results of MIBA.

Incineration Plants, and Cu was the most leached heavy metal element, followed by Ba. In addition, it was also found that the amount of each heavy metal leaching from the MIBA was far below the heavy metal element leaching standard for the reuse of general utility wastes when compared with the standard for reusing.

2.4 Mix Design

In this study, the MIBA from the three different incineration plants in Taichung, Taiwan, was applied to LRPC, and the MIBA was used to replace some of the natural aggregates. According to the study by Liu et al. [2], when the substitution amount of MIBA reaches 40%, the compressive strength of the previous concrete will significantly decrease. Therefore, this study set the substitution amount of MIBA produced by three different incineration plants at 0%, 10%, 20%, and 30%, respectively. The MIBA of the three different incineration plants was used with three different substitution quantities to investigate the effects of the MIBA on the engineering properties of LRPC and to propose the recommended usage of MIBA in LRPC by the test results. In this study, the proportioning method of LRPC is based on the volumetric method of ACI 211. The

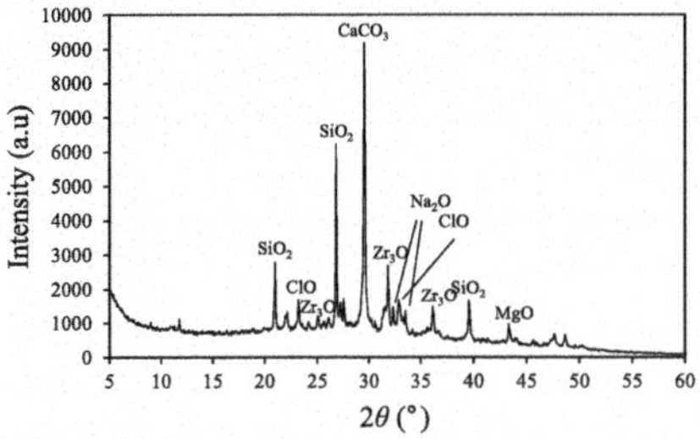

(a) Wenshan

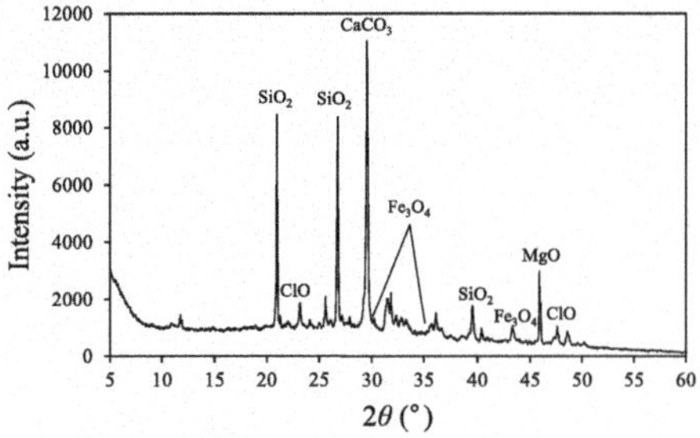

(b) Houli

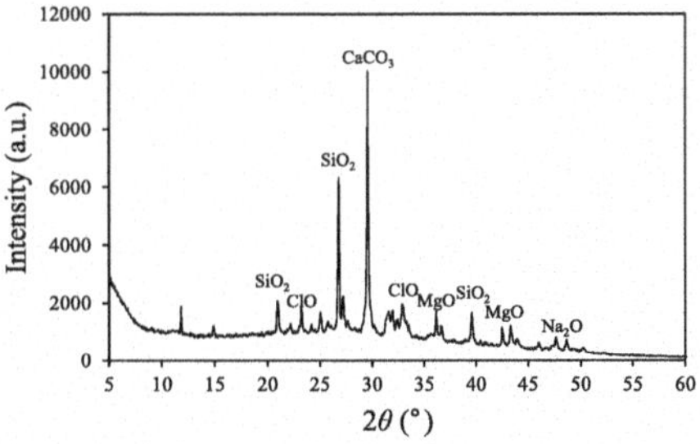

(c) Wuri

Fig. 3. X-ray diffractograms of MIBA.

Table 2. Chemical compositions of MIBA.

%	$CaCO_3$	SiO_2	MgO	ClO	Zr_3O	Na_2O	Fe_2O_3
Wenshan	61.8	24.3	3.6	3.6	1.4	5.3	-
Houli	52.2	34.5	12.2	0.4	-	-	0.7
Wuri	63.9	28.5	1.1	4.5	-	2.1	-

Table 3. TCLP leaching concentrations from MIBA.

mg/L	As	Pb	Cu	Cd	Cr	Hg	Cr6+	Se	Ba
Wenshan	ND	ND	2.29	< 0.100	ND	ND	ND	ND	0.585
Houli	ND	<0.100	2.92	< 0.100	ND	ND	ND	ND	0.620
Wuri	ND	ND	2.10	< 0.100	ND	ND	ND	ND	0.603
Standard	5.0	5.0	15.0	1.0	5.0	0.2	2.5	1.0	1.0

volumetric method is based on the porosity of the materials, and the target porosity and water-cement ratio are set for calculation. Since the volumetric method is suitable for controlling the porosity of the LRPC, it is suitable for LRPC, which emphasizes water permeability. In addition, the water-cement ratio of LRPC suggested by the American Concrete Institute is 0.35–0.45. Liu et al. [2] also claim that when the water-cement ratio is 0.45, the slump value of LRPC can comply with the standardized value of 0–50mm. Therefore, in this study, the water-cement ratio is set to 0.45, and the target porosity is 25%. The design of the proportion of LRPC is shown in Table 4. The experimental site of this study was in Kaohsiung. The local average temperature and relative humidity measured during the experimental process ranged from 28.9 to 29.4°C and 78.0 to 79.9%, respectively. The material preparation, mixing, and specimen production process for LRPC are as follows in Fig. 4.

Table 4. Mix design of low-density recycled pervious concrete.

MIBA content (% by weight)	Water-cement ratio	Water (kg/m^3)	Cement (kg/m^3)	MIBA (kg/m^3)	Coarse natural aggregate (kg/m^3)
0%	0.45	81	113	0	1658
10%	0.45	81	113	165	1484
20%	0.45	81	113	328	1312
30%	0.45	81	113	490	1142

3 Results and Discussion

3.1 Compressive Strength

The compressive strength test for LRPC was conducted in accordance with the ASTM D4832 [30] testing method. For each type of incineration bottom ash, amount of substitution, and curing age, three cylindrical specimens with a diameter of 15 cm and a height of 30 cm were prepared for 72 specimens. The average compressive strength for each proportion design of LRPC was then calculated. According to the Construction Manual of the Public Works Commission Chapter 03341 [31], the LRPC of 3 days and 28 days of compressive strength should meet the requirements of more than 10 kg/cm^2 and between 25–50 kg/cm^2. The test results of compressive strength of the LRPC are shown in Fig. 5 and Fig. 6. Based on the test results, when the age of LRPC reaches 3 days, the compressive strength of LRPC decreases with the increased use of MIBA as substitution, and when the substitution of MIBA reaches 30%, the compressive strength of LRPC with the age of 3 days is below 10 kg/cm2. When the age of LRPC reaches 28 days, its compressive strength decreases with the increase of MIBA substitution; the maximum value of MIBA substitution is 10%, which is because there is an upper limit

Fig. 4. Mixing and casting process of low-density recycled pervious concrete.

of compressive strength of LRPC at the age of 28 days. Therefore, substitution values between 0% and 20% can meet the requirement. Furthermore, the use of MIBA from the Wenshan, Houli, and Wuri Incineration Plant did not significantly affect the compressive strength of LRPC.

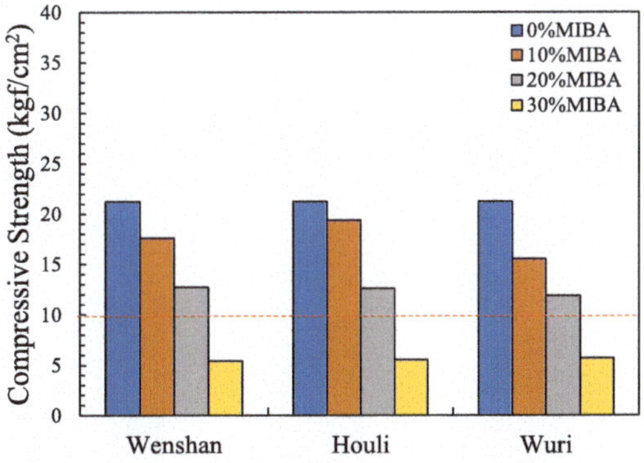

Fig. 5. 3-day compressive strength of all test series.

3.2 Porosity

Since LRPC is a porous material, choosing the test method to measure porosity using the volume method is necessary. Therefore, this study conducted the porosity of LRPC in

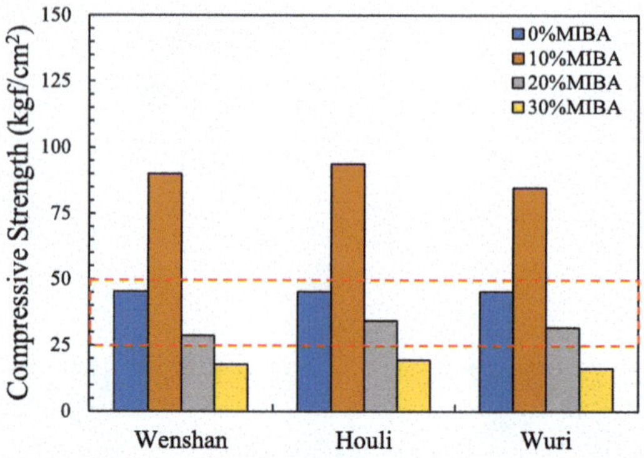

Fig. 6. 28-day compressive strength of all test series.

accordance with the ASTM C173 [32] test method. A total of 36 tests were performed and presented as average porosity. Figure 7 shows the porosity test results of MIBA applied to LRPC. The porosity test was conducted to investigate the effects of MIBA from three different incineration plants and different substitution amounts of MIBA on the porosity of LRPC. The test results show that the porosity of LRPC decreases gradually as the MIBA substitution increases because the MIBA used in this study contains a portion of fine aggregates, so when the amount of MIBA increases, the MIBA fine aggregates will fill a portion of the pores of LRPC. When the amount of MIBA substitution reaches 40%, the porosity shows a significant decline in trend. In addition, the porosity of LRPC using the MIBA from the Wuri Incinerator Plant showed a much higher porosity compared to MIBA used from the Wenshan Incinerator Plant and Houli Incinerator Plant because the proportion of coarse aggregates in the MIBA from the Wuri Incineration Plant was higher than that of the other two mentioned incineration plants, which led to a higher porosity.

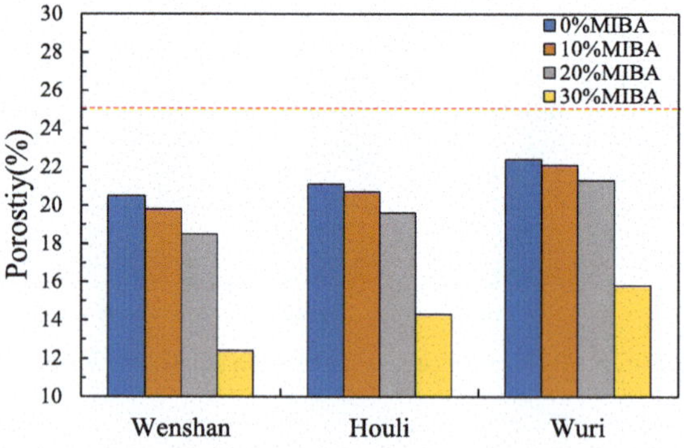

Fig. 7. Porosity of all test series.

3.3 Unit Weight

As Chapter 03341 of the Public Works Commission Building Program [31] specified, the unit weight of LRPC should be between 1500–2000 kg/cm^3. Therefore, this study conducted 36 unit weight tests of LRPC in accordance with ASTM C138 [33]. The average unit weight of LRPC is shown in Fig. 8. Figure 8 shows the unit weight test results of LRPC. The test results show that the unit weight of LRPC decreases with the increase of MIBA substitution because the specific gravity of MIBA is significantly lower than that of natural coarse aggregates, which leads to the decrease of the unit weight of LRPC with the increase of substitution. However, even with 40% MIBA substitution, the unit weight of LRPC met the specification requirements, and the use of MIBA from all three incineration plants did not significantly affect the unit weight of LRPC.

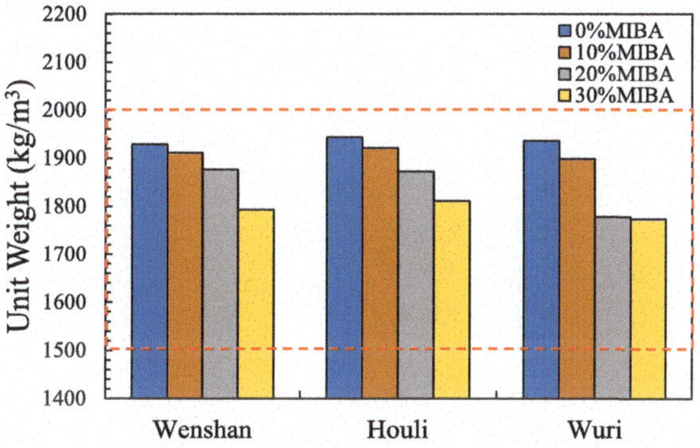

Fig. 8. Unit Weight of all test series.

3.4 Permeability

In this study, 36 permeability tests of LRPC were performed by referring to the field permeability test method mentioned in Japan's "Drained Asphalt Pavement Technical Guidelines" [34]. Figure 9 shows the average water permeability of LRPC. According to the test results, it is known that when the substitution amount of MIBA increases, the water permeability of LRPC tends to decrease because MIBA contains a part of the fine aggregate material. Furthermore, when comparing the water permeability of LRPC using MIBA from three different incineration plants, it was found that LRPC using MIBA from Wuri Incineration Plant had the best water permeability because of the proportion of coarse aggregates in MIBA obtained from Wuri Incineration Plant was higher than that of the coarse aggregates of MIBA obtained from Wenshan Incineration Plant and Houli Incineration Plant. As the porosity of LRPC increased, its water permeability improved, which is consistent with the results of the porosity test in this study. According to the standards of the Public Works Commission Building Specifications Chapter 03341 [31], the permeability of LRPC must be greater than 600 ml/15s. Therefore, regardless of the use of MIBA from the three incineration plants mentioned above, if the substitution amount of MIBA is greater than 20%, the permeability of LRPC will be less than 600ml/15s.

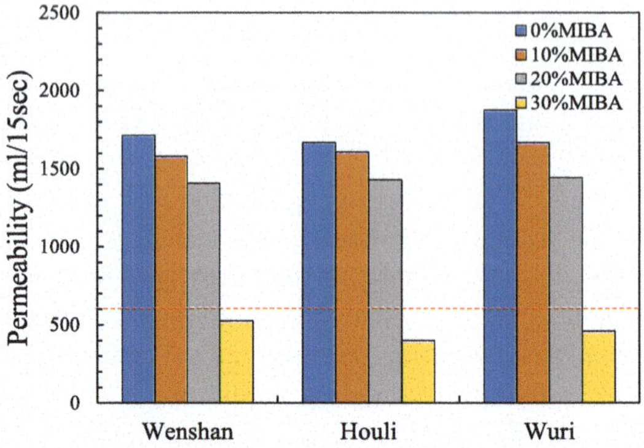

Fig. 9. Permeability of all test series.

3.5 Chloride

This study conducted the chloride ion content test of LRPC in accordance with the CNS 13465 [35] test method. The purpose was to investigate the effect of MIBA on the chloride ion content in LRPC. Therefore, this study scraped the cement mortar from the surface of coarse aggregates in freshly mixed LRPC, and the chloride ion content was tested in the form of cement mortar. Figure 10 shows the chlorine ion content test results of LRPC, which shows that with the increase of MIBA substitution, the chlorine ion content increases accordingly because MIBA from the Wenshan Incineration Plant, Houli Incineration Plant, and Wuri Incineration Plant have ClO. However, the effects of MIBA from the Houli Incineration Plant were highest on the chloride ion content of LRPC, followed by the Wenshan Incineration Plant and then the Wuri Incineration Plant. In addition, Gong et al. [36] indicated that excessive chloride ion content in materials can lead to steel reinforcement corrosion. Therefore, when incineration bottom ash is used in construction concrete, it should be carefully evaluated whether the chloride ion content will cause corrosion of the steel reinforcement.

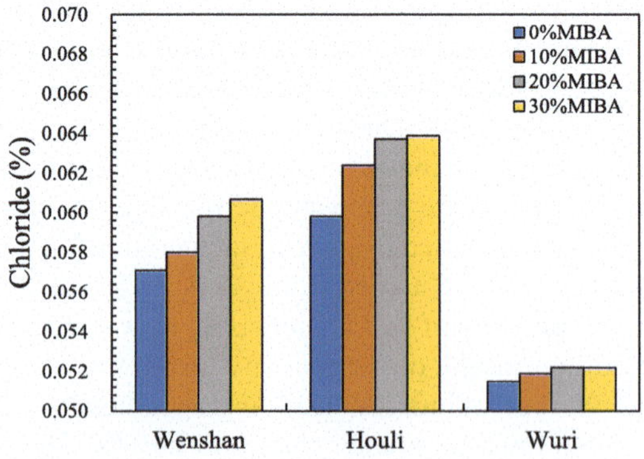

Fig. 10. Chlorine ion content of all test series.

4 Conclusions

This study investigates the effects of MIBA from the three different incineration plants in Taichung City, Taiwan, and different MIBA substitutions on the engineering properties of LRPC proposed recommended MIBA substitution. According to the results of the study, the following conclusions were drawn:

- The water absorption rate of MIBA is significantly higher than that of natural coarse aggregate. The Wuri Incineration Plant has the highest MIBA water absorption rate among the three incineration plants in Taichung City, Taiwan.
- The anti-abrasion ability of MIBA is lower than that of natural coarse aggregate. Among the three types of MIBA used in this study, the anti-abrasion ability of MIBA from the Wenshan Incineration Plant is relatively better.
- The amount of heavy metal elements dissolved in MIBA is lower than that of the general standards for reusing industrial waste, which indicates that MIBA has the potential to be reused.
- The increase of MIBA substitution reduces the water permeability of LRPC, especially when the amount of replacement exceeds 20%, which is less than that of the specification standard. However, the water permeability of MIBA from the Wuri Incineration Plant was relatively better, which was related to the higher proportion of coarse aggregates.
- With increasing MIBA substitution, the chlorine ion content in LRPC increased accordingly. The effect of MIBA on the chlorine ion content of three incineration plants in Taichung City, Taiwan, varied. Among them, the MIBA from the Houli Incineration Plant has the most chlorine ions, and the MIBA from the Wuri Incineration Plant has the least.
- Since construction steel reinforcement is prone to corrosion due to the influence of chloride ion content, the use of incineration bottom ash in building concrete should be cautiously approached. Assessing the chloride ion content of incineration bottom ash is recommended when considering its use in concrete.
- After considering the engineering properties of compressive strength and water permeability, this study recommends that the substitution of different MIBA in LRPC should be 20%.

References

1. Yildirim, I.Z., Prezzi, M.: Chemical, mineralogical, and morphological properties of steel slag. Adv. Civil Eng. **2011**, 1–13 (2011)
2. Liu, P., Lin, D.F., Wang, W.: Suitability of using pervious concrete containing incinerator bottom ashes as a roadway base course. J. Perform. Constr. Facil. **37**(1), 04022064 (2023)
3. Kundariya, N., et al.: A review on integrated approaches for municipal solid waste for environmental and economical relevance: Monitoring tools, technologies, and strategic innovations. Bioresour. Technol. **342**, 125982 (2021)
4. Chen, Z., Liu, Y., Zhu, W., Yang, E.: Incinerator bottom ash (IBA) aerated geopolymer. Constr. Build. Mater. **112**, 1025–1031 (2016)

5. Kumar, S.S., Singh, D.: From waste to resource: Evaluating the possibility of incinerator bottom ash composites for geotechnical applications. Int. J. Environ. Sci. Technol. 1–12 (2023)
6. Phoungthong, K., Zhang, H., Shao, L., He, P.: Variation of the phytotoxicity of municipal solid waste incinerator bottom ash on wheat (triticum aestivum l.) seed germination with leaching conditions. Chemosphere **146**: 547–554 (2016)
7. Silva, R., de Brito, J., Lynn, C.J., Dhir, R.K.: Environmental impacts of the use of bottom ashes from municipal solid waste incineration: A review. Resour. Conserv. Recycl. **140**, 23–35 (2019)
8. Xu, Y.: Unconfined compressive strength of municipal solid waste incineration bottom ashes. Geotech. Geol. Eng. **37**, 1373–1382 (2018)
9. Zhang, Z., Wang, J.Q., Liu, L., Ma, J., Shen, B.: Preparation of additive-free glass-ceramics from MSW incineration bottom ash and coal fly ash. Constr. Build. Mater. **254**, 119345 (2020)
10. Baino, F., Ferraris, M.: Production and characterization of ceramic foams derived from vitrified bottom ashes. Mater. Lett. **236**, 281–284 (2019)
11. Lin, D., Wang, W., Chen, C., Lin, K.: Applying mixture of municipal incinerator bottom ash and sewage sludge ash for ceramic tile manufacturing. Materials **14**(14), 3863 (2021)
12. Yang, G., Gang, Y., Wang, H., Jin, Q.: Test and analysis of influencing factors on the permeability coefficient of steel slag permeable concrete. Baosteel Tech. Res. **5**(2), 24–26 (2011)
13. He, B.J., Zhu, J., Zhao, D.X., Gou, Z.H., Qi, J.D., Wang, J.: Co-benefits approach: Opportunities for implementing sponge city and urban heat island mitigation. Land Use Policy **86**, 147–157 (2019)
14. Okada, K.: Lotus ceramics for counteracting urban heat island effects. In Eco-efficient Materials for Mitigating Building Cooling Needs, 195213 (2015)
15. Zheng, M., Chen, S.F., Wang, B.: Mix design method for permeable base of porous concrete. Int. J. Pavement Res. Technol. **5**(2), 102–107 (2012)
16. Cheng, A., Hsu, H., Chao, S., Lin, K.: Experimental study on properties of pervious concrete made with recycled aggregate. Int. J. Pavement Res. Technol. **4**, 104–110 (2011)
17. Hesami, S., Ahmadi, S., Sadeghi, V., Hassanabadi, M.E.: Evaluation of polyphenylene sulfide and steel fibers on mechanical properties of pervious concrete pavement. International Journal of Pavement Research and Technology **7**, 218–222 (2014)
18. The bureau of standards, metrology and inspection (BSNI): Method of test for sieve analysis of fine and coarse aggregates, CNS 486, Taiwan (2001)
19. The bureau of standards, metrology and inspection (BSNI): Method of test for density, relative density (specific gravity), and absorption of fine aggregate, CNS 487, Taiwan (2013)
20. The bureau of standards, metrology and inspection (BSNI): Method of test for density, relative density (specific gravity), and absorption of coarse aggregate, CNS 488, Taiwan (2008)
21. The bureau of standards, metrology and inspection (BSNI): Method of test for resistance to degradation of coarse aggregate (smaller then 37.5 mm) by abrasion and impact in the los angeles machine, CNS 490, Taiwan (2009)
22. Wang, C., et al.: Study of waste incineration bottom ash as fine aggregate applied to green alkali-activated bottom ash-slag concrete: Mechanical properties, microstructure, durability. Constr. Build. Mater. **449**, 138484 (2024)
23. Wang, D., et al.: Mechanical performance of recycled aggregate concrete in green civil engineering: review. Case Stud. Constr. Mater. **19**, e02384 (2023)
24. Li, S., et al.: Fate of sulfur and chlorine during co-incineration of municipal solid waste and industrial organic solid waste. Sci. Total. Environ. **920**, 171040 (2024)
25. Wu, B., Wang, D., Chai, X., Takahashi, F., Shimaoka, T.: Characterization of chlorine and heavy metals for the potential recycling of bottom ash from municipal solid waste incinerators as cement additives. Front. Environ. Sci. Eng. **10**, 1–9 (2016)

26. He, Y., Kasina, M.: The sequential extraction of municipal solid waste incineration bottom ash: heavy metals mobility and sustainable application of ashes. Sustainability **15**(19), 14638 (2023)
27. Quek, A., Xu, W.Y., Guo, L., Wu, D.: Heavy metal removal from incineration bottom ash through washing with rainwater and Seawater. Int. J. Waste Resour. **6**, 1–9 (2016)
28. National environmental research academy: Toxicity characteristic leaching procedure of business waste, NIEA R201.15C (2017)
29. Yeo, R.J., Sng, A., Wang, C., Tao, L., Zhu, Q., Bu, J.: Strategies for heavy metals immobilization in municipal solid waste incineration bottom ash: A critical review. Rev. Environ. Sci. Biotechnol. **23**, 503–568 (2024)
30. American Society for Testing and Materials: Standard test method for preparation and testing of controlled low strength material (clsm) cylindrical test specimens, ASTM D4832, West Conshohocken, PA: ASTM (2023)
31. Public Works Commission Building Specifications Homepage, https://pcic.pcc.gov.tw/pwc-web/service/tec0301, last accessed 2024/11/19
32. American Society for Testing and Materials: Standard test method for air content of freshly mixed concrete by the volumetric method, ASTM C173, ASTM, West Conshohocken, PA (2023)
33. American Society for Testing and Materials: Standard test method for density (unit weight), yield, and air content (gravimetric) of concrete, ASTM C138, ASTM, West Conshohocken, PA (2024)
34. Japan Road Association: Drained asphalt pavement technical guidelines. Japan Road Association, Japan (1996)
35. The Bureau of Standards, Metrology and Inspection (BSNI): Method of test for the water-soluble chloride ion content of fresh concrete, CNS 15308, Taiwan (2014)
36. Gong, W., Wang, N., Zhang, N., Chen, G.: Time-dependent model and life prediction for reinforcement corrosion in magnesium oxychloride cement concrete. Struct. Concr. **21**, 1865–1879 (2020)

Study on Five-Channel Online Analysis Method of Seawater Nutrient Salts for Unmanned Ship System

Ran Ma[1,2], Yuying Liu[1](✉), Yan Liu[1,2], Xiangfeng Kong[1,2], Shuwei Zhang[1,2], and Fengqing Liu[1,2]

[1] Institute of Oceanographic Instrumentation, Qilu University of Technology (Shandong Academy of Sciences), Qingdao 266061, People's Republic of China
694880817@qq.com, kongxiangfeng@qlu.edu.cn

[2] Laoshan Laboratory, Qingdao, Shandong 266237, People's Republic of China

Abstract. Nutrient monitoring in the marine environment is essential for protecting marine ecosystems. Monitoring nutrient salts in the marine environment is essential for the protection of marine ecosystem. However, traditional monitoring methods have the disadvantages of complicated operation, slow response, and inability to monitor in real time. In order to improve the monitoring efficiency and accuracy, this paper proposes a multi-channel seawater nutrient salt monitoring system based on unmanned shipboard. The system designs the hardware circuit with an STM32F103 microcontroller as the core to realize the real-time and synchronous monitoring of nutrient salts such as nitrate, phosphate, silicate, nitrite, and ammonia nitrogen in seawater. The monitoring system consists of a central control node and five nutrient salts, each independent collection node, the control node and each collection node between the ZigBee wireless communication. Compared with other communication methods, wireless communication is easy to replace and maintain, can reduce the data interruption time, and ensures the validity of data. Experimental results show that the system can efficiently and accurately monitor nutrient salt concentrations in seawater with good monitoring efficiency and accuracy. Compared with the existing technology, this system shows higher data accuracy and better real-time performance, providing an effective technical means for marine environment monitoring.

Keywords: Unmanned Shipboard · Multichannel Monitoring · Seawater Nutrient Salinity · Filtering Algorithm

1 Introduction

The health and balance of marine ecosystems is an important topic in global environmental science [1]. Changes in the concentration of nutrients in seawater, especially key elements such as nitrogen, phosphorus, and silicon, are critical for the growth and reproduction of marine organisms and the overall productivity of the ecosystem [2]. In recent years, with the increase of human activities, ocean eutrophication has become

C. Weng (Ed.): WRE 2024, LNCE 634, pp. 486–500, 2025.
https://doi.org/10.1007/978-981-96-5095-8_40

increasingly serious, leading to the frequent occurrence of red tides and other ecological disasters, which pose a threat to the marine environment and human health [3]. Therefore, real-time monitoring of nutrient salt concentration in seawater is of great scientific significance and practical application value for understanding the dynamic changes of marine ecosystems, timely prevention, and effective control of marine ecological disasters.

In the study of seawater nutrient salt concentration monitoring technology, spectrophotometry, as a mature seawater nutrient salt detection technology, has been widely used in marine environmental monitoring. This method is based on the absorption characteristics of substances to specific wavelengths of light, and quantitatively analyzes the concentration of nutrient salts by measuring the absorbance [4]. However, with the deepening of marine scientific research, the traditional single-channel monitoring system has gradually exposed limitations such as complex operation, slow detection speed, and the inability to realize simultaneous multi-parameter detection [5]. These limitations are especially obvious in real-time monitoring and large-scale marine surveys, limiting the responsiveness to the marine environment. In order to overcome these limitations, scholars in the field have carried out a lot of innovative research: Ma Ran et al. proposed a nutrient salt in situ analysis method based on microfluidic technology, using microfluidic chips and micro three-way valve manifolds to form a valve island system, to realize the in situ monitoring of nutrient salts, and this method has a significant advantage in improving the detection accuracy and reducing the consumption of reagents; Qi Shengbo et al. developed a multi-scale seawater nutrient salt in situ analyzer based on spectrophotometric method [6]. Qi Shengbo et al. developed a multi-range seawater nutrient salts in situ sensor detection system based on spectrophotometry, which improved the sensitivity and stability of detection by optimizing the optical design and electronic signal processing [7]; furthermore, Liu J et al. designed a multi-channel in situ nutrient salts monitoring system based on STM32F401, which significantly improved the sensitivity and accuracy of the signal extraction by using the microfluidic chip technology and analog phase-locked amplification circuits [8].

Although significant progress has been made in the research of seawater nutrient salts monitoring technology, there are still some key issues in the existing online monitoring system: the complexity of operation leads to a time-consuming monitoring process and a high technological threshold; the slow detection speed affects the timeliness of the monitoring; and at the same time, the lack of simultaneous multi-parameter detection capability limits the comprehensive assessment of the marine environment [9]. In addition, the real-time and response speed of the existing system is insufficient, making it difficult to reflect the rapid changes in the marine environment in time, while the lack of data accuracy and repeatability directly affects the scientific and effectiveness of the monitoring results [10]. As an emerging marine monitoring platform with high flexibility, wide coverage, and good cost-effectiveness, unmanned vessels can easily reach areas difficult to reach by traditional monitoring means for continuous data collection, and the maintenance cost is relatively low. Therefore, developing a multi-channel seawater nutrient salt monitoring system based on unmanned shipboard will significantly improve the extensiveness, real-time and accuracy of monitoring. Moreover, by integrating advanced sensor technology and efficient data processing algorithms, the system improves the response speed and data quality of monitoring, thus providing more efficient and reliable data support for marine environmental monitoring, which is of great academic value and practical application significance.

2 Technical Framework

2.1 Detection Principle

Spectrophotometry. Spectrophotometry quantitatively analyzes the concentration of a substance by measuring its absorption characteristics of light at a specific wavelength. The method is based on the Lambert-Beer law, which states that the absorbance of a solution is directly proportional to the concentration of the substance in the solution [11]. The specific formula is:

$$A = \log\left(\frac{I_0}{I}\right) = \varepsilon bc \tag{1}$$

where A is the absorbance and I_0 is the incident light intensity, I is the transmitted light intensity, ε is the molar absorbance coefficient, b is the optical range, and c is the solution concentration. In this system, the concentrations of four nutrient salts, nitrate, nitrite, phosphate, and silicate, can be accurately calculated by measuring the change in absorbance of the seawater samples after the addition of color-developing reagents.

Fluorescence. The fluorescence method quantitatively analyzes the concentration of a substance by measuring the intensity of the fluorescent signal emitted by the substance when irradiated with light of a specific wavelength. The method has the advantages of high sensitivity and good selectivity, and is suitable for the detection of trace or micro-substances. The method is suitable for the detection of trace or micro-volume substances [12]. In this system, the fluorescence method is used to detect the concentration of ammonia nitrogen by mixing and reacting with o-phthalaldehyde (OPA) sufficiently, and then heating it up to 40 °C in a circulating system through a thermostat device in a cycle in order to excite the fluorescence in the wavelength range of 400–480 nm. During this process, the fluorescence signal will be kept orthogonal to the excitation light. This system can accurately calculate the ammonia nitrogen content by detecting the intensity of the fluorescence signal and the LED reference signal. The structure of the flow path system used in this method is shown in Fig. 1.

2.2 System Design

The monitoring system developed in this paper is a highly integrated multi-channel seawater nutrient salt online monitoring platform designed to realize real-time and accurate monitoring of key nutrient salt elements in the marine environment. As shown in Fig. 2, the system includes a central control node and various nutrient salt concentration collection nodes, and the hardware part adopts a modularized design concept, which consists of a main control node and five single-channel control nodes, that are connected to each other through the wireless communication protocol. Each single control node is

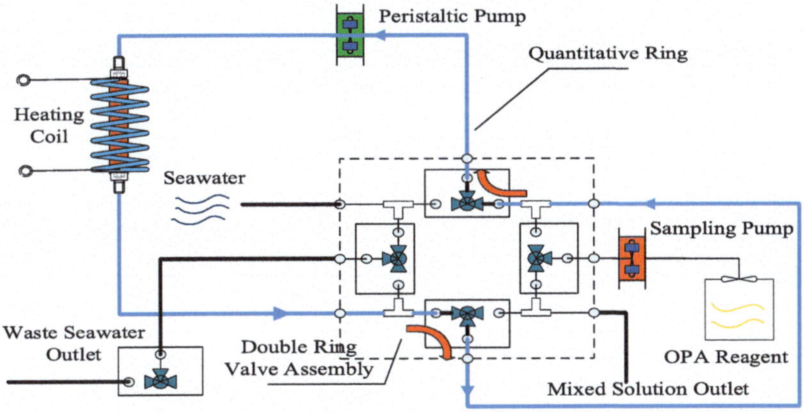

Fig. 1. System flow path architecture diagram.

equipped with a corresponding power management module, driver module, signal processing module, communication module, leakage detection module, and storage module. These modules work together and are controlled and deployed by a microprocessor to improve detection efficiency and enhance system stability. The design of the power management module, which cuts off the power supply of high-power modules in real time during their non-operating state, will significantly reduce the standby power consumption of the system, effectively solving the problem of unmanned boat platforms in small space and limited power consumption.

As the core of the system, the control node integrates key components such as the power management module, communication, storage, etc. It is responsible for the overall coordination and control of the system, including key tasks such as system configuration, parameter download, data reading, and arithmetic analysis.

Each single control node is responsible for an independent nutrient parameter acquisition channel covering the detection of nitrate, nitrite, phosphate, silicate, and ammonia nitrogen. These control nodes are equipped with the STM32F103 as the core of the minimum system, and are equipped with relevant modules for signal acquisition and processing. The power management module has a multi-level protection mechanism, including fuses, voltage regulator diodes, and current limiting resistors for over-voltage and over-current protection. The module integrates 12V and 5V power management circuits and accurately controls the on-off of the field effect tubes through the MCU to realize intelligent power management and save energy consumption. In addition, the DC-DC power supply module is applied to isolate the analog power supply voltage from the digital power supply voltage to provide a stable power supply for the signal processing module.

The solenoid valve driver module adopts an MCU-controlled NPN-type Darlington tube array, and realizes the smooth switching of Darlington tubes between saturation and cut-off regions through the co-emitter connection and resistance configuration. The UV lamp and heater drive modules use relay control to quickly drive the high-power devices by controlling NPN-type transistors through the MCU. The constant-current drive design of the LED light source realizes precise control of the current through resistors of different

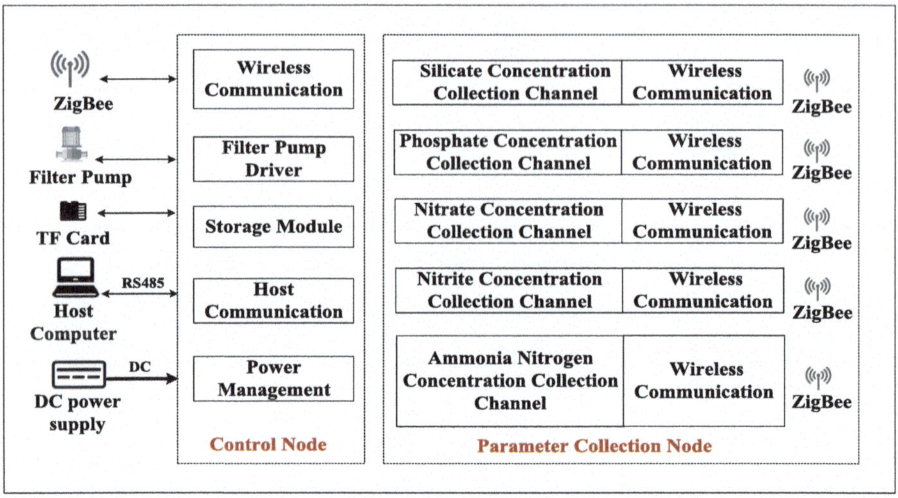

Fig. 2. Overall system architecture.

resistance values to meet the specific requirements of light wavelengths of different elemental detections. The mixing and filtering pump drive module adopts a stepping motor drive circuit, and the microprocessor generates 1~10 kHz pulse signals through a timer for fine speed regulation and commutation by changing the pulse timing to improve the accuracy of fluid control.

In the signal processing module, the signals collected by the photodiode are amplified by the process controller amplifier module and then converted into digital signals by the analog-to-digital converter module, which is transmitted to the microcontroller unit for further analysis and processing. In order to improve the reliability of the system, a leakage detection module is designed, which consists of a voltage regulator tube and a resistor, and can cause a voltage change in the event of a leakage, triggering the MCU to start the alarm program. The communication module adopts Zigbee wireless communication and RS485 communication protocols to ensure stable communication between the main control node and the single control node, as well as between the main control node and the host computer. The control node receives control commands from the host computer via RS485 and sends the commands to the parameter acquisition node via Zigbee wireless communication. The low power consumption advantage of Zigbee communication significantly reduces energy consumption and can effectively extend the range of unmanned vessels.

The system adopts a modular design to realize real-time, synchronous, and high-precision monitoring of key nutrient elements in seawater, such as nitrate, nitrite, phosphate, silicate, and ammonium salt. The system realizes stable data transmission and remote sharing through RS485 and Zigbee wireless communication protocols. The intelligent power management module combines over-voltage and over-current protection mechanisms to guarantee the stable operation of the system. In contrast, the leakage detection module enhances the safety of the system. The system module design is shown in Fig. 3.

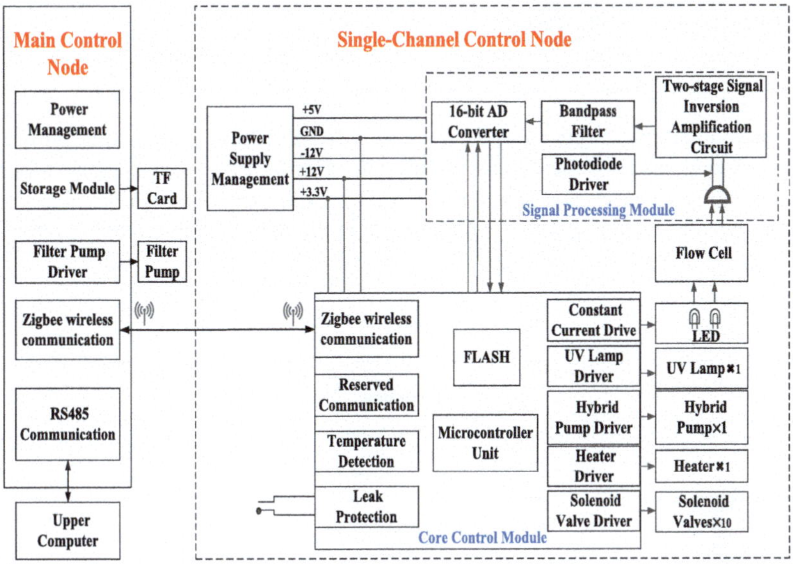

Fig. 3. System module design diagram.

2.3 Dynamic Adaptive RLS Filtering Algorithm

In order to improve the accuracy of the monitoring data, the recursive least squares (RLS) algorithm is introduced in this system. As an efficient recursive parameter estimation method, the RLS algorithm can quickly update the parameter estimates by recursive formulas each time new observations are acquired without recalculating the entire data set. The algorithm not only dramatically reduces the computational burden and time cost, but is also particularly suitable for real-time signal processing. In this study, an innovative adaptive recursive least squares (RLS) filtering algorithm is proposed, with special emphasis on a novel dynamic forgetting factor adjustment method to optimize the real-time monitoring accuracy of nutrient concentrations in a multichannel seawater monitoring system. The core of the method lies in the adaptive tuning formula of the forgetting factor λ.

$$\lambda = \lambda 0 \cdot (1 + \gamma \cdot |(n)|) \quad (2)$$

The innovation of this paper lies in the dynamic adjustment mechanism of the forgetting factor λ, which breaks through the limitation of the fixed forgetting factor in the traditional RLS algorithm. By introducing a constant of proportionality γ and the absolute value of the current prediction error $|(n)|$, the algorithm can adaptively adjust the degree of dependence on historical data according to the accuracy and stability requirements of real-time data. This adaptivity significantly improves the responsiveness of the algorithm in the face of environmental changes and data anomalies, thus enhancing the robustness of the system while maintaining high accuracy.

3 Experimental Procedure

3.1 Experimental Step

In this paper, an adaptive recursive least squares (RLS) filtering algorithm with dynamically adjusted forgetting factor λ is applied to a multi-channel seawater monitoring system according to the design steps to ensure the accuracy of real-time monitoring data of nutrient salts in seawater. The realized flow is shown in Fig. 4.

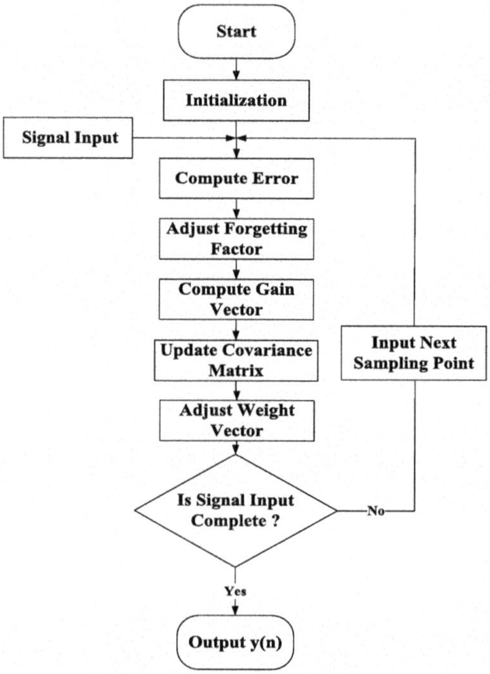

Fig. 4. Algorithm flowchart.

Step 1, data preprocessing and initialization: at each time point n, the system obtains the concentration data of seawater samples in real time through the parameter acquisition node, which are used as input signals (n). At the same time, the historical monitoring data are used to generate the desired signal (n). The algorithm initializes the weight vector (0) to zero, while the inverse of the covariance matrix (0) is initialized as δ^{-1} I, where δ is a positive number close to zero and I denotes the unit matrix.

Step 2, calculate the error and adjust the forgetting factor: in each iteration of the RLS algorithm, the system first calculates the current prediction error ϵ(n). Then, based on the magnitude of the prediction error, the forgetting factor λ is dynamically adjusted. This process ensures that the filter can adaptively adjust the old and new data weights to cope with data changes under different environmental conditions.

$$\varepsilon(n) = d(n) - w^T(n-1)x(n) \tag{3}$$

$$\lambda = \lambda_0 \cdot (1 + \gamma \cdot |\varepsilon(n)|) \tag{4}$$

Step 3, gain vector and weight update: The system calculates a gain vector (n) based on the adjusted forgetting factor, which is used to correct the current weight vector and update the covariance matrix $P(n)$ that ensures the stability of the system over a long period. The weight vector $\omega(n)$ is adjusted according to the gain vector (n) and the a priori error (n) to optimize the parameter configuration of the filter.

$$k(n) = P(n-1)x(n)/\left(\lambda + x^{n}T(n)P(n-1)x(n)\right) \tag{5}$$

$$P(n) = 1/\lambda\left[P(n-1) - k(n)x^{\wedge}T(n)P(n-1)\right] \tag{6}$$

$$\omega(n) = \omega(n-1) + k(n)\varepsilon(n) \tag{7}$$

Step 4, OUTPUT SIGNAL CALCULATION: The system calculates the final output signal (n), which is used as the monitoring result of the system at the current point in time to provide real-time monitoring data of nutrient salts.

$$y(n) = w^{T}(n-1)x(n) \tag{8}$$

Step 5, model evaluation and performance optimization: evaluate the performance of the RLS algorithm and adjust the algorithm parameters according to the evaluation results, such as the initial value of the forgetting factor λ_0 and the adjustment factor γ, to improve the accuracy and robustness of the model.

3.2 Analysis of Experimental Results

Tn order to comprehensively evaluate the performance of the proposed dynamic adaptive recursive least squares (RLS) filtering algorithm, a comparison experiment with the traditional fixed forgetting factor RLS model is designed in this study. In the experiment, the uncertainty in the ocean monitoring is simulated by adding random noise to the seawater monitoring signal. After setting the parameters of the RLS algorithm, including the number of weight orders (Ord = 48), the number of iterations (N = 500), and the length of data intercepts (K = 100), the two RLS models were tested in this experiment.

Figure 5 shows the filtering effect of the fixed forgetting factor RLS algorithm, there is a significant deviation between the filtered signal and the ideal signal, indicating that the consistency between the filtered signal and the ideal signal is poor; although the mean square error is higher in the initial stage, it shows a decreasing trend with the increase in the number of iterations, reflecting the gradual reduction of the prediction error as the algorithm is iterated. However, the final level of error is still high after many iterations, which implies that the performance of the filter in terms of noise suppression taking is not satisfactory.

As shown in Fig. 6, after applying the RLS algorithm with a dynamically adjusted forgetting factor to filter the raw seawater monitoring signals, the obtained signals are more consistent with the ideal signals, especially in the latter half of the signals, and the

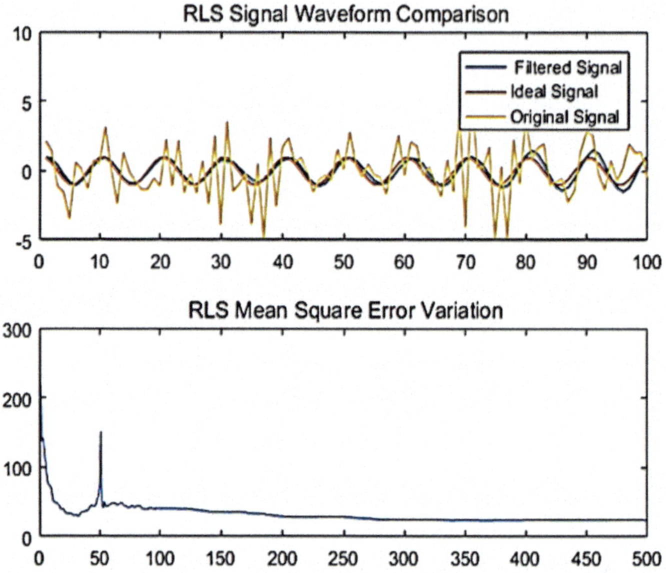

Fig. 5. Filtering effect of RLS algorithm with fixed forgetting factor.

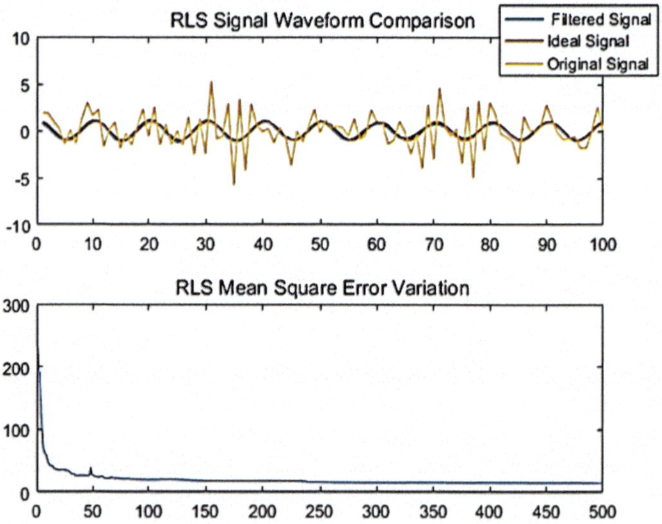

Fig. 6. Filtering effect of RLS algorithm with dynamically adjusted forgetting factor.

difference between the two is significantly reduced. This result indicates that the filter can effectively extract the useful signal from the noise; the mean square error is higher in the initial stage but gradually decreases with the increase of iterations and stabilizes at a lower level. This indicates that the prediction accuracy of the algorithm gradually improves with the iteration of the algorithm and the filtering performance is optimized.

In contrast, the filtered signal of the dynamic adaptive RLS model is closer to the ideal signal, showing better tracking performance and lower steady-state error; although the traditional RLS model can also filter out some noise, its filtering effect is relatively poor. In terms of the mean square error (MSE), the dynamic adaptive RLS model significantly reduces the MSE with the increase of the iteration number and eventually stabilizes at a lower level, a trend that indicates that the algorithm can quickly adapt to the signal changes and improve the detection accuracy; whereas, the MSE of the traditional RLS model decreases at a slower rate, reflecting its insufficient adaptability to the signal changes. In summary, the dynamic adaptive RLS model not only outperforms the traditional model in terms of filtering effect but also shows obvious advantages in terms of the fast adaptability of the algorithm and detection accuracy, and this result verifies the necessity of dynamically adjusting the forgetting factor in improving the performance of the RLS algorithm.

4 Long-Term Sea Trials Based on Unmanned Vessels

The long-term test experiment aims to verify the long-term stability of the multi-channel seawater nutrient salt monitoring method based on unmanned shipborne proposed in this paper, and to verify the consistency of the nutrient salt analyzing instrument. The sea test is proposed to select the outside of the Huangsha Harbor outlet in Jiangsu Province, such as Fig. 7①, ② and ③ three sea areas, three sea areas are affected by land source input, can effectively analyze the land source input of seaweed outbreak influencing factors, therefore, through the long period of unmanned ship real-time observation of seaweed outbreak proliferation degree and nutrient salts and other water quality environment relationship research and analysis, through the integration of the three sets of self-developed nutrient salts analytical instruments, to carry out the comparative experiments, to Verify the consistency of the method.

Fig. 7. Geographic location of the unmanned ship sea trial area in the Yellow Sea area.

The experimental methods are as follows: from May 2024 to July 2024, the real-time data of the three sets of seawater nutrient in situ analysis systems were collected and recorded on the experimental platform of the unmanned ship at sea, and water samples were collected at the corresponding time, and the water samples were tested in the laboratory according to the testing method stipulated by the national standard GB 17378.4-2007 with the use of comparative equipment, and the obtained data were compared with the data of seawater nutrient in situ sensors at the same time to calculate the difference and examine the accuracy of the data. The data will be compared with the data of seawater nutrient in-situ sensor at the same time, and the difference will be calculated to examine the accuracy of the data. The specific experimental steps are as follows:

(1) For each parameter, 2 sets of identical water samples of 20mL each were selected.
(2) Each group of water samples is filtered by pumping, and the filter pore size is uniformly selected as 0.45µm.
(3) The duration of the specific measurements was 3 months, with 2 regular measurements per day for a total of 180 measurements.
(4) The test methods used for the comparative analysis are shown in Table 1:

Table 1. Methods for analysis of seawater nutrient salts.

Nutrient salt parameters	Sensor Detection Methods	Expected Measurement Range
sphates	phosphomolybdenum blue spectrophotometry	3 ~ 300µg/L
ammonia nitrogen	fluorescence photometry	6 ~ 2000µg/L
nitrite	Naphthalene ethylenediamine spectrophotometric method	6 ~ 300µg/L
nitrate	Cadmium column reduction + naphthalene ethylenediamine spectrophotometry	15 ~ 1000µg/L
silicate	silicon molybdenum blue method	10 ~ 3000µg/L

The system calibration procedure is as follows:

- Zero calibration: Zero calibration is performed using pure water that does not contain the substance to be measured.
- Standard Solution Calibration: Standard solution calibration using a standard solution that covers the expected measurement range.
- Multi-point calibration: The system tests different concentrations of standard solutions and records their responses to obtain multiple data points.
- Calibration Curve Establishment: A calibration curve is established through a linear regression mathematical model using the concentrations of the standard solutions and the corresponding system response values.
- Repeatability test: Measure the same standard solution several times to check the consistency and repeatability of the readings.

The results of the data are analyzed below:

After three months of continuous monitoring at three sites, the data obtained were visualized using a horizontally stacked plot (shown in Fig. 8), where the horizontal coordinates represent the nutrient concentrations (in ug/L), while the vertical coordinates represent the number of measurements. Lines of different colors in the graph tracked the real-time data at the three monitoring sites: yellow for site 1, green for site 2, and purple for site 3. Through comparative analysis, it can be observed that the data from the three stations show highly consistent trends, a phenomenon that not only confirms the reliability of the nutrient salt analyzer in terms of measurement consistency but also demonstrates that the monitoring system maintains excellent stability and small drifts during long-term operation.

Further, the calculated correlation coefficients of the key nutrient salt parameters-nitrate, nitrite, ammonia nitrogen, silicate, and phosphate-were 0.9995, 0.9998, 0.9999, 0.9997, and 0.9992, respectively, which satisfy the requirement of correlation coefficient ($\geq$0.999), indicating that these parameters are well correlated with the measurement results. The relative standard deviations (RSDs) of phosphate, nitrate, nitrite and ammonia nitrogen were 0.95%, 1.17%, 0.78% and 0.45%, respectively, which were in compliance with the performance indexes of measurement precision, thus verifying the accuracy and reproducibility of the measurements of these parameters.

In addition, a generally significant decrease in the concentrations of the five key nutrients was observed between the 90th and 120th measurements of the monitoring cycle, i.e., within 15 days of seagrass growth. This phenomenon reveals the high demand for nutrient salts by seagrass during growth, which in turn leads to a decrease in nutrient salt concentrations in the water column. This not only provides us with direct evidence of the nutrient salt demand for seagrass growth, but also emphasizes the significant impact of biological activities on marine nutrient cycling.

5 Conclusions

In this paper, a multi-channel seawater nutrient salt monitoring system based on an unmanned shipboard is developed, and a dynamic adaptive recursive least squares (RLS) filtering algorithm is developed, to improve the efficiency and accuracy of nutrient salt concentration monitoring in the marine environment. The hardware circuit of the system is based on the STM32F103 microcontroller as the core, and real-time and synchronized monitoring of key nutrient salts in seawater, such as nitrate, phosphate, silicate, nitrite, and ammonia nitrogen, is realized through modular design. Meanwhile, the application of intelligent power management module realizes the optimization of energy consumption and enhances the stability of the system.

The dynamic adaptive RLS filtering algorithm proposed in this paper significantly improves the system's ability to respond to environmental changes and data anomalies by introducing a dynamic adjustment mechanism of the forgetting factor. Compared with the traditional fixed forgetting factor RLS model, the dynamic adaptive RLS model shows significant improvement in both filtering effect and mean square error, which can adapt to signal changes faster and improve detection accuracy. The experimental results verify the superior performance of the proposed algorithm in dealing with noisy data and outliers.

In summary, this system can effectively monitor the nutrient salt concentration in seawater, showing high data accuracy and excellent real-time performance, which is

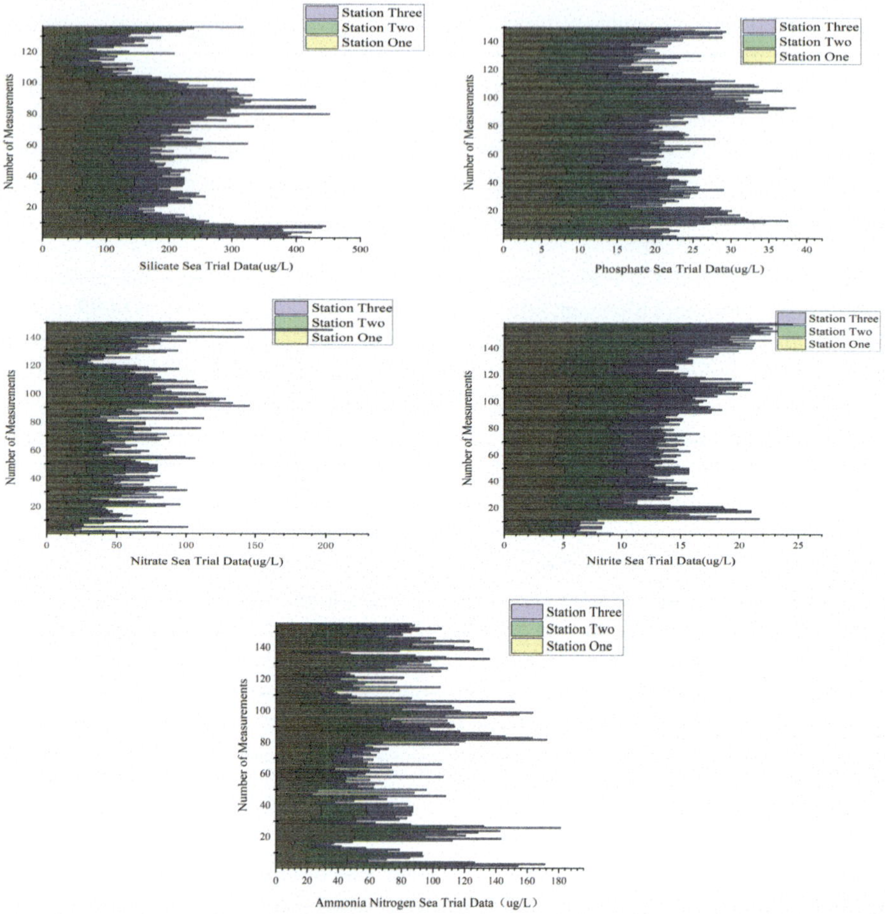

Fig. 8. Unmanned ship sea trial in the Yellow Sea: sea trial data of five nutrients at three stations.

of great academic value and practical application for marine environment monitoring. Future work in this study will further optimize the performance of the algorithm, expand the functions of the monitoring system, and conduct field tests and applications in a wider range of marine environments. Continuous technological innovation and system improvement will provide a more reliable and effective technical means for monitoring marine environment.

6 Research Limitations and Outlook

While this study provides an in-depth analysis of an unmanned shipboard multichannel seawater nutrient salinity monitoring system, some limitations remain. First, due to budget and resource constraints, our experiments were conducted only in specific sea areas, which may limit the generalizability of the results. Second, although we used an advanced adaptive recursive least square filtering algorithm to improve the accuracy of the data, the performance of the algorithm under extreme environmental conditions has not been fully tested. In addition, the sensors used in this study still have some measurement errors that affect the accuracy of the data. Future research could test the performance of the system in a wider range of marine environments to verify its stability and reliability. Meanwhile, further optimization of the algorithm to improve its robustness under extreme conditions is also an important direction for future work. In addition, exploring the use of higher-precision sensor technologies to reduce measurement errors is also key to improving the performance of the monitoring system.

Acknowledgments. This work was supported by the Major Innovation Project of Science and Education Integration Pilot Project of Qilu University of Technology (Shandong Academy of Sciences)(2023JBZ02), Major Innovation Project of Science and Education Integration Pilot Project of Qilu University of Technology (Shandong Academy of Sciences)(2023JBZ03), Science Education Industry Integration Pilot Major Innovation Project (2023HTZX01) of Qilu University of Technology (Shandong Academy of Sciences), Open Research Fund of Key Laboratory of Marine Ecological Monitoring and Restoration Technology, Ministry of Natural Resources, (MEMRT202301).

References

1. Oduor, N.A., Munga, C.N., Ong'anda, H.O., et al.: Nutrients and harmful algal blooms in Kenya's coastal and marine waters: a review. Ocean Coast. Manag. **233**, 106454 (2023)
2. Glibert, P.M., Heil, C.A., Madden, C.J., et al.: Dissolved organic nutrients at the interface of fresh and marine waters: flow regime changes, biogeochemical cascades and picocyanobacterial blooms-the example of Florida Bay, USA. Biogeochemistry **164**(1), 229–255 (2023)
3. Howard, M.D.A., Smith, J., Caron, D.A., et al.: Integrative monitoring strategy for marine and freshwater harmful algal blooms and toxins across the freshwater-to-marine continuum. Integr. Environ. Assess. Manag. **19**(3), 586–604 (2023)
4. Altahan, M.F., Esposito, M., Achterberg, E.P.: Improvement of on-site sensor for simultaneous determination of phosphate, silicic acid, nitrate plus nitrite in seawater. Sensors **22**, 3479 (2022)
5. Briciu-Burghina, C. et al.: Sensors for coastal and ocean monitoring. Ann. Rev. Anal. Chem. **16**(1), 451–469 (2023)
6. Ma, R. et al.: Research on signal processing method for total organic carbon of water quality online monitor. IOP Conf. Ser. Earth Environ. Sci. **82**(1), (2017)
7. Liu, Y., Lu, H., Cui, Y.: A review of marine in situ sensors and biosensors. J. Mar. Sci. Eng. **11**(7), 1469 (2023)
8. Zhang, Y., et al.: Marine ecological environment monitoring and management system based on sensor technology. In 2024 International Conference on Integrated Circuits and Communication Systems (ICICACS). IEEE, 2024: 1–6 (2024)
9. Mahmud, MA Parvez, et al.: Recent progress in sensing nitrate, nitrite, phosphate, and ammonium in aquatic environment. Chemosphere **259**, 127492 (2020)

10. Zhu, X., Yu, K., Zhu, X., Su, J., Wu, C.: An improved algorithm for measuring nitrate concentrations in seawater based on deep-ultraviolet spectrophotometry: a case study of the Aoshan Bay seawater and western pacific seawater. Sensors **21**, 965 (2021)
11. Wang, H., Ju, A., Wang, L.: Ultraviolet spectroscopic detection of nitrate and nitrite in seawater simultaneously based on partial least squares. Molecules **26**, 3685 (2021)
12. Yang, Z., et al.: An in-situ analyzer for long-term monitoring of nitrite in seawater with versatile liquid waveguide capillary cells. Development, optimization and application. Mar. Chem. **245**, 104149 (2022)

Experimental Study on the Plant-Growing Performance of Porous Concrete

Bin Yan[1], Chenglin Li[2], Jing Zhang[1], Xianchong Liu[1], and Zhenwei Gao[1,1](✉)

[1] College of Water Conservancy, Shenyang Agricultural University, Shenyang 110866, China
{yanb2020,2008500006,gzw2020}@syau.edu.cn, 2975095994@qq.com
[2] Tianjin Zhongshui Xinhua Project Planning and Design Limited Company, Tianjin 300220, China
1318259671@qq.com

Abstract. The study used the orthogonal experimental method to analyze the significant impact of various factors on the vegetation performance of porous concrete. The results showed that the plant growth was good during the experiment, and at 70 days, the height of grass plants in each experimental group could reach about 40cm, indicating that porous concrete has good plant growth performance. There are significant differences in the thickness of the test specimens at different levels, with the plant height of the 6cm thick test group being significantly higher than that of other levels, indicating that the thickness of porous concrete specimens is an important factor affecting grass height. The smaller the thickness of the specimen, the better the plant growth performance. The porosity of porous concrete, specimen thickness, and coarse aggregate particle size all significantly affect the weight of vegetation. The larger the coarse aggregate particle size and porosity, the more suitable it is for plant growth. Among them, porosity has the most significant impact on the weight of grass planted.

Keywords: Porous Concrete · Vegetation Performance · Porosity · Specimen Thickness · Coarse Aggregate Particle Size · Filler

1 Introduction

For many years, to effectively prevent soil erosion and ensure the safety of embankments, the slopes of rivers and lakes were usually covered by masonry or ordinary reinforced concrete structures. These structures blocked the connection between river water and land soil and prevented the necessary ecological environment for the growth of animals and plants [1]. This resulted in the destruction of the natural ecosystem between river banks and coastal land, and the destruction of biological species and decrease in the biodiversity index.

Porous concrete is an ecological concrete composed of coarse aggregate, cement, water, and admixture. It has certain strength and anti-scour ability and a more continuous pore structure that can provide space for plant growth and animal habitat. France began experimental work on porous vegetation concrete for riverbank slope protection in the 1980s [2]. Anti-erosion vegetation concrete for riverbank slope protection has been

C. Weng (Ed.): WRE 2024, LNCE 634, pp. 501–516, 2025.
https://doi.org/10.1007/978-981-96-5095-8_41

developed through proper material selection and mix design [3, 4]. Porous vegetation eco-concrete forms a sturdy surface layer that helps slow down water flow and prevent erosion of slope soil by river or ocean currents. Plant roots can maintain soil stability and promote the activity of microorganisms in the soil, contributing to soil formation and ecosystem restoration [5]. Plant roots absorb water from the soil, hence increasing the soil suction and improving the shear strength of the soil through soil-root reinforcement [6]. The biodiversity on the surface of porous concrete is higher than that of soil matrix, indicating that porous concrete matrix is more conducive to the attachment and growth of microorganisms [7]. Therefore, compared with traditional river slope protection technology, the vegetation porous concrete ecological slope protection could not only enhance slope stability, meet the requirements for flood control and drainage of urban rivers, and prevent soil erosion, but also could solve the contradiction between "greening and hardening," improve water quality, and maintain biodiversity, thus playing a positive role in creating the ecological landscape of urban rivers and improving the living environment.

At present, the studies on the plant-growing performance of porous concrete mainly focus on the following two aspects: (1) to identify the methods of reducing the pH in porous concrete to meet the requirements of plant root growth and improve the plant-growing performance and (2) to investigate the effect of porosity, coarse aggregate gradation, the thickness of soil covering, pore filler substances and nutrients, and the thickness of porous concrete on the plant-growing performance.

Studies by Xu et al. [8], Chen et al. [9], and Tian et al. [10] show that changing the mix ratio of porous concrete can change its porosity and pH value, thus influencing its plant-growing performance. The study by Liao et al. [11] shows that adding the admixture not only reduces the alkalinity of the porous concrete but also improves its strength. To a certain extent, the use of specific curing methods also plays a role in reducing the pH. Researches by Wang et al. [12] and Li et al. [13] show that water immersion was an effective method to improve the alkaline environment of the porous concrete. The pH value of the porous concrete could be reduced by dissolving and precipitating alkaline matter, increasing the water content, and reducing the concentration of alkaline substances. Wang et al. [14] summarize in detail the methods to control the pH of the porous concrete, which play an important role in guiding the preparation of the porous concrete.

Studies by Chen et al. [15] and Zhu et al. [16] show that filling the porous concrete pores with nutritious soil slurry could provide nutrients for early plant growth. The chemical properties of the filling materials and the various nutrients required by plants are important factors in determining whether plants can take root, sprout, and grow in porous concrete for a long time. The study by Horiguchi et al. [17] shows that porous concrete with crushed oyster shell aggregates has good vegetation growth performance because oyster shell aggregates are lighter, have smaller particle sizes, and have a porous microstructure that makes them highly water retaining and suitable for plant growth.

Jiang et al. [18] report that the plant-growing performance of porous concrete could be affected by its water–cement ratio, aggregate gradation, and porosity. In general, within a certain range, the higher the porosity, the larger the coarse aggregate particle size, and the higher the water cement ratio, the better the vegetation performance of porous concrete.

Studies by Li [19], Zhou et al. [20], Li et al. [21], and Hu et al. [22] show that porosity has a significant effect on the survival and healthy growth of grass seeds. For plant roots grown in concrete pores, the living space provided by the pores and the chemical characteristics inside the pores directly determined the planting performance of the porous concrete. On the one hand, the pores need to store sufficient nutrients for plant growth, and plants also need some space to develop their roots. On the other hand, plants can only survive in environments with specific acidity and alkalinity ranges. For vegetation type porous concrete used for slope protection, the porosity needs to be controlled between 20% and 30%. Research by Zhang et al. has shown that porous concrete with an aggregate grading of 20–25mm is more suitable for vegetation growth than other grading [23]. A study by Quan [24] shows that the coarse aggregate gradation of porous concrete affects its pore structure, which in turn affects its symbiotic performance with plants. Meanwhile, the plant growth performance of porous concrete is related to the thickness of the porous concrete test block and the covering soil layer. The smaller the thickness of the porous concrete and the larger the thickness of the covering soil, the better the plant growth performance.

The above research has played an important role in improving the vegetation performance of porous concrete. However, research on the significant impact of different factors on the vegetation performance of porous concrete is still rare, which limits the engineering application of porous concrete. Considering that tall fescue is a perennial herbaceous plant with the developed fibrous root system, strong stress resistance, high germination rate, and ability to adapt to various soil and climatic conditions, tall fescue was selected as the experimental plant. Planting tall fescue on porous concrete, after a 70 day growth cycle, observing the growth status of tall fescue, analyzing the significant impact of thickness, porosity, coarse aggregate particle size, and pore filling on the vegetation performance of porous concrete, in order to provide theoretical basis for the preparation and application of porous concrete.

2 Materials and Methods

2.1 Test Materials

The test materials include the following: cement (P.O42.5 ordinary Portland Cement), silica fume (92% grade special Microsilica powder for high-strength concrete application), admixture (high-performance polycarboxylic acid water-reducing agent), coarse aggregate (high-quality pebbles of 10–25 mm diameter), and mixing water (tap water).

2.2 Preparation of Porous Concrete and Materials Suitable for Growth

Preparation of Porous Concrete. By comparing the compressive strength and frost resistance of porous concrete under different mix ratios, the optimum mix ratio comprising 0.27 water-cement ratio, 7% silica fume and 0.16% water reducing agent was selected to study the grass planting performance. Under these conditions, the compressive strength of the porous concrete was 12 MPa, and it could meet the demand for slope protection strength.

The thickness, porosity, and diameter of the coarse aggregate of porous concrete were selected as the test factors and each factor was tested at four levels. The diameter of the coarse aggregate for the four levels was 10–15, 15–20, 20–25, and 10–25 mm; porosity was 22%, 24%, 26%, and 28%; and the specimen thickness was 6, 9, 12, and 15 cm.

The porous concrete specimen was made using a self-designed concrete mold with a diameter of 25 cm. The porosity of the concrete was controlled by the ratio of the aggregate to cement. The experiment was divided into 16 groups, and the material dosage of each group's porous concrete test block is shown in Table 1. The test blocks were demolded 24 h later and cured for 28 days by immersing them in clean water.

Preparation of materials suitable for growth. The materials suitable for growth used in the experiment could be divided into filling materials that were used to fill in the concrete pores and soil materials that were used to cover the surface of the porous concrete.

The filling materials comprised tillage soil, compound fertilizer, fly ash, and a water retention agent, which were combined in a mass ratio of 1000:10:20:3. The tillage soil provided the basic conditions for plant growth; fly ash improved the alkaline environment in the pores and enabled plants to grow under a suitable pH [25]; compound fertilizer increased the soil fertility and the N, P, and K content in the soil [26]; and the water retention agent improved the soil structure and maintained the water and fertilizer content, preserving heat and increasing the ability of drought resistance.

The filling materials were filled using the non-dynamic mixing method. The materials suitable for growth and water were mixed in a mass ratio of 1:1.5 to form a slurry and filled into the pores. The slurry was sprinkled on the porous concrete surface until the slurry seeped into the bottom of the concrete specimen [27]. The covering material was the tillage soil, which directly covered the surface of the porous concrete after sowing at a thickness of 10 mm.

2.3 Orthogonal Experimental Design

In order to reduce the number of test groups, the orthogonal testing method was selected under the condition of determining the mix ratio of porous concrete; four factors of coarse aggregate particle size (A), porosity (B), specimen thickness (C), and pore filler (D) were studied at four levels. The level of each factor is shown in Table 2. Based on the test factors and levels, the L16(45) level of the orthogonal test table was selected to carry out the experiment. The orthogonal test design is shown in Table 3.

2.4 Grass Selection and Conservation

The selection of grass seeds should meet the requirements of fast germination, high sprouting rate, well-developed roots, and strong resistance to stress. As far as possible, use local species of grass to avoid ecological disasters [28]. Through preliminary plant selection experiments, it was found that the fibrous root system of tall fescue is well-developed, with strong stress resistance and a high germination rate. Therefore, tall fescue was selected as the experimental plant. Spread 30g/m^2 of tall fescue grass seeds

Table 1. Material dosage of the porous concrete test block.

Test No	Diameter of Aggregate/Mm	Porosity/%	Specimen Thickness/Cm	Coarse Aggregate /G	Ratio of Aggregate to Cement	Cementitious Material/G	Cement/G	Silica Fume /G (7%)	Water/G (Water Cement Ratio 0.27)	Water Reducing Agent/G (0.16%)
1	10–15	22	6	1600.8	4	400	374	26	104	0.64
2		24	9	2401.2	4.5	534	499	35	139	0.85
3		26	12	3201.6	5	640	598	42	166	1.02
4		28	15	4002	5.5	728	680	48	189	1.16
5	15–20	22	9	2401.2	4	600	561	39	156	0.96
6		24	6	1600.8	4.5	356	332	23	92	0.57
7		26	15	4002	5	800	748	52	208	1.28
8		28	12	3201.6	5.5	582	544	38	151	0.93
9	20–25	22	12	3201.6	4	800	748	52	208	1.28
10		24	15	4002	4.5	889	831	58	231	1.42
11		26	6	1600.8	5	320	299	21	83	0.51
12		28	9	2401.2	5.5	437	408	29	114	0.70
13	10–25	22	15	4002	4	1001	935	65	260	1.60
14		24	12	3201.6	4.5	711	665	47	185	1.14
15		26	9	2401.2	5	480	449	31	125	0.77
16		28	6	1600.8	5.5	291	272	19	76	0.47

Table 2. Factors and levels of the orthogonal experimental design of the porous concrete.

Level	Factors			
	A (coarse aggregate particle size/mm)	B (porosity/%)	C (specimen thickness/cm)	D (filler)
1	10–15	22	6	Soil
2	15–20	24	9	Soil + fertilizer
3	20–25	26	12	Soil + fertilizer + fly ash
4	10–25	28	15	Soil + fertilizer + fly ash + SPA

Table 3. Orthogonal test design.

Number	Factors			
	A (coarse aggregate particle size /mm)	B (porosity /%)	C (specimen thickness /cm)	D (filler)
1	1 (10–15)	1 (22)	1 (6)	1 (Soil)
2	1	2 (24)	2 (9)	2 (Soil + fertilizer)
3	1	3 (26)	3 (12)	3 (Soil + fertilizer + fly ash)
4	1	4 (28)	4 (15)	4 (Soil + fertilizer + fly ash + SPA)
5	2 (15–20)	1	2	3
6	2	2	1	4
7	2	3	4	1
8	2	4	3	2
9	3 (20–25)	1	3	4
10	3	2	4	3
11	3	3	1	2
12	3	4	2	1
13	4 (10–25)	1	4	2
14	4	2	3	1
15	4	3	2	4
16	4	4	1	3

evenly on the surface of porous concrete filled with suitable materials, and then cover with 10mm thick soil material.

One week after sowing is the critical period for grass germination, during which water should be sprayed daily to keep it moist. After the grass seeds grow into seedlings, water and maintain them every other day. After one month, water twice a week for maintenance. The frequency of watering per day is determined based on climate conditions and soil foundation. In addition, appropriate pruning and weed control should be carried out during the germination and growth period of the plant [15].

2.5 Plant-Growing Performance Evaluation

The plant-growth performance of porous concrete was evaluated by measuring changes in plant height and weight and the growth status of plant roots.

The plant height was characterized by measuring the length from the soil surface at the top of the test block to the longest leaf of the plant, with a measurement interval of 7 days. IBM SPSS Statistics 22 software was used to calculate the estimated marginal mean of plant height under various influencing factors and to analyze the variation process of the estimated marginal value of plant height over time. The marginal mean is the average value calculated after excluding the influence of other variables. By analyzing the changes in the marginal mean, the main factors influencing the grass planting performance of porous concrete can be identified.

Because plant height cannot reflect grass coverage, while plant weight comprehensively reflects the influence of plant height and coverage, the plant growth density of porous concrete test blocks can be fed back by measuring plant weight. After three months of experimentation, the porous concrete test blocks were taken out, and the plants above the top soil surface were cut off to measure their weight. Porous concrete test blocks were crushed to observe whether plant roots could grow within the pores of porous concrete and penetrate the specimens. IBM SPSS Statistics 22 software was used for variance analysis of grass planting weight to determine the significant impact of various experimental factors on the grass planting performance of porous concrete, and to evaluate the vegetation performance of porous concrete.

3 Results and Discussion

After four days of planting, the seeds began to germinate. During the experimental period, the plants grew well, and there was no death in a large part of the area. The growth status of the seeds in each group at 7, 35, and 63 days is shown in Fig. 1.

3.1 Determination of Plant Height

Grass Height Determination Results. The grass heights recorded during the experimental period are shown in Table 4.

The results show that the grass height increased continuously with time. At two weeks, grass height was 7–14 cm. At four weeks, 75% of grass had a height exceeding 20cm. At six weeks, 69% of grass had a height exceeding 30cm. At 70 days, the grass height in each experimental group was measured to be around 40cm, which indicated that the porous concrete had a good plant-growing performance.

7d Plant Growth Condition 35d Plant Growth Condition 63d Plant Growth Condition

Fig. 1. Grass growth status at 7, 35, and 63 days.

A similar conclusion was drawn from an experimental study on the vegetation performance of porous concrete made with by-product materials [29]. This study used 13 kinds of seeds (11 wild flowers and two lawn grasses), and the experimental results showed that seeds began to sprout one week after planting. After two weeks, the grass was 4.9–7.1 cm long in the quadrat zones. After four weeks, the length was about 20 cm; after six weeks, it was at least 30 cm.

Table 4. Grass heights (cm) recorded during the experimental period.

Test number	Time /d								
	14	21	28	35	42	49	56	63	70
1	10	16	35	36	38	46	49	49	53
2	10	14	27	32	33	39	39	41	43
3	14	16	22	30	28	36	38	39	46
4	12	19	18	26	27	30	29	32	35
5	11	16	25	31	32	41	37	40	44
6	10	16	25	34	36	43	46	44	48
7	8	17	26	35	32	39	43	39	45
8	9	14	23	29	31	39	41	40	45
9	10	15	9	28	29	34	36	38	44
10	7	13	17	24	26	31	32	34	39
11	9	18	29	33	37	42	45	42	47
12	7	13 ara>ara>	23	28	32	37	41	44	45
13	10	19	27	36	34	40	41	40	46
14	7	12	21	27	29	38	34	38	43
15	10	17	15	28	31	40	38	41	49
16	14	20	34	37	43	47	46	45	49

Analysis of the Estimated Marginal Value of Plant Height over Time. The changes in the estimated marginal average value of plant height over time under various influencing factors are shown in Fig. 2–5.

Figure 2–5 show that the plant height of different levels of influencing factors was almost the same in the 10 days after germination of grass seeds. This indicates that the influence of various factors on plant height was not significant at the beginning of the experiment. The mulching soil provided the conditions for plant germination and growth at this stage.

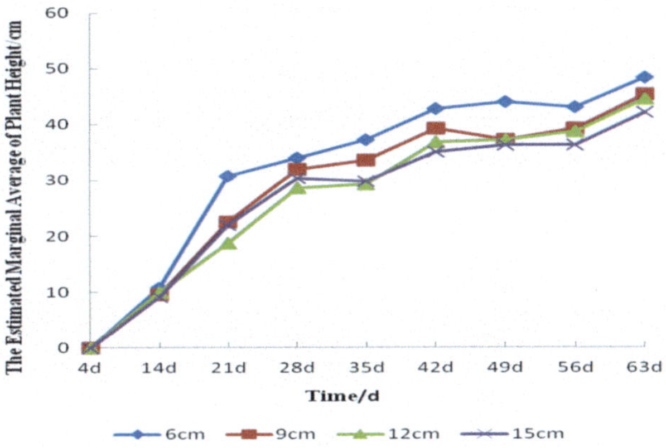

Fig. 2. The estimated marginal average of plant height with time for different specimen thicknesses.

As can be seen in Fig. 2, the plant height of the test group with a porous concrete thickness of 6 cm was higher than that of the other levels with the change in time, which indicated that there was a significant difference between the experimental level and other experimental levels. The observation also shows that the thickness of the porous concrete specimen had a significant effect on the height of the grass, which was mainly because that the smaller the thickness of the porous concrete specimen, the lower the alkali content in the specimen, the easier is it for the plant root system to survive, and the easier is it for the plant root system to grow through the porous concrete to the natural soil layer, so the better the plant growth performance.

It could be seen from Figs. 3–5 that there was little difference in the height of grass planting at different levels of test factors and that the broken lines basically coincided, indicating that the porosity of the porous concrete, composition of the pore filler, and the coarse aggregate size had little effect on plant height. This was mainly because the porosity selected in the experiment was between 22% and 28%, which was more suitable for plant growth. The alkaline content of the porous concrete was relatively low when it was soaked in clear water during the curing stage of the concrete, and, therefore, there was no significant difference in the plant-growing height among different test factors.

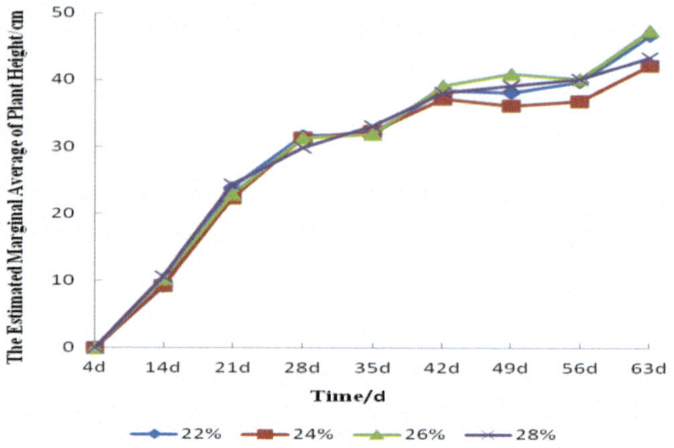

Fig. 3. The estimated marginal average of plant height with time for different porosity values.

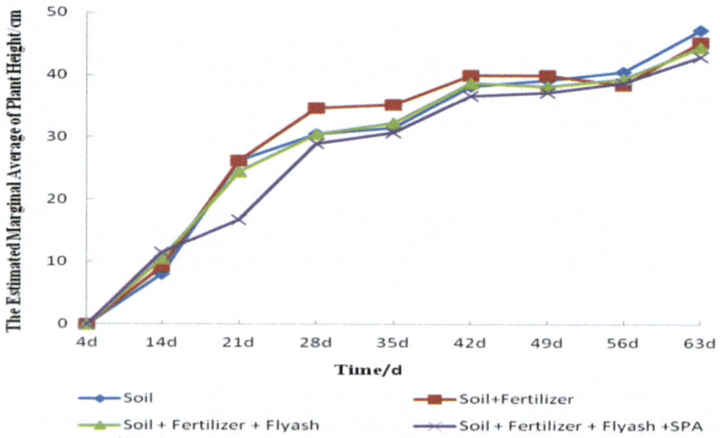

Fig. 4. The estimated marginal average of plant height with time for different filling level.

3.2 Plant Weight Determination and Evaluation of Root-Growth Status

The results of the plant weight measurement are shown in Table 5. The growth status of the plant roots is shown in Fig. 6.

The results of the variance analysis of plant weight are shown in Table 6. It could be seen from Table 6 that the Sig. Value tested by the variance model is 0.002, less than 0.05, which indicates that the model was applicable. The Sig. Value of porosity was 0.000 and that of sample thickness was 0.005, both of which were less than 0.01, indicating that the effect of both on plant weight was highly significant. The Sig. Value of the coarse aggregate size was 0.042, less than 0.05, which indicated that the effect of the coarse aggregate size on plant weight reached a significant level. The Sig. Value of the filler on plant weight was 0.166, more than 0.05, indicating that the effect of filler on plant weight was not significant.

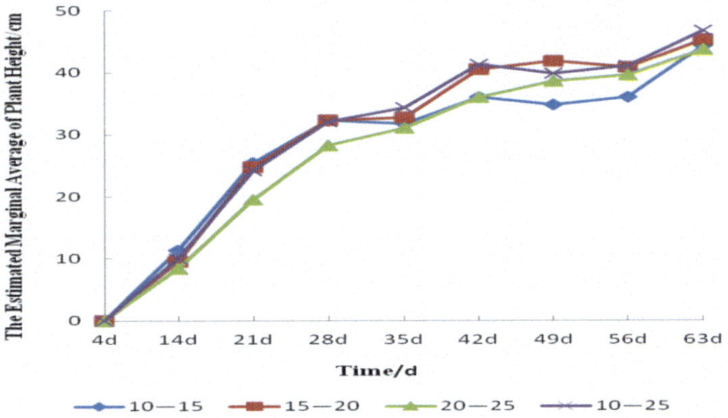

Fig. 5. The estimated marginal average of plant height with time for different coarse aggregate particle size.

Fig. 6. Root-growth status of plants.

As can be seen from the sum of the squared deviations contributed by the various factors, the porosity of porous concrete had the greatest effect on the plant weight (sum of squares = 6135.907), followed by the thickness of porous concrete specimen on the plant weight (sum of squares = 1127.472). Next was the effect of coarse aggregate size on plant weight (sum of squares = 263.350), and the fillers had the least effect on plant weight (sum of squares = 86.819).

The relationship between each factor level and the estimated marginal average of plant weight is shown in Fig. 7. The data in Fig. 7 shows the following results: (1) The plant weight increased with the increase in the coarse aggregate size, and when compared with better graded coarse aggregates, the single-particle coarse aggregate had a better plant-growing performance. These were mainly due to the fact that under the same porosity, the larger the coarse aggregate size, the larger the average pore diameter, which was more suitable for plant root growth. Therefore, the single-particle coarse aggregates with larger particle size should be used as far as possible in the preparation of porous concrete for plant growing. (2) With the increase in the porosity of the concrete, the grass planting weight increased, which indicates that the larger the porosity of the concrete, the better its plant-growing performance. However, with the increase in the porosity of the concrete, the compressive strength of porous concrete decreased. Under the condition of guaranteeing the strength to the porous concrete, its porosity should be

Table 5. Grass planting weight and root condition.

Test number	Plant weight/g	Root growth status	Penetrate concrete specimens or not
1	34.24	1	No
2	36.51	2	Yes
3	45.9	2	No
4	65.56	3	Yes
5	26.86	1	No
6	50.35	1	Yes
7	41.50	2	No
8	78.99	3	Yes
9	30.97	1	No
10	28.60	1	No
11	75.63	3	Yes
12	86.19	3	Yes
13	20.73	1	No
14	42.39	1	No
15	67.42	2	Yes
16	89.03	3	Yes

Notes: the growth status of plant roots in porous concrete specimens was categorized into three grades based on the observation. Level 1 denotes a very small number of roots, level 2 represents relatively small roots, and level 3 expresses a large number of roots

as large as possible. (3) With the decrease in the thickness of porous concrete specimen, the grass planting weight increased, which indicated that the smaller the thickness of porous concrete, the better its plant-growing performance. However, when the thickness of the porous concrete decreases, its strength and scour resistance will be reduced. To ensure the strength of concrete, its laying thickness should be as small as possible. (4) There was no significant difference in the marginal average value of plant weight prediction under various levels of pore fillers, which indicated that the effect of the pore fillers on the plant-growing properties of the porous concrete was not significant. It was possible that the pore fillers were mainly composed of ploughed soil, which contained more nutrients for plant growth and was suitable for plant growth. Therefore, adding other materials had little effect in improving the plant-growing performance.

Based on the variance analysis of plant weight data and the analysis of the relationship between the level of factors and the marginal average value of plant weight prediction, it can be observed that the porosity of the concrete, the thickness of specimen, and the diameter of the coarse aggregate were the important factors that influenced the plant-growing performance of porous concrete, which should be taken into account in the application of porous concrete in practical engineering.

Table 6. Results of the variance analysis.

Source	Type III sum of squares	Df	Mean square	F	Sig
Corrected model	7613.548[a]	12	634.462	76.383	.002
Intercept	42114.222	1	42114.222	5070.140	.000
coarse aggregate size	263.350	3	87.783	10.568	.042
porosity	6135.907	3	2045.302	246.234	.000
sample thickness	1127.472	3	375.824	45.246	.005
filler	86.819	3	28.940	3.484	.166
Error	24.919	3	8.306		
Total	49752.690	16			
Corrected total	7638.467	15			

Notes: Dependent variable: plant weight. [a] R Squared = .997 (adjusted R squared = .984)

Ganapathy et al. [30] discovered that Coarse Aggregate Size Ratio (CASR) had a substantial impact on the void content, strength and planting properties of Vegetation Porous Concrete (VPC). Despite an increase in CASR improved the VPC's strength, the increased interlocking property caused the internal concrete pores to become tortuous and the root travelling through the concrete to narrow, which further slowed down the growth of the root system. In this sense, a lower CASR increased the root's ability to penetrate the concrete and enhanced plant growth. The research results of this article are consistent with this finding.

4 Conclusions

This study investigated the effects of porous concrete thickness, porosity, coarse aggregate particle size, and pore fillers on the vegetation performance of porous concrete through orthogonal experiments and analyzed the significance level of each factor's influence on the vegetation performance of porous concrete. The main conclusions of this study are as follows:

(1) During the experiment, plants grew well on the porous concrete, and the plant height of each test group could reach up to 40 cm in 70 days. The results showed that porous concrete had good plant-growing performance.
(2) There was a significant difference in plant height between the experimental group with a porous concrete specimen thickness of 6cm and other thickness levels, indicating that the thickness of porous concrete was an important factor affecting grass growth height. There was no significant difference in the porosity, pore filler, and coarse aggregate particle size of porous concrete at different levels, indicating that these factors had little effect on the plant growth height.

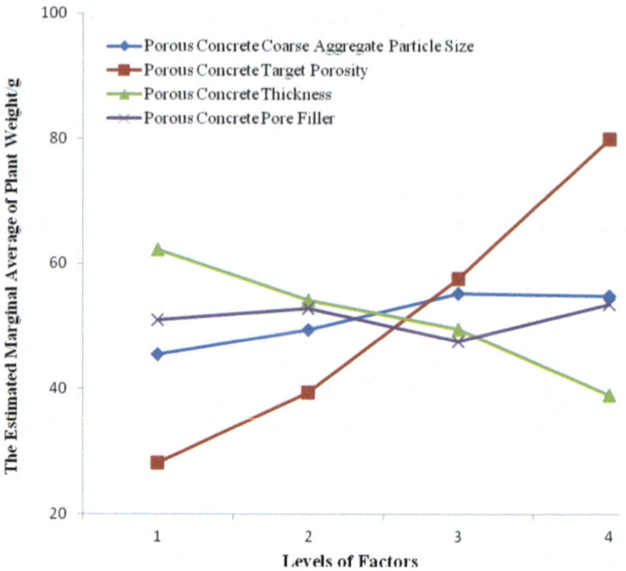

Fig. 7. Relationship between plant weight and factor levels.

(3) The porosity, specimen thickness, and coarse aggregate particle size were important factors influencing the growth performance of porous concrete, which need to be carefully considered in practical engineering applications of porous concrete. The effect of pore filler on the growth performance of porous concrete was not significant.

This study only conducted preliminary experimental research on the vegetation performance of porous concrete, and there are still some issues that need further research and resolution, such as how to control the selection of raw materials and basic performance parameters, how to further improve the mechanical and durability properties of porous concrete, and how to establish an evaluation system for vegetation effects. At the same time, it is necessary to actively promote the engineering application of plant-based porous concrete.

References

1. Feng, X.X., Nian, F.X., Yao, X.M., et al.: Application of ecological technology of mossy riverbed in river ecological remediation projects. Water Resour. Plan. Des. **7**, 84–88 (2024)
2. Bachawati, M.E., Manneh, R., Belarbi, R., et al.: Cradle-to-gate life cycle assessment of traditional gravel ballasted, white reflective, and vegetative roofs: A Lebanese case study. J. Clean. Prod. **137**, 833–842 (2016)
3. Zhang, H.L., Zhao, Z.F., Ma, G.F., et al.: Quantitative evaluation of soil anti-erodibility in riverbank slope remediated with nature-based soil bioengineering in Liaohe River, Northeast China. Ecol. Eng.Eng **151**, 105840 (2020)
4. Jiang, H., Liu, Y.: Performance evaluation and field application of prefabricated grass-planting concrete blocks with concave-convex construction for mid-sized and small river revetment projects. Pol. J. Environ. Stud. **31**(1), 99–110 (2022)

5. Zheng, C.W., Zhang, Z.H., Huang, Z.Y., et al.: Review of porous vegetation eco-concrete (PVEC) technology: From engineering requirements to material design. Compos. B Eng. **279**, 111442 (2024)
6. Nguyen, T.S., Likitlersuang, S., Jotisankasa, A.: Influence of the spatial variability of the root cohesion on a slope-scale stability model: a case study of residual soil slope in Thailand. Bull. Eng. Geol. Environ. **78**(5), 3337–3351 (2019)
7. Gao, S.: Mechanism and engineering demonstration for removal of nitrogen and phosphorus from farmland drainage water by porous concrete-based ecological ditch. Anhui Agricultural University, Anhui (2022)
8. Xu, Y., Chen, F.Q.: Effects of concrete content in vegetation concrete matrix on seed germination and seeding establishment of Cynodon Dactylon. Proc. Eng. **28**, 105–109 (2012)
9. Chen, F., Xu, Y., Wang, C., et al.: Effects of concrete content on seed germination and seeding establishment in vegetation concrete matrix in slop restoration. Ecol. Eng. **58**, 99–104 (2013)
10. Tian, L., Pang, Z.M., Quan, H.Z., et al.: Physical properties and vegetative adaptation of eco-porous concrete. Bullet. Chinese Ceram. Soc. **35**(10), 3381–3386 (2016)
11. Liao, W.Y., Shi, X., Huang, Z.F., et al.: Study on decreasing alkalinity of planting concrete and the resulting planting effect. Concrete **7**, 155–158 (2013)
12. Wang, Y.J., Zhai, A.L., Zhao, J., et al.: Experimental studies on planting tests of recycled brick aggregate vegetation-type concrete. Hydro-Sci. Eng. **3**, 90–95 (2016)
13. Li, X.D., Li, K., Weng, S.F.: Research on lowering the pH and EC of porous concrete and plant growth test. New Build. Mater. **01**(44), 49–51+56 (2017)
14. Wang, G.L., Wang, Z.L., Zhang, H.X., et al.: Mixture design, alkalinity control, selection of planting soil and plant of planting eco-concrete. Concrete **2**, 102–109 (2013)
15. Chen, J., Lu, J.L., Xu, F.L., et al.: Preparation of concrete and study on planting experiment for plants-growing. Beton Chinese Edition-Ready-mixed Concr. **5**, 56–59 (2015)
16. Zhu, J., Xue, D.J., Liu, D., et al.: Construction technology of plant ecological concrete for slope protection. Constr. Technol. **43**(18), 109–112 (2014)
17. Horiguchi, I., Mimura, Y., Monteiro, P.J.M.: Plant-growing performance of pervious concrete containing crushed oyster shell aggregate. Clean. Mater. **2**, 100027 (2021)
18. Jiang, C.B., Liu, Y.Z., Xiang, T.S., et al.: Studies on the suitable of porous concrete for planting. Concrete **5**, 152–156 (2014)
19. Li, Q.G.: Study on the water holding performance and water-supply measures for Eco-concrete. Nanjing Hydraulic Research Institute, Nanjing (2007)
20. Zhou, X.L., Zhang, S.: Preparation and application of greening concrete. Coal Ash **2**, 42–44 (2010)
21. Li, Z.Q., Xiang, Y., Li, J.M., et al.: Properties of porous Eco-concrete for plant-growing. Build. Energy Effic. **4**, 50–53 (2012)
22. Hu, Y.Y., Hu, C.M., Xie, L., et al.: Effect of void status in eco-concrete for planting on plant growth. J. South China Univ. Technol. (Nat. Sci. Ed.) **34**(12), 5–9 (2006)
23. Zhang, W.L., Yuan, Z.J., Li, D.Q., et al.: Mechanical and vegetation performance of porous concrete with recycled aggregate in riparian buffer area. J. Clean. Prod. **332**, 130015 (2022)
24. Quan, H.Z.: Coexistence of porous ecological concrete with plants. Bull. Chin. Ceram. Soc. **7**, 1985–1988 (2015)
25. Pathan, S.M., Aylmore, L.A., Colmer, T.D.: Properties of several fly ash materials in relation to use as soil amendments. J. Environ. Qual. **32**(2), 687–693 (2003)
26. Zhang, W.: Research on planting structure of planting concrete and formulating nutrition soil. Concrete **11**, 124–127 (2012)
27. Huang, J.P.: Research on preparation and properties of planting concrete of the slope protection adaptive to urban rivers in South China. South China University of Technology, Guangzhou (2011)

28. Xu, W.N., Ye, J.J., Zhou, M.G., et al.: Several problems of vegetation technology for protecting slopes using vegetation-growing concrete. Water Resour. Hydrop. Eng. **35**(10), 50–52 (2004)
29. Kim, H.H., Park, C.G.: Performance evaluation and field application of porous vegetation concrete made with by-product materials for ecological restoration projects. Sustainability **8**, 294 (2016)
30. Ganapathy, G.P., Alagu, A., Ramachandran, S., et al.: Effects of fly ash and silica fume on alkalinity, strength and planting characteristics of vegetation porous concrete. J. Market. Res. **24**, 5347–5360 (2023)

Impacts of Mitigation Scenarios on Nitrogen in Surface Runoff Under Drought-Flood Abrupt Alternation in the Huaibei Plain, China

Wuxia Bi[1,2](✉), Cheng Zhang[1,2], Zhaoyu Dong[3], and Dawei Zhang[1,2]

[1] State Key Laboratory of Simulation and Regulation of Water Cycle in River Basin, China Institute of Water Resources and Hydropower Research, Beijing 100038, China
{biwx,zhangc,zhangdw}@iwhr.com

[2] Research Center On Flood & Drought Disaster Prevention and Reduction of the Ministry of Water Resources, Beijing 100038, China

[3] Institute of Environment and Ecology, Tsinghua Shenzhen International Graduate School, Tsinghua University, Shenzhen 518055, China
dongzy24@mails.tsinghua.edu.cn

Abstract. Global compound extreme events have increased, especially drought-flood abrupt alternation (DFAA), one of the new types of compound extreme events, has occurred more and more frequently. The compound extreme events mainly have negative impacts on the environment and ecosystems. Huaibei Plain is prone to DFAA events, with the frequency of about every 3 to 4 years. According to our previous studies, the soil phosphorus loss in surface runoff increased facing to DFAA based on field experiments and numerical simulations. To adapt to climate change, this study set mitigation scenarios, including drought/flood level degradation scenarios, to explore their effects on soil nitrogen (N) loss in surface runoff. The results reveal that the N loss in surface runoff decreased by 2.9% to 3.6% in history under drought level degradation scenarios, and decreased by –2.6% to 10.4% in the future. Under flood level degradation scenarios, the highest value of N loss in surface runoff decreased by 1.5% in the future, while decreased by 28.4% in history. The findings could help quantitively evaluate the effects of mitigation measures on extreme events. Further, it would guide the development of targeted measures.

Keywords: Climate Change · Drought-Flood Abrupt Alternation · Huaibei Plain · Mitigation Measures · Soil Nitrogen Loss · Surface Runoff

1 Introduction

In recent years, compound (extreme) events have increased [1]. Of them, drought-flood abrupt alternation (DFAA) has occurred frequently, especially in China [2]. Previous studies have reported that DFAA events significantly affected the environment, grain yield, and so on [3–5]. At the same time, more and more attention has been paid to water quality, especially the nutrient pollutants, such as phosphorus, nitrogen (N), etc., which are related to human health [6, 7]. Thus, it is significant to explore the effects of compound extreme events on water quality indicators.

C. Weng (Ed.): WRE 2024, LNCE 634, pp. 517–523, 2025.
https://doi.org/10.1007/978-981-96-5095-8_42

This study searched the published articles in the core collection of the Web of Science, with keywords such as drought, flood, and N. There are 15,060 papers published until now. According to the research knowledge map, previous studies mainly focus on growth, response, yield, soil, water, management, etc. (Fig. 1). Thus, N has significant effects on soil, water, and plant growth. Several studies reported the eco-environmental effects of droughts and floods on the N cycle. For the effects of floods on N in runoff, Ghidey and Alberts found that floods could erode the soil and cause 5% of the soil N loss into the runoff. And with the increase in precipitation intensity, the replenishment effect became stronger [8]. Allen et al. reported that nitrous oxide emissions increased due to soil waterlogging [9]. Xiao et al. reported that during floods, a significant decrease can be found in the concentration of ammonia N and nitrate N in surface runoff [10]. Hong and Li found that in surface runoff, the concentrations of ammonia N and nitrate N decreased increased rainfall intensity [11]. Wang et al. found that the change of soluble N was mainly related to nitrate N during rainfall in purple soil hilly areas [12]. Yang et al. revealed that the N concentration was negatively affected by the amount, duration, and intensity during the rainfall-runoff process [13]. Chang et al. found that total N contents decreased nonlinearly with the decrease in precipitation intensity [14]. Kizuka et al. reported that NO_3-N was mainly supplied by flood waters [15]. In comparison, few studies were reported on the effects of DFAA for N, thus even less on mitigation measures on N loss in surface runoff under DFAA.

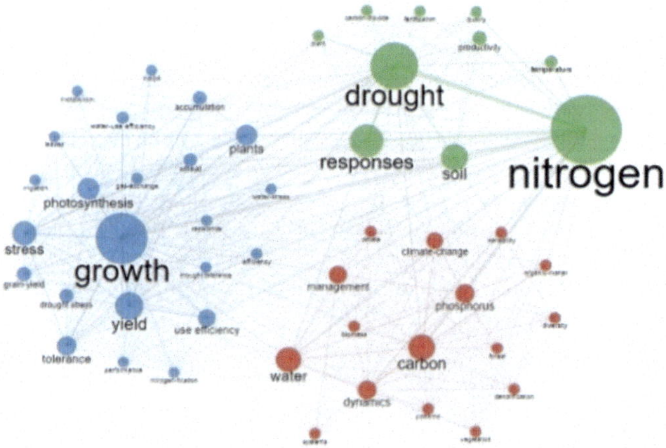

Fig. 1. Knowledge map of keywords.

The main objective is to evaluate the mitigation effects of drought/flood level degradation scenarios on DFAA. Huaibei Plain was chosen as the study area, which is an important grain base. The DFAA occurred every 3 to 4 years there, with over 85% in summer [16]. In addition, N is the main indicator of surface water pollution here. Therefore, this paper calculated the mitigation effects of DFAA on N loss in surface runoff in the summer maize planting areas. The study first determined the DFAA events in history (from 1964 to 2017) and in the future (from 2020 to 2050) then calculated the N loss in surface runoff in the Huaibei Plain based on field experiment; further quantitively estimated the N loss under mitigation scenarios (Fig. 2). The main novelty is that the mitigation effects for DFAA were quantitatively assessed, which would guide mitigating the negative effects of extreme events on the water environment for local government.

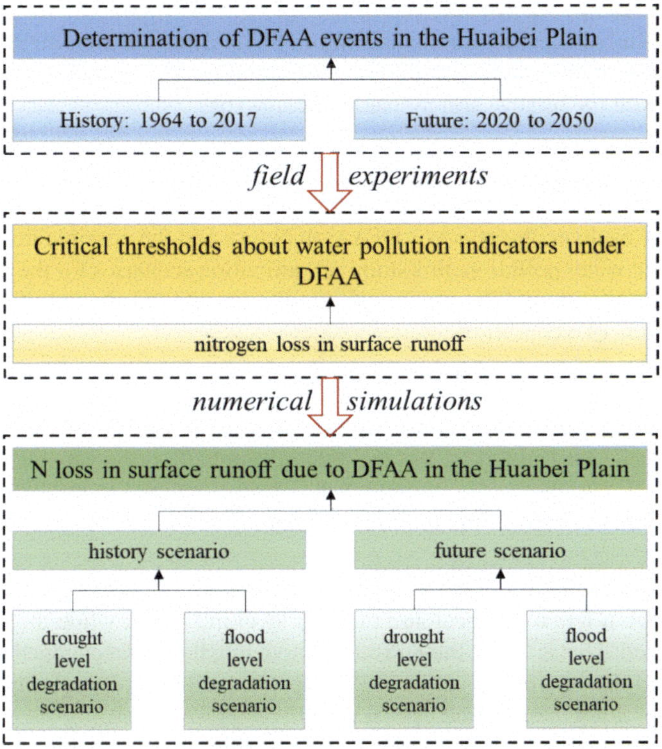

Fig. 2. Research outlines.

2 Materials and Methods

2.1 Data

The data contains three parts. The first part is the historical meteorological data. A dataset of measured daily data of 16 meteorological stations in or surrounding the Huaibei Plain was used (http://data.cma.cn/en). The second part is the future predicted data. Based on previous studies, this study chose the climate scenario HADGEM2-ES. It was developed by the Coupled Model Intercomparison Project (CMIP) 5, which reported that its fitting effect on precipitation data at different scales in the study area was relatively better than other scenarios [17]. The third part is the measured critical thresholds obtained by the field experiments with different DFAA scenarios. The soil N loss in surface runoff (water part) was 0.120%, 0.147%, 0.0474%, and 0.0510% in the light drought-light flood, moderate drought-light flood, light drought-moderate flood, and moderate drought-moderate flood treatments.

2.2 DFAA Determination Method

Our previous study proposed a multi-indicator DFAA determination method on a daily scale. The method considers the duration of drought, the duration and amount of precipitation, and the relative soil moisture, etc. [4, 16].

2.3 Mitigation Scenarios

To quantitively evaluate the effect of DFAA on N loss and the possible mitigation measures on surface water pollution, this study set mitigation scenarios for the DFAA events in history and the future. The numerical simulation was conducted. The mitigation scenarios include two conditions:

(1) Drought level degradation (DLD) scenario

The scenario refers to the decrease of drought level in DFAA, while the flood level keeps unchanged. That is to say, the moderate drought-X flood was changed to light drought-X flood, and the severe drought-X flood was changed to moderate drought-X flood ('X' represents 'light', 'moderate', or 'severe').

(2) Flood level degradation (FLD) scenario

The scenario refers to the decrease of flood level in DFAA, while the drought level keeps unchanged. That is to say, the X drought-moderate flood was changed to X drought-light flood, and the X drought-severe flood was changed to X drought-moderate flood ('X' represents 'light', 'moderate', or 'severe').

The above-mentioned mitigation scenarios were simulated in history and the future. The history scenarios calculated the regional N loss in surface runoff from 1964 to 2017 [17]. The future scenarios calculated the regional N loss from 2020 to 2050 with the climate scenario HADGEM2-ES.

3 Results and Discussion

3.1 Effects of DFAA on N Loss in Surface Runoff

The field experiments show that the average annual N loss in surface runoff was 0.0747% to 0.102% for the history scenario, with the highest center of N loss in the center of the Huaibei Plain (Fig. 3a). While in the future scenario, the N loss was 0.0607% to 0.134%. The highest N loss center was in the center and east of the Huaibei Plain (Fig. 3b). The spatial distribution of annual N loss due to DFAA changed slightly over time, so more attention should be paid to the highest N pollution center. At the same time, the highest value of N loss in surface runoff would increase by 31.4% in the future, which indicates that the surface water pollution caused by DFAA will increase in the next 30 years.

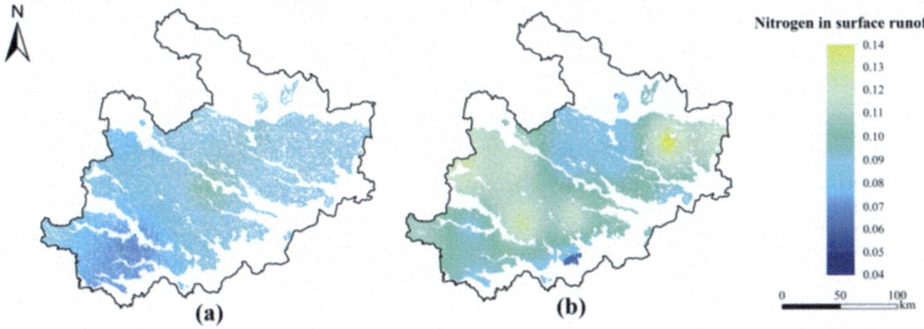

Fig. 3. The nitrogen loss in surface runoff due to DFAA in (a) history scenario; (b) future scenario.

The N loss into surface runoff due to DFAA would cause eutrophication and health risks, further threatens the living environment. In addition, DFAA increased soil nutrient loss into groundwater, air, etc. [16]. It implies that except for N input in surface runoff, DFAA may cause more serious water pollution. Our previous studies reveal that DFAA increased phosphorus concentration in surface runoff in the Huaibei Plain, which verified that the DFAA events have negative effects on the water environment [4, 16]. Thus, it is necessary to propose mitigation measures and estimate their effects, which benefit the local water environment management.

3.2 Effects of Mitigation Scenarios on N Loss in Surface Runoff

For the DLD scenarios, the mean annual N loss in surface runoff was 0.0720% to 0.0990% in the history scenario (Fig. 4a), and was 0.0623% to 0.129% in the future scenario (Fig. 4c). Compared with the history and future scenarios, the average annual N loss in surface runoff under DLD scenarios decreased by 2.9% to 3.6% from 1964 to 2017, and –2.6% to 10.4% from 2020 to 2050, respectively. The highest center of N loss in surface runoff did not shift under DLD scenarios.

For the FLD scenarios, the mean annual N loss in surface runoff was 0.122% to 0.131% (Fig. 4b) and 0.120% to 0.132% (Fig. 4d) in the history and future scenarios, respectively. The FLD scenario increased the highest value of N loss in surface runoff by about 28.4% compared to that in the history scenario, while it decreased by 1.5% compared to that in a future scenario. The lowest value of N loss in the FLD scenarios increased by 63.3% and 97.7% to that in the history and future scenarios, respectively. The highest center of N loss from 1964 to 2017 shifted from the center to the southwest of the study area and shifted from the center and east to the north from 2020 to 2050.

The highest value of N loss decreased in the DLD scenarios, while the lowest value of N loss increased slightly in the future scenario. The main possible reasons are that the N loss in surface runoff decreased from 0.147% to 0.120% and from 0.0510% to 0.0474% when moderate drought-light flood and moderate drought-moderate flood changed to light drought-light flood, respectively. Thus, the mitigation measures on DLD in the DFAA events would decrease the N loss in surface runoff, further mitigating the DFAA effects on the surface water environment.

The FLD scenarios increased the N loss in both history and future scenarios, while the highest value of N loss decreased slightly in the future scenario. The main reason is that the N loss into surface runoff was low in the DFAA with high flood levels, according to our field experiments. In a word, the degradation effects of these scenarios depend on the DFAA frequency in different levels and the N loss data obtained by field experiments. The detailed impact mechanism should be further studied.

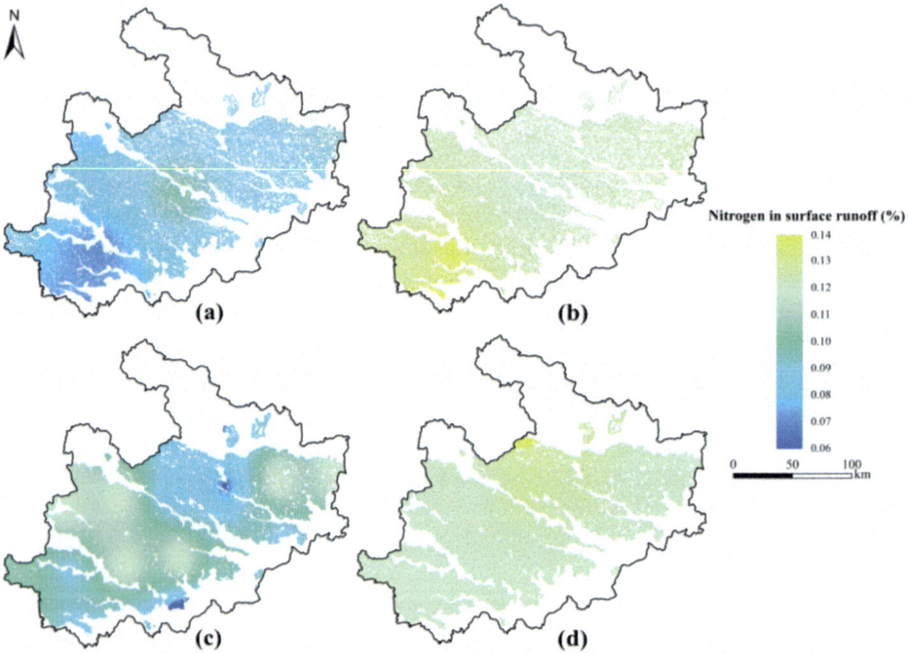

Fig. 4. The average annual nitrogen loss in surface runoff under (a), drought; (b) flood level degradation scenario in the history scenario; under (c), drought; (d), flood level degradation scenario in the future scenario.

4 Conclusions

This study set mitigation scenarios for DFAA in history and in the future. The main findings include i) Based on the field experiments, the highest value of N loss into surface runoff increased by 31.4% in the future. ii) Under the DLD scenario, the N loss in surface runoff decreased by 2.9% to 3.6%, and –2.6% to 10.4% in history (1964 to 2017) and in the future (2020 to 2050), respectively. iii) The N loss in surface runoff increased by 28.4% to 63.3%, and –1.5% to 97.7% in the history and future scenarios under the FLD scenarios, and iv) The highest center of N loss didn't shift under the DLD scenarios, while it shifted under the FLD scenarios. Further studies may consider more complicated conditions, which could guide the targeted countermeasures.

Acknowledgements. This study was supported by the National Natural Science Foundation of China (No. 52209043), the National Key Research and Development Program of China (No. 2022YFC3080300, 2021YFC2001404), the MWR Major Science & Technology Program (SKS-2022007), and the IWHR Research and Development Support Program (WH0145B022021).

References

1. IPCC. Climate change 2021: the physical science basis. Contribution of Working Group I to the Sixth Assessment Report of the Intergovernmental Panel on Climate Change. New York (2021)
2. Bi, W., et al.: Drought-flood abrupt alternation events over China. Sci. Total. Environ. **875**, 162529 (2023)
3. Shi, W., et al.: Drought-flood abrupt alternation dynamics and their potential driving forces in a changing environment. J. Hydrol. **597**, 126179 (2021)
4. Bi, W., et al.: Effects of drought-flood abrupt alternation on phosphorus in summer maize farmland systems. Geoderma **363**, 114147 (2020)
5. Bi, W., et al.: Response of summer maize growth to drought-flood abrupt alternation. Front. Earth Sci. **11**, 1086769 (2023)
6. Finlay, J.C., Small, G.E., Sterner, R.W.: Human influences on nitrogen removal in lakes. Science **342**(6155), 247–250 (2013)
7. Saturday, A., Kangume, S., Bamwerinde, W.: Content and dynamics of nutrients in the surface water of shallow Lake Mulehe in Kisoro District, South-western Uganda. Appl Water Sci **13**(7), 150 (2023)
8. Ghidey, F., Alberts, E.E.: Temporal and spatial patterns of nitrate in a Claypan soil. J. Environ. Qual. **28**(2), 584–594 (1999)
9. Allen, D.E., Kingston, G., Rennenberg, H., Dalal, R.C., Schmidt, S.: Effect of nitrogen fertilizer management and waterlogging on nitrous oxide emission from subtropical sugarcane soils. Agr Ecosyst Environ **136**(3–4), 209–217 (2010)
10. Xiao, M., Shuang-En, Y., Wang, Y., Huang, R.: Nitrogen and phosphorus changes and optimal drainage time of flooded paddy field based on environmental factors. Water Sci. Eng. **6**(2), 164–177 (2013)
11. Hong, L., Li, H.: Characteristics of nitrogen and phosphorus losses in surface runoff from farmland in a typical irrigation district in southern China. Geogr. Res. **30**(1), 115–124 (2011)
12. Wang, C., Zhao, P., Gao, M.: Characteristics of nitrogen and phosphorus transportation through runoff in a typical ecological-hydrological unit of hilly area of purple soil. J. Hydraul. Eng. **44**(6), 748–755 (2013)
13. Yang, F., Jiang, Y., Wang, C., Huang, X., Wu, Z., Chen, L.: Characteristics of nitrogen and phosphorus losses in Longhong Ravine Basin of westlake in rainstorm runoff. Environ. Sci. **37**(1), 141–147 (2016)
14. Chang, D., Li, S., Lai, Z.: Effects of extreme precipitation intensity and duration on the runoff and nutrient yields. J. Hydrol. **626**, 130281 (2023)
15. Kizuka, T., Mikami, H., Kameyama, S., Ono, S., Suzuki, H.: Hydrological environment affects the nutrient retention and runoff function of naturally re-wetted agricultural peatland in lowland river floodplain. Sci. Total. Environ. **857**, 159483 (2023)
16. Bi, W., et al.: Soil phosphorus loss increases under drought-flood abrupt alternation in summer maize planting area. Agric. Water Manag. **262**, 107426 (2022)
17. Gao, C., Zhang, Z., Chen, S., Liu, Q.: The high-resolution simulation of climate change model under RCP4.5 scenarios in the Huaihe River Basin. Geograph. Res. **33**, 467–477 (2014)

A Low-Power Seawater Nitrate Profile Measurement Method Based on Ultraviolet Spectroscopy

Yan Liu[1,2], Guoliang He[1], Ran Ma[1,2](✉), Yang Wang[1,2], Haikuan Ma[1,2], Dongzhi Chu[1,2], Xiangfeng Kong[1,2], and Qian Shi[1,2]

[1] Institute of Oceanographic Instrumentation, Qilu University of Technology (Shandong Academy of Sciences), Qingdao 266061, Shandong, P.R. China
mr_sdioi@126.com, yangwangphy@139.com, kongxiangfeng@qlu.edu.cn
[2] Laoshan Laboratory, Qingdao 266237, Shandong, P.R. China

Abstract. The deep-sea environment, an underexplored and mysterious realm, is not only home to exotic creatures, but also key to studying Earth's history, climate change, and marine ecosystems. Its extreme conditions pose challenges for scientific investigations, and the distribution and transformation of nitrates are particularly important for understanding ocean productivity and global climate change. To this end, a low-power seawater nitrate profile measurement method based on ultraviolet spectroscopy technology is proposed for the detection of nitrate concentration in the marine environment. The sensor directly measures the characteristic absorption peaks of nitrate by ultraviolet spectroscopy, avoiding the dependence of traditional chemical analysis methods on reagents. The hardware design adopts a series of low-power modules, such as a low-power STM32L1 microcontroller and ultraviolet LED light source, to ensure the stability and low energy consumption of the sensor in long-term monitoring of the deep sea. In addition, the sensor integrates temperature compensation and organic matter compensation algorithms, which can effectively reduce the influence of ambient temperature fluctuations and dissolved organic matter in seawater on measurement accuracy. At the same time, the precision and recovery of the sensor experimented, the precision of the sensor was measured many times at the nitrate concentration of 500ug/L, the RSD was only 4.52%, the recovery rate of the 15 spike experimental samples prepared by the standard preparation of wheat island seawater was between 90%~110%, and the recovery rate of the 15 samples was 109.24%, and the lowest was 97.73%, which were within the requirements of the national standard. Nitrate sensors provide highly accurate nitrate concentration measurements at different depths and environmental conditions, in good agreement with traditional chemical analysis data. Its good power consumption and profiling function provide reliable data support for the study of deep-sea nitrogen cycle and carbon sink, which is of great significance for global climate change prediction.

Keywords: Deep-sea Profile · UV Spectroscopy · Nitrate Sensor · Low-power Consumption

C. Weng (Ed.): WRE 2024, LNCE 634, pp. 524–540, 2025.
https://doi.org/10.1007/978-981-96-5095-8_43

1 Introduction

The deep sea is one of the least affected areas of the planet by human activities and is an important part of the global carbon cycle, which has an important impact on the global climate. As an important part of the nitrogen cycle, the concentration of nitrate is closely related to deep-sea carbon sinks [1]. Since deep-sea ecosystems are sensitive to external changes, monitoring deep-sea nitrate concentrations is helpful to study the role of deep-sea in global climate change and predict the impact of climate change on marine ecosystems. Nitrate concentration measurements can be made with high-resolution data obtained by spectral sensing technology, which can reveal more nuanced dynamics in these processes and facilitate a deeper understanding of deep-sea science [2].

Common methods for measuring nitrate content include wet chemical analysis [3], ion-selective electrode method [4], and ultraviolet absorption spectroscopy [5].

The chemical analysis method is cumbersome, complex in structure, long in measurement time. It requires self-filling to prepare cadmium columns, which makes the reagents highly toxic and has the problem of secondary contamination of chemical reagents [6]. Due to interfering substances, the electrodes need to be calibrated frequently. These problems also make the application of the ion-selective electrode method in seawater relatively rare [7]. Ultraviolet spectroscopy has a wide dynamic range and can cover nitrate detection from low to high concentrations, adapting to the needs of different marine environments. The ultraviolet method uses the strong absorption characteristics of nitrate in the ultraviolet band of 200 nm–400 nm to correct the influence of other interfering substances in the water body by measuring the absorbance of the sample at different wavelengths to obtain the concentration of nitrate [8]. The significant advantage of this method is that compared with wet chemical analysis, ultraviolet spectroscopy can complete the analysis in a shorter time, without using chemical reagents, direct optical measurement, fast detection speed, and avoid secondary pollution of the marine environment. The measurement process is simple, fast, and accurate and is particularly suitable for in-situ measurement of seawater nitrate, which can be used for long-term automatic detection unattended [9].

Sensors in deep-sea environments are often battery-powered, so low-power design is important. The low-power design of spectral nitrate sensors allows for smaller, lighter sensors, increasing the portability of the device, making it simpler, more efficient, and less costly for users to carry, deploy, and recycle sensors [10]. This design not only significantly extends the operating time of the sensor but also reduces the need for frequent battery replacement or maintenance, thereby reducing the high cost of deployment and recycling. The low-power design can reduce power problems such as overheating, reduce the frequency of maintenance, improve the overall reliability and durability of the equipment, and further reduce the operating cost. Long-term, continuous nitrate monitoring is essential for studying changes in the marine environment and ecosystem health [11]. The low-power design ensures that the sensor can continuously monitor changes in the deep-sea environment with limited resources, providing a stable stream of data that facilitates long-term trend analysis and scientific research.

Hach Corporation in the United States developed the NITRATAXsc online nitrate analyzer based on dual-wavelength detection. The light source on the analyzer's probe emits 210 nm ultraviolet light, which is partially absorbed by the sample flowing in the slit and reaches the detector on the other side of the probe to measure the absorbance value to calculate the nitrate concentration. The main application scenarios of this product are freshwater fields such as drinking water, surface water, and sewage treatment. Considering that there may be turbidity interference in the actual detection that may affect the detection accuracy, the analyzer uses a dual-beam measurement system to simultaneously measure the absorbance at 350 nm to compensate for turbidity, with a measurement accuracy of ±3% [12].

The ISUS nitrate sensor, developed by Kenneth S. Johnson et al. [13], can operate at depths of up to 2 km and is suitable for long-term monitoring of nitrate, disulfide, and bromide concentrations. The study found that in areas with water depths greater than 400 m, changes in seawater temperature could cause nitrate measurements to deviate by about 10% while also affecting the estimation of salinity by measuring bromide concentrations [14].

SUNA (Submersible UV Nitrate Sensor) is a shallow water sensor developed by Satlantic based on ISUS technology. Although its structure and calibration algorithms are like those of ISUS, the main difference of the SUNA is the design of the optical measuring cell. ISUS uses a reflection probe design, while SUNA uses a straight-through measuring cell primarily used in nearshore and freshwater environments [15].

The submersible multiparameter ultraviolet spectrophotometer (spectro::lyser) is a device developed by the Austria company S::CAN to measure COD, biological oxygen demand (BOD), total organic carbon (TOC), nitrate, benzene, toluene, xylene and phenol in wastewater to ultrapure water [16–18]. The deep-sea in-situ UV spectroscopy sensor OPUS, developed by the German company Trios, is a multi-parameter measuring instrument capable of measuring nitrates, BOD, COD, TOC, humic acids, and certain phenols [19]. The sensor uses a pass-through cell coated with a "nano-coating" on the optical window to prevent biofouling. Meyer. D et al. [20] demonstrated the sensor's ability to resolve nitrate profiles by deploying OPUS for the first time in coastal waters to monitor nitrate and hydrogen sulfide levels.

2 Low-Power Hardware Design

Low-power design at the hardware level is crucial, as power consumption occurs if the electronic components are powered on and the current is flowing. The spectral nitrate sensor is composed of the main control unit single-chip microcomputer and the peripheral functional circuit of the data operation module. The peripheral functional circuit is composed of active devices and passive components (see Fig. 1). The more electronic components in the system, the greater the power consumption, so while doing circuit design, it is necessary to choose low-power devices and functional composite devices.

2.1 The Low-Power Design of the Main Control Microcontroller and Communication Module

The microcontroller is the core of the nitrate sensor. Its performance will directly determine the working ability of the sensor; in addition, the power consumption it generates is an important source of system power consumption, so choosing a suitable low-power

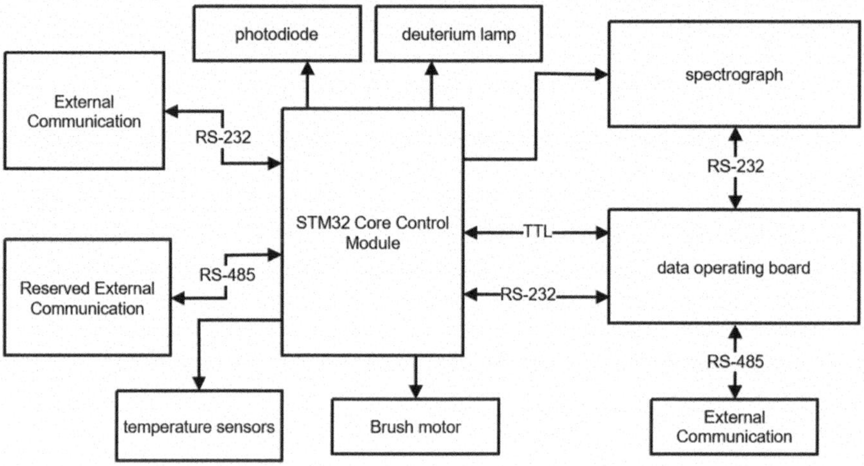

Fig. 1. Block diagram of the sensor hardware.

microcontroller will help improve the efficiency of the system, and greatly reduce the power consumption of the system. Considering that in addition to low power consumption, nitrate sensor also has a variety of serial communication needs such as debugging, sensor information collection, and communication system information transmission, the main control of the spectroscopic nitrate sensor uses STM32L1, which not only has ultra-low power consumption but also has a wealth of peripherals, thereby reducing the design of peripheral circuits. As a result, it is suitable for battery-powered applications that require very low power consumption and long-term operation, such as the deep-sea planning scenario described in this article.

There are two ways for the main control to communicate with the host computer, namely RS485 communication and RS232 communication. RS232 is a standard for serial communication, and the circuit uses a serial port MAX3232 chip. It can convert the TTL level of the main controller output to RS232 level, so that the microcontroller master can communicate with the host computer for 232 communication. MAX3232 needs low power consumption; its supply current is usually as low as 0.3 mA in the idle state, which makes it very energy-efficient when not transmitting data, and it can automatically enter a low-power state when no valid RS232 signal is detected, which can significantly reduce power consumption. In this sensor, the overall power consumption of the system can be further reduced when it enters stop mode. MAX3485 is a low-power, half-duplex RS-485 transceiver with a very low quiescent current of only 200 μA, which matches the sensor's low power consumption needs in the deep sea. The chip has an enable pin (EN) and a receiver enable pin (RE). With the combination of these two pins, the chip can be put into low-power shutdown mode, and the current consumption can be reduced to 1 nA. The DE and RE pins can be turned off before the host controller goes into stop mode, further reducing power consumption.

2.2 Design of High-Efficiency and Low-Power Data Computing Module

TTL level signal connects the main sensor controller to the data operation module. Equipped with a 600 MHz Cortex-M7 core and a floating-point unit (FPU), it can perform complex mathematical operations and data processing tasks at extremely high speeds. Its high-speed processing and multitasking capabilities ensure real-time and low latency of the system, which is particularly important for real-time monitoring and feedback from spectral nitrate sensors. With 1MB of RAM and 8MB of flash memory, the data computing module can store and processing large amounts of spectral data, ensuring data integrity and processing continuity when processing high-resolution spectral data. Despite its robust performance, it operates with relatively low power consumption, which is necessary for spectral nitrate sensors that require long operating periods. The main controller will directly cut off the power supply of the data computing module when the sensor is not working to achieve low-power operation during operation and no power consumption when it is in stop mode.

2.3 UV Optical Path Design

The UV light in the sensor is provided by a Miniature UV-Vis Light Source FiberLight module, whose UV light source emitting source is a deuterium lamp. This miniature deuterium lamp is powered and switched on and off by a master controller. It has an emission spectrum ranging from 185 nm to 700 nm. It is suitable for the detection of substances with absorption characteristics in the UV region, such as nitrate. The UV light source is used with a decrease in light intensity. The light source of this sensor can maintain a consistent light output to a certain extent, and the power supply of the micro deuterium lamp will be disconnected when not necessary. It not only saves energy consumption but also prolongs the life of the deuterium lamp. The light emitted by the UV light source is measured and captured by the Ocean ST spectrometer after transmitting the seawater, which not only has the characteristics of high sensitivity and high resolution but also can accurately and quickly detect the absorption peak of nitrate in the ultraviolet spectrum range, and then directly transmit the collected spectral data to the data computing module for direct data processing through an RS232 communication. Their combination is small in addition to meeting the requirements for accurate measurements and low power consumption, and miniaturized designs are especially important for deep-sea environments, where equipment needs to accommodate tight spaces and complex installation conditions.

2.4 Passive Component Optimization and Low-Power Strategies

Passive components, such as resistors, capacitors, and inductors, are the most common and widely used electronic components in circuits, and they are also the main source of energy consumption in circuits. Resistors are energy-dissipating components that consume electrical energy as long as an electric current passes through them. Capacitors and inductors are energy storage components that, due to their characteristics, also generate power consumption during charging and discharging. In circuit design, these passive components should be avoided as much as possible. Resistors are commonly used for current-limiting, pull-up, and pull-down resistors. If the requirements are met, the current can be reduced by increasing its resistance value, thereby reducing the power consumption.

2.5 Software and Data Processing Process of Nitrate Sensor

There is an inseparable relationship between the hardware and embedded parts, which are both important parts of the nitrate sensor. The software design of the sensor is mainly divided into two parts, one part is the main control program design, and the other part is the algorithm data processing of the data operation module. The main controller is mainly responsible for controlling a series of different working modes, including communication, power management, and working together between various peripherals to measure nitrate, and the data computing module is mainly responsible for processing the data measured by the spectrometer and calculating the nitrate concentration value and returning it to the main controller.

Two communication modes between the sensor and the host computer are communicated through RS485 serial port and RS232 serial port, respectively. The sensor supports various working modes, which can detect and transmit back data at regular intervals or can be actively requested by the host computer for nitrate detection. The flow chart of the main control software program for the nitrate sensor is shown below. After turning on the power supply, the main controller turns on the power supply of the data operation module, then controls the temperature sensor for temperature detection. The sensor is equipped with two temperature sensors, each is measured ten times and averaged. When calculating nitrate concentrations, these two temperature values are transmitted to the data operation module through the serial port for deep-sea temperature compensation. The density of the solution changes with the temperature of the seawater, it also changes the refractive index of the solution. Next, the main controller turns on the deuterium light source and waits for the light source to stabilize. Then, the spectrometer is turned on to measure the spectral intensity and transmit the data to the data calculation module. The data operation module processes the spectral data compensates and calculates the temperature data transmitted by the main controller, and finally obtains the nitrate concentration. The data operation module transmits the nitrate concentration data to the main control board. Then, the main control board communicates with the host computer to transmit the nitrate concentration data to the host computer (see Fig. 2). In addition, the sensor has a data storage function, which stores spectral data and nitrate concentration information each time a nitrate concentration is detected to study changes in nitrate concentration in the deep-sea environment.

3 Nitrate Sensor Spectral Analysis Algorithm

3.1 Temperature Compensation Algorithm

When the data operation module receives the spectral data information from the spectrometer, it converts the spectral data into a second-order spectral derivative sequence. Then, according to the temperature information transmitted by the main control, the spectral compensation series obtained by the following second-order spectral compensation

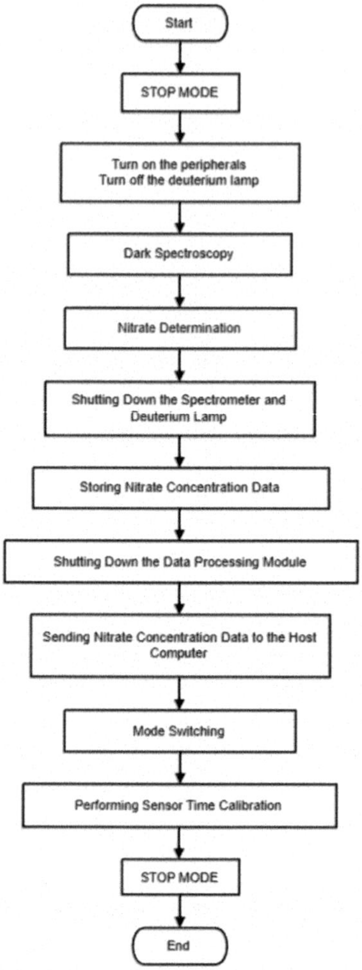

Fig. 2. Flowchart of sensor operation.

formula is substituted:

$$\begin{bmatrix} cp_1 \\ cp_2 \\ \vdots \\ cp_m \end{bmatrix} = \left(j_0 + j_1T + j_2T^2\right)S_1 + \left(k_0 + k_1T + k_2T^2\right)S_2 \tag{1}$$

where *cp* is the second derivative value of the temperature compensation spectrum, and the subscript shows the number of spectral data. *J, K* are the temperature fitting coefficients. *T* is the current temperature, and the value is in degrees Celsius. *S* stands for Fingerprint Spectrum.

Finally, the temperature-compensated second-order derivative spectrum of the seawater sample can be obtained by subtracting the corresponding elements in the spectral temperature compensation series. In this way, the temperature compensation of the initial data of nitrate spectra is completed, and the second derivative spectra of seawater samples after temperature compensation can continue to be used for nitrate spectral modeling and concentration measurement, but the temperature fitting coefficient and fingerprint spectra should be modeled and calibrated before practical application.

Under laboratory conditions, the water sample temperature is T_1, the UV absorption spectra of standard seawater and pure water samples are measured using a combination of a microspectrometer and a deuterium lamp. The standard seawater UV absorbance curve is calculated according to the Beer–Lambert law using pure water as a reference, and the obtained absorbance values are recorded as a series in order of wavelength from small to large. The second derivative of the absorbance curve is then calculated and recorded as a series of second derivatives. In this step, the UV absorption spectra of standard seawater and pure water samples at different temperatures T_2, T_3, T_4, T_5, and T_6 are repeatedly measured, and multiple sets of second derivative sequences are obtained and arranged as matrices:

$$MD_{m\times 6} = \begin{bmatrix} sd_1^{T1} & sd_1^{T2} & \cdots & sd_1^{T6} \\ sd_2^{T1} & sd_2^{T2} & \cdots & sd_2^{T6} \\ \vdots & \vdots & \ddots & \vdots \\ sd_m^{T1} & sd_m^{T2} & \cdots & sd_m^{T6} \end{bmatrix} \tag{2}$$

The T_n corner mark is only used to describe the arrangement of data and does not participate in numerical operations.

The second derivative matrix of the spectrum is decomposed by singular values as follows:

$$MD_{m\times 6} = U_{m\times m} S_{m\times 6} V_{6\times 6}^T \tag{3}$$

The U-matrix is the basic spectrum, the S-matrix is the diagonal array of singular values, and the V-matrix is the intensity change of each elementary spectrum.

After the singular value decomposition operation, the value of each element in the U-matrix, including m row and m column, can be expressed as:

$$U_{m\times m} = \begin{bmatrix} u_{1,1} & u_{1,2} & \cdots & u_{1,m} \\ u_{2,1} & u_{2,2} & \cdots & u_{2,m} \\ \vdots & \vdots & \ddots & \vdots \\ u_{m,1} & u_{m,2} & \cdots & u_{m,m} \end{bmatrix} \tag{4}$$

The first two columns of the U-matrix, S_1 and S_2, are extracted as the two parameters of the second-order spectral compensation formula, respectively:

$$S_1 = \begin{bmatrix} u_{1,1} \\ u_{2,1} \\ \vdots \\ u_{m,1} \end{bmatrix} \tag{5}$$

$$S_2 = \begin{bmatrix} u_{1,2} \\ u_{2,2} \\ \vdots \\ u_{m,2} \end{bmatrix} \tag{6}$$

The linear equation is established by using the linear combination method of S_1 and S_2, and the second-order spectral derivative sequence is reconstructed, which can be expressed as:

$$b_1 \begin{bmatrix} u_{1,1} \\ u_{2,1} \\ \vdots \\ u_{m,1} \end{bmatrix} + b_2 \begin{bmatrix} u_{1,2} \\ u_{2,2} \\ \vdots \\ u_{m,2} \end{bmatrix} = \begin{bmatrix} sd_1^T \\ sd_2^T \\ \vdots \\ sd_m^T \end{bmatrix} \tag{7}$$

where, b_1, b_2 represent the intensity parameters of the fingerprint spectrum and are temperature-dependent.

At this time, b_1, b_2 are unknown, and the second derivative series at the temperature of $T_1, T_2, \ldots\ldots, T_6$ are brought into the superdefinite equation, and the parameter values of b_1, b_2 are solved respectively. For each temperature T, one equation can be established, and one set of b_1, b_2 parameters can be obtained after solving, and six sets of b_1, b_2 parameters can be obtained corresponding to the temperatures $T_1, T_2, \ldots\ldots, T_6$. Then, the polynomial function is established with b_1, b_2 and the temperature value parameter:

$$\mathrm{b_1(T) = j_0 + j_1T + j_2T^2} \tag{8}$$

$$\mathrm{b_2(T) = k_0 + k_1T + k_2T^2} \tag{9}$$

The coefficients j_0, j_1, j_2 and k_0, k_1 *and* k_2 can be obtained by substituting the parameters of b_1, b_2 and the corresponding temperature value parameters into the above two polynomial functions, and fitting by the least squares method. Spectral temperature compensation can be performed with coefficients, temperature fitting coefficients, and fingerprint spectra.

3.2 Organic Compensation and Nitrate Concentration Calculation Algorithm

The data calculation module processes the second derivative of the spectral data transmitted by the spectrometer and then uses the temperature compensation method to compensate for the temperature. Then, the concentration can be calculated. Firstly, the temperature-compensated second derivative sequence of the spectrum is substituted into the following spectral reconstruction equation, and the parameters a_3-a_5 can be obtained by fitting by the least squares method.

$$\mathrm{a_0H_0 + a_1H_1 + a_2H_2 + a_3H_3 + a_4H_4 + a_5H_5} = \begin{bmatrix} sd_1 \\ sd_2 \\ \vdots \\ sd_m \end{bmatrix} \tag{10}$$

Among them, a_0-a_5 is the fitting parameter, H is the nitrate spectral fingerprint, and sd is the second-order guide of the spectrum.

Substituting the fitting parameters a_3-a_5 into the following multiple regression equation is obtained to obtain the nitrate concentration value C of the seawater.

$$C = k_0 + k_3 a_3 + k_4 a_4 + k_5 a_5 \tag{11}$$

where k is the calibration coefficient, and C is the nitrate concentration value, the unit is mg/L.

The calculation of the nitrate concentration still needs to be calibrated before application. It is necessary to construct various nitrate-artificial seawater solutions and second-derivative spectra of artificial seawater at different concentrations under laboratory conditions in the same way as temperature compensation. In addition, the second derivative spectra of a variety of chlorobromide-artificial seawater solutions with different concentrations should be constructed. Because the high concentration of chlorine and bromine plasma in seawater also has a high absorption peak in the wavelength range of 200 to 400 nm, which interferes with nitrate measurements, it is necessary to exclude the influence of interference factors.

The second derivative spectra of nitrate-artificial seawater solution and chlorobromide-artificial seawater solution with different concentrations can be arranged into two matrices, and the singular value decomposition is carried out according to the same steps in the nitrate temperature compensation method to extract the number series in the U matrix. The difference is that the first column of the U matrix extraction of chlorobromide-artificial seawater solution is used as spectral fingerprint H_2, and the first three columns of the U matrix extraction of nitrate-artificial seawater solution are used as spectral fingerprints H_3, H_4, H_5.

So far, the linear combination method of the above spectral fingerprints is used to establish the spectral reconstruction equation, which can be expressed as:

$$a_0 H_0 + a_1 H_1 + a_2 H_2 + a_3 H_3 + a_4 H_4 + a_5 H_5 = \begin{bmatrix} sd_1 \\ sd_2 \\ \vdots \\ sd_m \end{bmatrix} \tag{12}$$

Among them, H_0 is the constant spectral fingerprint, the series with all values of 1, H_1 is the second derivative spectrum of artificial seawater, and H_0 is the constant spectral fingerprint.

The second derivative spectra of multiple nitrate-artificial seawater solutions with different concentrations from the previous ones were constructed, and the multiple second derivative spectra were substituted into the spectral reconstruction equation, and the fitting parameters a_0-a_5 could be solved by the least squares method. The a_3-a_5 parameters are correlated with the nitrate spectral fingerprint intensity, and multiple regression equations are constructed according to the a_3, a_4, a_5, and the corresponding known nitrate concentrations:

$$C = k_0 + k_3 a_3 + k_4 a_4 + k_5 a_5 \tag{13}$$

where k is the calibration coefficient.

The calibration coefficients can be obtained by solving the multiple regression equations. Recalibration is performed before each instrument installation to ensure instrument accuracy. With these parameters, the program can be imported into the data calculation module, and when the spectral data from the spectrometer is received, the nitrate concentration is calculated.

4 Experimental Process and Data

4.1 Sensor Precision Experiment

After the hardware construction, software program construction, initial calibration of the sensor, and compensation of interference factors are completed, the sensor performance test is further carried out to determine the precision, recovery rate, and detection limit of the sensor, and to verify the accuracy of the measurement data of the nitrate sensor.

High precision ensures data reliability, reduces measurement errors, and enhances comparability. First, high precision indicates that the sensor can produce consistent results when measured repeatedly, which is critical for data reliability in scientific research and industrial applications, ensuring that data in an experiment or during production is not affected by random errors. Secondly, high precision helps reduce measurement errors, ensures the stability and repeatability of the nitrate concentration value, and makes measurement results more accurate and reliable. Finally, when comparing data at different times or between different laboratories, precision can ensure data comparability and consistency so that all parties can effectively compare and analyze based on consistent standards, thereby improving the overall value and application effect of the data.

Nitrate mixtures at concentrations of 100 μg/L, 500 μg/L, and 1500 μg/L were prepared, and five measurements per sample were taken using a sensor. The RSDs were 16.63% for the 100 μg/L nitrate mixture, 4.52% for the 500 μg/L nitrate mixture, and 3.33% for the 1500 μg/L nitrate mixture (Table 1).

The sensor performs particularly well in high (1500 μg/L) and medium (500 μg/L) nitrate mixtures. In high-concentration measurements, its RSD is only 3.33%, showing extremely high measurement stability and reliability, which makes it able to provide accurate data in complex environments, fully demonstrating its great value as an analytical tool. In the medium concentration range, the sensor also exhibits good sensitivity, with an RSD of 4.52% at 500 μg/L, which is close to the national standard, further verifying its applicability. This excellent performance not only increases the confidence of the experimental results but also significantly improves the efficiency of data analysis and reduces the potential for error. Therefore, the excellent performance of the sensor in the measurement of high and medium concentrations makes it a reliable choice in environmental monitoring and water quality analysis and has a wide range of application prospects and promotion value. In short, this sensor deserves more attention and application in related fields due to its superior performance and efficient data processing capabilities.

Table 1. Experimental data for RSD.

Number	Concentration of The Sample (μg/L)	Nitrate Detection Concentration (μg/L)	RSD
1	100	80.23	16.63%
2	100	115.67	
3	100	98.51	
4	100	92.18	
5	100	107.93	
6	500	465.87	4.52%
7	500	487.42	
8	500	511.68	
9	500	522.33	
10	500	534.99	
11	1500	1476.23	3.33%
12	1500	1550.45	
13	1500	1493.88	
14	1500	1582.12	
15	1500	1524.67	

4.2 Sensor Recovery Rate Experiment

Recovery testing plays a key role in verifying the measurement accuracy of sensors, compensating for interferences, and standardizing applications. First, a recovery test allows you to evaluate whether the sensor can accurately measure the amount of nitrate added to the sample. If the recovery rate of the test result is close to 100%, it indicates that the sensor accurately reflects the actual measured value, thus guaranteeing the accuracy of the measurement. Second, recovery testing helps to detect and correct for possible interferences, ensuring that the sensor response matches the actual concentration of the analyte of interest, increasing the reliability of the data. Finally, by ensuring consistent measurement results across different sample matrices, recovery testing enhances the standardization and universal applicability of the sensor, allowing it to provide consistent measurement performance across a wide range of environments and conditions.

The recovery test results should meet the requirements of GB 17378.2-2007 Marine Monitoring Code Part 2: Data Processing and Analysis Quality Control. The Marine Monitoring Code stipulates that the allowable recovery rate of natural samples is 60%~110% when the concentration of the measured substance in seawater is less than 100μg/L, 80%~110% when the concentration of the measured substance is between 100μg/L~1000μg/L, and 90%~110% when the concentration of the measured substance exceeds 1000μg/L.

As can be seen from the charts, the recoveries of the 15 spiked experimental samples prepared for the seawater of Wheat Island were all between 90%~110%. The recovery

Table 2. Experimental data on seawater recovery rate in Wheat Island.

Number	Spiked Concentration (μg/L)	Nitrate Detection concentration(μg/L)	Recovery Rate
1	0	7.74	-
2	25	27.31	109.24%
3	50	53.24	106.48%
4	75	79.27	105.69%
5	100	104.82	104.82%
6	280	278.45	99.16%
7	460	452.78	98.43%
8	640	623.19	97.73%
9	820	844.57	103.0%
10	1000	1012.34	101.2%
11	1400	1389.76	99.26
12	1800	1799.45	99.97%
13	2200	2245.32	102.05%
14	2600	2601.47	100.06%
15	3000	3064.12	102.14%

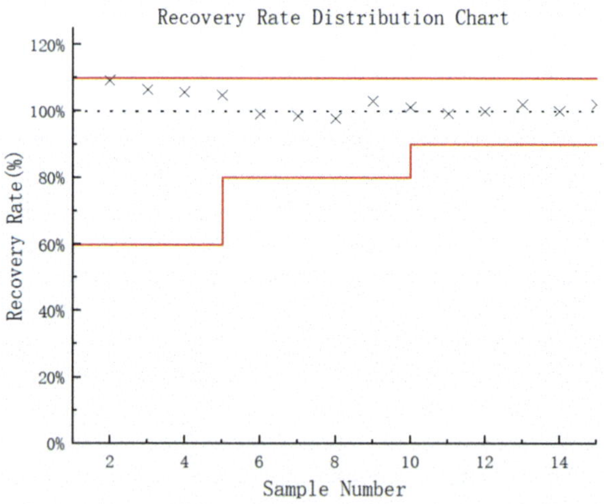

Fig. 3. Map of seawater recovery rate in Wheat Island.

rates of the 15 samples were the highest and the lowest was 109.24% and 97.73%, which met the relevant requirements of the Marine Monitoring Code (see Table 2 and Fig. 3). This high recovery rate not only demonstrates the accuracy and reliability of the sensor in sample analysis but also reflects its adaptability to complex sample environments. This excellent recovery performance compared to national standards means that the sensor can effectively capture and quantify nitrate concentrations, reducing errors due to sample loss.

The performance of the nitrate sensor not only enhances the confidence of the experimental results but also improves the efficiency of the laboratory and ensures the quality of data in environmental monitoring and water quality analysis. This efficient recycling capability makes the sensor an indispensable tool in scientific and practical fields, supporting water quality monitoring strongly. Therefore, with its superior recovery rate and stable measurement performance, this sensor is worthy of promotion and use in various applications. It is ideal for improving the accuracy and reliability of environmental monitoring.

4.3 South China Sea Nitrate Planning Experiment

After evaluating the performance of the sensor, we installed and precisely calibrated the nitrate sensor in the South China Sea. At the beginning of the experiment, the sensor was slowly lowered to a depth of about 700 m to perform a vertical profile measurement of nitrate concentrations. During this time, the sensor continuously measures and records the nitrate concentration in the water. The main purpose of this decentralization process is to obtain nitrate concentration data at different depth levels of the water body to characterize nitrate distribution in the vertical direction. After the decommissioning, we gradually retract the sensor at the same speed, and during the ascent, the sensor again measures and records the concentration of nitrate in the water body. This bi-directional measurement provides both possible trends in nitrate concentrations during dips and rises and helps verify data consistency while helping us rule out data bias due to operational or instrumental errors. By comparing the data from the up-and-down process, we can analyze the trend of nitrate concentration between different depths and explore its potential spatial distribution pattern.

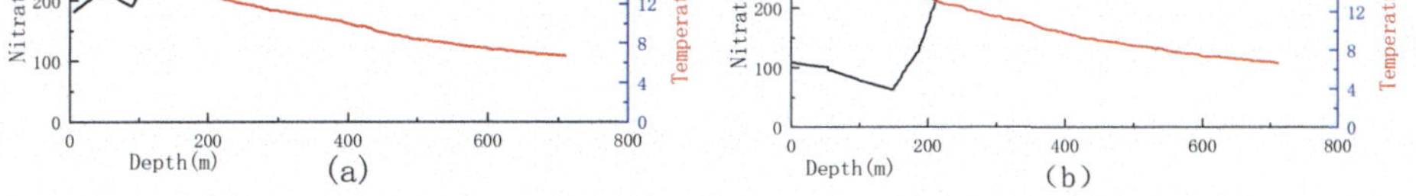

Fig. 4. Diagram of nitrate concentrations in the South China Sea.

As can be seen from the images, relatively low nitrate concentrations can be observed at the surface of the ocean during the gradual descent from sea level to a depth of 700 m.

As we explore deeper waters, nitrate concentrations show a gradual upward trend. This process is not instantaneous but rather a slow but steady increase in nitrate concentration with each meter of depth. When it reaches a depth of about 650 m, i.e. into the deep seawater, the nitrate concentration has reached a high level. As the image shows, from this depth, the rate of nitrate concentration growth slows down significantly, and it has remained relatively stable within a certain range, although it continues to advance to depth. This means that in a deep-sea environment below 650 m, the change in nitrate concentration with increasing depth becomes very small and almost negligible. This nitrate distribution pattern reflects the complex interplay of physical, chemical, and biological processes between water bodies at different ocean depths (see Fig. 4).

On the surface of the ocean, phytoplankton can grow and reproduce through photosynthesis due to abundant light, which consumes large amounts of nutrients such as nitrates, resulting in low nitrate concentrations in surface water [21]. However, with the increase of depth, the growth of phytoplankton is limited, and the sedimentation of dead organic matter in the surface layer into the deep sea and the release of nutrients such as nitrate through microbial decomposition led to a gradual increase in nitrate concentration in the deep sea [22]. When a depth of about 650 m is reached, the change in nitrate concentration tends to be flat due to the relatively closed deep-sea environment; the water exchange is less, and the replenishment and consumption of nitrate reach a state of equilibrium [23]. In addition, changes in temperature, pressure, and other factors can also affect the distribution of nitrate in the deep ocean, for example, nitrate concentrations are higher in deeper water bodies with lower temperatures because lower temperatures slow the rate of biodegradation and allow nitrate accumulation, while lower temperatures help dissolve more gases and substances, which indirectly affects nitrate concentrations [24]. These factors combine to create a unique nitrate distribution pattern in the deep ocean.

5 Conclusion

The application of low-power spectroscopic nitrate sensors in deep-sea environments is of great significance. Due to the complexity and extreme conditions of the deep-sea environment, traditional monitoring methods often face challenges in energy supply and equipment maintenance. The low-power spectroscopy nitrate sensor achieves efficient, continuous water quality monitoring through advanced spectroscopy, low-power electronics, and an optimized light source design while significantly reducing energy consumption, eliminating the need for frequent battery replacements or external power sources during long-term operation, especially in demanding applications such as the deep sea. The sensor not only extends the life of the equipment and reduces maintenance costs but also improves monitoring sustainability, providing scientists with a new tool for long-term, high-resolution chemical monitoring in extreme environments. Monitoring nitrate concentrations in the deep ocean can also help fill gaps in existing ocean data, reveal the distribution and dynamics of nitrate in marine waters, and gain insight into the impact of nitrate on marine ecosystems. By analyzing these data, we can not only promote a comprehensive understanding of the deep-sea environment and ecosystems, but also provide a scientific basis for marine environmental protection and resource management. These detailed experimental methods and data analysis strategies will provide important references for future marine scientific research and promote the development of overall scientific research.

Acknowledgement. This work was supported by the Major Innovation Project of Science and Education Integration Pilot Project of Qilu University of Technology (Shandong Academy of Sciences) (2023JBZ02), Major Innovation Project of Science and Education Integration Pilot Project of Qilu University of Technology (Shandong Academy of Sciences) (2023JBZ03), Science Education Industry Integration Pilot Major Innovation Project (2023HTZX01) of Qilu University of Technology (Shandong Academy of Sciences), Open Research Fund of Key Laboratory of Marine Ecological Monitoring and Restoration Technology, Ministry of Natural Resources (MEMRT202301).

References

1. Wankel, S., Buchwald, C., Ziebis, W., et al.: Nitrogen cycling in the deep sedimentary biosphere: nitrate isotopes in porewaters underlying the oligotrophic North Atlantic. Biogeosciences **12**, 7483–7502 (2015). https://doi.org/10.5194/BG-12-7483-2015
2. Nehir, M., Esposito, M., Begler, C., et al.: Improved calibration and data processing procedures of opus optical sensor for high-resolution in situ monitoring of nitrate in seawater. Front. Mar. Sci. **8**, 663800 (2021)
3. Yang, Z., Zhang, J., Zhao, J., et al.: A high-sensitivity lab-on-a-chip analyzer for online monitoring of nitrite and nitrate in seawater based on liquid waveguide capillary cells. Lab Chip **24**(22), 3528–3535 (2024)
4. Morawska, K., Malinowski, S., Krawczyk, J., et al.: Multi-walled carbon nanotubes and copper oxide nanoparticles composite used as transducer medias in nitrate ion-selective electrodes. J. Electrochem. Soc. **171**(8), 087511 (2024)
5. Dong, J., Tang, J., Wu, G., et al.: Effective correction of dissolved organic carbon interference in nitrate detection using ultraviolet spectroscopy combined with the equivalent concentration offset method. RSC Adv. **14**(8), 5370–5379 (2024)
6. Niu, W., Li, H., Guo, X., Chen, J., et al.: Field determination of nitrate in seawater using a novel on-line coppered cadmium column. Front. Mar. Sci. **7**(11), 1–8 (2023)
7. Mai, Z., Xiao, S., Zhang, W., Wang, K.: A Multi-ion interference decoupling model based on ion-selective electrode arrays. Anal. Methods **15**, 5038–5049 (2023)
8. Geng, Y., Zhang, Y., Yuan, D., Feng, X., et al.: Study on nitrate detection method in water using uv full-spectrum combined with PCA-BPNN. Proc. SPIE **12757**, 1275732–8 (2023). https://doi.org/10.1117/12.2690289
9. Wang, H., Ju, A., Wang, L.: Ultraviolet spectroscopic detection of nitrate and nitrite in seawater simultaneously based on partial least squares. Molecules **24**(23), 4361–4375 (2021)
10. Beaton, A., Schaap, A., Pascal, R., et al.: Lab-on-chip for in situ analysis of nutrients in the deep sea. ACS Sens. **6**(3), 808–815 (2022). https://doi.org/10.1021/acssensors.1c01685
11. Forrest, T., et al.: A submersible probe with in-line calibration and a symmetrical reference element for continuous direct nitrate concentration measurements. Environ. Sci. Process. Impacts **24**(9), 2099–2111 (2023)
12. Egli, P.J., et al.: Sustained, autonomous coastal nutrient observations aboard moorings and vertical profilers. Oceans 2009 Conference. Biloxi: IEEE, 1–9 (2009)
13. Johnson, K.S., Coletti, L.J., et al.: In situ ultraviolet spectrophotometry for high resolution and long-term monitoring of nitrate, bromide and bisulfide in the ocean. Deep-Sea Res. Part **I**(49), 1291–1305 (2002)

14. Sakamoto, C.M. et al.: Hourly in situ nitrate on a coastal mooring: A 15-year record and insights into new production. Oceanography **30**(4): 114–127 (2017)
15. Macintyre, G., et al.: ISUS/SUNA nitrate measurements in networked ocean observing systems. Proceedings of the Oceans 2009 Conference (pp. 1–7). IEEE, Biloxi, MS (2009)
16. Langergraber, G., Fleischmann, N., Hofstaedter, F., et al.: A multivariate calibration procedure for UV/VIS spectrometric quantification of organic matter and nitrate in wastewater. Water Sci. Technol. **47**(2), 63–71 (2003)
17. Rieger. L. et al.: Spectral in-situ analysis of NO_2, NO_3, COD, DOC and TSS in the effluent of a WWTP. Water Sci. Technol. **50**(11), 143–152 (2004)
18. Broeke, J.V.D., Langergraber, G., Weingartner, A., et al.: On-line and in-situ UV/vis spectroscopy for multi-parameter measurements: A brief review. Spectrosc. Eur. **18**(4), 15–18 (2006)
19. Pellerin, B.A., Downing, B.D., Kendall, C., et al.: Assessing the sources and magnitude of diurnal nitrate variability in the San Joaquin River (California)with an in situ optical nitrate sensor and dual nitrate isotopes. Freshw. Biol. **54**(2), 376–387 (2009)
20. Meyer, D., Prien, R.D., Rautmann, L., Wurst, M., Seifert, T.: In situ determination of nitrate and hydrogen sulfide in the Baltic Sea using an ultraviolet spectrophotometer. Front. Mar. Sci. **5**, 431 (2018). https://doi.org/10.3389/fmars.2018.00431
21. Guitao, S., et al.: Using stable isotopes to distinguish atmospheric nitrate production and its contribution to the surface ocean across hemispheres. Earth Planet. Sci. Lett. 564 (2021)
22. D'Ortenzio, F., Lavigne, H., Besson, F., et al.: Observing mixed layer depth, nitrate, and chlorophyll concentrations in the northwestern Mediterranean. Geophys. Res. Lett. **41**, 6443–6451 (2014). https://doi.org/10.1002/2014GL061020
23. Matsunaga, A., Kurokawa, H., Shozen, K., et al.: Temporal changes on nitrate concentrations in deep seawater pumped up at 321m depth in Toyama Bay, Japan Sea. Deep Ocean Water Appl. Soc. **4**, 67–72 (2003)
24. Yin, J.-P., Lin, L., Wang, Y.-S., Du, J.: Temporal variability of temperature-nitrate relationship in a coastal region. Chin. J. Oceanol. Limnol. **32**, 879–885 (2014). https://doi.org/10.1007/s00343-014-3136-0

A Method of Seawater TOC Signal Processing Based on Improved Threshold Function

Ran Ma[1,2], Yi Liu[1], Yan Liu[1,2](✉), Xiangfeng Kong[1,2], Fengqing Liu[1,2], and Qian Shi[1,2]

[1] Institute of Oceanographic Instrumentation, Qilu University of Technology (Shandong Academy of Sciences), Qingdao 266061, China
sdqdliuyan@126.com
[2] Laoshan Laboratory, Qingdao 266237, Shandong, China

Abstract. This paper presents a wavelet denoising method utilizing an enhanced threshold function to enhance the real-time monitoring of total organic carbon (TOC) in seawater. The conventional threshold denoising method has limitations when handling complex signals, often resulting in errors and residual noise during data reconstruction. This paper introduces an enhanced threshold function that addresses the limitations of the conventional method and reduces noise in high-frequency signals. The simulation results show that the signal-to-noise ratio (SNR) and minimum mean square error (MSE) of the signal with noise measured by the analyzer are 23.0370 and 0.042075. The SNR and MSE of the signal after noise reduction by the enhanced wavelet transform are 34.7860 and 0.0028127. In contrast, the values of the SNR and MSE of the traditional threshold noise reduction are between the noisy signal and the enhanced wavelet transform. Experimental data show that the enhanced threshold method retains useful signals more effectively than the traditional threshold method. In the long-term sea test data, the truth values of TOC monitoring results before and after denoising show a high consistency, and the Pearson correlation coefficient reaches 0.954, which verifies the stability and reliability of the method.

Keywords: Total organic carbon in seawater · Wavelet Transform · Signal denoising · Improved threshold function · Chemiluminescent immunoassay

1 Introduction

As science and technology advance rapidly, energy consumption has surged, leading to an increasingly severe water pollution problem. With the advancement of industrialization and urbanization, a large number of wastewater discharged by factories, pesticides and fertilizers produced by agricultural activities, and nutrients and harmful substances in urban solid waste all enter water bodies directly or indirectly, resulting in water pollution. The presence of organic matter in water will increase the oxygen consumption of dissolved oxygen, which will cause the deterioration of water quality. Therefore, the problem of water pollution has increasingly attracted the attention of the whole society.

C. Weng (Ed.): WRE 2024, LNCE 634, pp. 541–553, 2025.
https://doi.org/10.1007/978-981-96-5095-8_44

The total organic carbon (TOC) of seawater is the sum of dissolved organic matter and suspended organic matter in water, which is an important indicator to characterize the pollution level of organic pollutants and can directly reflect the status of water polluted by organic matter [1, 2]. Therefore, the academic community has widely concerned about the accurate measurement of TOC content.

At present, the main measurement methods for TOC are high temperature combustion (HTCO) and wet oxidation. HTCO is mainly catalyzed by high temperatures, and its oxidation is completely suitable for more polluted waters. Still, the temperature control in the oxidation process is difficult, which may lead to a decrease in accuracy [3]. The wet oxidation method can provide higher detection precision and accuracy, especially for the determination of low concentrations of organic carbon, wet oxidation process requires the use of chemical reagents, persulfate is a common oxidizer, may bring reagent cost and environmental problems in the treatment of waste liquid, at the same time, the reaction time of wet oxidation is longer, may affect the detection efficiency [4, 5]. In addition, most of the existing test methods are conducted in the laboratory, so it is difficult to implement in situ, real-time water tests. On this basis, Liu Yan et al. created an innovative method to measure organic carbon content and its component characteristics through the chemiluminescence technique [6]. This method uses ozone as an oxidizing agent to react with organic substances in seawater. Most organic substances can be oxidized by ozone, resulting in luminescence during the oxidation reaction [7]. This method has a high reaction speed and does not require other chemical reagents to produce secondary pollution, which greatly improves the reaction efficiency [8]. However, the luminous signal generated during the reaction is too weak, and it needs to be amplified and analyzed, which inevitably produces various complex noise pollution during the experiment.

When the total carbon in the ocean is monitored on site, the signal obtained contains the sharp signal sent by large-grain seaweed, which requires the study of high resolution and low time resolution, that is, the spectrum analysis of it, the decomposition of it by time-frequency domain conversion, and the elimination of noise [9]. Wavelet analysis is an efficient time-frequency analysis technique, especially suited to the treatment of non-stationary signals and has a solid theoretical foundation. Wavelet packet transform is a kind of extension of wavelet, which breaks down the signal into the time domain and frequency domain, transforming the original signal into a superposition of multiple levels. On this basis, the threshold method is employed to establish an appropriate threshold for each coefficient, allowing for the elimination of noise signals within the signal components [10]. This method shows a good application prospect.

In 1992, Donoho et al. first proposed wavelet threshold contraction to reduce noise [11]. In 2008, Mohamad Bagher Rahnama et al., using wavelet transform and neural network, established a joint model based on an artificial neural network to forecast the precipitation of Halil River runoff [12]. In 2019, Yu Binbin proposed a new threshold function. The regulator has a special value, but it is discontinuous at the threshold point [13]. In 2020, Xie Bo proposed a threshold function based on exponential function, but only considering the noise reduction effect of Gaussian noise [14]. Dong Xin et al. introduced a wavelet thresholding technique for diminishing noise in one-dimensional vibration signals with high smoothness and low noise reduction effect [15]. In 2022,

Zhang used two adjusting parameters to construct a threshold function that is derivable everywhere and continuous, which has a good noise reduction effect, but its effect is limited to the objective evaluation of Gaussian noise reduction results [16].

On this basis, the threshold value of the wavelet transform is improved, and the data obtained by the TOC analyzer is processed by this method. This method addresses the limitations of the traditional threshold function by preserving valuable signals while effectively eliminating noise. Finally, through the experimental verification, it has been demonstrated that this method provides effective noise reduction, and can be used to reduce total organic carbon signal in seawater.

2 Method and Principle

Wavelet filter is an algorithm with multiple basis functions to choose from. This method has the advantage of multi-resolution, can be analyzed in both the time and frequency domains, and preserves signal mutability while eliminating noise. The process is shown in Fig. 1.

The basic principle of the wavelet transform algorithm is that after the wavelet transform, the effective signal shows different energy distribution and noise characteristics. When the wavelet coefficient's amplitude is significant, it is usually considered an effective signal, and when the wavelet coefficient has a low amplitude, noise predominates. Therefore, based on wavelet coefficient, the data is processed by the threshold method, minimize the noise coefficient while preserving the signal coefficient. This method has the advantages of less computation, simple principle and effective noise elimination. Finally, the wavelet base is flexibly selected to make the wavelet transform more applicable [17].

First, according to different signal characteristics, select the appropriate basis function, select the basis function with orthogonality, compact support, and regularity, and analyze it. Next, the selected signal basis function is discretized and the number of layers N to be decomposed is determined. The original signal is decomposed into N layers in different frequency ranges to obtain the signal coefficient of each layer. The decomposed signal is divided into low frequency and high frequency parts. In the decomposed signal of each layer, the high-frequency part is retained, and related processing is carried out, while the low-frequency part is decomposed and transformed in the next layer. This process repeats the decomposition of the low-frequency portion until the desired number of decomposition layers is achieved. The high-frequency coefficient (D) derived from the decomposition at each level is processed through the enhanced threshold function, and the high-frequency coefficient D1 is obtained. A signal is generated by merging the high-frequency coefficient D1 from the top layer with the low-frequency coefficient A from the last layer. Following this, the high-frequency coefficients from each layer are combined sequentially to produce a smooth and clean data signal.

After further decomposition of the low-frequency coefficient, when the decomposition scale increases, the effective signal exhibits a large wavelet coefficient that remains unchanged as the decomposition scale increases, while the wavelet coefficient of the noisy signal has a smaller value and more dispersion. Therefore, to eliminate most of the noise, we must choose the appropriate threshold value to manipulate the signal. The

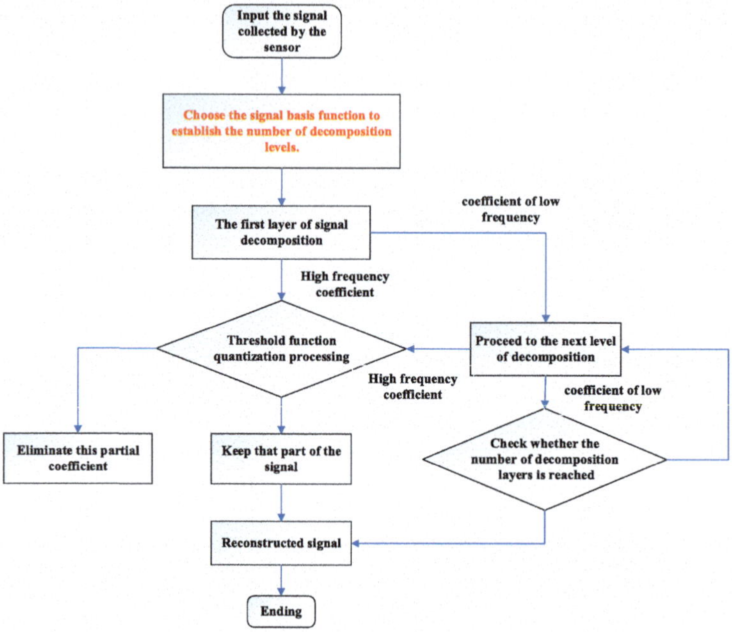

Fig. 1. Data signal processing flow chart.

crucial aspect of the wavelet thresholding method is determining how to choose the threshold.

3 Research on Denoising Techniques Utilizing an Enhanced Threshold Function

3.1 Existing Threshold Function

Currently, there are two typical threshold functions: hard and soft threshold functions [18].

Hard threshold function expression.

$$\hat{\zeta}_{\alpha,\beta} = \begin{cases} \zeta_{\alpha,\beta} & \left|\zeta_{\alpha,\beta}\right| \geq \chi \\ 0 & \left|\zeta_{\alpha,\beta}\right| < \chi \end{cases} \tag{1}$$

where $\zeta_{\alpha,\beta}$ is the high frequency coefficient after wavelet transformation, $\hat{\zeta}_{\alpha,\beta}$ is the high frequency coefficient after threshold processing, and χ is the threshold.

$$\chi = \kappa\sqrt{2\ln(\varpi)} \tag{2}$$

In the formula, κ indicates the noise standard deviation, which is determined by dividing 0.6745 by the average absolute value of the high-frequency coefficient, and ϖ signifies the length of the high-frequency coefficient for each level.

The size of the processed signal coefficient $\zeta_{\alpha,\beta}$ and the threshold point is judged, the signal coefficient exceeding the threshold value is directly retained, and if it does not exceed the threshold, it is zeroed. Although the hard threshold function can more effectively maintain the signal's original characteristics, the wavelet coefficients at the threshold points are not continuous, which leads to the oscillation of the wavelet reconstruction [19–21].

Expression of the soft threshold function.

$$\hat{\zeta}_{\alpha,\beta} = \begin{cases} \operatorname{sgn}(\zeta_{\alpha,\beta})(|\zeta_{\alpha,\beta}| - \chi) & |\zeta_{\alpha,\beta}| \geq \chi \\ 0 & |\zeta_{\alpha,\beta}| < \chi \end{cases} \tag{3}$$

When the signal coefficient exceeds the threshold value of $\zeta_{\alpha,\beta}$, a constant value of χ is applied, and when it falls below the threshold value, it is assigned a value of zero. The threshold function maintains continuity at the threshold, addressing the limitations of the hard threshold function. However, the coefficient after wavelet transformation and the real value has certain errors, which will make the reconstructed signal too smooth, resulting in the loss of some parts [22].

3.2 Improved Threshold Function

To mitigate the drawbacks of the conventional soft and hard threshold functions, this paper proposes a new threshold function and provides its calculation formula.

$$\hat{\zeta}_{\alpha,\beta} = \begin{cases} \zeta_{\alpha,\beta} - \operatorname{sgn}(\zeta_{\alpha,\beta}) * \dfrac{-\operatorname{arccot}(|\zeta_{\alpha,\beta}| - \chi) + \frac{\pi}{2} + \chi}{e^{\left(\sqrt[7]{|\zeta_{\alpha,\beta}|^7 - \chi^7}/\chi\right)}} & |\zeta_{\alpha,\beta}| \geq \chi \\ 0 & |\zeta_{\alpha,\beta}| < \chi \end{cases} \tag{4}$$

This novel threshold function overcomes the limitations of the traditional threshold function.

Continuity:

$$\lim_{\zeta_{\alpha,\beta} \to \chi^+} \hat{\zeta}_{\alpha,\beta} = \chi - \frac{-\operatorname{arccot}(\chi - \chi) + \frac{\pi}{2} + \chi}{e^{\left(\sqrt[7]{\chi^7 - \chi^7}/\chi\right)}} = \chi - \frac{\chi}{1} = 0 \tag{5}$$

$$\lim_{\zeta_{\alpha,\beta} \to \chi^+} \hat{\zeta}_{\alpha,\beta} = \lim_{\zeta_{\alpha,\beta} \to \chi^-} \hat{\zeta}_{\alpha,\beta} \tag{6}$$

$$\lim_{\zeta_{\alpha,\beta} \to \chi^-} \hat{\zeta}_{\alpha,\beta} = 0 \tag{7}$$

According to the formula, the improved threshold function is continuous at the threshold.

Gradualness:

$$\lim_{\zeta_{\alpha,\beta} \to \infty} \hat{\zeta}_{\alpha,\beta} = \zeta_{\alpha,\beta} - \frac{-\operatorname{arccot}(\infty - \chi) + \frac{\pi}{2} + \chi}{e^{\left(\sqrt[7]{\infty^7 - \chi^7}/\chi\right)}} = \zeta_{\alpha,\beta} - 0 = \zeta_{\alpha,\beta} \tag{8}$$

It is evident that when the absolute value of the signal coefficient exceeds the threshold, the output signal of the enhanced threshold function quickly and infinitely approaches the signal coefficient before threshold processing, thereby minimizing the error. The comparison of related improvements is shown in Fig. 2.

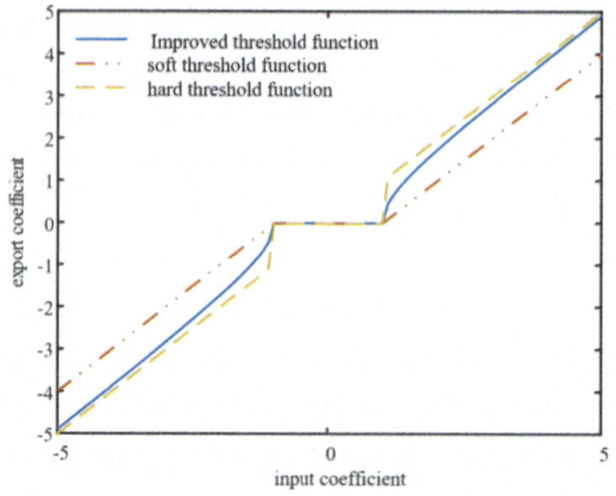

Fig. 2. Improve the comparison of threshold, hard and soft threshold functions.

3.3 Optimal Wavelet Base and Number of Decomposition Levels

To further evaluate the efficacy of the enhanced threshold function, the simulated TOC signal is designed by simulation software, and random white noise is introduced to mimic the real signal.

The denoising effect also depends on the setting of the wavelet coefficient and the number of decomposition layers, among which the wavelet basis is the key to the decomposition and reconstruction of the signal, and it also has some effect on the size of the coefficient. Firstly, the suitable wavelet basis function is chosen. The wavelet basis is crucial for wavelet decomposition and reconstruction, and it significantly affects the magnitude of the wavelet coefficients, thereby impacting the results of noise reduction. Common wavelet basis functions include the Haar function, Daubechies wavelet system, Biorthogonal wavelet system, Coiflet wavelet system, and Symlets wavelet system. The signal-to-noise ratio of several wavelet bases after denoising is compared through simulation tests, and the results are shown in Table 1 below. In order to get a good noise reduction result, the wavelet base must satisfy the orthogonal, symmetrical, and regular conditions. Orthogonality is used to ensure the accuracy of signal reconstruction, symmetry can reduce the distortion of signal reconstruction, and regularity can obtain good frequency domain resolution and smoothness [20]. Through the simulation and comparison of various wavelet bases, this paper selects various db wavelets with good performance, analyzes the SNR and the MSE after noise cancellation, and selects the same threshold. The simulation effect is shown in Table 2. Using db3 wavelet, it has the highest SNR and the best noise reduction effect.

Table 1. Comparison of denoising effect of wavelet basis function.

Wavelet basis	Haar	Db3	Biro2.6	Coif2	Sym5
SNR (db)	21.3233	34.7860	34.5759	34.5462	34.6996

Table 2. Comparison of denoising effect of each DB wavelet.

Wavelet basis	Db2	Db3	Db4	Db5	Db6
SNR (db)	34.1714	34.7860	34.7463	34.6746	33.3787

Based on selecting the optimal wavelet base, the ideal number of decomposition layers should be determined when denoising the signal. As seen from Table 3, the signal-to-noise ratio increases first and then decreases with the increase of decomposition times. This is mainly due to too few layers, and the removal of noise is not thorough enough, resulting in a less than ideal noise reduction effect. A greater number of decomposition layers improves noise reduction. Still, a greater number of decomposition layers can result in the loss of valuable signal data, leading to an overall decline in the effectiveness of denoising. When the signal is decomposed into five layers, the SNR is the highest, and the denoising effect is the best. Therefore, the subsequent experiment adopts a 5-layer decomposition.

Table 3. Comparison of wavelet denoising effect of each layer.

Decomposition level	2	3	4	5	6	7
SNR (db)	29.3335	31.8500	33.7850	34.7860	32.4035	19.8472

3.4 Simulated Analysis

Using simulated TOC data signal, wavelet base DB3 and decomposition layer 5 were selected for simulation analysis, as shown in Fig. 3a and Fig. 3b. The enhanced threshold function for noise reduction greatly improved accuracy and smoothness of the output signal.

For the effect evaluation of signal denoising, SNR, MSE and waveform similarity parameters (NCC) are generally calculated. When comparing the denoising outcomes of traditional threshold functions with those of enhanced threshold functions, the larger SNR, the larger NCC and the smaller MSE, the denoising effect is superior. Three threshold functions were used to process the data, and the SNR, MSE and NCC values were obtained respectively. The results are shown in the following Table 4. As can be seen from the table, the SNR, MSE, and NCC of the signal with noise measured by the analyzer are 23.0370, 0.042075, and 0.99755. The SNR, MSE, and NCC of the signal denoised by the improved threshold function are 34.7860, 0.0028127, and 0.99984 after the signal denoised by the analyzer is denoised by the improved threshold function Well, compared with the traditional method, the SNR and NCC values of this method are the highest, and the MSE is the smallest, which proves that the improved threshold function has the best noise reduction effect. Simulation results indicate that the proposed method successfully reduces noise, retains the characteristics of the original signal, and at the same time effectively suppress the oscillation of the reconstructed signal.

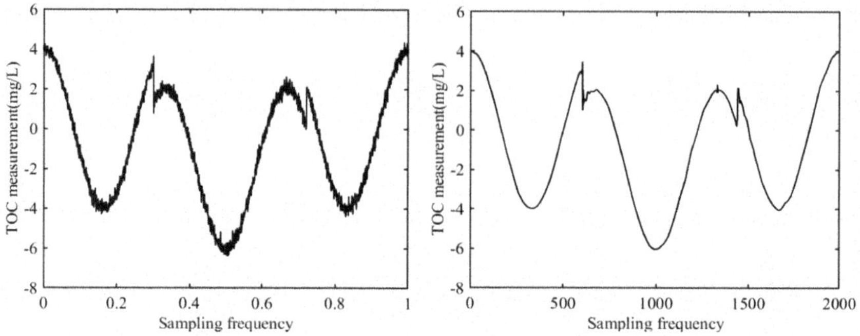

(a) Noisy signal. **(b)** Improved threshold function denoising.

Fig. 3. Simulation signal analysis.

Table 4. Three kinds of threshold function data processing analysis.

Denoising method	Noisy signal	Hard threshold function	Soft threshold function	Improved threshold function
SNR	23.0370	34.5017	32.9216	34.7860
MSE	0.042075	0.0030030	0.0043207	0.0028127
NCC	0.99755	0.99983	0.99975	0.99984

4 Experimental Verification

In order to verify the feasibility of the TOC analyzer prototype, a series of continuous sea test studies were carried out in the offshore waters of Qingdao. Five different concentrations were set for the on-site seawater samples, the sewage outlet and the artificial seawater, and the accuracy, accuracy, applicability, stability, continuous operation ability and automation degree of the instrument were emphasized. The results were compared with those obtained by high temperature combustion method in laboratory.

Sample configuration: on-site seawater sample: on-site seawater supply. Sewage outfall sample: Sewage flowing into the sea sewage outfall provided (at low tide), mainly domestic wastewater. Artificial seawater sample: Dissolve 31 g of sodium chloride (NaCl, superior pure), 10 g of magnesium sulfate ($7H_2O$, superior pure), and 0.05 g of sodium bicarbonate ($NaHCO_3 \bullet H_2O$, superior pure) in water and dilute to 1000 mL. Create artificial seawater with a salinity of 35. Five different concentrations of seawater samples were prepared for the above three samples. Concentration point 1: bathing place: artificial seawater was 2:1. Concentration point 2: bath. Concentration point 3: bath: sewage outlet is 5:1. Concentration point 4: bath: sewage outlet is 5:2. Concentration point 5: Bath: sewage outlet is 1:1.

According to the relevant provisions of GB17378.2 Marine Monitoring Code Part 2: Data Processing and Analysis quality Control, various performance indexes of the analyzer prototype were calculated. Accuracy refers to the comprehensive evaluation of systematic and random errors contained in various detection methods or detection systems, which directly affects the reliability of detection results.

In this experiment, the precision of the analyzer is analyzed with a known concentration of standard solution, and the average of the measured results is drawn, to indirectly evaluate the accuracy of the analyzer prototype. As can be seen from Fig. 4, the relationship between the results obtained from the instrumental method and those measured by the combustion method is approximately linear, which shows the accuracy of the measurement results of the analyzer prototype.

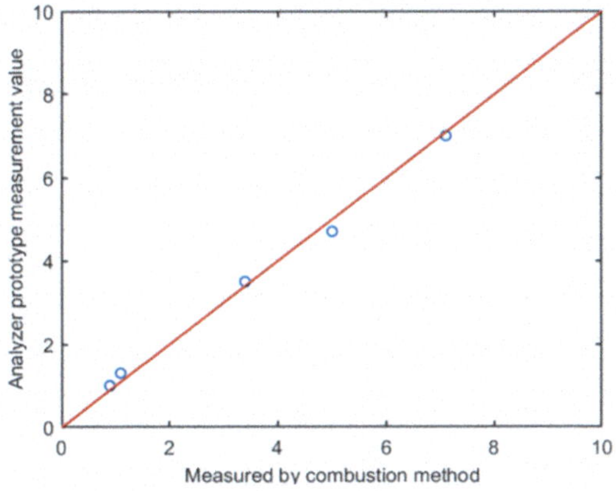

Fig. 4. Comparison of results of analyzer measurement and combustion method.

Precision means that the same sample is tested multiple times under the same experimental conditions, and the relative standard deviation (RSD%) is used to express the dispersion of the obtained results.

In this study, Repeatability/Intra-assay precision was used to measure the precision of the instrument, that is, 5 parallel samples were measured on one sample with the analyzer prototype and the corresponding precision was calculated. The relative standard deviation of field seawater samples was measured by instruments to evaluate the precision, and a total of 24 groups of data were measured by the two methods. The analysis results of the relative standard deviation are shown in Table 5. The results showed that the data with a relative standard deviation of less than 5% accounted for 100% of all groups in 5 parallel tests, indicating that the method has good precision and can meet the needs of Marine sample analysis.

Table 5. The precision of the two methods was compared.

RSD (%) Range	0%–5%	5%–10%	10%–15%	>15%
Prototype analyzer	24	0	0	0
combustion method	24	0	0	0

Sensitivity means that with a certain degree of reliability, a special analytical method can detect the lowest concentration or minimum of the substance under test from the sample. This is usually expressed as the detection limit (DL).

In this experiment, artificial seawater was used to dilute a field seawater sample with very low TOC content as a blank solution, and the detection limit of the instrument was obtained through 11 consecutive instrumental analyses. Artificial seawater dilution field seawater until the instrument just responds, at this time, the ratio of artificial seawater to field seawater configuration is 10:1. It was found that the limit of detection was 0.05 mg/L, which is low and sensitive, and can meet the needs of Marine samples, which relevent data can be shown in Table 6.

Table 6. Analyzer prototype detection limit data (mg/L).

TOC measurement	0.01	0.02	0.03	0.01	0.04	0.05	0.01	0.04	0.01	0.02	0.04
standard deviation	0.013										
LOD (limit of detection)	0.05										

It was found that the results of the two methods were close to each other, and the results of the two methods were in good agreement. The detection limit was 0.05 mg/L, and the precision (RSD) was <5%. The results indicate that the model analyzer is accurate and accurate, stable performance, and can meet the needs of seawater sample analysis.

To evaluate the noise reduction effectiveness of the enhanced wavelet threshold method in this study, this paper utilizes a residual analysis approach for assessment, that is, calculate the difference between the signal obtained by the total organic carbon analyzer and the data after noise removal, and the residual histogram is drawn by comparing the data directly collected by the experimental platform with the denoising processing results. Simulation software was used to conduct simulation analysis on the collected data, and the following comparison chart was obtained.

From Fig. 5 and Fig. 6, the shape of the residuals of the method is close to the center symmetry, which is close to the normal state, indicating that the method can effectively eliminate noise. Mean and Median almost coincide and are near line 0, indicating that the wavelet denoising method will not destroy the data structure of the effective signal and obtain the best effect for signal denoising.

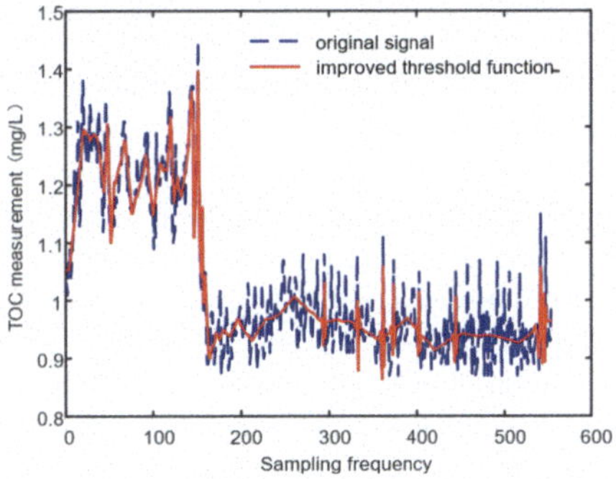

Fig. 5. Improved threshold function denoising.

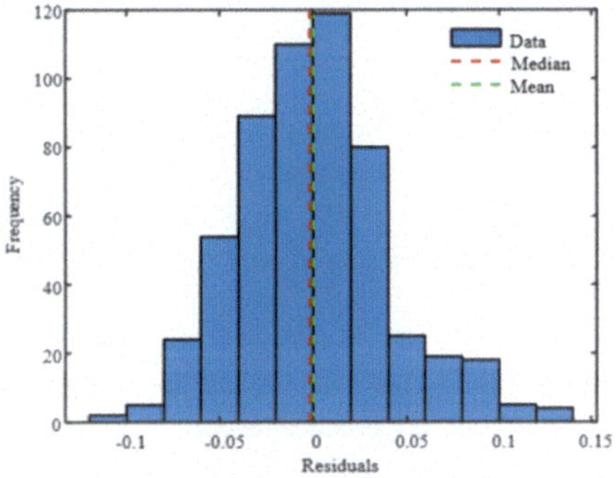

Fig. 6. Improved residual histogram of threshold function before and after denoising.

5 Conclusion

This paper introduces a novel threshold function that utilizes wavelet multi-scale and multi-frequency techniques for denoising seawater total organic carbon monitoring data. The enhanced threshold function is built upon the conventional soft and hard threshold functions, which can avoid the defects of the soft and hard threshold functions at the same time and realize efficient noise reduction. Based on the available experimental data, a simulation analysis is conducted using the improved wavelet transform, and the obtained Pearson correlation coefficient is 0.954, this demonstrates a very strong linear correlation between the two. The denoising process can significantly diminish noise while preserving the key characteristics of the main signal, which proves the feasibility of the denoising function. Additionally, the enhanced threshold algorithm is combined with the traditional threshold algorithm and improved. Simulation analysis shows that the SNR of the traditional threshold function is 32.9216 and 34.5017, and the MSE is 0.0043207 and 0.0030030, while the SNR of the improved threshold function is 34.7860 and the MSE is 0.0028127. The SNR of the improved threshold function is higher than the traditional threshold denoising, and the MSE is lower than the traditional threshold denoising. By comparing the noise reduction effectiveness of the two methods, the method is feasible. The application of this method to the long-term monitoring of

Marine total organic carbon can provide an effective early warning for the pollution of organic compounds in China's Marine areas, and provide a scientific basis for the safe operation of Marine ecosystems.

Acknowledgement. This work was supported by the Major Innovation Project of Science and Education Integration Pilot Project of Qilu University of Technology (Shandong Academy of Sciences) (2023JBZ02), Major Innovation Project of Science and Education Integration Pilot Project of Qilu University of Technology (Shandong Academy of Sciences) (2023JBZ03), Science Education Industry Integration Pilot Major Innovation Project (2023HTZX01) of Qilu University of Technology (Shandong Academy of Sciences), Open Research Fund of Key Laboratory of Marine Ecological Monitoring and Restoration Technology, Ministry of Natural Resources, (MEMRT202301).

References

1. Yang, H.Y.: Study on the experimental method for determination of total organic carbon by TOC-L total organic carbon analyzer. Tot. Corros. Control **35**(10), 34–38 (2021)
2. Bi, W., Li, Y., Sun, J.C., Jia, Y.J., Huang, P.J., Xu, F.: Research review and development trend of marine total organic carbon detection based on spectral technology. J. Yanshan Univ. **46**(1), 1–8 (2022)
3. Maestre, S.E., Mora, J., Hernandis, V., Todoli, J.L.: A system for the direct determination of the nonvolatile organic carbon, dissolved organic carbon, and inorganic carbon in water samples through inductively coupled plasma atomic emission spectrometry. Anal. Chem. **75**(1), 111–7 (2003)
4. Yang, H., Graham, N.J.D., Wang, W., Liu, M., Yu, W.: Evaluating and improving the reliability of the UV-persulfate method for the determination of TOC/DOC in surface water. Water Res. 9: 116918 (2021)
5. Zhang, S., Wang, Z., Yan, X., Wang, J., Liu, Y.: Measurement methods of total organic carbon in seawater. E3S Web Conf. **233**, 01078 (2021)
6. Liu, Y., et al.: Study of analysis technology of organic carbon in seawater with ozonation chemiluminescence method. Mar. Technol. **29**(04), 96–101 (2010)
7. Shen, J.M., Li, X.Y., Chen, Z.L., Qi, F., Ye, M.M., Xu, Z.Z.: Ozonation degradation of nitrochlorobenzene isomers in aqueous solution: i. discuss of kinetics and processes. J. Harbin Inst. Technol. **4**, 540–545 (2008)
8. Ren, G., Ma, R., Zhang, S.W., Liu, Y.: Research and implementation of total organic carbon of seawater using the chemiluminescence dynamic method. Mar. Technol. **30**(03), 21–24 (2011)
9. Li, M., Liu, Y., Ma, R., Wang, Z.Y.: Research on wavelet multi threshold algorithm for denoising weak signal of DOC in seawater. J. Sens. Technol. **34**(01), 75–79 (2021)
10. Lan, Z.Q., Gao, P., Wang, P., Wang, Y.J., Liang, J.D., Hu, G.M.: Improved wavelet packet noise reduction for microseismic data via fuzzy partition. IEEE Geosci. Remote Sens. Lett. **19**, 1–5 (2022)
11. Donoho, D.L., Johnstone, I.M.: Ideal spatial adaptation by wavelet shrinkage. Biometrika **81**(3), 425–455 (1994)
12. Rahnama, M.B., Noury, M.: Developing of Halil river rainfall-runoff model, using conjunction of wavelet transform and artificial neural networks. Res. J. Environ. Sci. **2**(5), 385–392 (2008)
13. Binbin, Y.: An improved infrared image processing method based on adaptive threshold denoising. EURASIP J. Image Video Process. **2019**(1), 1–12 (2019). https://doi.org/10.1186/s13640-018-0401-8

14. Xie, B., Xiong, Z., Wang, Z., Zhang, L., Zhang, D., Li, F.: Gamma spectrum denoising method based on improved wavelet threshold. Nucl. Eng. Technol. **52**(8), 1771–1776 (2020)
15. Dong, X., Li, G.L., He, K., Jia, Y.C., Xu, K., Li, B.: Spectral wavelet threshold denoising and its application in vibration signal analysis of hob spindle. Chinese J. Mech. Eng. **56**(11), 96–107 (2020)
16. Zhang, Z.R., Deng, C.X., Yue, X.H.: Fingerprint image denoising method base on improved threshold function. J. Harbin Univ. Sci. Technol. **27**(1), 55 (2022)
17. Liu, J., Chen, J., Han, B., Ma, X.F., An, J.: Deep learning underwater target classification based on improved wavelet threshold denoising. Tech. Acoust. **42**(1), 25–33 (2023)
18. Donoho, D.L.: De-noising by soft-thresholding. IEEE Trans. Inf. Theory **41**(3), 613–627 (1995)
19. Wu, X.J.: Improved wavelet threshold based on lifting wavelet in application of signal denoising. Laser J. **35**(08), 15–18 (2014)
20. Dong, W., Chen, C., Li, Z., Wang, Y.: A wide range measurement of methane using single saturated absorption spectral line concentration measurement method. IEEE Trans. Instrum. Measur. (2023)
21. Lin, X., Dong, M., Nie, W., Rao, G., Lu, J.: Time-to-frequency conversion method for tunable diode laser absorption spectrum. Optik **270**, 170049 (2022)
22. Wang, W.: Analysis of denoising effect of wavelet threshold denoising method in radar signal detection. Shandong Ind. Technol. **04**, 40–45 (2020)

Research on High-Quality Development of Urban Economy Under Dual Constraints of "Carbon" and "Water": A Case Study of Xi'an

Dongping Xu[1,2], Keying Li[1], Yingxue Shi[1], Yuping Yang[1], and Qiuling Lu[3](✉)

[1] School of Civil Engineering, Xi'an University of Technology, Xi'an Shaanxi 710000, China
[2] State Key Laboratory of Eco-Hydraulic Engineering in Northwest Arid Area, Xi'an, Shaanxi 710048, China
[3] Northwest Investigation & Design Institute, Key Laboratory of Eco-Hydrology and Disaster Prevention of Arid Areas, National Forestry and Grassland Administration, National Forestry and Grassland Administration, Beijing, China
253087478@qq.com

Abstract. In the current complex international environment, promoting high-quality development (HQD) of the urban economy plays a crucial role in enhancing the ability of China's urban economies to respond to international challenges. Water resources (WRE) and carbon emissions (CEM) are two important subsystems that constrain the HQD of the urban economy. Studying the coupled, coordinated, and synergistic relationships between urban economic quality development and these subsystems is crucial for promoting the HQD in the urban economy. Taking Xi'an as an example, this thesis uses the entropy weight TOPSIS method and the coupling coordination evaluation model to assess the development levels of Xi'an's urban economy, WRE support, and CEM, and the coupling, coordination, and systemic relationships among them. The findings are as follows: (1) The comprehensive development levels of the subsystems of urban economic development quality, water resource support, and CEM in Xi'an show a continually improving situation. (2) The coupling state between Xi'an's urban economic development quality subsystem and other two subsystems show a trend of initial improvement, followed by a decrease, and then stabilization. (3) The coordination state between Xi'an's urban economic development quality subsystem and other two subsystems remain highly consistent, displaying a wave-like rising trend. (4) The synchronous development pace of Xi'an's urban economic development quality subsystem with other two subsystems is generally aligned. By studying the issues related to HQD of urban economy under the dual constraints of "carbon" and "water", this research not only enriches the theoretical understanding of HQD of urban economy but also provides useful references for promoting high-quality economic development in Xi'an and other cities.

Keywords: HQD of Urban Economy · WRE · CEM · Coupling · Coordination · Synergy

C. Weng (Ed.): WRE 2024, LNCE 634, pp. 554–573, 2025.
https://doi.org/10.1007/978-981-96-5095-8_45

1 Introduction

In the face of long-term challenges from the complex international economic environment, China's economy is currently undergoing a pivotal shift from speed-oriented to quality-oriented development. This transition plays a crucial role in comprehensively enhancing the country's economic development level and bolstering its capacity to tackle international challenges [1]. The 19th National Congress of the Communist Party of China emphasized this shift, stating that China's economy has shifted from high-speed growth to HQD. The transformation of economic development into the pursuit of high-quality changes in production methods, economic structure, and growth dynamics. In the 20th Party Congress report, General Secretary Xi Jinping stressed that "HQD is the primary task of building a modern socialist country in an all-round way". Given its pivotal role as an important carrier of China's economic advancement, research on the level and influencing factors of HQD in the urban economy assumes the utmost significance.

The concept of "HQD" has garnered high attention from the leaders of the Party and the country and has become a research hotspot in the academic community. Foreign scholars believe that HQD should pay attention to the role of economic growth in poverty eradication, thereby promoting economic growth and improving the welfare levels of the entire population [2, 3]. Domestically, scholars hold roughly two viewpoints on the connotation of HQD. Scholars such as Cheng [4], Ren [5], and Li [6] argue that HQD should advocate a value of national sharing, with innovation as the core and coordinated development as the fundamental requirement, thus significantly promoting the sustainable development of society, economy, and ecology. Scholars represented by Jin [7] and Ma [8], another aspect, believe that high quality, with economic HQD at its core, should be an economic development pattern, framework, and mechanical condition in the new era of China that meets the growing real needs of the people, a development model that is more efficient, of higher quality. On this basis, some scholars have focused their research on cities, conducting extensive studies on the HQD of urban economies. Firstly, scholars conduct research on evaluation indicator systems and evaluation methods. Scholars have mainly focused on two perspectives: the "New Development Idea (NDI)" and "meeting the people's aspiration for a better life." For instance, Fan et al. [9] approached from the perspective of the NDI, constructing a high-quality evaluation index system of urban economy that includes characteristics such as effectiveness, sustainability, stability, and coordination. Meanwhile, Li et al. [10] built an evaluation system from the perspectives of economic vitality, innovation efficiency, and social harmony. Regarding evaluation methods, Zeng et al. [11] simulated the achievement level of carbon peak and carbon neutrality in the process of HQD of Chinese urban clusters under the dual constraints of "double carbon" targets using a system dynamics model. Wang et al. [12] analyzed the spatial and temporal changes in the HQD level of 61 prefecture-level cities in the Yellow River Basin using the entropy weight TOPSIS and K-mean clustering methods. In terms of empirical research on HQD evaluation, scholars have conducted research at different scales, such as cities [13], river basins [14, 15], provinces [16], and urban agglomerations [17, 18], bringing research on urban HQD to a new height. In order to further explore the influencing factors of promoting HQD of cities, scholars have studied the impact of ecological conservation [19], industrial agglomeration [20], spatial relations [21], digital economy [22], technological innovation [23], resource conditions [24], talent agglomeration [25], and environmental legislation [26] on urban HQD.

Among the various factors influencing the HQD of cities, research on the constraints of WRE and CEM has become a recent hot topic. WRE, as an indispensable foundational resource for urban development, determines the quality and progress of high-quality urban development. According to Wang et al. [27], the loading capacity of WRE is a core constraint factor for HQD and ecological environmental protection. They suggest establishing dynamic connections between natural environmental resources and socioeconomic growth under the rigid constraint of WRE to promote HQD. Jiang et al. [28] studied the relationship between water resource utilization efficiency and HQD of the urban economy using a decoupling model. They argue that identifying and utilizing the "decoupling-connection" fluctuation relationship between the two can promote HQD under the constraint of WRE in the Yellow River Basin. Li et al. [29] proposed a framework for optimizing water resource allocation in northern water-scarce regions for HQD. They emphasized that scientifically reasonable water resource allocation schemes are an important basis for achieving HQD in water-deficient areas of northern China. Under the dual carbon strategy, how to achieve HQD of the urban economy under the constraint of carbon emission targets has become a more academically concerned issue. Chen et al. [30] explored the synergistic mechanism between pollution reduction, carbon reduction, and high-quality economic development, and investigated the relationship and driving factors between the synergy of pollution and carbon reduction and the coordinated development of high-quality economy. Kuang et al. [31] and Liu et al. [32] investigated the spatial-temporal differentiation and driving mechanisms of construction of coupling model and harmonization of carbon emission intensity, low-carbon economy, and HQD through coupling models. Such research is important in leveraging the dual interaction between a low-carbon economy and HQD, creating distinctive development paths, and optimizing regional cooperation and coordination relationships. While the academic community has conducted sufficient study on the relationship and driving mechanisms between water resource constraints, the current studies all show the relationship between a single element and HQD, and the problems related to HQD and coupled coordination of urban economy under the dual constraints of carbon and water are still unclear. Given that cities frequently encounter multiple constraints, this thesis seeks to delve into the HQD of urban economies under the dual constraints of carbon and water. Its objective is to enhance the promotion of HQD in the urban economy within the pattern of ecological civilization.

Xi'an, positioned as a national-level central city in the semi-arid western region, faces developmental constraints due to the scarcity of WRE. Simultaneously, as a leading city in the western region, it actively advocates for low-carbon-oriented HQD of the urban economy, assuming a significant driving and exemplary role in the western region. Hence, this thesis chooses Xi'an City as a case study to investigate the HQD of the urban economy under the dual constraints of "carbon" and "water."

2 Research Design

2.1 Evaluation System

As a fundamental resource supporting urban development, the quantity and quality of WRE restrict the pace and caliber of urban growth. CEM in the process of urban development reflects the health and quality problems of urban industrial structure, energy utilization, energy efficiency and other development. To reveal the constraining roles played by WRE and CEM in the process of urban economic development, an evaluation indicator system composed of three subsystems - urban economic development quality, water resource utilization, and CEM - has been established. This system comprehensively evaluates the development status of Xi'an's high-quality urban development under the dual restrictions of WRE and CEM. The aim is to promote its HQD while reducing the constraints of WRE and CEM, and to further enrich the theory and practice of high-quality urban development.

The evaluation indicator system is constructed following the principles of scientificity, systematicness, dynamism, accessibility, and operability. Firstly, the main subsystems of the evaluation are established, and the subsystems of urban economic HQD, WRE support and carbon emission are taken as the system layers of the evaluation system. Secondly, based on the fundamental theories of HQD, WRE, and CEM, criteria such as economic progression, innovative progression, coordinated progression, green progression, open progression, shared progression, water resource utilization, and CEM targets are established as the criteria layer. Thirdly, referring to relevant literature [8–10], economic growth, industrial structure, and development quality are represented as elements layer of economic development; innovative development is depicted from aspects such as innovation input, innovation force, and innovation output; coordinated development is elucidated from aspects of urban-rural coordination, urban-rural structure, and industrial coordination; green development is analyzed from the perspectives of environmental pollution, energy-water consumption, and environmental governance; open development is clarified from aspects of foreign economic development, factor circulation, and public services; shared development is described from the viewpoint of social security; water resource utilization is expounded from aspects such as water resource quantity, water use structure, and water resource efficiency; and CEM constraints are interpreted from aspects such as CEM volume, carbon emission structure, and carbon emission intensity. Finally, specific indicators for each element layer are determined through repeated comparisons. Based on the indicator system construction approach and in conjunction with relevant studies [11–13], the evaluation indicator system for the HQD level of Xi'an under the dual constraints of "carbon" and "water" is determined (see Table 1).

2.2 Data Sources and Processing

The basic data of the empirical analysis are all from the Xi'an Statistical Yearbook from 2011 to 2023. In order to eliminate the influence of price changes on each variable index, GDP price index and consumer price index were used to deflate the related variable index with 2010 as the base period.

Table 1. The evaluation index system of Xi'an City's high-quality economic development level under the double constraints of "carbon" and "water".

System Layer A	Criterion Layer B	Element Layer C	Index Layer X	Calculation formula	Index weights	Attribute
Urban economy high-quality development subsystem	Economy	Economic growth	Per capita GDP growth rate	Per capita GDP growth rate	3.784	+
		Industrial structure	Upgrading of industrial structure	Industrial structure upgrading index	5.655	+
		Development quality	Fiscal revenue as a share of GDP	Fiscal revenue/Total GDP	2.313	+
			GDP per employed population	GDP total/Number of employees	2.966	+
	Innovation	Innovation input	R&D investment intensity	R&D expenditure/Total GDP	1.366	+
			The proportion of educational expenditure	Total educational expenditure/Total GDP	2.42	+
		Innovative force	Proportion of scientific and technological personnel	Number of technology personnel/Entire population	4.486	+
			Number of full-time teachers in higher education	Number of full-time teachers in higher education/Entire population	4.605	+
		Innovative output	Patents per capita	Number of students in higher education *100,000/Total population	4.644	+
			High-tech industry revenue share	Technology market turnover/Total GDP	4.637	+
	Coordination	Urban-rural coordination	Urbanization level	Urbanization level	4.13	+
			Engel coefficient ratio of urban and rural residents	Engel coefficient ratio of urban and rural residents	2.433	-

(continued)

Table 1. *(continued)*

System Layer A	Criterion Layer B	Element Layer C	Index Layer X	Calculation formula	Index weights	Attribute
		Urban-rural structure	Urban and rural binary comparison coefficient	(Proportion of the value added of the primary industry/Proportion of employment in primary industry)/(Proportion of the added value of the secondary and tertiary industries/Proportion of employment in secondary and tertiary industries)	2.361	+
			Urban-rural dual contrast coefficient	Proportion of the value added of the primary industry-Proportion of employment in primary industry	2.753	-
		Industrial coordination	Industrial upgrading index	Tertiary industry value/Secondary industry value	3.898	+
			Proportion of cultural and sports expenditures	Proportion of cultural and sports expenditures/Fiscal expenditure	2.676	+
	Greenness	Environmental pollution	Wastewater discharge per unit of GDP	Wastewater discharge/Total GDP	1.754	-
			Sulfur dioxide emissions per unit of GDP	Sulfur dioxide emissions/Total GDP	2.726	-
			PM2.5 concentration	Average PM2.5 concentration	1.862	-
		Energy water consumption	Electricity consumption per 10,000 yuan GDP	Electricity consumption of the whole society/Total GDP	1.71	-

(continued)

Table 1. *(continued)*

System Layer A	Criterion Layer B	Element Layer C	Index Layer X	Calculation formula	Index weights	Attribute
			Water consumption for 10,000 yuan of GDP	Water consumption of the whole society/Total GDP	2.259	-
		Environmental governance	Green coverage rate of built-up area	Green coverage rate of built-up areas	2.38	+
			Comprehensive utilization rate of industrial solid waste	Industrial solid waste generation/Total solid waste	1.264	+
			Expenditure on energy conservation and environmental protection	Expenditure on energy conservation and environmental protection/Total GDP	3.56	+
	Openness	Foreign economy	Proportion of total imports and exports	Total amount of import and export/Total GDP	4.59	+
			Utilization ratio of foreign capital	Foreign direct investment/total GDP	1.238	+
		Factor circulation	Per capita highway mileage	Highway mileage/Entire population	2.533	+
			Per capita road freight volume	Road freight volume/Entire population	7.485	+
	Sharing	Public service	Perfection of cultural facilities	Library collections/Entire population	2.871	+
			Number of beds in medical and health institutions per unit of population	Number of medical and health beds/Entire population	3.008	+
		Social security	Social security expenditure	Social security expenditure/Total GDP	2.2	+

(continued)

Table 1. *(continued)*

System Layer A	Criterion Layer B	Element Layer C	Index Layer X	Calculation formula	Index weights	Attribute
			Number of people participating in basic medical insurance	Number of people participating in basic medical insurance/Entire population	3.434	+
Water resource subsystem	Water resources utilization	Amount of water resources	Total water resources	Total water resources	27.644	-
			Water resources per capita	Total water resources/Entire population	22.99	+
		Water use structure	Non-agricultural water use ratio	Total water consumption of secondary and tertiary industries/Total water resources	21.932	+
		Water use efficiency	Water consumption per unit GDP	Total water consumption/GDP	27.435	-
Carbon emission subsystem	Carbon emission constraints	Carbon emissions	Total carbon emissions	Total carbon emissions	27.914	-
			Carbon emissions per capita	Total carbon emissions/Entire population	34.203	-
		Carbon emission structure	Carbon emission structure	Carbon emissions of secondary and tertiary industries/Total carbon emissions	20.921	+
		Carbon emission intensity	Carbon emissions per unit of GDP	Carbon emissions/Total GDP	16.962	-

2.3 Entropy Weight TOPSIS Method

Entropy weight TOPSIS method is to use entropy weight method to determine the evaluation index weights, and then use TOPSIS method to rank the evaluation objects through the technique of approximating the ideal solution. The entropy weight method can objectively reflect the importance of a certain indicator when making decisions. Combining the TOPSIS method based on the entropy weight method avoids the loss of information about the indicator system in the process of data processing and makes the evaluation results more objective and accurate. Here, we exemplify the evaluation process by focusing on the urban economic development quality subsystem.

(1) Construct the evaluation index matrix. Construct the evaluation index matrix for the urban economic HQD level of Xi'an based on the 40 indicators Xij of the city from 2010 to 2022:

$$X = \begin{bmatrix} X_{11} \; ... \; X_{1j} \\ ... \; ... \; ... \\ X_{i1} \; ... \; X_{ij} \end{bmatrix} \tag{1}$$

(2) Indicator standardization. The calculation formula for the standardization of each indicator is:

$$Y_{ij} = \begin{cases} \dfrac{X_{ij} - \min X_j}{\max X_j - \min X_j} (\text{Forward indicator}) \\ \dfrac{\max X_j - X_{ij}}{\max X_j - \min X_j} (\text{Negative indicator}) \end{cases} \tag{2}$$

where Y_{ij} is the standardized result of X_{ij}; min X_j is the minimum value of the j-th indicator; Max X_j is the maximum value of the j-th indicator.

(3) Calculate the weights of each indicator. The calculation formula for the weight of each indicator is:

$$P_{ij} = \frac{Y_{ij}}{\sum\limits_{j=1}^{n} Y_{ij}} \tag{3}$$

The indicator data weight matrix is:

$$P = \begin{bmatrix} P_{11} \; ... \; P_{1j} \\ ... \; ... \; ... \\ P_{i1} \; ... \; P_{ij} \end{bmatrix} \tag{4}$$

where Pij is the weight of the indicator Xij.

(4) Indicator entropy weight calculation. The calculation formula for the indicator entropy weight is:

$$e_j = -\frac{1}{\ln m}\sum_{i=1}^{m} P_{ij}\ln P_{ij}, d_j = 1 - e_j, W_j = \frac{d_j}{\sum_{j=1}^{n} d_j} \quad (5)$$

(5) Determine the positive and negative ideal solutions of each indicator. The formula is:

$$\begin{cases} Y_j^+ = \left(\max Y_{ij} | i = 1, 2, \cdots, m\right) = (Y_1^+, Y_2^+, \cdots Y_n^+) \\ Y_j^- = \left(\max Y_{ij} | i = 1, 2, \cdots, m\right) = (Y_1^-, Y_2^-, \cdots Y_n^-) \end{cases} \quad (6)$$

where, Yj + represents the positive ideal solution formed by the maximum value of each indicator; Yj- represents the negative ideal solution formed by the minimum value of each indicator.

(6) Determine the Euclidean distance Di+ and Di- between the evaluated object and the positive and negative ideal solutions. The formula is:

$$D_i^+ = \sqrt{\sum_{i=1}^{n} W_j(Y_j^+ - Y_{ij})^2}, D_i^- = \sqrt{\sum_{i=1}^{n} W_j(Y_{ij} - Y_j^-)^2} \quad (7)$$

(7) Calculate the proximity, i.e., the evaluation value of the evaluated object, based on the Euclidean distance. The formula is:

$$F_i = \frac{D_i^-}{D_i^+ + D_i^-} \quad (8)$$

where Fi represents the evaluation value of the evaluated object. The larger the value, the better the evaluation and the higher the development level.

2.4 Coupling Coordination Degree Model

The coupling coordination degree analysis method is utilized to examine the interplay between the three subsystems outlined earlier. In particular, the WRE support and carbon emission subsystem is coupled and coordinated with the urban economic HQD subsystem, respectively. The coupling degree of each subsystem is computed individually. The coupling model is calculated as follows:

$$C = 2\sqrt{\frac{F_i \times F_j}{\left(F_i + F_j\right)^2}} \quad (9)$$

where C represents the coupling degree of the system; Fi represents the evaluation value of the WRE support system or carbon emission system, and Fj represents the evaluation value of the urban economic HQD. The criteria for determining the coupling degree are specified in Table 2, which divides it into four levels. A larger C value indicates a stronger correlation between the two systems and a higher level of coupling, while a lower value indicates the opposite.

Table 2. Coupling degree C, Coordination degree D judgment table.

Coupling degree	(0, 0.2]	(0.2, 0.4]	(0.4, 0.6]	(0.6, 0.8]	(0.8, 1.0]
Type C	Low level coupling	Antagonistic stage	Run-in stage	Fundamental coupling	High level coupling
Type D	Dissonance (I)	Borderline disorder (II)	Low coordination (III)	Moderate coordination (IV)	High coordination (V)

Table 3. Judgment Table for Collaborative Development Coefficient U.

Judging condition	(0, 0.8]	(0.8, 1.2]	(1.2, ∞]
Cooperative state	"Quality of urban economic development" leading type	Synchronized development	"Water resources" or "carbon emission" leading type
Sort	U*u*	U	*Uw or Uc*

In order to assess the overall coordination level of the subsystems, the coordination degree and synergy development coefficient are further introduced to reflect the degree of coordination between the "urban economic development quality subsystem" and the "WRE support subsystem and CEM subsystem" comprehensively and accurately. The calculation formulas are as follows:

$$T = \alpha F_i + \beta F_j \tag{10}$$

$$D = \sqrt{C \times T} \tag{11}$$

$$U = F_i / F_j \tag{12}$$

where, C represents the coupling degree, D represents the coordination degree, T represents the comprehensive coordination index of the two systems, and U represents the coefficient of coordinated development, indicating the degree of coordinated development between the two systems. α and β are weight coefficients, where $\alpha > 0$, $\beta > 0$, and $\alpha + \beta = 1$. Referring to the studies of Yang [33] and Weng et al. [34], the coefficients of both are taken as 1/2. Similarly, when there are three systems, the coefficients of all three are set to 1/3. Liu et al. [35] divides the level of coordination degree into 5 types, and the level of collaborative development into 3 types, and combines the level of coordination and coordination into a unified classification, for example, "moderate coordination - leading quality of urban economic development" is recorded as "IV-Uc" type, and the rest is also the same.

3 Result Analysis of Coupling Coordination Degree

3.1 Evaluation Results Analysis of Comprehensive Development Level

Equations (1)–(8) is used to conduct a comprehensive evaluation of the development level of urban economic HQD subsystem, WRE subsystem and carbon emission subsystem, and the evaluation results are shown in Table 4.

(1) Evaluation results of the urban economic subsystem development level. The evaluation results in Table 4 indicate that the urban economic development quality subsystem of Xi'an City maintained a stable and positive development trajectory from 2010 to 2022. Specifically, it can be divided into three stages: a rapid development stage (2010–2015), a development adjustment stage (2016–2018), and a stable development stage (2019–2022). During the rapid development stage, the quality of urban economic development in Xi'an underwent a rapid increase from a low level to a relatively high level, with its development level rising swiftly from 0.2819 to 0.5445, with an average growth rate of 14.07%. This period coincided with China's 12th Five-Year Plan, during which the country vigorously promoted industrial restructuring, innovation, and advocated for green and shared development. Aligning with national policies, Xi'an actively transformed backward production capacity, improved industrial structure, continuously coordinated urban and rural development, and promoted integrated construction. Concurrently, it vigorously promoted innovative development methods, advocating for an economy dominated by green development, thus significantly enhancing the HQD level of Xi'an's urban economy. In the development adjustment stage, the quality of urban economic development in Xi'an gradually declined to 0.4519 in 2018, with an average annual decline of 6.77%. The development correction is the result of the adjustment of the parts that are not adapted to the development state of the previous stage, mainly the correction of the parts that are not adapted to the economic, social, ecological and cultural development conditions of Xi'an in the previous development stage, which will inevitably lead to the decline of the quality of urban economic development to a certain extent. Entering the stable development stage, various sectors of Xi'an's urban economy underwent comprehensive adjustments, with the degree of coordination and cooperation continuously strengthened, leading to a rapid increase to 0.6416 in 2021, with an average annual increase of 8.84%. It is worth mentioning that despite the severe impact of COVID-19 on the national economy, Xi'an's urban economy did not experience a catastrophic decline, only adjusting by 6.51%. This indicates its strong resilience.

(2) Evaluation results of the WRE support subsystem. The evaluation results in Table 4 indicate that from 2010 to 2022, the comprehensive development level of Xi'an's WRE support subsystem experienced an initial increase followed by a subsequent decline. This can be divided into two phases: a period of steady improvement (2010–2017) and a phase of gradual decline (2018–2022). During the phase of steady improvement, indicators such as water usage structure, water use efficiency, and per capita WRE in Xi'an showed continuous improvement, contributing to the steady improvement of its WRE support subsystem. In the phase of gradual decline, indicators such as total water consumption, water usage structure, and per capita WRE

Table 4. Evaluation results of the comprehensive development level, coupling and coordination state of the three systems.

	Comprehensive development level of high quality of urban economy	Comprehensive development level of water resources	Comprehensive development level of carbon emission	Coupling state between high quality development of urban economy and water resources	Coupling state between high quality development of urban economy and carbon emission	Harmonious state between HQD of urban economy and water resources	Harmonious state between HQD of urban economy and carbon emission
2010	0.2914	0.4833	0.6273	0.6043	0.6347	0.4056	0.4605
2011	0.3869	0.5269	0.5829	0.7564	0.6874	0.5420	0.5682
2012	0.4770	0.5728	0.6400	0.8694	0.7394	0.6510	0.6670
2013	0.5132	0.6140	0.6276	0.8788	0.7591	0.6784	0.6926
2014	0.5299	0.6479	0.5350	0.9346	0.7789	0.7252	0.7107
2015	0.5445	0.6597	0.4564	0.9617	0.8062	0.7710	0.7338
2016	0.5072	0.7230	0.4127	0.9042	0.7895	0.7276	0.6752
2017	0.5044	0.7348	0.4670	0.8830	0.7846	0.7130	0.6816
2018	0.4733	0.6815	0.4393	0.7950	0.7614	0.6369	0.6092
2019	0.4951	0.6688	0.4408	0.8216	0.7860	0.6711	0.6403
2020	0.5463	0.6385	0.4277	0.8105	0.8091	0.6677	0.6671
2021	0.6103	0.6260	0.3825	0.8832	0.8266	0.7320	0.6929
2022	0.5706	0.6183	0.4423	0.7584	0.7985	0.6169	0.6759

exhibited a downward trend, leading to a decrease in the development level of the urban WRE support subsystem of Xi'an.

(3) (Evaluation results of the CEM subsystem. The evaluation results reveal a declining trend in Xi'an's CEM subsystem, broadly divided into two phases: a rapid decline stage (2010–2016) and a slow decline stage (2017–2022). During the rapid decline stage, the CEM decreased from 0.6273 to 0.4127, with an average annual decline of 6.74%. This period suggests rapid integration of green development principles into urban economic strategies, resulting in a swift reduction in the CEM subsystem. In the slow decline stage, all departments of the urban economy of Xi'an experienced rapid carbon emission reduction, and the carbon emission reduction potential was fully released. In the follow-up period, although the carbon reduction effect was still obvious, the speed slowed down, as shown by the slow decline trend of the urban carbon emission subsystem, with an average annual decline of only 1.40%, and the speed was only 20.8% of the rapid decline stage.

3.2 Analysis of Coupling Degree Evaluation Results

Using Eq. (9), the coupling relationships between Xi'an's urban economic development quality, WRE support, and CEM, are calculated separately. The results are shown in Table 4.

(1) Coupling relationship between urban economic development quality and WRE support subsystem. Except for a few years, such as 2010, 2011 and 2022, the coupling relationship between the quality of urban economic development and the WRE support subsystem of Xi'an is lower than 0.8, and the rest are in the range of 0.8–1.0, basically showing a high-level coupling state. This indicates that the quality of urban economic development and the subsystem of WRE support show a very strong interdependence, indicating that the interaction between the HQD of Xi'an urban economy and the subsystem of WRE support is obvious.

(2) Coupling relationship between urban economic development quality and the CEM subsystem. The coupling relationship between these two subsystems in Xi'an ranges between 0.6 and 0.8, indicating a basic coupling state. This suggests that although the interdependence between urban economic development quality and CEM is slightly weaker than that of the WRE support, there is still a relatively strong mutual dependence. It indicates that the interaction between the HQD of Xi'an's urban economy and the CEM subsystem is also quite evident.

3.3 Analysis of Coordination Degree Evaluation Results

Using Eqs. (10) to (11), the coordination relationships between urban economic development quality and WRE support, as well as CEM in Xi'an, are calculated separately. The results are shown in Table 4.

The coordination relationships between urban economic development quality and the other two subsystems exhibit a high level of consistency in Xi'an, characterized by three stages: rapid improvement stage (2010–2015), slow decline stage (2016–2018), and slow improvement stage (2019–2022). This progression transitions from a low coordination to a moderate coordination state. During the rapid improvement stage, the coordination relationships between the urban economic development quality subsystem and the two other subsystems both rapidly progress from a low coordination stage to a moderate coordination stage, reaching the peak in the study period their peak. This indicates a rapid enhancement in the positive interactive development status between the urban economic development quality subsystem and two subsystems. In the slow decline stage, the coordination status between urban economic development quality and the other subsystems shows a gradual decline trend. Still, it remains at a moderate coordination level throughout, indicating that the benign interaction indicators among the three systems always show a good benign interaction state. However, there is a certain degree of correction. In the slow improvement stage, the positive interaction relationships among the three systems experience rapid improvement and adjustments, demonstrating a rational slow improvement trend. This indicates that the positive interaction relationships among the three systems are becoming more mature, and although the improvement is slow, the quality has improved. It is worth noting that towards the end of the study period, influenced by the COVID-19 pandemic, there is a certain decline in the coordination among the three systems, indicating that the pandemic has, to some extent, disrupted the positive interaction relationship between the subsystems.

3.4 Analysis of Synergy Evaluation Results

Based on the evaluation of the degree of coordination, the synergistic relationship between "water support" and "CEM" and the quality of urban economic development is analyzed separately. Using Eq. (12), the synergistic relationships between Xi'an's urban economic development quality and WRE support and CEM are calculated separately, and the results are shown in Table 5. The synergistic relationship between the urban economic development quality subsystem and the WRE support subsystem indicates that during the study period, the WRE support subsystem was predominantly at an advanced development level for most of the time, while the urban economic development quality subsystem only maintained a certain development lead at the end of the study period. Xi'an is primarily a water-deficient city, with WRE being a crucial determinant for supporting the HQD of Xi'an's urban economy. To ensure the HQD of Xi'an's urban economy, the Xi'an municipal government has made significant efforts to construct the urban water resource support system, continuously improve water use efficiency and water structure, and sustainably develop WRE. This has led to the maintenance of a high level of development quality in the water resource support system, as demonstrated by its leadership over the urban economic development quality subsystem for most of the study period.

The synergistic relationship between the urban economic development quality and CEM subsystem is characterized by three stages. In the early stage of the study period, the CEM subsystem primarily led development. In the mid-term, there was generally a coordinated development between the two subsystems, and in the later stage, the urban economic development quality subsystem predominantly led in development. The synergistic relationship between the CEM system and urban high-quality economic development largely aligns with the broader economic environment in China. In the early stage of the study period, China vigorously developed economic construction.

Although much attention was paid to green environmental protection development in this period, the control of carbon emission was relatively loose, which was reflected in the slow development of carbon reduction, which was consistent with the situation that the carbon emission subsystem of Xi'an was ahead of the urban economic development subsystem. As international attention to climate change intensified, China, as a major country responsible of climate change, placed significant emphasis on carbon emission management, continuously introducing strategies and concepts for green development. This facilitated a gradual reduction in CEM nationwide and noticeable improvements in Xi'an's CEM, reflecting a coordinated development between the CEM and urban economic development quality subsystem. With further development of the green economy, towards the end of the study period, Xi'an's urban economic development quality subsystem gradually surpassed the CEM subsystem.

Table 5. Collaborative evaluation results of three systems.

	Fw/Fu	Synergistic state between high quality urban economic development and water resources	Fc/Fu	Synergistic state between high quality urban economic development and carbon emission
2010	1.6587	III-Uw	2.1529	III-Uc
2011	1.3618	III-Uw	1.5065	III-Uc
2012	1.2006	IV-Uw	1.4244	IV-Uc
2013	1.1964	IV-U	1.2231	IV-Uc
2014	1.2228	IV-Uw	1.0097	IV-U
2015	1.0913	IV-U	0.7549	IV-Uu
2016	1.4254	IV-Uw	0.8137	IV-U
2017	1.4569	IV-Uw	0.9259	IV-U
2018	1.5035	IV-Uw	0.9692	IV-U
2019	1.3508	IV-Uw	0.8903	IV-U
2020	0.9674	IV-Uu	0.7830	IV-Uu
2021	0.7996	IV-Uu	0.6167	IV-Uu
2022	0.7330	IV-Uu	1.0206	IV-U

4 Conclusion

(1) The comprehensive development level of the three subsystems in Xi'an, including urban economic development quality, WRE support, and CEM, exhibits a consistent upward trend. Through detailed studies on the quality assurance systems for each subsystem, distinct internal driving forces for improvement of the integrated development level of the three subsystems. In the urban economic development

quality subsystem, innovative and green development hold weights of 22.158% and 17.515%, respectively, playing pivotal roles in determining its quality. Within the WRE support subsystem, water resource quantity and water use efficiency both carry weights of 27.644%, underscoring their importance in determining the quality of WRE development. Similarly, in the CEM subsystem, total CEM and per capita CEM have weights of 27.914% and 34.203%, respectively, which is crucial for determining the carbon emission subsystem.

(2) The coupling relationships between the urban economic development quality subsystem and the subsystems of WRE support and CEM in Xi'an demonstrate a trend of initial improvement, followed by a decline and then stabilization. They have both transitioned from a stage of adaptation to one of high-level coupling. Although there have been some downward adjustments, they have essentially stabilized at a high level of coupling, with their interdependence gradually deepening. In terms of the degree of coupling change, the coupling status between the urban economic development quality subsystem and WRE shows greater variability compared to that with the CEM subsystem, indicating a relatively stable level of interdependence between the urban economic development quality subsystem and the CEM subsystem.

(3) The coordination relationships between the urban economic development quality subsystem and the subsystems of WRE support and CEM in Xi'an remain highly consistent, showing a trend of wave-like ascension. The comprehensive upward trend in coordination continues to show a positive trend of improvement in the coordination status between the urban economic development, quality subsystem and the other two subsystems in Xi'an. When combined with the trend of coupling, it suggests a favorable developmental situation where the relationship between the urban economic development quality subsystem and the WRE support and CEM subsystems demonstrates synchronous progress in both coupling and coordination.

(4) The urban economic development quality subsystem in Xi'an is basically in accordance with and aligned with the coordinated pace of the subsystems of WRE support and CEM. In the initial phase of the research period, there was a tendency for the WRE support and CEM subsystems to take precedence, highlighting their substantial influence on the development quality subsystem of the urban economy. They compelled the urban economic development quality subsystem to adapt to their changes. However, in the latter phase of the research period, the urban economic development quality subsystem began to lead both the WRE support and CEM subsystems. This transition suggests that urban economic development has gradually overcome the constraints imposed by these two subsystems and is progressing towards a path of virtuous circle development.

Acknowledgements. Upon the completion of this research, I am filled with gratitude. I would like to express my sincere appreciation to the Key Scientific Research Project of Shaanxi Provincial Department of Education (21JT031) "Research on the High-Quality Development Path of Xi'an City under the Constraint of Water Resources" fund and the Soft Science Research Program of Forestry and Grassland (2023131020) "Research on the Connection and Development between Nature Reserves and Communities" for their financial support, which has provided crucial assistance for the research. In addition, I am especially grateful to my team member Yan Jiaxing and

Zheng Jiaojiao for their help in data collection. It is the close collaboration among every team member that has enabled the successful completion of this research.

References

1. Shen, K.R., Zhao, Q.: Promoting high-quality economic development during the 14th Five-year Plan Period by adopting new development pattern of dual cycle. Econ. Rev. J. **2020**(10), 18–25 (2020)
2. Martinez, M., Mlachila, M.: The quality of the recent high-growth episode in Sub-Saharan Africa. IMF Working Papers, **13**(53), 1–35 (2013)
3. Mlachila, M., Tapsoba, R., Tapsoba, S.J.: A quality of growth index for developing countries: a proposal. IMF Working Papers, **14**(172), 1–32 (2014)
4. Cheng, C.P.: How to implement the fundamental requirements of high-quality development. Governance **2018**(05), 27–33 (2018)
5. Ren, B.P., Zhao, T.: The core essence and policy orientation of high-quality development. Red Flag Manuscr. **2019**(13), 23–25 (2019)
6. Li, M.X., Ren, B.P.: Comprehensive evaluation and path choice of China's high-quality development in the new era. Financ. Econ. **2019**(05), 26–40 (2019)
7. Jin, B.: Study on the "High-Quality Development" economics. China Ind. Econ. **2018**(4), 5–18 (2018)
8. Ma, R., Luo, H., Wang, H.W., et al.: Study of evaluating high-quality economic development in Chinese regions. China Soft Sci. **2019**(07), 60–67 (2019)
9. Fan, J., Zhang, Q., Luo, C.: Evolution trend of economic development quality of Yangtze River Delta urban agglomeration and its countermeasures. J. Ind. Technol. Econ. **37**(12), 70–77 (2018)
10. Li, J.C., Shi, L.M., Xu, A.T.: Probe into the assessment indicator system on high-quality development. Stat. Res. **2019**(1), 4–14 (2019)
11. Zeng, P.: SD model and simulation of high-quality development of China's urban agglomeration under "Dual Carbon" goals. J. Hehai Univ. (Philos. Soc. Sci.), **24**(05): 52–61+129 (2022)
12. Wang, H.L., Xu, Y.J., Zhang, J.N., et al.: Study on the measurement of urban high-quality development in the Yellow River Basin based on the perspective of "Sansheng." Yellow River **46**(01), 25–31 (2024)
13. Ou, J.F., Xu, C.J., Liu, Y.Q.: The measurement of high-quality development level from five development concepts: Empirical analysis of 21 prefecture-level cities in Guangdong Province. Econ. Geogr. **40**(06), 77–86 (2020)
14. Lu, C.Y., et al.: Spatio-temporal comprehensive measurement and influencing factors of high-quality development in the Yellow River Basin. J. Lanzhou Univ. (Nat. Sci.), **59**(04): 476–483+488 (2023)
15. Fan, J., Wang, Y.F., Wang, Y.X.: High quality regional development research based on geographical units: discuss on the difference in development conditions and priorities of the Yellow River Basin compared to the Yangtze River Basin. Econ. Geogr. **40**(01), 1–11 (2020). https://doi.org/10.15957/j.cnki.jjdl.2020.01.001
16. Wang, Y.N., Zhang, X.L., Zhao, Y.J., et al.: Dynamic evaluation of high-quality development in Henan Province. Areal Res. Dev. **42**(06), 33–39 (2023)
17. Liu, Y., Li, F.T., Sun, S.Y.: Measurement and obstacle factor analysis of high-quality economic development in Chengdu-Chongqing urban agglomeration. J. Southwest Univ. Sci. Technol. (Philos. Soc. Sci. Ed.) **41**(01), 43–51 (2024)

18. Li, H., Cao, L.: Measurement of high-quality economic development of urban agglomeration in the middle reaches of the Yangtze River. Stat. Decis. **37**(24), 101–105 (2021). https://doi.org/10.13546/j.cnki.tjyjc.2021.24.022
19. Li, F.Z., Miao, Q.: The spatial network characteristics of the coupling between urban ecological protection and high-quality economic development in the Yellow River Basin. Stat. Decis. **38**(05), 80–84 (2022). https://doi.org/10.13546/j.cnki.tjyjc.2022.05.015
20. Li, T., Xue, L., Li, G.P.: The evolution of spatial pattern of industrial agglomeration and its impact on the high-quality economic development: Empirical analysis based on the data of 278 cities in China. Geogr. Res. **41**(04), 1092–1106 (2022)
21. Zhu, Y.M., Jia, Z.Y.: Research on the spatial connections and characteristics of high-quality economic development of cities: A case study of the Central Plains Urban Agglomeration. Ecol. Econ. **38**(12), 82–88 (2022)
22. Zhou, L., Gong, Z.M.: The impact of digital economy level on green high-quality development in Chinese Cities. Econ. Geogr. **42**(11), 133–141 (2022). https://doi.org/10.15957/j.cnki.jjdl.2022.11.015
23. Zhang, Y.T., Zhang, S.R., He, S.K.: Study on the mechanism of synergistic promotion of high-quality economic development by environmental regulation and scientific and technological environmental regulation, technological innovation and high-quality economic development-A case study of 16 prefecture-level cities in Cheng-du-Chongqing Economic Circle. Resour. Dev. Market (2024). https://link.cnki.net/urlid/51.1448.N.20240305.1111.004
24. Zhang, G.X., Zhang, J.Y.: Spatiotemporal evolution of high-quality development of resource-based cities in the Yellow River Basin. China Popul. Resour. Environ. **33**(2), 124–133 (2023)
25. Sun, J.S., Li, Q.S.: Research on the impact of talent agglomeration on high quality economy development-a double test of spatial metrology and threshold regression based on major cities in China along the Silk Road. J. Xinjiang Normal Univ. (Nat. Sci. Ed.), **42**(04): 65–76+94 (2023)
26. Wang, J.M., Ding, H.: Environmental regulation of resource-based cities and high-quality economic development. Stat. Decis. **38**(15), 108–112 (2022). https://doi.org/10.13546/j.cnki.tjyjc.2022.15.020
27. Wang, X.F., Jiang, C.H.: High-quality development progress in the Yellow River Basin under the rigid constraint of water resources. J. Econ. Water Res. **41**(02): 18–24+32+93–94 (2023)
28. Jiang, X.C., Liu, J.H.: Decoupling analysis between water resource use efficiency and high-quality economic development in Yellow River Streaming. Res. Ind. **25**(2), 65–75 (2023)
29. Li, M.H., Zhao, X.P., Zhang, Z.Z., et al.: Development model and application of optimal allocation of water resources in water-deficient areas for high-quality. Yellow River **46**(2), 73–79 (2024)
30. Chen, X.L., et al.: The mechanism and empirical on the "Pollution reduction and carbon reduction" and high-quality economic development under the "Dual-Carbon" target: A Case of the China's coastal urban agglomerations. World Reg. Stud. 1–16 [2024–03–14] (2024)
31. Kuang, C.E., Li, W.Y., Huang, X.S.: Spatial-temporal evolution and driving factors of coupling coordination between carbon emission intensity and high-quality economic development in urban agglomerations in the middle Reaches of the Yangtze River. Econ. Geogr. **42**(08), 30–40 (2022). https://doi.org/10.15957/j.cnki.jjdl.2022.08.004
32. Liu, Z.H. et al.: Research on the synergistic effect of low-carbon economy and high-quality development under the "Dual-Carbon" strategy: A Case Study of Beijing-Tianjin Hebei Urban Agglomeration. Environ. Sci. 1–20 [2024–03–14]. https://doi.org/10.13227/j.hjkx.202312013 (2024)
33. Yang, J., Cheng, Y.L.: Research on the spatiotemporal differentiation and influencing factors of the coupling coordination degree of digital economy, environ-mental regulation, and sustainable development. Ecol. Econ. **40**(05), 154–163 (2024)

34. Weng, G.M., Tang, Y.B., Pan, Y., et al.: Spatiotemporal evolution and spatial difference of tourism-ecology-urbanization coupling coordination in Beijing-Tianjin-Hebei Urban Agglomeration. Econ. Geogr. **41**(12), 196–204 (2021). https://doi.org/10.15957/j.cnki.jjdl.2021.12.021
35. Liu, N.R.S., Wu, R.G., Jin, L.: Coupling coordinated evaluation of sustainable livelihoods and ecological livability of farming and herding households in semi-agricultural and semi-pastoralist areas of northern China. J. Arid Land Res. Environ. **37**(11), 74–81 (2023)

The Impact of Urbanization Development on Carbon Emissions: A Study Based on the STIRPAT Model

Yifen Xia[1], Yuanzhuo Wu[1(✉)], Jiatang Luo[1], and Chun Fu[1,2]

[1] School of Public Policy and Administration, Nanchang University, Nanchang 330000, Jiangxi, China
17807065007@163.com

[2] Jiangxi Regional Economic Research Institute, Jiangxi 330000, China

Abstract. This study investigated the dynamics and mechanisms between urban development and carbon emissions using the STIRPAT model. The IPCC calculation method and ridge regression were used to analyze panel data from Jiangxi Province spanning 2003 to 2021. The findings reveal that carbon emissions are positively influenced by per capita disposable income, outputs from the secondary and tertiary industries, urban population proportion, and built-up areas, with the secondary industry exhibiting the most substantial effect. In contrast, energy intensity demonstrates a negative relationship with carbon emissions, suggesting that improved energy intensity can reduce emissions. However, this reduction is insufficient to offset the carbon emissions generated by other urbanization factors. Additionally, an inverted U-shaped relationship between urbanization and carbon emissions was identified, indicating that a 1% increase in urbanization initially results in a 0.083% rise in carbon emissions, which subsequently declines after surpassing a specific threshold. These results underscore the necessity of promoting sustainable urbanization strategies to mitigate carbon emissions.

Keywords: Urbanization · Carbon Emission · STIRPAT Model

1 Introduction

Urbanization has been pivotal in China's socio-economic transformation since 1950. By 2021, China had achieved an urbanization rate of 63.9% within a span of three decades, a transformation that took centuries to accomplish in Western countries. Exploring the success of China's urbanization, Guan et al. [1] attribute the nation's achievements to a diversified strategy. Economic, social, and environmental factors shape urbanization's trajectory toward sustainable development. Huang et al. [2] contend that industrial configuration and urban-rural synergy are vital economic drivers. Building on this, Turok et al. [3] highlight urban attraction and rural propulsion as crucial forces driving the process in developing nations. Using spatial panel econometrics, Zhou et al. [4] found that industrial structuring helps reduce regional disparities, although its effects vary across time and space. Polyzos et al. [5] concluded that social factors, such as demographic

C. Weng (Ed.): WRE 2024, LNCE 634, pp. 574–589, 2025.
https://doi.org/10.1007/978-981-96-5095-8_46

magnitude, are influenced by government policies and market forces, with the latter playing a more significant role. Additionally, urbanization is a multifaceted process shaped by demographic, economic, and social factors that significantly contribute to carbon emissions, making it crucial to decouple these dimensions from carbon emissions to mitigate climate change [6].

Traditional urban development models have relied heavily on fossil fuels for construction, transportation, and energy provision. This dependence on carbon-intensive energy sources has significantly increased greenhouse gas emissions, particularly carbon dioxide, exacerbating global warming. The relationship between urbanization and carbon emissions is complex and non-linear [7], shaped by various social, economic, and technological factors. This complexity is firstly reflected in the research on the interplay between carbon emissions and economic growth, much of which supports the Environmental Kuznets Curve (EKC) hypothesis [8]. Song et al. [9] extended this concept by identifying a threshold effect between economic growth and environmental quality, while Sikdar et al. [10] examined this relationship in OPEC countries. They revealed deviations from the typical "inverted U-shape" of the EKC, with specific sectors exhibiting a "U-shaped" trend. Beyond economic growth, carbon emissions are influenced by population size and energy utilization [11]. Furthermore, Liu et al. [12] found the mediating role of energy efficiency in reducing emissions, again underscoring the multifaceted nature of the determinants of carbon emissions.

The 20th National Congress of the Communist Party of China (CPC) emphasized the value of ecological assets and harmonious coexistence between humans and nature. Achieving China's "Dual-Carbon" target demands a sustainable, low-carbon urban model that integrates social, economic, and ecological systems. However, reducing carbon emissions poses challenges at different stages of urbanization [13]. Studies reveal that the relationship between urbanization and carbon dioxide emissions in China varies by region, with urbanized, urbanizing, and under-urbanized areas displaying distinct patterns [14]. In response to these challenges, "Jiangxi Province Urban System Planning (2015–2030)" aims to improve urban planning, management, and industry development for coordinated urban growth. Despite high GDP growth, Jiangxi's total GDP ranks low nationally, indicating the need to accelerate urbanization, leading to increased energy demands. To address this, Jiangxi's "14th Five-Year Plan" targets reductions in energy consumption and carbon emissions to balance economic development with environmental protection [15].

While numerous scholars have studied the factors influencing carbon emissions and urbanization separately, there is a gap in research that integrates both dimensions. Furthermore, the debate remains unresolved regarding whether many of these factors positively or negatively influence carbon emissions, with some studies reaching conflicting conclusions. A potential reason for these discrepancies could be the inability of traditional Ordinary Least Squares (OLS) regression models to address multicollinearity issues between variables, leading to biased results.

This study addresses these gaps by examining factors influencing carbon emissions from an urbanization perspective, integrating economic, social, and environmental elements. Additionally, the study introduces quadratic terms to investigate potential nonlinear relationships often overlooked in conventional models. This approach aims to provide a deeper understanding of the relationship between urbanization and carbon emissions, contributing to developing more effective low-carbon urban strategies.

2 Theoretical Analysis and Research Hypotheses

2.1 Urbanization Rate and Carbon Emissions

Rapid urbanization is a crucial feature of contemporary economic development. The intensification of urbanization is closely linked with a rise in carbon emissions within urban environments [16]. At lower levels of urbanization, urbanization trajectory stimulates the economy through capital-intensive processes such as the development of infrastructure, residential construction, and the expansion of transportation networks—all predicated on the consumption of energy-dense resources [17]. However, as urbanization progresses and reaches higher levels, adopting more efficient technologies, shifting towards service-oriented economies, and improved urban planning and management can help mitigate carbon emissions [18].

Hypothesis 1: There is an inverted U-shaped relationship between the rate of urbanization and carbon emissions.

2.2 Industrial Output Value and Carbon Emissions

The association between industrial output and carbon dioxide emissions is significant and multifaceted. Industry expansion typically goes hand in hand with an escalation in energy use and material conversion, culminating in increased carbon emissions [19]. Furthermore, the emissions are intricately linked to both the structural composition and the production practices of the industrial sectors [20]. The secondary sector—including manufacturing, construction, and electric power production—primarily relies on the combustion of fossil fuels such as coal, oil, and gas, which are primary sources of carbon dioxide [21]. The growth of coal-fired power stations, metallurgical operations, and cement manufacturing are energy-demanding. Growth in the output of these industries will lead to a significant increase in carbon dioxide emissions [22].

Hypothesis 2: There is a positive relationship between the output value of the secondary industry and carbon emissions.

The tertiary sector, encompassing the service industry, plays a vital role in the modern economy and includes sectors such as retail, finance, education, and healthcare [23]. As urbanization and economic development progress, the demand for tertiary services swells, precipitating an uptick in urban energy needs. The activities in the tertiary sector frequently involve transportation-related services such as roads, water, and aviation, which contribute to increased carbon emissions [24]. Furthermore, the energy required for heating, cooling, and lighting large office buildings and commercial centers within the service industry contributes significantly to carbon emissions [25]. In addition, the service sector's improper treatment of municipal solid and electronic waste can produce methane and other greenhouse gas emissions [26].

Hypothesis 3: There is a positive relationship between the output value of the tertiary industry and carbon emissions.

2.3 Per Capita Disposable Income and Carbon Emissions

Empirical evidence indicates that residential income impacts carbon emissions growth with a marginal contribution rate of around 0.87% [27]. As urban residents shift from bare subsistence to higher living standards, the increased demand for housing, transportation, and consumer goods like home appliances leads to higher carbon emissions [28]. This trend is driven by the correlation between rising per capita incomes and greater household energy consumption and emissions. Additionally, entrenched consumption patterns and lack of awareness about energy conservation and environmental protection among some urban residents lead to resource wastage and indirect carbon emissions [29]. Studies have found a parallel rise in economic output, carbon dioxide emissions, and per capita disposable income across urban areas [30].

Hypothesis 4: A positive relationship exists between per capita disposable income and carbon emissions.

2.4 Urban Built-Up Areas and Carbon Emissions

Accelerating urbanization leads to an expansion of built-up areas, which inherently increases carbon emissions from energy use in buildings and transportation [31]. Furthermore, the expansion of built-up urban areas can lead to changes in land use, reducing agricultural and forested lands, which can significantly impact surrounding natural ecosystems and impair the carbon sequestration system [32]. The augmentation of built-up areas correlates with an elevated demand for building energy, thereby increasing carbon emissions [33]. However, regional structural changes in built-up areas can reduce total building carbon emissions by up to 3.7 million tons by shifting the concentration of construction activities to locations with lower energy intensity and carbon emission factors [34].

Hypothesis 5: A positive relationship exists between urban built-up areas and carbon emissions.

2.5 Energy Intensity and Carbon Emissions

Energy intensity exerts profound influence on carbon emissions during the urbanization process [35]. A concerted push to improve energy efficiency will be crucial to avoid persistent high energy intensity and elevated carbon emissions [36]. It has been demonstrated that a 1% increment in energy consumption correlates with a 0.15% rise in carbon emissions in the short term and a more pronounced 1.03% increase in the long term [37]. Reduced energy intensity coupled with green technological innovations can improve utilization efficiency and facilitate effective carbon emission reductions [38]. Energy intensity is a crucial driver for reducing carbon emissions, particularly in the context of rapid urbanization.

Hypothesis 6: There is a negative relationship between energy intensity and carbon emissions.

Based on the above hypotheses, the research framework of this study is shown in Fig. 1.

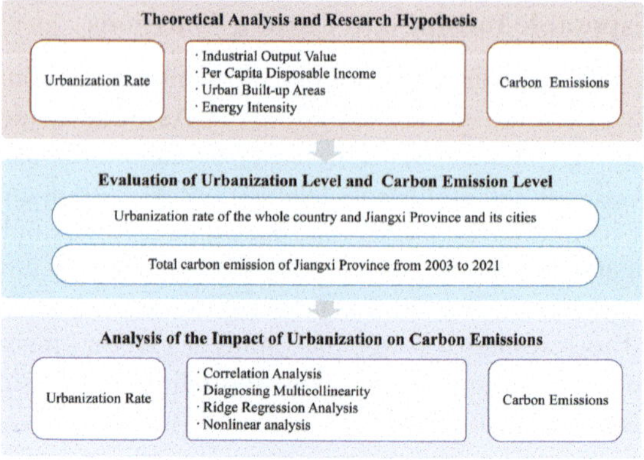

Fig. 1. Research framework.

3 Research Design

3.1 Variable

The Kaya Identity outlines key factors affecting carbon emissions: per capita energy use, population size, and energy infrastructure. Yang et al. [39] expanded on this by examining how carbon emissions relate to demographic trends, per capita economic output, energy efficiency, and carbon intensity. Urbanization is crucial in increasing carbon emissions through expanded spatial footprints, lifestyle changes, and higher energy demands. Its effect is influenced by economic size, industrial structure, and energy use. This paper analyzes the impact of urban development on carbon emissions by using calculated carbon emissions as the dependent variable, with urbanization rate, secondary industry output, tertiary industry output, urban built-up areas, urban residents' per capita disposable income, and energy intensity as independent variables representing urbanization.

3.2 Data Source

The dataset used in this study was sourced from the IPCC Guidelines for National Greenhouse Gas Inventories, the Jiangxi Statistical Yearbook, and the China Energy Statistical Yearbook. Before analysis, the raw data were standardized to address inconsistencies in units and magnitudes, eliminating biases from heterogeneous data and improving the accuracy and reliability of comparative evaluations.

3.3 Model

The IPAT framework, introduced by Ehrlich et al. [40], represents environmental impact as a product of population (P), affluence (A), and technology (T). The Stochastic Impacts by Regression on Population, Affluence, and Technology (STIRPAT) model, developed by York et al. [41]. Refined this by incorporating stochastic elements, providing a more nuanced analysis. The model is formulated as:

$$I_i = aP_i^b \times A_i^c \times T_i^d \times \varepsilon_i \tag{1}$$

where I quantifies the environmental impact, P encapsulates the population size, and A indicates the level of affluence, T assesses the technology level, a is the model coefficient, b is the index of the population size variable, c is the index of the affluence level variable, d is the index of the technology level variable, i marks the data point in a time series, ε denotes the random disturbance term.

For this study, per capita disposable income, secondary and tertiary industry output, urban population proportion, and built-up areas are proxies for affluence and population size, while energy intensity represents technology. The carbon emission function based on the STIRPAT model is:

$$C = aU^{b1} \times I^{b2} \times S^{b3} \times W^{b4} \times F^{b5} \times E^{b6} \times \varepsilon \tag{2}$$

Taking the logarithm:

$$InC_i = a + bInU_i + cInS_i + dInT_i + eInG_i + fInB_i + gInE_i + \varepsilon_i \tag{3}$$

The study also integrates the environmental Kuznets curve, adding the square of the urbanization rate to explore the relationship between urbanization and carbon emissions:

$$InC_i = a + bInU_i + cInS_i + dInT_i + eInG_i + fInB_i + gInE_i + h(InU_i)^2 + \varepsilon_i \tag{4}$$

where C represents the total carbon emissions, a is the constant term of the model, U denotes the proportion of the urban population to the total population, S reflects the output value of the secondary industry, T is the output value of the tertiary industry, G indicates the per capita disposable income of urban residents, B measures the area of the urban built-up areas, E captures the energy intensity, Coefficients b, c, d, e, f, g represent the elasticities of the respective variables within the model, and ε is the stochastic error term.

4 Empirical Analysis

4.1 Evaluation of Urbanization Level

By the end of 2021, Jiangxi's urbanization rate had increased by 27% compared to 2003, reflecting significant progress in the province's urban development. Despite steady progress, the urbanization trajectory in Jiangxi Province still needs to catch up. It lags the national average and falls short of the benchmarks established by more economically advanced regions. As illustrated in Fig. 2, from 2003 to 2017, Jiangxi's urbanization rate consistently lagged 4–5 percentage points behind the national average, though its growth rate was mainly in line with the national trend. After 2017, while Jiangxi's urbanization rate remained below the national average, its growth rate surpassed the national level.

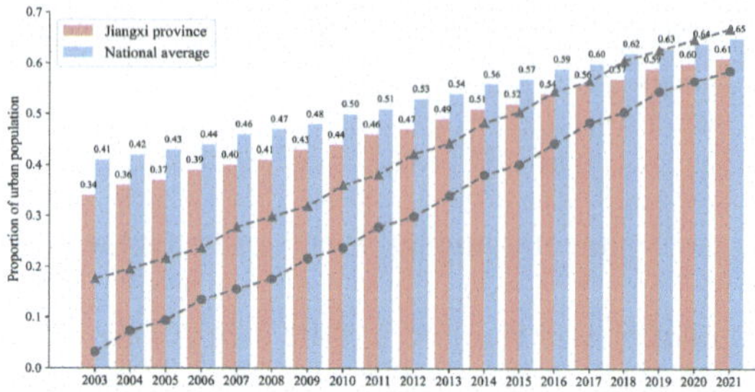

Fig. 2. Urbanization rate of Jiangxi province and national average.

As shown in Fig. 3, Jiangxi's average urbanization rate in 2021 was 61.46%, with Nanchang at 78% and Ji'an at 53%. The standard deviation in urbanization rates among the province's districts was 7.28%, with a maximum divergence of 25.23%, highlighting substantial regional development disparities. Nanchang's high urbanization rate contrasts with the slower rates in other districts, which are influenced by natural resources, economic conditions, policy directions, and infrastructure development.

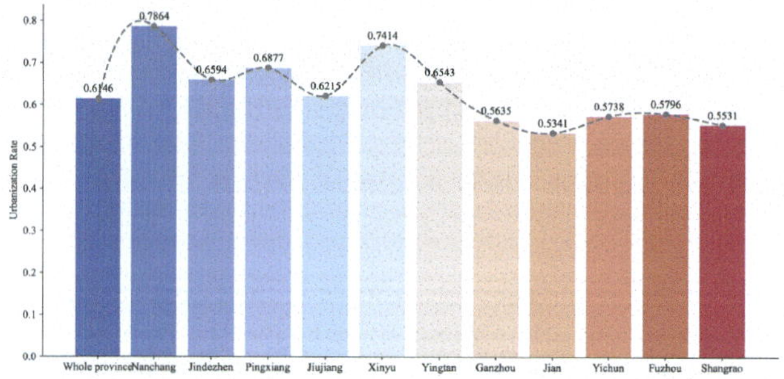

Fig. 3. Urbanization rate by city in Jiangxi Province in 2021.

4.2 Evaluation of Carbon Emission Level

This study utilizes the IPCC method, referencing data from the IPCC Guidelines for National Greenhouse Gas Inventories, the Jiangxi Statistical Yearbook, and the China Energy Statistical Yearbook. The estimation model is:

$$C = \Sigma_i E_i \times CV_i \times CCF_i \times COF_i \times (44/12) \tag{5}$$

Energy consumption data must be adjusted to a common basis using energy conversion coefficients to standard coal equivalent:

$$C = \Sigma_i E_i \times e_i \times p_i \times (44/12) \tag{6}$$

where C is total carbon emissions, E represents energy consumption, CV indicates the mean low-level calorific value of the energy source, COF signifies the carbon oxidation factor for the energy, and 44/12 is the molecular weight ratio of carbon dioxide to elemental carbon. Ei quantifies the usage of the ith fossil energy type, ei provides the standard coal equivalent conversion coefficient for the ith energy type, and pi presents the carbon emission coefficient.

Figure 4 illustrates a significant rise in carbon emissions in Jiangxi Province from 67.19 million tons in 2003 to 218.82 million tons in 2021, with an average annual growth rate of 7.05%. After a decade of swift increase, a deceleration became apparent, with a notable decline recorded between 2011–2012, indicating the potential for a low-carbon future. The slowdown, attributed to successful energy conservation and emission reduction strategies post-2012, shows an initial reversal of the emission increase. Nevertheless, the reduction has been inconsistent, with emissions rising again in 2017, signaling the necessity for Jiangxi to intensify efforts in clean energy adoption and energy consumption decarbonization to curtail emissions persistently.

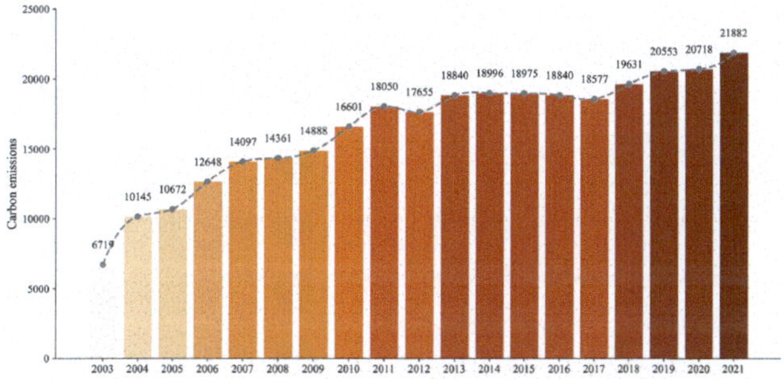

Fig. 4. Total carbon emission of Jiangxi Province from 2003 to 2021.

4.3 Correlation Analysis

Table 1 reveals substantial positive correlations between the natural logarithm of carbon emissions (LnC) and other variables: LnU (urban population proportion), LnS (output value of the secondary industry), LnT (output value of the tertiary industry), and LnB (urban built-up areas) with correlation coefficients of 0.924, 0.896, 0.822, and 0.897, respectively. Conversely, the natural logarithm of LnE (energy intensity) exhibits strong negative correlations with the abovementioned variables.

As depicted in Fig. 5, the correlation coefficient between LnU and its square (LnU)^2 is exceptionally high at 0.997, suggesting potential collinearity issues. This is further exemplified by the correlation coefficients observed between LnG and LnB (0.998), as well as LnG and (LnU)^2 (0.999). Although the correlation matrix identifies statistically significant relationships among these variables, it does not imply causation and may complicate the interpretative process due to multicollinearity.

4.4 Diagnosing Multicollinearity

An OLS regression analysis produced a coefficient of determination (R^2) of 0.978, with an F-statistic of 68.441 and a corresponding p-value below 0.001, demonstrating

Table 1. Correlation analysis matrix.

Variable	InC	InU	InS	InT	InG	InB	InE	$(InU)^2$
InC	1	–	–	–	–	–	–	–
InU	0.924**	1	–	–	–	–	–	–
InS	0.896**	0.991**	1	–	–	–	–	–
InT	0.822**	0.969**	0.958**	1	–	–	–	–
InG	0.822**	0.993**	0.985**	0.990**	1	–	–	–
InB	0.897**	0.996**	0.987**	0.983**	0.998**	1	–	–
InE	0.990**	0.948**	0.922**	0.815**	0.911**	0.923**	1	–
$(InU)^2$	0.896**	0.997**	0.991**	0.985	0.999**	0.999**	–0.992	1

Note: * $p < 0.05$, ** $p < 0.01$, *** $p < 0.01$

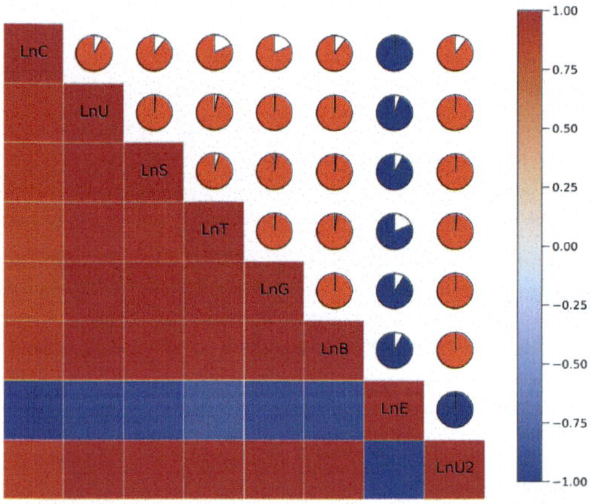

Fig. 5. Heat map of the correlation coefficient.

the overall statistical significance of the model. Nevertheless, the inability of individual regressors to meet the significance criteria as per the t-test necessitates a multicollinearity assessment within the regression equation.

Analysis of Table 2 reveals that the VIF for explanatory variables—the output value of the secondary and tertiary industries, urban population proportion, built-up areas, per capita disposable income, and energy intensity—is significantly higher than the threshold of 10. This indicates severe multicollinearity, which compromises the validity of the OLS regression results. To address this analytical deficiency, ridge regression analysis is applied to provide a more robust examination of the data.

4.5 Ridge Regression Analysis

This study employs ridge regression to mitigate the distortions of the high intercorrelation among independent variables. By introducing a bias term k (>0) to the diagonal of

Table 2. OLS regression and multicollinearity test.

Model	Unstandardized coefficient		Standardized coefficient	t	Significance	Covariance statistic	
	B	Standard error				Tolerance	VIF
Constant	4.902	1.394	–	3.517	0.005	–	–
LnU	1.411	2.060	1.699	0.685	0.508	0.000	3014.021
LnS	–0.328	0.591	–0.392	–0.555	0.590	0.004	245.135
LnT	–4.059	2.000	–4.972	–2.030	0.067	0.000	2941.648
LnG	–1.305	1.284	–1.595	–1.016	0.331	0.001	1207.706
LnB	0.344	0.710	0.427	0.485	0.637	0.003	379.994
LnE	–2.377	0.920	–2.832	–2.585	0.25	0.002	588.290
		1.249	–2.921	–1.965	0.075	0.001	1083.748

the independent variable matrix, ridge regression enhances the accuracy of regression analyses. Variation in the ridge parameter k alters the regression coefficients. The corresponding regression coefficients can be identified upon stabilizing the variables on ridge trace plots constructed using various k values. As indicated in Fig. 6, the linearization coefficients for different explanatory variables tend to stabilize when ($k > 0.35$), the five ridge trace curves approach stability.

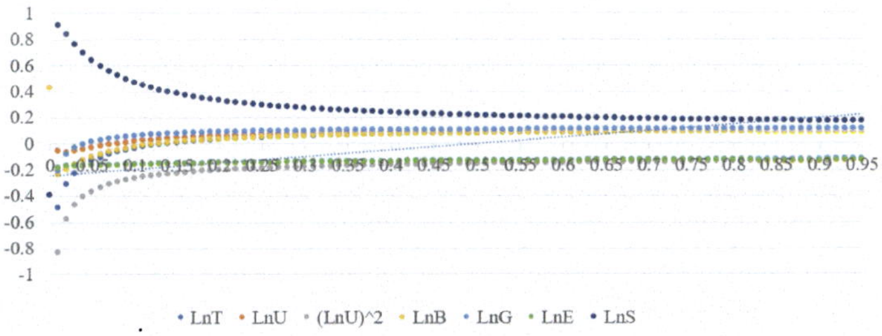

Fig. 6. Ridge trace plot.

The regression coefficients for the explanatory variables were estimated using a ridge parameter of $k = 0.35$, yielding the regression equation as follows:

$$\begin{aligned} \mathrm{InC} &= 0.083\mathrm{InU_i} + 0.251\mathrm{InS_i} + 0.068\mathrm{InT_i} + 0.103\mathrm{InG_i} \\ &+ 0.069\mathrm{InB_i} - 0.138\mathrm{InE_i} - 1.171(\mathrm{InU_i})^2 + \varepsilon_i \end{aligned} \tag{7}$$

Analysis of Table 3 reveals that a confluence of factors—namely, the urban population proportion, output value of the secondary and tertiary industries, per capita disposable income, urban built-up areas, and energy intensity—collectively explain 86.6% of

Table 3. Ridge regression results.

Model	Unstandardized coefficient		Standardized coefficient	t	p	VIF
	B	Standard error	Beta			
Constant	0.000	0.108	-	0.000	1.000	–
LnU	0.083	0.025	0.083	3.251	0.008**	0.053
LnS	0.251	0.051	0.251	4.920	0.000**	0.214
LnT	0.068	0.025	0.068	2.758	0.019*	0.050
LnG	0.103	0.019	0.103	5.533	0.000**	0.028
LnB	0.069	0.030	0.069	2.318	0.041*	0.072
LnE	–0.138	0.026	–0.138	–5.269	0.000**	0.056
$(\text{LnU})^2$	–0.171	0.024	–0.171	–7.142	0.000**	0.047
R^2	0.866					
Adjusted R^2	0.780					
F	$F(7,11) = 10.119, p = 0.001$					

Note: * p < 0.05, ** p < 0.01, *** p < 0.01

the variance in total carbon emissions. Positive regression coefficients for the urban population proportion, output value of the secondary and tertiary industries, per capita disposable income, and urban built-up areas indicate that these factors positively influence carbon emissions during urbanization. Conversely, the negative coefficient for energy intensity implies that improvements in energy efficiency are linked to reduced carbon emissions.

The regression results show that the coefficient for the natural logarithm of urbanization (LnU) is 0.083, and the coefficient for its square ((LnU)^2) is -0.171, indicating an inverted U-shaped relationship between urbanization and carbon emissions, validating Hypothesis 1. Initially, carbon emissions rise with increasing urbanization but decline after reaching a certain threshold.

Urbanization significantly affects carbon emissions, with the secondary sector having a substantial positive impact, shown by an elasticity coefficient of 0.251%. This result supports Hypothesis 2, showing that the secondary sector (industries such as manufacturing) contributes more to carbon emissions than other sectors. In contrast, the tertiary sector has a minor effect on emissions, with an elasticity coefficient of 0.068%, confirming Hypothesis 3. This indicates that the secondary sector contributes more to emissions than the tertiary sector. The elasticity coefficient of per capita disposable income is 0.103%, reflecting its positive relationship with carbon emission, confirming Hypothesis 4. This reflects the significant impact of economic well-being on the province's carbon footprint. The expansion of urban built-up areas contributes significantly to carbon emissions, with an elasticity coefficient of 0.069%, supporting Hypothesis 5. This indicates that expanding urban built-up areas exacerbates carbon emissions, contributing to environmental degradation. The elasticity coefficient of energy intensity concerning carbon emissions is –0.138%, validating Hypothesis 6. This negative relationship indicates that improvements in energy efficiency contribute to reducing carbon emissions.

5 Discussion

The inverted U-shaped relationship between urbanization and carbon emissions aligns with the Environmental Kuznets Curve [42]. This suggests that while urbanization initially increases emissions, they adopt sustainable technologies and practices as regions advance, leading to emission reductions. The negative coefficient of (LnU2) supports this pattern, indicating that surpassing the turning point in urbanization is crucial for achieving environmental benefits [43]. However, Jiangxi Province remains in a relatively low stage of urbanization, leading to increased carbon emissions as urbanization progresses, with significant disparities among cities.

For the relationship between the secondary and tertiary industries and carbon emissions, both sectors contribute to increased emissions, but the secondary industry has a greater and more significant impact [44]. This is consistent with studies showing that manufacturing and heavy industries are major drivers of emissions [45]. In contrast, the tertiary sector's more minor impact aligns with findings that service economies as energy consumers contribute less to emissions than industrial sectors [46].

The positive elasticity of per capita disposable income suggests that rising income correlates with higher carbon emissions, emphasizing the need for policies promoting energy-efficient lifestyles in affluent regions [47]. Urban expansion also significantly drives carbon emissions, as growing built-up areas increase demand for energy-intensive sectors like construction. This underscores the environmental costs of unchecked sprawl and the need for green urban planning [48].

Finally, the negative relationship between energy intensity and emissions highlights the critical role of renewable energy [49]. However, while energy intensity can reduce carbon emissions, this effect is insufficient to offset the increase in emissions driven by factors such as the growth of secondary industries. Continued investment in energy-saving technologies and incentivizing their adoption remain essential [50].

6 Conclusions and Policy Recommendations

6.1 Conclusions

This study using the STIRPAT model for Jiangxi Province (2003–2021) shows that carbon emissions are positively influenced by per capita disposable income, secondary and tertiary industry output, urban population proportion, and built-up areas. Among these, the secondary industry exerts the most substantial impact on emissions. Energy intensity has a negative relationship with carbon emissions, demonstrating that improved energy intensity reduces emissions. However, the resulting reduction is insufficient to offset the carbon emissions generated by other factors of urbanization development.

Additionally, an inverted U-shaped relationship between urbanization and carbon emissions is identified, indicating that as urbanization increases, carbon emissions initially rise before beginning to decline after reaching a certain threshold. Currently, Jiangxi Province's urbanization remains in the pre-turning point phase, where a 1% increase in urbanization initially results in a 0.083% rise in emissions. This underscores the need for continued urbanization progress to achieve sustainable development.

This study fills existing research gaps through the following contributions. First, it examines the factors influencing carbon emissions from an urbanization perspective, integrating economic and social elements. Second, using ridge regression instead of ordinary least squares allows for mitigating neglected multicollinearity issues. Third, by introducing quadratic terms, this study captures potential non-linear relationships between urbanization and carbon emissions, providing a more nuanced understanding of dynamics. This study ultimately contributes to the development of more effective low-carbon urban strategies.

6.2 Policy Recommendations

In light of the analysis presented, this study offers policy recommendations on four fronts to facilitate Jiangxi Province's transition to a sustainable, low-carbon economy.

Firstly, promote high-quality development. Accelerating industrial restructuring and integrating digital and traditional economies is essential. Enhance the modern service sector, promote low-carbon practices in traditional industries, and decouple economic growth from energy consumption to achieve carbon-neutral goals.

Secondly, commit to sustainable development. Follow the "Beautiful Jiangxi" initiative by integrating sustainable development principles into urbanization. Align policies with rural rejuvenation and scientific innovation to replace outdated models with innovative, low-carbon growth strategies.

Thirdly, promote low-carbon ethos and education. Raise public awareness about the environmental impacts of carbon emissions. Governments should promote low-carbon principles, advocate eco-friendly transportation, and expand green spaces to guide citizens toward sustainable lifestyles.

Fourthly, refine urban spatial organization. Improve land governance by regulating land use and urban housing development. Enact strict land management measures, establish protective legislation for arable land, curb new construction expansion, and balance development with conservation.

References

1. Guan, X.L., Wei, H.K., Lu, S.S., Dai, Q., Su, H.J.: Assessment on the urbanization strategy in China: Achievements, challenges and reflections. Habitat Int. **71**, 97–109 (2018)
2. Huang, D., Zheng, Q.: Analysis of regional differences in the influence of China's urbanization modes on rural sustainable development. Front. Environ. Sci. **10**, 938897 (2022)
3. Turok, I., McGranahan, G.: Urbanization and economic growth: The arguments and evidence for Africa and Asia. Environ. Urban. **25**(2), 465–482 (2013)
4. Zhou, Q.Y., Li, Z.Q.: The impact of industrial structure upgrades on the urban-rural income gap: An empirical study based on China's provincial panel data. Growth Chang. **52**(3), 1761–1782 (2021)

5. Polyzos, E., Kuck, S., Abdulrahman, K.: Demographic change and economic growth: The role of natural resources in the MENA region. Res. Econ. **76**(1), 1–13 (2022)
6. Duan, X.L., Li, X., Tan, W.F., Xiao, R.: Decoupling relationship analysis between urbanization and carbon emissions in 33 African countries. Heliyon **8**(9), 11 (2022)
7. Muhammad, S., Long, X.L., Salman, M., Dauda, L.: Effect of urbanization and international trade on CO_2 emissions across 65 belt and road initiative countries. Energy **196**, 15 (2020)
8. Grossman, G.M., Krueger, A.B.: Economic growth and the environment. Q. J. Econ. **110**(2), 353–377 (1995)
9. Song, Z.G.: Economic growth and carbon emissions: Estimation of a panel threshold model for the transition process in China. J. Clean. Prod. **278**, 12 (2021)
10. Sikdar, C., Mukhopadhyay, K.: The nexus between carbon emission, energy consumption, economic growth and changing economic structure in India. J. Dev. Areas **52**(4), 67–83 (2018)
11. Bétila, R.R.: Economic freedom and carbon emissions across the globe: the mediating effect of renewable energy consumption. Environ. Sci. Pollut. Res. **30**(36), 86300–86327 (2023)
12. Liu, X., Niu, Q., Dong, S., Zhong, S.: How does renewable energy consumption affect carbon emission intensity? Temporal-spatial impact analysis in China. Energy **284**, 128690 (2023)
13. Shi, X.C., Li, X.Y.: Research on three-stage dynamic relationship between carbon emission and urbanization rate in different city groups. Ecol. Ind. **91**, 195–202 (2018)
14. Zhang, S.X., Li, Z.F., Ning, X., Li, L.: Gauging the impacts of urbanization on CO_2 emissions from the construction industry: Evidence from China. J. Environ. Manage. **288**, 12 (2021)
15. Xinfa, T., Xue, L.: Research on energy policies of Jiangxi province under the dual-carbon constraints. Front. Environ. Sci. **10**, 13 (2022)
16. Guo, X.M., Fang, C.L.: How does urbanization affect energy carbon emissions under the background of carbon neutrality? J. Environ. Manage. **327**, 10 (2023)
17. Wu, Y.Z., Shen, J.H., Zhang, X.L., Skitmore, M., Lu, W.S.: The impact of urbanization on carbon emissions in developing countries: A Chinese study based on the U-Kaya method. J. Clean. Prod. **163**, S284–S298 (2017)
18. Sun, W., Huang, C.C.: How does urbanization affect carbon emission efficiency? Evidence from China. J. Clean. Prod. **272**, 9 (2020)
19. Yan, M., Sun, H., Gu, K.Y.: Driving factors and key emission reduction paths of Xinjiang industries carbon emissions: An industry chain perspective. J. Clean. Prod. **374**, 12 (2022)
20. Gao, P., Yue, S.J., Chen, H.T.: Carbon emission efficiency of China's industry sectors: From the perspective of embodied carbon emissions. J. Clean. Prod. **283**, 9 (2021)
21. Chen, W.D., Wu, F.Y., Geng, W.X., Yu, G.Y.: Carbon emissions in China's industrial sectors. Resour. Conserv. Recycl. **117**, 264–273 (2017)
22. Hou, H.T., Xie, B., Cheng, Y.Y.: Analysis of carbon emissions and emission reduction from coal-fired power plants based on dual carbon targets. Sustainability **15**(9), 14 (2023)
23. Muhammad, S., Pan, Y.C., Agha, M.H., Umar, M., Chen, S.Y.: Industrial structure, energy intensity and environmental efficiency across developed and developing economies: The intermediary role of primary, secondary and tertiary industry. Energy **247**, 16 (2022)
24. Li, F.Y., et al.: Changing patterns and determinants of transportation carbon emissions in Chinese cities. Energy **174**, 562–575 (2019)
25. Lu, M.X., Lai, J.: Review on carbon emissions of commercial buildings. Renew. Sustain. Energy Rev. **119**, 15 (2020)
26. Wang, M.X., et al.: Energy conservation and carbon reduction potential for solid waste in China. Environ. Sci. Pollut. Res. **27**(33), 42158–42167 (2020)
27. Adeleye, B.N., Osabohien, R., Lawal, A.I., De Alwis, T.: Energy use and the role of per capita income on carbon emissions in African countries. PLoS ONE **16**(11), e0259488 (2021)
28. Zhang, S., Shi, B., Ji, H.: How to decouple income growth from household carbon emissions: A perspective based on urban-rural differences in China. *Energy Economics* 106816 (2023)

29. Wang, Z., Cui, C., Peng, S.: How do urbanization and consumption patterns affect carbon emissions in China? A decomposition analysis. J. Clean. Prod. **211**, 1201–1208 (2019)
30. Khalfaoui, R., Tiwari, A.K., Khalid, U., Shahbaz, M.: Nexus between carbon dioxide emissions and economic growth in G7 countries: Fresh insights via wavelet coherence analysis. J. Environ. Planning Manage. **66**(1), 31–66 (2023)
31. Chuai, X.W., Huang, X.J., Lu, Q.L., Zhang, M., Zhao, R.Q., Lu, J.Y.: Spatiotemporal changes of built-up land expansion and carbon emissions caused by the Chinese construction industry. Environ. Sci. Technol. **49**(21), 13021–13030 (2015)
32. Zhu, E.Y., et al.: Carbon emissions induced by land-use and land-cover change from 1970 to 2010 in Zhejiang, China. Sci. Total. Environ. **646**, 930–939 (2019)
33. Son, J., Chang, S.: Interrelations between Greenhouse Gas (GHG) emission and total floor area of buildings: With the case study of public facilities in Ontario, Canada. Arch. Res. **19**(4), 95–99 (2017)
34. Li, R., You, K., Cai, W., Wang, J., Liu, Y., Yu, Y.: Will the southward center of gravity migration of population, floor area, and building energy consumption facilitate building carbon emission reduction in China? Build. Environ. **242**, 110576 (2023)
35. Du, G., Sun, C.W., Ouyang, X.L., Zhang, C.: A decomposition analysis of energy-related CO_2 emissions in Chinese six high-energy intensive industries. J. Clean. Prod. **184**, 1102–1112 (2018)
36. Rahman, M.M., Sultana, N., Velayutham, E.: Renewable energy, energy intensity and carbon reduction: Experience of large emerging economies. Renew. Energy **184**, 252–265 (2022)
37. Afriyie, D., Wang, Z., Hu, S., Ampofo, G.K.M., Asante, D.A.: Exploring the dynamic nexus between urbanization and industrialization with carbon emissions in sub-Saharan Africa: evidence from panel PMG-ARDL estimation. Environ. Sci. Pollut. Res. **30**(3), 6373–6389 (2023)
38. Pang, G., Ding, Z., Shen, X.: Spillover effect of energy intensity reduction targets on carbon emissions in China. Front. Environ. Sci. **11**, 203 (2023)
39. Yang, J., et al.: Driving forces of China's CO_2 emissions from energy consumption based on Kaya-LMDI methods. Sci. Total. Environ. **711**, 15 (2020)
40. Ehrlich, P.R., Holdren, J.P.: Impact of population growth. Science **171**(3977), 1212–1217 (1971)
41. York, R., Rosa, E.A., Dietz, T.: Bridging environmental science with environmental policy: Plasticity of population, affluence, and technology. Soc. Sci. Q. **83**(1), 18–34 (2002)
42. Verbič, M., Satrovic, E., Muslija, A.: Environmental Kuznets curve in Southeastern Europe: the role of urbanization and energy consumption. Environ. Sci. Pollut. Res. **28**(41), 57807–57817 (2021). https://doi.org/10.1007/s11356-021-14732-6
43. Fan, J.J., Wang, J.L., Qiu, J.X., Li, N.: Stage effects of energy consumption and carbon emissions in the process of urbanization: Evidence from 30 provinces in China. Energy **276**, 10 (2023)
44. Zheng, Y., Tang, J., Huang, F.B.: The impact of industrial structure adjustment on the spatial industrial linkage of carbon emission: From the perspective of climate change mitigation. J. Environ. Manage. **345**, 17 (2023)
45. Dong, B.Y., et al.: Carbon emissions, the industrial structure and economic growth: Evidence from heterogeneous industries in China. Environ. Pollut. **262**, 12 (2020)
46. Zhao, J., Dong, X.C., Dong, K.Y.: How does producer services' agglomeration promote carbon reduction? The case of China. Econ. Modell. **104**, 13 (2021)
47. Zhou, Y., Chen, M.X., Tang, Z.P., Mei, Z.A.: Urbanization, land use change, and carbon emissions: Quantitative assessments for city-level carbon emissions in Beijing-Tianjin-Hebei region. Sustain. Cities Soc. **66**, 12 (2021)

48. Wu, Y., Zong, T., Shuai, C.Y., Jiao, L.D.: How does new-type urbanization affect total carbon emissions, per capita carbon emissions, and carbon emission intensity? An empirical analysis of the Yangtze River economic belt, China. J. Environ. Manage. **349**, 16 (2024)
49. Grøntoft, T.: Historical dry deposition of air pollution in the urban background in Oslo, Norway, compared to Western European data. Atmos. Environ. **267**, 118777 (2021)
50. Chen, R., Meng, Q., Yu, J.J.: Optimal government incentives to improve the new technology adoption: Subsidizing infrastructure investment or usage? Omega-Int. J. Manage. Sci. **114**, 16 (2023)

Author Index

C. Weng (Ed.): WRE 2024, LNCE 634, pp. 591–593, 2025.
https://doi.org/10.1007/978-981-96-5095-8

The manufacturer's authorised representative in the EU is Springer Nature Customer Service Centre GmbH, Europaplatz 3, 69115 Heidelberg, Germany. If you have any concerns regarding our products, please contact ProductSafety@springernature.com

Printed and bound by CPI Group (UK) Ltd, Croydon, CR0 4YY
01/10/2026
02225935-0019